PC UW
LLYFRGELL
LIBRARY
ABERYSTWYTH

Applied Mathematical Sciences
Volume 118

Springer
New York
Berlin
Heidelberg
Barcelona
Budapest
Hong Kong
London
Milan
Paris
Santa Clara
Singapore
Tokyo

Applied Mathematical Sciences

1. *John:* Partial Differential Equations, 4th ed.
2. *Sirovich:* Techniques of Asymptotic Analysis.
3. *Hale:* Theory of Functional Differential Equations, 2nd ed.
4. *Percus:* Combinatorial Methods.
5. *von Mises/Friedrichs:* Fluid Dynamics.
6. *Freiberger/Grenander:* A Short Course in Computational Probability and Statistics.
7. *Pipkin:* Lectures on Viscoelasticity Theory.
8. *Giacoglia:* Perturbation Methods in Non-linear Systems.
9. *Friedrichs:* Spectral Theory of Operators in Hilbert Space.
10. *Stroud:* Numerical Quadrature and Solution of Ordinary Differential Equations.
11. *Wolovich:* Linear Multivariable Systems.
12. *Berkovitz:* Optimal Control Theory.
13. *Bluman/Cole:* Similarity Methods for Differential Equations.
14. *Yoshizawa:* Stability Theory and the Existence of Periodic Solution and Almost Periodic Solutions.
15. *Braun:* Differential Equations and Their Applications, 3rd ed.
16. *Lefschetz:* Applications of Algebraic Topology.
17. *Collatz/Wetterling:* Optimization Problems.
18. *Grenander:* Pattern Synthesis: Lectures in Pattern Theory, Vol. I.
19. *Marsden/McCracken:* Hopf Bifurcation and Its Applications.
20. *Driver:* Ordinary and Delay Differential Equations.
21. *Courant/Friedrichs:* Supersonic Flow and Shock Waves.
22. *Rouche/Habets/Laloy:* Stability Theory by Liapunov's Direct Method.
23. *Lamperti:* Stochastic Processes: A Survey of the Mathematical Theory.
24. *Grenander:* Pattern Analysis: Lectures in Pattern Theory, Vol. II.
25. *Davies:* Integral Transforms and Their Applications, 2nd ed.
26. *Kushner/Clark:* Stochastic Approximation Methods for Constrained and Unconstrained Systems.
27. *de Boor:* A Practical Guide to Splines.
28. *Keilson:* Markov Chain Models—Rarity and Exponentiality.
29. *de Veubeke:* A Course in Elasticity.
30. *Shiatycki:* Geometric Quantization and Quantum Mechanics.
31. *Reid:* Sturmian Theory for Ordinary Differential Equations.
32. *Meis/Markowitz:* Numerical Solution of Partial Differential Equations.
33. *Grenander:* Regular Structures: Lectures in Pattern Theory, Vol. III.
34. *Kevorkian/Cole:* Perturbation Methods in Applied Mathematics.
35. *Carr:* Applications of Centre Manifold Theory.
36. *Bengtsson/Ghil/Källén:* Dynamic Meteorology: Data Assimilation Methods.
37. *Saperstone:* Semidynamical Systems in Infinite Dimensional Spaces.
38. *Lichtenberg/Lieberman:* Regular and Chaotic Dynamics, 2nd ed.
39. *Piccini/Stampacchia/Vidossich:* Ordinary Differential Equations in $\mathbf{R}^n$.
40. *Naylor/Sell:* Linear Operator Theory in Engineering and Science.
41. *Sparrow:* The Lorenz Equations: Bifurcations, Chaos, and Strange Attractors.
42. *Guckenheimer/Holmes:* Nonlinear Oscillations, Dynamical Systems and Bifurcations of Vector Fields.
43. *Ockendon/Taylor:* Inviscid Fluid Flows.
44. *Pazy:* Semigroups of Linear Operators and Applications to Partial Differential Equations.
45. *Glashoff/Gustafson:* Linear Operations and Approximation: An Introduction to the Theoretical Analysis and Numerical Treatment of Semi-Infinite Programs.
46. *Wilcox:* Scattering Theory for Diffraction Gratings.
47. *Hale et al:* An Introduction to Infinite Dimensional Dynamical Systems—Geometric Theory.
48. *Murray:* Asymptotic Analysis.
49. *Ladyzhenskaya:* The Boundary-Value Problems of Mathematical Physics.
50. *Wilcox:* Sound Propagation in Stratified Fluids.
51. *Golubitsky/Schaeffer:* Bifurcation and Groups in Bifurcation Theory, Vol. I.
52. *Chipot:* Variational Inequalities and Flow in Porous Media.
53. *Majda:* Compressible Fluid Flow and System of Conservation Laws in Several Space Variables.
54. *Wasow:* Linear Turning Point Theory.
55. *Yosida:* Operational Calculus: A Theory of Hyperfunctions.
56. *Chang/Howes:* Nonlinear Singular Perturbation Phenomena: Theory and Applications.
57. *Reinhardt:* Analysis of Approximation Methods for Differential and Integral Equations.
58. *Dwoyer/Hussaini/Voigt (eds):* Theoretical Approaches to Turbulence.
59. *Sanders/Verhulst:* Averaging Methods in Nonlinear Dynamical Systems.
60. *Ghil/Childress:* Topics in Geophysical Dynamics: Atmospheric Dynamics, Dynamo Theory and Climate Dynamics.

(continued following index)

Edwige Godlewski   Pierre-Arnaud Raviart

Numerical Approximation of Hyperbolic Systems of Conservation Laws

With 75 Illustrations

Edwige Godlewski
Laboratoire d'Analyse Numérique
Université Pierre et Marie Curie
4 Place Jussieu
75252 Paris Cedex 05
France

Pierre-Arnaud Raviart
Ecole Polytechnique
Centre de Mathématiques Appliquées
91128 Palaiseau Cedex
France

Editors

Mathematics Subject Classification (1991): 76N15, 76N20, 70G05

Library of Congress Cataloging-in-Publication Data
Godlewski, Edwige.
Numerical approximation of hyperbolic systems of conservation laws/
Edwige Godlewski, Pierre-Arnaud Raviart.
p. cm. — (Applied mathematical sciences; 118)
Includes bibliographical references and index.
ISBN 0-387-94529-6 (hard:alk. paper)
1. Gas dynamics. 2. Conservation laws (Mathematics).
3. Differential equations, Hyperbolic—Numerical solutions.
I. Raviart, Pierre-Arnaud, 1939- . II. Title. III. Series: Applied
mathematical sciences (Springer-Verlag New York Inc.); v. 118.
QA1.A647 vol. 118
[QA930]
510 s—dc20
[533.2′01515353] 96-13585

Printed on acid-free paper.

Production managed by Hal Henglein; manufacturing supervised by Joe Quatela.
Camera-ready copy prepared from the authors' TeX file.
Printed and bound by Edwards Brothers, Inc., Ann Arbor, MI.
Printed in the United States of America.

9 8 7 6 5 4 3 2 1

ISBN 0-387-94529-6 Springer-Verlag New York Berlin Heidelberg SPIN 10501626

Preface

This work is devoted to the theory and approximation of nonlinear hyperbolic systems of conservation laws in one or two space variables. It follows directly a previous publication on hyperbolic systems of conservation laws by the same authors, and we shall make frequent references to Godlewski and Raviart (1991) (hereafter noted G.R.), though the present volume can be read independently. This earlier publication, apart from a first chapter, especially covered the scalar case. Thus, we shall detail here neither the mathematical theory of multidimensional *scalar* conservation laws nor their approximation in the one-dimensional case by finite-difference conservative schemes, both of which were treated in G.R., but we shall mostly consider systems. The theory for systems is in fact much more difficult and not at all completed. This explains why we shall mainly concentrate on some theoretical aspects that are needed in the applications, such as the solution of the Riemann problem, with occasional insights into more sophisticated problems.

The present book is divided into six chapters, including an introductory chapter. For the reader's convenience, we shall resume in this Introduction the notions that are necessary for a self-sufficient understanding of this book —the main definitions of hyperbolicity, weak solutions, and entropy— present the practical examples that will be thoroughly developed in the following chapters, and recall the main results concerning the scalar case.

Chapter I is devoted to the resolution of the Riemann problem for a general hyperbolic system in one space dimension, introducing the classical notions of Riemann invariants and simple waves, the rarefaction and shock curves, and characteristics and entropy conditions. The theory is then applied to the p-system.

In Chapter II, we make a closer study of the one-dimensional system of gas dynamics. We solve the Riemann problem in detail and then present the simplest models of reacting flow, first the Chapman–Jouguet theory and then the Z.N.D. model for detonation.

After this theoretical approach, we go into the numerical approximation of hyperbolic systems by conservative finite-difference methods. The most usual schemes for one-dimensional systems are developed in Chapter III, with special emphasis on the application to gas dynamics. The last section begins with a short account on the kinetic theory so as to introduce kinetic schemes.

Chapter IV is devoted to the study of finite volume methods for bidimensional systems, preceded by some theoretical considerations on multidimensional systems.

For the sake of completeness, we could not avoid the problem of boundary conditions. Chapter V is but an introduction to the complex theory and presents some numerical boundary treatment.

The authors wish to thank R. Abgrall, F. Coquel, F. Dubois, and particularly T. Gallouët, B. Perthame, and D. Serre, from whom they learned a great deal and who answered willingly and most amiably their many questions.

They owe thanks to the SMAI reading committee and to the reviewers, who made very valuable suggestions.

The first author is grateful to all her colleagues who encouraged her in completing this huge work, especially to H. Le Dret and F. Murat for so often giving her their time, and to L. Ruprecht for her kind and competent assistance in the retyping of the final manuscript; such friendly help was invaluable.

Paris, France
September 1995

E. Godlewski and P.-A. Raviart

Contents

Introduction

1 Definitions and examples

In this section, we present the general form of systems of conservation laws in several space variables and we give some important examples of such systems that arise in continuum physics.

Let Ω be an open subset of $\mathbb{R}^p$, and let $\mathbf{f}_j, 1 \leq j \leq d$, be d smooth functions from Ω into $\mathbb{R}^p$; the general form of a system of conservation laws in several space variables is

$$\frac{\partial \mathbf{u}}{\partial t} + \sum_{j=1}^{d} \frac{\partial}{\partial x_j} \mathbf{f}_j(\mathbf{u}) = \mathbf{0}, \quad \mathbf{x} = (x_1, ..., x_d) \in \mathbb{R}^d, \quad t > 0, \tag{1.1}$$

where

$$\mathbf{u} = \begin{pmatrix} u_1 \\ \vdots \\ u_p \end{pmatrix}$$

is a vector-valued function from $\mathbb{R}^d \times [0, +\infty[$ into Ω. The set Ω is called the set of states and the functions

$$\mathbf{f}_j = \begin{pmatrix} f_{1j} \\ \vdots \\ f_{pj} \end{pmatrix}$$

are called the flux-functions. One says that system (1.1) is written in *conservative form*.

Formally, the system (1.1) expresses the conservation of the p quantities $u_1, ..., u_p$. In fact, let D be an arbitrary domain of $\mathbb{R}^d$, and let $\mathbf{n} = (n_1, ..., n_d)^T$ be the outward unit normal to the boundary ∂D of D. Then, it follows from (1.1) that

$$\frac{d}{dt} \int_D \mathbf{u}\, d\mathbf{x} + \sum_{j=1}^{d} \int_{\partial D} \mathbf{f}_j(\mathbf{u})\, n_j\, dS = \mathbf{0}.$$

This balance equation has now a very natural meaning: the time variation of $\int_D \mathbf{u}\, d\mathbf{x}$ is equal to the losses through the boundary ∂D.

In all the following, we shall be concerned with the study of *hyperbolic* systems of conservation laws, which we define in the following way. For all $j = 1, ..., d$, let

$$\mathbf{A}_j(\mathbf{u}) = \Big(\frac{\partial f_{ij}}{\partial u_k}(\mathbf{u})\Big)_{1\le i,k\le p}$$

be the Jacobian matrix of $\mathbf{f}_j(\mathbf{u})$; the system (1.1) is called *hyperbolic* if, for any $\mathbf{u} \in \Omega$ and any $\boldsymbol{\omega} = (\omega_1, ..., \omega_d) \in \mathbb{R}^d, \boldsymbol{\omega} \neq \mathbf{0}$, the matrix

$$\mathbf{A}(\mathbf{u}, \boldsymbol{\omega}) = \sum_{j=1}^{d} \omega_j\, \mathbf{A}_j(\mathbf{u})$$

has p real eigenvalues $\lambda_1(\mathbf{u}, \boldsymbol{\omega}) \le \lambda_. 1(\mathbf{u}, \boldsymbol{\omega}) \le ... \le \lambda_p(\mathbf{u}, \boldsymbol{\omega})$ and p linearly independent corresponding eigenvectors $\mathbf{r}_1(\mathbf{u}, \boldsymbol{\omega}), ..., \mathbf{r}_p(\mathbf{u}, \boldsymbol{\omega})$, i.e.,

$$\mathbf{A}(\mathbf{u}, \boldsymbol{\omega})\, \mathbf{r}_k(\mathbf{u}, \boldsymbol{\omega}) = \lambda_k(\mathbf{u}, \boldsymbol{\omega})\, \mathbf{r}_k(\mathbf{u}, \boldsymbol{\omega}), \quad 1 \le k \le p.$$

If, in addition, the eigenvalues $\lambda_k(\mathbf{u}, \boldsymbol{\omega})$ are all distinct, the system (1.1) is called *strictly hyperbolic.*

In fact, little is known about systems in more than one space variable unless they are *symmetrizable*, i.e., there exists for all $\mathbf{u} \in \Omega$ a positive-definite symmetric matrix $\mathbf{A}_0(\mathbf{u})$ smoothly varying with $\mathbf{u}$ such that the matrices

$$\mathbf{A}_0(\mathbf{u})\mathbf{A}_j(\mathbf{u}), \quad 1 \le j \le d$$

are symmetric. Symmetrizable systems of conservation laws are clearly hyperbolic. Note that most of the systems of conservation laws that arise in practice are symmetrizable; this is a consequence of the existence of an entropy function (Godunov–Mock theorem, see Theorem 3.1 below).

For such systems, we shall study the *Cauchy problem*, or initial value problem (IVP): Find a function $\mathbf{u} : (\mathbf{x}, t) \in \mathbb{R}^d \times [0, \infty[\to \mathbf{u}(\mathbf{x}, t) \in \Omega$ that is a solution of (1.1) satisfying the initial condition

$$\mathbf{u}(\mathbf{x}, 0) = \mathbf{u}_0(\mathbf{x}), \quad \mathbf{x} \in \mathbb{R}^d, \tag{1.2}$$

where $\mathbf{u}_0 : \mathbb{R}^d \to \Omega$ is a given function. The initial boundary value problem (IBVP) will be considered in Chapter V. One aim of this introduction is to make precise in which sense (1.1), (1.2) is to be taken.

In the one-dimensional case, when $\mathbf{u}_0$ has the following particular form,

$$\mathbf{u}_0(x) = \begin{cases} \mathbf{u}_\ell, & x < 0 \\ \mathbf{u}_r, & x > 0, \end{cases} \tag{1.3}$$

this Cauchy problem is called the (one-dimensional) *Riemann problem* (see Chapter IV, Remark 2.10 for the definition of a 2-D Riemann problem).

In the scalar case (i.e. $p = 1$), the simplest example of a nonlinear conservation law is given by Burgers' equation.

Example 1.1. The Burgers equation. The scalar parabolic equation

$$\frac{\partial u}{\partial t} + u\,\frac{\partial u}{\partial x} - \nu\,\frac{\partial^2 u}{\partial x^2} = 0 \tag{1.4a}$$

was introduced in particular by Burgers as the simplest differential model for a fluid flow and is therefore often called the (viscous) Burgers equation. Though very simple, this equation can be regarded as a model for decaying free turbulence (see Cole 1951). A number of authors have developed the asymptotic theory of Navier–Stokes equations in terms of Burgers' equation, and it is thus often used in numerical tests (Shu and Osher 1989). Burgers studied the limit equation when ν tends to zero, which we write in conservation form

$$\frac{\partial u}{\partial t} + \frac{\partial}{\partial x}\left(\frac{u^2}{2}\right) = 0. \tag{1.4b}$$

(1.4b) is the inviscid Burgers equation (or Burgers' equation without viscosity), which, for brevity, we shall simply call from now on Burgers' equation. It occurs in particular in wave theory to depict the distortion of waveform in simple waves (see Lighthill 1978, Sec. 2.9, Whitman 1974, Sec. 2.8).

We shall see that Burgers' equation possesses all the features of a scalar convex equation

$$\frac{\partial u}{\partial t} + \frac{\partial}{\partial x}\, f(u) = 0, \tag{1.5}$$

where $f : \mathbb{R} \to \mathbb{R}$ is a *convex* smooth function. In particular, the Cauchy problem for Burgers' equation may have discontinuous weak solutions even for a smooth initial function u_0, and the solution of the Riemann problem is either a shock propagating or a rarefaction wave (see Figure 2.2), both being kinds of waves that are involved in the solution of the Riemann problem for a system.

Finally, it is worth mentioning that the Cauchy problem for (1.4a) has an explicit solution, obtained using the Cole–Hopf transform

$$u = -2\nu\,\frac{\varphi_x}{\varphi}.$$

It eliminates the nonlinear term and transforms (1.4a) into the heat equation

$$\frac{\partial \varphi}{\partial t} = \nu\,\frac{\partial^2 \varphi}{\partial x^2},$$

for which explicit expressions of the solution are known. For details, we refer to the original papers of Hopf (1950), Cole (1951), and to Whitman, (1974, Chapter 4) where a thorough study of equation (1.4a) (including limit as $\nu \to 0$ and shock structure) can be found. □

Let us next introduce some fundamental examples of the systems of conservation laws that we shall study in the following chapters.

Example 1.2. The p-system. A model for one-dimensional isentropic gas dynamics in Lagrangian coordinates is given by the following system of two equations:

$$\begin{cases} \dfrac{\partial v}{\partial t} - \dfrac{\partial u}{\partial x} = 0 \\ \dfrac{\partial u}{\partial t} + \dfrac{\partial}{\partial x}\, p(v) = 0, \end{cases} \tag{1.6}$$

where v is the specific volume, u is the velocity, and the pressure $p = p(v)$ is a given function of v. For a polytropic isentropic ideal gas, we have

$$p(v) = Av^{-\gamma}$$

for some constants $A = A(s) > 0$ (depending on the entropy) and $\gamma \geq 1$.

System (1.6) is of the form

$$\frac{\partial \mathbf{w}}{\partial t} + \frac{\partial}{\partial x}\, \mathbf{f}(\mathbf{w}) = \mathbf{0},$$

where

$$\mathbf{w} = \begin{pmatrix} v \\ u \end{pmatrix}, \quad \mathbf{f}(\mathbf{w}) = \begin{pmatrix} -u \\ p(v) \end{pmatrix}, \quad \Omega = \{(v,u) \in \mathbb{R}^2; v > 0\}.$$

This system is strictly hyperbolic provided that we assume $p'(v) < 0$. In that case, the Jacobian matrix of $\mathbf{f}$

$$\mathbf{A}(\mathbf{w}) = \begin{pmatrix} 0 & -1 \\ p'(v) & 0 \end{pmatrix}$$

has indeed two real distinct eigenvalues

$$\lambda_1 = -\sqrt{(-p'(v))} < \lambda_2 = \sqrt{(-p'(v))}.$$

Besides the scalar case, the p-system (1.6) is the simplest nontrivial example of a nonlinear system of conservation laws. Note that any nonlinear wave equation

$$\frac{\partial^2 g}{\partial t^2} - \frac{\partial}{\partial x}\left(\sigma\left(\frac{\partial g}{\partial x}\right)\right) = 0$$

may be put in the form (1.6) by setting

$$u = \frac{\partial g}{\partial t}, \quad v = \frac{\partial g}{\partial x}, \quad p(v) = -\sigma(v),$$

as one can easily check. □

Example 1.3. The equations of gas dynamics in Eulerian coordinates. An example that appears to be fundamental in the applications is the system of gas dynamics. In Eulerian coordinates, the Euler equations for a compressible inviscid fluid (where we neglect heat conduction) can be written in the following conservative form:

$$(1.7)\qquad \begin{cases} \dfrac{\partial \rho}{\partial t} + \displaystyle\sum_{j=1}^{3} \frac{\partial}{\partial x_j}(\rho u_j) = 0, \\ \dfrac{\partial}{\partial t}(\rho u_i) + \displaystyle\sum_{j=1}^{3} \frac{\partial}{\partial x_j}(\rho u_i u_j + p\,\delta_{ij}) = 0, \quad 1 \le i \le 3, \\ \dfrac{\partial}{\partial t}(\rho e) + \displaystyle\sum_{j=1}^{3} \frac{\partial}{\partial x_j}\Big((\rho e + p)u_j\Big) = 0. \end{cases}$$

In (1.7), ρ is the density of the fluid, $\mathbf{u} = (u_1, u_2, u_3)$ the velocity, p the pressure, ε the specific (i.e., per unit mass) internal energy, and $e = \varepsilon + \frac{|\mathbf{u}|^2}{2}$ the specific total energy. Equations (1.7) express respectively the laws of conservation of mass, momentum, and total energy for the fluid.

We need to add to (1.7) an equation for p. We shall see in Chapter IV that Galilean invariance requires that p does not depend on $\mathbf{u}$. The "equation of state" can be taken in the form

$$p = p(\rho, \varepsilon),$$

e.g., for a polytropic ideal gas, the equation of state is given by

$$p = (\gamma - 1)\rho\,\varepsilon, \quad \gamma > 1.$$

Now, setting

$$q_i = \rho\, u_i, \quad 1 \le i \le 3, \quad E = \rho\, e,$$

the system (1.7) can be put into the general framework of system (1.1) if we take

$$(1.8a)\qquad \mathbf{U} = \begin{pmatrix} \rho \\ q_1 \\ q_2 \\ q_3 \\ E \end{pmatrix}, \quad \mathbf{f}_1(\mathbf{U}) = \begin{pmatrix} q_1 \\ p + q_1^2/\rho \\ q_1 q_2/\rho \\ q_1 q_3/\rho \\ (E+p)q_1/\rho \end{pmatrix},$$

$$(1.8b)\qquad \mathbf{f}_2(\mathbf{U}) = \begin{pmatrix} q_2 \\ q_1 q_2/\rho \\ p + q_2^2/\rho \\ q_2 q_3/\rho \\ (E+p)q_2/\rho \end{pmatrix}, \quad \mathbf{f}_3(\mathbf{U}) = \begin{pmatrix} q_3 \\ q_1 q_3/\rho \\ q_2 q_3/\rho \\ p + q_3^2/\rho \\ (E+p)q_3/\rho \end{pmatrix}$$

with

$$p = p\Big(\rho, \frac{E}{\rho} - \frac{|\mathbf{q}|^2}{2\rho^2}\Big)$$

and if the set of states is

$$\Omega = \Big\{(\rho, \mathbf{q} = (q_1, q_2, q_3), E);\ \rho > 0,\ \mathbf{q} \in \mathbb{R}^3, E - \frac{|\mathbf{q}|^2}{2\rho} > 0\Big\}.$$

We shall see that in the one-dimensional case (resp. bidimensional), the system (1.8) of the gas dynamics equations with usual equation of state is indeed a strictly hyperbolic (resp. hyperbolic) symmetrizable nonlinear system of conservation laws.

In many applications we do not have to solve the full system but only a somewhat reduced system of equations. On the one hand, one can often suppose, for instance, that the flow is isentropic so that the equation of state reduces to

$$p = p(\rho),$$

and it suffices to solve the system of equations for conservation of mass and momentum. On the other hand, if we assume that the flow has some symmetry, one can reduce the number of space variables. For instance, assuming slab symmetry, the gas dynamics equations become (with obvious notations)

$$\text{(1.9)} \qquad \begin{cases} \dfrac{\partial \rho}{\partial t} + \dfrac{\partial}{\partial x}(\rho\, u) = 0, \\ \dfrac{\partial}{\partial t}(\rho\, u) + \dfrac{\partial}{\partial x}(\rho\, u^2 + p) = 0, \\ \dfrac{\partial}{\partial t}(\rho\, e) + \dfrac{\partial}{\partial x}((\rho\, e + p)u) = 0. \end{cases}$$

Again, choosing $\rho, q = \rho\, u$, and $E = \rho\, e$ as dependent variables, (1.9) fits into our framework with $d = 1, p = 3$. □

Example 1.4. The equations of gas dynamics in Lagrangian coordinates. Let us rewrite equations (1.6) in Lagrangian coordinates. Denote again by $\mathbf{u} = \mathbf{u}(\mathbf{x}, t)$ the velocity field of the fluid flow; we consider the differential system

$$\text{(1.10)} \qquad \frac{d\mathbf{x}}{dt} = \mathbf{u}(\mathbf{x}, t)$$

and, for all $\boldsymbol{\xi} \in \mathbb{R}^3$, we denote by $t \to \mathbf{x}(\boldsymbol{\xi}, t)$ the solution of (1.10) that satisfies the initial condition

$$\text{(1.11)} \qquad \mathbf{x}(\mathbf{0}) = \boldsymbol{\xi}.$$

Then $(\boldsymbol{\xi} = (\xi_1, \xi_2, \xi_3), t)$ are the Lagrangian coordinates associated with the velocity field $\mathbf{u}$. If we set

$$\text{(1.12)} \qquad J(\boldsymbol{\xi}, t) = \det\Big(\frac{\partial x_i}{\partial \xi_j}(\boldsymbol{\xi}, t)\Big),$$

it is a standard exercise to show that

$$\text{(1.13)} \qquad \frac{\partial J}{\partial t}(\boldsymbol{\xi}, t) = J(\xi, t)(\operatorname{div} \mathbf{u})(\mathbf{x}(\boldsymbol{\xi}, t), t),$$

where

$$\mathrm{div}\,\mathbf{u} = \sum_{j=1}^{3} \frac{\partial u_j}{\partial x_j}.$$

Now, given a function $\varphi = \varphi(\mathbf{x}, t)$ expressed in Eulerian coordinates, we denote by $\overline{\varphi} = \overline{\varphi}(\boldsymbol{\xi}, t)$ this function expressed in Lagrangian coordinates, i.e.,

$$\overline{\varphi}(\boldsymbol{\xi}, t) = \varphi(\mathbf{x}(\boldsymbol{\xi}, t), t).$$

Then, using (1.13), an easy computation shows that

$$\frac{\partial}{\partial t}(\overline{\varphi}\, J) = J\Big(\overline{\frac{\partial \varphi}{\partial t}} + \overline{\mathrm{div}(\varphi \mathbf{u})}\Big) = J\Big(\overline{\frac{\partial \varphi}{\partial t}} + \overline{\sum_{j=1}^{3} \frac{\partial}{\partial x_j}(\varphi u_j)}\Big).$$

Hence the gas dynamics equations become

(1.14)
$$\begin{cases} \dfrac{\partial}{\partial t}(\overline{\rho}\, J) = 0 & \text{(conservation of mass)} \\ \dfrac{\partial}{\partial t}(\overline{\rho u_i} J) + J\,\overline{\dfrac{\partial p}{\partial x_i}} = 0, \ 1 \leq i \leq 3, & \text{(conservation of momentum),} \\ \dfrac{\partial}{\partial t}(\overline{\rho e}\, J) + J\,\overline{\mathrm{div}(u)} = 0 & \text{(conservation of energy).} \end{cases}$$

Let us next assume slab symmetry so that equations (1.14) become

$$\begin{cases} \dfrac{\partial}{\partial t}(\overline{\rho}\, J) = 0, \\ \dfrac{\partial}{\partial t}(\overline{\rho u}\, J) + J\,\overline{\dfrac{\partial p}{\partial x}} = 0, \\ \dfrac{\partial}{\partial t}(\overline{\rho e}\, J) + J\,\overline{\dfrac{\partial}{\partial x}(pu)} = 0. \end{cases}$$

Since

$$\frac{\partial}{\partial \xi} = \Big(\frac{\partial x}{\partial \xi}\Big)\frac{\partial}{\partial x} = J\,\frac{\partial}{\partial x},$$

we obtain, suppressing the bars for simplicity,

$$\begin{cases} \dfrac{\partial}{\partial t}(\rho\, J) = 0, \\ \dfrac{\partial}{\partial t}(\rho\, u\, J) + \dfrac{\partial p}{\partial \xi} = 0, \\ \dfrac{\partial}{\partial t}(\rho\, e\, J) + \dfrac{\partial}{\partial \xi}(pu) = 0. \end{cases}$$

Let us put these equations in a more classical form. The first equation (of conservation of mass) gives

$$\rho J = \rho_0, \tag{1.15}$$

where $\rho_0(\xi) = \rho(\xi, 0)$. If we introduce the specific volume

$$\tau = \frac{1}{\rho},$$

we get

$$J = \rho_0 \, \tau$$

so that the equation (see (1.13))

$$\frac{\partial J}{\partial t} = J \, \frac{\partial u}{\partial x} = \frac{\partial u}{\partial \xi}$$

becomes

$$\rho_0 \, \frac{\partial \tau}{\partial t} - \frac{\partial u}{\partial \xi} = 0.$$

Moreover, using (1.15), the equations of conservation of momentum and energy become respectively

$$\rho_0 \, \frac{\partial u}{\partial t} + \frac{\partial p}{\partial \xi} = 0,$$

$$\rho_0 \, \frac{\partial e}{\partial t} + \frac{\partial}{\partial \xi} (pu) = 0.$$

Finally, we introduce a mass variable m such that

$$dm = \rho_0 \, d\xi.$$

Then, the equations of gas dynamics in slab symmetry and in Lagrangian coordinates can be written

$$\begin{cases} \dfrac{\partial \tau}{\partial t} - \dfrac{\partial u}{\partial m} = 0, \\ \dfrac{\partial u}{\partial t} + \dfrac{\partial p}{\partial m} = 0, \\ \dfrac{\partial e}{\partial t} + \dfrac{\partial}{\partial m} (pu) = 0, \end{cases} \tag{1.16}$$

with $p = p(\tau, \varepsilon) = p(\tau, e - \frac{u^2}{2})$. By setting

$$\mathbf{V} = \begin{pmatrix} \tau \\ u \\ e \end{pmatrix}, \quad \mathbf{f}(\mathbf{V}) = \begin{pmatrix} -u \\ p \\ pu \end{pmatrix}, \quad \Omega = \Big\{ \mathbf{V}; \tau > 0, u \in \mathbb{R}, e - \frac{u^2}{2} > 0 \Big\},$$

we obtain a nonlinear system of conservation laws

$$\frac{\partial \mathbf{V}}{\partial t} + \frac{\partial}{\partial m} \, \mathbf{f}(\mathbf{V}) = \mathbf{0}$$

that again will be seen to be strictly hyperbolic and symmetrizable under sensible physical assumptions.

If we assume in addition that the flow is isentropic, the equation of state reduces to $p = p(\tau)$. Again, it suffices to solve the system of equations of conservation of mass and momentum

$$\begin{cases} \dfrac{\partial \tau}{\partial t} - \dfrac{\partial u}{\partial m} = 0, \\ \dfrac{\partial u}{\partial t} + \dfrac{\partial p}{\partial m} = 0. \end{cases}$$

We obtain the p-system (1.6) but with different notations. □

In fact, in a number of applications, we need to consider systems of conservation laws with source terms

$$\frac{\partial \mathbf{u}}{\partial t} + \sum_{j=1}^{d} \frac{\partial}{\partial x_j} \mathbf{f}_j(\mathbf{u}) = \mathbf{g}(\mathbf{u}); \quad \mathbf{x} = (x_1, ..., x_d) \in \mathbb{R}^d, \quad t > 0, \tag{1.17}$$

where $\mathbf{g}$ is a smooth function from Ω into $\mathbb{R}^p$. Let us give an example of such a situation.

Example 1.5. The equations of reacting gas flows. Consider the more complex example of a system of reacting gas flows. For simplicity, we assume that the system consists of two species, the unburnt gas and the burnt gas, and we denote by z the mass fraction of the burnt gas, so that $1 - z$ is the mass fraction of the unburnt gas. Again, if we neglect viscosity and heat conduction, the equations of the system are obtained by adjoining to the gas dynamics equations (1.7) the chemical reaction equation

$$\frac{\partial}{\partial t}(\rho\, z) + \sum_{j=1}^{d} \frac{\partial}{\partial x_j}(\rho\, z\, u_j) = \rho\, r, \tag{1.18}$$

where

$$r = r(\rho, p, z) \tag{1.19}$$

is the reaction rate. To close the system of equations, we need to add an equation of state, which can be taken in the form

$$\varepsilon = \varepsilon(\rho, p, z). \tag{1.20}$$

The simplest model is obtained by assuming that the unburnt gas is converted to burnt gas through an irreversible exothermic chemical reaction. If we assume that the kinetics of the chemical reaction obeys an Arrhenius mechanism, we have

$$r = K(1 - z) \exp\Big(-\frac{E^*}{RT}\Big),$$

where E^* is the activation energy and T is the temperature. In practice, a simpler expression for r is often used, for instance

$$r = K(1-z)^\alpha, \quad \alpha = \frac{1}{2} \text{ or } 1.$$

As regards the equation of state, one usually defines the internal energy ε of the mixture of burnt and unburnt gases up to an additive constant by

$$\varepsilon = z\,\varepsilon_b(\rho, p) + (1-z)\,\varepsilon_u(\rho, p),$$

where ε_b and ε_u are the internal energies of the burnt and unburnt gas, respectively. If we assume that both gases are ideal polytropic gases with the same γ law, we have

$$\varepsilon_b(\rho, p) = (\gamma-1)^{-1}\frac{p}{\rho} + Q_b, \quad \varepsilon_u(\rho, p) = (\gamma-1)^{-1}\frac{p}{\rho} + Q_u,$$

where Q_b and Q_u are the energies of formation of the burnt and unburnt gases. Observe that

$$Q_b < Q_u$$

for an exothermic reaction. Hence, in that case we may write

$$\varepsilon(\rho, p, z) = (\gamma-1)^{-1}\frac{p}{\rho} + (1-z)(Q_u - Q_b) + Q_b. \tag{1.21}$$

Setting

$$Z = \rho\, z,$$

and, using notations (1.8),

$$\mathbf{V} = \begin{pmatrix} \mathbf{U} \\ Z \end{pmatrix}, \quad \mathbf{g}_i = \begin{pmatrix} \mathbf{f}_i \\ Zq_i/\rho \end{pmatrix}, \quad i = 1, 2, 3,$$

$$\mathbf{g}(\mathbf{V}) = \begin{pmatrix} \mathbf{0} \\ \rho r \end{pmatrix},$$

which are all column vectors in $\mathbb{R}^6$, with

$$p = p\Big(\rho, \frac{E}{\rho} - \frac{|\mathbf{q}|^2}{2\rho^2}, \frac{Z}{\rho}\Big) \text{ and } r = r\Big(\rho, p, \frac{Z}{\rho}\Big)$$

we get a system of the form (1.17).

In Lagrangian coordinates, equation (1.18) becomes

$$\frac{\partial}{\partial t}(\overline{\rho\, z}\, J) = \overline{\rho}\, J\, r$$

or equivalently, by the equation of conservation of mass,

$$\frac{\partial}{\partial t}\,\overline{z} = r.$$

Now, assuming slab symmetry, the equations of reacting gas flows become in Eulerian coordinates

$$
(1.22)\qquad \begin{cases} \dfrac{\partial \rho}{\partial t} + \dfrac{\partial}{\partial x}(\rho\, u) = 0 \\ \dfrac{\partial}{\partial t}(\rho\, u) + \dfrac{\partial}{\partial x}(\rho\, u^2 + p) = 0 \\ \dfrac{\partial}{\partial t}(\rho\, e) + \dfrac{\partial}{\partial x}((\rho\, e + p)u) = 0 \\ \dfrac{\partial}{\partial t}(\rho\, z) + \dfrac{\partial}{\partial x}(\rho\, z\, u) = \rho r, \end{cases}
$$

and

$$
(1.23)\qquad \begin{cases} \dfrac{\partial \tau}{\partial t} - \dfrac{\partial u}{\partial m} = 0 \\ \dfrac{\partial u}{\partial t} + \dfrac{\partial p}{\partial m} = 0 \\ \dfrac{\partial e}{\partial t} + \dfrac{\partial}{\partial m}(pu) = 0 \\ \dfrac{\partial z}{\partial t} = r \end{cases}
$$

is the corresponding system in Lagrangian coordinates. □

One can even consider systems of conservation laws more general than (1.17), namely systems of the form

$$
(1.24)\qquad \frac{\partial \mathbf{u}}{\partial t} + \sum_{j=1}^{d} \frac{\partial}{\partial x_j}\mathbf{f}_j(\mathbf{x}, t, \mathbf{u}) = \mathbf{g}(\mathbf{x}, t, \mathbf{u}), \quad \mathbf{x} \in \mathbb{R}^d,\ t > 0,
$$

where $\mathbf{f}_j, 1 \le j \le d$, and the source term $\mathbf{g}$ are smooth functions from $\mathbb{R}^d \times [0, +\infty[\times \Omega$ into $\mathbb{R}^p$.

However, for the sake of simplicity, we shall essentially concentrate in the following on systems of the form (1.1) and occasionally on systems of the form (1.17) since they cover a large range of physical situations.

2 Weak solutions of systems of conservation laws

Let us go back to the Cauchy problem (1.1), (1.2) for a general system of conservation laws. We shall say that a function $\mathbf{u} : \mathbb{R}^d \times [0, +\infty[\to \Omega$ is a *classical solution* of (1.1), (1.2) if $\mathbf{u}$ is a C^1 function that satisfies the equations (1.1), (1.2) pointwise.

An essential feature of this problem is that there do not exist in general classical solutions of (1.1), (1.2) beyond some finite time interval, even when the initial condition $\mathbf{u}_0$ is a very smooth function. Let us illustrate this fact by studying the simple case of a one-dimensional scalar equation.

2.1 Characteristics in the scalar one-dimensional case

Thus assume $p = d = 1$ and let $f : \mathbb{R} \to \mathbb{R}$ be a C^1 function. We consider the problem

$$\frac{\partial u}{\partial t} + \frac{\partial}{\partial x} f(u) = 0, \quad x \in \mathbb{R},\ t > 0, \tag{2.1}$$

$$u(x, 0) = u_0(x), \quad x \in \mathbb{R}, \tag{2.2}$$

and set

$$a(u) = f'(u). \tag{2.3}$$

Let u be a classical solution of (2.1) so that we can write (2.1) in *nonconservative* form

$$\frac{\partial u}{\partial t} + a(u)\frac{\partial u}{\partial x} = 0,$$

and let us introduce the *characteristic* curves associated with (2.1). They are defined as the integral curves of the differential equation

$$\frac{dx}{dt} = a(u(x(t), t)). \tag{2.4}$$

Proposition 2.1

Assume that u is a smooth solution of (2.1). The characteristic curves (2.4) are straight lines along which u is constant.

Proof. Consider a characteristic curve passing through the point $(x_0, 0)$, i.e., a solution of the ordinary differential system

$$\begin{cases} \dfrac{dx}{dt} = a(u(x(t), t)) \\ x(0) = x_0. \end{cases}$$

It exists at least on a small time interval $[0, t_0[$. Along such a curve, u is constant since

$$\begin{aligned} \frac{d}{dt} u(x(t), t) &= \frac{\partial u}{\partial t}(x(t), t) + \frac{\partial u}{\partial x}(x(t), t)\,\frac{dx}{dt}(t) \\ &= \Big(\frac{\partial u}{\partial t} + a(u)\frac{\partial u}{\partial x}\Big)(x(t), t) = 0. \end{aligned}$$

Therefore, it follows from (2.4) that the characteristic curves are straight lines whose constant slopes depend on the initial data, and the characteristic straight line passing through the point $(x_0, 0)$ is defined by the equation

$$x = x_0 + t\,a(u_0(x_0)). \tag{2.5}$$

This important property gives a way to construct smooth solutions.

One sets

$$u(x,t) = u_0(x_0), \tag{2.6}$$

where x_0 is solution of (2.5). This is the so-called *method of characteristics* (see example 2.1). □

Let us next assume that there exist two points $x_1 < x_2$ such that

$$m_1 = \frac{1}{a(u_0(x_1))} < m_2 = \frac{1}{a(u_0(x_2))}.$$

Then, the characteristics C_1 and C_2 drawn from the points $(x_1, 0)$ and $(x_2, 0)$, respectively, have slopes m_1 and m_2 and intersect necessarily at some point P.

At this point P, the solution u should take both values $u_0(x_1)$ and $u_0(x_2)$, which is clearly impossible. Hence, the solution u cannot be continuous at the point P. Note that this phenomenon is independent of the smoothness of the functions u_0 and f. Indeed, using (2.5), we see that the two characteristics intersect at time t if

$$t(a(u_0(x_1)) - a(u_0(x_2))) = x_2 - x_1.$$

Thus, unless the function $x \to a(u_0(x))$ is monotone increasing, in which case this equation has no positive solution t, we cannot define a classical solution u for all time $t > 0$ (see Figure 2.1). Moreover, one can determine the critical time T^* up to which a classical solution exists and can be constructed by the method of characteristics; T^* is given by

$$T^* = -\frac{1}{\min(\alpha, 0)}, \qquad \alpha = \min_{y \in \mathbb{R}} \frac{d}{dy}\, a(u_0(y)).$$

The multidimensional case will be considered in Chapter IV, Section 1.2.

Example 2.1. We want to solve the following Cauchy problem for Burgers' equation (1.4) with the initial condition

$$u(x,0) = u_0(x) = \begin{cases} 1, & x \le 0, \\ 1 - x, & 0 \le x \le 1, \\ 0, & x > 1. \end{cases}$$

By using the method of characteristics, we can solve up to the time when the characteristics intersect. We already know by (2.5) that the characteristic passing through the point $(x_0, 0)$ is given by

$$x = x(x_0, t) = x_0 + t u_0(x_0)$$

so that

$$x(x_0, t) = \begin{cases} x_0 + t, & x_0 \le 0, \\ x_0 + t(1 - x_0), & 0 \le x_0 \le 1, \\ x_0, & x_0 \ge 1. \end{cases}$$

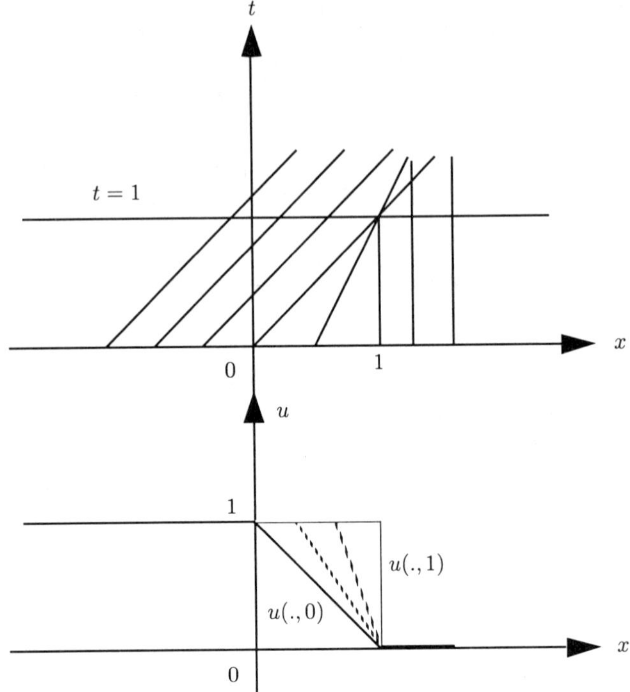

FIGURE 2.1. Method of characteristics for Burgers' equation.

For $t < 1$, the characteristics do not intersect (see Figure 2.1). Hence, given a point (x, t) with $t < 1$, we draw the (backward) characteristic passing through this point, and we determine the corresponding point x_0

$$x_0 = \begin{cases} x - t, & \text{if } x \leq t < 1, \\ \dfrac{1-x}{1-t}, & \text{if } t \leq x \leq 1, \\ 0, & \text{if } x \geq 1. \end{cases}$$

It consists of a front moving to the right and steepening until it becomes a "shock" (see Figure 2.1). This discontinuity corresponds to the fact that at time $t = 1$ the characteristics intersect. □

In short, using the method of characteristics (involving the implicit function theorem), one can prove that for u smooth enough a classical solution of (2.1), (2.2) exists in a small time interval. On the other hand, we have seen that in the nonlinear case $a'(u) \neq 0$ discontinuities may develop after a finite time. The above considerations lead us to introduce *weak solutions* (which are indeed weaker that the classical solutions!).

2.2 Weak solutions: the Rankine–Hugoniot condition

Consider the Cauchy problem (1.1), (1.2), and assume $\mathbf{u}_0 \in \mathbf{L}^\infty_{\text{loc}}(\mathbb{R}^d)^p$, where $\mathbf{L}^\infty_{\text{loc}}$ is the space of locally bounded measurable functions; we want to state precisely in which sense (1.1), (1.2) is to be taken. Let $\mathbf{C}^1_0(\mathbb{R}^d \times [0,+\infty[)$ denote the space of C^1 functions φ with compact support in $\mathbb{R}^d \times [0,+\infty[$ (which means that φ is the restriction to $\mathbb{R}^d \times [0,+\infty[$ of a C^1 function with compact support in an open set containing $\mathbb{R}^d \times [0,+\infty[$). We begin by noticing that if $\mathbf{u}$ is C^1 and $\varphi \in \mathbf{C}^1_0(\mathbb{R}^d \times [0,+\infty[)^p$, we obtain by Green's theorem (or integration by parts)

$$-\int_0^\infty \int_{\mathbb{R}^d} \Big\{ \frac{\partial \mathbf{u}}{\partial t} + \sum_{j=1}^d \frac{\partial}{\partial x_j} \mathbf{f}_j(\mathbf{u}) \Big\} \cdot \varphi \, d\mathbf{x}\, dt$$

$$= \int_0^\infty \int_{\mathbb{R}^d} \Big\{ \mathbf{u} \cdot \frac{\partial \varphi}{\partial t} + \sum_{j=1}^d \mathbf{f}_j(\mathbf{u}) \cdot \frac{\partial \varphi}{\partial x_j} \Big\} d\mathbf{x}\, dt + \int_{\mathbb{R}^d} \mathbf{u}(\mathbf{x},0) \cdot \varphi(\mathbf{x},0)\, d\mathbf{x},$$

where the dot $\cdot$ denotes the Euclidean inner product on $\mathbb{R}^p$. Thus, any *classical* solution $\mathbf{u}$ of (1.1), (1.2) satisfies $\forall \varphi \in \mathbf{C}^1_0(\mathbb{R}^d \times [0,+\infty[)^p$

$$\text{(2.7)} \quad \int_0^\infty \int_{\mathbb{R}^d} \Big\{ \mathbf{u} \cdot \frac{\partial \varphi}{\partial t} + \sum_{j=1}^d \mathbf{f}_j(\mathbf{u}) \cdot \frac{\partial \varphi}{\partial x_j} \Big\} + \int_{\mathbb{R}^d} \mathbf{u}_0(\mathbf{x}) \cdot \varphi(\mathbf{x},0)\, d\mathbf{x} = 0.$$

Next, we remark that (2.7) makes sense if $\mathbf{u} \in \mathbf{L}^\infty_{\text{loc}}(\mathbb{R}^d \times [0,+\infty[)^p$. Hence, we introduce the following definition.

Definition 2.1

Assume that $\mathbf{u}_0 \in \mathbf{L}^\infty_{\text{loc}}(\mathbb{R}^d)^p$. A function $\mathbf{u} \in \mathbf{L}^\infty_{\text{loc}}(\mathbb{R}^d \times [0,+\infty[)^p$ is called a weak solution of the Cauchy problem (1.1), (1.2) if $\mathbf{u}(\mathbf{x},t) \in \Omega$ a.e. and satisfies (2.7) for any function $\varphi \in \mathbf{C}^1_0(\mathbb{R}^d \times [0,+\infty[)^p$.

By construction, a classical solution of problem (1.1), (1.2) is also a weak solution. Conversely, by choosing φ in $\mathbf{C}^\infty_0(\mathbb{R}^d \times]0,\infty[)^p$, where $\mathbf{C}^\infty_0(\mathbb{R}^d \times]0,+\infty[)$ is the space of C^∞ functions with compact support in $\mathbb{R}^d \times]0,\infty[$, we obtain that any weak solution $\mathbf{u}$ satisfies (1.1) in the sense of distributions on $\mathbb{R}^d \times]0,\infty[$. Moreover, if $\mathbf{u}$ happens to be a C^1 function, then it is a classical solution. Indeed, let $\varphi \in \mathbf{C}^1_0(\mathbb{R}^d \times]0,+\infty[)^p$; integrating (2.7) by parts gives

$$\int_0^\infty \int_{\mathbb{R}^d} \Big\{ \frac{\partial \mathbf{u}}{\partial t} + \sum_{j=1}^d \frac{\partial}{\partial x_j} \mathbf{f}_j(\mathbf{u}) \Big\} \cdot \varphi \, d\mathbf{x}\, dt = 0,$$

so that (1.1) holds pointwise.

Next, if we multiply (1.1) by a test function $\varphi \in \mathbf{C}^1_0(\mathbb{R}^d \times [0,+\infty[)^p$, integrate by parts, and compare with (2.7), we obtain

$$\int_{\mathbb{R}^d} (\mathbf{u}(\mathbf{x},0) - \mathbf{u}_0(\mathbf{x})) \cdot \varphi(\mathbf{x},0)\, d\mathbf{x} = 0,$$

which yields (1.2) pointwise.

We shall now consider solutions of (1.1) in the sense of distributions that are only piecewise smooth and therefore admit discontinuities.

First, we notice that the above argument shows that any distributional solution $\mathbf{u}$ is a classical solution of (1.1) in any domain where $\mathbf{u}$ is C^1. We restrict the study to a particular type of discontinuous function; in fact, this is not restrictive for the examples that we shall encounter (if one regards the structure of BV functions, this simplification is not too unreasonable, see for instance Di Perna (1979)). For the sake of brevity, we say that a function $\mathbf{u}$ is "piecewise C^1" if there exists a finite number of smooth orientable surfaces Σ in the $(t, \mathbf{x})$-space outside of which $\mathbf{u}$ is a C^1 function and across which $\mathbf{u}$ has a jump discontinuity. Given a surface of discontinuity Σ, we denote by $\mathbf{n} = (n_t, n_{x_1}, n_{x_2}, ..., n_{x_d})^T (\neq \mathbf{0})$ a normal vector to Σ and by $\mathbf{u}_+$ and $\mathbf{u}_-$ the limits of $\mathbf{u}$ on each side of Σ,

$$\mathbf{u}_\pm(\mathbf{x}, t) = \lim_{\varepsilon \to 0, \varepsilon > 0} \mathbf{u}((\mathbf{x}, t) \pm \varepsilon \mathbf{n}).$$

In the one-dimensional case, we assume in addition that any line of discontinuity Σ has a parametrization of the form $(t, \xi(t))$, where $\xi : t \to \xi(t)$ is a C^1 function from some time interval (t_1, t_2) into $\mathbb{R}$, and we set in the same way

$$\mathbf{u}_\pm(\xi(t), t) = \lim_{\varepsilon \to 0} \mathbf{u}((\xi(t), t) \pm \varepsilon \mathbf{n}).$$

We now show that even in the frame of "piecewise C^1" functions not every discontinuity is admissible. The values of $\mathbf{u}$ and $\mathbf{f}(\mathbf{u})$ on each side of Σ are linked by a relation that is hidden in the equations, as results from the following theorem.

Theorem 2.1

Let $\mathbf{u} : \mathbb{R}^d \times [0, +\infty[\to \Omega$ *be a piecewise* C^1 *function (in the above sense). Then,* $\mathbf{u}$ *is a solution of (1.1) in the sense of distributions on* $\mathbb{R}^d \times]0, +\infty[$ *if and only if the two following conditions are satisfied:*
(i) $\mathbf{u}$ *is a classical solution of (1.1) in the domains where* $\mathbf{u}$ *is* C^1*;*
(ii) $\mathbf{u}$ *satisfies the jump condition*

$$(\mathbf{u}_+ - \mathbf{u}_-) n_t + \sum_{j=1}^{d} (\mathbf{f}_j(\mathbf{u}_+) - \mathbf{f}_j(\mathbf{u}_-)) \, n_{x_j} = \mathbf{0} \tag{2.8}$$

along the surfaces of discontinuity.

The jump relation (2.8) is known as the *Rankine–Hugoniot condition.*

Proof. Suppose that $\mathbf{u}$ is a piecewise C^1 function solution of (1.1) in the sense of distributions on $\mathbb{R}^d \times]0, +\infty[$. We have already observed that $\mathbf{u}$ satisfies property (i).

Now, let Σ be a surface of discontinuity of $\mathbf{u}$, M a point of Σ, and D a small ball centered at M (for simplicity, we assume that $\Sigma \cap D$ is the only surface of discontinuity of $\mathbf{u}$ in D). We denote by $D_\pm$ the two open components of D on each side of Σ. Let $\varphi \in \mathbf{C}_0^\infty(D)^p$. We write

$$0 = \int_D \left\{\mathbf{u} \cdot \frac{\partial \varphi}{\partial t} + \sum_{j=1}^d \mathbf{f}_j(\mathbf{u}) \cdot \frac{\partial \varphi}{\partial x_j}\right\} d\mathbf{x}\, dt = \int_{D_+} + \int_{D_-}.$$

Suppose for instance that the normal vector $\mathbf{n}$ to the surface Σ points in the direction of D_+. Then, applying Green's formula in D_+ and D_- gives

$$\begin{aligned} 0 = &- \int_{D_+} \left\{\frac{\partial \mathbf{u}}{\partial t} + \sum_{j=1}^d \frac{\partial}{\partial x_j} \mathbf{f}_j(\mathbf{u})\right\} \cdot \varphi \, d\mathbf{x}\, dt \\ &- \int_{\Sigma \cap D} \{n_t\, \mathbf{u}_+ + \sum_{j=1}^d n_{n_j}\, \mathbf{f}_j(\mathbf{u}_+)\} \cdot \varphi \, dS \\ &- \int_{D_-} \left\{\frac{\partial \mathbf{u}}{\partial t} + \sum_{j=1}^d \frac{\partial}{\partial x_j} \mathbf{f}_j(\mathbf{u})\right\} \cdot \varphi \, d\mathbf{x}\, dt \\ &+ \int_{\Sigma \cap D} \{n_t\, \mathbf{u}_- + \sum_{j=1}^d n_{x_j}\, \mathbf{f}_j(\mathbf{u}_-\} \cdot \varphi \, dS. \end{aligned}$$

Since $\mathbf{u}$ is a classical solution of (1.1) in D_+ and D_-, the first and third integrals vanish; we obtain

$$\int_{\Sigma \cap D} \{-n_t(\mathbf{u}_+ - \mathbf{u}_-) - \sum_{j=1}^d n_{x_j}(\mathbf{f}_j(\mathbf{u}_+) - \mathbf{f}_j(\mathbf{u}_-))\} \cdot \varphi \, dS = 0.$$

This holds for arbitrary φ, and hence we obtain the jump relation (2.8) at the point M.

On the other hand, if $\mathbf{u}$ is a piecewise C^1 function that satisfies properties (i) and (ii), it is a simple matter to check that $\mathbf{u}$ is indeed a distributional solution of (1.1). □

Denote by

$$[\mathbf{u}] = \mathbf{u}_+ - \mathbf{u}_- \tag{2.9}$$

the *jump* of $\mathbf{u}$ across Σ and similarly by

$$[\mathbf{f}_j(\mathbf{u})] = \mathbf{f}_j(\mathbf{u}_+) - \mathbf{f}_j(\mathbf{u}_-)$$

the jump of $\mathbf{f}_j(\mathbf{u})$, $1 \le j \le d$; then (2.8) can be written

$$n_t[\mathbf{u}] + \sum_{j=1}^d n_{x_j}[\mathbf{f}_j(\mathbf{u})] = \mathbf{0}. \tag{2.8}$$

If $(n_{x_1}, ..., n_{x_d}) \neq (0, ..., 0)$, we can take the normal vector in the form

$$\mathbf{n} = \begin{pmatrix} -s \\ \boldsymbol{\nu} \end{pmatrix},$$

where $s \in \mathbb{R}$ and $\boldsymbol{\nu} = (\nu_1, ..., \nu_d)^T$ is a unit vector in $\mathbb{R}^d$. Then (2.8) can be equivalently written

$$s[\mathbf{u}] = \sum_{j=1}^{d} \nu_j [\mathbf{f}_j(\mathbf{u})].$$

We recall here that Σ has a standard orientation determined by the choice of the canonical volume form on its tangent space and thus a canonical normal vector field associated to this orientation. Now, if Σ is oriented and $\frac{\mathbf{n}}{|\mathbf{n}|}$ is the outward unit normal vector to Σ, $\boldsymbol{\nu}$ and s may be interpreted respectively as the direction and the speed of propagation of the discontinuity. For instance, suppose that in the two-dimensional case ($d = 2$) Σ is a smooth surface in $\mathbb{R}^3$ that has a parametrization of the form $(t, x_1, x_2 = \xi(t, x_1))$. We have

$$\mathbf{n} = (n_t, n_{x_1}, n_{x_2})^T = \Big(1 + \Big(\frac{\partial \xi}{\partial x_1}\Big)^2\Big)^{-1/2} \Big(-\frac{\partial \xi}{\partial t}, -\frac{\partial \xi}{\partial x_1}, 1\Big)^T,$$

and $\boldsymbol{\nu}$ points in the direction of the positive x_2-axis.

In the one-dimensional case ($d = 1$) we have assumed that Σ is a smooth curve with parametrization $(t, \xi(t))$, and we have

$$\mathbf{n} = (-s, 1)^T, \quad s = \frac{d\xi}{dt}, \tag{2.10}$$

so that the Rankine–Hugoniot jump condition (2.8) becomes

$$s[\mathbf{u}] = [\mathbf{f}(\mathbf{u})]. \tag{2.11}$$

For a scalar equation, we obtain

$$s = \frac{[f(u)]}{[u]},$$

whereas for a system, (2.11) represents in fact p equations

$$s[u_i] = [f_i(u)], \quad 1 \leq i \leq p$$

in which the same s appears.

Remark 2.1. Note that Σ is traditionally represented in the (x,t)-space. Thus, in the figures, Figure 3.1 for instance, s is the reciprocal of the slope. The wave propagation is such that s is always finite. This remark is valid for systems too. □

Remark 2.2. If the function $\mathbf{u}$ is continuous, the Rankine–Hugoniot jump condition is automatically satisfied. In that case, it suffices to check that $\mathbf{u}$ satisfies (1.1) in the domains where $\mathbf{u}$ is C^1 to prove that $\mathbf{u}$ is indeed a distributional solution of (1.1). □

Next, we want to point out, again with a simple example, that a weak solution of the Cauchy problem (1.1), (1.2) is not necessarily unique.

2.3 Example of nonuniqueness of weak solutions

We consider the Riemann problem (1.3) for Burgers' equation (1.4), i.e.,

$$\tag{2.12} \begin{cases} \dfrac{\partial u}{\partial t} + \dfrac{\partial}{\partial x}\dfrac{u^2}{2} = 0 \\ u_0(x) = \begin{cases} u_\ell, & x < 0, \\ u_r, & x > 0. \end{cases} \end{cases}$$

If $u_\ell \neq u_r$, the Rankine–Hugoniot condition (2.11) shows that we obtain a weak solution of the problem by propagating the discontinuity at speed $s = \frac{1}{2}(u_\ell + u_r)$, which gives

$$\tag{2.13} u(x,t) = \begin{cases} u_\ell, & x < st \\ u_r, & x > st. \end{cases}$$

Now, let us check that there are many other weak solutions. In fact, let a be a constant such that $a \geq \max(u_\ell, -u_r)$. The function u defined by

$$u(x,t) = \begin{cases} u_\ell, & x < s_1 t \\ -a, & s_1 t < x < 0 \\ a, & 0 < x < s_2 t \\ u_r, & x > s_2 t \end{cases}$$

is also a weak solution if $s_1 = \frac{u_\ell - a}{2}$, $s_2 = \frac{u_r + a}{2}$, so that the Rankine–Hugoniot jump condition (2.11) is satisfied along each line of discontinuity of u. We thus obtain a one-parameter family of discontinuous weak solutions.

On the other hand, if we suppose $u_\ell \leq u_r$, we can also exhibit a continuous solution. In that case, the characteristics do not intersect (see Figure 2.2). Clearly, the method of characteristics enables us to determine the solution everywhere except in the region $u_\ell \leq \frac{x}{t} \leq u_r$. However, we notice that the function $v(x,t) = \frac{x}{t}$ is a classical solution of Burgers' equation since

$$\frac{\partial v}{\partial t} + v\frac{\partial v}{\partial t} = -\frac{x}{t^2} + \Big(\frac{x}{t}\Big)\Big(\frac{1}{t}\Big) = 0.$$

Hence, the continuous function

$$\tag{2.14} u(x,t) = \begin{cases} u_\ell, & x \leq u_\ell\, t \\ x/t, & u_\ell\, t \leq x \leq u_r\, t \\ u_r, & x \geq u_r\, t \end{cases}$$

is also a weak solution of problem (2.12) (see Figure 2.2).

Note that the function $\frac{x}{t}$ was not found by chance. For more general strictly convex fluxes f, assuming $u_\ell \leq u_r$, we can still find a self-similar

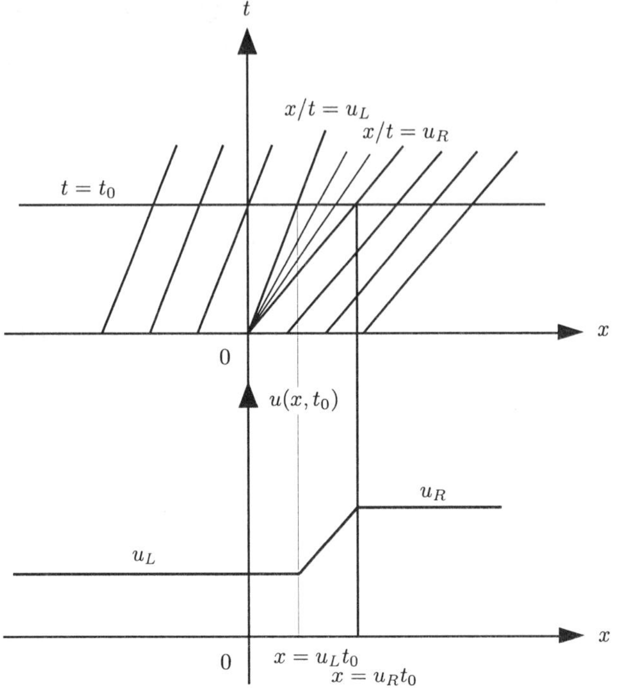

FIGURE 2.2. Example of rarefaction for Burgers' equation.

continuous solution of the Riemann problem (see an illustration in Figure 3.1(ii))

$$u(x,t) = v(\frac{x}{t}), \quad f'(u_\ell) \le \frac{x}{t} \le f'(u_r), \text{ where } f'(v(\xi)) = \xi.$$

Let us also note the obvious property that in the "rarefaction fan" $f'(u_\ell) \le \frac{x}{t} \le f'(u_r)$, where we cannot use the method of characteristics from the data at time 0, this solution (2.14) is smooth and constant on the lines $\frac{x}{t} = Cte$, which are characteristics for $t > 0$.

We have just noticed that a weak solution of the Cauchy problem (1.1), (1.2) is not necessarily unique. Hence, we need to find some criterion that enables us to choose the "physically relevant" solution among all the weak solutions of (1.1), (1.2). This criterion is based on the concept of *entropy* that we introduce now. In the above example, when $u_\ell \le u_r$, it happens that the "relevant" solution is the continuous one. One heuristic reason is that for the other discontinuous solutions we have added some arbitrary extra information a that is contained neither in the data u_0 nor in the equation itself.

3 Entropy solutions

3.1 A mathematical notion of entropy

Let us consider the following problem: given any smooth solution $\mathbf{u}$ of (1.1), we wonder whether $\mathbf{u}$ would satisfy an additional conservation law of the form

$$\frac{\partial}{\partial t} U(\mathbf{u}) + \sum_{j=1}^{d} \frac{\partial}{\partial x_j} F_j(\mathbf{u}) = 0, \tag{3.1}$$

where U and $F_j, 1 \leq j \leq d$, are sufficiently smooth functions from Ω into $\mathbb{R}$. We want to check that this is indeed the case if

$$U'(\mathbf{u})\mathbf{f}_j'(\mathbf{u}) = F_j'(\mathbf{u}), \quad 1 \leq j \leq d, \tag{3.2}$$

where, for ease of notation, we identify the linear forms $U'(\mathbf{u}), F_j'(\mathbf{u}) : \mathbb{R}^p \to \mathbb{R}$ with the corresponding row vectors

$$U' = \nabla U^T = \Big(\frac{\partial U}{\partial u_1}, \ldots, \frac{\partial U}{\partial u_p}\Big), \quad F_j' = \nabla F_j^T = \Big(\frac{\partial F_j}{\partial u_1}, \ldots, \frac{\partial F_j}{\partial u_p}\Big)$$

and the linear mapping $\mathbf{f}_j' : \mathbb{R}^p \to \mathbb{R}^p$ with the matrix

$$\mathbf{f}_j' = \mathbf{A}_j = \Big(\frac{\partial f_{ij}}{\partial u_k}\Big) \qquad 1 \leq i, k \leq p.$$

In fact, assuming that $\mathbf{u}$ is a classical solution of (1.1) and carrying out the differentiation, we obtain

$$U'(\mathbf{u})\Big(\frac{\partial \mathbf{u}}{\partial t} + \sum_{j=1}^{d} \mathbf{f}_j'(\mathbf{u}) \frac{\partial \mathbf{u}}{\partial x_j}\Big) = 0$$

and by (3.2)

$$U'(\mathbf{u}) \frac{\partial \mathbf{u}}{\partial t} + \sum_{j=1}^{d} F_j'(\mathbf{u}) \frac{\partial \mathbf{u}}{\partial x_j} = 0, \tag{3.3}$$

so that (3.1) follows.

Let us now define the notion of *entropy*. We shall restrict ourselves to convex entropy functions though one may consider a more general framework.

Definition 3.1

Assume that Ω is convex. Then, a convex function $U : \Omega \to \mathbb{R}$ is called an entropy for the system of conservation laws (1.1) if there exist d functions $F_j : \Omega \to \mathbb{R}, 1 \leq j \leq d$, called entropy fluxes, such that the relations (3.2) hold.

Hence, if (3.2) holds, any classical solution of (1.1) satisfies the additional conservation law (3.1). On the other hand, this is not true in general of a weak solution and in particular of a piecewise C^1 weak solution. Such a weak solution $\mathbf{u}$ must satisfy the Rankine–Hugoniot condition (2.8) along the surfaces of discontinuity, whereas a solution of (3.1) should satisfy the corresponding jump condition

$$n_t[U(\mathbf{u})] + \sum_{j=1}^{d} n_{x_j}[F_j(\mathbf{u})] = 0, \tag{3.4}$$

which is in general incompatible with (2.8). We shall see below that, for an entropy solution, this last jump condition should be replaced by a jump inequality.

The problem is to find entropy functions and if possible all entropy functions associated with the nonlinear system of conservation laws (1.1). This is easy for a scalar conservation law ($p = 1$) since in that case any convex function U is an entropy. In fact, it suffices to take for F_j a primitive of the function $U' \, f_j'$.

In the general case $p > 1$, finding the entropy functions is a much more difficult problem. Note that equations (3.2) can be written in the form of a system of $p \times d$ linear partial differential equations of the first order in the $(d+1)$ unknown functions $U, F_j, 1 \leq j \leq d$, namely

$$\sum_{i=1}^{p} \frac{\partial f_{ij}}{\partial u_k} \frac{\partial U}{\partial u_i} - \frac{\partial F_j}{\partial u_k} = 0, \quad 1 \leq j \leq d, \ 1 \leq k \leq p. \tag{3.5}$$

Except in the case $p = 2, d = 1$ (which is treated for instance in Serre (1987)), the existence of entropy functions is a special property of the system. However, in all the practical examples derived from mechanics or physics, we are able to find an entropy function that has a physical meaning. A classification of hyperbolic systems ($d = 1$) with respect to their entropies is given in Serre (1989).

Example 3.1. The p-system. We consider again the p-system (1.6), where we assume $p'(v) < 0$. Let P be a primitive of p. By multiplying the first equation (1.6) by $-p(v)$, the second equation (1.6) by u, and adding, we obtain

$$\frac{\partial}{\partial t}\left(\frac{u^2}{2} - P(v)\right) + \frac{\partial}{\partial x}(p(v)u) = 0.$$

Therefore, the pair (U, F) defined by

$$U(v,u) = \frac{u^2}{2} - P(v), \quad F(v,u) = p(v)u$$

is a pair of an entropy function and an entropy flux for the p-system. Note that the Hessian matrix of U, given by

$$U''(v,u) = \begin{pmatrix} -p'(v) & 0 \\ 0 & 1 \end{pmatrix},$$

is positive-definite so that U is strictly convex. □

Example 3.2. Symmetric systems. Let us go back to the general situation (1.1) but assume that the $p \times p$ matrices $\mathbf{A}_j = \mathbf{f}'_j(\mathbf{u})$ are symmetric, i.e.,

$$\frac{\partial f_{ij}}{\partial u_k} = \frac{\partial f_{kj}}{\partial u_i}, \quad 1 \leq i, k \leq p.$$

The above relations are exactly the compatibility conditions for the existence of a function $g_j = g_j(\mathbf{u})$ such that

$$\frac{\partial g_j}{\partial u_i} = f_{ij}.$$

Next, we observe that the following strictly convex function

$$U(\mathbf{u}) = \frac{1}{2} \sum_{i=1}^{p} u_i^2$$

is an entropy for the symmetric system (1.1) associated with the entropy fluxes

$$F_j(\mathbf{u}) = \sum_{i=1}^{p} u_i \, f_{ij}(\mathbf{u}) - g_j(\mathbf{u}).$$

Indeed, since

$$\frac{\partial}{\partial x_j} g_j(\mathbf{u}) = \sum_{i=1}^{p} \frac{\partial g_j}{\partial u_i}(\mathbf{u}) \frac{\partial u_i}{\partial x_j} = \sum_{i=1}^{p} f_{ij}(\mathbf{u}) \frac{\partial u_i}{\partial x_j},$$

we have

$$\begin{aligned} \frac{\partial}{\partial x_j} F_j(\mathbf{u}) &= \sum_{i=1}^{p} u_i \frac{\partial}{x_j} (f_{ij}(\mathbf{u})) + \sum_{i=1}^{p} f_{ij}(u) \frac{\partial u_i}{\partial x_j} - \sum_{i=1}^{p} f_{ij}(\mathbf{u}) \frac{\partial u_i}{\partial x_j} \\ &= \sum_{i=1}^{p} u_i \frac{\partial}{\partial x_j} (f_{ij}(\mathbf{u})) \end{aligned}$$

so that

$$\frac{\partial}{\partial t} U(\mathbf{u}) + \sum_{j=1}^{d} \frac{\partial}{\partial x_j} F_j(\mathbf{u}) = \sum_{i=1}^{p} u_i \Big\{ \frac{\partial u_i}{\partial t} + \sum_{j=1}^{d} \frac{\partial}{\partial x_j} (f_{ij}(\mathbf{u})) \Big\} = 0.$$

We shall now prove that the situation of Example 3.2 is in fact almost general. □

Let us see that a nonlinear system of conservation laws that admits a strictly convex entropy is *symmetrizable.* More precisely, we can state the following theorem.

Theorem 3.1

Let $U : \Omega \to \mathbb{R}$ be a strictly convex function. A necessary and sufficient condition for U to be an entropy for the system (1.1) is that the $(p \times p)$ matrices $U''(\mathbf{u})\mathbf{f}'_j(\mathbf{u}), 1 \leq j \leq d$, are symmetric.

Proof. Since the function U is strictly convex, its Hessian $U''(\mathbf{u})$ is a symmetric positive-definite matrix. Now, we assume that U is an entropy so that the conditions (3.5) hold. Then, differentiating (3.5) with respect to u_ℓ gives

$$\frac{\partial^2 F_j}{\partial u_k \partial u_\ell} - \sum_{i=1}^{p} \frac{\partial^2 f_{ij}}{\partial u_k \partial u_\ell} \frac{\partial U}{\partial u_i} = \sum_{i=1}^{p} \frac{\partial f_{ij}}{\partial u_k} \frac{\partial^2 U}{\partial u_i \partial u_\ell}, \quad 1 \leq k, \ell \leq p.$$

Since the left-hand side member of the above equation is symmetric in k and ℓ, the same is true of the right-hand side, i.e.,

$$\sum_{i=1}^{p} \frac{\partial f_{ij}}{\partial u_k} \frac{\partial^2 U}{\partial u_i \partial u_\ell} = \sum_{i=1}^{p} \frac{\partial f_{ij}}{\partial u_\ell} \frac{\partial^2 U}{\partial u_i \partial u_k}, \quad 1 \leq k, \ell \leq p. \tag{3.6}$$

This means exactly that the matrix $U''(\mathbf{u})\mathbf{f}'_j(\mathbf{u})$ is symmetric, $1 \leq j \leq d$.

Conversely, assume that conditions (3.6) hold. Using

$$\frac{\partial}{\partial u_\ell}\Big(\sum_{i=1}^{p} \frac{\partial f_{ij}}{\partial u_k} \frac{\partial U}{\partial u_i}\Big) = \sum_{i=1}^{p}\Big(\frac{\partial f_{ij}}{\partial u_k} \frac{\partial^2 U}{\partial u_i \partial u_\ell} + \frac{\partial^2 f_{ij}}{\partial u_k \partial u_\ell} \frac{\partial U}{\partial u_i}\Big),$$

we obtain

$$\frac{\partial}{\partial u_\ell}\Big(\sum_{i=1}^{p} \frac{\partial f_{ij}}{\partial u_k} \frac{\partial U}{\partial u_i}\Big) = \frac{\partial}{\partial u_k}\Big(\sum_{i=1}^{p} \frac{\partial f_{ij}}{\partial u_\ell} \frac{\partial U}{\partial u_i}\Big), \quad 1 \leq k, \ell \leq p.$$

These relations are the compatibility conditions that ensure the existence of a function F_j such that

$$\frac{\partial F_j}{\partial u_k} = \sum_{i=1}^{p} \frac{\partial f_{ij}}{\partial u_k} \frac{\partial U}{\partial u_i}.$$

Hence, it follows from (3.5) that U is an entropy function associated with the entropy fluxes $F_j, 1 \leq j \leq d$. □

As a corollary of Theorem 3.1, the existence of a strictly convex entropy U implies that the system (1.1) is symmetrizable. In fact, premultiplication by $U''(u)$ gives the system

$$U''(\mathbf{u}) \frac{\partial \mathbf{u}}{\partial t} + \sum_{j=1}^{d} U''(\mathbf{u})\mathbf{f}'_j(\mathbf{u}) \frac{\partial \mathbf{u}}{\partial x_j} = \mathbf{0},$$

where the matrix $U''(\mathbf{u})$ is symmetric and positive-definite, and the matrices $U''(\mathbf{u})\mathbf{f}'_j(\mathbf{u})$ are symmetric.

The symmetrization of (1.1) can also be accomplished by introducing new dependent variables $\mathbf{v}$, i.e., by setting $\mathbf{u} = \mathbf{u}(\mathbf{v})$. Equation (1.1) then becomes

$$\mathbf{u}'(\mathbf{v}) \frac{\partial \mathbf{v}}{\partial t} + \sum_{j=1}^{d} \mathbf{f}_j'(\mathbf{u}) \, \mathbf{u}'(\mathbf{v}) \frac{\partial \mathbf{v}}{\partial x_j} = \mathbf{0}$$

so that (1.1) is symmetrized if again $\mathbf{u}'(\mathbf{v})$ is a symmetric positive-definite matrix and the matrices $\mathbf{f}_j'(\mathbf{u}) \, \mathbf{u}'(\mathbf{v})$ are symmetric.

Theorem 3.2

A necessary and sufficient condition for the system (1.1) to possess a strictly convex entropy U is that there exists a change of dependent variables $\mathbf{u} = \mathbf{u}(\mathbf{v})$ that symmetrizes (1.1).

Proof. Suppose that (1.1) is symmetrized by the change of variables $\mathbf{u} = \mathbf{u}(\mathbf{v})$. The symmetry of the matrix

$$\mathbf{u}'(\mathbf{v}) = \left(\frac{\partial u_i}{\partial v_k} \right)_{1 \le i,k \le p}$$

implies the existence of a function $q(\mathbf{v})$ such that

$$q'(\mathbf{v}) = \mathbf{u}(\mathbf{v})^T.$$

Similarly, the symmetry of the matrices $\mathbf{f}_j'(\mathbf{u}(\mathbf{v})) \, \mathbf{u}'(\mathbf{v})$, $1 \le j \le d$, implies the existence of functions $p_j(\mathbf{v})$ such that

$$p_j'(\mathbf{v}) = \mathbf{f}_j(\mathbf{u}(\mathbf{v}))^T, \quad 1 \le j \le d.$$

The positive-definiteness of $\mathbf{u}'(\mathbf{v})$ is equivalent to the strict convexity of the function $q(\mathbf{v})$, which implies in turn that the mapping $\mathbf{v} \to q'(\mathbf{v})$ is one-to-one. Hence $\mathbf{v}$ can be regarded as a function of $\mathbf{u}$. We set

$$U(\mathbf{u}) = \mathbf{v}(\mathbf{u})^T \, \mathbf{u} - q(\mathbf{v}(\mathbf{u})), \quad F_j(\mathbf{u}) = \mathbf{v}(\mathbf{u})^T \, \mathbf{f}_j(\mathbf{u}) - p_j(v(\mathbf{u})), \ 1 \le j \le p.$$

By differentiating U and F_j, we obtain

$$U'(\mathbf{u}) = \mathbf{v}(\mathbf{u})^T + \mathbf{u}^T \, \mathbf{v}'(\mathbf{u}) - q'(\mathbf{v}(\mathbf{u}))\mathbf{v}'(\mathbf{u}) = \mathbf{v}(\mathbf{u})^T$$

and

$$F_j'(\mathbf{u}) = \mathbf{v}(\mathbf{u})^T \, \mathbf{f}_j'(\mathbf{u}) + \mathbf{f}_j(\mathbf{u})^T \, \mathbf{v}'(\mathbf{u}) - p_j'(\mathbf{v}(\mathbf{u}))\mathbf{v}'(\mathbf{u}) = \mathbf{v}(\mathbf{u})^T \, \mathbf{f}_j'(\mathbf{u})$$

so that the relations (3.2) hold. Moreover, we have

$$U''(\mathbf{u}(\mathbf{v})) \cdot \mathbf{u}'(\mathbf{v}) = \mathbf{I},$$

and therefore the matrix

$$U''(\mathbf{u}(\mathbf{v})) = \mathbf{u}'(\mathbf{v})^{-1}$$

is positive-definite. This proves that U is a strictly convex entropy.

Conversely, if U is a strictly convex entropy, the mapping $\mathbf{u} \to U'(\mathbf{u})$ is one-to-one. Hence, we can define the change of variables

$$\mathbf{v}^T = U'(\mathbf{u})$$

(*entropy* variables). Next, we define the conjugate, or polar, functions $U^\star$ and $F_j^\star$ of U and F_j, $1 \le j \le d$ by

$$U^\star(\mathbf{v}) = \mathbf{v}^T\,\mathbf{u}(\mathbf{v}) - U(\mathbf{u}(\mathbf{v})),\ \ F_j^\star(\mathbf{v}) = \mathbf{v}^T\,\mathbf{f}_j(\mathbf{u}(\mathbf{v})) - F_j(\mathbf{u}(\mathbf{v})).$$

Differentiating with respect to $\mathbf{v}$ gives

$$U^{\star'}(\mathbf{v}) = \mathbf{u}(\mathbf{v})^T - \mathbf{v}^T\,\mathbf{u}'(\mathbf{v}) - U'(\mathbf{u}(\mathbf{v}))\mathbf{u}'(\mathbf{v}) = \mathbf{u}(\mathbf{v})^T$$

and

$$F_j^{\star'}(\mathbf{v}) = \mathbf{f}_j(\mathbf{u}(\mathbf{v}))^T + \mathbf{v}^T\,\mathbf{f}_j'(\mathbf{u}(\mathbf{v}))\mathbf{u}'(\mathbf{v}) - F_j'(\mathbf{u}(\mathbf{v}))\mathbf{u}'(\mathbf{v}) = \mathbf{f}_j(\mathbf{u}(\mathbf{v})).$$

Hence, the matrices $\mathbf{u}'(\mathbf{v}) = U^{\star''}(\mathbf{v})$ and $\mathbf{f}_j'(\mathbf{u}(\mathbf{v}))\mathbf{u}'(\mathbf{v}) = F_j^{\star''}(\mathbf{v}), 1 \le j \le d$, are symmetric. Moreover, the matrix

$$\mathbf{u}'(\mathbf{v}) = U''(\mathbf{u}(\mathbf{v}))^{-1}$$

is positive-definite, which proves that the change of variables $\mathbf{v}^T = U'(\mathbf{u})$ symmetrizes (1.1). □

Example 3.3. Gas dynamics equations. Let us consider again the gas dynamics equations in Eulerian coordinates (Example 1.3). We define the (thermodynamic) specific entropy s by

$$T\,ds = d\varepsilon + p\,d\tau, \quad \tau = \frac{1}{\rho},$$

where $T = T(\rho, \varepsilon)$ is the temperature. We shall check in Chapter II, Section 1.1, that $U = -\rho s$ is indeed a strictly convex entropy function (of the conservative variables (ρ, q_1, q_2, q_3, E)) associated with the entropy fluxes

$$F_i = -\rho\,u_i\,s, \quad 1 \le i \le 3.$$

In fact, for a polytropic ideal gas, we have

$$T = \frac{\varepsilon}{C_v},$$

where the constant C_v is the specific heat at constant volume (see Chapter II, Section 1.2). Since $p = (\gamma - 1)\rho\,\varepsilon$, we obtain in this case

$$ds = C_v\Big(\frac{d\varepsilon}{\varepsilon} - \frac{p}{\varepsilon}\frac{d\rho}{\rho^2}\Big) = C_v\Big(\frac{d\varepsilon}{\varepsilon} - (\gamma - 1)\frac{d\rho}{\rho}\Big),$$

so that

$$s = s_0 + C_v\,\mathrm{Log}\Big(\frac{\varepsilon}{\rho^{\gamma-1}}\Big),$$

where s_0 is an arbitrary constant. This can be equivalently written

$$s = s_0 + C_v\Big(\mathrm{Log}(E - \frac{|\mathbf{q}|^2}{2\rho}) - \gamma \,\mathrm{Log}\,\rho\Big),$$

where $|\mathbf{q}|^2 = q_1^2 + q_2^2 + q_3^2$, and it is an easy but lengthy matter to check directly that the function

$$(\rho, q_1, q_2, q_3, E) \to -\rho\Big(\mathrm{Log}(E - \frac{|\mathbf{q}|^2}{2\rho}) - \gamma \,\mathrm{Log}\,\rho\Big)$$

is strictly convex. □

3.2 The vanishing viscosity method

The concept of entropy will enable us to select among the weak solutions of (1.1), (1.2) the physically relevant solution. To make this point clear, we begin by introducing a viscous perturbation of the nonlinear system of conservation laws. Given a small parameter $\varepsilon > 0$, we associate with (1.1) the parabolic system

$$\frac{\partial \mathbf{u}_\varepsilon}{\partial t} + \sum_{j=1}^{d} \frac{\partial}{\partial x_j} \mathbf{f}_j(\mathbf{u}_\varepsilon) - \varepsilon\, \Delta \mathbf{u}_\varepsilon = \mathbf{0}, \tag{3.7}$$

where $-\varepsilon\, \Delta \mathbf{u}_\varepsilon$ can be viewed as a viscosity term. Now, assuming that the systems (3.7) (with $\mathbf{u}_\varepsilon(\mathbf{x}, 0) = \mathbf{u}_{0\varepsilon}(\mathbf{x}) \to \mathbf{u}_0(\mathbf{x})$ as $\varepsilon \to 0$) have sufficiently smooth solutions $\mathbf{u}_\varepsilon$ (see Goodman and Xin (1992) and Gisclon (1994) for the existence of $\mathbf{u}_\varepsilon$), we want to recover solutions of (1.1) as the limits as $\varepsilon \to 0$ of solutions of (3.7). In that direction, we can state the following theorem.

Theorem 3.3

Assume that (1.1) admits an entropy U with entropy fluxes $F_j, 1 \le j \le d$. Let $(\mathbf{u}_\varepsilon)_\varepsilon$ be a sequence of sufficiently smooth solutions of (3.7) such that

$$\|\mathbf{u}_\varepsilon\|_{\mathbf{L}^\infty(\mathbb{R}^d \times [0,+\infty[)^p} \le C, \tag{3.8}$$

$$\mathbf{u}_\varepsilon \to \mathbf{u} \quad \text{as } \varepsilon \to 0 \quad \text{a.e. in} \quad \mathbb{R}^d \times [0, +\infty[, \tag{3.9}$$

where $C > 0$ is a constant independent of ε. Then $\mathbf{u}$ is a weak solution of (1.1) and satisfies the entropy condition

$$\frac{\partial}{\partial t} U(\mathbf{u}) + \sum_{j=1}^{d} \frac{\partial}{\partial x_j} F_j(\mathbf{u}) \le 0 \tag{3.10}$$

in the sense of distributions on $\mathbb{R}^d \times]0, +\infty[$.

The inequality (3.10) means that for any function $\varphi \in \mathbf{C}_0^\infty(\mathbb{R}^d\times]0,+\infty[)$, $\varphi \geq 0$, we have

$$\int_0^\infty \int_{\mathbb{R}^d} \Big(U(\mathbf{u}) \frac{\partial \varphi}{\partial t} + \sum_{j=1}^d F_j(\mathbf{u}) \frac{\partial \varphi}{\partial x_j} \Big) d\mathbf{x}\, dt \geq 0. \tag{3.11}$$

Proof. Let U be a C^2 entropy function; by applying $U'(\mathbf{u}_\varepsilon)$ to (3.7) and taking into account the relations (3.2), we obtain

$$\frac{\partial}{\partial t} U(\mathbf{u}_\varepsilon) + \sum_{j=1}^d \frac{\partial}{\partial x_j} F_j(\mathbf{u}_\varepsilon) = \varepsilon\, U'(\mathbf{u}_\varepsilon)\, \Delta \mathbf{u}_\varepsilon.$$

We rewrite the right-hand side of the above equation in the following way:

$$\varepsilon\, U'(\mathbf{u}_\varepsilon)\, \Delta \mathbf{u}_\varepsilon = \varepsilon\, \Delta U(\mathbf{u}_\varepsilon) - \varepsilon \sum_{j=1}^d \Big(\frac{\partial \mathbf{u}_\varepsilon}{\partial x_j} \Big)^T U''(\mathbf{u}_\varepsilon) \frac{\partial \mathbf{u}_\varepsilon}{\partial x_j},$$

so that by the convexity of U

$$\varepsilon\, U'(\mathbf{u}_\varepsilon) \Delta \mathbf{u}_\varepsilon \leq \varepsilon\, \Delta U(\mathbf{u}_\varepsilon)$$

and therefore

$$\frac{\partial}{\partial t} U(\mathbf{u}_\varepsilon) + \sum_{j=1}^d \frac{\partial}{\partial x_j} F_j(\mathbf{u}_\varepsilon) \leq \varepsilon\, \Delta U(\mathbf{u}_\varepsilon). \tag{3.12}$$

Now, suppose that $(\mathbf{u}_\varepsilon)_\varepsilon$ is a sequence of smooth solutions of (3.7) that satisfies the properties (3.8) and (3.9). Then

$$\mathbf{u}_\varepsilon \to \mathbf{u} \text{ as } \varepsilon \to 0 \quad \text{in } \mathcal{D}'(\mathbb{R}^d\times]0,+\infty[)^p$$

i.e., in the sense of distributions on $\mathbb{R}^d\times]0,+\infty[$ so that

$$\frac{\partial \mathbf{u}_\varepsilon}{\partial t} \to \frac{\partial \mathbf{u}}{\partial t}, \quad \varepsilon\, \Delta \mathbf{u}_\varepsilon \to 0 \quad \text{in } \mathcal{D}'(\mathbb{R}^d\times]0,+\infty[)^p.$$

Next, using (3.8) and the Lebesgue dominated convergence theorem, we have

$$\mathbf{f}_j(\mathbf{u}_\varepsilon) \to \mathbf{f}_j(\mathbf{u}) \quad \text{in } \mathbf{L}^1_{\text{loc}}(\mathbb{R}^d\times]0,+\infty[)^p.$$

Hence, by passing to the limit in (3.7), we obtain that $\mathbf{u}$ is a solution of (1.1) in the sense of distributions on $\mathbb{R}^d\times]0,+\infty[$.

Similarly, we have

$$\frac{\partial}{\partial t} U(\mathbf{u}_\varepsilon) \to \frac{\partial}{\partial t} U(\mathbf{u}), \quad \frac{\partial}{\partial x_j} F_j(\mathbf{u}_\varepsilon) \to \frac{\partial}{\partial x_j} F_j(\mathbf{u}), \quad \varepsilon \Delta U(\mathbf{u}_\varepsilon) \to 0$$

in $\mathcal{D}'(\mathbb{R}^d\times]0,+\infty[)$.

Therefore, passing to the limit in (3.12) gives

$$\frac{\partial}{\partial t} U(\mathbf{u}) + \sum_{j=1}^d \frac{\partial}{\partial x_j} F_j(\mathbf{u}) \leq 0 \quad \text{in } \mathcal{D}'(\mathbb{R}^d\times]0,+\infty[),$$

which gives the result. □

Therefore, the distributional solutions of (1.1) constructed by the method of vanishing viscosity satisfy the entropy condition (3.10) for all entropy functions U. This leads us to introduce the definition of an *entropy solution.*

Definition 3.2

A weak solution $\mathbf{u}$ *of (1.1), (1.2) is called an entropy solution if* $\mathbf{u}$ *satisfies, for all entropy functions* U *of (1.1) and for all test functions* $\varphi \in \mathbf{C}_0^1(\mathbb{R}^d \times [0,\infty[)$, $\varphi \geq 0$,

$$(3.13) \qquad \int_0^\infty \int_{\mathbb{R}^d} \Big(U(\mathbf{u}) \frac{\partial \varphi}{\partial t} + \sum_{j=1}^d F_j(\mathbf{u}) \frac{\partial \varphi}{\partial x_j} \Big) d\mathbf{x}\, dt + \int_{\mathbb{R}^d} U(\mathbf{u}_0(\mathbf{x}))\varphi(\mathbf{x},0) d\mathbf{x} \geq 0.$$

By arguing as in the proof of Theorem 2.1, it is a simple matter to check that a piecewise C^1 function $\mathbf{u}$ is an entropy solution of (1.1), (1.2) if
(i) $\mathbf{u}$ is a classical solution of (1.1) in the domains where $\mathbf{u}$ is C^1 and satisfies (1.2) a.e.,
(ii) $\mathbf{u}$ satisfies the Rankine–Hugoniot condition (2.8),
(iii) $\mathbf{u}$ satisfies the jump inequality

$$(3.14) \qquad n_t[U(\mathbf{u})] + \sum_{j=1}^d n_{x_j}[F_j(\mathbf{u})] \leq 0$$

along the surfaces of discontinuity, if $\frac{\mathbf{n}}{|\mathbf{n}|}$ is the outward (unit) normal vector pointing in the D_+ direction.

Note that in the one-dimensional case, (3.14) becomes

$$(3.15) \qquad s[U(\mathbf{u})] \geq [F(\mathbf{u})],$$

where s is again given by (2.10).

We shall encounter in Chapter I (Section 5) other criteria (which, however, coincide in the most usual cases) for selecting admissible solutions.

Remark 3.1. We can easily extend the definitions of weak and entropy solutions to systems with source terms (1.17): the Rankine–Hugoniot condition (2.8) remains unchanged, while inequality (3.10) should be replaced by

$$\frac{\partial}{\partial t} U(u) + \sum_{j=1}^d \frac{\partial}{\partial x_j} F_j(u) \leq U'(u) g(u) \quad \text{in } \mathcal{D}',$$

where U and F_j satisfy the same conditions (3.2). □

Remark 3.2. It would be of interest to consider more general viscous perturbations of the system (1.1). In fact, for a compressible fluid, if we take into account the effect of viscosity together with heat conduction, the Euler equations of gas dynamics (1.7) are replaced by the Navier–Stokes equations

$$(3.16)\quad \begin{cases} \dfrac{\partial \rho}{\partial t} + \displaystyle\sum_{j=1}^{3} \dfrac{\partial}{\partial x_j}(\rho u_j) = 0, \\ \dfrac{\partial}{\partial t}(\rho u_i) + \displaystyle\sum_{j=1}^{3} \dfrac{\partial}{\partial x_j}(\rho u_i u_j + p\,\delta_{ij} - \tau_{ij}) = 0, \quad 1 \le i \le 3, \\ \dfrac{\partial}{\partial t}(\rho e) + \displaystyle\sum_{j=1}^{3} \dfrac{\partial}{\partial x_j}\Big((\rho e + p)u_j - \sum_{\ell=1}^{3} \tau_{j\ell} u_\ell - k \dfrac{\partial T}{\partial x_j}\Big) = 0, \end{cases}$$

with

$$(3.17)\quad \begin{aligned} \tau_{ij} &= \gamma(\operatorname{div} \mathbf{u})\delta_{ij} + 2\mu\, D_{ij}(\mathbf{u}), \\ D_{ij}(\mathbf{u}) &= \frac{1}{2}\Big(\frac{\partial u_i}{\partial x_j} + \frac{\partial u_j}{\partial x_i}\Big), \quad 1 \le i, j \le 3. \end{aligned}$$

In (3.16) and (3.17), T is the temperature, $D(\mathbf{u}) = (D_{ij}(\mathbf{u}))$ is the deformation tensor, γ and μ are Lamé coefficients of viscosity, and k is the coefficient of thermal conductivity. Usually, the coefficients γ, μ, and k depend only on T. Now, we need to add to (3.16), (3.17) two equations of state of the form

$$(3.18)\quad p = p(\rho, \varepsilon), \quad T = T(\rho, \varepsilon).$$

Note that it may be easier to use ρ and T as the independent thermodynamic variables so that the equations of state become

$$p = p(\rho, T), \quad \varepsilon = \varepsilon(\rho, T).$$

The simplest model is obtained by taking

$$p = (\gamma - 1)\rho\,\varepsilon, \quad \varepsilon = C_v\, T$$

and assuming that the coefficients γ, μ, and k are constant.

By using again the notations (1.8) and setting

$$\mathbf{Q}_j\Big(u, \frac{\partial u}{\partial x_1}, \frac{\partial u}{\partial x_2}, \frac{\partial u}{\partial x_3}\Big) = \begin{pmatrix} 0 \\ \tau_{j1} \\ \tau_{j2} \\ \tau_{j3} \\ \sum_{\ell=1}^{3} \tau_{j\ell} u_\ell + k \frac{\partial T}{\partial x_j} \end{pmatrix}, \quad 1 \le j \le 3,$$

the Navier–Stokes equations can be written in the form

$$\frac{\partial \mathbf{u}}{\partial t} + \sum_{j=1}^{3} \frac{\partial}{\partial x_j} \mathbf{f}_j(\mathbf{u}) - \sum_{j=1}^{3} \frac{\partial}{\partial x_j} \mathbf{Q}_j = \mathbf{0}.$$

Observe that we have

$$\mathbf{Q}_j = \sum_{j=1}^{3} \mathbf{A}_{j\ell}(\mathbf{u}) \frac{\partial \mathbf{u}}{\partial x_\ell}$$

for some $p \times p$ matrices $\mathbf{A}_{j\ell}(\mathbf{u})$. Hence (3.16) becomes

$$\frac{\partial \mathbf{u}}{\partial t} + \sum_{j=1}^{3} \frac{\partial}{\partial x_j} \mathbf{f}_j(\mathbf{u}) - \sum_{j,\ell=1}^{3} \frac{\partial}{\partial x_j} \left(\mathbf{A}_{j\ell}(\mathbf{u}) \frac{\partial \mathbf{u}}{\partial x_\ell} \right) = \mathbf{0}. \tag{3.19}$$

Therefore, in the general case, a realistic situation consists in introducing instead of (3.7) the following viscous perturbation of the system (1.1):

$$\frac{\partial \mathbf{u}_\varepsilon}{\partial t} + \sum_{j=1}^{d} \frac{\partial}{\partial x_j} \mathbf{f}_j(\mathbf{u}_\varepsilon) - \varepsilon \sum_{j,\ell=1}^{d} \frac{\partial}{\partial x_j} \left(\mathbf{A}_{j\ell}(\mathbf{u}_\varepsilon) \frac{\partial \mathbf{u}_\varepsilon}{\partial x_\ell} \right) = \mathbf{0}, \tag{3.20}$$

where the $\mathbf{A}_{j\ell}$ are $p \times p$ matrix-valued functions. Now, applying $U'(\mathbf{u}_\varepsilon)$ to (3.20), we obtain

$$\frac{\partial}{\partial t} U(\mathbf{u}_\varepsilon) + \sum_{j=1}^{d} \frac{\partial}{\partial x_j} F_j(\mathbf{u}_\varepsilon) - \varepsilon \sum_{j,\ell=1}^{d} U'(\mathbf{u}_\varepsilon) \frac{\partial}{\partial x_j} \left(\mathbf{A}_{j\ell}(\mathbf{u}_\varepsilon) \frac{\partial \mathbf{u}_\varepsilon}{\partial x_\ell} \right) = 0.$$

We have

$$U'(\mathbf{u}_\varepsilon) \frac{\partial}{\partial x_j} \left(\mathbf{A}_{j\ell}(\mathbf{u}_\varepsilon) \frac{\partial \mathbf{u}_\varepsilon}{\partial x_\ell} \right) = \frac{\partial}{\partial x_j} \left(U'(\mathbf{u}_\varepsilon)\, \mathbf{A}_{j\ell}(\mathbf{u}_\varepsilon) \frac{\partial \mathbf{u}_\varepsilon}{\partial x_\ell} \right) - \left(\frac{\partial \mathbf{u}_\varepsilon}{\partial x_j} \right)^T U''(\mathbf{u}_\varepsilon) \mathbf{A}_{j\ell}(\mathbf{u}_\varepsilon) \frac{\partial \mathbf{u}_\varepsilon}{\partial x_\ell}.$$

Hence, if we assume that the augmented $dp \times dp$ matrix

$$\begin{pmatrix} U''(\mathbf{u})\mathbf{A}_{11} & \cdots & U''(\mathbf{u})\mathbf{A}_{1d}(\mathbf{u}) \\ \vdots & & \vdots \\ U''(\mathbf{u})\mathbf{A}_{d1} & \cdots & U''(\mathbf{u})\mathbf{A}_{dd}(\mathbf{u}) \end{pmatrix}$$

is symmetric and semi-positive-definite, we get

$$\frac{\partial}{\partial t} U(\mathbf{u}_\varepsilon) + \sum_{j=1}^{d} \frac{\partial}{\partial x_j} F_j(\mathbf{u}_\varepsilon) \le \varepsilon \sum_{j,\ell=1}^{d} \frac{\partial}{\partial x_j} \left(U'(\mathbf{u}_\varepsilon) \mathbf{A}_{j\ell}(\mathbf{u}_\varepsilon) \frac{\partial \mathbf{u}_\varepsilon}{\partial x_\ell} \right). \tag{3.21}$$

This property holds indeed in the case of the gas dynamics equations (see Dutt (1988)). The problem is then to pass to the limit in the inequality (3.21) as ε tends to zero. This can be done in the following simple case:
(i) $\mathbf{A}_{j\ell} = \mathbf{D}\, \delta_{j\ell}$, where $\mathbf{D}$ is a $p \times p$ symmetric positive definite constant matrix (and $\delta_{j\ell}$ the Kronecker symbol),
(ii) U is a strictly convex positive entropy with

$$U''\mathbf{D} \ge \alpha \quad \text{for some } \alpha > 0. \tag{3.22}$$

Then one can prove that the conclusion of Theorem 3.3 holds (see Harten 1983a). □

Remark 3.3. For more general systems, without assuming that $\mathbf{D}$ is constant, the positive definiteness (3.22) of the matrix $U''\mathbf{D}$ is Mock's assumption of *admissibility*. It is also involved in the study of boundary conditions for the initial boundary value problem (see Chapter V, Section 2.2), obtained with the vanishing viscosity method (see Gisclon and Serre 1994 and Gisclon 1994) i.e., by studying the limit as $\varepsilon \to 0$ of the solution $\mathbf{u}_\varepsilon$ of the viscous system

$$\frac{\partial \mathbf{u}_\varepsilon}{\partial t} + \sum_{j=1}^{3} \frac{\partial}{\partial x_j} \mathbf{f}_j(\mathbf{u}_\varepsilon) - \varepsilon \sum_{j=1}^{3} \frac{\partial}{\partial x_j} \left(\mathbf{D}(\mathbf{u}_\varepsilon) \frac{\partial \mathbf{u}_\varepsilon}{\partial x_j} \right) = \mathbf{0} \tag{3.23}$$

together with initial and boundary conditions. When the matrix $\mathbf{D}$ satisfies (3.22), which is more precisely stated as: for any compact K in $\mathbb{R}^p$, there exists $\alpha_K > 0$ such that for any $\mathbf{u} \in K$

$$(U''(\mathbf{u})\mathbf{v}), \mathbf{D}(\mathbf{u})\mathbf{v}) \geq \alpha_K |\mathbf{v}|^2, \forall \mathbf{v} \in \mathbb{R}^p,$$

it is called dissipative. Note that the so-called residual boundary conditions for the limit solution depend on the matrix $\mathbf{D}$.

We shall also meet a more general related assumption that is linked to the well-posedness of the viscous problem (3.23) linearized at a state $\mathbf{u}_0$ and to the existence of viscous shock profiles (see Chapter I, Remark 5.2). It consists of the requirement that there exists a positive-definite symmetric matrix $\mathbf{M}(\mathbf{u}_0)$ such that $\mathbf{M}(\mathbf{u}_0)\mathbf{A}(\mathbf{u}_0)$ is symmetric and such that $\mathbf{M}(\mathbf{u}_0)\mathbf{D}(\mathbf{u}_0)$ is positive-definite. This condition guarantees that the viscosity matrix is *strictly stable* (see Majda and Pego 1985). □

Remark 3.4. Assumptions (3.8), (3.9) are satisfied in the scalar case (Godlewski and Raviart 1991, hereafter noted G.R., Chapter II, Section 3). For general systems, it may be that only an $\mathbf{L}^\infty$ estimate on $\mathbf{u}_\varepsilon$ is available, which does not allow us to pass to the limit in the nonlinear term $\mathbf{f}(\mathbf{u}^\varepsilon)$ when ε tends to 0, so that we cannot use Theorem 3.3. More powerful tools are provided by compensated compactness and measure-valued solutions (see Chapter IV, Section 4.2.3 for the definition). We refer for instance to Murat (1978), Tartar (1979, 1983), DiPerna (1983a, 1983b), Serre (1987a), Chen (1992). □

It is conjectured that an entropy solution of (1.1), (1.2) is necessarily unique. This is indeed proved in the case of a scalar equation ($p = 1$), as stated in Theorem 3.4 below. The conjecture is still widely open in the case of systems even in one dimension ($d = 1, p \geq 2$). We shall only prove, in Chapter I, Section 6, a uniqueness result for the Riemann problem (1.1), (1.3) in the case of a "convex" system and for nearby states. (We shall not consider more general data, for which we refer to the notes at the end of Chapter I.) The Riemann problem is involved in many approximation schemes, which justifies a thorough study of both the theoretical aspects and some specific examples (such as the Euler system in Chapter II).

3.3 Existence and uniqueness of the entropy solution in the scalar case

A thorough proof of the following theorem can be found in G.R., Chapter II (Section 5). Existence is obtained by the vanishing viscosity method while uniqueness follows from Kruzkov's result, which states moreover that the solution has a finite domain of dependence corresponding to a finite speed of propagation.

Theorem 3.4

Assume that the function u_0 belongs to $\mathbf{L}^\infty(\mathbb{R}^d)$. Then, the Problem (1.1), (1.2) has a unique entropy solution $u \in \mathbf{L}^\infty(\mathbb{R}^d \times (0,T))$. This solution satisfies for almost all $t \geq 0$

$$\|u(\cdot,t)\|_{\mathbf{L}^\infty(\mathbb{R}^d)} \leq \|u_0\|_{\mathbf{L}^\infty(\mathbb{R}^d)}. \tag{3.24}$$

Moreover, if u and v are the entropy solutions of (1.1) associated with the initial conditions u_0 and v_0, respectively, we have

$$u_0 \geq v_0 \Longrightarrow u(\cdot,t) \geq v(\cdot,t) \ a.e. \tag{3.25}$$

Finally, if u_0 belongs to $\mathbf{L}^\infty(\mathbb{R}^d) \cap \mathbf{BV}(\mathbb{R}^d)$, then $u(\cdot,t)$ belongs to $\mathbf{BV}(\mathbb{R}^d)$ with

$$TV(u(\cdot,t)) \leq TV(u_0). \tag{3.26}$$

Remark 3.5. The proof of uniqueness uses the well-known Kruzkov's entropy pair

$$U(u) = |u-k|, \quad F_j(u) = \operatorname{sgn}(u-k)f_j(k)), \quad j = 1, ..., d.$$

This important result extends to equations with source term (1.24) (see Kruzkov 1970). □

For what concerns the Riemann problem in the scalar one-dimensional case, the unique solution can be constructed explicitly in the general case (G.R., Chapter II, Section 6); it is piecewise smooth, self-similar, monotonic, and is usually written in the form

$$u(x,t) = w_R\Big(\frac{x}{t}; u_L, u_R\Big). \tag{3.27}$$

We recall the formula for w_R only when the function f is strictly convex, since in the general case it is difficult to exhibit a compact formula (moreover, in the case of a system, we shall assume in Chapter I, Section 6, that the sytem is *convex*, in which case the elementary waves that are involved in the solution of the Riemann problem are also rarefactions and shocks). In this case ($f'' > 0$), it is easy to prove that only decreasing shocks are admissible; moreover, the characteristics are going into the shock (see Figure 3.1(i)). Any increasing discontinuity introduced in the initial data is immediately spread out. Then the entropy solution consists of either a rarefaction when $u < v$ (Figure 3.1 (ii)), which is found by looking for a smooth selfsimilar solution of (1.5) connecting u and v, or a shock when $u > v$ (Figure 3.1(i)), which propagates at a constant velocity s given by the Rankine-Hugoniot condition (2.11).

The function $\xi \to w_R(\xi; u, v)$ is precisely defined by
if $u < v$

$$(3.28) \qquad w_R(\xi; u, v) = \begin{cases} u, & \xi \leq a(u) \\ a^{-1}(\xi), & a(u) \leq \dfrac{x}{t} \leq a(v), \quad a(u) = f'(u) \\ v, & \xi \geq a(v), \end{cases}$$

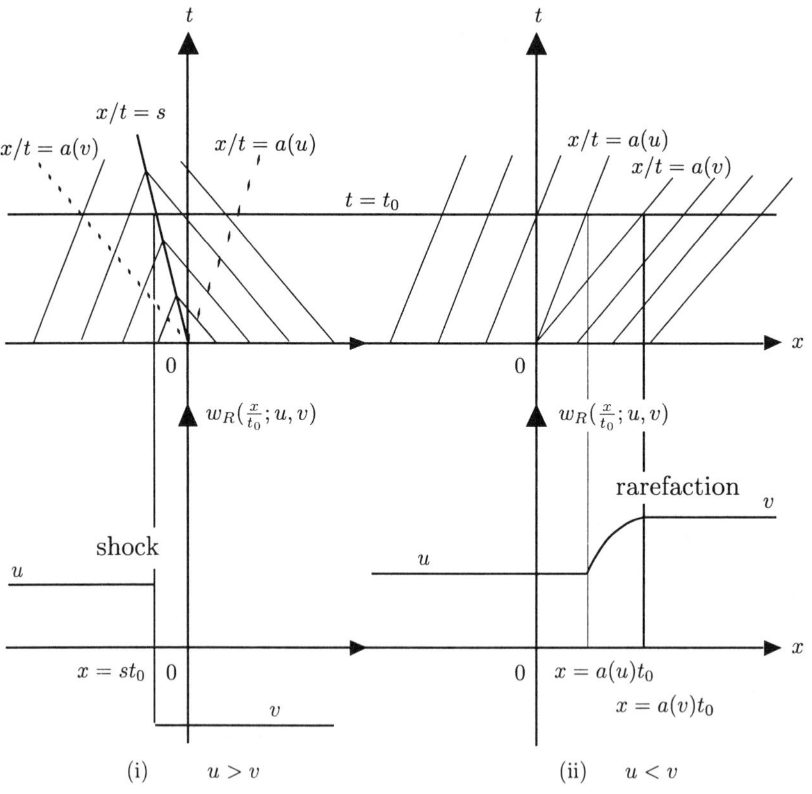

FIGURE 3.1. Solution of the scalar Riemann problem (convex case).

if $u = v$ $\quad w_R(\xi; u, v) = u$,
if $u > v$

$$w_R(\xi; u, v) = \begin{cases} u & \text{if } \xi < s \\ v & \text{if } \xi > s, \end{cases}$$

where

$$s = \frac{f(u) - f(v)}{u - v}$$

is the speed of propagation of the discontinuity.

Notes

The examples presented in Section 1 will be examined in the next chapters, at the end of which the reader will find bibliographical notes. Let us just mention some other examples: Buckley–Leverett equations and mathematical models for oil recovery (Temple 1982, Schaeffer and Shearer 1987; Shearer 1988; Holden 1991; Tveito and Winther 1991); traffic flow (Whitham 1974, Sec. 3.1; Holden and Risbro 1995 and the references therein); propagation of waves in elasticity theory (Keyfitz and Kranzer 1980); magnetohydrodynamics (Jeffrey and Taniuti 1964). Further applications are mentionned in the Notes at the end of Chapters III and IV. For examples as well as for most theoretical results, we refer particularly to the recent book of Serre (1996).

The formation of singularities was shown with a simple example; for more general situations, see John (1974). We did not mention the "equal area" rule, which, in a single conservation law, gives a geometric way of constructing the correct position for the discontinuity (Whitham 1974 section 2.8). In general, definitions or results concerning scalar equations and characteristics can be found in John's book, in the recent books of Taylor (1996) and as we already mentionned of Serre (1996), also specifically in Jeffrey (1976), Taniuti and Nishihara (1983). In Smoller's book (1994, Chapter 16), one finds a thoroughly different proof of Theorem 3.4 (concerning the convex case $f'' > 0$ only). This proof involves the convergence of a finite difference scheme and uses another way of writing the entropy condition (the "spreading estimate" or one-sided Lipschitz condition) due to Oleinik and a nonlinear version of the Holmgren method via the "adjoint" equation. Concerning the smoothness of the entropy solution, see Tadmor and Tassa (1993), and the references therein.

Some important papers concerned with the notion of entropy are those of Lax (1971), Dafermos (1973), T.-P. Liu (1976) and Bardos (1982); see also the references in Remark 5.2, Section 5 in Chapter I. Note that we have presented here a mathematical notion of entropy associated with a pair of entropy–entropy flux. We shall consider more geometrical conditions for systems in Chapter I, Section 5. Concerning the vanishing viscosity method, besides the scalar case, which is thoroughly studied in Godlewski and Raviart (1991), the reader will find further references at the end of Chapter I.

The scalar Riemann problem is also treated in Chang and Hsiao (1989), where one finds, moreover, the study of wave interaction.

I

Nonlinear hyperbolic systems in one space dimension

The main goal of this chapter is to study the Riemann problem for a general nonlinear hyperbolic system of conservation laws in one space dimension. We begin by considering the case of a linear hyperbolic system with constant coefficients for which the Riemann problem is easily solved. Next, in the nonlinear case we introduce the notions of rarefaction waves, shock waves, and contact discontinuities, which play an essential role in the explicit construction of the solution of the Riemann problem. These notions are illustrated on the examples of the p-system and the gas dynamics equations. Then, we prove the local existence of an entropy solution of the Riemann problem for a general system in the sense that the initial states are sufficiently close. In fact, in the case of the p-system, we are able to prove that the Riemann problem is always solvable. We shall show in the next chapter that this is also true for the gas dynamics equations. These two basic chapters present the results with detailed proofs, which could be easily shortened for a reader already familiar with the subject.

1 Linear hyperbolic systems with constant coefficients

We begin by considering the first-order linear system

$$\frac{\partial \mathbf{u}}{\partial t} + \mathbf{A}\frac{\partial \mathbf{u}}{\partial x} = \mathbf{0}, \quad x \in \mathbb{R}, \ t > 0, \tag{1.1}$$

where $\mathbf{u} = (u_1, ..., u_p)^T$ is a p-vector (all the vectors of $\mathbb{R}^p$ will be considered as column vectors) and $\mathbf{A}$ is a $p \times p$ constant matrix. We assume that the system is strictly hyperbolic, i.e., the matrix $\mathbf{A}$ has p distinct real eigenvalues arranged in increasing order

$$\lambda_1 < \lambda_2 < ... < \lambda_p.$$

With each eigenvalue λ_k we associate a right eigenvector $\mathbf{r}_k \in \mathbb{R}^p$

$$\mathbf{A}\mathbf{r}_k = \lambda_k \mathbf{r}_k \tag{1.2}$$

and a "left" eigenvector $\mathbf{l}_k^T$

$$\mathbf{l}_k^T \mathbf{A} = \lambda_k \mathbf{l}_k^T \tag{1.3}$$

i.e., $\mathbf{l}_k$ is an eigenvector of $\mathbf{A}^T$. Since the eigenvalues are distinct, the eigenvectors $\mathbf{r}_k$, $1 \le k \le p$, form a basis of $\mathbb{R}^p$ and we have

$$\mathbf{l}_j^T \mathbf{r}_k = \mathbf{l}_j \cdot \mathbf{r}_k = 0, \quad j \neq k$$

(we use either the dot $\cdot$ for the scalar product of the two vectors or $\mathbf{l}_j^T \mathbf{r}_k$, which denotes the product of the $1 \times p$ and $p \times 1$ matrices). Moreover, we can normalize the vectors $\mathbf{l}_k^T$ in such a way that

$$\mathbf{l}_k^T \mathbf{r}_k = 1.$$

Hence, using the Kronecker delta symbol, we obtain

$$\mathbf{l}_j^T \mathbf{r}_k = \delta_j^k, \quad 1 \le j,\ k \le p. \tag{1.4}$$

Now, setting

$$\mathbf{u} = \sum_{k=1}^{p} \alpha_k \mathbf{r}_k, \quad \alpha_k = \mathbf{l}_k^T \mathbf{u},$$

we have

$$\frac{\partial \mathbf{u}}{\partial t} + \mathbf{A} \frac{\partial \mathbf{u}}{\partial x} = \sum_{k=1}^{p} \Big(\frac{\partial \alpha_k}{\partial t} + \lambda_k \frac{\partial \alpha_k}{\partial x} \Big) \mathbf{r}_k$$

so that the first-order system (1.1) is equivalent to the p independent scalar first-order equations

$$\frac{\partial \alpha_k}{\partial t} + \lambda_k \frac{\partial \alpha_k}{\partial x} = 0, \quad 1 \le k \le p.$$

Thus, we can derive an explicit expression for the solution $\mathbf{u}$ of the Cauchy problem

$$\begin{cases} \dfrac{\partial \mathbf{u}}{\partial t} + \mathbf{A} \dfrac{\partial \mathbf{u}}{\partial x} = \mathbf{0}, & x \in \mathbb{R},\ t > 0, \\ \mathbf{u}(x, 0) = \mathbf{u}_0(x). \end{cases} \tag{1.5}$$

Indeed, setting

$$\alpha_{k0}(x) = \mathbf{l}_k^T \mathbf{u}_0(x),$$

we obtain

$$\alpha_k(x, t) = \alpha_{k0}(x - \lambda_k t) = \mathbf{l}_k^T \mathbf{u}_0(x - \lambda_k t), \ 1 \le k \le p,$$

and therefore

$$\mathbf{u}(x, t) = \sum_{k=1}^{p} \mathbf{l}_k^T \mathbf{u}_0(x - \lambda_k t) \mathbf{r}_k. \tag{1.6}$$

Consider in particular the Riemann problem for the system (1.1) corresponding to the initial condition

$$\mathbf{u}_0(x) = \begin{cases} \mathbf{u}_L, & x < 0, \\ \mathbf{u}_R, & x > 0. \end{cases} \tag{1.7}$$

If we define α_{kR} and α_{kL}, $1 \leq k \leq p$, by

$$\mathbf{u}_L = \sum_{k=1}^{p} \alpha_{kL} \mathbf{r}_k, \qquad \mathbf{u}_R = \sum_{k=1}^{p} \alpha_{kR} \mathbf{r}_k,$$

we obtain

$$\alpha_k(x,t) = \begin{cases} \alpha_{kL}, & x < \lambda_k t, \\ \alpha_{kR}, & x > \lambda_k t, \end{cases}$$

so that the solution $\mathbf{u}$ of the Riemann problem (1.5), (1.7) is self-similar, i.e., of the form

$$\mathbf{u}(x,t) = \mathbf{w}_R\Big(\frac{x}{t}; \mathbf{u}_L, \mathbf{u}_R\Big). \tag{1.8}$$

Moreover, for $\lambda_m < \frac{x}{t} < \lambda_{m+1}$, $\mathbf{u}$ takes the constant value

$$\mathbf{w}_m = \sum_{k=1}^{m} \alpha_{kR} \mathbf{r}_k + \sum_{k=m+1}^{p} \alpha_{kL} \mathbf{r}_k, \quad 0 \leq m \leq p \tag{1.9}$$

with the convention $\lambda_0 = -\infty$, $\lambda_{p+1} = +\infty$. Hence, we have

$$\mathbf{w}_R\Big(\frac{x}{t}; \mathbf{u}_L, \mathbf{u}_R\Big) = \begin{cases} \mathbf{w}_0 = \mathbf{u}_L, & \frac{x}{t} < \lambda_1, \\ \mathbf{w}_1, & \lambda_1 < \frac{x}{t} < \lambda_2, \\ \vdots & \\ \mathbf{w}_{p-1}, & \lambda_{p-1} < \frac{x}{t} < \lambda_p, \\ \mathbf{w}_p = \mathbf{u}_R, & \frac{x}{t} > \lambda_p, \end{cases} \tag{1.10}$$

which shows that, in general, the initial discontinuity breaks up into p discontinuity waves, which propagate with the characteristic speeds λ_k, $1 \leq k \leq p$ (see Figure 1.1). Note that the intermediate states $\mathbf{w}_m$ satisfy

$$\mathbf{w}_m - \mathbf{w}_{m-1} = (\alpha_{mR} - \alpha_{mL})\mathbf{r}_m$$

and therefore

$$\mathbf{A}(\mathbf{w}_m - \mathbf{w}_{m-1}) - \lambda_m(\mathbf{w}_m - \mathbf{w}_{m-1}).$$

Thus, across the line of discontinuity $x = \lambda_m t$, the Rankine–Hugoniot jump condition (2.11) of the Introduction is indeed satisfied.

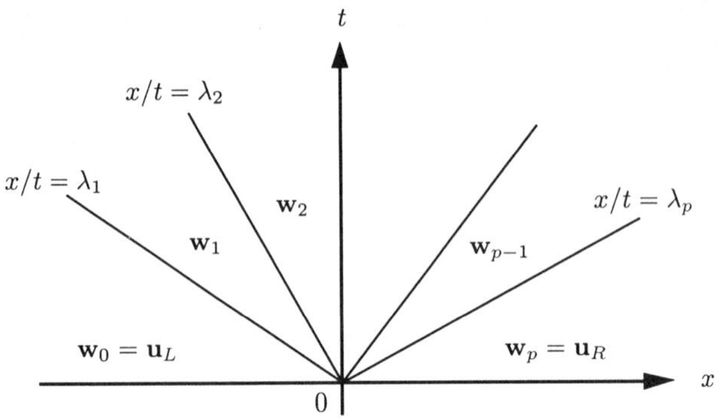

FIGURE 1.1. Solution of the Riemann problem for a linear system.

2 The nonlinear case. Definitions and examples

We turn now to nonlinear hyperbolic systems. Let Ω be an open subset of $\mathbb{R}^p$ and $f : \Omega \to \mathbb{R}^p$ be a sufficiently smooth function (of class C^2 at least). We consider the nonlinear system of conservation laws

$$\frac{\partial \mathbf{u}}{\partial t} + \frac{\partial}{\partial x}\mathbf{f}(\mathbf{u}) = \mathbf{0}, \quad x \in \mathbb{R},\ t > 0, \tag{2.1}$$

where $\mathbf{u} = (u_1, ..., u_p)^T$ and $\mathbf{f}(\mathbf{u}) = \Big(f_1(\mathbf{u}), ..., f_p(\mathbf{u})\Big)^T$. Let us assume for simplicity that the system (2.1) is strictly hyperbolic, i.e., for any state $\mathbf{u} \in \Omega$, the $p \times p$ Jacobian matrix

$$\mathbf{A}(\mathbf{u}) = \Big(\frac{\partial f_i}{\partial u_j}(\mathbf{u})\Big)_{1 \le i,j \le p}$$

has p distinct real eigenvalues

$$\lambda_1(\mathbf{u}) < \lambda_2(\mathbf{u}) < \cdots < \lambda_p(\mathbf{u}) \cdot$$

With each eigenvalue $\lambda_k(\mathbf{u})$ we associate a right eigenvector $\mathbf{r}_k(\mathbf{u})$

$$\mathbf{A}(\mathbf{u})\mathbf{r}_k(\mathbf{u}) = \lambda_k(\mathbf{u})\mathbf{r}_k(\mathbf{u}) \tag{2.2}$$

and a "left eigenvector" $\mathbf{l}_k(\mathbf{u})$

$$\mathbf{l}_k(\mathbf{u})^T\mathbf{A}(\mathbf{u}) = \lambda_k(\mathbf{u})\mathbf{l}_k(\mathbf{u})^T \tag{2.3}$$

i.e., $\mathbf{l}_k(\mathbf{u})$ is an eigenvector of $\mathbf{A}(\mathbf{u})^T$. Since the eigenvalues are distinct, $\Big(\mathbf{l}_k(\mathbf{u})\Big)_k$ is a dual basis of $\Big(\mathbf{r}_k(\mathbf{u})\Big)_k$ and

$$\mathbf{l}_j(\mathbf{u})^T\mathbf{r}_k(\mathbf{u}) = \mathbf{l}_j(\mathbf{u}) \cdot \mathbf{r}_k(\mathbf{u}) = 0, \qquad k \neq j. \tag{2.4}$$

In fact, most of the results of the following sections hold for a hyperbolic system, provided the eigenvectors are complete and the eigenvalues have constant multiplicity (see Remark 6.1 below). If we assume that $\mathbf{f}$ is a C^m function, we obtain that $\lambda_k, \mathbf{r}_k, \mathbf{l}_k$ are C^{m-1} functions of $\mathbf{u}$ (C^1 functions at least).

Definition 2.1

The kth *characteristic field is said to be genuinely nonlinear if*

$$D\lambda_k(\mathbf{u}) \cdot \mathbf{r}_k(\mathbf{u}) \neq 0, \quad \forall \mathbf{u} \in \Omega. \tag{2.5}$$

The kth *characteristic field is said to be linearly degenerate if*

$$D\lambda_k(\mathbf{u}) \cdot \mathbf{r}_k(\mathbf{u}) = 0, \quad \forall \mathbf{u} \in \Omega. \tag{2.6}$$

In (2.5) and (2.6), $D\lambda_k(\mathbf{u}) = \lambda_k'(\mathbf{u}) \in \mathcal{L}(\mathbb{R}^p; \mathbb{R})$ denotes the derivative of $\lambda_k(\mathbf{u})$ and can be identified with a vector $\nabla\lambda_k(\mathbf{u})$ of $\mathbb{R}^p$.

Example 2.1. Consider first the case $p = 1$, i.e., (2.1) is a *scalar* equation. Then, we have in the (x, t)-space

$$\lambda(u) = a(u) = f'(u).$$

Hence, we are in the genuinely nonlinear case if and only if $a(u)$ does not vanish, i.e., if and only if f is a strictly convex or a strictly concave function. On the other hand, we are in the linearly degenerate case if and only if $a(u) = a$ does not depend on u, which corresponds to the case of a linear hyperbolic equation with constant coefficients. □

Example 2.2. The p-system. Let us consider again the p-system introduced in the Introduction, Example 1.2:

$$\frac{\partial \mathbf{w}}{\partial t} + \frac{\partial}{\partial x}\mathbf{f}(\mathbf{w}) = \mathbf{0}, \quad \mathbf{w} = \begin{pmatrix} v \\ u \end{pmatrix}, \quad \mathbf{f}(\mathbf{w}) = \begin{pmatrix} -u \\ p(v) \end{pmatrix}. \tag{2.7}$$

We have already noticed that the p-system is strictly hyperbolic provided that we assume $p'(v) < 0$. In that case, since the Jacobian matrix of $\mathbf{f}$ depends only on v, we can note $\mathbf{A}(\mathbf{w}) = \mathbf{A}(v)$, and

$$\mathbf{A}(v) = \begin{pmatrix} 0 & -1 \\ p'(v) & 0 \end{pmatrix}$$

has two real distinct eigenvalues

$$\lambda_1(v) = -\sqrt{-p'(v)} < 0 < \lambda_2(v) = \sqrt{-p'(v)}. \tag{2.8}$$

Moreover, the corresponding right eigenvectors can be chosen as

$$\mathbf{r}_1(v) = \begin{pmatrix} 1 \\ \sqrt{-p'(v)} \end{pmatrix}, \quad \mathbf{r}_2(v) = \begin{pmatrix} 1 \\ -\sqrt{-p'(v)} \end{pmatrix}. \tag{2.9}$$

Hence, we obtain

$$\nabla\lambda_1(v) = \begin{pmatrix} \frac{p''(v)}{2\sqrt{-p'(v)}} \\ 0 \end{pmatrix}, \quad \nabla\lambda_2(v) = \begin{pmatrix} -\frac{p''(v)}{2\sqrt{-p'(v)}} \\ 0 \end{pmatrix}$$

so that

$$D\lambda_1(v) \cdot \mathbf{r}_1(v) = \nabla\lambda_1(v)^T \mathbf{r}_1(v) = \frac{p''(v)}{2\sqrt{-p'(v)}},$$

$$D\lambda_2(v) \cdot \mathbf{r}_2(v) = \nabla\lambda_2(v)^T \mathbf{r}_2(v) = -\frac{p''(v)}{2\sqrt{-p'(v)}}.$$

If we suppose $p''(v) > 0$, we obtain that the two characteristic fields are genuinely nonlinear. □

In fact, in many applications, it can be easier when studying the characteristic fields to work on a nonconservative form of the nonlinear system (2.1). Indeed, let $\boldsymbol{\theta}$ be a C^1 diffeomorphism of an open subset $\vartheta \subset \mathbb{R}^p$ onto Ω. By making the change of dependent variables

$$\mathbf{u} = \boldsymbol{\theta}(\mathbf{v}), \tag{2.10}$$

the differential system

$$\frac{\partial \mathbf{u}}{\partial t} + \mathbf{A}(\mathbf{u})\frac{\partial \mathbf{u}}{\partial x} = \mathbf{0} \tag{2.11}$$

becomes

$$D\boldsymbol{\theta}(\mathbf{v})\,\frac{\partial \mathbf{v}}{\partial t} + \mathbf{A}(\boldsymbol{\theta}(\mathbf{v}))\;D\boldsymbol{\theta}(\mathbf{v})\,\frac{\partial \mathbf{v}}{\partial \mathbf{x}} = \mathbf{0},$$

where $D\boldsymbol{\theta}(\mathbf{v}) = \boldsymbol{\theta}'(\mathbf{v})$ denotes the Jacobian matrix of $\boldsymbol{\theta}(\mathbf{v})$, and therefore

$$\frac{\partial \mathbf{v}}{\partial t} + \mathbf{B}(\mathbf{v})\frac{\partial \mathbf{v}}{\partial x} = \mathbf{0} \tag{2.12}$$

with

$$\mathbf{B}(\mathbf{v}) = \Big(D\boldsymbol{\theta}(\mathbf{v})\Big)^{-1}\;\mathbf{A}(\boldsymbol{\theta}(\mathbf{v}))\;D\boldsymbol{\theta}(\mathbf{v}).$$

Denote by $\mu_k(\mathbf{v})$ and $\mathbf{s}_k(\mathbf{v})$, $1 \le k \le p$, the eigenvalues and the corresponding right eigenvectors of the $p \times p$ matrix $\mathbf{B}(\mathbf{v})$

$$\mathbf{B}(\mathbf{v})\mathbf{s}_k(\mathbf{v}) = \mu_k(\mathbf{v})\mathbf{s}_k(\mathbf{v}).$$

By the similarity of the matrices $\mathbf{A}(\boldsymbol{\theta}(\mathbf{v}))$ and $\mathbf{B}(\mathbf{v})$, we obtain

$$\mu_k(\mathbf{v}) = \lambda_k(\boldsymbol{\theta}(\mathbf{v})). \tag{2.13}$$

Moreover, we can take

$$\mathbf{s}_k(\mathbf{v}) = (D\boldsymbol{\theta}(\mathbf{v}))^{-1}\mathbf{r}_k(\boldsymbol{\theta}(\mathbf{v})); \tag{2.14}$$

indeed

$$\begin{aligned}
\mathbf{B}(\mathbf{v})\mathbf{s}_k(\mathbf{v}) &= \Big(D\boldsymbol{\theta}(\mathbf{v})\Big)^{-1}\mathbf{A}(\boldsymbol{\theta}(\mathbf{v}))D\boldsymbol{\theta}(\mathbf{v})\Big(D\boldsymbol{\theta}(\mathbf{v})\Big)^{-1}\mathbf{r}_k(\boldsymbol{\theta}(\mathbf{v}))\\
&= \Big(D\boldsymbol{\theta}(\mathbf{v})\Big)^{-1}\mathbf{A}(\boldsymbol{\theta}(\mathbf{v}))\mathbf{r}_k(\boldsymbol{\theta}(\mathbf{v}))\\
&= \lambda_k(\boldsymbol{\theta}(\mathbf{v}))\Big(D\boldsymbol{\theta}(\mathbf{v})\Big)^{-1}\mathbf{r}_k(\boldsymbol{\theta}(\mathbf{v})) = \mu_k(\mathbf{v})\mathbf{s}_k(\mathbf{v}).
\end{aligned}$$

Next, we note that

$$D\mu_k(\mathbf{v}) \cdot \mathbf{s}_k(\mathbf{v}) = D\lambda_k(\boldsymbol{\theta}(\mathbf{v})) \cdot D\boldsymbol{\theta}(\mathbf{v}) \cdot \mathbf{s}_k(\mathbf{v}) = D\lambda_k(\boldsymbol{\theta}(\mathbf{v})) \cdot \mathbf{r}_k(\boldsymbol{\theta}(\mathbf{v})),$$

i.e.,

$$D\mu_k(\mathbf{v}) \cdot \mathbf{s}_k(\mathbf{v}) = D\lambda_k(\mathbf{u}) \cdot \mathbf{r}_k(\mathbf{u}), \ \ \mathbf{u} = \boldsymbol{\theta}(\mathbf{v}). \tag{2.15}$$

Therefore, the notions of genuinely nonlinear or linearly degenerate characteristic fields do not depend on the chosen conservative or nonconservative form of the nonlinear hyperbolic system (2.1). This remark will enable us to simplify the analysis in the next two examples.

Example 2.3. The gas dynamics equations in Lagrangian coordinates. Let us consider again the gas dynamics equations written in Lagrangian coordinates that were introduced in the Introduction, Example 1.4:

$$\begin{cases} \dfrac{\partial \tau}{\partial t} - \dfrac{\partial u}{\partial x} = 0, \\ \dfrac{\partial u}{\partial t} + \dfrac{\partial p}{\partial x} = 0, \\ \dfrac{\partial e}{\partial t} + \dfrac{\partial}{\partial x}(pu) = 0. \end{cases} \tag{2.16}$$

Again $\tau = \frac{1}{\rho}$ is the specific volume, u the velocity, p the pressure, ε the specific internal energy, and $e = \varepsilon + \frac{u^2}{2}$ the specific total energy; the independent variable x here stands for the mass variable. We supplement (2.16) with an equation of state, which can be chosen in the (incomplete) form

$$p = p(\tau, \varepsilon). \tag{2.17}$$

This equation of state must satisfy various conditions of a thermodynamic nature which will be explicitly given later on. Here we just recall some results from thermodynamics that will be made precise in Chapter II. In order to simplify the computations, it is convenient to make a change of dependent variables. We introduce the specific entropy s defined via the second law of thermodynamics

$$T ds = d\varepsilon + p d\tau, \tag{2.18}$$

where T denotes the temperature. We have

$$T\frac{\partial s}{\partial t} = \frac{\partial \varepsilon}{\partial t} + p\frac{\partial \tau}{\partial t}.$$

If we assume that the dependent variables are sufficiently smooth functions of x and t, it follows from (2.16) that

$$T\frac{\partial s}{\partial t} = \frac{\partial e}{\partial t} - u\frac{\partial u}{\partial t} + p\frac{\partial \tau}{\partial t} = 0.$$

Since T is > 0, we obtain

$$\frac{\partial s}{\partial t} = 0,$$

which expresses the conversation of entropy for smooth flows. Now, we observe that the mapping

$$\boldsymbol{\theta} = \begin{pmatrix} \tau \\ u \\ s \end{pmatrix} \rightarrow \begin{pmatrix} \tau \\ u \\ e \end{pmatrix}$$

is smooth and one-to-one. Indeed, from

$$e = \varepsilon(\tau, s) + \frac{u^2}{2}$$

we get

$$D\boldsymbol{\theta} = \begin{pmatrix} 1 & 0 & 0 \\ 0 & 1 & 0 \\ -p & u & T \end{pmatrix},$$

and T is > 0. Hence, in the case of smooth flows, we may equivalently write the system (2.16) in the nonconservative form

$$\begin{cases} \dfrac{\partial \tau}{\partial t} - \dfrac{\partial u}{\partial x} = 0, \\ \dfrac{\partial u}{\partial t} + \dfrac{\partial p}{\partial x} = 0, \\ \dfrac{\partial s}{\partial t} = 0. \end{cases} \tag{2.19}$$

On the other hand, we know from thermodynamics that given any two of the thermodynamic variables $\tau, p, \varepsilon, T, s$, we can determine the remaining three variables. Thus, we may assume that p, ε, and T are given functions of τ and s so that the equation of state becomes

$$p = p(\tau, s) \tag{2.20}$$

with the following properties:

$$\frac{\partial p}{\partial \tau}(\tau, s) = -\frac{c^2}{\tau^2} < 0, \quad \frac{\partial^2 p}{\partial \tau^2}(\tau, s) > 0. \tag{2.21}$$

In (2.21) c is the local sound speed. Recall that for a polytropic ideal gas, we have (see Introduction, in Example 3.3, $p = (\gamma - 1)\exp(\frac{(s-s_0)}{C_v})\rho^\gamma$)

$$p = C\exp(\frac{s}{C_v})\tau^{-\gamma}, \tag{2.22}$$

where C_v is the specific heat at constant volume. The system (2.19), (2.20) now has a form suitable for the study of the characterisitic fields. Indeed, the Jacobian matrix of the system (2.19) is given by

$$\begin{pmatrix} 0 & -1 & 0 \\ \frac{\partial p}{\partial \tau} & 0 & \frac{\partial p}{\partial s} \\ 0 & 0 & 0 \end{pmatrix}.$$

Since the corresponding characteristic polynomial is

$$\lambda(\lambda^2 + \frac{\partial p}{\partial \tau}) = 0,$$

we obtain the distinct real eigenvalues

$$\lambda_1 = -\sqrt{-\frac{\partial p}{\partial \tau}} = -\frac{c}{\tau} < \lambda_2 = 0 < \lambda_3 = \sqrt{-\frac{\partial p}{\partial \tau}} = \frac{c}{\tau}, \tag{2.23}$$

with $p = p(\tau, s)$. The associated eigenvectors can be taken as

$$\mathbf{r}_1 = \begin{pmatrix} \tau \\ c \\ 0 \end{pmatrix}, \quad \mathbf{r}_2 = \begin{pmatrix} p_s \\ 0 \\ -p_\tau \end{pmatrix}, \quad \mathbf{r}_3 = \begin{pmatrix} \tau \\ -c \\ 0 \end{pmatrix}, \tag{2.24}$$

where

$$p_\tau = \frac{\partial p(\tau, s)}{\partial \tau}, \qquad p_s = \frac{\partial p(\tau, s)}{\partial s}.$$

Then, we have

$$\nabla\lambda_1 = \begin{pmatrix} (\frac{\tau}{2c})p_{\tau\tau} \\ 0 \\ (\frac{\tau}{2c})p_{\tau s} \end{pmatrix}, \quad \nabla\lambda_2 = \mathbf{0}, \quad \nabla\lambda_3 = \begin{pmatrix} (-\frac{\tau}{2c})p_{\tau\tau} \\ 0 \\ (-\frac{\tau}{2c})p_{\tau s} \end{pmatrix}$$

so that

$$\begin{aligned} \nabla\lambda_1^T \mathbf{r}_1 &= (\frac{\tau^2}{2c})\frac{\partial^2 p}{\partial \tau^2}, \\ \nabla\lambda_2^T \mathbf{r}_2 &= 0, \\ \nabla\lambda_3^T \mathbf{r}_3 &= -(\frac{\tau^2}{2c})\frac{\partial^2 p}{\partial \tau^2}. \end{aligned}$$

Using (2.21), we find that the first and the third characteristic fields are genuinely nonlinear while the second characteristic field is linearly degenerate. □

Example 2.4. The gas dynamics equations in Eulerian coordinates. We turn to the gas dynamics equations written in Eulerian coordinates that were introduced in the Introduction, Example 1.3:

$$\begin{cases} \frac{\partial \rho}{\partial t} + \frac{\partial}{\partial x}(\rho u) = 0, \\ \frac{\partial}{\partial t}(\rho u) + \frac{\partial}{\partial x}(\rho u^2 + p) = 0, \\ \frac{\partial}{\partial t}(\rho e) + \frac{\partial}{\partial x}\Big((\rho e + p)u\Big) = 0, \end{cases} \tag{2.25}$$

with $e = \varepsilon + \frac{u^2}{2}$ (total specific energy), and the equation of state

$$p = p(\rho, \varepsilon). \tag{2.26}$$

Now we notice that in the case of smooth flows, the system (2.25) can be equivalently written in the nonconservative form

$$\begin{cases} \dfrac{\partial \rho}{\partial t} + u\dfrac{\partial \rho}{\partial x} + \rho\dfrac{\partial u}{\partial x} = 0, \\ \dfrac{\partial u}{\partial t} + u\dfrac{\partial u}{\partial x} + \dfrac{1}{\rho}\dfrac{\partial p}{\partial x} = 0, \\ \dfrac{\partial \varepsilon}{\partial t} + u\dfrac{\partial \varepsilon}{\partial x} + \dfrac{p}{\rho}\dfrac{\partial u}{\partial x} = 0. \end{cases} \tag{2.27}$$

If we again introduce the specific entropy s, which by (2.18) satisfies

$$T ds = d\varepsilon - \frac{p}{\rho^2}\, d\rho,$$

we obtain

$$T\Big(\frac{\partial s}{\partial t} + u\frac{\partial s}{\partial x}\Big) = \frac{\partial \varepsilon}{\partial t} + u\frac{\partial \varepsilon}{\partial x} - \frac{p}{\rho^2}\Big(\frac{\partial \rho}{\partial t} + u\frac{\partial \rho}{\partial x}\Big) = 0$$

and therefore

$$\frac{\partial s}{\partial t} + u\frac{\partial s}{\partial x} = 0.$$

Next, we observe that the mapping

$$\boldsymbol{\theta} : \begin{pmatrix} \rho \\ u \\ s \end{pmatrix} \to \begin{pmatrix} \rho \\ \rho u \\ \rho e \end{pmatrix}$$

is smooth and one-to-one with Jacobian

$$D\boldsymbol{\theta} = \begin{pmatrix} 1 & 0 & 0 \\ u & \rho & 0 \\ e + \frac{p}{\rho} & \rho u & \rho T \end{pmatrix},$$

so that for smooth flows an equivalent form of (2.27) is given by

$$\begin{cases} \dfrac{\partial \rho}{\partial t} + u\dfrac{\partial \rho}{\partial x} + \rho\dfrac{\partial u}{\partial x} = 0, \\ \dfrac{\partial u}{\partial t} + u\dfrac{\partial u}{\partial x} + \dfrac{1}{\rho}\dfrac{\partial p}{\partial x} = 0, \\ \dfrac{\partial s}{\partial t} + u\dfrac{\partial s}{\partial x} = 0. \end{cases} \tag{2.28}$$

Choosing ρ and s as the two independent thermodynamic variables, we supplement the system (2.27) with the equation of state

$$p = p(\rho, s). \tag{2.29}$$

Note that the properties (2.21) become here

$$
\begin{cases}
\dfrac{\partial p}{\partial \rho}(\rho, s) = c^2 > 0, \\
\dfrac{\partial^2 p}{\partial \rho^2} + \dfrac{2}{\rho}\dfrac{\partial p}{\partial \rho} = \dfrac{2c}{\rho}(\rho \dfrac{\partial c}{\partial \rho} + c) > 0.
\end{cases}
\tag{2.30}
$$

The Jacobian matrix of the system (2.28) now has the simple form

$$
\mathbf{B} = \begin{pmatrix} u & \rho & 0 \\ \frac{1}{\rho}\frac{\partial p}{\partial \rho} & u & \frac{1}{\rho}\frac{\partial p}{\partial s} \\ 0 & 0 & u \end{pmatrix}.
$$

Since the characteristic equation is given by

$$
(u-\lambda)\Big((u-\lambda)^2 - \frac{\partial p}{\partial \rho}\Big) = 0,
$$

we obtain the real distinct eigenvalues

$$
\lambda_1 = u - c < \lambda_2 = u < \lambda_3 = u + c. \tag{2.31}
$$

The associated eigenvectors of $\mathbf{B}$ can be chosen as

$$
\mathbf{r}_1 = \begin{pmatrix} \rho \\ -c \\ 0 \end{pmatrix}, \quad \mathbf{r}_2 = \begin{pmatrix} \frac{\partial p}{\partial s} \\ 0 \\ -c^2 \end{pmatrix}, \quad \mathbf{r}_3 = \begin{pmatrix} \rho \\ c \\ 0 \end{pmatrix}. \tag{2.32}
$$

Thus, we have

$$
\nabla\lambda_1 = \begin{pmatrix} -\frac{\partial c}{\partial \rho} \\ 1 \\ -\frac{\partial c}{\partial s} \end{pmatrix}, \quad \nabla\lambda_2 = \begin{pmatrix} 0 \\ 1 \\ 0 \end{pmatrix}, \quad \nabla\lambda_3 = \begin{pmatrix} \frac{\partial c}{\partial \rho} \\ 1 \\ \frac{\partial c}{\partial s} \end{pmatrix}
$$

and

$$
\nabla\lambda_1^T \mathbf{r}_1 = -(c + \rho\frac{\partial c}{\partial \rho}), \quad \nabla\lambda_2{}^T \mathbf{r}_2 = 0, \quad \nabla\lambda_3{}^T \mathbf{r}_3 = c + \rho\frac{\partial c}{\partial \rho}.
$$

Using (2.30), we obtain again that the first and the third characteristic fields are genuinely nonlinear while the second characteristic field is linearly degenerate.

As expected, we get the same results in Lagrangian and Eulerian coordinates. □

Remark 2.1. For the sake of completness, let us also give the expression of the matrices that "diagonalize" the Jacobian matrix $\mathbf{B}$. Considering the matrix $\mathbf{T}$ with columns $(\mathbf{r}_1, \mathbf{r}_2, \mathbf{r}_3)$, then $\mathbf{T}^{-1}$ has for rows the "left eigenvectors" $\mathbf{l}_k(\mathbf{u})^T$ (see (2.3)), which gives, setting $p_s = \frac{\partial p(\rho,s)}{\partial s}$,

$$
\mathbf{T}^{-1} = \begin{pmatrix} \frac{1}{2\rho} & -\frac{1}{2c} & \frac{p_s}{2\rho c^2} \\ 0 & 0 & -\frac{1}{c^2} \\ \frac{1}{2\rho} & \frac{1}{2c} & \frac{p_s}{2\rho c^2} \end{pmatrix},
$$

and $\mathbf{T}^{-1}\mathbf{B}\mathbf{T} = \text{diag}\,(\lambda_i)$. These expressions are used in many numerical schemes and also when one linearizes the system (in the variables (ρ, u, s)) about a constant state $\overline{\mathbf{V}}$. Then, setting

$$\mathbf{V} = (\rho, u, s)^T, \mathbf{W} = \mathbf{T}^{-1}(\overline{\mathbf{V}})\mathbf{V},$$

we get a system of p decoupled equations in the components α_k of $\mathbf{W}$, which are called "characteristic variables". Note that we can equivalently write

$$\mathbf{V} = \sum \alpha_k \mathbf{r}_k(\overline{\mathbf{V}}),$$

and if we introduce a linearized pressure by

$$p = \frac{\partial p}{\partial \rho}(\overline{\mathbf{V}})\rho + \frac{\partial p}{\partial s}(\overline{\mathbf{V}})s = \overline{c}^2\rho + \frac{\partial p}{\partial s}(\overline{\mathbf{V}})s,$$

the characteristic variables α_k take the simple form

$$\begin{aligned} \alpha_1 &= \frac{(p - \overline{\rho}\,\overline{c}\,u)}{2\overline{c}^2\,\overline{\rho}}, \\ \alpha_2 &= -\frac{s}{\overline{c}^2}, \\ \alpha_3 &= \frac{(p + \overline{\rho}\,\overline{c}\,u)}{2\overline{c}^2\,\overline{\rho}}, \end{aligned}$$

which will be used in numerical schemes. □

Remark 2.2. We can also write the system in a still different nonconservative form using the "primitive" variables $(\rho, u, p)^T$. The mapping $\boldsymbol{\varphi} : (\rho, u, p)^T \longrightarrow (\rho, \rho u, \rho e)^T$ is one-to-one, with, for a γ-law $p = (\gamma - 1)\rho\varepsilon$,

$$D\boldsymbol{\varphi} = \begin{pmatrix} 1 & 0 & 0 \\ u & \rho & 0 \\ \frac{u^2}{2} & \rho u & \frac{1}{(\gamma-1)} \end{pmatrix},$$

and

$$(D\boldsymbol{\varphi})^{-1} = \begin{pmatrix} 1 & 0 & 0 \\ \frac{-u}{\rho} & \frac{1}{\rho} & 0 \\ \frac{(\gamma-1)u^2}{2} & -(\gamma-1)u & \gamma - 1 \end{pmatrix}.$$

Then the Jacobian matrix of the corresponding system is

$$\mathbf{B}' = \begin{pmatrix} u & \rho & 0 \\ 0 & u & \frac{1}{\rho} \\ 0 & \rho c^2 & u \end{pmatrix};$$

the eigenvectors associated to $\lambda_1 = u - c$, $\lambda_2 = u$, $\lambda_3 = u + c$ can be chosen as

$$\mathbf{r}'_1 = \begin{pmatrix} 1 \\ -c/\rho \\ c^2 \end{pmatrix}, \quad \mathbf{r}'_2 = \begin{pmatrix} 1 \\ 0 \\ 0 \end{pmatrix}, \quad \mathbf{r}'_3 = \begin{pmatrix} 1 \\ c/\rho \\ c^2 \end{pmatrix}.$$

If $\mathbf{T}'$ is the matrix with columns $(\mathbf{r}'_1, \mathbf{r}'_2, \mathbf{r}'_3)$, then $\mathbf{T}'^{-1}\mathbf{B}'\mathbf{T}' = \text{diag}\,(\lambda_i)$ and $\mathbf{T}'^{-1}$ is given by

$$\mathbf{T}'^{-1} = \begin{pmatrix} 0 & -\frac{\rho}{2c} & \frac{1}{2c^2} \\ 1 & 0 & -\frac{1}{c^2} \\ 0 & \frac{\rho}{2c} & \frac{1}{2c^2} \end{pmatrix}.$$

In fact, the corresponding nonconservative system is not equivalent to (2.25) for discontinuous solutions, and neither is (2.28); one should take the variables (τ, u, p), $\tau = \frac{1}{\rho}$ instead, in order to obtain the same jump conditions. □

Let us go back to the general system of conservation laws (2.1). If the kth characteristic field is genuinely nonlinear, it is convenient to normalize the right and left eigenvectors $\mathbf{r}_k(\mathbf{u})$ and $\mathbf{l}_k(\mathbf{u})^T$ in such a way that

$$\begin{cases} D\lambda_k(\mathbf{u}) \cdot \mathbf{r}_k(\mathbf{u}) = 1, \\ \mathbf{l}_k(\mathbf{u})^T \mathbf{r}_k(\mathbf{u}) = 1. \end{cases} \tag{2.33}$$

Similarly, if the kth characteristic field is linearly degenerate, i.e.,

$$D\lambda_k(\mathbf{u}) \cdot \mathbf{r}_k(\mathbf{u}) = 0,$$

one normalizes the vectors $\mathbf{r}_k(u)$ and $\mathbf{l}_k(u)$ so that

$$\mathbf{l}_k(\mathbf{u})^T \mathbf{r}_k(\mathbf{u}) = 1. \tag{2.34}$$

3 Simple waves and Riemann invariants

3.1 Rarefaction waves

Let $\mathbf{u}_L$ and $\mathbf{u}_R$ be two states of $\Omega \subset \mathbb{R}^p$; in this section, we are looking for piecewise smooth continuous functions $\mathbf{u} : (x, t) \to \mathbf{u}(x, t)$, solutions of (2.1) that connect $\mathbf{u}_L$ and $\mathbf{u}_R$:

$$\begin{cases} \dfrac{\partial \mathbf{u}}{\partial t} + \dfrac{\partial}{\partial x}\mathbf{f}(\mathbf{u}) = \mathbf{0}, \\ \mathbf{u}(x, 0) = \begin{cases} \mathbf{u}_L, & x < 0, \\ \mathbf{u}_R, & x > 0. \end{cases} \end{cases} \tag{3.1}$$

At first, we restrict ourselves to *self-similar solutions* of (3.1), i.e., solutions of the form

$$\mathbf{u}(x, t) = \mathbf{v}\left(\frac{x}{t}\right). \tag{3.2}$$

We begin by considering *classical* self-similar solutions of (3.1); these solutions satisfy the equation

$$\frac{\partial \mathbf{u}}{\partial t} + \mathbf{A}(\mathbf{u})\frac{\partial \mathbf{u}}{\partial x} = \mathbf{0}$$

in the classical sense. We must have

$$-\left(\frac{x}{t^2}\right)\mathbf{v}'\left(\frac{x}{t}\right)+\left(\frac{1}{t}\right)\mathbf{A}\left(\mathbf{v}\left(\frac{x}{t}\right)\right)\mathbf{v}'\left(\frac{x}{t}\right)=\mathbf{0},$$

so that by setting $\xi = \frac{x}{t}$

$$\Big(\mathbf{A}(\mathbf{v}(\xi)) - \xi\,\mathbf{I}\Big)\,\mathbf{v}'(\xi) = \mathbf{0}.$$

Hence, either we obtain

$$\mathbf{v}'(\xi) = \mathbf{0}$$

or there exists an index $k \ \in 1, ..., p$ such that

$$\mathbf{v}'(\xi) = \alpha(\xi)\mathbf{r}_k(\mathbf{v}(\xi)), \quad \lambda_k(\mathbf{v}(\xi)) = \xi.$$

If $\mathbf{v}'(\xi)$ is nonzero on an interval, since the eigenvalues are distinct, the index k does not depend on ξ in that interval. If we differentiate the second equation with respect to ξ, we get

$$D\lambda_k(\mathbf{v}(\xi)) \cdot \mathbf{v}'(\xi) = 1,$$

and using the first equation

$$\alpha(\xi)D\lambda_k(\mathbf{v}(\xi)) \cdot \mathbf{r}_k(\mathbf{v}(\xi)) = 1. \tag{3.3}$$

Equation (3.3) cannot be solved if the kth characteristic field is linearly degenerate. But if the kth field is genuinely nonlinear, we get, with the normalization (2.33),

$$\alpha(\xi) = 1.$$

Hence, we find either

$$\mathbf{v}'(\xi) = \mathbf{0}$$

or

$$\begin{cases} \mathbf{v}'(\xi) = \mathbf{r}_k(\mathbf{v}(\xi)), \\ \lambda_k(\mathbf{v}(\xi)) = \xi, \end{cases} \tag{3.4}$$

and $\mathbf{v}$ is therefore an integral curve of the field $\mathbf{r}_k$. Thus, assume that the kth characteristic field is genuinely nonlinear and that the function $\mathbf{v}$ is a solution of (3.3) with

$$\mathbf{v}(\lambda_k(\mathbf{u}_L)) = \mathbf{u}_L, \quad \mathbf{v}(\lambda_k(\mathbf{u}_R)) = \mathbf{u}_R$$

(this presupposes that $\mathbf{u}_L$ and $\mathbf{u}_R$ are on the same integral curve of $\mathbf{r}_k$ and that λ_k increases from $\mathbf{u}_L$ to $\mathbf{u}_R$ along this curve). Then, it follows from the above analysis that the function

$$\mathbf{u}(x,t)=\begin{cases}\mathbf{u}_L, & \frac{x}{t}\le\lambda_k(\mathbf{u}_L),\\ \mathbf{v}(\frac{x}{t}), & \lambda_k(\mathbf{u}_L)\le\frac{x}{t}\le\lambda_k(\mathbf{u}_R),\\ \mathbf{u}_R, & \frac{x}{t}\ge\lambda_k(\mathbf{u}_R)\end{cases}\tag{3.5}$$

is a continuous self-similar weak solution of (2.1).

Definition 3.1

Such a self-similar weak solution (3.5) of (2.1) is called a k-centered simple wave or a k-rarefaction wave connecting the states $\mathbf{u}_L$ and $\mathbf{u}_R$.

We shall see in Section 3.2 that the straight lines that form the rarefaction fan in Figure 3.1 are the characteristic curves of the kth field.

Concerning the existence of k-simple waves connecting two states, we have the following local result:

Theorem 3.1

Assume that the kth characteristic field is genuinely nonlinear with the normalization (2.33). Given a state $\mathbf{u}_L\in\Omega$, there exists a curve $\mathcal{R}_k(\mathbf{u}_L)$ of states of Ω that can be connected to $\mathbf{u}_L$ on the right by a k-simple wave. Moreover, there exists a parametrization of $\mathcal{R}_k(\mathbf{u}_L)$: $\varepsilon\to\mathbf{\Phi}_k(\varepsilon)$ defined for $0\le\varepsilon\le\varepsilon_0, \varepsilon_0$ small enough, such that

$$\mathbf{\Phi}_k(\varepsilon)=\mathbf{u}_L+\varepsilon\mathbf{r}_k(\mathbf{u}_L)+\frac{\varepsilon^2}{2}D\mathbf{r}_k(\mathbf{u}_L)\cdot\mathbf{r}_k(\mathbf{u}_L)+O(\varepsilon^3).\tag{3.6}$$

Proof. Let $\mathbf{v}:\xi\to\mathbf{v}(\xi)$ be the solution of the differential system

$$\mathbf{v}'(\xi)=\mathbf{r}_k(\mathbf{v}(\xi)),\quad \xi>\lambda_k(\mathbf{u}_L),\tag{3.7}$$

$$\mathbf{v}(\lambda_k(\mathbf{u}_L))=\mathbf{u}_L.\tag{3.8}$$

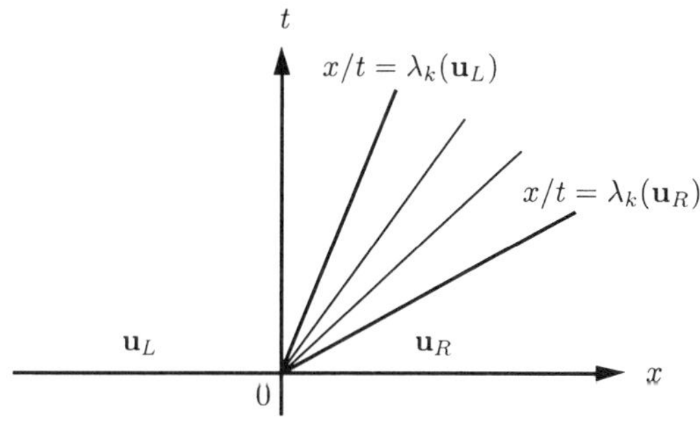

FIGURE 3.1. Rarefaction fan.

The function $\mathbf{v}$ exists for $\lambda_k(\mathbf{u}_L) \le \xi \le \lambda_k(\mathbf{u}_L) + \varepsilon_0$, $\varepsilon_0 > 0$ small enough. Using (3.7), we have

$$\frac{d}{d\xi}\lambda_k(\mathbf{v}(\xi)) = D\lambda_k(\mathbf{v}(\xi)) \cdot \mathbf{v}'(\xi) = D\lambda_k(\mathbf{v}(\xi)) \cdot \mathbf{r}_k(\mathbf{v}(\xi))$$

and by (2.33)

$$\frac{d}{d\xi}\lambda_k(\mathbf{v}(\xi)) = 1.$$

Thus $\mathbf{v}$ satisfies

$$\lambda_k(\mathbf{v}(\xi)) - \lambda_k(\mathbf{v}(\lambda_k(\mathbf{u}_L))) = \xi - \lambda_k(\mathbf{u}_L)$$

so that by (3.8)

$$\lambda_k(\mathbf{v}(\xi)) = \xi.$$

Hence, the function $\mathbf{v}$ is indeed the solution of (3.4) for which (3.8) holds.

Next, we define

$$\mathcal{R}_k(\mathbf{u}_L) = \{\mathbf{v}(\xi)\ ;\ \lambda_k(\mathbf{u}_L) \le \xi \le \lambda_k(\mathbf{u}_L) + \varepsilon_0\}.$$

The curve $\mathcal{R}_k(\mathbf{u}_L)$ is therefore the set of all states of Ω that can be connected to $\mathbf{u}_L$ on the right by a k-simple wave. Setting

$$\mathbf{\Phi}_k(\varepsilon) = \mathbf{v}\Big(\lambda_k(\mathbf{u}_L) + \varepsilon\Big)\ ,\ 0 \le \varepsilon \le \varepsilon_0\ ,$$

we have

$$\mathbf{\Phi}_k(0) = \mathbf{u}_L$$

and

$$\mathbf{\Phi}_k'(0) = \mathbf{v}'(\lambda_k(\mathbf{u}_L)) = \mathbf{r}_k(\mathbf{v}(\lambda_k(\mathbf{u}_L))) = \mathbf{r}_k(\mathbf{u}_L).$$

Finally, since

$$\mathbf{v}''(\varepsilon) = D\mathbf{r}_k(\mathbf{v}(\xi)) \cdot \mathbf{v}'(\xi) = D\mathbf{r}_k(\mathbf{v}(\xi)) \cdot \mathbf{r}_k(\mathbf{v}(\xi)),$$

we find

$$\mathbf{\Phi}_k''(0) = D\mathbf{r}_k(\mathbf{u}_L) \cdot \mathbf{r}_k(\mathbf{u}_L).$$

This proves the expansion (3.6). □

The curve $\mathcal{R}_k(\mathbf{u}_L)$ is called a *k-rarefaction curve*. It is an *integral curve of* $\mathbf{r}_k$, which is thus tangent to $\mathbf{r}_k(\mathbf{u}_L)$ at the point $\mathbf{u}_L$.

To go on with our study of elementary waves associated with one specific characteristic family (for which we no longer assume that it is genuinely nonlinear), let us introduce the Riemann invariants.

3.2 Riemann invariants

Definition 3.2

A smooth function $w : \Omega \to \mathbb{R}$ is called a k-Riemann invariant if it satisfies

$$Dw(\mathbf{u}) \cdot \mathbf{r}_k(\mathbf{u}) = 0, \quad \forall \mathbf{u} \in \Omega. \tag{3.9}$$

A k-Riemann invariant w is constant on a curve $\mathbf{v} : \xi \in \mathbb{R} \to \mathbf{v}(\xi) \in \mathbb{R}^p$ iff

$$\frac{d}{d\xi} w(\mathbf{v}(\xi)) = Dw(\mathbf{v}(\xi)) \cdot \mathbf{v}'(\xi) = 0,$$

which holds if $\mathbf{v}$ is an integral curve of $\mathbf{r}_k$

$$\mathbf{v}'(\xi) = \mathbf{r}_k(\mathbf{v}(\xi)). \tag{3.10}$$

This means that a *k-Riemann invariant is constant along the trajectories of the vector field* $\mathbf{r}_k$. Note that (3.9) is a first-order linear differential equation in $\mathbb{R}^p$. It can usually be integrated explicitly, as we shall see.

Remark 3.1. When the kth field is linearly degenerate, λ_k is a k-Riemann invariant (see (2.6)). □

Let us show that there exist locally $(p-1)$ k-Riemann invariants whose gradients are linearly independent. In general, we can solve (3.9) locally by the method of characteristics as soon as w is given on some initial surface that is not characteristic for the vector field $\mathbf{r}_k$. Recall that a surface $\mathcal{S} = \{\mathbf{u} \in \mathbb{R}^p, \varphi(\mathbf{u}) = 0\}$ is characteristic if

$$D\varphi(\mathbf{u}) \cdot \mathbf{r}_k(\mathbf{u}) = 0 \quad \forall \mathbf{u} \in \mathcal{S},$$

i.e., if $\mathbf{r}_k(\mathbf{u})$ is tangent to $\mathcal{S}$ for $\mathbf{u} \in \mathcal{S}$. For $p = 2$, we thus take a curve that is not tangent to $\mathbf{r}_k(\mathbf{u})$ at any point, and we assign arbitrary values for w along it. For simplicity, we prove the result in the case where the hyperplane

$$\mathcal{S}_p = \{\mathbf{u} \in \mathbb{R}^p, u_p = 0\}$$

is not characteristic, i.e.,

$$\mathbf{r}_k(\mathbf{u}) \notin \mathcal{S}_p \text{ for } \mathbf{u} \in \mathcal{S}_p \cap \Omega.$$

Lemma 3.1

Assume that the hyperplane $\mathcal{S}_p = \{\mathbf{u} \in \mathbb{R}^p, u_p = 0\}$ is not characteristic for $\mathbf{r}_k$. Then, there exists a smooth change of variables $\mathbf{u} = \boldsymbol{\theta}(\mathbf{v})$ defined in a neighborhood of $\mathcal{S}_p$ such that (3.9) is equivalent to

$$\frac{\partial z}{\partial v_p} = 0, \quad z = w \circ \boldsymbol{\theta}. \tag{3.11}$$

Moreover, the $(p-1)$ functions

$$w_j = z_j \circ \boldsymbol{\theta}^{-1}(\mathbf{v}), \; 1 \leq j \leq p-1,$$

where

$$z_j(\mathbf{v}) = v_j, \; 1 \leq j \leq p-1,$$

are k-Riemann invariants whose gradients are linearly independent.

Proof. Let $\mathbf{e}_p = (0, 0, ..., 1)^T \in \mathbb{R}^p$ be the pth canonical basis vector. Since by definition

$$\frac{\partial z}{\partial v_p}(\mathbf{v}) = Dz(\mathbf{v}) \cdot \mathbf{e}_p$$

and

$$Dz(\mathbf{v}) \cdot \mathbf{e}_p = D(w \circ \boldsymbol{\theta})(\mathbf{v}) \cdot \mathbf{e}_p = Dw(\boldsymbol{\theta}(\mathbf{v})) \cdot D\boldsymbol{\theta}(\mathbf{v}) \cdot \mathbf{e}_p = Dw(\mathbf{u}) \cdot \frac{\partial \theta}{\partial v_p}(\mathbf{v}),$$

we see that (3.11) is equivalent to (3.9) iff $\boldsymbol{\theta}$ satisfies

$$\frac{\partial \boldsymbol{\theta}}{\partial v_p}(\mathbf{v}) = \mathbf{r}_k(\boldsymbol{\theta}(\mathbf{v})). \tag{3.12}$$

Given the values of $\boldsymbol{\theta}$ on the surface $\mathcal{S}_p$, for instance

$$\boldsymbol{\theta}(v_1, ..., v_{p-1}, 0) = (v_1, ..., v_{p-1}, 0),$$

for each $(v_1, ..., v_{p-1})$, (3.12) has a smooth solution that is an integral curve of $\mathbf{r}_k$ defined for v_p small enough. Since by assumption

$$\frac{\partial \boldsymbol{\theta}}{\partial v_p}(v_1, ..., v_{p-1}, 0) = \mathbf{r}_k(v_1, ..., v_{p-1}, 0) \notin \mathcal{S}_p$$

and

$$\frac{\partial \boldsymbol{\theta}}{\partial v_i}(v_1, ..., v_{p-1}, 0) = \mathbf{e}_i, \quad i = 1, ..., p-1,$$

this enables us to define a mapping $\mathbf{v} \to \boldsymbol{\theta}(\mathbf{v})$ which is one-to-one in a neighborhood of $\mathcal{S}_p$. Then, the coordinate functions $z_j(\mathbf{v}) = v_j, 1 \leq j \leq p-1$ are solutions of (3.11), and their gradients are obviously linearly independent. The result follows easily. □

We have thus proven that locally there exist $(p-1)$ k-Riemann invariants whose gradients are linearly independent. Now, given $w_1, ..., w_j$, a set of j k-Riemann invariants, and $Z : \mathbb{R}^j \to \mathbb{R}$, $Z(w_1, ..., w_j)$ is clearly a k-Riemann invariant and vice versa. We have the following result:

Lemma 3.2

Given $(p-1)$ k-Riemann invariants $w_1, ..., w_{p-1}$ whose gradients are linearly independent, there exists (locally), for any k-Riemann invariant w, a function $Z = \mathbb{R}^{p-1} \to \mathbb{R}$ such that

$$w = Z(w_1, ..., w_{p-1}).$$

Proof. By Lemma 3.1, the proof is obvious since, after a smooth change of variables, the $(p-1)$ k-Riemann invariants are the coordinates $z_j(\mathbf{v}) =$

v_j, $1 \leq j \leq p-1$. The general case follows in a similar way. Let us complete the set $w_1, ..., w_{p-1}$ by a smooth function $w_p : \Omega \to \mathbb{R}$ so that the p gradients $\nabla w_1, ..., \nabla w_{p-1}, \nabla w_p$ are linearly independent. For instance, we can find locally a function w_p such that $\nabla w_p(\mathbf{u})$ is collinear to $\mathbf{r}_k(\mathbf{u})$, since by (3.9) $\nabla w_1, ..., \nabla w_{p-1}$ are all orthogonal to $\mathbf{r}_k$. Then, the mapping $\mathbb{R}^p \to \mathbb{R}^p$ defined by

$$\mathbf{u} = (u_1, ..., u_p)^T \to (w_1, ..., w_p)^T(\mathbf{u}) = \mathbf{v} \tag{3.13}$$

is a diffeomorphism of $\mathbb{R}^p$. Denote by $\boldsymbol{\theta}$ the inverse diffeomorphism so that $\mathbf{u} = \boldsymbol{\theta}(\mathbf{v})$, and we can write

$$w(\mathbf{u}) = w \circ \boldsymbol{\theta} \circ \boldsymbol{\theta}^{-1}(\mathbf{u}).$$

If we prove that

$$\frac{\partial(w \circ \boldsymbol{\theta})}{\partial w_p}(\mathbf{v}) = 0, \tag{3.14}$$

the result will follow by setting $Z = w \circ \boldsymbol{\theta}$ since then

$$w(\mathbf{u}) = w \circ \boldsymbol{\theta}(w_1, ..., w_{p-1})(\mathbf{u}) = Z(w_1, ..., w_{p-1})(\mathbf{u}).$$

It remains to check (3.14). We write

$$\frac{\partial(w \circ \boldsymbol{\theta})}{\partial w_p}(\mathbf{v}) = D(w \circ \boldsymbol{\theta})(\mathbf{v}) \cdot \mathbf{e}_p = Dw(\mathbf{u}) \cdot \boldsymbol{\theta}'(\mathbf{v}) \cdot \mathbf{e}_p,$$

where $\mathbf{u} = \boldsymbol{\theta}(\mathbf{v})$. The vector

$$\mathbf{f}_p = \boldsymbol{\theta}'(\mathbf{v}) \cdot \mathbf{e}_p = \Big((\boldsymbol{\theta}^{-1})'(\mathbf{u})\Big)^{-1} \cdot \mathbf{e}_p$$

obviously satisfies

$$(\boldsymbol{\theta}^{-1})'(\mathbf{u}) \cdot \mathbf{f}_p = \mathbf{e}_p.$$

By the definition of $\boldsymbol{\theta}$, this yields

$$Dw_k(\mathbf{u}) \cdot \mathbf{f}_p = 0, \quad k = 1, ..., p-1.$$

Now, since w is a k-Riemann invariant, $\nabla w(\mathbf{u})$ is orthogonal to $\mathbf{r}_k(\mathbf{u})$ so that in the basis $\nabla w_1, ..., \nabla w_{p-1}, \nabla w_p$, we have

$$\nabla w(\mathbf{u}) = \sum_{k=1}^{p-1} \alpha_k(\mathbf{u}) \nabla w_k(\mathbf{u}),$$

which implies

$$Dw(\mathbf{u}) \cdot \mathbf{f}_p = 0$$

and proves the desired result. □

Example 3.1. The p-system. Consider once more the p-system (2.7). Since we have a system of two equations, we are looking for one 1-Riemann invariant w_1 and one 2-Riemann invariant w_2. By (2.9) we have

$$\nabla w_1(\mathbf{w}) \cdot \mathbf{r}_1(v) = \frac{\partial w_1}{\partial v} + \sqrt{-p'(v)}\, \frac{\partial w_1}{\partial u} = 0.$$

Hence $w_1(\mathbf{w}) = w_1(v, u)$ is given by

$$w_1(v, u) = u - \int^v \sqrt{-p'(y)}\, dy. \tag{3.15a}$$

Similarly, we get

$$\nabla w_2(\mathbf{w}) \cdot \mathbf{r}_2(v) = \frac{\partial w_2}{\partial v} - \sqrt{-p'(v)}\, \frac{\partial w_2}{\partial u} = 0,$$

so that w_2 is given by

$$w_2(v, u) = u + \int^v \sqrt{-p'(y)}\, dy. \tag{3.15b}$$

We have thus obtained global Riemann invariants. As a consequence, any k-Riemann invariant is globally defined as a function of w_k, for $k = 1, 2$.□

Again, it may be more convenient to work on a nonconservative form of the system of conservation laws (2.1). The above computations for Lemma 3.2 show that the notion of Riemann invariant is independent of the chosen conservative or nonconservative form of (2.1). Indeed, by using the change of dependent variables (2.10) and setting

$$z(\mathbf{v}) = w(\boldsymbol{\theta}(\mathbf{v})),$$

we obtain

$$Dz(\mathbf{v}) \cdot \mathbf{s}_k(\mathbf{v}) = Dw(\boldsymbol{\theta}(\mathbf{v})) \cdot \boldsymbol{\theta}'(\mathbf{v})^{-1}\mathbf{r}_k(\boldsymbol{\theta}(\mathbf{v})) = Dw(\boldsymbol{\theta}(\mathbf{v})) \cdot \mathbf{r}_k(\boldsymbol{\theta}(\mathbf{v})).$$

Hence, a k-Riemann invariant may be equivalently defined by

$$Dz(\mathbf{v}) \cdot \mathbf{s}_k(\mathbf{v}) = 0.$$

Example 3.2. The gas dynamics equations in Lagrangian coordinates. We consider again the equations (2.16). Let us check that the three pairs of Riemann invariants can be taken as

$$(u + \ell, s), (u, p), (u - \ell, s), \tag{3.16}$$

where the function $\ell = \ell(\tau, s)$ is defined up to an additive function of s by

$$\frac{\partial \ell}{\partial \tau} = -\frac{c}{\tau}. \tag{3.17}$$

As we have just noticed, we may work on the nonconservative form (2.19) of the gas dynamics equations. Since $\mathbf{r}_1 = (1, \frac{c}{\tau}, 0)^T$, the 1-Riemann invariants $w = w(\tau, u, s)$ are solutions of the equation

$$\frac{\partial w}{\partial \tau} + \frac{c}{\tau}\frac{\partial w}{\partial u} = 0.$$

Hence $u+\ell$ and s are indeed two 1-Riemann invariants whose derivatives

$$\nabla(u+\ell) = \left(\frac{\partial \ell}{\partial \tau}, 1, \frac{\partial \ell}{\partial s}\right)^T, \quad \nabla s = (0,0,1)^T$$

are clearly linearly independent. Here again, any 1-Riemann invariant is globally of the form $Z(u+\ell, s)$. We proceed in a similar way for the 2- and 3-Riemann invariants. □

Example 3.3. The gas dynamics equations in Eulerian coordinates. Let us consider finally the equations (2.25). Observe that the function $\ell = \ell(\rho, s)$ defined by (3.17) may be equivalently defined by

$$\frac{\partial \ell}{\partial \rho}(\rho, s) = \frac{c}{\rho}. \tag{3.18}$$

Then using the nonconservative form (2.28) of the equations (2.25), it is an easy matter to check that the three pairs of Riemann invariants can be again taken as in (3.16). □

Let us check the following simple useful property of Riemann invariants.

Theorem 3.2

On a k-rarefaction wave, all k-Riemann invariants are constant.

Proof. Let $\mathbf{u}$ be a k-rarefaction wave of the form (3.5), and let w be a k-Riemann invariant. The function $w(\mathbf{u}) : (x,t) \to w(\mathbf{u}(x,t))$ is continuous for $t > 0$. First, $w(\mathbf{u})$ is constant for $\frac{x}{t} \le \lambda_k(\mathbf{u}_L)$ and $\frac{x}{t} \ge \lambda_k(\mathbf{u}_R)$. Next, for $\lambda_k(\mathbf{u}_L) \le \frac{x}{t} \le \lambda_k(\mathbf{u}_R)$, $\mathbf{u}$ is an integral curve of $\mathbf{r}_k$, which proves the result . □

More generally, let us define a wider class of smooth k-waves.

Definition 3.3

A smooth solution $\mathbf{u}(x,t)$ of (2.1) defined on a domain D of $\mathbb{R} \times \mathbb{R}_+$ is called a k-simple wave if $w(\mathbf{u}(x,t))$ is constant in D for any k-Riemann invariant w.

Example 3.3 (revisited). For the gas dynamics equations, a 1-simple wave is one for which $u+\ell$ and s are constant. We shall use this fact in Chapter II, Section 3, and also in Chapter III for Osher's scheme. □

Proposition 3.1

If $\mathbf{u}$ is a k-simple wave, the field of values of $\mathbf{u}$ is restricted to only one integral curve of $\mathbf{r}_k$ in $\mathbb{R}^p$.

Proof. We can find at least one k-Riemann invariant whose constant values on two integral curves are distinct. Indeed, in the case $p=2$, we have seen that given a curve $\mathcal{C}$ that is not tangent to $\mathbf{r}_k$, we can assign arbitrary values for w, for instance strictly increasing along $\mathcal{C}$. The value on each integral curve of $\mathbf{r}_k$ intersecting $\mathcal{C}$ is then determined, and w takes distinct values along distinct integral curves. The general case follows similarly. □

Hence, we look for a function

$$\mathbf{u}(x,t) = \mathbf{v}(\varphi(x,t)),$$

where $\mathbf{v}$ is solution of (3.10) with some initial value $\mathbf{v}_0$:

$$\begin{cases} \mathbf{v}'(\xi) = \mathbf{r}_k(\mathbf{v}(\xi)) \\ \mathbf{v}(\xi_0) = \mathbf{v}_0. \end{cases} \tag{3.19}$$

Such a solution $\mathbf{v}$ exists at least locally (i.e., for ξ close to ξ_0) as we have already observed. The equation for φ is obtained by writing

$$\begin{aligned} \mathbf{0} &= \frac{\partial \mathbf{u}}{\partial t} + \frac{\partial}{\partial x}\mathbf{f}(\mathbf{u}) \\ &= \mathbf{v}'(\varphi)\frac{\partial \varphi}{\partial t} + \mathbf{A}(\mathbf{u})\mathbf{v}'(\varphi)\frac{\partial \varphi}{\partial x} \\ &= \mathbf{r}_k(\mathbf{v}(\varphi))\frac{\partial \varphi}{\partial t} + \mathbf{A}(\mathbf{u})\mathbf{r}_k(\mathbf{v}(\varphi))\frac{\partial \varphi}{\partial x} \\ &= \mathbf{r}_k(\mathbf{v}(\varphi))\left\{\frac{\partial \varphi}{\partial t} + \lambda_k(\mathbf{v}(\varphi))\frac{\partial \varphi}{\partial x}\right\}. \end{aligned}$$

Thus, we look for φ as a smooth solution of a quasilinear scalar equation

$$\frac{\partial \varphi}{\partial t} + \lambda_k(\mathbf{v}(\varphi))\frac{\partial \varphi}{\partial x} = 0. \tag{3.20}$$

We have already studied such equations in the Introduction, Section 2. The characteristics, i.e., the integral curves of the differential equation

$$\frac{dx}{dt} = \lambda_k(\mathbf{v}(\varphi(x,t))), \tag{3.21}$$

are straight lines along which φ is constant. We deduce the implicit formula for φ by integrating along the characteristics

$$\varphi(x,t) = \varphi_0\Big(x - \lambda_k(\mathbf{v}(\varphi(x,t)))t\Big) \quad \text{if } \varphi(\cdot,0) = \varphi_0, \tag{3.22}$$

which is valid as long as φ remains smooth. This gives the expression of a k-simple wave

$$u(x,t) = \mathbf{v}\Big(\varphi_0\Big(x - \lambda_k(\mathbf{v}(\varphi))t\Big)\Big).$$

Let us introduce the *characteristic curves* C_k of the kth field. They are defined as the integral curves of the differential system

$$\frac{dx}{dt} = \lambda_k(\mathbf{u}(x,t)). \tag{3.23}$$

When $\mathbf{u}$ is a k-simple wave, we have thus proved the following theorem.

Theorem 3.3

Let **u** *be a k-simple wave. Then, the characteristics of the* kth *field are straight lines along which* **u** *is constant.*

We can determine some of the k-simple waves when the kth field is either genuinely nonlinear or linearly degenerate.

Example 3.4. Simple waves for a genuinely nonlinear field. Assume that the kth field is genuinely nonlinear. With the normalization (2.33) and (3.19), we have

$$\frac{d}{d\varphi}\Big(\lambda_k(\mathbf{v}(\varphi))\Big) = D\lambda_k(\mathbf{v}(\varphi)) \cdot \mathbf{v}'(\varphi) = 1.$$

Hence, after possibly changing the function φ_0 by an additive constant, we get

$$\lambda_k(\mathbf{v}(\varphi)) = \varphi,$$

and (3.20) becomes

$$\frac{\partial \varphi}{\partial t} + \varphi \frac{\partial \varphi}{\partial x} = 0,$$

which means that φ is a solution of the Burgers equation. As a first example, we find the self-similar k-centered simple waves previously defined in (3.6) that correspond to $\varphi(x,t) = \frac{x}{t}$, or more generally to $\frac{(x-x_0)}{(t-t_0)}$. They are called rarefaction (centered) waves because the characteristics form a fan centered at point (x_0, t_0) that diverges in opposition to compression centered waves, for which the fan converges and that generate singularities (see Figure 3.2). We shall also find compression waves for $t \geq t_0$ (λ_k decreases from right to left in the fan) in Chapter III, when studying Osher's scheme. We have also encountered noncentered simple waves when resolving the G.R.P. (generalized Riemann problem) in G.R., Chapter IV, Section

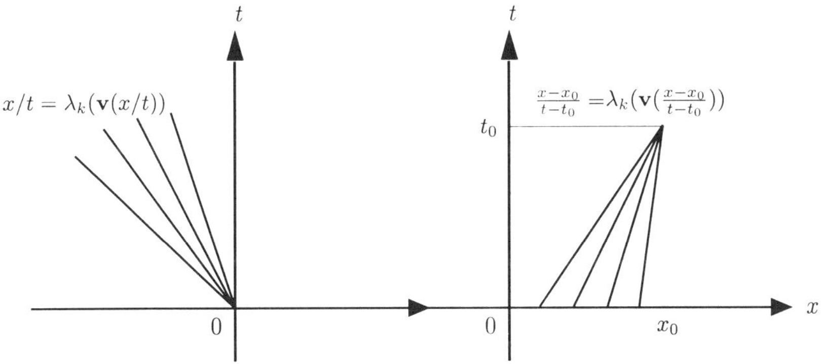

FIGURE 3.2. Rarefaction and compression centered waves.

3.3. In the case of a k-rarefaction wave (3.6) connecting two states $\mathbf{u}_L$ and $\mathbf{u}_R$, the fan that we have already depicted in Section 3.1 (see Figure 3.1) is thus composed of the characteristics C_k which are straight lines and is bordered by the lines $\frac{x}{t} = \lambda_k(\mathbf{u}_L)$ and $\frac{x}{t} = \lambda_k(\mathbf{u}_R)$.

Example 3.5. Simple waves for a linearly degenerate field. Assume now that the kth field is linearly degenerate. We cannot find centered simple waves, as we have already observed, because (3.3) never holds in that case. Nonetheless, we know (see Remark 3.1) that λ_k is a k-Riemann invariant, so that λ_k is constant on a k-simple wave. Thus, the functions $\mathbf{u}$ of the form

$$\mathbf{u}(x,t) = \mathbf{v}(\varphi_0(x - \overline{\lambda}_k t)), \quad \overline{\lambda}_k = \lambda_k(\mathbf{v}), \tag{3.24}$$

with $\mathbf{v}$ satisfying (3.19), are k-simple waves, and the characteristics C_k are now the parallel lines $x - \overline{\lambda}_k t = C$. We cannot connect two states $\mathbf{u}_L$ and $\mathbf{u}_R$ by a continuous k-wave, and we must consider a discontinuity wave. This will be the object of the next section. □

4 Shock waves and contact discontinuities

Given two states $\mathbf{u}_L$ and $\mathbf{u}_R \in \Omega$, we are now looking for piecewise constant discontinuous solutions of (2.1) that connect $\mathbf{u}_L$ and $\mathbf{u}_R$. Let us recall that along a line of discontinuity $x = \xi(t)$ of a weak solution $\mathbf{u}$ of (2.1), $\mathbf{u}$ satisfies the Rankine–Hugoniot jump conditions

$$\sigma[\mathbf{u}] = [\mathbf{f}(\mathbf{u})], \tag{4.1}$$

where $\sigma = \xi'(t)$ is the speed of propagation of the discontinuity (see Introduction, Section 2). Therefore, the function

$$\mathbf{u}(x,t) = \begin{cases} \mathbf{u}_L, & x < \sigma t, \\ \mathbf{u}_R, & x > \sigma t \end{cases} \tag{4.2}$$

is a weak solution of (2.1) provided that the real number σ satisfies

$$\mathbf{f}(\mathbf{u}_R) - \mathbf{f}(\mathbf{u}_L) = \sigma(\mathbf{u}_R - \mathbf{u}_L). \tag{4.3}$$

Such a solution (4.1), (4.2) of the nonlinear hyperbolic system (2.1) is called a *discontinuity wave*. Given a state $\mathbf{u}_L \in \Omega$, we want to determine all the states $\mathbf{u}_R \in \Omega$ to which $\mathbf{u}_L$ can be connected on the right by a discontinuity wave. Thus, we introduce the following definition.

Definition 4.1

The Rankine–Hugoniot set of $\mathbf{u}_0$ *is the set of all states* $\mathbf{u} \in \Omega$ *such that there exists* $\sigma(\mathbf{u}_0, \mathbf{u}) \in \mathbb{R}$ *with*

$$\sigma(\mathbf{u}_0, \mathbf{u})(\mathbf{u} - \mathbf{u}_0) = \mathbf{f}(\mathbf{u}) - \mathbf{f}(\mathbf{u}_0). \tag{4.4}$$

The structure of the Rankine–Hugoniot set of $\mathbf{u}_0$ is given by the following theorem.

Theorem 4.1

Let $\mathbf{u}_0$ *be in* Ω. *The Rankine–Hugoniot set of* $\mathbf{u}_0$ *is locally made of* p *smooth curves* $\mathcal{S}_k(\mathbf{u}_0), 1 \leq k \leq p$. *Moreover, for all* k, *there exists a parametrization of* $\mathcal{S}_k(\mathbf{u}_0) : \varepsilon \to \boldsymbol{\Psi}_k(\varepsilon)$ *defined for* $|\varepsilon| \leq \varepsilon_1, \varepsilon_1$ *small enough, such that*

$$\boldsymbol{\Psi}_k(\varepsilon) = \mathbf{u}_0 + \varepsilon\, \mathbf{r}_k(\mathbf{u}_0) + \frac{\varepsilon^2}{2} D\mathbf{r}_k(\mathbf{u}_0) \cdot \mathbf{r}_k(\mathbf{u}_0) + O(\varepsilon^3) \tag{4.5}$$

and

$$\sigma(\mathbf{u}_0, \boldsymbol{\Psi}_k(\varepsilon)) = \lambda_k(\mathbf{u}_0) + \frac{\varepsilon}{2} D\lambda_k(\mathbf{u}_0) \cdot \mathbf{r}_k(\mathbf{u}_0) + O(\varepsilon^2). \tag{4.6}$$

Proof. We write

$$\begin{aligned} \mathbf{f}(\mathbf{u}) - \mathbf{f}(\mathbf{u}_0) &= \int_0^1 \frac{d}{ds} \mathbf{f}(\mathbf{u}_0 + s(\mathbf{u} - \mathbf{u}_0))ds \\ &= \Big(\int_0^1 \mathbf{A}(\mathbf{u}_0 + s(\mathbf{u} - \mathbf{u}_0))ds\Big)(\mathbf{u} - \mathbf{u}_0). \end{aligned}$$

Hence, setting

$$\mathbf{A}(\mathbf{u}, \mathbf{v}) = \int_0^1 \mathbf{A}(\mathbf{u} + s(\mathbf{v} - \mathbf{u}))ds,$$

the jump condition (4.4) becomes

$$\Big(\mathbf{A}(\mathbf{u}_0, \mathbf{u}) - \sigma(\mathbf{u}_0, \mathbf{u})\Big)(\mathbf{u} - \mathbf{u}_0) = \mathbf{0}. \tag{4.7}$$

Note that the $p \times p$ matrix $\mathbf{A}(\mathbf{u}_0, \mathbf{u}_0) = \mathbf{A}(\mathbf{u}_0)$ has p distinct eigenvalues $\lambda_1(\mathbf{u}_0) < ... < \lambda_p(\mathbf{u}_0)$, and the function $\mathbf{u} \to \mathbf{A}(\mathbf{u}_0, \mathbf{u})$ is continuous. Thus, using a continuity argument, there exists a neighborhood $\mathcal{N}$ of $\mathbf{u}_0$ in Ω and p real functions $\mathbf{u} \to \lambda_k(\mathbf{u}_0, \mathbf{u}), 1 \leq k \leq p$, defined in $\mathcal{N}$ such that $\lambda_k(\mathbf{u}_0, \mathbf{u}), 1 \leq k \leq p$, are the p distinct real eigenvalues of $\mathbf{A}(\mathbf{u}_0, \mathbf{u})$ with

$$\lambda_k(\mathbf{u}_0, \mathbf{u}_0) = \lambda_k(\mathbf{u}_0).$$

We denote by $\mathbf{l}_k(\mathbf{u}_0, \mathbf{u})^T, 1 \leq k \leq p$, the "left eigenvectors" of the matrix $\mathbf{A}(\mathbf{u}_0, \mathbf{u})$, i.e.,

$$\mathbf{l}_k(\mathbf{u}_0, \mathbf{u})^T \mathbf{A}(\mathbf{u}_0, \mathbf{u}) = \lambda_k(\mathbf{u}_0, \mathbf{u}) \mathbf{l}_k(\mathbf{u}_0, \mathbf{u})^T.$$

Now, using (4.7), a state $\mathbf{u} \in \mathcal{N}$ belongs to the Rankine–Hugoniot set of $\mathbf{u}_0$ if and only if there exists an index $k \in \{1, ..., p\}$ such that

$$\sigma(\mathbf{u}_0, \mathbf{u}) = \lambda_k(\mathbf{u}_0, \mathbf{u}) \tag{4.8}$$

and

(4.9) $\mathbf{u} - \mathbf{u}_0$ is a right eigenvector of $\mathbf{A}(\mathbf{u}_0, \mathbf{u})$ associated with $\lambda_k(\mathbf{u}_0, \mathbf{u})$.

On the other hand, (4.9) holds if and only if

$$\mathbf{l}_j(\mathbf{u}_0, \mathbf{u})^T(\mathbf{u} - \mathbf{u}_0) = 0, \quad j \neq k. \tag{4.10}$$

This gives a system of $(p-1)$ equations in p unknowns which can be written in the form

$$\mathbf{G}_k(\mathbf{u}) = \mathbf{M}_k(\mathbf{u})(\mathbf{u} - \mathbf{u}_0) = \mathbf{0},$$

where

$$\mathbf{M}_k(\mathbf{u}) = \begin{pmatrix} \mathbf{l}_1(\mathbf{u}_0, \mathbf{u})^T \\ \vdots \\ \mathbf{l}_{k-1}(\mathbf{u}_0, \mathbf{u})^T \\ \mathbf{l}_{k+1}(\mathbf{u}_0, \mathbf{u})^T \\ \vdots \\ \mathbf{l}_p(\mathbf{u}_0, \mathbf{u})^T \end{pmatrix}.$$

We have

$$\mathbf{G}_k(\mathbf{u}_0) = 0, \ D\mathbf{G}_k(\mathbf{u}_0) = \mathbf{M}_k(\mathbf{u}_0).$$

Moreover, since the vectors $\mathbf{l}_j(\mathbf{u}_0, \mathbf{u}_0) = \mathbf{l}_j(\mathbf{u}_0)$, $1 \le j \le p$, are linearly independent, the $(p-1) \times p$ matrix $\mathbf{M}_k(\mathbf{u}_0)$ has rank $p-1$. Therefore, by the implicit function theorem, there exists a one-parameter family $\mathcal{S}_k(\mathbf{u}_0)$ of solutions of (4.10), $\theta \to \boldsymbol{\Psi}_k(\theta)$, $|\theta| \le \theta_1$ small enough, with

$$\boldsymbol{\Psi}_k(0) = \mathbf{u}_0. \tag{4.11}$$

In addition, we have

$$\begin{cases} \sigma(\mathbf{u}_0, \boldsymbol{\Psi}_k(\theta)) = \sigma_k(\mathbf{u}_0, \boldsymbol{\Psi}_k(\theta)), \\ \sigma(\mathbf{u}_0, \boldsymbol{\Psi}_k(0)) = \lambda_k(\mathbf{u}_0). \end{cases} \tag{4.12}$$

Hence we have proved that the Rankine–Hugoniot set of $\mathbf{u}_0$ is locally made of p curves $\mathcal{S}_k(\mathbf{u}_0)$. It remains to check the expansions (4.5) and (4.6). First, it follows from (4.10) and (4.11) that

$$\begin{aligned} 0 &= \lim_{\theta \to 0} \mathbf{l}_j(\mathbf{u}_0, \boldsymbol{\Psi}_k(\theta))^T \Big(\frac{\boldsymbol{\Psi}_k(\theta) - \boldsymbol{\Psi}_k(0)}{\theta} \Big) \\ &= \mathbf{l}_j(\mathbf{u}_0)^T \boldsymbol{\Psi}_k'(0), \qquad j \neq k, \end{aligned}$$

so that $\boldsymbol{\Psi}_k'(0)$ is collinear to $\mathbf{r}_k(\mathbf{u}_0)$. Hence, we can change our parametrization in order to get

$$\boldsymbol{\Psi}_k'(0) = \mathbf{r}_k(\mathbf{u}_0). \tag{4.13}$$

Next, for the sake of simplicity, we set

$$\begin{aligned} \mathbf{A}_k(\theta) &= \mathbf{A}(\boldsymbol{\Psi}_k(\theta)), \\ \sigma_k(\theta) &= \lambda_k(\mathbf{u}_0, \boldsymbol{\Psi}_k(\theta)). \end{aligned}$$

Then, differentiating the condition (4.4)

$$\sigma_k(\boldsymbol{\Psi}_k - \mathbf{u}_0) = \mathbf{f}(\boldsymbol{\Psi}_k) - \mathbf{f}(\mathbf{u}_0),$$

we obtain

$$\sigma_k'(\boldsymbol{\Psi}_k - \mathbf{u}_0) + \sigma_k \boldsymbol{\Psi}_k' = \mathbf{A}_k \boldsymbol{\Psi}_k'$$

and

$$\sigma_k''(\boldsymbol{\Psi}_k - \mathbf{u}_0) + 2\sigma_k' \boldsymbol{\Psi}_k' + \sigma_k \boldsymbol{\Psi}_k'' = \mathbf{A}_k' \boldsymbol{\Psi}_k' + \mathbf{A}_k \boldsymbol{\Psi}_k'',$$

which gives for $\theta = 0$

$$2\sigma_k'(0)\mathbf{r}_k(\mathbf{u}_0) + \lambda_k(\mathbf{u}_0)\boldsymbol{\Psi}_k''(0) = \mathbf{A}_k'(0)\mathbf{r}_k(\mathbf{u}_0) + \mathbf{A}(\mathbf{u}_0)\boldsymbol{\Psi}_k''(0)$$

or equivalently

$$\Big(\mathbf{A}(\mathbf{u}_0) - \lambda_k(\mathbf{u}_0)\Big)\boldsymbol{\Psi}_k''(0) + \mathbf{A}_k'(0)\mathbf{r}_k(\mathbf{u}_0) = 2\sigma_k'(0)\mathbf{r}_k(\mathbf{u}_0). \tag{4.14}$$

On the other hand, differentiating

$$\mathbf{A}_k \mathbf{r}_k(\boldsymbol{\Psi}_k) = \lambda_k(\boldsymbol{\Psi}_k)\mathbf{r}_k(\boldsymbol{\Psi}_k)$$

gives

$$\mathbf{A}_k'\mathbf{r}_k(\boldsymbol{\Psi}_k) + \mathbf{A}_k D\mathbf{r}_k(\boldsymbol{\Psi}_k)\cdot\boldsymbol{\Psi}_k' = D\lambda_k(\boldsymbol{\Psi}_k)\cdot\boldsymbol{\Psi}_k'\mathbf{r}_k(\boldsymbol{\Psi}_k) + \lambda_k(\boldsymbol{\Psi}_k)D\mathbf{r}_k(\boldsymbol{\Psi}_k)\cdot\boldsymbol{\Psi}_k'$$

and for $\theta = 0$

$$\begin{cases} \Big(\mathbf{A}(\mathbf{u}_0) - \lambda_k(\mathbf{u}_0)\Big)D\mathbf{r}_k(\mathbf{u}_0)\cdot\mathbf{r}_k(\mathbf{u}_0) \\ \quad + \mathbf{A}_k'(0)\mathbf{r}_k(\mathbf{u}_0) - D\lambda_k(\mathbf{u}_0)\cdot\mathbf{r}_k(\mathbf{u}_0)\cdot\mathbf{r}_k(\mathbf{u}_0) = 0. \end{cases} \tag{4.15}$$

Hence, by subtracting (4.15) from (4.14), we get

$$\begin{cases} \Big(\mathbf{A}(\mathbf{u}_0) - \lambda_k(\mathbf{u}_0)\Big)\Big(\boldsymbol{\Psi}_k''(0) - D\mathbf{r}_k(\mathbf{u}_0)\cdot\mathbf{r}_k(\mathbf{u}_0)\Big) \\ \quad + D\lambda_k(\mathbf{u}_0)\cdot\mathbf{r}_k(\mathbf{u}_0)\cdot\mathbf{r}_k(\mathbf{u}_0) = 2\sigma_k(0)\mathbf{r}_k(\mathbf{u}_0). \end{cases} \tag{4.16}$$

Now, multiplying (4.16) by $\mathbf{l}_k(\mathbf{u}_0)^T$ on the left and using the normalization

$$\mathbf{l}_k(\mathbf{u}_o)^T\mathbf{r}_k(\mathbf{u}_0) = 1,$$

we find

$$\sigma_k'(0) = \frac{1}{2}D\lambda_k(\mathbf{u}_o)\cdot\mathbf{r}_k(\mathbf{u}_0). \tag{4.17}$$

Substituting $\sigma_k'(0)$ in (4.16) by its expression (4.17) gives

$$\Big(\mathbf{A}(\mathbf{u}_0) - \lambda_k(\mathbf{u}_0)\Big)\Big(\boldsymbol{\Psi}_k''(0) - D\mathbf{r}_k(\mathbf{u}_0)\cdot\mathbf{r}_k(\mathbf{u}_0)\Big) = 0.$$

Therefore, there exists a real number β such that

$$\boldsymbol{\Psi}_k''(0) = D\mathbf{r}_k(\mathbf{u}_0)\cdot\mathbf{r}_k(\mathbf{u}_0) + \beta\mathbf{r}_k(\mathbf{u}_0). \tag{4.18}$$

Again, we change our parametrization by setting

$$\theta = \varepsilon - \frac{1}{2}\beta\varepsilon^2.$$

Then, we have by (4.12) and (4.17)

$$\begin{aligned}\sigma(\mathbf{u}_0, \boldsymbol{\Psi}_k(\varepsilon)) &= \lambda_k(\mathbf{u}_0) + \frac{\theta}{2} D\lambda_k(\mathbf{u}_0)\cdot \mathbf{r}_k(\mathbf{u}_0) + O(\theta^2)\\ &= \lambda_k(\mathbf{u}_0) + \frac{\varepsilon}{2} D\lambda_k(\mathbf{u}_0)\cdot \mathbf{r}_k(\mathbf{u}_0) + O(\varepsilon^2),\end{aligned}$$

which gives (4.6). Next, using (4.11), (4.13), and (4.18), we obtain

$$\begin{aligned}\boldsymbol{\Psi}_k(\varepsilon) &= \mathbf{u}_0 + \theta \mathbf{r}_k(\mathbf{u}_0) + \frac{\theta^2}{2}\Big(D\mathbf{r}_k(\mathbf{u}_0)\cdot \mathbf{r}_k(\mathbf{u}_0) + \beta \mathbf{r}_k(\mathbf{u}_0)\Big) + O(\theta^3)\\ &= \mathbf{u}_0 + \varepsilon \mathbf{r}_k(\mathbf{u}_0) + \frac{\varepsilon^2}{2} D\mathbf{r}_k(\mathbf{u}_0)\cdot \mathbf{r}_k(\mathbf{u}_0) + O(\varepsilon^3)\end{aligned}$$

i.e., the expansion (4.5). □

Remark 4.1. From (4.5), we have

$$\lambda_k(\boldsymbol{\Psi}_k(\varepsilon)) = \lambda_k(\mathbf{u}_0) + \varepsilon D\lambda_k(\mathbf{u}_0)\cdot \mathbf{r}_k(\mathbf{u}_0) + O(\varepsilon^2),$$

and together with (4.6), we get

$$\sigma(\mathbf{u}_0, \boldsymbol{\Psi}_k(\varepsilon)) = \frac{1}{2}\Big(\lambda_k(\mathbf{u}_0) + \lambda_k(\boldsymbol{\Psi}_k(\varepsilon))\Big) + O(\varepsilon^2).$$

Thus, the shock speed is approximated at the order $O(\varepsilon^2)$ by the mean value of the characteristic speeds on both sides. □

Corollary 4.1

For any k-Riemann invariant w, we have

$$w(\boldsymbol{\Psi}_k(\varepsilon)) = w(\mathbf{u}_0) + O(\varepsilon^3). \tag{4.19}$$

Proof. Let w be a k-Riemann invariant. By differentiating the relation (3.9), we find

$$D^2 w(\mathbf{u})\cdot\Big(\mathbf{r}_k(\mathbf{u}), \mathbf{v}\Big) + Dw(\mathbf{u})\cdot D\mathbf{r}_k(\mathbf{u})\cdot \mathbf{v} = 0, \tag{4.20}$$

where $D^j w(\mathbf{u}) \in \mathcal{L}_j(\mathbb{R}^p;\mathbb{R})$ denotes the jth Frechet derivative of w at the point $\mathbf{u}$. Now, using (4.5) gives

$$\begin{aligned}w(\boldsymbol{\Psi}_k(\varepsilon)) &= w\Big(\mathbf{u}_0 + \varepsilon \mathbf{r}_k(\mathbf{u}_0) + \frac{\varepsilon^2}{2} D\mathbf{r}_k(\mathbf{u}_0)\cdot \mathbf{r}_k(\mathbf{u}_0) + O(\varepsilon^3)\Big)\\ &= w(\mathbf{u}_0) + \varepsilon Dw(\mathbf{u}_0)\cdot \mathbf{r}_k(\mathbf{u}_0) + \frac{\varepsilon^2}{2}\Big\{Dw(\mathbf{u}_0)\cdot D\mathbf{r}_k(\mathbf{u}_0)\cdot \mathbf{r}_k(\mathbf{u}_0)\\ &\quad + D^2 w(\mathbf{u}_0)\cdot\Big(\mathbf{r}_k(\mathbf{u}_0), \mathbf{r}_k(\mathbf{u}_0)\Big)\Big\} + O(\varepsilon^3).\end{aligned}$$

Hence (4.19) follows from (3.9) and (4.20). □

Consider the case where the kth characteristic field is *genuinely nonlinear*. The curve $\mathcal{S}_k(\mathbf{u}_0)$ is then called a *k-shock curve*. Moreover, using the normalization (2.33), (4.6) can be written

$$\sigma(\mathbf{u}_0, \boldsymbol{\Psi}_k(\varepsilon)) = \lambda_k(\mathbf{u}_0) + \frac{\varepsilon}{2} + O(\varepsilon^2). \tag{4.21}$$

If $\mathbf{u}_R$ belongs to the k-shock curve $\mathcal{S}_k(\mathbf{u}_L)$, or equivalently if $\mathbf{u}_L$ belongs to the k-shock curve $\mathcal{S}_k(\mathbf{u}_R)$, a weak solution of (2.1) of the form (4.2), (4.3) is called a *k-shock wave*. In fact, we shall see in the next section that not all the states $\mathbf{u}$ of the k-shock curve $\mathcal{S}_k(\mathbf{u}_0)$ are admissible but only those that correspond to $\varepsilon < 0$ (for the normalization (2.33)).

Now, if we consider $|\varepsilon|$ as a measure of the strengh of the k-shock connecting $\mathbf{u}_0$ and $\mathbf{u} = \boldsymbol{\Psi}_k(\varepsilon)$, it follows from the Corollary of Theorem 4.1 that *across a weak k-shock, the change in any k-Riemann invariant is of order 3 in ε.*

Let us next turn to the case where the kth characteristic field is *linearly degenerate*. Then, we can state the following result.

Theorem 4.2

If the kth characteristic field is linearly degenerate, the curve $\mathcal{S}_k(\mathbf{u}_0)$ given by Theorem 4.1 is an integral curve of the vector-field $\mathbf{r}_k$, and we have

$$\sigma(\mathbf{u}_0, \boldsymbol{\Psi}_k(\varepsilon)) = \lambda_k(\boldsymbol{\Psi}_k(\varepsilon)) = \lambda_k(\mathbf{u}_0). \tag{4.22}$$

Moreover, we have for any k-Riemann invariant w

$$w(\boldsymbol{\Psi}_k(\varepsilon)) = w(\mathbf{u}_0). \tag{4.23}$$

Proof. Let us consider the integral curve of the vector field $\mathbf{r}_k$ passing through the point $\mathbf{u}_0$, i.e., the solution $\xi \to \mathbf{v}(\xi)$ of

$$\begin{cases} \mathbf{v}'(\xi) = \mathbf{r}_k(\mathbf{v}(\xi)), \\ \mathbf{v}(0) = \mathbf{u}_0. \end{cases}$$

Let us check that the Rankine–Hugoniot jump condition (4.4) holds along this integral curve with constant speed $\sigma(\mathbf{u}_0, \mathbf{u}) = \lambda_k(\mathbf{u}_0)$ if the kth characteristic field is linearly degenerate. Indeed, we have

$$\begin{aligned} &\frac{d}{d\xi}\{\mathbf{f}(\mathbf{v}(\xi)) - \mathbf{f}(\mathbf{u}_0) - \lambda_k(\mathbf{v}(\xi))(\mathbf{v}(\xi) - \mathbf{u}_0)\} \\ &\quad = \Big(\mathbf{A}(\mathbf{v}(\xi)) - \lambda_k(\mathbf{v}(\xi))\Big)\mathbf{v}'(\xi) - D\lambda_k(\mathbf{v}(\xi)) \cdot \mathbf{v}'(\xi)(\mathbf{v}(\xi) - \mathbf{u}_0) \\ &\quad = \Big(\mathbf{A}(\mathbf{v}(\xi)) - \lambda_k(\mathbf{v}(\xi))\Big)\mathbf{r}_k(\mathbf{v}(\xi)) - D\lambda_k(\mathbf{v}(\xi)) \cdot \mathbf{r}_k(\mathbf{v}(\xi))(\mathbf{v}(\xi) - \mathbf{u}_0) \\ &\quad = \mathbf{0} \end{aligned}$$

and therefore

$$\mathbf{f}(\mathbf{v}(\xi)) - \mathbf{f}(\mathbf{u}_0) = \lambda_k(\mathbf{v}(\xi))(\mathbf{v}(\xi) - \mathbf{u}_0).$$

Hence, the integral curve coincides with $\mathcal{S}_k(\mathbf{u}_0)$, and furthermore

$$\sigma(\mathbf{u}_0, \mathbf{v}(\xi)) = \lambda_k(\mathbf{v}(\xi)).$$

On the other hand, let w be a k-Riemann invariant. As we have already observed, w is constant on an integral curve of $\mathbf{r}_k$; indeed, we have by (3.9)

$$\frac{d}{d\xi} w(\mathbf{v}(\xi)) = Dw(\mathbf{v}(\xi)) \cdot \mathbf{v}'(\xi) = Dw(\mathbf{v}(\xi)) \cdot \mathbf{r}_k(\mathbf{v}(\xi)) = 0$$

so that

$$w(\mathbf{v}(\xi)) = w(\mathbf{u}_0).$$

The theorem follows since λ_k is a k-Riemann invariant when the kth characteristic field is linearly degenerate. □

Thus, assume that the kth characteristic field is linearly degenerate and that $\mathbf{u}_R \in \mathcal{S}_k(\mathbf{u}_L)$ or, equivalently, $\mathbf{u}_L \in \mathcal{S}_k(\mathbf{u}_R)$. Then, a weak solution $\mathbf{u}$ of (2.1) of the form (4.2), (4.3), where

$$\sigma = \lambda_k(\mathbf{u}_L) = \lambda_k(\mathbf{u}_R) = \overline{\lambda}_k$$

i.e.,

$$\mathbf{u}(x,t) = \begin{cases} \mathbf{u}_L, & x < \overline{\lambda}_k t, \\ \mathbf{u}_R, & x > \overline{\lambda}_k t, \end{cases} \tag{4.24}$$

is called a *k-contact discontinuity* (see Figure 4.1).

Remark 4.2. It is noteworthy that the function (4.24) is the limit of k-simple (noncentered) waves (3.24). Indeed, let φ_0^ε, $\varepsilon > 0$ be a smooth increasing function such that

$$\varphi_0^\varepsilon(x) = \begin{cases} \xi_0, & x \le 0, \\ \xi_R, & x > \varepsilon, \end{cases}$$

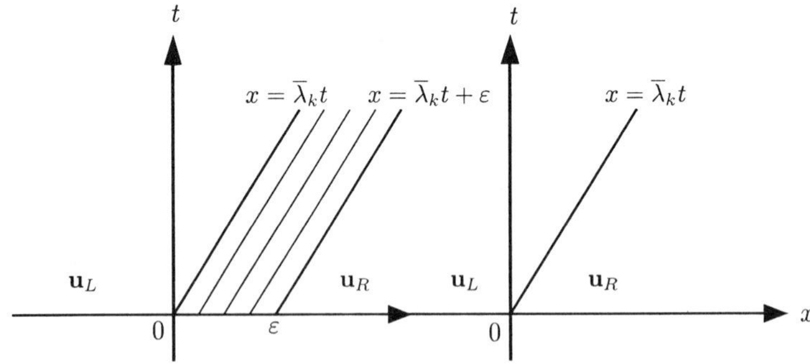

FIGURE 4.1. Contact discontinuity.

where the solution $\mathbf{v}$ of (3.19) with $\mathbf{u}_0 = \mathbf{u}_L$ satisfies $\mathbf{v}(\xi_R) = \mathbf{u}_R$. The corresponding k-simple wave (3.24) is such that

$$\mathbf{u}^\varepsilon(x,t) = \begin{cases} \mathbf{u}_L, & x \le \overline{\lambda}_k t, \\ \mathbf{u}_R, & x > \overline{\lambda}_k t + \varepsilon \end{cases}$$

and thus approaches (4.24) as $\varepsilon \to 0$.

As we have noticed in Example 3.5, the characteristics of the k-simple wave are the parallel lines $x = \overline{\lambda}_k t + C$. □

The above results can be readily applied to the p-system; this will be done in detail in Section 7 of this chapter, where we shall indeed obtain a global description of the rarefaction curves and shock curves. In many applications, however, we have already noticed that it could be more convenient to use nonconservative variables. Let us check that the results of Theorems 4.1 and 4.2 are still valid when we use a nonconservative form (2.12) of the nonlinear hyperbolic system (2.1). We introduce new dependent variables $\mathbf{v}$ defined by (2.10), and we look for a parametrization of the curve $\mathcal{S}_k(\mathbf{u}_0)$ of the form

$$\mathbf{v}(\varepsilon) = \mathbf{v}_0 + \varepsilon\mathbf{v}_1 + \varepsilon^2\mathbf{v}_2 + O(\varepsilon^3).$$

Setting

$$\mathbf{u}(\varepsilon) = \boldsymbol{\Psi}_k(\varepsilon),$$

we have

$$\begin{aligned} \mathbf{u}(\varepsilon) = \boldsymbol{\theta}(\mathbf{v}(\varepsilon)) = \boldsymbol{\theta}(\mathbf{v}_0) + \varepsilon D\boldsymbol{\theta}(\mathbf{v}_0)\cdot\mathbf{v}_1 + \varepsilon^2\{D\boldsymbol{\theta}(\mathbf{v}_0)\cdot\mathbf{v}_2 \\ + \frac{1}{2}D^2\boldsymbol{\theta}(\mathbf{v}_0)\cdot(\mathbf{v}_1,\mathbf{v}_1)\} + O(\varepsilon^3). \end{aligned}$$

Comparing this with the expansion (4.5) gives

$$\begin{aligned} \mathbf{u}_0 &= \boldsymbol{\theta}(\mathbf{v}_0), \\ \mathbf{r}_k(\mathbf{u}_0) &= D\boldsymbol{\theta}(\mathbf{v}_0)\cdot\mathbf{v}_1, \end{aligned}$$

$$D\mathbf{r}_k(\mathbf{u}_0) = 2D\boldsymbol{\theta}(\mathbf{v}_0)\cdot\mathbf{v}_2 + D^2\boldsymbol{\theta}(\mathbf{v}_0)\cdot(\mathbf{v}_1,\mathbf{v}_1), \tag{4.25}$$

and so on. Hence, we find

$$\mathbf{v}_0 = \boldsymbol{\theta}^{-1}(\mathbf{u}_0)$$

and, by (2.14),

$$\mathbf{v}_1 = D\boldsymbol{\theta}(\mathbf{v}_0)^{-1}\mathbf{r}_k(\boldsymbol{\theta}(\mathbf{v}_0)) = \mathbf{s}_k(\mathbf{v}_0).$$

Moreover, by differentiating

$$\mathbf{r}_k(\boldsymbol{\theta}(\mathbf{v})) = D\boldsymbol{\theta}(\mathbf{v})\cdot\mathbf{s}_k(\mathbf{v}),$$

we obtain

$$D\mathbf{r}_k(\boldsymbol{\theta}(\mathbf{v}))\cdot D\boldsymbol{\theta}(\mathbf{v})\cdot\mathbf{s}_k(\mathbf{v}) = D^2\boldsymbol{\theta}(\mathbf{v})\cdot(\mathbf{s}_k(\mathbf{v}),\mathbf{s}_k(\mathbf{v})) + D\boldsymbol{\theta}(\mathbf{v})\cdot D\mathbf{s}_k(\mathbf{v})\cdot\mathbf{s}_k(\mathbf{v}).$$

Thus, we have

$$D\mathbf{r}_k(\mathbf{u})\cdot\mathbf{r}_k(\mathbf{u}) = D^2\boldsymbol{\theta}(\mathbf{v})\cdot\Big(\mathbf{s}_k(\mathbf{v}),\mathbf{s}_k(\mathbf{v})\Big) + D\boldsymbol{\theta}(\mathbf{v})\cdot D\mathbf{s}_k(\mathbf{v})\cdot\mathbf{s}_k(\mathbf{v})$$

and, since $\mathbf{v}_1 = \mathbf{s}_k(\mathbf{v}_0)$,

$$D\mathbf{r}_k(\mathbf{u}_0)\cdot\mathbf{r}_k(\mathbf{u}_0) = D^2\boldsymbol{\theta}(\mathbf{v}_0)\cdot(\mathbf{v}_1,\mathbf{v}_1) + D\boldsymbol{\theta}(\mathbf{v}_0)\cdot D\mathbf{s}_k(\mathbf{v}_0)\cdot\mathbf{s}_k(\mathbf{v}_0),$$

so that by (4.25)

$$\mathbf{v}_2 = \frac{1}{2}D\mathbf{s}_k(\mathbf{v}_0)\cdot\mathbf{s}_k(\mathbf{v}_0).$$

Therefore, we get

$$\mathbf{v}(\varepsilon) = \mathbf{v}_0 + \varepsilon\mathbf{s}_k(\mathbf{v}_0) + \frac{\varepsilon^2}{2}D\mathbf{s}_k(\mathbf{v}_0)\cdot\mathbf{s}_k(\mathbf{v}_0) + O(\varepsilon^3). \tag{4.26}$$

On the other hand, setting $\sigma(\varepsilon) = \sigma\Big(\mathbf{u},\boldsymbol{\Psi}_k(\varepsilon)\Big)$ and using (2.13), (2.15) together with (4.6), we have

$$\sigma(\varepsilon) = \mu_k(\mathbf{v}_0) + \frac{\varepsilon}{2}D\mu_k(\mathbf{v}_0)\cdot s_k(\mathbf{v}_0) + O(\varepsilon^2). \tag{4.27}$$

Finally, when the kth characteristic field is linearly degenerate, we know that

$$\mathbf{u}'(\varepsilon) = \mathbf{r}_k(\mathbf{u}(\varepsilon)),$$

so that

$$D\boldsymbol{\theta}(\mathbf{v}(\varepsilon))\cdot\mathbf{v}'(\varepsilon) = D\boldsymbol{\theta}(\mathbf{v}(\varepsilon))\cdot\mathbf{s}_k(\mathbf{v}(\varepsilon)).$$

Hence

$$\begin{aligned}\mathbf{v}'(\varepsilon) &= \mathbf{s}_k(\mathbf{v}(\varepsilon)),\\ \mathbf{v}(0) &= \mathbf{v}_0,\end{aligned}$$

$\mathcal{S}_k(\mathbf{u}_0)$ is an integral curve of the vector field $\mathbf{s}_k(\mathbf{v})$, and moreover,

$$\mu_k(\mathbf{v}(\varepsilon)) = \mu_k(\mathbf{v}_0) = \lambda_k(\mathbf{u}_0). \tag{4.28}$$

Example 4.1. Consider the gas dynamics equations in Lagrangian coordinates written in the nonconservative form (2.19), i.e., using the nonconservative variables (τ, u, s). It follows from (2.24) that

$$D\mathbf{r}_1\cdot\mathbf{r}_1 = \begin{pmatrix}\tau\\ \tau\frac{\partial c}{\partial\tau}\\ 0\end{pmatrix}, \qquad D\mathbf{r}_3\cdot\mathbf{r}_3 = \begin{pmatrix}\tau\\ -\tau\frac{\partial c}{\partial\tau}\\ 0\end{pmatrix}.$$

Hence, the 1-shock curve passing through the point (τ_0, u_0, s_0) has a parametrization of the form

$$(4.29) \qquad \begin{cases} \tau(\varepsilon) = \tau_0(1 + \varepsilon + \dfrac{\varepsilon^2}{2}) + O(\varepsilon^3), \\ u(\varepsilon) = u_0 + \varepsilon c_0 + \dfrac{\varepsilon^2}{2}\tau_0(\dfrac{\partial c}{\partial \tau})_0 + O(\varepsilon^3), \\ s(\varepsilon) = s_0 + O(\varepsilon^3). \end{cases}$$

We find that the change in entropy s is of third order in ε, as was expected from Corollary 4.1 of Theorem 4.1 since s is a 1-Riemann invariant (see Example 3.2). On the other hand, for such a 1-shock, we have

$$\begin{aligned} p(\varepsilon) = p\Big(\tau(\varepsilon), s(\varepsilon)\Big) &= p(\tau_0 + \varepsilon\tau_0 + ..., s_0 + ...) \\ &= p(\tau_0, s_0) + \varepsilon\tau_0 \frac{\partial p}{\partial \tau}(\tau_0, s_0) + O(\varepsilon^2) \end{aligned}$$

and therefore, by (2.21),

$$(4.30) \qquad p(\varepsilon) = p_0 - \varepsilon\frac{c_0^2}{\tau_0} + O(\varepsilon^2).$$

Similarly, the 3-shock curve passing through the point (τ_0, u_0, s_0) has a parametrization of the form

$$(4.31) \qquad \begin{cases} \tau(\varepsilon) = \tau_0(1 + \varepsilon + \dfrac{\varepsilon^2}{2}) + O(\varepsilon^3), \\ u(\varepsilon) = u_0 - \varepsilon c_0 - \dfrac{\varepsilon^2}{2}\tau_0(\dfrac{\partial c}{\partial \tau})_0 + O(\varepsilon^3), \\ s(\varepsilon) = s_0 + O(\varepsilon^3) \end{cases}$$

and

$$(4.32) \qquad p(\varepsilon) = p_0 - \varepsilon\,\frac{c_0^2}{\tau_0} + O(\varepsilon^2).$$

Finally, since u, p are 2-Riemann invariants, the 2-contact discontinuity curve passing through (τ_0, u_0, s_0) satisfies, by Theorem 4.2,

$$(4.33) \qquad u(\varepsilon) = u_0\ ,\ p(\varepsilon) = p_0,$$

and we have

$$(4.34) \qquad \sigma(\varepsilon) = 0$$

for this 2-contact discontinuity. □

Example 4.2. We turn to the gas dynamics equations in Eulerian coordinates written in the nonconservative form (2.28). Using (2.32), we have

$$D\mathbf{r}_1 \cdot \mathbf{r}_1 = \begin{pmatrix} \rho \\ -\rho\frac{\partial c}{\partial \rho} \\ 0 \end{pmatrix}, \qquad D\mathbf{r}_3 \cdot \mathbf{r}_3 = \begin{pmatrix} \rho \\ \rho\frac{\partial c}{\partial \rho} \\ 0 \end{pmatrix}.$$

and the 1-shock curve passing through the point (ρ_0, u_0, s_0) has a parametrization of the form

$$(4.35)\qquad \begin{cases} \rho(\varepsilon) = \rho_0(1+\varepsilon+\dfrac{\varepsilon^2}{2}) + O(\varepsilon^3), \\ u(\varepsilon) = u_0 - \varepsilon c_0 - \dfrac{\varepsilon^2}{2}\rho_0(\dfrac{\partial c}{\partial \rho})_0 + O(\varepsilon^3), \\ s(\varepsilon) = s_0 + O(\varepsilon^3), \end{cases}$$

while the 3-shock curve passing through this point is given by

$$(4.36)\qquad \begin{cases} \rho(\varepsilon) = \rho_0(1+\varepsilon+\dfrac{\varepsilon^2}{2}) + O(\varepsilon^3), \\ u(\varepsilon) = u_0 + \varepsilon c_0 + \dfrac{\varepsilon^2}{2}\rho_0(\dfrac{\partial c}{\partial \rho})_0 + O(\varepsilon^3), \\ s(\varepsilon) = s_0 + O(\varepsilon^3). \end{cases}$$

Note that for both shock curves, we have

$$(4.37)\qquad p(\varepsilon) = p_0 + \varepsilon \rho_0 c_0^2 + O(\varepsilon^2).$$

On the other hand, the 2-contact discontinuity curve passing through the point (ρ_0, u_0, s_0) again satisfies

$$(4.38)\qquad u(\varepsilon) = u_0 \ , \ p(\varepsilon) = p_0$$

with

$$(4.39)\qquad \sigma(\varepsilon) = u_0,$$

which characterizes the physical contact discontinuities. □

5 Characteristic curves and entropy conditions

Assume that $\mathbf{u}$ is a classical solution of the system (2.1) in a domain D so that

$$\frac{\partial \mathbf{u}}{\partial t} + \mathbf{A}(\mathbf{u})\frac{\partial \mathbf{u}}{\partial x} = \mathbf{0} \quad \text{in } D$$

or equivalently, by (2.3),

$$(5.1)\qquad \mathbf{l}_k(\mathbf{u})^T\Big(\frac{\partial \mathbf{u}}{\partial t} + \lambda_k(\mathbf{u})\frac{\partial \mathbf{u}}{\partial x}\Big) = 0 \quad , \quad 1 \le k \le p.$$

Each equation (5.1) involves differentiation in only one direction. Introducing as in (3.21) the characteristic curves C_k as the integral curves of the differential system

$$(5.2)\qquad \frac{dx}{dt} = \lambda_k(\mathbf{u}(x,t)),$$

and denoting by $s_k \to \big(x(s_k), t(s_k)\big)$ a parametric representation of C_k, each equation (5.1) becomes

$$\mathbf{l}_k(\mathbf{u})^T \frac{d\mathbf{u}}{ds_k} = 0, \quad 1 \le k \le p.$$

The equations (5.1) are therefore called the *characteristic equations.*

Note that the change (2.10) of dependent variables does not affect the eigenvalues λ_k and thus the characteristic curves.

Example 5.1. Consider first the case $p = 2$. For ease of notation, we denote by λ and μ the eigenvalues of the 2×2 matrix $\mathbf{A}(\mathbf{u})$, by $\mathbf{r}_\lambda$ and $\mathbf{r}_\mu$ (resp. $\mathbf{l}_\lambda^T$ and $\mathbf{l}_\mu^T$) the corresponding right (resp. left) eigenvectors. Next, we introduce a Riemann invariant w (resp. z) associated with the eigenvalue λ (resp. μ) which satisfies by definition

$$\nabla w^T \mathbf{r}_\lambda = 0 \text{ (resp. } \nabla z^T \mathbf{r}_\mu = 0).$$

Recall that we have obtained in Example 3.1 the global existence of such Riemann invariants. Since $\mathbf{l}_\mu^T \mathbf{r}_\lambda = 0$, it follows that ∇w is collinear to $\mathbf{l}_\mu$ so that

$$\nabla w^T \, \mathbf{A} = \mu \, \nabla w^T.$$

Hence, we obtain

$$\frac{\partial w}{\partial t} + \mu \frac{\partial w}{\partial x} = \nabla w^T \Big(\frac{\partial \mathbf{u}}{\partial t} + \mu \frac{\partial \mathbf{u}}{\partial x} \Big) = \nabla w^T \Big(\frac{\partial \mathbf{u}}{\partial t} + \mathbf{A} \frac{\partial \mathbf{u}}{\partial x} \Big) = 0.$$

Similarly, we find

$$\frac{\partial z}{\partial t} + \lambda \frac{\partial z}{\partial x} = \nabla z^T \Big(\frac{\partial \mathbf{u}}{\partial t} + \lambda \frac{\partial \mathbf{u}}{\partial x} \Big) = \nabla z^T \Big(\frac{\partial \mathbf{u}}{\partial t} + \mathbf{A} \frac{\partial \mathbf{u}}{\partial x} \Big) = 0.$$

Hence, the system of characteristic equations is given here by

$$\begin{cases} \dfrac{\partial z}{\partial t} + \lambda \dfrac{\partial z}{\partial x} = 0, \\ \dfrac{\partial w}{\partial t} + \mu \dfrac{\partial w}{\partial x} = 0. \end{cases} \tag{5.3}$$

If we denote by C_λ (resp. C_μ) the characteristic curves associated with the eigenvalue λ (resp. μ), we find that w is constant along the characteristics C_μ while z is constant along the characteristics C_λ.

If we turn to the p-system (2.7), we have (see Example 3.1)

$$\lambda = -\sqrt{-p'(v)}, \quad \mu = \sqrt{-p'(v)},$$
$$w = u - \int^v \sqrt{-p'(y)} dy, \quad z = u + \int^v \sqrt{-p'(y)} dy.$$

As a consequence, if instead of C_λ and C_μ we denote by C_- and C_+, respectively, the characteristic curves, we find that $u \pm \int^v \sqrt{-p'(y)} dy$ is constant on the $C\mp$ characteristics. Then, if by some device we have determined the characteristic curves, we can obtain the solution $(v,u)^T$ of the p-system provided that this solution is smooth.

The theory for systems of two equations is made simpler by the use of the two Riemann invariants and the simple form of the characteristic equations (see Lax 1972); more details concerning "reducible" hyperbolic systems can be found, for instance, in Li Ta-tsien (1994).

Example 5.2. Again, we consider the gas dynamics equations in Lagrangian coordinates (Example 3.2 revisited). Let us derive the system of characteristic equations. We start from the nonconservative form (2.19). Since

$$dp = -\Big(\frac{c^2}{\tau^2}\Big)d\tau + \frac{\partial p}{\partial s}\,ds,$$

it follows from the third equation (2.19) that

$$\frac{\partial p}{\partial t} = -\frac{c^2}{\tau^2}\frac{\partial \tau}{\partial t},$$

so that by the first equation (2.19)

$$\frac{\partial p}{\partial t} + \frac{c^2}{\tau^2}\frac{\partial u}{\partial x} = 0.$$

Hence, another nonconservative form of the gas dynamics equations in Lagrangian coordinates is given by

$$\begin{cases} \dfrac{\partial p}{\partial t} + \dfrac{c^2}{\tau^2}\dfrac{\partial u}{\partial x} = 0, \\ \dfrac{\partial u}{\partial t} + \dfrac{\partial p}{\partial x} = 0, \\ \dfrac{\partial s}{\partial t} = 0. \end{cases} \tag{5.4}$$

By multiplying the second equation (5.4) by $\pm\frac{c}{\tau}$ and adding to the first equation, we obtain the characteristic equations

$$\begin{cases} \Big\{\dfrac{\partial p}{\partial t} - \dfrac{c}{\tau}\dfrac{\partial p}{\partial x}\Big\} - \dfrac{c}{\tau}\Big\{\dfrac{\partial u}{\partial t} - \dfrac{c}{\tau}\dfrac{\partial u}{\partial x}\Big\} = 0, \\ \dfrac{\partial s}{\partial t} = 0, \\ \Big\{\dfrac{\partial p}{\partial t} + \dfrac{c}{\tau}\dfrac{\partial p}{\partial x}\Big\} + \dfrac{c}{\tau}\Big\{\dfrac{\partial u}{\partial t} + \dfrac{c}{\tau}\dfrac{\partial u}{\partial x}\Big\} = 0. \end{cases} \tag{5.5}$$

Note that the second equation (5.5) may be equivalently replaced by

$$\frac{\partial p}{\partial t} + \frac{c^2}{\tau^2}\frac{\partial \tau}{\partial t} = 0.$$

Therefore, the corresponding characteristic curves are given by

$$
\tag{5.6}
\begin{cases}
\dfrac{dx}{dt} = -\dfrac{c}{\tau} & (C_-), \\
\dfrac{dx}{dt} = 0 & (C_0), \\
\dfrac{dx}{dt} = \dfrac{c}{\tau} & (C_+).
\end{cases}
$$

We find that the characteristics C_0 are straight lines along which the entropy s is constant.

We have

$$
\tag{5.7}
\begin{cases}
dp - \dfrac{c}{\tau}\, du = 0 & \text{along the } C_- \text{ characteristics}, \\
ds = 0 \quad \text{or } dp + (\dfrac{c^2}{\tau^2})\, d\tau = 0 & \text{along the } C_0 \text{ characteristics}, \\
dp + \dfrac{c}{\tau}\, du = 0 & \text{along the } C_+ \text{ characteristics},
\end{cases}
$$

which is another way of writing the characteristic equations. □

Example 5.3. We turn to the gas dynamics equations written in Eulerian coordinates (Example 3.3 revisited). To derive the characteristic equations, we use the nonconservative equations (2.23) together with the equation of state (2.29). Since

$$
dp = c^2 d\rho + \frac{\partial p}{\partial s}\, ds,
$$

we obtain from the third equation (2.28)

$$
\Big(\frac{\partial p}{\partial t} + u\,\frac{\partial p}{\partial x}\Big) - c^2\Big(\frac{\partial \rho}{\partial t} + u\,\frac{\partial \rho}{\partial x}\Big) = 0.
$$

Together with the first equation (2.28), this gives

$$
\tag{5.8}
\frac{\partial p}{\partial t} + u\,\frac{\partial p}{\partial x} + \rho c^2\,\frac{\partial u}{\partial x} = 0.
$$

Multiplying the second equation (2.28) by $\pm\rho c$ and adding to the above equation, we find

$$
\tag{5.9}
\begin{cases}
\Big\{\dfrac{\partial p}{\partial t} + (u-c)\,\dfrac{\partial p}{\partial x}\Big\} - \rho c\Big\{\dfrac{\partial u}{\partial t} + (u-c)\,\dfrac{\partial u}{\partial x}\Big\} = 0, \\
\dfrac{\partial s}{\partial t} + u\,\dfrac{\partial s}{\partial x} = 0, \\
\Big\{\dfrac{\partial p}{\partial t} + (u+c)\,\dfrac{\partial p}{\partial x}\Big\} + \rho c\Big\{\dfrac{\partial u}{\partial t} + (u+c)\,\dfrac{\partial u}{\partial x}\Big\} = 0,
\end{cases}
$$

which is the desired characteristic system, while the associated characteristic curves are

$$\left\{\begin{aligned} \frac{dx}{dt} &= u - c \quad (C_-),\\ \frac{dx}{dt} &= u \qquad\quad (C_0),\\ \frac{dx}{dt} &= u + c \quad (C_+). \end{aligned}\right. \tag{5.10}$$

Again, the characteristic equations (5.9) may be equivalently written

$$\left\{\begin{array}{ll} dp - \rho c\, du = 0 & \text{along the } C_- \text{ characteristics,}\\ ds = 0 \quad \text{or } dp - c^2 d\rho = 0 & \text{along the } C_0 \text{ characteristics,}\\ dp + \rho c\, du = 0 & \text{along the } C_+ \text{ characteristics} \end{array}\right. \tag{5.11}$$

(compare with (5.7)). □

In the general case, we obtain the following result, which appears to be a consequence of Theorem 3.3.

Theorem 5.1

Assume that $\mathbf{u}$ *is a* k*-rarefaction wave in* D*. Then the characteristic curves* C_k *are straight lines along which* $\mathbf{u}$ *is constant.*

Proof. We give a direct proof (which is not valid for a linearly degenerate field). Let $w_j, 1 \le j \le p-1$, be $(p-1)$ k-Riemann invariants, with derivatives Dw_j linearly independent in D. Since by Theorem 3.2, $w_j(\mathbf{u})$ is constant in D, we have

$$\nabla w_j(\mathbf{u})^T \Big(\frac{\partial \mathbf{u}}{\partial t} + \lambda_k \frac{\partial \mathbf{u}}{\partial x} \Big) = \frac{d}{ds_k} w(\mathbf{u}) = 0.$$

Together with (5.1), this gives

$$\mathbf{N}_k(\mathbf{u}) \Big(\frac{\partial \mathbf{u}}{\partial t} + \lambda_k \frac{\partial \mathbf{u}}{\partial x} \Big) = \mathbf{0},$$

where $\mathbf{N}_k(\mathbf{u})$ is the $p \times p$ matrix

$$\mathbf{N}_k(\mathbf{u}) = \begin{pmatrix} \nabla w_1(\mathbf{u})^T \\ \vdots \\ \nabla w_{p-1}(\mathbf{u})^T \\ \mathbf{l}_k(\mathbf{u})^T \end{pmatrix}.$$

Since $\nabla w_j(\mathbf{u})^T \mathbf{r}_k(\mathbf{u}) = 0,\ 1 \le j \le p-1$, and $\mathbf{l}_k(\mathbf{u})^T \mathbf{r}_k(\mathbf{u}) = 1$, the matrix $\mathbf{N}_k(\mathbf{u})$ is nonsingular so that

$$\frac{\partial \mathbf{u}}{\partial t} + \lambda_k(\mathbf{u}) \frac{\partial \mathbf{u}}{\partial x} = 0.$$

Hence $\mathbf{u}$ is constant along each C_k-characteristic, and C_k is indeed a straight line. □

Let us consider again the problem of determining all the states $\mathbf{u}_R \in \Omega$ that can be connected to a given state $\mathbf{u}_L \in \Omega$ by a discontinuity wave. We have already noticed in the scalar case that not any shock discontinuity was admissible. In the case of systems, we want to show how heuristic arguments based on characteristics lead us to exclude a part of the k-shock curve $\mathcal{S}_k(\mathbf{u}_L)$. Observe first that in the smoothness regions of a weak solution $\mathbf{u}$ of (2.1), the characteristic curves propagate the information from the boundary data. In particular, if Σ is a curve $x = \xi(t)$ in the (x,t)-plane with $\sigma = \xi'(t)$ and if

$$\lambda_1(\mathbf{u}) < \dots < \lambda_k(\mathbf{u}) < \sigma < \lambda_{k+1}(\mathbf{u}) < \dots < \lambda_p(\mathbf{u}) \text{ on } \Sigma$$

(see Figure 5.1), we need to give $(p-k)$ boundary conditions on Σ (resp. k boundary conditions on Σ) in order to specify the solution $\mathbf{u}$ in the region $\{(x,t), x > \sigma t\}$ (resp. in the region $x < \sigma t$).

Now, if a piecewise C^1 weak solution $\mathbf{u}$ of (2.1) is discontinuous across the curve Σ and satisfies

$$\lambda_k(\mathbf{u}_+) < \sigma < \lambda_{k+1}(\mathbf{u}_+), \tag{5.12}$$

we need to specify $(p-k)$ conditions on the right boundary of the discontinuity. Similarly, if

$$\lambda_j(\mathbf{u}_-) < \sigma < \lambda_{j+1}(\mathbf{u}_-), \tag{5.13}$$

we have to give j conditions on the left boundary of the discontinuity, which yields in total $p-k+j$ conditions. On the other hand, eliminating σ from the Rankine–Hugoniot jump relations

$$\sigma(\mathbf{u}_+ - \mathbf{u}_-) = \mathbf{f}(\mathbf{u}_+) - \mathbf{f}(\mathbf{u}_-)$$

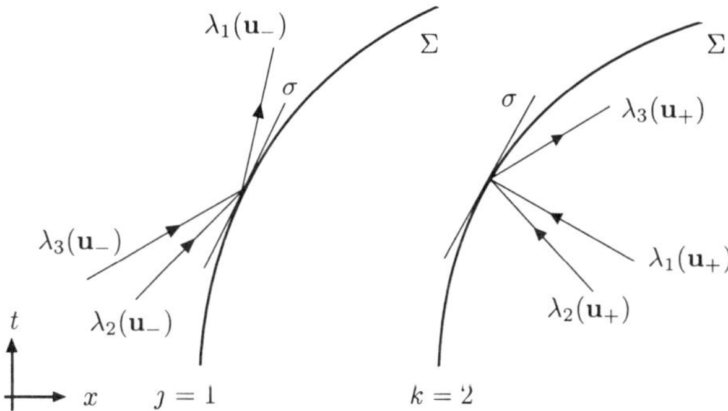

FIGURE 5.1. Characteristics and discontinuity curve.

provides $(p-1)$ relations between $\mathbf{u}_+$ and $\mathbf{u}_-$. Hence, assuming that (5.12) holds, it seems natural to require that (5.13) holds with $j = k-1$.

If $\mathbf{u}$ satisfies $\lambda_k(\mathbf{u}_+) = \sigma$, the same argument requires that $\lambda_k(\mathbf{u}_-) = \sigma$. Thus, we introduce the following definition.

Definition 5.1

We shall say that the discontinuity satisfies the Lax entropy conditions if there exists an index $k \in \{1, 2, ..., p\}$ such that we have either
(i)

$$\begin{cases} \lambda_k(\mathbf{u}_+) < \sigma < \lambda_{k+1}(\mathbf{u}_+), \\ \lambda_{k-1}(\mathbf{u}_-) < \sigma < \lambda_k(\mathbf{u}_-) \end{cases} \tag{5.14}$$

if the kth *characteristic field is genuinely nonlinear; or*
(ii)

$$\lambda_k(\mathbf{u}_-) = \sigma = \lambda_k(\mathbf{u}_+) \tag{5.15}$$

if the kth *characteristic field is linearly degenerate.*

Then, using the parametrization of Theorem 4.1, we define $\mathcal{S}_k^a(\mathbf{u}_L)$ as the set of states $\boldsymbol{\Psi}_k(\varepsilon) \in \mathcal{S}_k(\mathbf{u}_L)$ that can be connected to $\mathbf{u}_L$ (on the right of $\mathbf{u}_L$) by a k-discontinuity wave that satisfies the Lax entropy conditions.

Theorem 5.2

If the kth *characteristic field is genuinely nonlinear, the curve $\mathcal{S}_k^a(\mathbf{u}_L)$ consists of the states $\boldsymbol{\Psi}_k(\varepsilon) \in \mathcal{S}_k(\mathbf{u}_L)$ that satisfy*

$$\varepsilon \leq 0, |\varepsilon| \leq \varepsilon_1 \text{ small enough} \tag{5.16}$$

(for the normalization (2.33)).

If the kth *characteristic field is linearly degenerate, $\mathcal{S}_k^a(\mathbf{u}_L)$ coincides with the whole curve $\mathcal{S}_k(\mathbf{u}_L)$.*

Proof. Assume that the kth characteristic field is genuinely nonlinear. Then, setting

$$\mathbf{u}(\varepsilon) = \boldsymbol{\Psi}_k(\varepsilon), \quad \sigma(\varepsilon) = \sigma(\mathbf{u}_L, \boldsymbol{\Psi}_k(\varepsilon)), \tag{5.17}$$

we have by (4.5) and (4.21)

$$\mathbf{u}(\varepsilon) = \mathbf{u}_L + \varepsilon\, \mathbf{r}_k(\mathbf{u}_L) + O(\varepsilon^2),$$
$$\sigma(\varepsilon) = \lambda_k(\mathbf{u}_L) + \frac{\varepsilon}{2} + O(\varepsilon^2),$$

so that by (2.33)

$$\begin{aligned} \lambda_k(\mathbf{u}(\varepsilon)) &= \lambda_k(\mathbf{u}_L) + \varepsilon D\lambda_k(\mathbf{u}_L) \cdot \mathbf{r}_k(\mathbf{u}_L) + O(\varepsilon^2) \\ &= \lambda_k(\mathbf{u}_L) + \varepsilon + 0(\varepsilon^2) = \sigma(\varepsilon) + \frac{\varepsilon}{2} + O(\varepsilon^2). \end{aligned}$$

Now, we need to find the conditions that ensure

$$\lambda_k(\mathbf{u}(\varepsilon)) < \sigma(\varepsilon) < \lambda_{k+1}(\mathbf{u}(\varepsilon)),$$
$$\lambda_{k-1}(\mathbf{u}_L) < \sigma(\varepsilon) < \lambda_k(\mathbf{u}_L).$$

First, we have $\lambda_k(\mathbf{u}(\varepsilon)) < \sigma(\varepsilon)$ for ε small enough if and only if ε is < 0. Next, since $\lambda_k(\mathbf{u}_L) < \lambda_{k+1}(\mathbf{u}_L)$ and

$$\sigma(\varepsilon) \to \lambda_k(\mathbf{u}_L)\ ,\ \lambda_{k+1}(\mathbf{u}(\varepsilon)) \to \lambda_{k+1}(\mathbf{u}_L) \text{ as } \varepsilon \to 0,$$

we obtain $\sigma(\varepsilon) < \lambda_{k+1}(\mathbf{u}_L)$ for $|\varepsilon|$ small enough.

On the other hand, we have $\sigma(\varepsilon) < \lambda_k(\mathbf{u}_L)$ for $|\varepsilon|$ small enough if and only if ε is < 0. Finally, since $\lambda_{k-1}(\mathbf{u}_L) < \lambda_k(\mathbf{u}_L)$, we get $\lambda_{k-1}(\mathbf{u}_L) < \sigma(\varepsilon)$ for $|\varepsilon|$ small enough. This proves the first part of the theorem.

When the kth characteristic field is linearly degenerate, the result of Theorem 4.2 ensures that the whole curve $\mathcal{S}_k(\mathbf{u}_L)$ satisfies the Lax entropy conditions. □

Another way of selecting the relevant part of the curve $\mathcal{S}_k(\mathbf{u}_L)$ is based on entropy considerations (see the Introduction, Section 3). Let us recall that a convex smooth function $U : \Omega \to \mathbb{R}$ is an entropy if there exists a smooth function $F : \Omega \to \mathbb{R}$, called entropy flux, such that

$$U'(\mathbf{u})\mathbf{A}(\mathbf{u}) = F'(\mathbf{u}),\ \forall \mathbf{u} \in \Omega.$$

Then, a piecewise C^1 weak solution of (2.1) is an entropy solution if, across each discontinuity line, it satisfies together with the Rankine–Hugoniot jump condition (4.1) the following entropy condition:

$$\sigma[U(\mathbf{u})] \geq [F(\mathbf{u})] \tag{5.18}$$

for all entropy pairs (U, F).

Hence, using again the parametrization of Theorem 4.1, we want to determine the states $\boldsymbol{\Psi}_k(\varepsilon) \in \mathcal{S}_k(\mathbf{u}_L)$ that satisfy the inequality

$$\sigma(\mathbf{u}_L, \boldsymbol{\Psi}_k(\varepsilon))\Big(U(\boldsymbol{\Psi}_k(\varepsilon)) - U(\mathbf{u}_L)\Big) \geq F(\boldsymbol{\Psi}_k(\varepsilon)) - F(\mathbf{u}_L). \tag{5.19}$$

Theorem 5.3

Let (U, F) be an entropy pair. If the kth characteristic field is genuinely nonlinear and if U is strictly convex, the inequality (5.19) holds for $|\varepsilon|$ small enough if and only if $\varepsilon \leq 0$.

If the kth characteristic field is linearly degenerate, we have

$$\sigma(\mathbf{u}_L, \mathbf{u})\Big(U(\mathbf{u}) - U(\mathbf{u}_L)\Big) = F(\mathbf{u}) - F(\mathbf{u}_L)\ ,\ \forall \mathbf{u} \in \mathcal{S}_k(\mathbf{u}_L). \tag{5.20}$$

Proof. Given an entropy pair (U, F), we set

$$E(\varepsilon) = \sigma(\varepsilon)\Big(U(\mathbf{u}(\varepsilon)) - U(\mathbf{u}_L)\Big) - \Big(F(\mathbf{u}(\varepsilon)) - F(\mathbf{u}_L)\Big) \tag{5.21}$$

where $\mathbf{u}(\varepsilon)$ and $\sigma(\varepsilon)$ are defined as in (5.17). We want to determine the values of ε for which

$$E(\varepsilon) \geq 0\,, \quad |\varepsilon| \text{ small enough.}$$

Clearly, we have

$$E(0) = 0.$$

Next, differentiating (5.21) gives

$$E'(\varepsilon) = \sigma'(\varepsilon)\Big(U(\mathbf{u}(\varepsilon)) - U(\mathbf{u}_L)\Big) + \sigma(\varepsilon) DU(\mathbf{u}(\varepsilon)) \cdot \mathbf{u}'(\varepsilon) - DF(\mathbf{u}(\varepsilon)) \cdot \mathbf{u}'(\varepsilon),$$

and by the condition $U'\mathbf{A} = F'$

(5.22)
$$E'(\varepsilon) = \sigma'(\varepsilon)\Big(U(\mathbf{u}(\varepsilon)) - U(\mathbf{u}_L)\Big) - DU(\mathbf{u}(\varepsilon)) \cdot \Big(\mathbf{A}(\mathbf{u}(\varepsilon)) - \sigma(\varepsilon)\Big)\mathbf{u}'(\varepsilon).$$

On the other hand, by differentiating the Rankine–Hugoniot jump condition

$$\sigma(\varepsilon)\Big(\mathbf{u}(\varepsilon) - \mathbf{u}_L\Big) = f(\mathbf{u}(\varepsilon)) - f(\mathbf{u}_L),$$

we obtain

$$\sigma'(\varepsilon)\Big(\mathbf{u}(\varepsilon) - \mathbf{u}_L\Big) = \Big(\mathbf{A}(\mathbf{u}(\varepsilon)) - \sigma(\varepsilon)\Big)\mathbf{u}'(\varepsilon). \tag{5.23}$$

Hence, by combining (5.22) and (5.23), we get

$$E'(\varepsilon) = \sigma'(\varepsilon)\{U(\mathbf{u}(\varepsilon)) - U(\mathbf{u}_L) - DU(\mathbf{u}(\varepsilon)) \cdot \Big(\mathbf{u}(\varepsilon) - \mathbf{u}_L\Big)\} \tag{5.24}$$

and therefore

$$E'(0) = 0.$$

Again, differentiating (5.24) gives

$$\left\{\begin{aligned} E''(\varepsilon) =\, & \sigma''(\varepsilon)\Big(U(\mathbf{u}(\varepsilon)) - U(\mathbf{u}_L) - DU(\mathbf{u}(\varepsilon)) \cdot \Big(\mathbf{u}(\varepsilon) - \mathbf{u}_L\Big)\\ & - \sigma'(\varepsilon) D^2U(\mathbf{u}(\varepsilon)) \cdot \Big(\mathbf{u}'(\varepsilon), \mathbf{u}(\varepsilon) - \mathbf{u}_L\Big)\end{aligned}\right. \tag{5.25}$$

so that

$$E''(0) = 0.$$

Differentiating (5.25) once more, we find

$$\begin{aligned} E'''(\varepsilon) &= \sigma'''(\varepsilon)\{U(\mathbf{u}(\varepsilon)) - U(\mathbf{u}_L) - DU(\mathbf{u}(\varepsilon)) \cdot \Big(\mathbf{u}(\varepsilon) - \mathbf{u}_L\Big)\}\\ &- 2\sigma''(\varepsilon) D^2U(\mathbf{u}(\varepsilon)) \cdot \Big(\mathbf{u}'(\varepsilon), \mathbf{u}(\varepsilon) - \mathbf{u}_L\Big)\\ &- \sigma'(\varepsilon)\{D^3U(\mathbf{u}(\varepsilon)) \cdot \Big(\mathbf{u}'(\varepsilon), \mathbf{u}'(\varepsilon), \mathbf{u}(\varepsilon) - \mathbf{u}_L\Big)\\ &+ D^2U(\mathbf{u}(\varepsilon)) \cdot \Big(\mathbf{u}''(\varepsilon), \mathbf{u}(\varepsilon) - \mathbf{u}_L\Big) + D^2U(\mathbf{u}(\varepsilon)) \cdot \Big(\mathbf{u}'(\varepsilon), \mathbf{u}'(\varepsilon)\Big)\}\end{aligned}$$

and

$$E'''(0) = -\sigma'(0)D^2U(\mathbf{u}_L)\Big(\mathbf{u}'(0), \mathbf{u}'(0)\Big).$$

Assume now that the kth characteristic field is genuinely nonlinear. We have by (4.5) and (4.21)

$$\mathbf{u}'(0) = \mathbf{r}_k(\mathbf{u}_L), \ \sigma'(0) = \frac{1}{2},$$

and therefore

$$E(\varepsilon) = -\frac{\varepsilon^3}{12} D^2U(\mathbf{u}_L) \cdot \Big(\mathbf{r}_k(\mathbf{u}_L), \mathbf{r}_k(\mathbf{u}_L)\Big) + O(\varepsilon^4). \tag{5.26}$$

If we suppose also that the entropy function U is strictly convex, we obtain that $E(\varepsilon)$ is ≥ 0 for $|\varepsilon|$ small enough if and only if ε is ≤ 0.

Assume next that the kth characteristic field is linearly degenerate. Then by Theorem 4.2, $\mathcal{S}_k(\mathbf{u}_L)$ is an integral curve of the vector field $\mathbf{r}_k$ so that $\mathbf{u}'(\varepsilon)$ is proportional to $\mathbf{r}_k(\mathbf{u}(\varepsilon))$. Moreover, $\sigma(\varepsilon) = \lambda_k(\mathbf{u}_L) = \lambda_k(\mathbf{u}(\varepsilon))$. Hence, it follows from (5.22) that

$$E'(\varepsilon) = 0.$$

Since $E(0) = 0$, we obtain $E(\varepsilon) = 0$ for all ε, which proves (5.20). □

Hence, we have proven that, for sufficiently weak shocks, the Lax entropy conditions (5.14) are equivalent to the condition (5.18) associated with a strictly convex entropy U. In the general case, however, it is not yet clear whether the above entropy conditions are equivalent or not. In practice, it is often more convenient to use the Lax entropy conditions to determine the admissible part $\mathcal{S}_k^a(\mathbf{u}_L)$ of the shock curve $\mathcal{S}_k(\mathbf{u}_L)$.

Remark 5.1. Assume that the kth characteristic field is genuinely nonlinear. Then, arguing exactly as in the proof of Theorem 5.2 or Theorem 5.3, one can check that the set of left states $\boldsymbol{\Psi}_k(\varepsilon) \in \mathcal{S}_k(\mathbf{u}_R)$ that can be connected to a given right state $\mathbf{u}_R$ by an admissible k-shock wave consists of the states $\boldsymbol{\Psi}_k(\varepsilon)$ that satisfy

$$\varepsilon > 0, |\varepsilon| \text{ small enough}$$

(for the normalization (2.33)). □

Remark 5.2. Liu has introduced another entropy condition, noted (E), which extends Oleinik's condition (see G.R., Chapter II, Lemma 6.1) to a system. It is written

$$\sigma(\mathbf{u}_L, \mathbf{u}_R) \leq \sigma(\mathbf{u}_L, \mathbf{u}) \text{ for any } \mathbf{u} \in \mathcal{S}_k(\mathbf{u}_L) \text{ between } \mathbf{u}_L \text{ and } \mathbf{u}_R, \tag{5.27}$$

(or $\sigma(\mathbf{u}_L, \mathbf{u}_R) < \sigma(\mathbf{u}_L, \mathbf{u})$ for Liu's strict condition). Condition (E) implies the shock inequality

$$\lambda_k(\mathbf{u}_R) < \sigma(\mathbf{u}_L, \mathbf{u}_R) < \lambda_k(\mathbf{u}_L),$$

and, for a characteristic field that is genuinely nonlinear, condition (E) is equivalent to Lax's entropy inequalities (5.14) (Liu 1975). However, if $D\lambda_k(\mathbf{u}) \cdot \mathbf{r}_k(\mathbf{u})$ vanishes at some state on $\mathcal{S}_k(\mathbf{u}_L)$, one uses condition (E) instead of Lax's entropy condition to solve the Riemann problem.

Moreover, Majda and Pego (1985) have proven that, for weak shocks, Liu's strict entropy condition is equivalent to the existence of a viscous shock profile. We define a *viscous profile* as follows: it is a (smooth) *traveling wave* $\mathbf{u}_\varepsilon$ i.e.,

$$\mathbf{u}_\varepsilon(x,t) = \mathbf{v}\Big(\frac{x-\sigma t}{\varepsilon}\Big), \tag{5.28}$$

that is solution of a parabolic system

$$\frac{\partial \mathbf{u}_\varepsilon}{\partial t} + \frac{\partial}{\partial x}\mathbf{f}(\mathbf{u}_\varepsilon) = \varepsilon \frac{\partial}{\partial x}\Big(\mathbf{D}(\mathbf{u}_\varepsilon)\frac{\partial \mathbf{u}_\varepsilon}{\partial x}\Big), \tag{5.29}$$

where $\mathbf{D}$ is some "admissible" viscosity matrix (see Introduction, Remark 3.3), and such that the vanishing viscosity limit of $\mathbf{u}_\varepsilon$ is an admissible shock (4.2), which reads

$$\mathbf{u}_\varepsilon(x,t) \longrightarrow \mathbf{u}(x,t) = \begin{cases} \mathbf{u}_L, & x < \sigma t, \\ \mathbf{u}_R, & x > \sigma t \end{cases} \tag{5.30}$$

as $\varepsilon \to 0$. The problem of the existence of a viscous profile is called the shock-structure problem and was introduced by Gelfand. Majda and Pego have characterized the admissible matrices $\mathbf{D}$ in terms of linear uniform $\mathbf{L}^2$-stability (or well-posedness) of the viscous system linearized about a constant state. Note that this linear stability criterion can be characterized in terms of the algebraic structure of $\mathbf{D}$. In turn, this structure of $\mathbf{D}$ is implied by the existence of an entropy pair (U, F) such that $U''\mathbf{D}$ is positive definite (Mock's criterion) (see Remark 3.1 in the Introduction).

Conversely, the existence of viscous profiles can also be used to derive a viscosity criterion ("chord condition") for selecting admissible shocks, for instance in the case of the system of one-dimensional elasticity (which corresponds to a p-system that is not strictly hyperbolic since p'' vanishes at one point and p' twice). In this example, the viscosity matrix is diagonal and is obtained from physical considerations (viscoelastic system), and the "admissible" shocks are thus selected as the limit of viscous profiles (see Pego 1987 and the references therein). Freistühler (1990) has considered the same problem for hyperbolic systems with rotational invariance (which are not strictly hyperbolic).

All these criteria for selecting reasonable weak solutions are thus linked but not always equivalent (depending on the problems), strictly hyperbolic or not, weak or strong shocks (we refer to the above-mentionned papers for details, see also Dafermos 1989, Menikoff and Plohr 1989, Brio 1988, Azevedo and Marchesin 1991, Holden et al. 1991, Shearer and Schecter

1991, Warnecke 1991, Temple 1982, Čanic and Plohr 1995, Kawashima and Matsumura 1994, Serre 1996).

Before concluding this remark, let us find the conditions satisfied by a viscous profile, i.e., a solution of (5.28)–(5.30). First, we notice that in order that $\mathbf{u}_\varepsilon$ be a solution of (5.29), $\mathbf{v}(\xi)$ in (5.28) should satisfy the differential system

$$-\sigma\mathbf{v}' + \mathbf{A}(\mathbf{v})\mathbf{v}' = \Big(\mathbf{D}(\mathbf{v})\mathbf{v}'\Big)', \tag{5.31}$$

which no longer depends on the viscosity coefficient ε. Also, if $\mathbf{u}_\varepsilon$ is a viscous profile, by assumption the limit of $\mathbf{u}_\varepsilon$ must be an admissible shock (5.30). Hence, on the one hand the speed σ of the traveling wave is given by the Rankine–Hugoniot condition

$$\sigma(\mathbf{u}_R - \mathbf{u}_L) = \mathbf{f}(\mathbf{u}_R) - \mathbf{f}(\mathbf{u}_L)$$

and thus depends on the conservation form but not on the viscosity matrix $\mathbf{D}$. On the other hand, we see that $\mathbf{v}$ should satisfy the "boundary conditions"

$$\mathbf{v}(\xi) \to \mathbf{u}_L \text{ as } \xi \to -\infty, \quad \mathbf{v}(\xi) \to \mathbf{u}_R \text{ as } \xi \to +\infty. \tag{5.32}$$

Indeed, the family $\mathbf{v}_\varepsilon(\xi) = \mathbf{v}(\frac{\xi}{\varepsilon})$ converges a.e. as $\varepsilon \to 0, \varepsilon > 0$, to $\mathbf{v}_0(\xi)$

$$\mathbf{v}(\frac{\xi}{\varepsilon}) \longrightarrow \mathbf{v}_0(\xi) = \begin{cases} \lim\limits_{\xi\to-\infty} \mathbf{v}(\xi), \xi < 0, \\ \lim\limits_{\xi\to+\infty} \mathbf{v}(\xi), \xi > 0. \end{cases} \tag{5.33}$$

Thus, setting $\xi = x - \sigma t$ and identifying the limits yields (5.32).

Now, integrating the system (5.31), we get

$$-\sigma\mathbf{v} + \mathbf{f}(\mathbf{v}) = \mathbf{D}(\mathbf{v})\mathbf{v}' + C, \tag{5.34}$$

and taking the limit of (5.32) as $\xi \to \pm\infty$, we obtain

$$C = -\sigma\mathbf{u}_L + \mathbf{f}(\mathbf{u}_L) = -\sigma\mathbf{u}_R + \mathbf{f}(\mathbf{u}_R).$$

Thus, a viscous profile, if it exists (which is the case if the shock connecting $\mathbf{u}_R$ to $\mathbf{u}_L$ satisfies Liu's strict entropy condition and $\mathbf{D}$ is admissible), is a solution of the nonlinear system of ordinary differential equations

$$\mathbf{D}(\mathbf{v})\mathbf{v}' = \mathbf{f}(\mathbf{v}) - \mathbf{f}(\mathbf{u}_L) - \sigma(\mathbf{v} - \mathbf{u}_L) \tag{5.35}$$

together with (5.32); i.e., $\mathbf{v}$ is an orbit (or trajectory) connecting the critical (or rest) points $\mathbf{u}_L$ to $\mathbf{u}_R$. □

Remark 5.3. In many practical problems, we encounter nonlinear hyperbolic systems in *nonconservative* form,

$$\frac{\partial\mathbf{u}}{\partial t} + \mathbf{A}(\mathbf{u})\frac{\partial\mathbf{u}}{\partial x} = \mathbf{0}. \tag{5.36}$$

In order to solve the Riemann problem for such a system, we need to define what we mean by a shock wave. Following the lines of the previous remark, we consider a parabolic regularization of the nonconservative system

$$\frac{\partial \mathbf{u}_\varepsilon}{\partial t} + \mathbf{A}(\mathbf{u}_\varepsilon)\frac{\partial \mathbf{u}_\varepsilon}{\partial x} = \varepsilon\frac{\partial}{\partial x}\Big(\mathbf{D}(\mathbf{u}_\varepsilon)\frac{\partial \mathbf{u}_\varepsilon}{\partial x}\Big), \tag{5.37}$$

where $\mathbf{D}$ is a given viscosity matrix. A traveling wave solution of (5.37) is a smooth function $\mathbf{v}$ with, say, $\mathbf{v}' \in \mathbf{L}^1(\mathbb{R})$, of the form

$$\mathbf{u}_\varepsilon(x,t) = \mathbf{v}\Big(\frac{x-\sigma t}{\varepsilon}\Big). \tag{5.38}$$

We check that the corresponding differential system for $\mathbf{v}$ is still (5.31), which does not depend on ε, but it cannot be integrated in the form (5.34).

If there exists a solution $\mathbf{v}$ to (5.31) with the "boundary conditions" (5.32), this gives a way of defining a shock wave solution $\mathbf{u}$ of the nonconservative system (5.36) that connects $\mathbf{u}_L, \mathbf{u}_R$ by setting

$$\mathbf{u}(x,t) = \begin{cases} \mathbf{u}_L, & x < \sigma t, \\ \mathbf{u}_R, & x > \sigma t. \end{cases}$$

This solution, i.e., the triple $(\mathbf{u}_L, \mathbf{u}_R, \sigma)$, now depends on the diffusion matrix $\mathbf{D}$ (we shall say that it is consistent with $\mathbf{D}$) and is the limit of viscous profiles when the viscosity tends to 0. Indeed, set $\mathbf{u}_\varepsilon(x,t) = \mathbf{v}(\frac{x-\sigma t}{\varepsilon})$, where $\mathbf{v}$ satisfies (5.31), (5.32). Then $\mathbf{u}_\varepsilon$ satisfies (5.37) and, using (5.32), (5.33), we have a.e. as $\varepsilon \to 0$

$$\mathbf{u}_\varepsilon(x,t) \longrightarrow \begin{cases} \mathbf{u}_L, & x < \sigma t, \\ \mathbf{u}_R, & x > \sigma t. \end{cases} \tag{5.39}$$

We emphasize again that when the system is conservative ($\mathbf{A}(\mathbf{u}) = \mathbf{f}'(\mathbf{u})$), a necessary condition for the existence of a solution $\mathbf{v}$ to (5.31),(5.32) is the Rankine–Hugoniot condition, which links the triple $(\mathbf{u}_L, \mathbf{u}_R, \sigma)$, and the shock wave solutions do not depend on $\mathbf{D}$.

Other approaches for defining discontinuous solutions to (5.36) suppose that one can give a meaning in some way or other to the "nonconservative product" $\mathbf{A}(\mathbf{u})\frac{\partial \mathbf{u}}{\partial x}$ when $\mathbf{u}$ is, say, a Heaviside function. This can be done following the theory of Dal Maso, Le Floch, and Murat (1995), who extend the work of Volpert; the definition of the jump $[\mathbf{A}(\mathbf{u})\frac{\partial \mathbf{u}}{\partial x}]_\Phi$ then depends on a path Φ connecting $\mathbf{u}_L$ and $\mathbf{u}_R$ in the states space (a straight line in the case of Volpert's product). A different approach, using the generalized functions, consists roughly in defining the product HH', where H is the Heaviside function, by $HH' \sim aH'$; here a is some parameter that is obtained in a unique way when the equations have a conservative form. For instance, in the case of Burgers' equation written in nonconservation form

$$\frac{\partial u}{\partial t} + u\frac{\partial u}{\partial x} = 0,$$

substituting a shock $u(x,t) = (u_R - u_L)H(x - \sigma t) + u_L$ gives formally

$$-\sigma H'\Delta u + (\Delta u H + u_L)\Delta u H' = 0,$$

and thus

$$\Delta u H H' = (\sigma - u_L)H'.$$

Since

$$\sigma = \frac{u_L + u_R}{2}$$

for the corresponding conservative equation, this yields $HH' = \frac{1}{2}H'$ (i.e., $a = \frac{1}{2}$). Otherwise, one must invoke physical considerations when the equations are derived from physics (see Colombeau et al. 1989).

In any case, some extra information (choice of Φ or a) is needed in order to define discontinuous solutions of (5.36). The above considerations show that this information can also be taken from a diffusion matrix $\mathbf{D}$, which leads to a definition of weak shocks as the limit of viscous profiles (see Raviart and Sainsaulieu 1992 and Sainsaulieu 1993). The choice of $\mathbf{D}$ is usually guided by physical considerations. □

6 Solution of the Riemann problem

We are now able to solve the Riemann problem for the system (2.1)

$$\text{(6.1)} \qquad \begin{cases} \dfrac{\partial \mathbf{u}}{\partial t} + \dfrac{\partial}{\partial x}\mathbf{f}(\mathbf{u}) = \mathbf{0}, & x \in \mathbb{R}, \quad t > 0, \\ \mathbf{u}(x,0) = \begin{cases} \mathbf{u}_L, & x < 0, \\ \mathbf{u}_R, & x > 0 \end{cases} \end{cases}$$

when the data $\mathbf{u}_L$ and $\mathbf{u}_R$ are sufficiently close ($\mathbf{u}_L - \mathbf{u}_R$ is "small"). To do so, we begin by summarizing some of the results of the previous sections. Assume first that the kth characteristic field is genuinely nonlinear and normalized by (2.33). It follows from (3.6) and (4.5) that the curves $\mathcal{R}_k(\mathbf{u}_L)$ and $\mathcal{S}_k(\mathbf{u}_L)$ are osculatory at $\varepsilon = 0$. Hence, the function $\chi_k : (\varepsilon, \mathbf{u}_L) \to \chi_k(\varepsilon; \mathbf{u}_L)$ defined for $|\varepsilon|$ small enough by

$$\text{(6.2)} \qquad \chi_k(\varepsilon; \mathbf{u}_L) = \begin{cases} \mathbf{\Phi}_k(\varepsilon), & \varepsilon \geq 0, \\ \mathbf{\Psi}_k(\varepsilon), & \varepsilon \leq 0 \end{cases}$$

is of class C^2. Moreover, using Theorems 3.1 and 5.2, we obtain that the set

$$\text{(6.3)} \qquad \{\chi_k(\varepsilon; \mathbf{u}_L), |\varepsilon| \text{ small enough}\}$$

is exactly the set of all neighboring states $\mathbf{u}$ that can be connected to $\mathbf{u}_L$ either by a k-rarefaction wave or by an admissible k-shock wave.

Next, when the kth characteristic field is linearly degenerate, setting

$$\chi_k(\varepsilon; \mathbf{u}_L) = \boldsymbol{\Psi}_k(\varepsilon) \tag{6.4}$$

we obtain that (6.3) is the set of all neighboring states $\mathbf{u}$ that can be connected to $\mathbf{u}_L$ by a k-contact discontinuity.

Now, we can state the main result of this chapter.

Theorem 6.1

Assume that for all $k \in 1, ..., p$, the kth *characteristic field is either genuinely nonlinear or linearly degenerate. Then for all $\mathbf{u}_L \in \Omega$ there exists a neighborhood ϑ of $\mathbf{u}_L$ in Ω with the following property: If $\mathbf{u}_R$ belongs to ϑ, the Riemann problem (6.1) has a weak solution that consists of at most $(p+1)$ constant states separated by rarefaction waves, admissible shock waves, or contact discontinuities. Moreover, a weak solution of this kind is unique.*

A solution of this kind is depicted in Figure 6.1.

Proof. Let $\mathbf{u}_L$ be in Ω. We consider the mapping

$$\boldsymbol{\chi} : \boldsymbol{\varepsilon} = (\varepsilon_1, ..., \varepsilon_p)^T \to \boldsymbol{\chi}(\boldsymbol{\varepsilon}) = \chi_p\Big(\varepsilon_p; \chi_{p-1}\Big(\varepsilon_{p-1}; ...; \chi_1(\varepsilon_1; u_L)...\Big)\Big)$$

defined in a neighborhood of $\mathbf{0}$ in $\mathbb{R}^p$ with values in $\Omega \in \mathbb{R}^p$. In other words, the left state $\mathbf{u}_L$ is connected on the right to $\chi_1(\varepsilon_1; \mathbf{u}_L) = \mathbf{u}_1$ by a 1-wave, then $\mathbf{u}_1$ to $\mathbf{u}_2 = \chi_2(\varepsilon_2; \mathbf{u}_1)$ on the right by a 2-wave, ... , and $\mathbf{u}_{p-1}$ to $\mathbf{u}_p = \chi_p(\varepsilon_p; \mathbf{u}_{p-1})$ on the right by a p-wave. We want to check whether we can reach in that way any state $\mathbf{u}_R \in \Omega$ located in a neighborhood of $\mathbf{u}_L$ or, equivalently, solve the equation

$$\boldsymbol{\chi}(\boldsymbol{\varepsilon}) = \mathbf{u}_R. \tag{6.5}$$

We begin by noticing that $\boldsymbol{\chi}$ is a mapping of class C^2 and

$$\boldsymbol{\chi}(0) = \chi_p\Big(0; \chi_{p-1}\Big(0; ...; \chi_1(0, \mathbf{u}_L)...\Big)\Big) = \mathbf{u}_L.$$

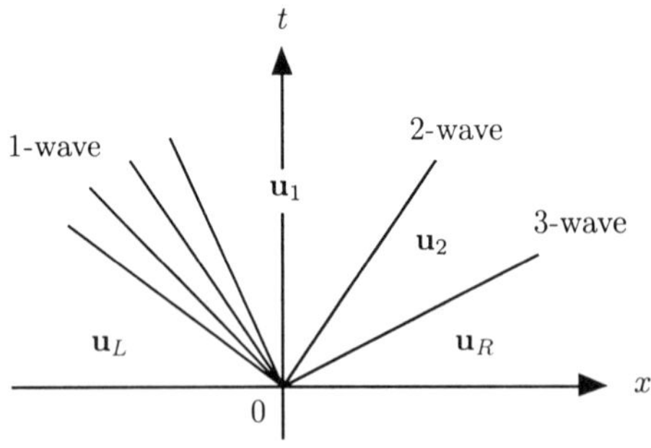

FIGURE 6.1. Solution of the Riemann problem in (x, t)-space.

On the other hand, we have by (3.6) and (4.5)

$$\boldsymbol{\chi}_k(\varepsilon_k; \mathbf{u}) = \mathbf{u} + \varepsilon_k \mathbf{r}_k(\mathbf{u}) + O(\varepsilon_k^2),$$

so that

$$\begin{aligned}\boldsymbol{\chi}_2\Big(\varepsilon_2; \boldsymbol{\chi}_1(\varepsilon_1; \mathbf{u}_L)\Big) &= \boldsymbol{\chi}_2\Big(\varepsilon_2; \mathbf{u}_L + \varepsilon_1 \mathbf{r}_1(\mathbf{u}_L) + O(\varepsilon_1^2)\Big) \\ &= \mathbf{u}_L + \varepsilon_1 \mathbf{r}_1(\mathbf{u}_L) + O(\varepsilon_1^2) + \varepsilon_2 \mathbf{r}_2\Big(\mathbf{u}_L + \varepsilon_1 \mathbf{r}_1(u_L) + O(\varepsilon_1^2)\Big) + O(\varepsilon_2^2) \\ &= \mathbf{u}_L + \varepsilon_1 \mathbf{r}_1(\mathbf{u}_L) + \varepsilon_2 \mathbf{r}_2(\mathbf{u}_L) + O(\varepsilon_1^2 + \varepsilon_2^2)\end{aligned}$$

and by induction

$$\boldsymbol{\chi}(\boldsymbol{\varepsilon}) = \mathbf{u}_L + \sum_{k=1}^{p} \varepsilon_k \mathbf{r}_k(\mathbf{u}_L) + O(|\varepsilon|^2).$$

This means exactly that the derivative $D\boldsymbol{\chi}(\mathbf{0}) \in \mathcal{L}(\mathbb{R}^p)$ of $\boldsymbol{\chi}$ at the origin is given by

$$D\boldsymbol{\chi}(0) \cdot \boldsymbol{\eta} = \sum_{k=1}^{p} \eta_k \mathbf{r}_k(\mathbf{u}_L) \qquad \forall \boldsymbol{\eta} = (\eta_1, ..., \eta_p)^T \in \mathbb{R}^p.$$

Since the vectors $\mathbf{r}_k(\mathbf{u}_L), 1 \le k \le p$, are linearly independent, the linear operator $D\boldsymbol{\chi}(\mathbf{0})$ is invertible. By the local inversion theorem, there exists a neighborhood ϑ of $\mathbf{u}_L$ in Ω such that, for all $\mathbf{u}_R \in \vartheta$, the equation (6.5) has a unique solution $\boldsymbol{\varepsilon} \in \mathbb{R}^p$. We thus obtain a solution $\mathbf{u}$ consisting of $(p+1)$ constant states $\mathbf{u}_0 = \mathbf{u}_L, \mathbf{u}_1, ..., \mathbf{u}_p = \mathbf{u}_R$ separated by k-waves, $1 \le k \le p$, that satisfy the Lax entropy conditions. □

If $\boldsymbol{\varepsilon} = (\varepsilon_1, ..., \varepsilon_p)^T$ is the solution of (6.5), ε_k is called the strength of the kth wave in the solution of the Riemann problem (6.1).

Remark 6.1. Consider now the case where the system is hyperbolic but not strictly hyperbolic, with a complete set of eigenvectors and eigenvalues having constant multiplicity. More precisely, assume that at least one eigenvalue has multiplicity greater than one but that each characteristic field is still either genuinely nonlinear or linearly degenerate. Then the results of Theorem 6.1 extend since first the vectors $\mathbf{r}_k(\mathbf{u}), 1 \le k \le p$, are linearly independent and, moreover, one can prove that there is just one possibility for the corresponding wave, i.e., a contact discontinuity, because the multiple fields are linearly degenerate (see Boillat 1972). Indeed, if, say, $\lambda_k(\mathbf{u}) = \lambda_{k+1}(\mathbf{u})$, differentiating the identity $\Big(\mathbf{A}(\mathbf{u}) - \lambda(\mathbf{u})\Big)\mathbf{r}_k(\mathbf{u}) = \mathbf{0}$ and taking the scalar product of the resulting equations with $\mathbf{l}_k(\mathbf{u})^T$ and $\mathbf{l}_{k+1}(\mathbf{u})^T$ yields for any vector $\mathbf{v} \in \mathbb{R}^p$

$$\mathbf{l}_k(\mathbf{u})^T \{\mathbf{A}'(\mathbf{u})\Big(\mathbf{v}, \mathbf{r}_k(\mathbf{u})\Big) - \Big(D\lambda(\mathbf{u}) \cdot \mathbf{v}\Big)\mathbf{r}_k(\mathbf{u})\} = 0,$$

$$\mathbf{l}_{k+1}(\mathbf{u})^T \{\mathbf{A}'(\mathbf{u})\Big(\mathbf{v}, \mathbf{r}_k(\mathbf{u})\Big) - \Big(D\lambda(\mathbf{u}) \cdot \mathbf{v}\Big)\mathbf{r}_k(\mathbf{u})\} = 0,$$

hence by (2.4)

$$\text{(6.6)} \qquad \begin{cases} \mathbf{l}_k(\mathbf{u})^T \mathbf{A}'(\mathbf{u})\Big(\mathbf{v}, \mathbf{r}_k(\mathbf{u})\Big) = D\lambda(\mathbf{u}) \cdot \mathbf{v}, \\ \mathbf{l}_{k+1}(\mathbf{u})^T \mathbf{A}'(\mathbf{u})\Big(\mathbf{v}, \mathbf{r}_k(\mathbf{u})\Big) = 0. \end{cases}$$

Similarly, for $\mathbf{r}_{k+1}(\mathbf{u})$

$$\text{(6.7)} \qquad \begin{cases} \mathbf{l}_{k+1}(\mathbf{u})^T \mathbf{A}'(\mathbf{u})\Big(\mathbf{v}, \mathbf{r}_{k+1}(\mathbf{u})\Big) = D\lambda(\mathbf{u}) \cdot \mathbf{v}, \\ \mathbf{l}_k(\mathbf{u})^T \mathbf{A}'(\mathbf{u})\Big(\mathbf{v}, \mathbf{r}_{k+1}(\mathbf{u})\Big) = 0. \end{cases}$$

Taking $\mathbf{v} = \mathbf{r}_{k+1}(\mathbf{u})$ in (6.5) (resp. $\mathbf{r}_k(\mathbf{u})$ in (6.6)) and using the symmetry of the Hessian gives

$$D\lambda(\mathbf{u}) \cdot \mathbf{r}_k(\mathbf{u}) = D\lambda(\mathbf{u}) \cdot \mathbf{r}_{k+1}(\mathbf{u}) = 0.$$

One encounters such a situation of multiple eigenvalues in the multicomponent Euler equations (see Chapter II, Remark 3.5) and for the one-dimensional system obtained by projecting the 2-D Euler system (see Chapter IV, Section 2, Remark 2.6). □

Remark 6.2. Theorem 6.1 and the above remark give the solution of the Riemann problem for nearby states when the system is "convex", i.e., when each characteristic field is either genuinely nonlinear or linearly degenerate. However, when the system is strictly hyperbolic but not convex, i.e., if $D\lambda_k(\mathbf{u}) \cdot \mathbf{r}_k(\mathbf{u})$ vanishes at some state on $\mathcal{S}_k(\mathbf{u}_L)$, then we may have $\sigma(\mathbf{u}_L, \mathbf{u}_1) = \lambda_k(\mathbf{u}_1)$ at some state $\mathbf{u}_1$. Hence, if $\mathbf{u}_R$ lies beyond $\mathbf{u}_1$ on $\mathcal{S}_k(\mathbf{u}_L)$, we cannot connect $\mathbf{u}_L$ to $\mathbf{u}_R$ by a unique shock satisfying Lax's condition. Instead, following the argument of the scalar case when f is nonconvex (see G.R., Chapter II, Section 6), we must consider compound k-waves, i.e., a succession of k-shocks (admissible in the sense of Liu; see Remark 5.2), and k-rarefactions. The states $\mathbf{u}_i$, where we switch from a shock to a rarefaction, satisfy $\sigma(\mathbf{u}_i, \mathbf{u}_{i+1}) = \lambda_k(\mathbf{u}_i)$. We refer to Liu (1975) for details, Menikoff and Plohr (1989) for the case of fluid dynamics.

The question of nonuniqueness (in the class described by Theorem 6.1) of the "entropy" solution for the Riemann problem appears for a system that is not strictly hyperbolic if there exists an "umbilic point", i.e., a state $\mathbf{u}_0$ such that two eigenvalues coincide at $\mathbf{u}_0$; the set of eigenvectors may be complete at $\mathbf{u}_0$ (hyperbolic degeneracy) or not (parabolic degeneracy) (see Shearer et al. 1987, Shearer and Schecter 1991, Isaacson et al. 1988, Glimm 1989, Freistühler and Pitman 1995, Liu and Xin 1995, Chen and Kan 1994). In that case, other criteria for selecting the "relevant" solution must be considered (see Remark 5.2). Related problems concern "resonant" systems (Isaacson and Temple 1992). Similar nonuniqueness problems also occur for systems of mixed type, i.e., such that the eigenvalues of $\mathbf{A}$ become complex in some region, called the elliptic region, (see for instance Azevedo and Marchesin 1991, Hsiao 1989).

The question of nonexistence of weak solutions can also arise for some nonstrictly hyperbolic systems, for which measure solutions (delta-shock waves) are needed (see Le Floch 1991, and Tan et al. 1994 and the references therein). □

7 The Riemann problem for the p-system

In the case of the p-system already introduced in Examples 2.2 and 3.1, one can carry out the computations and solve the Riemann problem for almost any two states.

Recall that the p-system is given by (2.7)

$$\begin{cases} \dfrac{\partial v}{\partial t} - \dfrac{\partial u}{\partial x} = 0, \\ \dfrac{\partial u}{\partial t} + \dfrac{\partial}{\partial x} p(v) = 0. \end{cases}$$

If $p'(v) < 0$, the two eigenvalues of $\mathbf{A} = \mathbf{f}'$ are

$$\lambda_1(v) = -\sqrt{-p'(v)} < 0 < \lambda_2(v) = +\sqrt{-p'(v)},$$

and the corresponding eigenvectors $\mathbf{r}_k$, $k = 1, 2$ are given by (2.9). Assuming, moreover, $p'' > 0$, the two characteristic fields are genuinely nonlinear. Observe that λ_k and $\mathbf{r}_k$ are functions of v only.

Let us construct first the curve $\mathcal{R}_1(\mathbf{w}_L)$ (noted $\mathcal{R}_1^L$ in the illustrations) of states that can be connected to a given state $\mathbf{w}_L = (v_L, u_L)$ by a 1-rarefaction wave. It is the integral curve of $\mathbf{r}_1$ issued from $\mathbf{w}_L = (v_L, u_L)$. By (2.9), we have

$$\begin{cases} v'(\zeta) = 1, u'(\zeta) = \sqrt{-p'(v(\xi))}, \\ v(\lambda_k(v_L)) = v_L, \quad u(\lambda_k(v_L)) = u_L. \end{cases}$$

Since $v'(\xi) \neq 0$, we can take v as the parameter and we get

$$u'(v) = \sqrt{-p'(v)},$$

and by integrating we obtain

$$u(v) = u_L + \int_{v_L}^{v} \sqrt{-p'(y)}dy.$$

Due to the assumption $p'' > 0$, the condition $\lambda_1(v) \geq \lambda_1(v_L)$ gives

$$v \geq v_L.$$

The 1 rarefaction curve is thus given in the (v, u) plane by

$$u(v) = u_L + \int_{v_L}^{v} \sqrt{-p'(y)}dy, \quad v \geq v_L. \tag{7.1}$$

It satisfies

$$u'(v) = \sqrt{-p'(v)} > 0,\ u''(v) = -\frac{p''(v)}{2}\sqrt{-p'(v)} < 0.$$

If $\mathbf{w}_R = (v_R, u_R)$ is any point on $\mathcal{R}_1(\mathbf{w}_L)$, the corresponding solution of the Riemann problem is depicted in Figure 7.1 (remember that $\lambda_1 < 0$).

In the same way, the 2-rarefaction curve $\mathcal{R}_2(\mathbf{w}_L)$ (noted $\mathcal{R}_2^L$ in the illustrations) is given in the (v, u) plane by

$$u(v) = u_L + \int_v^{v_L} \sqrt{-p'(y)}dy, v \le v_L. \tag{7.2}$$

It satisfies

$$u'(v) = -\sqrt{-p'(v)} < 0,\ u''(v) = \frac{p''(v)}{2}\sqrt{-p'(v)} > 0,$$

and is illustrated in Figure 7.1 together with the solution of the corresponding Riemann problem associated with $\mathbf{w}_R = (v_R, u_R) \in \mathcal{R}_2(\mathbf{w}_L)$.

Let us turn now to the description of the shock curves. The Rankine–Hugoniot jump condition is given in the present case by

$$\begin{cases} \sigma(v - v_L) = -(u - u_L), \\ \sigma(u - u_L) = p(v) - p(v_L). \end{cases} \tag{7.3}$$

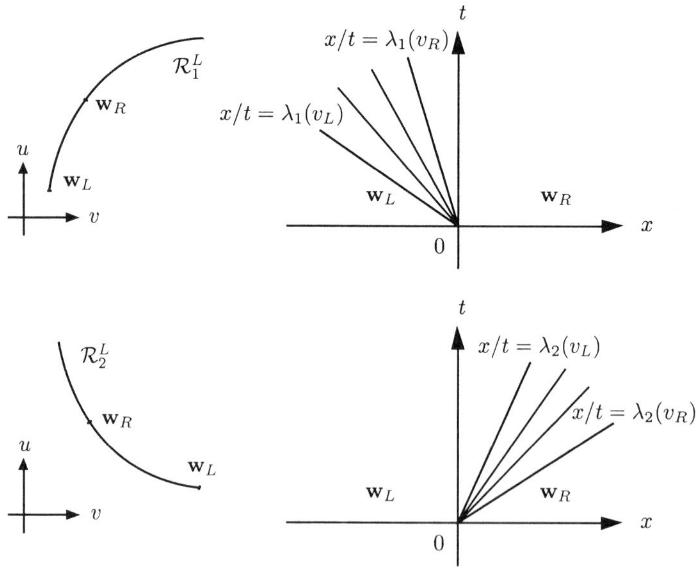

FIGURE 7.1. 1- and 2-rarefaction curves.

Eliminating σ from these two equations gives

$$(v - v_L)\Big(p(v) - p(v_L)\Big) = -(u - u_L)^2.$$

Since the function p is decreasing, we get

$$u - u_L = \pm\sqrt{(p(v) - p(v_L))(v_L - v)}. \tag{7.4}$$

Theorem 4.1 enables us to distinguish between the 1- and 2-shock curves. Due to (4.5), for a 1-shock we have

$$\frac{u - u_L}{v - v_L} = +\sqrt{-p'(v_L)} + O(\varepsilon).$$

Thus

$$u - u_L \text{ is } \begin{cases} \geq 0 & \text{for } v \geq v_L, \\ \leq 0 & \text{for } v \leq v_L, \end{cases}$$

and, from (7.4), we deduce that the 1-shock curve is given by

$$u - u_L = \begin{cases} -\sqrt{(p(v) - p(v_L))(v_L - v)}\ , & v \leq v_L, \\ +\sqrt{(p(v) - p(v_L))(v_L - v)}\ , & v \geq v_L. \end{cases} \tag{7.4a}$$

Notice that by (7.3) we have

$$\sigma_1 = -\frac{u - u_L}{v - v_L} < 0.$$

Similarly, for a 2-shock, we have by (4.5)

$$\frac{u - u_L}{v - v_L} = -\sqrt{-p'(v_L)} + O(\varepsilon).$$

It follows that the 2-shock curve is given by

$$u - u_L = \begin{cases} +\sqrt{(p(v) - p(v_L))(v_L - v)}, & v \leq v_L, \\ -\sqrt{(p(v) - p(v_L))(v_L - v)}, & v \geq v_L \end{cases} \tag{7.4b}$$

with

$$\sigma_2 = -\frac{u - u_L}{v - v_L} > 0.$$

Let us now select the admissible shocks, i.e., those that satisfy the Lax entropy conditions (5.14). For the 1-shock (5.14) yields

$$\begin{cases} \sigma_1 < \lambda_1(v_L), \\ \lambda_1(v) < \sigma_1 < \lambda_2(v). \end{cases}$$

Since $\lambda_1 < 0$ and $\lambda_2 > 0$, we have only to impose

$$\lambda_1(v) < \sigma_1 < \lambda_1(v_L). \tag{7.5}$$

First, since p' is monotonically increasing, $\lambda_1(v) < \lambda_1(v_L)$ implies

$$v < v_L.$$

Consider now the following part of the 1-shock curve:

$$u = u_L - \sqrt{(p(v) - p(v_L))(v_L - v)}\ , \ v \leq v_L. \tag{7.6}$$

By (7.3), the corresponding shock speed

$$\sigma_1 = -\frac{u - u_L}{v - v_L} = -\sqrt{(p(v) - p(v_L)/(v_L - v)}$$

readily satisfies (7.5) due to the convexity of p.

Thus $\mathcal{S}_1^a(\mathbf{w}_L)$ (also noted $\mathcal{S}_1^L$) is indeed given by (7.6). We have

$$u'(v) = \frac{p'(v)(v - v_L) + p(v) - p(v_L)}{2\sqrt{(p(v) - p(v_L)(v_L - v)}} > 0 \text{ for } v \leq v_L$$

(see Figure 7.2). With a similar analysis, we find that an admissible 2-shock satisfies

$$\lambda_2(v) < \sigma_2,$$
$$\lambda_1(v_L) < \sigma_2 < \lambda_2(v_L),$$

or equivalently

$$\lambda_2(v) < \sigma_2 < \lambda_2(v_L),$$

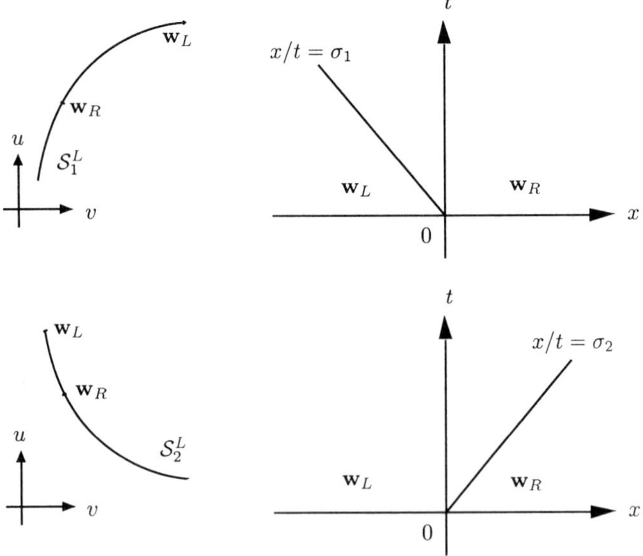

FIGURE 7.2. 1- and 2-shock curves.

which implies $v \geq v_L$. The curve $\mathcal{S}_2^a(\mathbf{w}_L)$ (also noted $\mathcal{S}_2^L$) is thus

$$u = u_L - \sqrt{(p(v) - p(v_L))(v_L - v)}\ ,\ v \geq v_L. \tag{7.7}$$

For a given right state $\mathbf{w}_R = (v_R, u_R)$, the speed of the shock connecting $\mathbf{w}_L = (v_L, u_L)$ and $\mathbf{w}_R = (v_R, u_R)$ is

$$\sigma_2 = -\frac{u_R - u_L}{v_R - v_L} = \sqrt{(p(v_L) - p(v_R))/(v_R - v_L)}.$$

Remark 7.1. We shall now see that the Lax entropy condition (5.14) is in the present case equivalent to the entropy condition (5.18) asssociated to the entropy pair (U, F) introduced in Chapter I, Example 3.1,

$$U(v, u) = \frac{u^2}{2} - P(v), F(v, u) = p(v)u,$$

where P is a primitive of p.

Indeed, with this choice of (U, F), (5.18) yields

$$\sigma\Big\{\frac{u^2 - u_L^2}{2} - \int_{v_L}^{v} p(y)dy\Big\} \geq up(v) - u_L p(v_L).$$

Using (7.3), we can eliminate σ and get

$$\frac{1}{2}(u + u_L)\Big(p(v) - p(v_L)\Big) + \Big(\frac{u - u_L}{v - v_L}\Big)\int_{v_L}^{v} p(y)dy \geq up(v) - u_L p(v_L),$$

or equivalently

$$(u - u_L)\Big\{\frac{1}{(v - v_L)}\int_{v_L}^{v} p(y)dy - \frac{1}{2}\Big(p(v) + p(v_L)\Big)\Big\} \geq 0.$$

Since, by the convexity of p, the term in brackets is always ≤ 0, we obtain the condition

$$u \leq u_L.$$

Comparing with (7.6), (7.7), we find the same admissible shock curves. □

Let us draw the curves $\mathcal{S}_i^a(\mathbf{w}_L)$ and $\mathcal{R}_i(\mathbf{w}_L), i = 1, 2$ together in the (v, u) plane. In conformity with the general case (see (6.2)), the curves $\mathcal{S}_i^a$ and $\mathcal{R}_i, i = 1, 2$ are osculatory so that we obtain Figure 7.3.

The curves above divide the (v, u)-plane into four regions (for a given left state (v_L, u_L)) labeled I to IV.

By Theorem 6.1, we can solve locally the Riemann problem for any right state $\mathbf{w}_R$ sufficiently close to $\mathbf{w}_L = (v_L, u_L)$. The solution consists of at most three constant states (including (v_L, u_L) and (v_R, u_R)) separated by a k-rarefaction or a k-admissible shock wave $k = 1, 2$. In fact, this result has a global character and holds for right states that are not necessarily in a neighborhood of (v_L, u_L).

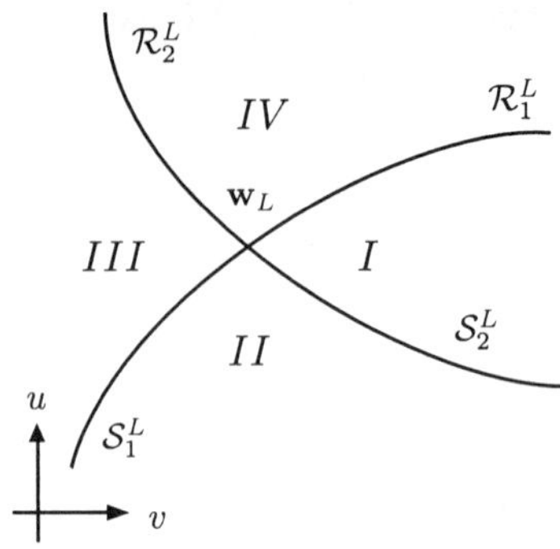

FIGURE 7.3. Regions delimited by the rarefaction and shock curves.

Theorem 7.1

(i) If $\mathbf{w}_R = (v_R, u_R)$ belongs to the regions I or III (see Figure 7.3) the Riemann problem (6.1) for the p-system always admits a solution.
(ii) If $\mathbf{w}_R = (v_R, u_R) \in II$, the Riemann problem always has a solution provided that p is defined on the whole real line or p is defined on $]0, +\infty[$ with $\lim_{v \to 0+} p(v) = +\infty$.
(iii) If $\mathbf{w}_R = (v_R, u_R) \in IV$, the Riemann problem may have no solution.

Proof. Let us consider the four different cases:

Case 1 : $(v_R, u_R) = \mathbf{w}_R \in I$ (see Figure 7.4). .
Let us prove that there exists a state $\overline{\mathbf{w}}_0 = (\overline{v}_0, \overline{u}_0)$ on $\mathcal{R}_1(w_L)$ such that $\mathbf{w}_R$ can be connected to $\overline{\mathbf{w}}_0$ by a 2-shock, i.e., $\mathbf{w}_R \in \mathcal{S}_2^a(\overline{\mathbf{w}}_0)$.

We begin by characterizing the states $\mathbf{w} = (v, u)$ that can be connected to $\mathbf{w}_R$ by a 1-rarefaction and a 2-shock by means of an intermediate state

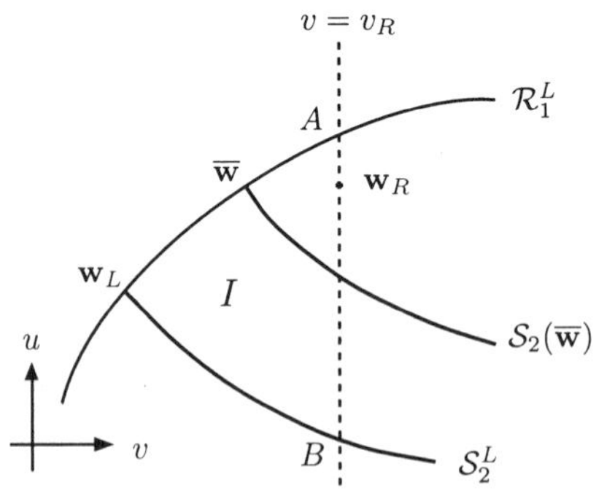

FIGURE 7.4. Region I.

$\overline{\mathbf{w}} = (\overline{v}, \overline{u}) \in \mathcal{R}_1(\mathbf{w}_L)$ First, we have by (7.1)

$$\overline{u} = u_L + \int_{v_L}^{\overline{v}} \sqrt{-p'(y)}dy \ , \ \overline{v} > v_L.$$

Next, by (7.7), $\mathcal{S}_2^a(\overline{\mathbf{w}})$ is the set of states $\mathbf{w} = (v, u)$ that satisfy

$$u = \overline{u} - \sqrt{(p(v) - p(\overline{v}))(\overline{v} - v)} \ , \ v > \overline{v}$$

so that $\mathbf{w}$ and $\overline{\mathbf{w}}$ are linked by the relation

$$u = u_L + \int_{v_L}^{\overline{v}} \sqrt{-p'(y)}dy - \sqrt{(p(v) - p(\overline{v}))(\overline{v} - v)} \ , \ v > \overline{v} > v_L.$$

Now $\mathcal{S}_2^a(\overline{\mathbf{w}})$ intersects the vertical line $v = v_R$ at a point $(v_R, G(\overline{v})) = \overline{\mathbf{w}}_R$, where G is defined by

$$G(v) = u_L + \int_{v_L}^{v} \sqrt{-p'(y)}dy - \sqrt{(p(v) - p(v_R))(v_R - v)} \ , v_R > v > v_L.$$

In order to prove that there exists a unique $\overline{v}_0$ such that $G(\overline{v}_0) = u_R$, hence such that $\overline{\mathbf{w}}_R = \mathbf{w}_R$, we first check that G is monotonically increasing. Indeed

$$G'(v) = \sqrt{-p'(v)} - \frac{\{p(v_R) - p(v) - p'(v)(v - v_R)\}}{2\sqrt{(p(v) - p(v_R))(v_R - v)}} > 0.$$

Then clearly the line $v = v_R$ intersects $\mathcal{R}_1(w_L)$ at a point $A = (v_R, u_A)$, and $\mathcal{S}_2^a(\mathbf{w}_L)$ at a point $B = (v_R, u_B)$ with

$$u_A = G(v_R) < u_R < u_B = G(v_L).$$

Hence, there exists a unique $\overline{v}_0 \in]v_L, v_R[$ such that $G(\overline{v}_0) = u_R$. Setting

$$u_0 = u_L + \int_{v_L}^{\overline{v}_0} \sqrt{-p'(y)}dy,$$

we get the point $\mathbf{w}_0 = (\overline{v}_0, \overline{u}_0)$ that we were looking for.

In short, $\mathbf{w}_L$ can be connected to $\mathbf{w}_R$ by a 1-rarefaction wave followed by a 2-shock propagating with speed $\sigma_2 = \sqrt{\frac{(p(v_R) - p(\overline{v}_0))}{(\overline{v}_0 - v_R)}}$ (see Figure 7.5).

Remark 7.2. Note that it is impossible in the preceding case to connect $\mathbf{w}_L$ and $\mathbf{w}_R$ by a 2-shock wave and a 1-rarefaction in this order (as Figure 7.4 might suggest), since the condition $\lambda_1 < 0 < \lambda_2$ cannot be violated. □

Case 2 : $\mathbf{w}_R \in III$.

By a similar argument, we can prove that $\mathbf{w}_L$ can be connected to $\mathbf{w}_R$ by a 1-shock followed by a 2-rarefaction wave (see Figure 7.6). The 1-shock propagates with speed $\sigma_1 = -\sqrt{\frac{(p(\overline{v}_0) - p(v_L))}{(v_L - \overline{v}_0)}}$.

Case 3 : $\mathbf{w}_R \in II$.

Let us check that the horizontal line $u = u_R$ meets $\mathcal{S}_1^a(\mathbf{w}_L)$ and $\mathcal{S}_2^a(\mathbf{w}_L)$, respectively, at two points A, B that are uniquely determined. To prove the

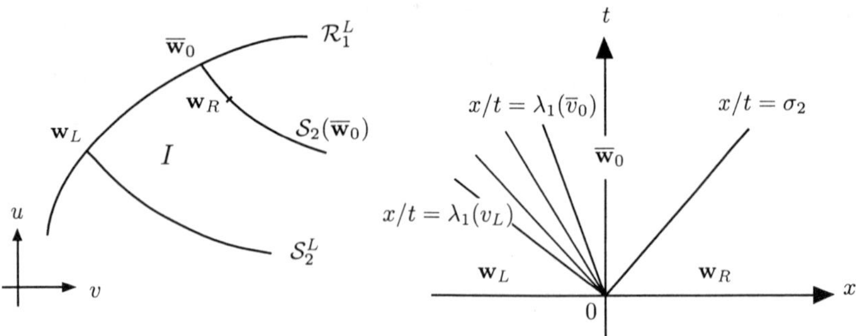

FIGURE 7.5. Solution of the Riemann problem: 1-rarefaction and 2-shock.

existence and uniqueness of B, we have to solve by (7.7)

$$u_L - \sqrt{(p(v) - p(v_L))(v_L - v)} = u_R, \quad v \geq v_L. \tag{7.8}$$

But the function

$$v \rightarrow u_L - \sqrt{(p(v) - p(v_L))(v_L - v)}$$

is easily seen to decrease for $v \in [v_L, +\infty[$ from u_L to $-\infty$. Hence, since $u_R < u_L$, there exists a unique $v = v_B$ solution of (7.8).

In the same way, the point where the line $u = u_R$ meets $\mathcal{S}_1^a(\mathbf{w}_L)$ satisfies by (7.6)

$$u_L - \sqrt{(p(v) - p(v_L))(v_L - v)} = u_R, \quad v \leq v_L. \tag{7.9}$$

The mapping $v \rightarrow (p(v) - p(v_L))(v_L - v)$ is decreasing for $v \leq v_L$. Moreover, on the one hand, if p is defined on $\mathbb{R}$, we have

$$\lim_{v \rightarrow -\infty} (p(v) - p(v_L))(v_L - v) = +\infty,$$

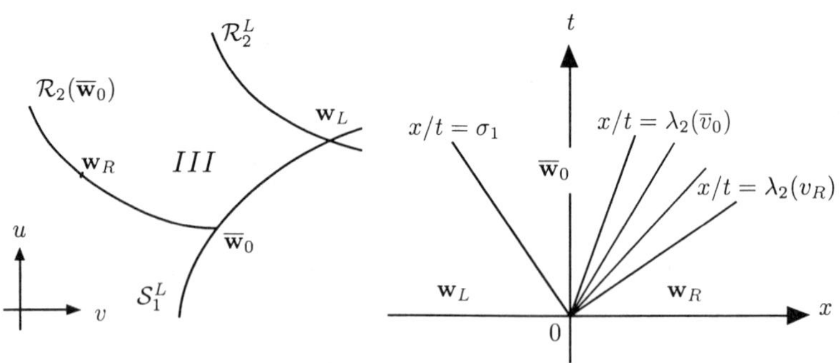

FIGURE 7.6. Solution of the Riemann problem : 1-shock and 2-rarefaction.

and, on the other hand, if p is only defined on $]0,+\infty[$ we have by assumption

$$\lim_{v\to 0+}(p(v)-p(v_L))(v_L-v)=+\infty.$$

Thus, in both cases we can find a unique $v=v_A<v_L$ solution of (7.9).

Then, we can conclude as in case 1 that there exists a unique state $\overline{\mathbf{w}}_0\in\mathcal{S}_a^1(\mathbf{w}_L)$ such that $\mathbf{w}_R$ belongs to $\mathcal{S}_2^a(\overline{\mathbf{w}}_0)$ (see Figure 7.7). Hence, we can connect $\mathbf{w}_L$ to $\mathbf{w}_R$ by a 1-shock followed by a 2-shock.

Case 4 : $\mathbf{w}_R\in IV$.

In this case, the Riemann problem may have no solution. Indeed, assume

$$u_\infty=\int_{v_L}^{+\infty}\sqrt{-p'(y)}dy<+\infty. \tag{7.10}$$

Then $\mathcal{R}_1(u_L)$ has a horizontal asymptote, namely the line $u=u_L+u_\infty$. Now take, for instance, a state $\mathbf{w}_R$ in IV with

$$v_R=v_L \text{ and } u_R>u_L+2u_\infty$$

(see Figure 7.8).

Then, for any $\overline{\mathbf{w}}\in\mathcal{R}_1(\mathbf{w}_L)$,

$$\overline{u}=u_L+\int_{v_L}^{\overline{u}}\sqrt{-p'(y)}dy<u_L+u_\infty,\ \overline{v}\geq v_L.$$

The 2-rarefaction curve passing by $\overline{\mathbf{w}}$ is given by

$$u=\overline{u}+\int_v^{\overline{v}}\sqrt{-p'(y)}dy,\quad v\leq\overline{v}$$

or

$$u=u_L+\int_{v_L}^{\overline{v}}\sqrt{-p'(y)}dy+\int_v^{\overline{v}}\sqrt{-p'(y)}dy.$$

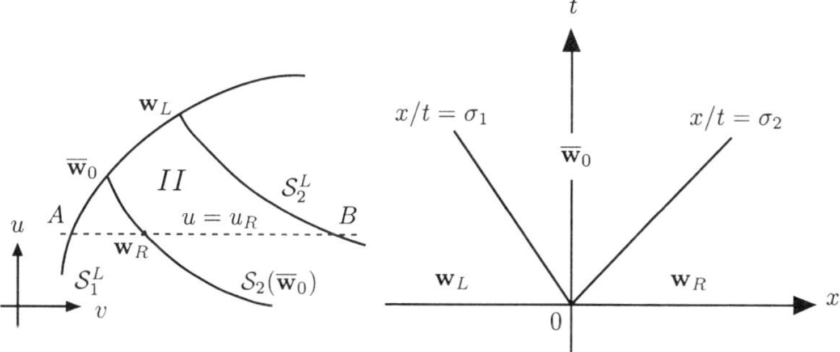

FIGURE 7.7. Solution of the Riemann problem: 1-shock and 2-shock.

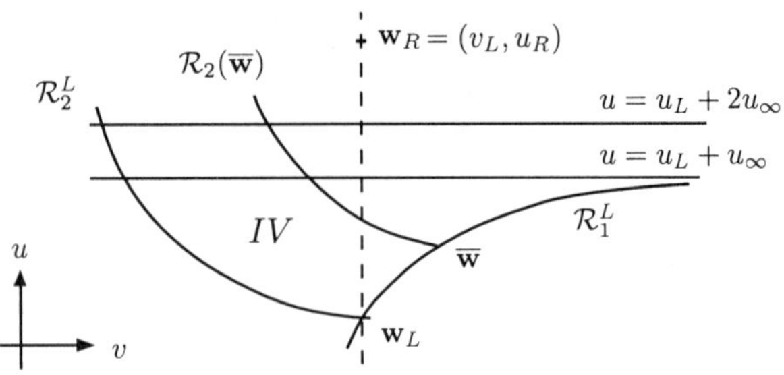

FIGURE 7.8. Region *IV*.

It meets the vertical line $v = v_L$ at a point (v_L, u) with

$$u = u_L + 2\int_{v_L}^{\overline{v}} \sqrt{-p'(y)}dy \leq u_L + 2u_\infty,$$

which proves that the point $\mathbf{w}_R$ cannot be reached by a 2-rarefaction wave (nor by a 2-shock of course!).

When the Riemann problem can be solved, which occurs at least for $\mathbf{w}_R$ close to $\mathbf{w}_L$, one can find $\overline{\mathbf{w}}_0 \in \mathcal{R}_1(\mathbf{w}_L)$ such that $\mathbf{w}_R \in \mathcal{R}_2(\overline{\mathbf{w}}_0)$ and the solution consists of two rarefaction waves (Figure 7.9).

Example 7.1. The vacuum problem.

To illustrate the situation in case 4, take $p(v) = Av^{-\gamma}$ with $\gamma > 1$, so that (7.10) holds. Also, we notice that as $v \to +\infty$

$$\lambda_i(v) = \pm\sqrt{-p'(v)} \quad \to \quad 0, \quad i = 1, 2.$$

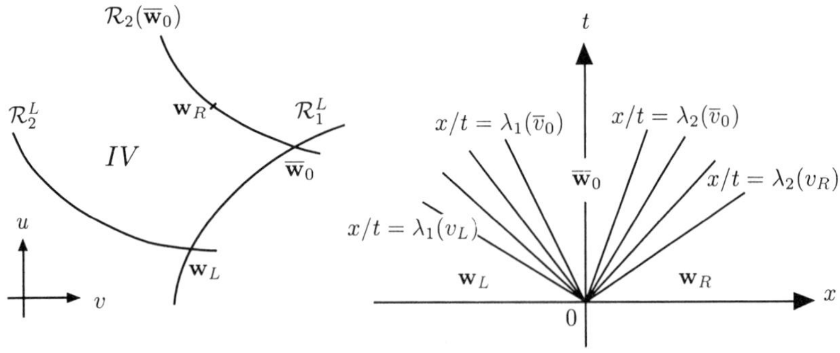

FIGURE 7.9. Solution of the Riemann problem: 1-and 2-rarefaction.

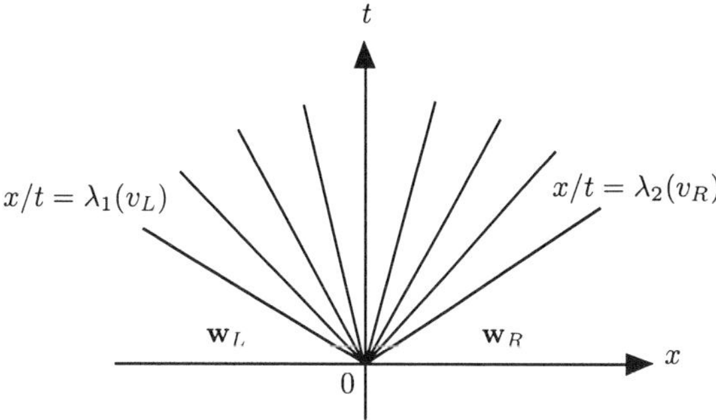

FIGURE 7.10. Vacuum.

To understand what happens, take the limit situation $\mathbf{w}_R = (v_L, u_L + 2u_\infty)$. Then $\mathbf{w}_R$ belongs to $\mathcal{R}_2(\overline{\mathbf{w}}_\infty)$, where $\overline{\mathbf{w}}_\infty = (+\infty, u_L + u_\infty)$ and the solution is given by two complete rarefaction waves (as depicted in Figure 7.10), which means that the fan of the 1-(resp. the 2-)rarefaction wave is bordered on the right (resp. the left) by the line $x = 0$. On this line, we have $v = +\infty$, which implies the formation of a vacuum since $\rho = \frac{1}{v} = 0$. □

Notes

We refer naturally to the important paper of Lax (1972), to the book of Smoller (1994), and particularly to the very complete one of Serre (1996); see also the illuminating introductory notes of Tartar (1989), the papers of Dafermos (1983) and Liu (1981), the books of LeVeque (1985), Chang and Hsiao (1989), and Li Ta-tsien (1994).

In Section 6, Theorem 6.1 ensures the existence of an entropy solution to the Riemann problem for nearby states; note that the solution may not exist for general Riemann data (Kranzer and Keyfitz 1991, Serre 1989 and the references therein, and Keyfitz and Kranzer 1995). The existence for more general small BV Cauchy data (near a constant state) is proven by Glimm (1965) and Schochet (1991) for small BV perturbation of a solvable Riemann problem (see the references therein for other results and also Young and Temple 1995) via the convergence of the random-choice method. The existence of solutions in the general case is mostly an open problem. For smooth local solutions see Kato (1975), for the use of convergent finite difference approximations see Nishida and Smoller (1990). In some cases global existence results are obtained by the vanishing viscosity method together with compensated compactness (see Tartar 1979), for instance for isentropic gas dynamics by DiPerna (1983), for the class of B. Temple systems by LeVeque and Temple (1985) and Serre (1987), and for "rich" systems by Serre (1991), see also Heibig (1994), Liu (1977), Rascle (1986), Peng (1992), Benzoni-Gavage and Serre (1993), and Rubino (1993). An alternative technique is wave-front tracking (Bressan and Colombo 1995 and the references therein). We also mention a result by Freistühler (1994) concerning a class of (non strictly) hyperbolic systems and related results to which he refers, and the papers of Isaacson and Temple (1992) concerning "resonant" systems. For the question of uniqueness, see Le Floch and Xin (1993), which contains a survey of the known results, and the references therein.

We have considered "convex" systems, i.e., strictly hyperbolic systems whose characteristic fields are either genuinely nonlinear or linearly degenerate. The existence of a solution to the Riemann problem for more general (nonconvex) systems is given in Liu (1975), see also Menikoff and Plohr (1989) and the interesting paper of Wendroff (1972) concerning the p-system when p is not convex. As in the scalar case, we must admit compound waves corresponding to a characteristic field that is non-genuinely nonlinear (see Remarks 5.3 and 6.2); otherwise, the conclusions of Theorem 6.1 are valid.

For the extension of these results to nonstrictly hyperbolic systems, the proof of Theorem 4.1 obviously fails since the rank of $\mathbf{M}(\mathbf{u}_0)$ is $< p - 1$ at an "umbilic point" (double eigenvalue, see Remark 6.2), which poses a bifurcation problem (see Shearer et al. 1987, and Chen and Kan 1994). Besides the references at the end of Chapter 17 of Smoller's book, see Keyfitz and Kranzer (1985), Tveito and Winther (1991), the papers of Keyfitz in the proceedings Keyfitz and Kranzer (1985) and Carasso, Raviart, and Serre (1986), those of Brio, Freisthüler, and Glimm in Ballmann and Jeltsch (1989), those of Azevedo and Marchesin, Hsiao, Liu and Xin, Le Floch, Shearer and Schecter, and Schaeffer and Shearer in Keyfitz and Shearer (1991), and Lindquist (1989). All those proceedings contain other interesting related papers.

For what concerns "convex" hyperbolic systems in nonconservative form, which are introduced in Remark 5.3, the existence of weak entropy BV solutions is proven in Le Floch and Liu (1992), for small BV data, by means of the convergence of Glimm's scheme.

Though it is very important in the applications, we did not mention the problem of invariant regions for system (2.1) (a subset S of the set of states is invariant if $(\mathbf{u}_0(x) \in S,\ \forall x \Rightarrow \mathbf{u}(x,t) \in S,\ \forall (x,t)))$; their characterization can be found in Chueh et al. (1977) and Hoff (1985).

We did not consider systems with source terms. However the case of reacting flows will be studied in the next chapter. Some other examples can be found in Leveque and Yee (1990), Leveque and Jinghua (1992), E (1992), Fan and Hale (1993), Chen et al. (1994), Isaacson and Temple (1995), Kevorkian et al. (1995), and Chen and Glimm (1994), Klingenberg and Lu (1994), Schroll et al. in Glimm et al. (1996).

II

Gas dynamics and reacting flows

1 Preliminaries

1.1 Properties of the physical entropy

Let us consider a fluid in a local thermodynamical equilibrium. Then we know from thermodynamics that the thermodynamical state of the fluid is completely determined by any two thermodynamic variables. Most often, we shall note by the same letter the corresponding mathematical functions, though they differ. For instance, we shall use

$$p = p(\tau, s) = p(\rho, \varepsilon), \quad \varepsilon = \varepsilon(\tau, s) = \varepsilon(\tau, p), \quad T = T(\tau, s) = T(\tau, p), \ldots$$

Let us choose in particular the specific volume τ and the specific (physical) entropy s and concentrate on the function $(\tau, s) \to \varepsilon(\tau, s)$. By the second law of thermodynamics, we have

$$d\varepsilon = T\,ds - p\,d\tau$$

so that

$$\frac{\partial \varepsilon}{\partial \tau} = -p < 0, \quad \frac{\partial \varepsilon}{\partial s} = T > 0. \tag{1.1}$$

Moreover, always assuming that the fluid is in a local thermodynamic equilibrium, we have the following important assumption:

$$\text{the function } (\tau, s) \to \varepsilon(\tau, s) \text{ is strictly convex.} \tag{1.2}$$

This means that the Hessian matrix of this function,

$$\begin{pmatrix} \dfrac{\partial^2 \varepsilon}{\partial \tau^2} & \dfrac{\partial^2 \varepsilon}{\partial \tau\,\partial s} \\ \dfrac{\partial^2 \varepsilon}{\partial \tau\,\partial s} & \dfrac{\partial^2 \varepsilon}{\partial s^2} \end{pmatrix},$$

is a (symmetric) positive-definite matrix, or equivalently that we have

$$\frac{\partial^2 \varepsilon}{\partial s^2} > 0, \quad \frac{\partial^2 \varepsilon}{\partial \tau^2}\frac{\partial^2 \varepsilon}{\partial s^2} - \Big(\frac{\partial^2 \varepsilon}{\partial \tau\,\partial s}\Big)^2 > 0. \tag{1.3}$$

As a consequence, we obtain

$$\frac{\partial p}{\partial \tau} = -\frac{\partial^2 \varepsilon}{\partial \tau^2} < 0,$$

which enables us to define the sound speed c by

$$c = \tau\sqrt{-\partial p/\partial \tau}. \tag{1.4}$$

Now, since $\frac{\partial \varepsilon}{\partial s} > 0$, one can invert the function $s \to \varepsilon(\tau, s)$ and introduce the function $(\tau, s) \to s(\tau, \varepsilon)$. The purpose of this section is to prove the following result.

Theorem 1.1

The three following assertions are equivalent:
(i) the function $(\tau, s) \to \varepsilon(\tau, s)$ is strictly convex;
(ii) the function $(\tau, \mathbf{u} = (u_1, ..., u_d), e) \to s(\tau, e - \frac{|\mathbf{u}|^2}{2})$ is strictly convex;
(iii) the function $(\rho, \mathbf{q} = (q_1, ..., q_d), E) \to -\rho\, s(\frac{1}{\rho}, \frac{E}{\rho} - \frac{|\mathbf{q}|^2}{2\rho^2})$ is strictly convex.

The proof of this result needs a few simple steps, all of which rely on the differentiation of a compound function. Let us recall that if f and g are two C^2 functions defined on an open subset U (respectively V) of a Banach space E (respectively F) with $f : U \to V$ and $g : V \to G$ (G a Banach space), we have the following formula expressing the second total derivative of $g \circ f$: for all $a \in U$, $(x_1, x_2) \in E \times E$

$$\begin{cases} D^2(g \circ f)(a)(x_1, x_2) \\ \qquad = Dg(f(a)) \cdot (D^2 f(a) \cdot (x_1, x_2)) \\ \qquad + D^2 g(f(a)) \cdot (Df(a) \cdot x_1, Df(a) \cdot x_2). \end{cases} \tag{1.5}$$

Lemma 1.1

The two following assertions are equivalent:
(i) the function $(\tau, s) \to \varepsilon(\tau, s)$ is strictly convex;
(ii) the function $(\tau, \varepsilon) \to -s(\tau, \varepsilon)$ is strictly convex.

Proof. We apply formula (1.5) to the functions $f(\tau, \varepsilon) = (\tau, s(\tau, \varepsilon))$ and $g(\tau, s) = \varepsilon(\tau, s)$. Hence $g \circ f(\tau, s) = \varepsilon(\tau, s(\tau, \varepsilon)) = \varepsilon$, and in that case

$$D^2(\varepsilon \circ f)(\tau, s) = 0. \tag{1.6}$$

Let us compute the corresponding right-hand side of (1.5) with $x_1 = x_2 = (\tau_1, \varepsilon_1)$. We have

$$\begin{cases} Df(\tau, \varepsilon)(\tau_1, \varepsilon_1) = (\tau_1, Ds(\tau, \varepsilon) \cdot (\tau_1, \varepsilon_1)), \\ D^2 f(\tau, \varepsilon)(\tau_1, \varepsilon_1)^{(2)} = (0, D^2 s(\tau, \varepsilon) \cdot (\tau_1, \varepsilon_1)^{(2)}), \end{cases} \tag{1.7}$$

and also for all (τ_1, s_1)

$$D\varepsilon(\tau, s) \cdot (\tau_1, s_1) = \frac{\partial \varepsilon}{\partial \tau}(\tau, s)\tau_1 + \frac{\partial \varepsilon}{\partial s}(\tau, s)s_1,$$

which together with (1.7) yields

$$(1.8) \quad \begin{cases} D\varepsilon(\tau, s(\tau, \varepsilon)) \cdot (D^2 f(\tau, \varepsilon)(\tau_1, \varepsilon_1)^{(2)}) \\ \qquad = \dfrac{\partial \varepsilon}{\partial s}(\tau, s(\tau, \varepsilon)) D^2 s(\tau, s) \cdot (\tau_1, \varepsilon_1)^{(2)} \end{cases}$$

and

$$(1.9) \quad \begin{cases} D^2\varepsilon(\tau, s(\tau, \varepsilon)) \cdot (Df(\tau, \varepsilon)(\tau_1, \varepsilon_1)^{(2)}) \\ \qquad = D^2\varepsilon(\tau, s(\tau, \varepsilon)) \cdot (\tau_1, Ds(\tau, \varepsilon) \cdot (\tau_1, \varepsilon_1))^{(2)}. \end{cases}$$

Combining (1.5)–(1.9), we get

$$D^2 s(\tau, \varepsilon) \cdot (\tau_1, \varepsilon_1)^{(2)} = -\Big(\frac{\partial \varepsilon}{\partial s}\Big)^{-1} D^2\varepsilon(\tau, s(\tau, \varepsilon)) \cdot (\tau_1 \cdot Ds(\tau, \varepsilon) \cdot (\tau_1, \varepsilon_1))^{(2)}.$$

Since $\frac{\partial \varepsilon}{\partial s} > 0$, this proves that (i) $\Longrightarrow$ (ii). The converse follows by exchanging the roles of the functions ε and s. □

The next step consists in proving another equivalent property in the case $d = 1$.

Lemma 1.2

The two following assertions are equivalent:
(i) the function $(\tau, \varepsilon) \to -s(\tau, \varepsilon)$ is strictly convex;
(ii) the function $(\tau, u, e) \to -s(\tau, e - \frac{u^2}{2})$ is strictly convex.

Proof. Let us now choose $f(\tau, u, e) = (\tau, e - \frac{u^2}{2})$ and $g(\tau, \varepsilon) = s(\tau, \varepsilon)$. Then, an easy computation shows that

$$Df(\tau, u, e) \cdot (\tau_1, u_1, e_1) = (\tau_1, e_1 - u\, u_1),$$
$$D^2 f(\tau, u, e) \cdot (\tau_1, u_1, e_1)^{(2)} = (0, -u_1^2),$$

and using (1.5) we get for all (τ_1, u_1, e_1)

$$(1.10) \quad \begin{cases} D^2(s \circ f)(\tau, u, e) \cdot (\tau_1, u_1, e_1)^{(2)} \\ \quad = -\dfrac{\partial s}{\partial \varepsilon}\Big(\tau, e - \dfrac{u^2}{2}\Big)u_1^2 + D^2 s\Big(\tau, e - \dfrac{u^2}{2}\Big) \cdot (\tau_1, e_1 - u\, u_1)^{(2)}. \end{cases}$$

Since $\frac{\partial s}{\partial \varepsilon} > 0$, this proves that (i) $\Longrightarrow$ (ii).

Now, if we take $u_1 = 0$ in (1.10), it follows that for all (τ_1, e_1)

$$D^2 s\Big(\tau, e - \frac{u^2}{2}\Big) \cdot (\tau_1, e_1)^{(2)} = D^2(s \circ f)(\tau, u, e) \cdot (\tau_1, 0, e_1)^{(2)},$$

which proves in turn that (ii) $\Longrightarrow$ (i). □

Now, let C be a convex cone of $\mathbb{R}^p$, and let $\pi : \mathbb{R}_+^\star \times C \to \mathbb{R}$ be a C^2 function. We define the function $\omega : \mathbb{R}_+^\star \times C \to \mathbb{R}$ by

$$\omega(\mu, \mathbf{v}) = \mu\, \pi\Big(\frac{1}{\mu}, \frac{\mathbf{v}}{\mu}\Big). \tag{1.11}$$

Lemma 1.3

The function ω is strictly convex (resp. convex) if and only if the function π is strictly convex (resp. convex).

Proof. Let us set

$$\hat{\pi}(\mu, \mathbf{v}) = \pi\Big(\frac{1}{\mu}, \frac{\mathbf{v}}{\mu}\Big)$$

so that

$$\omega(\mu, \mathbf{v}) = \mu\, \hat{\pi}(\mu, \mathbf{v}).$$

Then, simply by differentiating the bilinear form $(\mu, \mathbf{v}) \to \mu\, \mathbf{v}$, we get

$$D^2\omega(\mu, \mathbf{v}) \cdot (\mu_1, \mathbf{v}_1)^{(2)} = \mu\, D^2\hat{\pi}(\mu, \mathbf{v}) \cdot (\mu_1, \mathbf{v}_1)^{(2)} + 2\mu_1\, D\hat{\pi}(\mu, \mathbf{v}) \cdot (\mu_1, \mathbf{v}_1). \tag{1.12}$$

Let us next define on $\mathbb{R}_+^\star \times C$

$$r(\mu, \mathbf{v}) = \Big(\frac{1}{\mu}, \frac{\mathbf{v}}{\mu}\Big). \tag{1.13}$$

Since $\hat{\pi}(\mu, \mathbf{v}) = \pi \circ r(\mu, \mathbf{v})$, we have, again using (1.5),

$$\begin{cases} D\hat{\pi}(\mu, \mathbf{v})(\mu_1, \mathbf{v}_1) = D\pi(\frac{1}{\mu}, \frac{\mathbf{v}}{\mu}) \cdot (Dr(\mu, \mathbf{v})(\mu_1, \mathbf{v}_1)), \\ D^2\hat{\pi}(\mu, \mathbf{v})(\mu_1, \mathbf{v}_1)^{(2)} = D\pi(\frac{1}{\mu}, \frac{\mathbf{v}}{\mu}) \cdot (D^2 r(\mu, \mathbf{v})(\mu_1, \mathbf{v}_1)^{(2)}) \\ \qquad + D^2\pi(\frac{1}{\mu}, \frac{\mathbf{v}}{\mu}) \cdot (Dr(\mu, \mathbf{v})(\mu_1, \mathbf{v}_1))^{(2)}. \end{cases} \tag{1.14}$$

Now, we notice that differentiating two times the relation

$$\mu r(\mu, \mathbf{v}) = (1, \mathbf{v})$$

yields the identity

$$(0, 0) = 2\mu_1\, Dr(\mu, \mathbf{v}) \cdot (\mu_1, \mathbf{v}_1) + \mu\, D^2 r(\mu, \mathbf{v}) \cdot (\mu_1, \mathbf{v}_1)^{(2)}. \tag{1.15}$$

Substituting (1.14) in (1.12) and using (1.15), we get

$$\begin{aligned} D^2\, \omega(\mu, \mathbf{v}) \cdot (\mu_1, \mathbf{v}_1)^{(2)} &= \mu\, D^2\pi\Big(\frac{1}{\mu}, \frac{\mathbf{v}}{\mu}\Big) \cdot (Dr(\mu, \mathbf{v})(\mu_1, \mathbf{v}_1))^{(2)} \\ &\quad + D\pi\Big(\frac{1}{\mu}, \frac{\mathbf{v}}{\mu}\Big) \cdot \{2\mu_1\, Dr(\mu, \mathbf{v}) \cdot (\mu_1, \mathbf{v}_1) + \mu\, D^2 r(\mu, \mathbf{v}) \cdot (\mu_1, \mathbf{v}_1)^{(2)}\} \\ &= \mu\, D^2\pi\Big(\frac{1}{\mu}, \frac{\mathbf{v}}{\mu}\Big) \cdot Dr(\mu, \mathbf{v}) \cdot (\mu_1, \mathbf{v}_1))^{(2)}, \end{aligned}$$

and hence

$$D^2\,\omega(\mu,\mathbf{v})\cdot(\mu_1,\mathbf{v}_1)^{(2)} = \mu\, D^2\pi\Big(\frac{1}{\mu},\frac{\mathbf{v}}{\mu}\Big)\cdot\Big(-\frac{\mu_1}{\mu^2},-\Big(\frac{\mu_1}{\mu^2}\Big)\mathbf{v}+\Big(\frac{1}{\mu}\Big)\mathbf{v}_1\Big)^{(2)}.$$

This proves that ω is convex (resp. strictly convex) if π is convex (resp. strictly convex). The reverse property is clear since

$$\pi(\lambda,\mathbf{u}) = \lambda\,\omega\Big(\frac{1}{\lambda},\frac{\mathbf{u}}{\lambda}\Big)$$

enables us to exchange π and ω. □

Next, let C be a convex subset of $\mathbb{R}_+\times\mathbb{R}^p$, and let $\omega : C\to\mathbb{R}$ be C^2. We define a convex subset $\overline{C}$ of $\mathbb{R}^n\times\mathbb{R}^p$ by

$$\overline{C} : \{\mathbf{x},\mathbf{y})\in\mathbb{R}^n\times\mathbb{R}^p,\ (|\mathbf{x}|,\mathbf{y})\in C\},$$

and the function $\overline{\omega} : \overline{C}\to\mathbb{R}$ by

$$\overline{\omega}(\mathbf{u},\mathbf{v}) = \omega(|\mathbf{u}|,\mathbf{v}).$$

Lemma 1.4

Assume that the function ω is monotonically increasing with respect to the first variable. Then, the function $\overline{\omega}$ is strictly convex (resp. convex) if and only if the function ω is strictly convex (resp. convex).

Proof. Setting $n(\mathbf{x}) = |\mathbf{x}|$, and using once more (1.5), we have

$$\begin{aligned} &D^2\overline{\omega}(\mathbf{u},\mathbf{v})\cdot(\mathbf{u}_1,\mathbf{v}_1)^{(2)} \\ &\quad= \frac{\partial\omega}{\partial\mu}(|\mathbf{u}|,\mathbf{v})\,D^2n(\mathbf{u})\cdot(\mathbf{u}_1)^{(2)} + D^2\omega(|\mathbf{u}|,\mathbf{v})\cdot(Dn(\mathbf{u})\mathbf{u}_1,\mathbf{v}_1)^{(2)} \end{aligned}$$

with

$$Dn(\mathbf{u})\cdot\mathbf{u}_1 = \frac{\mathbf{u}^T\mathbf{u}_1}{|\mathbf{u}|}$$

so that

$$\begin{aligned} &D^2\overline{\omega}(\mathbf{u},\mathbf{v})\cdot(\mathbf{u}_1,\mathbf{v}_1)^{(2)} \\ &\quad= \frac{\partial\omega}{\partial\mu}(|\mathbf{u}|,\mathbf{v})D^2n(\mathbf{u})\cdot(\mathbf{u}_1)^{(2)} + D^2\omega(|\mathbf{u}|,\mathbf{v})\cdot\Big(\frac{\mathbf{u}^T\mathbf{u}^1}{|\mathbf{u}|},\mathbf{v}_1\Big)^{(2)}. \end{aligned}$$

By assumption, $\frac{\partial\omega}{\partial\mu} > 0$; hence, if ω is convex or strictly convex so is $\overline{\omega}$.

Conversely, we may write

$$\omega(\mu,\mathbf{v}) = \overline{\omega}(\mu\mathbf{e},\mathbf{v}),$$

where $\mathbf{e}$ is any fixed unit vector of $\mathbb{R}^n$. Then

$$D^2\omega(\mu,\mathbf{v})\cdot(\mu_1,\mathbf{v}_1)^{(2)} = D^2\overline{\omega}(\mu\mathbf{e},v)\cdot(\mu_1\mathbf{e},\mathbf{v}_1)^{(2)},$$

which proves the reverse property. □

Proof of Theorem 1.1. In the case $d = 1$, thanks to Lemmas 1.1 to 1.3, (1.2) implies that the function

$$(\rho, q, E) \to -\rho\, s\Big(\frac{1}{\rho}, \frac{E}{\rho} - \frac{q^2}{2\rho}\Big)$$

is strictly convex. In fact, taking $\pi(\tau, (u, e)) = -s(\tau, e - \frac{u^2}{2})$ in (1.11) gives

$$\omega(\rho, (q, E)) = -\rho s\Big(\frac{1}{\rho}, \frac{E}{\rho} - \frac{q^2}{2\rho}\Big) = -\rho s\Big(\tau, e - \frac{u^2}{2}\Big) = -\rho s(\tau, \varepsilon).$$

In the case $d > 1$, we may apply Lemma 1.4 to the function ω defined by

$$(\mu, \mathbf{v} = (\tau, e)) \to \omega(\mu, \mathbf{v}) = -s\Big(\tau, e - \frac{\mu^2}{2}\Big).$$

Since

$$\frac{\partial \omega}{\partial \mu}(\mu, \mathbf{v}) = +\mu \frac{\partial s}{\partial \varepsilon}\Big(\tau, e - \frac{\mu^2}{2}\Big) > 0,$$

Lemmas 1.1 to 1.3 imply the theorem. □

1.2 Ideal gases

Before concluding this introductory section, let us consider more closely the case of an *ideal gas*, or perfect gas, i.e., such that the equation of state satisfies Gay-Lussac and Boyle's law

$$p = \rho\, RT = \frac{RT}{\tau}, \tag{1.16}$$

where R is the specific gas constant; $R = \eta\mathcal{R}$, where η is the mole-mass fraction, and $\mathcal{R}$ the universal gas constant. From (1.1) and (1.16), one deduces

$$R\frac{\partial \varepsilon}{\partial s} + \tau \frac{\partial \varepsilon}{\partial \tau} = 0,$$

which implies that ε is a function of $\tau \exp(-\frac{s}{R})$, that is,

$$\varepsilon = \varphi\Big(\tau \exp\Big(-\frac{s}{R}\Big)\Big)$$

for some function φ. Computing $T = \frac{\partial \varepsilon}{\partial s}$ and taking hypothesis (1.3) into account, one can prove that ε is then a function of T only (see Courant and Friedrichs 1976, Chapter I, Section 4), $\varepsilon = \varepsilon(T)$. One defines the function $\gamma(T)$ by

$$\frac{d\varepsilon}{dT} = \frac{R}{\gamma(T) - 1}, \tag{1.17}$$

and $\frac{d\varepsilon}{dT}$ is the "specific heat at constant volume" C_v. Hence

$$p = \rho\, RT, \text{ where } R\frac{dT}{d\varepsilon} = \gamma(T) - 1$$

(which does not mean that $p = (\gamma(T) - 1)\rho\varepsilon$, unless γ is constant, see (1.19) below). Since $R > 0$ and $C_v > 0$, we have $\gamma(T) > 1$.

In the same way, introducing the specific enthalpy

$$h = \varepsilon + p\tau,$$

which satisfies

$$dh = T\, ds + \tau\, dp,$$

we obtain that for an ideal gas

$$h = \varepsilon + RT,$$

h is also a function of T only, and

$$\frac{dh}{dT} = \frac{R\gamma(T)}{(\gamma(T) - 1)}.$$

$\frac{dh}{dT}$ is the "specific heat at constant pressure" C_p. Hence

$$\gamma(T) = \frac{C_p}{C_v}$$

is the ratio of two specific heats. Note also that from (1.4) and (1.16)

$$c^2 = -\tau^2 \frac{\partial p}{\partial \tau}(\tau, s) = RT - R\tau\frac{\partial T}{\partial \tau}.$$

Now, since from (1.1, (1.16) and (1.17)

$$\tau R\frac{\partial T}{\partial \tau} = \tau\frac{dT}{d\varepsilon}\frac{\partial \varepsilon}{\partial \tau} = -p\tau(\gamma(T) - 1) = -RT(\gamma(T) - 1),$$

we get

$$c^2 = \gamma(T) RT = \frac{\gamma(T) p}{\rho}. \tag{1.18}$$

One sometimes says that the gas is thermally perfect (but calorically imperfect).

Let us now focus on the case of a *polytropic* ideal gas (thermally and calorically perfect) for which ε is proportional to T. Thus $\gamma, C_v = \frac{R}{(\gamma-1)}$, and $C_p = \frac{R\gamma}{(\gamma-1)}$ are constant, and

$$\varepsilon = C_v T = \frac{RT}{\gamma - 1}, \quad h = C_p T = \frac{\gamma RT}{\gamma - 1}, \tag{1.19}$$

which yields

$$p = (\gamma - 1)\rho\varepsilon. \tag{1.20}$$

Note that for usual cases $1 < \gamma \leq \frac{5}{3}$. We shall see in Chapter III, Section 7 (Remark 7.2 and (7.18)) that $\gamma = \frac{5}{3}$ corresponds to a monatomic gas in dimension $d = 3$ and $\gamma = \frac{7}{5} = 1.4$ for diatomic molecules. Now, we also obtain from (1.1) (see Introduction, Example 3.3)

$$ds = C_v\Big(\frac{d\varepsilon}{\varepsilon} - (\gamma - 1)\frac{d\rho}{\rho}\Big),$$

which gives

$$s - s_0 = C_v \operatorname{Log}\Big(\frac{\varepsilon}{\rho^{(\gamma-1)}}\Big)$$

or

$$\varepsilon = \rho^{(\gamma-1)} \exp\Big(\frac{(s - s_0)}{C_v}\Big).$$

Thus

$$p = (\gamma - 1) \exp\Big(\frac{(s - s_0)}{C_v}\Big)\rho^\gamma = A(s)\,\rho^\gamma.$$

Note also that

$$c^2 = \gamma R T = (\gamma - 1)h,$$

and the function ℓ such that $u \pm \ell$ are the 1- and 3-Riemann invariants is given by

$$\ell = \int \Big(\frac{c}{\rho}\Big) d\rho = \frac{2c}{\gamma - 1}.$$

Example 1.1. Air under normal conditions (p and T moderate enough) can be considered as a perfect gas with $\gamma = \frac{7}{5}$ (approximately a mixture of two diatomic molecular species: 20% of O_2, 80% N_2). When the temperature increases, the vibrational motion of oxygen and nitrogen molecules in air becomes important, and specific heats vary with temperature so that γ is no longer constant but depends on temperature: the air is thermally perfect. However, at even higher temperatures, the molecules of oxygen and nitrogen begin to dissociate, which is symbolized by the reactions

$$O_2 \to 2O, N_2 \to 2N.$$

The air becomes chemically reacting, and the specific heats are functions of both T and p. These facts are improperly termed *real gas effects.* □

Remark 1.1. A gas should rather be defined as *real* when intermolecular forces become important (at very cold temperatures and high pressures). The perfect gas equation of state must be replaced by more accurate relations such as the van der Waals equation

$$\Big(p + \frac{a}{\tau^2}\Big)(\tau - b) = RT,$$

where a and b are constants depending on the gas. □

Remark 1.2. A mixture of reacting gases such as air at high temperatures is said to be in chemical equilibrium if the forward and reverse reactions $AB \rightleftarrows A + B$ are balanced so that the species are present in fixed amount. The chemical composition is then uniquely determined by p and T. Most problems can be treated assuming a mixture of perfect gases, for which Dalton's law states that the total pressure is the sum of the partial pressures

$$p = \Sigma_\alpha p_\alpha,$$

where the summation is taken over all species, together with $p_\alpha = \rho \eta_\alpha \mathcal{R} T$ (η_α is the mole-mass fraction of species α). □

Let us now consider the more general case of a gas with an incomplete equation of state expressed in the form

$$p = p(\rho, \rho\varepsilon),$$

which will prove convenient in Chapter III. We set $\tilde{\varepsilon} = \rho\,\varepsilon$ and

$$\kappa = \frac{\partial p}{\partial \tilde{\varepsilon}}(\rho, \tilde{\varepsilon}), \quad \chi = \frac{\partial p}{\partial \rho}(\rho, \tilde{\varepsilon}), \tag{1.20}$$

and define

$$\gamma = \frac{\rho c^2}{p},$$

where c is the speed of sound. Then, for an ideal gas, this definition of γ and the previous one, $\gamma(T) = \frac{C_p}{C_v}$, coincide (which is not the case for a more general gas). Moreover, computing κ, and χ

$$\kappa = R\rho\Big(\frac{dT}{d\varepsilon}\Big)\frac{\partial \varepsilon}{\partial \tilde{\varepsilon}}\Big|_{\rho=\text{cte}} = RT'(\varepsilon) = \gamma(T) - 1$$

and

$$\chi = RT + \rho R\Big(\frac{dT}{d\varepsilon}\Big)\frac{\partial \varepsilon}{\partial \rho}\Big|_{\tilde{\varepsilon}=\text{cte}} = R\Big(T - \varepsilon\frac{dT}{d\varepsilon}\Big) = RT - (\gamma(T) - 1)\varepsilon,$$

we see that they depend only on T (or ε). Finally, we have

$$\chi\rho + \kappa\tilde{\varepsilon} = (RT - (\gamma(T) - 1)\varepsilon)\rho + (\gamma(T) - 1)\tilde{\varepsilon} = \rho RT.$$

Therefore, for an ideal gas, the function $p = p(\rho, \tilde{\varepsilon})$ satisfies the identity

$$p = \rho\frac{\partial p}{\partial \rho}(\rho, \tilde{\varepsilon}) + \tilde{\varepsilon}\frac{\partial p}{\partial \tilde{\varepsilon}}(\rho, \tilde{\varepsilon}). \tag{1.21}$$

Conversely, given a function $p = p(\rho, \varepsilon)$ such that (1.21) holds, we can show that it corresponds to an ideal gas. Indeed, assuming (1.21), an easy computation gives

$$\frac{\partial(p(\rho, \varepsilon)/\rho)}{\partial \rho} = 0.$$

Hence $\frac{p}{\rho}$ is a function of ε only that reads

$$\frac{p}{\rho} = T(\varepsilon),$$

where we have incorporated the constant R in the function T. Moreover, since

$$\frac{\partial(p(\rho,\varepsilon)/\rho)}{\partial \varepsilon} = \frac{\partial p}{\partial \tilde{\varepsilon}}(\rho, \tilde{\varepsilon})$$

and $\frac{\partial p}{\partial \tilde{\varepsilon}}(\rho, \tilde{\varepsilon}) = \kappa$ by definition (1.20) of κ, we find that κ is a function of ε only and that the function T satisfies

$$T'(\varepsilon) = \kappa.$$

Example 1.2. Consider a stiffened equation of Grüneisen type, which we write

$$p = (\gamma - 1)\rho\varepsilon + c_{\text{ref}}^2(\rho - \rho_{\text{ref}}), \quad \gamma > 1. \tag{1.22}$$

Such an equation is obtained by linearization from a Grüneisen equation of state for a metal (see Menikoff and Plohr 1989). We have $\kappa = \gamma - 1$ and $\chi = c_{\text{ref}}^2$. We shall see (Chapter III, Lemma 3.3) that

$$c^2 = \kappa h + \chi,$$

which gives

$$c^2 = \gamma(p + p_\infty)\tau,$$

where we have set $p_\infty = \frac{c_{\text{ref}}^2 \rho_{\text{ref}}}{\gamma}$ for convenience. □

We shall again use these notions in Chapter III, Section 3-5 for the application of the usual schemes to gas dynamics, and we shall also return to the subject in Chapter IV, Section 2.

2 Entropy satisfying shock conditions

Let us consider again the gas dynamics equations in slab symmetry written in conservation form and in Eulerian coordinates,

$$\begin{cases} \dfrac{\partial \rho}{\partial t} + \dfrac{\partial}{\partial x}(\rho u) = 0, \\ \dfrac{\partial}{\partial t}(\rho u) + \dfrac{\partial}{\partial x}(\rho u^2 + p) = 0, \\ \dfrac{\partial}{\partial t}(\rho e) + \dfrac{\partial}{\partial x}((\rho e + p)u) = 0. \end{cases} \tag{2.1}$$

We supplement equations (2.1) with an equation of state, which we can take to be of the form

$$\varepsilon = \varepsilon(\tau, p). \tag{2.2}$$

In this section, using the ideas of Chapter I, Section 4, we want to characterize the admissible shock discontinuities for the physical (or entropy) weak solutions of (2.1). In this particular case of fundamental importance, we will be able to give a global description of the shock curves, in contrast with the general situation of Chapter I where only a local description of shock curves was obtained. We start from the Rankine–Hugoniot jump conditions for the system (2.1):

$$\begin{cases} \sigma[\rho] = [\rho u], \\ \sigma[\rho u] = [\rho u^2 + p], \\ \sigma[\rho e] = [(\rho e + p)u], \end{cases} \tag{2.3}$$

where σ is the speed of the discontinuity.

In addition, since by Theorem 1.1 $-\rho s$ is a strictly convex function of the conservative variables $\rho, q = \rho u, E = \rho e$, we obtain that $-\rho s$ is a strictly convex (mathematical) entropy associated with the entropy flux $-\rho s u$ (the fact that $-\rho s u$ is indeed the associated entropy flux is not difficult to check). Therefore, the entropy considerations of the Introduction,Section 3 and Chapter I, Section 4 lead us to require that any physical weak solution of (2.1), (2.2) satisfies the entropy inequality (in the sense of distributions)

$$\frac{\partial}{\partial t}(\rho s) + \frac{\partial}{\partial x}(\rho s u) \geq 0. \tag{2.4}$$

This means that any jump discontinuity must satisfy the entropy jump condition

$$\sigma[\rho s] \leq [\rho s u]. \tag{2.5}$$

Let us denote by (0) and (1) the two states that are connected by the discontinuity. Later on, we shall give a more precise meaning to the indices 0 and 1, but for the moment this is just a way of discriminating between the two states under consideration. We set

$$v_i = u_i - \sigma, \quad i = 0, 1, \tag{2.6}$$

so that v_i is the flow velocity relative to the discontinuity.

Lemma 2.1

The Rankine–Hugoniot jump relations (2.3) can be equivalently written in the form

$$\begin{cases} \rho_0 v_0 = \rho_1 v_1, \\ \rho_0 v_0^2 + p_0 = \rho_1 v_1^2 + p_1, \\ \Big(\rho_0\Big(\varepsilon_0 + \frac{v_0^2}{2}\Big) + p_0\Big)v_0 = \Big(\rho_1\Big(\varepsilon_1 + \frac{v_1^2}{2}\Big) + p_1\Big)v_1, \end{cases} \tag{2.7}$$

with $\sigma = u_i - v_i$.

Proof. The first equation (2.7) coincides with the first equation (2.3). Next, we have

$$\rho_0 v_0^2 + p_0 = \rho_0(u_0 - \sigma)^2 + p_0 = \rho_0 u_0^2 + p_0 - \sigma\,\rho_0 u_0 - \sigma\,\rho_0 v_0.$$

Hence, the second equation (2.3) together with the first equation (2.7) imply the 2nd equation (2.7). Finally, we write

$$\Big(\rho_0\Big(\varepsilon_0 + \frac{v_0^2}{2}\Big) + p_0\Big)v_0 = \Big(\rho_0\Big(\varepsilon_0 + \frac{(u_0-\sigma)^2}{2}\Big) + p_0\Big)(u_0 - \sigma)$$
$$= \{(\rho_0 e_0 + p_0)u_0 - \sigma\,\rho_0 e_0\} - \sigma(\rho_0 v_0^2 + p_0) - \sigma^2 \rho_0 \frac{v_0}{2},$$

so that the third equation (2.3) together with the first two equations (2.7) implies the third equation (2.7). □

Next, we set

$$M = \rho_0 v_0 = \rho_1 v_1, \tag{2.8}$$

so that M is the mass flux through the curve of discontinuity. We have to distinguish between the two cases $M = 0$ and $M \neq 0$. In the case $M = 0$, we have necessarily $v_0 = v_1 = 0$, and therefore

$$\begin{cases} u_0 = u_1 = \sigma, \\ p_0 = p_1. \end{cases} \tag{2.9}$$

Since $\rho_0 \neq \rho_1$ (otherwise the discontinuity would not exist), we obtain a physical contact discontinuity (or slip line) that indeed corresponds to a mathematical 2-contact discontinuity with

$$\sigma = \lambda_2 = u.$$

When $M \neq 0$, we obtain a shock-discontinuity that may be a 1-shock or a 3-shock.

Lemma 2.2

For $M \neq 0$, the Rankine–Hugoniot jump conditions can be equivalently written

$$M = \frac{u_1 - u_0}{\tau_1 - \tau_0}, \tag{2.10}$$

$$M^2 = -\frac{p_1 - p_0}{\tau_1 - \tau_0}, \tag{2.11}$$

$$\varepsilon_1 - \varepsilon_0 + \frac{1}{2}(p_1 + p_0)(\tau_1 - \tau_0) = 0. \tag{2.12}$$

Proof. First using (2.8), we have

$$M\tau_i = v_i, \quad i = 0, 1 \tag{2.13}$$

and by (2.6)

$$\sigma = u_0 - M\tau_0 = u_1 - M\tau_1,$$

which gives (2.10). Next, again using (2.8), the second equation (2.7) yields

$$Mv_0 + p_0 = Mv_1 + p_1,$$

so that by (2.13)

$$\Big(\rho_i\Big(\varepsilon_i + \frac{v_i^2}{2}\Big) + p_i\Big)v_i = M\Big(\varepsilon_i + \frac{v_i^2}{2}\Big) + p_i v_i = M\Big(\varepsilon_i + \frac{v_i^2}{2} + p_i\tau_i\Big).$$

Hence, the third equation (2.7) becomes

$$\varepsilon_0 + \frac{v_0^2}{2} + p_0\tau_0 = \varepsilon_1 + \frac{v_1^2}{2} + p_1\tau_1,$$

or

$$\varepsilon_1 - \varepsilon_0 + \frac{(v_1^2 - v_0^2)}{2} + p_1\tau_1 - p_0\tau_0 = 0.$$

Since by (2.13) and (2.11)

$$v_1^2 - v_0^2 = M^2(\tau_1^2 - \tau_0^2) = -(p_1 - p_0)(\tau_1 + \tau_0),$$

we obtain

$$\varepsilon_1 - \varepsilon_0 - \frac{(p_1 - p_0)(\tau_1 + \tau_0)}{2} + p_1\tau_1 - p_0\tau_0 = 0,$$

and (2.12) follows.

Conversely, it is an easy matter to check that the equations (2.10) – (2.12) imply the Rankine–Hugoniot jump conditions (2.7) if we set $\sigma = u_i - M\tau_i, i = 0, 1$. □

Remark 2.1. We could have equivalently derived the relations (2.10) – (2.12) starting from the gas dynamics equations written in Lagrangian coordinates,

$$(2.14)\qquad \begin{cases} \dfrac{\partial \tau}{\partial t} - \dfrac{\partial u}{\partial m} = 0, \\ \dfrac{\partial u}{\partial t} + \dfrac{\partial p}{\partial m} = 0, \\ \dfrac{\partial e}{\partial t} + \dfrac{\partial}{\partial m}(pu) = 0. \end{cases}$$

Recall that in (2.14) m stands for a mass variable. Indeed, the Rankine–Hugoniot jump conditions associated with the system (2.14) are given by

$$(2.15)\qquad \begin{cases} \sigma_L[\tau] = -[u], \\ \sigma_L[u] = [p], \\ \sigma_L[e] = [pu], \end{cases} \quad \text{or} \quad \begin{cases} \sigma_L(\tau_1 - \tau_0) = -(u_1 - u_0), \\ \sigma_L(u_1 - u_0) = p_1 - p_0, \\ \sigma_L(e_1 - e_0) = p_1u_1 - p_0u, \end{cases}$$

where σ_L is the speed of propagation of the discontinuity with respect to the mass variable.

Comparing the first equation (2.15) with (2.10) gives

$$\sigma_L = -M. \tag{2.16}$$

In the case of a contact discontinuity, we have $\sigma_L = 0$, which implies

$$u_1 = u_0 \quad \text{and} \quad p_1 = p_0.$$

On the other hand, when $M \neq 0$, the first two equations (2.15) together with (2.16) imply trivially (2.11). Moreover, the third equation (2.15) can be written

$$M\Big(\varepsilon_1 - \varepsilon_0 + \frac{(u_1^2 - u_0^2)}{2}\Big) + p_1 u_1 - p_0 u_0 = 0,$$

or

$$\varepsilon_1 - \varepsilon_0 + \frac{(u_1 + u_0)}{2}\Big\{(u_1 - u_0) + \frac{(p_1 - p_0)}{M}\Big\} + \frac{(p_1 + p_0)(u_1 - u_0)}{2M} = 0,$$

which together with (2.10) and (2.11) gives (2.12). □

Several comments on equations (2.10) – (2.12) are now in order. First, combining (2.10) and (2.11) yields

$$M = -\frac{p_1 - p_0}{u_1 - u_0}. \tag{2.17}$$

Next, we observe that due to (2.11) the pressure p and the specific volume τ vary in opposite ways along a shock curve. Finally, we notice that the equation (2.12), called the *Hugoniot equation*, is of purely thermodynamic nature since by (2.2) it involves the thermodynamic variables τ and p. In other words, (2.12) is the projection onto the (τ, p)-plane of the shock relations. Let us then introduce the *Hugoniot function* $\mathcal{H} = \mathcal{H}(\tau, p)$ with *center* (τ_0, p_0) defined by

$$\mathcal{H}(\tau, p) = \varepsilon(\tau, p) - \varepsilon(\tau_0, p_0) + \frac{1}{2}(p + p_0)(\tau - \tau_0), \tag{2.18}$$

so that the Hugoniot equation (2.12) becomes

$$\mathcal{H}(\tau_1, p_1) = 0.$$

The graph $\mathcal{H}$ of the Hugoniot function with center (τ_0, p_0) in the (τ, p)-plane is called the *Hugoniot curve* or the *shock adiabatic with center* (τ_0, p_0). This is the set of all states (τ, p) that can be connected to the state (τ_0, p_0) by a shock (that is not necessarily admissible).

Example 2.1. Consider a polytropic ideal gas for which

$$\varepsilon = \frac{p\tau}{\gamma - 1}, \quad \gamma > 1.$$

One can write in this case

$$\begin{aligned}\mathcal{H}(\tau,p) &= \frac{p\tau - p_0\tau_0}{\gamma-1} + \frac{1}{2}(p+p_0)(\tau-\tau_0)\\ &= \frac{1}{2}\Big\{\Big(\tau\frac{(\gamma+1)}{(\gamma-1)} - \tau_0\Big)p + \Big(\tau - \tau_0\frac{(\gamma+1)}{(\gamma-1)}\Big)p_0\Big\}.\end{aligned}$$

Setting

$$\mu^2 = \frac{\gamma-1}{\gamma+1}, \tag{2.19}$$

we find

$$2\mu^2\mathcal{H}(\tau,p) = (\tau - \mu^2\tau_0)p - (\tau_0 - \mu^2\tau)p_0. \tag{2.20}$$

In this case, the Hugoniot curve is a rectangular hyperbola. Observe that for $\tau > \mu^{-2}\tau_0$ the pressure becomes negative so that the corresponding part of the Hugoniot curve has no physical meaning. Hence, along the Hugoniot curve, the values of τ may vary between two limits $\tau_{\min} = \mu^2\tau_0$ and $\tau_{\max} = \mu^{-2}\tau_0$ while the pressure p varies between 0 and $+\infty$ (see Figure 2.1). □

Example 2.2. If we take more generally a stiffened equation of state of Grüneisen type (Example 1.2), which we write

$$\varepsilon = \frac{1}{\gamma-1}p\tau + \frac{\gamma}{\gamma-1}p_\infty\tau - \varepsilon_\infty,$$

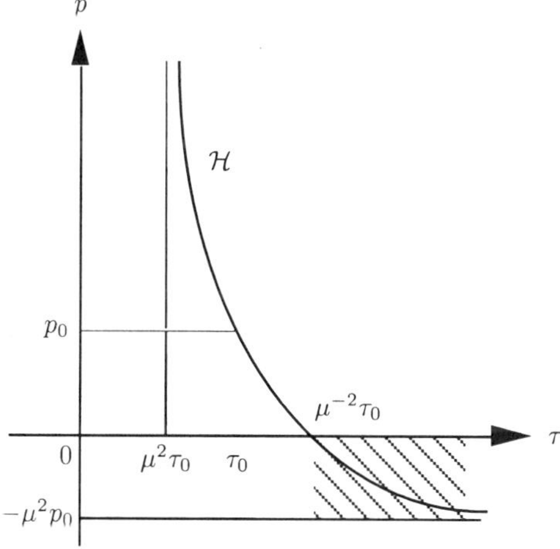

FIGURE 2.1. Hugoniot curve $\mathcal{H}$ for a polytropic ideal gas.

where $\varepsilon_\infty = \frac{c_{\mathrm{ref}}^2}{\gamma-1}$, we find that the Hugoniot curve is again a hyperbola

$$2\mu^2\,\mathcal{H}(\tau,p) = (\tau - \mu^2\tau_0)(p+p_\infty) - (\tau_0 - \mu^2\tau)(p_0+p_\infty).$$

We still have p varying between 0 and $+\infty$, $\tau_{\min} = \mu^2\tau_0$, but

$$\tau_{\max} = \tau_0\,\frac{p_\infty(1+\mu^2)+p_0}{p_\infty(1+\mu^2)+\mu^2 p_0} \le \mu^{-2}\tau_0$$

so that the upper limit of τ is smaller. □

Let us next study the properties of the Hugoniot curves. For the sake of brevity, we denote by $A_0 = (\tau_0, p_0)$ the center of the Hugoniot curve and we set

$$s_0 = s(\tau_0, p_0).$$

Lemma 2.3

Assume that the function $(\tau, s) \to p(\tau, s)$ *satisfies the conditions*

$$\frac{\partial p}{\partial \tau}(\tau,s) < 0, \qquad \frac{\partial p}{\partial s}(\tau,s) > 0. \tag{2.21}$$

Then, we have at point $A_0 = (\tau_0, p_0)$

$$\frac{\partial \mathcal{H}}{\partial \tau}(\tau_0,p_0) > 0, \qquad \frac{\partial \mathcal{H}}{\partial p}(\tau_0,p_0) > 0. \tag{2.22}$$

Proof. Differentiating (2.18), we have

$$d\mathcal{H} = d\varepsilon + \frac{1}{2}(p+p_0)d\tau + \frac{1}{2}(\tau-\tau_0)dp$$

and by (1.1)

$$d\mathcal{H} = T\,ds - \frac{1}{2}(p-p_0)d\tau + \frac{1}{2}(\tau-\tau_0)dp. \tag{2.23}$$

Thus

$$\begin{cases} \dfrac{\partial \mathcal{H}}{\partial \tau}(\tau,p_0) = T\,\dfrac{\partial s}{\partial \tau}(\tau,p_0), \\[2ex] \dfrac{\partial \mathcal{H}}{\partial p}(\tau_0,p) = T\,\dfrac{\partial s}{\partial p}(\tau_0,p). \end{cases} \tag{2.24}$$

Now it remains to compute the first partial derivatives of the function $s(\tau, p)$. From the identity

$$s = s(\tau, p(\tau, s)),$$

we deduce

$$\begin{cases} \dfrac{\partial s}{\partial \tau}(\tau,p) + \dfrac{\partial s}{\partial p}(\tau,p)\,\dfrac{\partial p}{\partial \tau}(\tau,s) = 0, \\[2ex] \dfrac{\partial s}{\partial p}(\tau,p)\,\dfrac{\partial p}{\partial s}(\tau,s) = 1. \end{cases}$$

Therefore, thanks to assumptions (2.21), it yields

$$\begin{cases} \dfrac{\partial s}{\partial \tau}(\tau, p) = -\Big(\dfrac{\partial p}{\partial s}(\tau, s)\Big)^{-1} \dfrac{\partial p}{\partial \tau}(\tau, s) > 0, \\ \dfrac{\partial s}{\partial p}(\tau, p) = \Big(\dfrac{\partial p}{\partial s}(\tau, s)\Big)^{-1} > 0, \end{cases} \tag{2.25}$$

which we substitute into (2.24) to get the desired result. □

Remark 2.2. As we have already noted, the first inequality (2.21) follows from the strict convexity of the function $\varepsilon(\tau, s)$ since (1.1) gives

$$\frac{\partial p}{\partial \tau}(\tau, s) = -\frac{\partial^2 \varepsilon}{\partial \tau^2}(\tau, s).$$

On the other hand, the second condition (2.21) holds in many physical cases and is indeed satisfied in the case of a polytropic ideal gas. □

Let us denote by a prime the derivative along the Hugoniot curve.

Lemma 2.4

Along the Hugoniot curve with center A_0, we have at point A_0

$$s' = 0, \quad s'' = 0, \quad s''' = -\frac{1}{2}\frac{\partial^2 p}{\partial \tau^2}(\tau_0, s_0)(\tau')^3. \tag{2.26}$$

Proof. From (2.23) and the relation $\mathcal{H}(\tau, p) = 0$, we get

$$T s' = (p - p_0)\frac{\tau'}{2} - (\tau - \tau_0)\frac{p'}{2}, \tag{2.27}$$

so that

$$s' = 0 \text{ at } A_0.$$

Next, differentiating (2.27), we write

$$(T s')' = (p - p_0)\frac{\tau''}{2} - (\tau - \tau_0)\frac{p''}{2},$$

which yields

$$(T s')' = 0 \text{ at } A_0.$$

But

$$(T s')' = T s'' + T' s' = T s'' \text{ at } A_0$$

and

$$s'' = 0 \text{ at } A_0.$$

Finally, differentiating once more (2.27) gives

$$(T s')'' = \frac{(p' \tau'' - \tau' p'')}{2} + \frac{(p - p_0)\tau'''}{2} - (\tau - \tau_0) p''',$$

so that

$$(Ts')'' = Ts''' = \frac{(p'\tau'' - \tau'p'')}{2} \quad \text{at } A_0.$$

On the other hand, using the equation of state in the form $p = p(\tau, s)$, we obtain

$$dp = \frac{\partial p}{\partial \tau}\, d\tau + \frac{\partial p}{\partial s}\, ds$$

and

$$d^2p = \frac{\partial^2 p}{\partial \tau^2}\,(d\tau)^2 + 2\Big(\frac{\partial^2 p}{\partial \tau\, \partial s}\Big) d\tau\, ds + \frac{\partial^2 p}{\partial s^2}\,(ds)^2 + \frac{\partial p}{\partial \tau}\, d^2\tau + \frac{\partial p}{\partial s}\, d^2 s.$$

Hence, along the Hugoniot curve $\mathcal{H} = 0$, we find

$$\text{(2.28)} \qquad \begin{cases} p' = \dfrac{\partial p}{\partial \tau}\,\tau' \quad \text{at } A_0, \\[2mm] p'' = \dfrac{\partial^2 p}{\partial \tau^2}\,(\tau')^2 + \dfrac{\partial p}{\partial \tau}\,\tau'' \quad \text{at } A_0, \end{cases}$$

and

$$T\,s''' = -\frac{1}{2}\,\frac{\partial^2 p}{\partial \tau^2}\,(\tau')^3 \text{ at } A_0,$$

which proves the result. □

As a first consequence of Lemma 2.4, we obtain that the Hugoniot curve with center A_0 and the isentropic curve $s = s_0$ passing through the point A_0 are osculatory at A_0. In fact, this property follows from the general theory of Chapter I. When the kth characteristic field is genuinely nonlinear, we know from Chapter I, Section 4 that the k-shock curve $\mathcal{S}_k(u_L)$ and the k-rarefaction curve $\mathcal{R}_k(u_L)$ are osculatory at the point u_L. This is exactly the situation here: the 1- and 3-shock curves are projected in the (τ, p)-plane onto the Hugoniot curve while the 1- and 3-rarefaction curves are projected onto the isentropic curve.

Let us point out another consequence of Lemmas 2.3 and 2.4. Since $\frac{\partial \mathcal{H}}{\partial p} > 0$ at A_0, we may parametrize the Hugoniot curve $\mathcal{H} = 0$ in the form $p = p(\tau)$ in a neighborhood of A_0. Moreover, if in addition to (2.21) we assume

$$\text{(2.29)} \qquad \frac{\partial^2 p}{\partial \tau^2}\,(\tau, s) > 0,$$

it follows from (2.23) that along the Hugoniot curve the physical entropy s is a decreasing function of τ in a neighborhood of A_0. More generally, we can state the following result.

Theorem 2.1

Assume that the function $p = p(\tau, s)$ satisfies the conditions (2.21) and (2.29). Then the entropy s is strictly monotone all along the Hugoniot curve with center A_0 and has a unique critical point at A_0.

Proof. By a critical point, we mean a point of the Hugoniot curve for which $s' = 0$. Let us prove that, along the Hugoniot curve, we have $s' \neq 0$ except at A_0. We begin by considering the straight lines Δ passing through A_0 (i.e., the Rayleigh lines) and parametrized by

$$\begin{cases} \tau = \tau_0 + a\alpha, \\ p = p_0 + b\alpha. \end{cases}$$

Observe that along such a line Δ we have

$$(p_0 - p)d\tau + (\tau - \tau_0)dp = 0.$$

Hence, using (2.23), we obtain

$$\frac{d\mathcal{H}}{d\alpha} = T\frac{ds}{d\alpha} \quad \text{along } \Delta. \tag{2.30}$$

Let us check that

(2.31) along Δ, s has at most one critical point, that is a maximum.

We first consider a straight line distinct from $\tau = \tau_0$. Along such a line, we have

$$\frac{dp}{d\alpha} = \frac{\partial p}{\partial \tau}\frac{d\tau}{d\alpha} + \frac{\partial p}{\partial s}\frac{ds}{d\alpha},$$

and since

$$\frac{d^2\tau}{d\alpha^2} = \frac{d^2 p}{d\alpha^2} = 0,$$

we find

$$0 = \frac{d^2 p}{d\alpha^2} = \frac{\partial^2 p}{\partial \tau^2}\left(\frac{d\tau}{d\alpha}\right)^2 + 2\left(\frac{\partial^2 p}{\partial \tau\,\partial s}\right)\frac{d\tau}{d\alpha}\frac{ds}{d\alpha} + \frac{\partial^2 p}{\partial s^2}\left(\frac{ds}{d\alpha}\right)^2 + \frac{\partial p}{\partial s}\frac{ds}{d\alpha}.$$

Now, we obtain at a critical point of s along Δ

$$\frac{\partial^2 p}{\partial \tau^2}\left(\frac{d\tau}{d\alpha}\right)^2 + \frac{\partial p}{\partial s}\frac{d^2 s}{d\alpha^2} = 0.$$

Since $\frac{d\tau}{d\alpha} \neq 0$, it follows from (2.21) and (2.29) that at such a critical point we have

$$\frac{d^2 s}{d\alpha^2} < 0.$$

In other words, a critical point of s along Δ is necessarily a local maximum. This implies the property (2.31).

It remains to consider the case of the line $\tau = \tau_0$. Since $\frac{\partial p}{\partial s} > 0$, the function $s \to p(\tau_0, s)$ is strictly increasing and the same is true of the reciprocal function $p \to s(\tau_0, p)$ so that s has no critical point along $\tau = \tau_0$. This proves our assertion (2.31).

Next, combining (2.30) and (2.31), we obtain that $\mathcal{H}$ has at most one critical point along Δ.

Then, let $A_1 = (\tau_1, p_1) \neq A_0$ be another critical point of s along the Hugoniot curve with center A_0. Since $\mathcal{H}' = 0$ along this Hugoniot curve, we use (2.27) to obtain

$$(p_0 - p_1)\tau' + (\tau_1 - \tau_0)p' = 0 \text{ at } A_1.$$

This means that the straight line $A_0\, A_1$ is tangent to the Hugoniot curve at the point A_1. Therefore, along $A_0\, A_1$, $\mathcal{H}$ is critical at A_1. On the other hand, $\mathcal{H}$ vanishes at the points A_0 and A_1 and must be critical along $A_0\, A_1$ at some intermediate point A_2 distinct from A_0 and A_1. Hence, along $A_0\, A_1$, $\mathcal{H}$ has two distinct critical points A_1 and A_2, which violates the above property.

Thus, along the Hugoniot curve with center A_0, the entropy s is critical at the point A_0 alone. □

As a consequence of Lemma 2.4 and Theorem 2.1, we obtain the following corollary.

Corollary 2.1

Assume, moreover, that the Hugoniot curve may be parametrized by τ. Then the entropy s is a strictly decreasing function of τ along the Hugoniot curve.

This is the case when $\frac{\partial \mathcal{H}}{\partial p} > 0$ and is indeed satisfied for a polytropic ideal gas (Example 2.1).

It remains to characterize the admissible shocks. We first notice that the sign of M enables us to recognize a 1-shock from a 3-shock. Indeed, since by (2.21), (2.24), and (2.25)

$$\frac{\partial \mathcal{H}}{\partial \tau}(\tau, p_0) = -T\Big(\frac{\partial p}{\partial s}\Big)^{-1}\frac{\partial p}{\partial \tau}(\tau, p_0) > 0,$$

the function $\tau \to \mathcal{H}(\tau, p_0)$ is strictly increasing, so that

$$M^2 = -\frac{p - p_0}{\tau - \tau_0}$$

cannot vanish when the point (τ, p) varies on the Hugoniot curve with center A_0. Hence M keeps a constant sign on a shock curve. In order to determine the sign of M, we use the parametrization of a shock curve derived in Chapter I, Example 4.2. We have (by (4.35) and (4.36))

$$\tau = \tau_0 - \varepsilon\tau_0 + O(\varepsilon^2)$$

and

$$\begin{aligned} u &= u_0 - \varepsilon\, c_0 + O(\varepsilon^2) \quad \text{for a 1-shock,} \\ u &= u_0 + \varepsilon\, c_0 + O(\varepsilon^2) \quad \text{for a 3-shock.} \end{aligned}$$

Hence

$$M = \frac{u - u_0}{\tau - \tau_0} = \begin{cases} c_0\, \rho_0 + O(\varepsilon) & \text{for a 1-shock,} \\ -c_0\, \rho_0 + O(\varepsilon) & \text{for a 3-shock,} \end{cases}$$

so that

$$\begin{cases} M > 0 & \text{for a 1-shock,} \\ M < 0 & \text{for a 3-shock.} \end{cases} \tag{2.32}$$

Let us now introduce some notations. Since $M \neq 0$, it follows from (2.8) that the gas crosses a shock. Then the side of the shock front through which the gas enters is called the *front side* (or the side ahead of the shock front) while the other side is called the *back side* (or the side behind). This means that the gas crosses the shock front from the front toward the back side, and one can say that the shock faces the front side. Hereafter, we shall use the following convention: the state (0) will refer to the state of the gas at *the front side* and the state (1) will refer to the state of the gas at the *back side*. In other words, we are in one of the two situations below (see Figure 2.2). Thus

for a 1-shock the state (0) is the left state

while

for a 3-shock the state (1) is the left state.

The left (resp. right) state is usually noted with the subscript $-$ or L (resp. $+$ or R).

Remark 2.3. The direction of propagation of the shock is given by the sign of σ and must be distinguished from the direction toward which the shock front faces, which depends on the sign of the relative velocities $u_i - \sigma$. Consider the situation illustrated in Figure 2.2 (left): the fluid particles are crossing the shock front from the left to the right since $u_0 > \sigma$. Hence, the left state is indeed the front state. This argument is of course independent of whether the front advances or recedes, and the conclusion holds with a negative σ. Note also that we might have taken $u_0 < u_1$. We shall see, however, below that this is excluded for an admissible shock. □

Now, a shock discontinuity is said to be *admissible* if it satisfies the entropy jump condition (2.5). Moreover, a shock is called a *compressive*

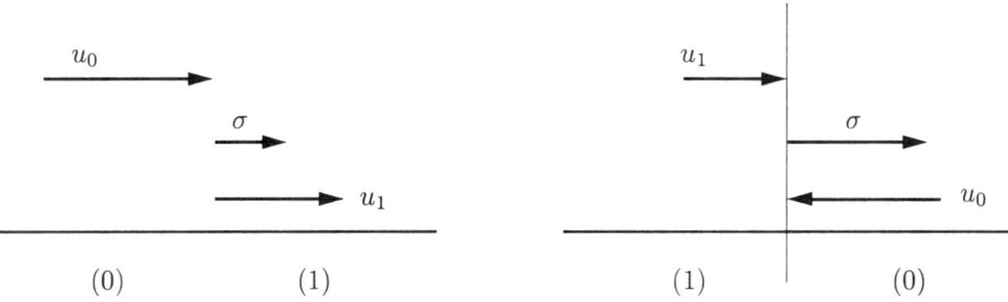

FIGURE 2.2. Relative velocities for a 1- and a 3-shock.

shock if it satisfies $p_1 > p_0$, i.e., if the pressure increases as the gas crosses the shock front.

Lemma 2.5

Assume the hypotheses of Corollary 2.1 of Theorem 2.1. Then, a shock is admissible if and only if it is a compressive shock.

Proof. Using (2.8), the entropy jump condition (2.5) can be written

$$M[s] = [\rho s(u - \sigma)] \geq 0$$

or, equivalently,

$$s_1 \geq s_0. \tag{2.33}$$

Indeed, when M is > 0, we have (see Figure 2.2)

$$M[s] = M(s_+ - s_-) = M(s_1 - s_0) \geq 0.$$

On the other hand, when M is < 0, we obtain

$$M[s] = M(s_0 - s_1) \geq 0,$$

so that the entropy increases across a shock and our assertion (2.33) follows.

Now, assuming the hypotheses of Corollary 2.1, we know that along the Hugoniot curve with center A_0, s is a strictly decreasing function of τ and therefore a strictly increasing function of p, so that

$$s_1 \geq s_0 \Longleftrightarrow p_1 \geq p_0$$

and the shock is indeed compressive. □

Remark 2.4. Note that the entropy jump condition $M[s] \geq 0$ is obtained in a more direct way if we use the Lagrangian variables since $-M$ is the shock velocity with respect to the mass variable (see Remark 2.1). □

As a simple consequence of Lemma 2.5, we obtain the following result.

Theorem 2.2

Assume the hypotheses of Corollary 2.1 of Theorem 2.1. A shock is admissible if and only if one of the four following equivalent properties holds:

$$\begin{cases} (i) & \rho_1 \geq \rho_0, \\ (ii) & p_1 \geq p_0, \\ (iii) & s_1 \geq s_0, \\ (iv) & u_- \geq u_+. \end{cases} \tag{2.34}$$

Proof. Since along the Hugoniot curve with center A_0 we have

$$\rho_1 \geq \rho_0 \Longleftrightarrow p_1 \geq p_0 \Longleftrightarrow s_1 \geq s_0,$$

the first three characterizations of an admissible shock are an obvious restatement of Lemma 2.5. They are also equivalent to $\tau_1 \leq \tau_0$. Let us check the fourth characterization (2.34). Using

$$\rho_0(u_0 - \sigma) = \rho_1(u_1 - \sigma) = M,$$

the condition $\rho_1 \geq \rho_0$ gives for $M > 0$

$$u_0 - \sigma \geq u_1 - \sigma$$

i.e., since the state (0) is the left state (see Figure 2.2)

$$u_0 = u_- \geq u_+ = u_1,$$

and for $M < 0$

$$u_0 - \sigma \leq u_1 - \sigma$$

i.e., (see Figure 2.2 again)

$$u_1 = u_- \geq u_+ = u_0.$$

Hence, for an admissible shock, the density, pressure, and entropy increase when the gas crosses the shock front. □

Theorem 2.3

Assume the hypotheses of the Corollary 2.1 of Theorem 2.1. Then, a shock is admissible if and only if

$$c_0 \leq |u_0 - \sigma|, \quad |u_1 - \sigma| \leq c_1, \tag{2.35}$$

i.e., if and only if the gas velocity relative to the shock front is supersonic at the front side and subsonic at the back side

Proof. Assume that $A_1 = (\tau_1, p_1)$ belongs to the Hugoniot curve with center A_0, and consider the straight line Δ joining the states A_0 and A_1, parametrized by α as in the proof of Theorem 2.1, with α_1 (corresponding to A_1) > 0, for instance. Since $\mathcal{H}(\tau_0, p_0) = \mathcal{H}(\tau_1, p_1) = 0$, along Δ we have $\frac{d\mathcal{H}}{d\alpha} = 0$ at some intermediate point A_2 of $A_0\, A_1$ distinct from A_0 and A_1. Then, it follows from (2.30) and (2.31) that along the line Δ, s has exactly one critical point A_2, which is a maximum. Hence, we obtain

$$\frac{ds}{d\alpha} > 0 \text{ at } A_0, \qquad \frac{ds}{d\alpha} < 0 \text{ at } A_1.$$

Since

$$\begin{aligned} \frac{ds}{d\alpha} &= \frac{\partial s}{\partial \tau}\frac{d\tau}{d\alpha} + \frac{\partial s}{\partial p}\frac{dp}{d\alpha} \\ &= \frac{(\tau_1 - \tau_0)}{\alpha_1}\frac{\partial s}{\partial \tau} + \frac{(p_1 - p_0)}{\alpha_1}\frac{\partial s}{\partial p}, \end{aligned}$$

(2.25) and (1.4) yield

$$\frac{ds}{d\alpha} = \frac{1}{\alpha_1}\{\rho^2c^2(\tau_1 - \tau_0) + p_1 - p_0\}\frac{\partial s}{\partial p}.$$

We find thus at A_0

$$\rho_0^2c_0^2(\tau_1 - \tau_0) + p_1 - p_0 > 0$$

and at A_1

$$\rho_1^2c_1^2(\tau_1 - \tau_0) + p_1 - p_0 < 0.$$

Since $\tau_1 < \tau_0$, this gives

$$\rho_0^2c_0^2 < -\frac{p_1 - p_0}{\tau_1 - \tau_0} < \rho_1^2c_1^2, \tag{2.36}$$

and by (2.11) and (2.8)

$$\rho_0^2c_0^2 < \rho_0^2v_0^2 = M^2 = \rho_1^2v_1^2 < \rho_1c_1^2.$$

This implies

$$c_0^2 < v_0^2, \quad v_1^2 < c_1^2,$$

which is equivalent to (2.35). □

Remark 2.5. For two given states, a front state A_0 and a back state A_1, that can be connected by an admissible shock, the Hugoniot curves $\mathcal{H}_0$ with center A_0 and $\mathcal{H}_1$ with center A_1 intersect at points A_0 and A_1, but they do not coincide (see Figure 2.3). If the left state is A_0, the shock connecting A_0 to A_1 is a 1-shock (see Figure 2.2), and a 3-shock if the left state is A_1. □

Remark 2.6. Let us prove that the inequalities (2.36) imply that a Rayleigh line through A_0 intersects the Hugoniot curve at (at most) one other point.

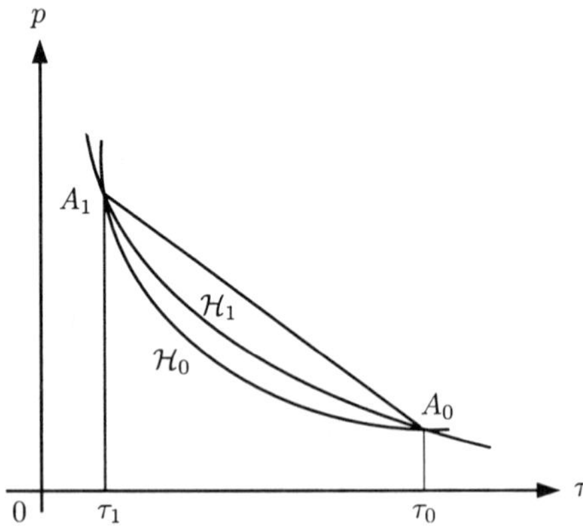

FIGURE 2.3. $\mathcal{H}_0$ with center A_0, $\mathcal{H}_1$ with center A_1.

First, we note that, due to assumption (2.28), we have by (2.29) (if the curve if parametrized by τ)

$$p''(\tau_0) = \frac{\partial^2 p}{\partial \tau^2}(\tau_0, s_0) > 0, \tag{2.37}$$

and the curve is convex in a neighborhood of the center A_0.

Now, by (2.36) and (1.4), we have

$$-\frac{\partial p}{\partial \tau}(\tau_0, s_0) = \rho_0^2 c_0^2 < -\frac{p_1 - p_0}{\tau_1 - \tau_0} = M^2 < \rho_1^2 c_1^2 = -\frac{\partial p}{\partial \tau}(\tau_1, s_1),$$

which we write

$$\frac{\partial p}{\partial \tau}(\tau_1, s_1) < \frac{p_1 - p_0}{\tau_1 - \tau_0} < \frac{\partial p}{\partial \tau}(\tau_0, s_0). \tag{2.38}$$

Consider the Hugoniot curve $\mathcal{H}$ with center A, and B, C two points on $\mathcal{H}$ with $\tau_B < \tau_A$ and $\tau_C > \tau_A$ (see Figure 2.4). We first consider the left side of $\mathcal{H}\,(\tau < \tau_A)$. Since the state B is such that $\tau_B < \tau_A$, the shock connecting A and B is admissible if A is the front side, $A = B_0 = (\tau_0, p_0)$, and B the back side $B = B_1 = (\tau_1, p_1)$, so that $\mathcal{H}$ is indeed the Hugoniot curve $\mathcal{H}_0$ with center $B_0 = A$. When parametrized by τ, it is given by the function $\tau \to p(\tau, s(\tau)) = p(\tau)$. Therefore, by (2.29), the slope of the tangent is

$$p'(\tau) = \frac{\partial p}{\partial \tau} + \frac{\partial p}{\partial s} s'(\tau),$$

and

$$p'(\tau_0) = \frac{\partial p}{\partial \tau}(\tau_0, s_0).$$

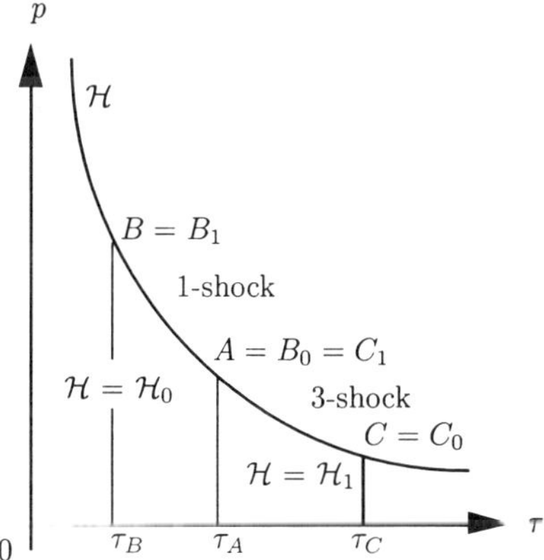

FIGURE 2.4. Hugoniot curve $\mathcal{H}$ with center A.

Now, since s is a strictly decreasing function of τ, $s' < 0$, we have by (2.21)

$$p'(\tau_1) = \frac{\partial p}{\partial \tau}(\tau_1, s_1) + \frac{\partial p}{\partial s}(\tau_1, s_1)s'(\tau_1) < \frac{\partial p}{\partial \tau}(\tau_1, s_1). \tag{2.39}$$

Thus, in this case (2.38) implies

$$p'(\tau_1) < \frac{p_1 - p_0}{\tau_1 - \tau_0} < p'(\tau_0), \tag{2.40}$$

which proves that the line $\Delta = A_0\, A_1$ crosses the left part of the Hugoniot curve $\{\tau < \tau_A\}$ of $\mathcal{H}_0$ at the point A_1 only.

Similarly, let us now consider the state C on $\mathcal{H}$ such that $\tau_C > \tau_A$. The shock connecting A and C is admissible if C is the front side $C = C_0 = (\tau_0, p_0)$, and $A = C_1 = (\tau_1, p_1)$. So $\mathcal{H}$ is the Hugoniot curve with center C_1, which we denote by $\mathcal{H}_1$ and parametrize by τ. To distinguish from the first case where the center corresponded to the index 0, we shall denote by a dot the differentiation along $\mathcal{H}_1$. We then have by (2.29)

$$\dot{p}(\tau_1) = \frac{\partial p}{\partial \tau}(\tau_1, s_1). \tag{2.41}$$

By using (2.27) applied to $\mathcal{H}_1$, we now observe that since s is a strictly decreasing function of τ,

$$T\,\dot{s}(\tau_0) = \frac{(p_0 - p_1)}{2} - \frac{(\tau_0 - \tau_1)}{2}\,\dot{p}(\tau_0) < 0.$$

Thus, for $\tau_0 > \tau_1$

$$\frac{p_0 - p_1}{\tau_0 - \tau_1} < \dot{p}(\tau_0).$$

Together with (2.41) and (2.37), this yields

$$\dot{p}(\tau_1) < \frac{p_0 - p_1}{\tau_0 - \tau_1} < \dot{p}(\tau_0), \tag{2.42}$$

which proves the analogous property for the right part of the Hugoniot curve $\mathcal{H}$. Properties (2.37),(2.40), and (2.42) give the desired result.

In Figure 2.3, we had represented for two given states A_0 (resp. A_1) the part of the Hugoniot curve $\mathcal{H}_0$ with center A_0 such that $\tau \leq \tau_0$ (resp. $\mathcal{H}_1$ with center A_1 such that $\tau \geq \tau_1$). Note that the slope of the tangent to $\mathcal{H}_0$ (resp. $\mathcal{H}_1$) at point A_1 is $p'(\tau_1)$ (resp. $\dot{p}(\tau_1)$) and by (2.39) and (2.41)

$$p'(\tau_1) < \dot{p}(\tau_1).$$

Similarly,

$$\dot{p}(\tau_0) = \frac{p_0 - p_1}{\tau_0 - \tau_1} - \frac{2T\,\dot{s}(\tau_0)}{\tau_0 - \tau_1} < p'(\tau_0),$$

so that $\mathcal{H}_1$ lies above $\mathcal{H}_0$. □

In the previous chapter, we introduced two different criteria for characterizing the admissible shocks, namely the Lax entropy conditions (Chapter

I, (5.14)) and the entropy condition based on a strictly convex entropy U (Chapter I, (5.18)). We have seen that these two criteria coincide for sufficiently weak shocks (see Chapter I, Theorem 5.3). Here, in the case of the gas dynamics equations, we want to check that the Lax-entropy conditions coincide with the entropy condition (2.5) (or equivalently (2.33)) for all shocks.

Theorem 2.4

Assume the hypotheses of the Corollary of Theorem 2.1. Then, for the gas dynamics equations, the Lax entropy inequalities are equivalent to the increase of the entropy across a shock.

Proof. Consider a 1-shock. In Eulerian coordinates, the Lax entropy conditions give in this case

$$u_+ - c_+ < \sigma < u_+, \quad \sigma < u_- - c_-,$$

or equivalently in terms of the states (0) and (1) (see Figure 2.2)

$$u_1 - c_1 < \sigma < u_1, \quad \sigma < u_0 - c_0.$$

These inequalities can be equivalently written

$$0 < u_1 - \sigma < c_1, \quad c_0 < u_0 - \sigma.$$

Since for a 1-shock we have $u_i - \sigma > 0$, $i = 0, 1$, Theorem 2.3 implies that the above inequalities characterize an admissible 1-shock. The case of a 3-shock is entirely similar. Note that since $(\lambda - u)^2 = -\frac{1}{\rho^2}\frac{\partial p}{\partial \tau}$ and $\rho_i(\sigma - u_i)^2 = -\frac{(p_1 - p_0)}{(\tau_1 - \tau_0)}$), inequalities (2.38) also express the Lax entropy criteria. □

Remark 2.7. Again, one could have proven Theorem 2.4 by working in Lagrangian coordinates. Indeed, consider a 3-shock; the Lax entropy conditions give here

$$\frac{c_+}{\tau_+} < -M = \sigma_L, \quad 0 < -M = \sigma_L < \frac{c_-}{\tau_-},$$

or equivalently by (2.13)

$$c_0 < -(u_0 - \sigma), \quad 0 < -(u_1 - \sigma) < c_1.$$

Using once more Theorem 2.3, this characterizes a 3-shock. □

Remark 2.8. Smoller, Temple, and Xin (1990) have proven that for a polytropic ideal gas with $1 < \gamma < \frac{5}{3}$, rarefaction shocks (of moderate strength), i.e., shocks that violate the Lax entropy condition (with $|p_+ - p_-|$ not too great) are unstable in the class of smooth solutions. This means that there exists a sequence of C^2 solutions (defined uniformly on $\mathbb{R} \times [0, T]$ for some $T > 0$) that converges in every $\mathbf{L}^p (p \geq 1)$ to the given discontinuous data at $t = 0$ but does not converge to the given (rarefaction) shock for any $t \in]0, T]$.

Thus, stability with respect to smoothing appears as another criterion for selecting "admissible" weak shocks (see Chapter I, Remark 5.2). □

3 Solution of the Riemann problem

In this section, we want to solve the Riemann problem for the gas dynamics equations either in Eulerian coordinates or in Lagrangian coordinates. In fact, it will be convenient in all the following to characterize the state of the gas by the three dependent variables (ρ, u, p). We look for an entropy solution of the system of equations (2.1), (2.2) satisfying the initial condition

$$(\rho, u, p)(x, 0) = \begin{cases} (\rho_L, u_L, p_L), & x < 0, \\ (\rho_R, u_R, p_R), & x > 0. \end{cases} \tag{3.1}$$

Now, it follows from the general theory of Chapter I that (at least for sufficiently close left and right states) we can find an entropy solution of the Riemann problem of the following form: the left state (ρ_L, u_L, p_L) is connected to the right state (ρ_R, u_R, p_R) by a 1-shock or a 1-rarefaction wave, a 2-contact discontinuity, and a 3-shock or a 3-rarefaction wave. The 2-contact discontinuity separates two constant states (ρ_I, u^*, p^*) and (ρ_{II}, u^*, p^*) so that u and p are continuous across the contact discontinuity. In Figure 3.1, the 1-wave is a rarefaction and the 3-wave a shock. Then, it is adequate to work in the (u, p)-plane in order to determine the types of the 1- and 3-waves (shock or rarefaction) and the unknown variables u^*, p^* at the contact discontinuity. It will be an easy matter afterwards to compute ρ_I and ρ_{II}.

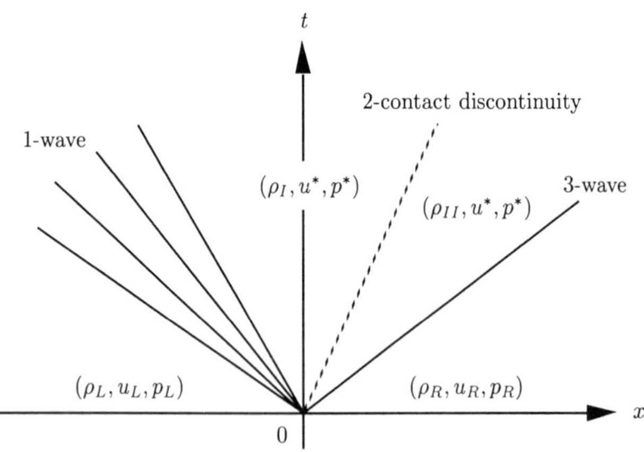

FIGURE 3.1. Solution of the Riemann problem in the (x, t)-plane.

Let us first determine the projections of the shock curves onto the (u, p)-plane. We consider two states $\mathbf{a}$ and $\mathbf{b}$ connected by a shock wave. Recall that by (2.11) and (2.17)

$$M = -\frac{p_a - p_b}{u_a - u_b},$$

$$M^2 = -\frac{p_a - p_b}{\tau_a - \tau_b},$$

and by (2.12)

$$\mathcal{H}_a(\tau_b, p_b) = 0,$$

where

$$\mathcal{H}_a(\tau, p) = \varepsilon(\tau, p) - \varepsilon(\tau_a, p_a) + \frac{1}{2}(p + p_a)(\tau - \tau_a).$$

Assume that the Hugoniot curve $\mathcal{H}_a(\tau, p) = 0$ may be parametrized by $p \in [0, +\infty[$, i.e., may be represented by an equation of the form

$$\tau = h_a(p), \quad \tau_a = h_a(p_a) \tag{3.2}$$

with the hypotheses

$$\lim_{p \to 0} h_a(p) = \tau_{\max}, \qquad \lim_{p \to +\infty} h_a(p) = \tau_{\min},$$

and

$$h_a'(p) < 0, \qquad \lim_{p \to +\infty} \sqrt{p}\, h_a'(p) = 0. \tag{3.3}$$

Remark 3.1. Let us assume that the Hugoniot curve may be parametrized by τ, and that $p(\tau) \to +\infty$ when $\tau \to \tau_{\min}$. Let us note that properties (3.3) hold if we assume that the Hugoniot curve is convex (decreasing). Indeed, h_a is also convex decreasing, $h_a(p) \to \tau_{\min}$ as $p \to +\infty$, and due to the convexity of h_a, we have for any p_0,

$$0 \le -h_a'(p) \le \frac{h_a(p_0) - h_a(p)}{p - p_0},$$

which implies the second inequality in (3.3). □

Note also that we have $\tau_b = h_a(p_b)$. Next, setting

$$M_a(p) = \begin{cases} \sqrt{(p - p_a)/(\tau_a - h_a(p))} & \text{for a 1-shock,} \\ -\sqrt{(p - p_a)/(\tau_a - h_a(p))} & \text{for a 3-shock,} \end{cases}$$

we have

$$u_b = u_a - \frac{p_b - p_a}{M_a(p_b)}.$$

Hence, defining the function

$$\Phi_a(p) = \frac{p - p_a}{|M_a(p)|} = (p - p_a)\sqrt{(\tau_a - h_a(p))/(p - p_a)},$$

or equivalently

$$\Phi_a(p) = \begin{cases} \sqrt{(p-p_a)(\tau_a - h_a(p))}, & p \geq p_a, \\ -\sqrt{(p-p_a)(\tau_a - h_a(p))}, & p \leq p_a, \end{cases} \tag{3.4}$$

we obtain

$$u_b = \begin{cases} u_a - \Phi_a(p_b) & \text{for a 1-shock,} \\ u_a + \Phi_a(p_b) & \text{for a 3-shock.} \end{cases} \tag{3.5}$$

Let us make out a list of the properties of the function Φ_a. First, obviously

$$\Phi_a(p_a) = 0.$$

Then, since

$$\lim_{p\to 0} h_a(p) = \tau_{\max}, \qquad \lim_{p\to+\infty} h_a(p) = \tau_{\min},$$

we have on the one hand

$$\Phi_a(0) = -\sqrt{p_a/(\tau_{\max} - \tau_a)}, \tag{3.6a}$$

$$\lim_{p\to+\infty} \Phi_a(p) = +\infty \tag{3.6b}$$

and, on the other hand,

$$\Phi_a'(p) = \frac{\tau_a - h_a(p) - (p-p_a)h_a'(p)}{2\Phi_a(p)}$$

so that by (3.3)

$$\Phi_a'(p) > 0, \tag{3.6c}$$

$$\lim_{p\to+\infty} \Phi_a'(p) = 0. \tag{3.6d}$$

Example 3.1. Consider again the polytropic ideal gas introduced in Example 2.1, for which

$$\mathcal{H}_a(\tau, p) = \frac{(\tau - \mu^2\tau_a)p - (\tau_a - \mu^2\tau)p_a}{2\mu^2},$$

with $\mu^2 = \frac{(\gamma-1)}{(\gamma+1)}$ given by (2.19) and $\tau_{\min} = \mu^2\tau_a$, $\tau_{\max} = \mu^{-2}\tau_a$. We have

$$h_a(p) = \frac{\tau_a(\mu^2 p + p_a)}{(p + \mu^2 p_a)},$$

which implies

$$\Phi_a(p) = (p - p_a)\sqrt{(1-\mu^2)\tau_a/(p + \mu^2 p_a)}.$$

Clearly, the function h_a satisfies the assumptions (3.3), and the properties (3.6) hold. Note also that

$$\Phi_a(0) = -\sqrt{p_a\tau_a(1-\mu^2)/\mu^2} = -\sqrt{2p_a\tau_a/(\gamma-1)},$$

and since $c^2 = \gamma p \tau$

$$\Phi_a(0) = -c_a \sqrt{2/\gamma(\gamma-1)}.$$

These results extend easily to a stiffened equation of state of Grüneisen type (Example 2.2). □

Let us consider the *shock curves*. The function

$$u = u_a \pm \Phi_a(p)$$

represents the states in the (u, p)-plane that can be connected to the state **a** on the right or on the left by a 1-shock or a 3-shock (see Figure 3.2). Now, let us select the *right* states that can be connected to the state **a** by an *admissible shock*. We use the characterization (2.34)(iv) of an admissible shock $u_- > u_+$. Since in that case $u_a = u_-$, we obtain diagram (i) of Figure 3.3.

Similarly, in order to determine the *left* states to which the state **a** can be connected by an admissible shock, we notice that $u_a = u_+$, so that we get diagram (ii) of Figure 3.3.

Let us next determine the projections of the *rarefaction curves* onto the (u, p)-plane. Defining a function $\ell = \ell(\rho, s)$ up to an additive constant by

$$\frac{\partial \ell}{\partial \rho}(\rho, s) = \frac{c(\rho, s)}{\rho},$$

we recall that we can choose $\{u - \ell, s\}$ as a pair of 1-Riemann invariants and $\{u + \ell, s\}$ as a pair of 3-Riemann invariants (see Chapter I, Example 2.2). In the following, it will be more convenient to use p, s as the thermodynamic independent variables. And we may equivalently define $\ell = \ell(p, s)$ by

$$\frac{\partial \ell}{\partial p}(p, s) = \frac{1}{(\rho c)}(p, s) \tag{3.7}$$

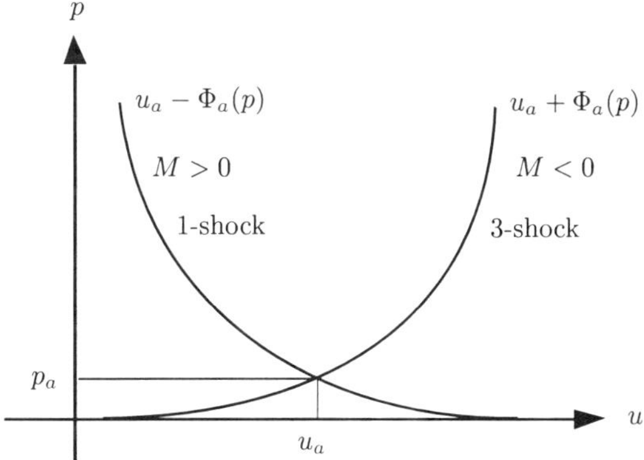

FIGURE 3.2. Right and left states that can be connected to **a** by a shock.

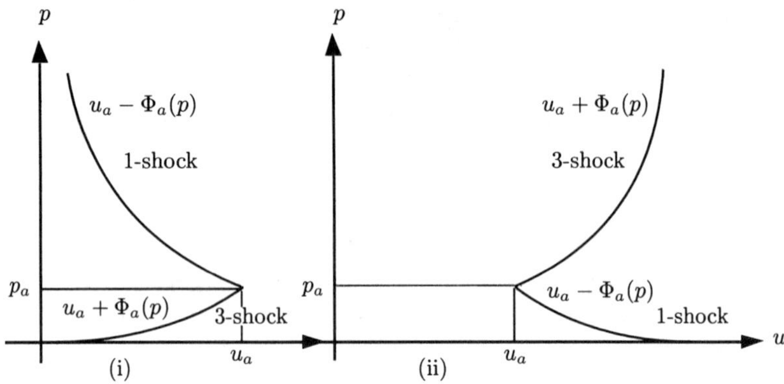

FIGURE 3.3. (i) Right and (ii) left states that can be connected to **a** by a shock.

since

$$\frac{c}{\rho} = \frac{\partial \ell}{\partial \rho}(p, s) = \frac{\partial \ell}{\partial p}(p, s)\frac{\partial p}{\partial \rho}(\rho, s) = \frac{\partial \ell}{\partial p}(p, s)c^2.$$

Now, let us consider two states **a** and **b** connected by a 1-rarefaction wave. By Theorem 3.2 of Chapter I, the 1-Riemann invariants are constant through a 1-rarefaction wave, so we have

$$\begin{cases} s_a = s_b, \\ u_a + \ell(p_a, s_a) = u_b + \ell(p_b, s_b). \end{cases} \tag{3.8}$$

Hence, setting

$$\Psi_a(p) = \ell(p, s_a) - \ell(p_a, s_a) \tag{3.9}$$

gives

$$u_b = u_a - \Psi_a(p_b).$$

Similarly, if the states **a** and **b** are connected by a 3-rarefaction wave, we have

$$\begin{cases} s_a = s_b, \\ u_a - \ell(p_a, s_a) = u_b - \ell(p_b, s_b), \end{cases}$$

so that

$$u_b = u_a + \Psi_a(p_b).$$

Thus, we obtain

$$u_b = \begin{cases} u_a - \Psi_a(p_b) & \text{for a 1-rarefaction,} \\ u_a + \Psi_a(p_b) & \text{for a 3-rarefaction.} \end{cases} \tag{3.10}$$

Let us study the properties of the function Ψ_a. Due to (3.7), (3.9) can be written

$$\Psi_a(p) = \int_{p_a}^{p} \left(\frac{1}{\rho c}\right) dp.$$

Since (see (2.21) of Chapter I)

$$\rho c = \sqrt{-\partial p/\partial\tau(\tau, s)}$$

and s is constant, we have

$$dp = \frac{\partial p}{\partial \tau}\, d\tau = -\rho^2 c^2 d\tau,$$

which yields

$$\Psi_a(p) = -\int_{\tau_a}^{\tau} \sqrt{-\partial p/\partial\tau}\; d\tau.$$

We note that

$$\Psi_a(p_b) = -\Psi_b(p_a), \tag{3.11a}$$

$$\Psi_a(0) = -\int_{\tau_a}^{+\infty} \sqrt{-\partial p/\partial\tau}\; d\tau < 0, \tag{3.11b}$$

and

$$\Psi_a'(p) = \frac{1}{\rho c} > 0. \tag{3.11c}$$

Now, we assume that the function $p = p(\tau, s)$ satisfies

$$\left|\frac{\partial p}{\partial \tau}(\tau, s)\right| \geq \frac{c}{\tau^{\alpha}}, \qquad \alpha \geq 2. \tag{3.12}$$

This implies for $p > p_a$

$$\Psi_a(p) = \int_{\tau}^{\tau_a} \sqrt{-\partial p/\partial\tau}\; d\tau \geq c\int_{\tau}^{\tau_a} \left(\frac{1}{\tau^{\alpha/2}}\right) d\tau,$$

hence

$$\lim_{p\to+\infty} \Psi_a(p) = +\infty \tag{3.11d}$$

and also

$$\lim_{p\to+\infty} \Psi_a'(p) = 0. \tag{3.11e}$$

Example 3.2. Let us go back to the case of a polytropic ideal gas (Example 2.1) for which

$$p = p(\rho, s) = A(s)\,\rho^{\gamma}, \qquad \gamma > 1.$$

We have

$$c^2 = \frac{\partial p}{\partial \rho} = \gamma A(s)\,\rho^{\gamma-1},$$

so that

$$\ell = \int \frac{c}{\rho}\, d\rho = \sqrt{\gamma A(s)} \int \rho^{(\gamma-3)/2} d\rho = \frac{2}{(\gamma-1)} \sqrt{\gamma A(s)}\, \rho^{(\gamma-1)/2}.$$

This yields

$$\ell = \frac{2}{\gamma - 1} \sqrt{\gamma p \tau} = \frac{2c}{\gamma - 1}$$

and

$$\Psi_a(p) = \frac{2\sqrt{\gamma}}{(\gamma-1)} \left(\sqrt{p\tau} - \sqrt{p_a \tau_a}\right).$$

It remains to express τ in terms of p. Since $s_a = s_b$, we can write

$$\frac{p}{p_a} = \left(\frac{\rho}{\rho_a}\right)^{\gamma}$$

or

$$\tau = \tau_a \left(\frac{p_a}{p}\right)^{1/\gamma}.$$

Hence, we find

$$\Psi_a(p) = \frac{2\sqrt{\gamma}}{(\gamma-1)} p_a^{1/2\gamma} \tau_a^{1/2} \{p^{(\gamma-1)/2\gamma} - p_a^{(\gamma-1)/2\gamma}\}.$$

Note that (3.12) holds with $\alpha = \gamma + 1$, which implies the properties (3.11). Moreover,

$$\Psi_a(0) = -2\sqrt{\gamma p_a \tau_a / (\gamma - 1)} = -\frac{2c_a}{\gamma - 1}.$$

The calculations are simple in this model case. □

The rarefaction curves defined by

$$u = u_a \pm \Psi_a(p)$$

represent the states in the (u, p)-plane that can be connected to the state **a** (on the right or on the left) by a 1-rarefaction or a 3-rarefaction wave (see Figure 3.4).

It remains to select the states that can be connected to the state **a** on the right (resp. on the left) by a rarefaction wave. We first state the following lemma.

Lemma 3.1

If a left state (with velocity u_-) is connected to a right state (with velocity u_+) by a rarefaction wave, we have

$$u_+ > u_-. \tag{3.13}$$

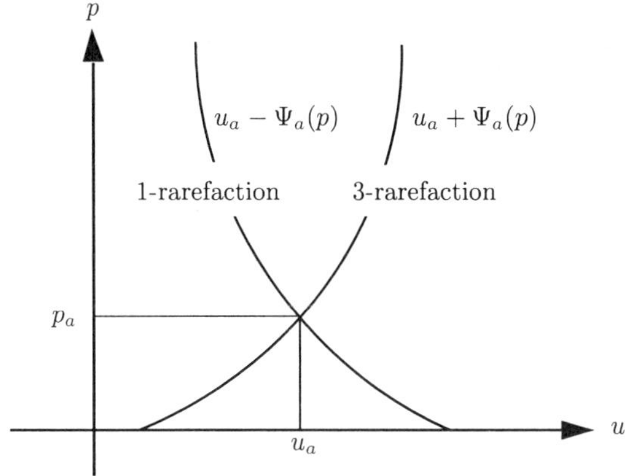

FIGURE 3.4. Right and left states that can be connected to **a** by a rarefaction wave.

Proof. Consider first a 1-rarefaction wave connecting a state **u** to the state **a**; we have by (3.8)

$$u + \ell = u_a + \ell_a, \quad s = s_a.$$

Hence, throughout the wave, we have

$$du = -d\ell \quad \text{and} \quad ds = 0.$$

Moreover,

$$\frac{d(u-c)}{du} = \frac{(d\ell + dc)}{d\ell} = \frac{\Big((c/\rho)d\rho + dc\Big)}{\Big((c/\rho)d\rho\Big)} = \frac{d(\rho c)}{c d\rho}.$$

But since

$$\rho c = \sqrt{-\partial p/\partial \tau},$$

we have

$$d(\rho c) = -\frac{\partial^2 p/\partial \tau^2}{2\sqrt{-\partial p/\partial \tau}} = \Big(\frac{1}{2}\Big)\rho^3 c\Big(\frac{\partial^2 p}{\partial \tau^2}\Big)d\rho$$

and

$$\frac{d(\rho c)}{c d\rho} > 0.$$

Hence, throughout the 1-rarefaction wave, we obtain

$$\frac{d(u-c)}{du} > 0,$$

so that $u - c$ increases with u. But for a 1-rarefaction wave, λ_1 increases from the left to the right and (with shorthand notations for $\lambda_1(\mathbf{U})$)

$$u_+ - c_+ = \lambda_1(u_+) > \lambda_1(u_-) = u_- - c_-,$$

which implies (3.13).

Now, for a 3-rarefaction wave, we have

$$u - \ell = u_a - \ell_a, \quad s = s_a,$$

so that $du = d\ell$ throughout the wave and

$$\frac{d(u+c)}{du} = \frac{(d\ell + dc)}{d\ell} = \frac{d(\rho c)}{c d\rho} > 0.$$

Hence $u + c$ increases with u. Since

$$\lambda_3(u_+) = u_+ + c_+ > u_- + c_- = \lambda_3(u_-),$$

we obtain again (3.13). □

Remark 3.2. The result of Lemma 3.1 is natural if we recall the characterization (2.34) (iv) of an admissible shock. Moreover, since in a rarefaction wave

$$dp = c^2 d\rho, \quad d\ell = \frac{c}{\rho}\, d\rho,$$

we have

$$\rho_+ < \rho_-, \quad p_+ < p_- \quad \text{for a 1-rarefaction wave,}$$
$$\rho_+ > \rho_-, \quad p_+ > p_- \quad \text{for a 3-rarefaction wave,}$$

which gives the analog of (2.34). □

We use the criterion (3.13), which characterizes a rarefaction wave, for obtaining the diagrams of Figure 3.5.

We are now in a position to solve the Riemann problem.

Theorem 3.1

Assume that the function $p(\tau, s)$ satisfies the conditions

$$\frac{\partial p}{\partial \tau} \leq -\frac{c}{\tau^\alpha}, \quad \alpha > 2, \quad \frac{\partial p}{\partial s} > 0, \quad \frac{\partial^2 p}{\partial \tau^2} > 0, \tag{3.14}$$

and that the Hugoniot curve may be parametrized as in (3.2) and (3.3). Then, the Riemann problem for the gas dynamics equations in Eulerian coordinates has a unique solution (in the class of admissible shock, contact discontinuities, and rarefaction waves separating constant states) if and only if the initial states satisfy the condition

$$u_R - u_L < -(\Psi_R(0) + \Psi_L(0)). \tag{3.15}$$

Proof. Let us denote by L (resp. R) the given left (resp. right) states. Consider the set of right states that can be connected to L by a 1-wave

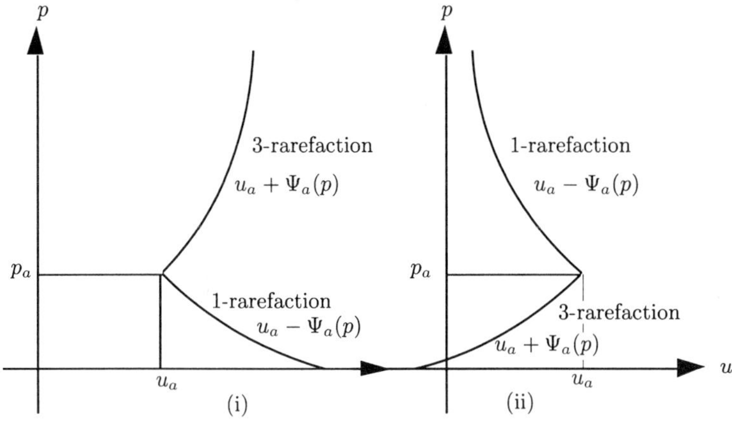

FIGURE 3.5. (i)Right and (ii)left states that can be connected to **a** by a rarefaction wave.

(shock or rarefaction). It can be defined by

$$
u = \begin{cases} u_L - \Phi_L(p), & p \geq p_L \qquad (\mathcal{S}_1^L), \\ u_L - \Psi_L(p), & p \leq p_L \qquad (\mathcal{R}_1^L). \end{cases} \tag{3.16}
$$

We have already observed that this curve is of class C^2 and is strictly decreasing from $u_L - \Psi_L(0)$ to $-\infty$ when p increases from 0 to $+\infty$ (Figure 3.6 (i)).

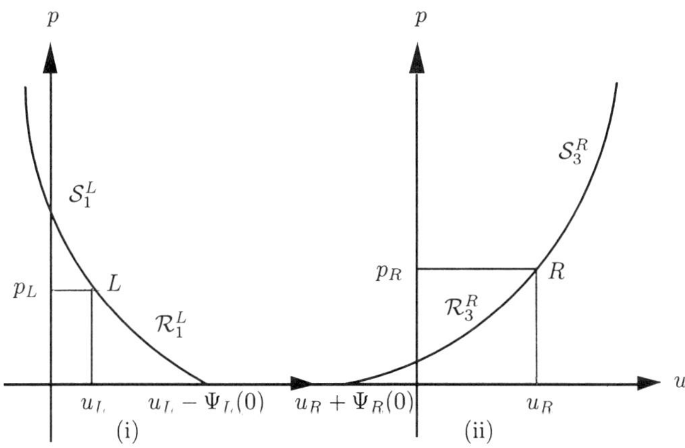

FIGURE 3.6. Right states that can be connected to L by a 1-wave. Left states to which R can be connected by a 3-wave.

Similarly, the set of left states to which R can be connected by a 3-wave (shock or rarefaction) is defined by

$$u = \begin{cases} u_R + \Phi_R(p), & p \geq p_R \qquad (\mathcal{S}_3^R), \\ u_R + \Psi_R(p), & p \leq p_R \qquad (\mathcal{R}_3^R). \end{cases}$$

This function is of class C^2 and is strictly increasing from $u_R + \Psi_R(0)$ to $+\infty$ when p increases from 0 to $+\infty$ (Figure 3.6 (ii)).

In order to solve the Riemann problem, we have to find the intersection of these two curves so as to determine (u^*, p^*).

Geometrically (see Figure 3.6), it is clear that a necessary and sufficient condition for (u^*, p^*) to be uniquely defined is given by

$$u_R + \Psi_R(0) < u_L - \Psi_L(0)$$

or, equivalently, by (3.15). Moreover, in that case, the nature of the 1-wave and the 3-wave is easily determined (a 1-rarefaction wave and a 3-shock wave in the case of Figures 3.1 and 3.7).

It remains to compute the densities ρ_I and ρ_{II} of the constant states I and II located on both sides of the contact discontinuity. Let us determine ρ_I. Assume first that the 1-wave is a shock wave. Then, we have by (3.2)

$$\tau_I = \frac{1}{\rho_I} = h_L(p^*).$$

On the other hand, if the 1-wave is a rarefaction wave, using the function $\rho = \rho(p, s)$, since in that case $s^* = s_L$, we obtain

$$\rho_I = \rho(p^*, s_L).$$

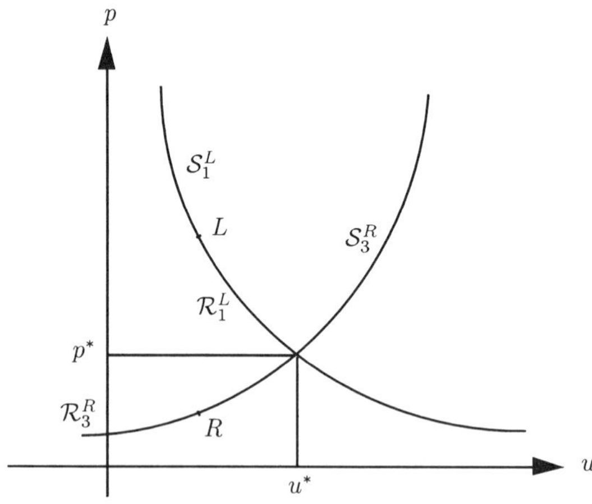

FIGURE 3.7. Intersection of the 1-wave (L) and 3-wave (R) curves.

Moreover, we have (see (2.30) in Example 2.4, Chapter I)

$$c_L^2 = \frac{\partial p}{\partial \rho}(\rho_L, s_L), \quad c_I^2 = \frac{\partial p}{\partial \rho}(\rho_I, s_L).$$

Remember that the 1-rarefaction wave is contained in the fan

$$\lambda_1(L) = u_L - c_L \leq \frac{x}{t} \leq u^* - c_I = \lambda_1(I).$$

e proceed in an analogous way for the 3-wave. □

Example 3.3. Let us go back to the case of a polytropic ideal gas (see Example 3.2). The condition (3.15) can be written

$$u_R - u_L < \frac{2(c_R + c_L)}{\gamma - 1}.$$

If the 1-wave is a shock wave, we obtain

$$\tau_I = \tau_L \frac{\mu^2 p^* + p_L}{p^* + \mu^2 p_L}.$$

On the other hand, if the 1-wave is a rarefaction wave, we may write

$$p_L = A(s_L)\rho_L^\gamma, \quad p^* = A(s_L)\rho_I^\gamma,$$

so that

$$\rho_I = \left(\frac{p^*}{p_L}\right)^{1/\gamma} \rho_L.$$

Again, in this model case the calculations are simple. □

In fact, we have four cases for the solution of the Riemann problem, depicted in Figure 3.8.

Remark 3.3. If the initial states L and R do not satisfy the condition (3.15), the Riemann problem has no solution in the above sense. However, one can yet define a solution by introducing a vacuum. The solution consists of two rarefaction waves separated by a vacuum where $\rho = 0$ and the other dependent variables are left undefined (see Figure 3.9). Then Theorem 3.1 is a global existence theorem in that the two initial states are not required to be close to each other. □

As an example, we consider the case of a shock tube that corresponds to the initial conditions of gas at rest with two different states on each side of $x = 0$: $u_L = u_R = 0$, $\rho_L > \rho_R$, $p_L > p_R$. The associated Riemann problem is easily solved by looking at Figure 3.8 (we are in case 3 of Figure 3.8). Now, the values of ρ, u, and p are as depicted in Figure 3.10, which gives the behavior but not the precise values of the density, velocity, and pressure. This shock tube problem serves very frequently as a test for numerical schemes (Sod 1978), and more precise values can be found in many references (Einfeldt 1988, for instance). Other tests corresponding to different initial conditions are often proposed (see, e.g., Montagné,

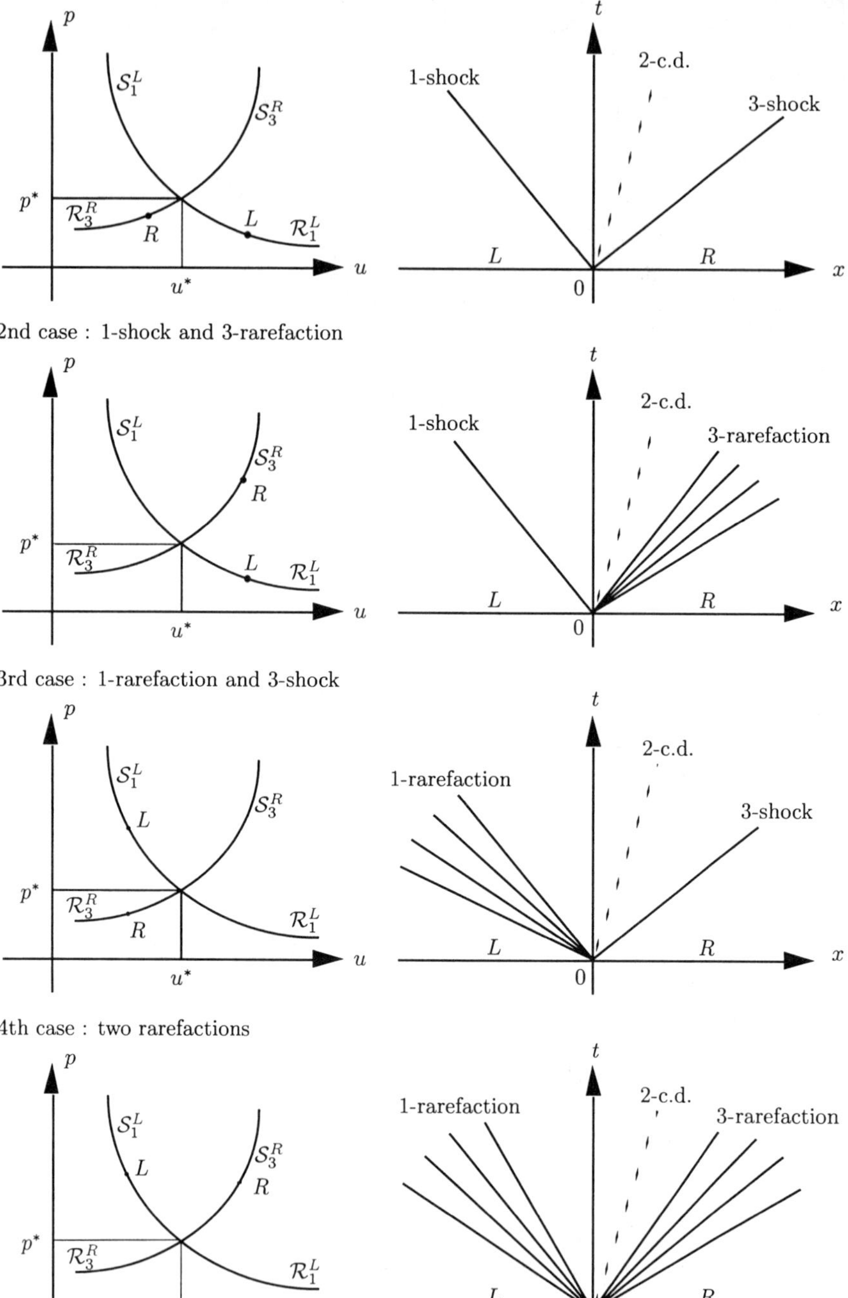

FIGURE 3.8. Solution of the Riemann problem.

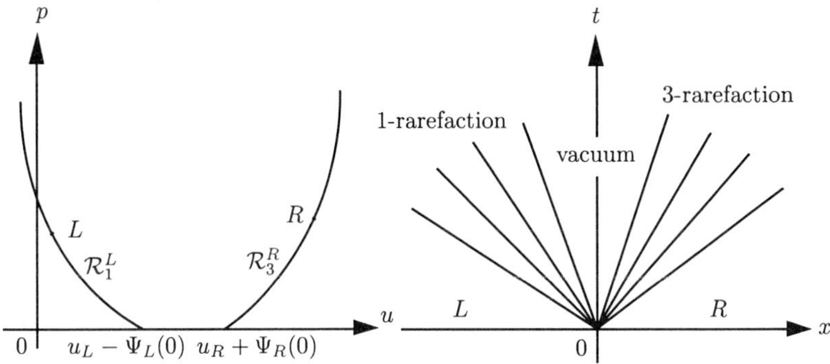

FIGURE 3.9. Vacuum.

Yee and Vinokur 1988). One also finds the example of two mirror states ($\mathbf{U}_L = (\rho, -u, p)$ and $\mathbf{U}_R = (\rho, u, p)$ (Einfeldt et al. 1991), which corresponds to case 4 if $u > 0$ or case 1 if $u < 0$, with a 2-contact discontinuity on $x = 0$.

We shall not detail the solution of the Riemann problem in Lagrangian coordinates: it has exactly the same structure as in Eulerian coordinates. The jump conditions are given in (2.15) (see Remark 2.1). If x is the mass variable, in the (x, t) plane the 2-contact discontinuity is the axis $x = 0$, across which u and p are constant.

Remark 3.4. Note the following property: on a 1-rarefaction wave, the C_- characteristics are the straight lines

$$\frac{x}{t} = -\frac{c}{\eta} = -g$$

(see Chapter I, (5.6)). Therefore, the C_+ characteristics satisfy at each point (x, t)

$$\frac{dx}{dt} = -\frac{x}{t}$$

which yields

$$xt = \text{ constant},$$

and the cross characteristics C_+ are thus hyperbolas (we have an analogous property for a 3-rarefaction). This property can be used to parametrize the rarefaction wave by characteristic coordinates (α, β) defined by

$$\alpha = -(-g_L x t)^{1/2}, \quad \beta = -g_L^{-1} \frac{x}{t}.$$

β is a normalized slope of the C_- line, while α is the x-coordinate of the intersection of the C_+-curve with the head characteristic (the slopes of the

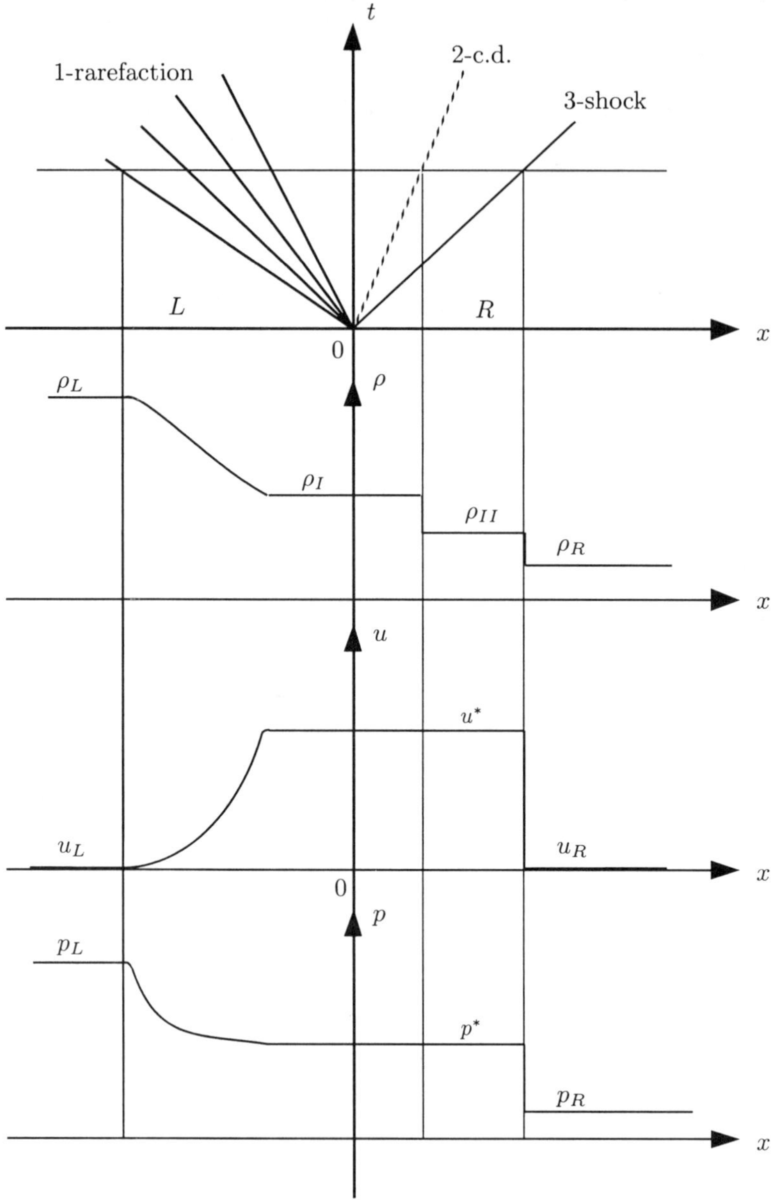

FIGURE 3.10. Solution of the shock tube problem.

rarefaction fan extend from $-g_L$ to $-g_L^*$, so that $\beta = 1$ at the head, and $\beta = \beta^* = \frac{g_L^*}{g_L}$ at the tail) (see Ben Artzi and Falcovitz 1984 for details). □

Remark 3.5. We can also solve the Riemann problem for a flow consisting of a mixture of reacting species (for simplicity, we consider two species). We add to the system of three equations (2.1) (which now states conservation of mass, momentum, and energy for the mixture) a fourth equation corresponding to the conservation of one species with mass fraction Y (the mass fraction of the other species is thus $1 - Y$)

$$\begin{cases} \dfrac{\partial \rho}{\partial t} + \dfrac{\partial}{\partial x}(\rho u) = 0, \\ \dfrac{\partial}{\partial t}(\rho u) + \dfrac{\partial}{\partial x}(\rho u^2 + p) = 0, \\ \dfrac{\partial}{\partial t}(\rho e) + \dfrac{\partial}{\partial x}((\rho e + p)u) = 0, \\ \dfrac{\partial}{\partial t}(\rho Y) + \dfrac{\partial}{\partial x}(\rho Y u) = 0. \end{cases}$$

In fact, the full system contains source terms from chemical production that we have not written in the right-hand side. For simplicity, we also assume for each species a perfect gas γ -law $p_i = (\gamma_i - 1)\rho Y_i \varepsilon_i$, $\varepsilon_i = C_{v_i} T$, and Dalton's law for the pressure of the mixture $p = p_1 + p_2$ (see Remark 1.2). The total energy of the mixture is $\rho e = \rho Y \varepsilon_1 + \rho(1 - Y)\varepsilon_2 + \frac{1}{2}\rho u^2$. We have $p = (\gamma - 1)(\rho e - \frac{1}{2}\rho u^2)$, where we find that γ is equal to

$$\gamma = \frac{Y C_{v_1}\gamma_1 + (1 - Y)C_{v_2}\gamma_2}{Y C_{v_1} + (1 - Y)C_{v_2}} = \frac{Y C_{p_1} + (1 - Y)C_{p_2}}{Y C_{v_1} + (1 - Y)C_{v_2}}.$$

We obtain a system in conservative variables $(\rho, \rho u, \rho e, \rho Y)$ that is not strictly hyperbolic, with eigenvalues $u - c, u, u, u + c$, where $c^2 = \frac{\partial p(\rho,s,Y)}{\partial \rho} = \frac{\gamma p}{\rho}$. The field corresponding to the double eigenvalue is linearly degenerate (see Remark 6.1, Chapter I), and the solution of the Riemann problem consists of four constant states separated by a 1-wave, a 2- (or 3-) contact discontinuity, and a 4-wave, as depicted in Figure 3.11.

It is worthwhile to note that the mass fraction Y is only discontinuous across a contact discontinuity (with the notations of Figure 3.11, $Y_I = Y_L$ and $Y_{II} = Y_R$). Indeed, Y is a 1- and 4-Riemann invariant and is thus constant on a rarefaction wave. Across a shock, it follows from the Rankine–Hugoniot jump conditions ((2.3) together with $\sigma[\rho Y] = [\rho u Y]$) that $[Y] = 0$ and Y is again constant. This justifies the fact that the last equation can also be written in nonconservative form

$$\frac{\partial Y}{\partial t} + u\frac{\partial Y}{\partial x} = 0,$$

which means that Y is purely convected by the flow. Indeed, since it is obtained by combining the first and fourth conservation equations, it holds

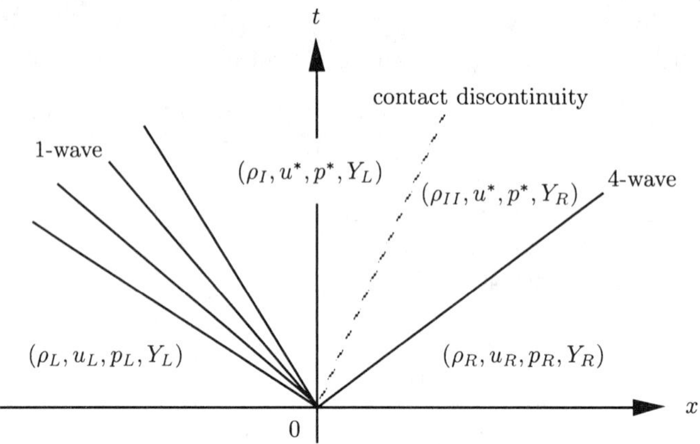

FIGURE 3.11. Solution of the Riemann problem for a mixture of two reacting gases.

for smooth solutions, while for a solution such that Y is discontinuous, the discontinuity must be a contact discontinuity across which u is constant, so that $u\,\frac{\partial Y}{\partial x}$ is well-defined. □

4 Reacting flows. The Chapman–Jouguet theory

Let us come now to the more complex example of reacting gas flow (Introduction, Example 1.5). We restrict ourselves to the plane one-dimensional flow involving a single exothermic reaction between two species, the unburnt gas and the burnt gas. Neglecting transport effects, the equations to be considered are the Euler equations plus a chemical reaction equation

$$\begin{cases} \dfrac{\partial \rho}{\partial t} + \dfrac{\partial}{\partial x}(\rho u) = 0, \\ \dfrac{\partial}{\partial t}(\rho u) + \dfrac{\partial}{\partial x}(\rho u^2 + p) = 0, \\ \dfrac{\partial}{\partial t}(\rho e) + \dfrac{\partial}{\partial x}((\rho e + p)u) = 0, \\ \dfrac{\partial}{\partial t}(\rho z) + \dfrac{\partial}{\partial x}(\rho z u) = \rho r, \end{cases} \tag{4.1}$$

where z denotes the mass fraction of the burnt gas (so that $1 - z$ is the mass fraction of the unburnt gas) and r is the reaction rate, which we assume to be of the form

$$r = r(\rho, p, z).$$

Note that now we must include in the internal energy ε a term corresponding to the heat of reaction of the mixture.

Example 4.1. Assuming that both gases are ideal polytropic with the same γ -law and with constant energy of formation Q_u (resp. Q_b) for the unburnt (resp. burnt) gas, we have

$$\varepsilon = \varepsilon(\rho, p, z) = (\gamma - 1)^{-1}\frac{p}{\rho} + (1-z)Q_u + z\,Q_b.$$

For an exothermic reaction, $Q_b < Q_u$. □

Let us begin by studying the even simpler case of an *infinite reaction rate.* This amounts to supposing that the reaction is completed instantaneously, so that we replace the chemical reaction equation by the following description: the reacting gas consists of two components, the unburnt gas and the burnt gas, each in thermodynamic equilibrium, and separated by an infinitely thin reaction front. The equations of state of the unburnt gas and the burnt gas, which are denoted respectively by $\varepsilon_u(\tau, p)$ and $\varepsilon_b(\tau, p)$, are assumed to be related by

$$\varepsilon_b(\tau, p) = \varepsilon_u(\tau, p) - Q, \tag{4.2}$$

where Q is the (constant) heat of complete reaction. In the case of an exothermic reaction, which we shall study later, we have

$$Q > 0.$$

In this model, we thus consider the usual gas dynamics equations, but with different equations of state for the unburnt and the burnt components of the gas. We now investigate the possible elementary waves that are allowed by this model. Indeed, in addition to the shock waves, the rarefaction waves, and the contact discontinuities, we shall have to introduce discontinuity waves that separate the unburnt gas from the burnt gas. These discontinuities will be referred to in the following as *combustion waves.*

If the reaction propagates with speed σ, the equations relating the properties on each side of the front are derived in the same way as the jump conditions for a discontinuity wave as in the Introduction, Section 2.2:

$$\begin{cases} \sigma[\rho] = [\rho u], \\ \sigma[\rho u] = [\rho u^2 + p], \\ \sigma[\rho e] = [(\rho e + p)u]. \end{cases} \tag{4.3}$$

We perform now the same analysis as in Section 2, with the only difference (with equations (2.3)) that, in the last relation (4.3), the equation of state differs on each side of the front.

Let us introduce the following convention: the state (0) will refer to the *unburnt* gas while the state (1) refers to the *burnt* gas. As before, we denote the velocity of the gases relative to the front by

$$v_i = u_i - \sigma, \quad i = 0, 1.$$

Then, relations (4.3) give after a computation similar to that of Lemma 2.1

$$(4.4)\qquad \begin{cases} \rho_0 v_0 = \rho_1 v_1, \\ \rho_0 v_0^2 + p_0 = \rho_1 v_1^2 + p_1, \\ \Big(\rho_0\Big(\varepsilon_0 + \frac{v_0^2}{2}\Big) + p_0\Big)v_0 = \Big(\rho_1\Big(\varepsilon_1 + \frac{v_1^2}{2}\Big) + p_1\Big)v_1. \end{cases}$$

Next, we set

$$M = \rho_0 v_0 = \rho_1 v_1$$

so that the second equation in (4.4) yields

$$M v_0 + p_0 = M v_1 + p_1$$

or

$$(4.5)\qquad M = -\frac{p_1 - p_0}{v_1 - v_0} = -\frac{p_1 - p_0}{u_1 - u_0}.$$

Also, since $v_i = M\tau_i$, we obtain

$$(4.6)\qquad M^2 = -\frac{p_1 - p_0}{\tau_1 - \tau_0}.$$

Now, eliminating the velocities in the third equation (4.2) gives

$$(4.7)\qquad \varepsilon_1 - \varepsilon_0 + \frac{1}{2}(p_1 + p_0)(\tau_1 - \tau_0) = 0.$$

This is the analog of the Hugoniot equation (2.12). Again, it is worthwhile to notice that $\varepsilon_0 = \varepsilon_u(\tau_0, p_0)$ and $\varepsilon_1 = \varepsilon_b(\tau_1, p_1)$ correspond to different equations of state.

In order to characterize all possible burnt states that can be connected to a given unburnt state (0) by a combustion wave, we define in the (τ, p)-plane the *Crussard curve* (or Hugoniot curve) $\mathcal{C}$ with *center* (τ_0, p_0) by

$$(4.8)\qquad \varepsilon(\tau, p) - \varepsilon_0 + \frac{1}{2}(p + p_0)(\tau - \tau_0) = 0,$$

where

$$\varepsilon(\tau, p) = \varepsilon_b(\tau, p)$$

denotes from now on the equation of state of the *burnt* gas. Due to the heat of reaction Q, the pole $A_0 = (\tau_0, p_0)$ does not belong to $\mathcal{C}$.

Example 4.2. Assuming as in Example 4.1 that both gases are ideal polytropic with the same γ-law, we find that the Crussard curve $\mathcal{C}$ is the rectangular hyperbola

$$\frac{1}{2\mu^2}\big((\tau - \mu^2\tau_0)p - (\tau_0 - \mu^2\tau)p_0\big) - Q = 0,$$

where $\mu^2 = \frac{(\gamma-1)}{(\gamma+1)}$ (see (2.19), Example 2.1). □

More generally, we can state the following result.

Lemma 4.1

Assume that the function $(\tau, s) \to p(\tau, s)$ satisfies the conditions (2.21),

$$\frac{\partial p}{\partial \tau} < 0, \quad \frac{\partial p}{\partial s}(\tau, s) > 0,$$

and that the reaction is exothermic,

$$\varepsilon(\tau_0, p_0) < \varepsilon_0.$$

Then, the Crussard curve with center $A_0 = (\tau_0, p_0)$ lies above A_0.

Proof. Let $B_0 = (\tau_0, p_1)$ be the point of $\mathcal{C}$ with abscissa τ_0 (see Figure 4.1). From (4.8) and (4.9), we deduce

$$\varepsilon(\tau_0, p_1) = \varepsilon_0 > \varepsilon(\tau_0, p_0).$$

Now, according to (2.21), the function $p \to \varepsilon(\tau_0, p)$ is strictly increasing since (see (1.1) and (2.25))

$$\frac{\partial \varepsilon}{\partial p}(\tau, p)_{|\tau} = \frac{\partial \varepsilon}{\partial s}(\tau, s)\frac{\partial s}{\partial p}(\tau, p) = T\Big(\frac{\partial p}{\partial s}(\tau, p)\Big)^{-1} > 0.$$

Thus

$$\varepsilon(\tau_0, p_1) = \varepsilon_0 > \varepsilon(\tau_0, p_0) \Longleftrightarrow p_1 > p_0,$$

and B_0 lies above A_0.

Notice that when (4.2) holds, (4.9) is equivalent to

$$\varepsilon(\tau_0, p_0) < \varepsilon_0 \Longleftrightarrow Q > 0.$$

In the same way, let $C_0 = (\tau_1, p_0)$ be the point on $\mathcal{C}$ with $p = p_0$. Arguing now on the specific enthalpy defined by

$$h = \varepsilon + p\tau,$$

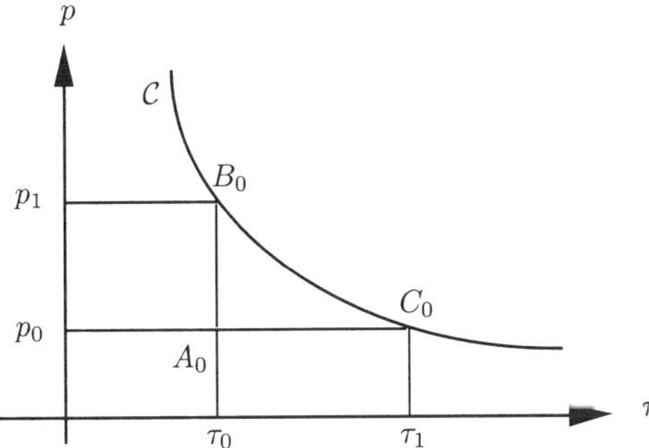

FIGURE 4.1. Crussard curve $\mathcal{C}$ with center A_0.

we have

$$h(\tau_0, p_0) = \varepsilon(\tau_0, p_0) + p_0\tau_0,$$
$$h(\tau_1, p_0) = \varepsilon(\tau_1, p_0) + p_0\tau_1 = \varepsilon_0 - p_0(\tau_1 - \tau_0) + p_0\tau_1 = \varepsilon_0 + p_0\tau_0,$$

which implies by (4.9)

$$h(\tau_1, p_0) > h(\tau_0, p_0).$$

Now (see (1.1) and (2.25))

$$\begin{aligned}\frac{\partial h}{\partial \tau}(\tau, p)_{|\tau} &= \frac{\partial \varepsilon}{\partial \tau}(\tau, p) + p \\ &= \frac{\partial \varepsilon}{\partial \tau}(\tau, s) + \frac{\partial \varepsilon}{\partial s}(\tau, s)\frac{\partial s}{\partial \tau}(\tau, p) + p = T\,\frac{\partial s}{\partial \tau}(\tau, p) > 0\end{aligned}$$

so that the function $\tau \to h(\tau, p_0)$ is strictly increasing and

$$h(\tau_1, p_0) > h(\tau_0, p_0) \Longleftrightarrow \tau_1 > \tau_0,$$

which ends the proof. □

From now on, we shall assume that (2.21) holds. In view of (4.5), the part of the Crussard curve between B_0 and C_0 is not admissible since (τ, p) must satisfy

$$\frac{p - p_0}{\tau - \tau_0} < 0.$$

The remaining part of the curve consists of two distinct branches: the upper branch $p > p_0$, called the *detonation branch*, and the lower one $\tau > \tau_0$, called the *deflagration branch.*

For a given final state $B = (\tau, p)$ of $\mathcal{C}$, the negative slope of the *Rayleigh line* joining A_0 and B

$$\frac{p - p_0}{\tau - \tau_0} = -M^2$$

enables us to determine the wave velocity $\sigma = u_i - M\,\tau_i$ through (4.4). Conversely, for a given combustion velocity σ, we see that the corresponding Rayleigh line may or may not intersect the Crussard curve. Let us assume moreover that (2.29) holds, which implies (see Remark 2.6) the property that in the (τ, p)-plane, the Rayleigh line drawn from a point $\overline{B} = (\overline{\tau}, \overline{p})$ with slope

$$-M^2 = \frac{p - \overline{p}}{\tau - \overline{\tau}}$$

intersects the Hugoniot curve $\mathcal{H}$ with pole $\overline{B}$ at one and only one other point. We can then prove the following lemma.

Lemma 4.2

Assume that the conditions (2.21) and (2.29) hold. For a given σ, there may exist 0 or 2 (possibly coalescing) burnt states that can be connected *to the (unburnt) state (0) by a combustion wave with speed σ. These two states belong to the same Hugoniot shock curve for the burnt gas.*

Proof. Consider the Rayleigh line Δ passing through A_0 with slope $-M^2$, and suppose that Δ intersects $\mathcal{C}$ at some point $B' = (\tau', p')$ (see Figure 4.2). Then, let us see that Δ intersects $\mathcal{C}$ at one and only one other point B'' (possibly coalescing with B'). Since B' belongs to $\mathcal{C}$, we have

$$\varepsilon(\tau', p') - \varepsilon_0 + \frac{1}{2}(p' + p_0)(\tau' - \tau_0) = 0. \tag{4.9}$$

Now, let us introduce the shock Hugoniot curve $\mathcal{H}'$ with center B' relative to the burnt gas,

$$\varepsilon(\tau, p) - \varepsilon(\tau', p') + \frac{1}{2}(p' + p)(\tau - \tau') = 0,$$

which we recall is the set of all (burnt) states that can be connected to the (burnt) state B' by a shock wave. If (2.29) holds and if the pressure may increase indefinitely along $\mathcal{H}'$, the Rayleigh line intersects $\mathcal{H}'$ at exactly one other point $B'' = (\tau'', p'')$ (possibly coalescing with B' if Δ is tangent to $\mathcal{H}'$), which thus satisfies

$$\varepsilon(\tau'', p'') - \varepsilon(\tau', p') + \frac{1}{2}(p' + p'')(\tau'' - \tau') = 0. \tag{4.10}$$

Let us check that B'' also belongs to the Crussard curve $\mathcal{C}$ with center A_0. Since A_0, B', and B'' lie on the same Rayleigh line, we can write

$$-M^2 = \frac{p' - p_0}{\tau' - \tau_0} = \frac{p'' - p_0}{\tau'' - \tau_0} = \frac{p'' - p'}{\tau'' - \tau'}. \tag{4.11}$$

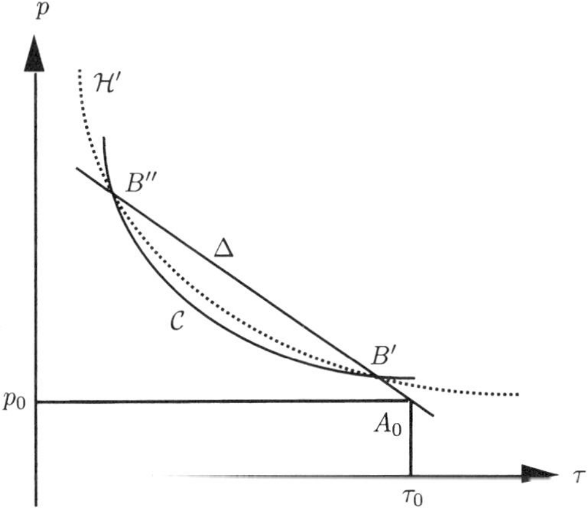

FIGURE 4.2. Rayleigh line Δ through A_0.

Setting

$$\varepsilon' = \varepsilon(\tau', p'), \quad \varepsilon'' = \varepsilon(\tau'', p''),$$

it follows from (4.9) and (4.11) that

$$\varepsilon' - \varepsilon_0 + \frac{(p'^2 - p_0^2)(\tau'' - \tau_0)}{2(p'' - p_0)} = 0, \tag{4.12}$$

and from (4.10) and (4.11) that

$$\varepsilon'' - \varepsilon' + \frac{(p''^2 - p'^2)(\tau'' - \tau_0)}{2(p'' - p_0)} = 0. \tag{4.13}$$

Adding (4.12) and (4.13), we get

$$\varepsilon' - \varepsilon_0 + \frac{1}{2}(p'' + p_0)(\tau'' - \tau_0) = 0,$$

which means exactly that B'' lies on the Crussard curve. □

According to Lemma 4.2, for M^2 large enough, the Rayleigh line with slope $-M^2$ intersects the Crussard curve at two points on the detonation branch. When M^2 decreases, it reaches a minimum value for which the two points coalesce when the corresponding Rayleigh line is tangent to $\mathcal{C}$. This point B^{CJ} represents the so-called *Chapman–Jouguet detonation* and separates the detonation branch of the Crussard curve into two parts: a detonation represented by a point on the upper part is called a *strong detonation*, while one represented by a point on the lower part is called a *weak detonation.*

For smaller values of M^2, the corresponding Rayleigh line Δ does not intersect $\mathcal{C}$ until M^2 reaches the value for which Δ is tangent to the deflagration branch. This point of tangency B_{CJ} corresponds to the *Chapman–Jouguet deflagration.* It separates the deflagration branch into two parts: the upper part represents *weak deflagrations*, the lower one *strong deflagrations* (see Figure 4.3).

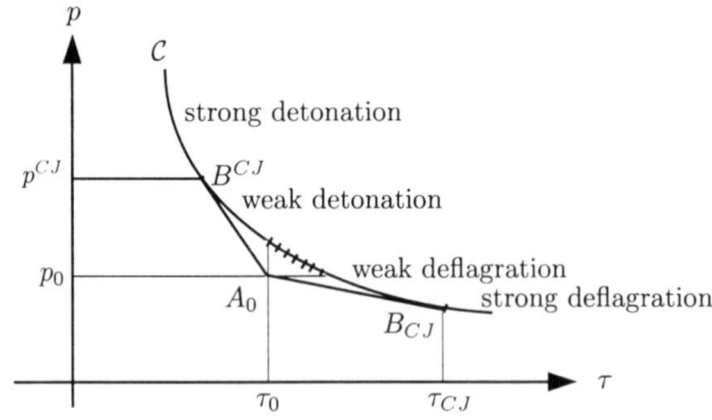

FIGURE 4.3. Detonation and deflagration branches on $\mathcal{C}$.

Remark 4.1. Let us show that a detonation wave may be viewed as a precompression shock propagating into the unburnt gas followed by a deflagration wave, both processes having the same velocity. Indeed, let a shock connecting the state (0) to a state (ρ^*, u^*, p^*) be followed by a deflagration connecting (ρ^*, u^*, p^*) to (ρ_1, u_1, p_1) in such a way that the fluxes through both discontinuities are the same,

$$M = \rho_0 v_0 = \rho^* v^* = \rho_1 v_1.$$

Then, writing the remaining conditions (2.11) and (2.12) for the shock, and (4.5) and (4.6) for the deflagration, we obtain the following system, which is equivalent to the Rankine–Hugoniot relations:

$$\begin{aligned} M^2 &= -\frac{p^* - p_0}{\tau^* - \tau_0} = -\frac{p_1 - p^*}{\tau_1 - \tau_0^*}, \\ \varepsilon^* - \varepsilon_0 &+ \frac{(p^* + p_0)(\tau^* - \tau_0)}{2} = \varepsilon_1 - \varepsilon^* + \frac{(p_1 + p^*)(\tau_1 - \tau^*)}{2} = 0. \end{aligned}$$

Using the same argument as in the proof of Lemma 4.2, we get, since the three states (0), (*), (1) lie on the same Rayleigh line,

$$\begin{aligned} M^2 &= -\frac{p_1 - p_0}{\tau_1 - \tau_0}, \\ \varepsilon_1 - \varepsilon_0 &+ \frac{1}{2}(p_1 + p_0)(\tau_1 - \tau_0) = 0. \end{aligned}$$

Thus, the process connecting the state (0) to the state (1) is indeed equivalent to a single process that is a detonation (see Figure 4.4(i)).

Note that the Crussard curve $\mathcal{C}^*$ with center $A^* = (\tau^*, p^*)$ lies above the shock Hugoniot curve $\mathcal{H}_0$ with center $A_0 = (\tau_0, p_0)$. Otherwise, the two curves would intersect at a point $A = (\tau, p)$ (see Figure 4.4 (ii)). By the above computations, A would belong to the same Rayleigh line as A_0 and A^*, which is obviously impossible.

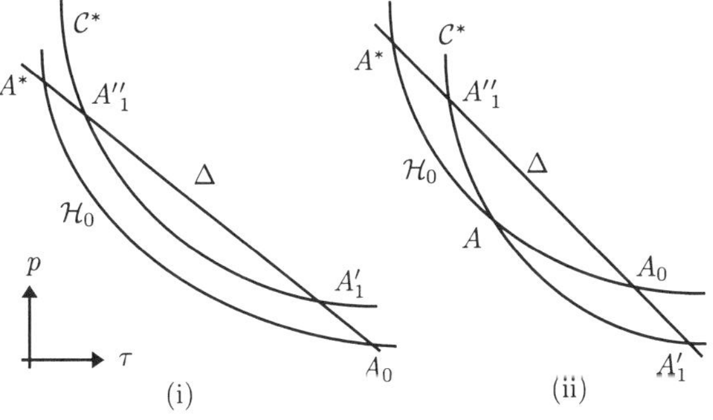

FIGURE 4.4. Case (ii) is not possible.

Moreover, we see in Figure 4.4 (i) that two cases are possible that correspond to the points A_1' and A_1''. The transition from A_0 to A_1'' (resp. from A_0 to A_1') is a strong (resp. weak) detonation representing a shock followed by a weak (resp. strong) deflagration. We shall come back to this interpretation in the next section when we introduce the Z.N.D. model. □

We shall now study in some detail the properties of the Chapman–Jouguet detonation or deflagration waves.

Lemma 4.3

Along the Crussard curve, the entropy (of the burnt gas) is stationary at the Chapman–Jouguet (C.J.) points and only at these points. For a C.J. detonation, the entropy s is a relative minimum, for the C.J. deflagration the entropy s is a relative maximum.

Proof. We follow the argument of the proof of Lemma 2.4, since the expressions defining the Crussard and the Hugoniot curves are identical. From (2.27), we deduce that along $\mathcal{C}$ (parametrized by τ)

$$T\dot{s} = \frac{(\tau - \tau_0)}{2}\left\{\frac{p - p_0}{\tau - \tau_0} - \dot{p}\right\},$$

where the dot $\cdot$ denotes differentiation along $\mathcal{C}$. Hence $\dot{s} = 0$ only at the points where the Rayleigh line is tangent to $\mathcal{C}$.

Now, if a Rayleigh line crosses the *detonation* branch at two points B' and B'' as in Figure 4.5, comparing the slopes gives

$$\dot{p}(\tau') > \frac{p' - p_0}{\tau' - \tau_0}, \quad \dot{p}(\tau'') < \frac{p'' - p_0}{\tau'' - \tau_0},$$

and we deduce that

$$\dot{s}(\tau') > 0 = \dot{s}(\tau^{CJ}) > \dot{s}(\tau'').$$

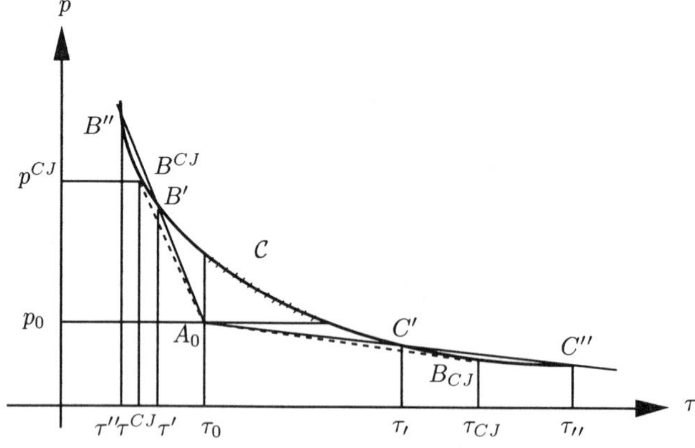

FIGURE 4.5. Chapman–Jouguet states on the Crussard curve with center A_0.

Hence, as τ decreases from τ_0, the entropy decreases from B_0 to B^{CJ} and then increases after B^{CJ}. Moreover, since B' and B'' are also on the same shock Hugoniot curve, we know from Corollary 2.1 that

$$s(\tau'') > s(\tau') > s(\tau^{CJ}). \tag{4.14a}$$

We check similarly the properties of the *deflagration* branch (see Figure 4.5) and get

$$s(\tau_{''}) < s(\tau_{,}) < s(\tau_{CJ}). \tag{4.14b}$$

Note that the isentrope $s = s^{CJ}$ is tangent to both the Crussard curve and the Rayleigh line at the point B^{CJ}. Indeed, the slope of the tangent to the isentrope $s(\tau, p) = s^{CJ}$ is given by

$$-\frac{\partial s}{\partial \tau}(\tau, p) / \frac{\partial s}{\partial p}(\tau, p) = \frac{\partial p}{\partial \tau}(\tau, s),$$

where we have used (2.25). Since $\dot{s}(\tau^{CJ}) = 0$, we get

$$\frac{\partial p}{\partial \tau}(\tau, s^{CJ}) = \dot{p}(\tau^{CJ}) = \frac{p^{CJ} - p_0}{\tau^{CJ} - \tau_0}. \tag{4.15}$$

We shall use this property below (Example 4.3; also in (4.22)). □

Lemma 4.4

Along the Crussard curve, the detonation (resp. the deflagration) speed $|v_0|$ is a local minimum (resp. maximum) for a C.J. detonation (resp. deflagration). Moreover, for a C.J. process the speed $|v_1|$ of the burnt gas relative to the front is equal to the local sound speed, i.e.,

$$|v_1| = c_1 \text{ at C.J. points.} \tag{4.16}$$

Proof. We have characterized the Chapman–Jouguet detonation as the detonation point on the Crussard curve with center $A_0 = (\tau_0, p_0)$ for which the Rayleigh line

$$M^2 = -\frac{p - p_0}{\tau - \tau_0}$$

has minimal slope M^2 (in absolute value). Since

$$M^2 = \frac{v_0^2}{\tau_0^2},$$

this proves that v_0^2 is also minimum. In particular, if $u_0 = 0$, the detonation speed $|\sigma|$ is minimum at B^{CJ}. The argument for a deflagration is similar.

Now, the local sound speed of the burnt gas satisfies (see (1.4))

$$-\rho^2 c^2 = \frac{\partial p}{\partial \tau}(\tau, s).$$

By (4.15), we have at the point $\tau^{CJ} = \tau$

$$-\rho^2 c^2 = \dot{p}(\tau) = \frac{p - p_0}{\tau - \tau_0} = -M^2 = -\rho^2 v^2,$$

which implies

$$v^2 = c^2 \text{ at a C.J. point}$$

and gives the result. □

Theorem 4.1

The gas flow relative to the reaction front satisfies the following properties:
(i) For a strong detonation $|v_0| > c_0,\ |v_1| < c_1,$
(ii) For a weak detonation $|v_0| > c_0,\ |v_1| > c_1,$
(iii) For a weak deflagration $|v_0| < c_0,\ |v_1| < c_1,$
(iv) For a strong deflagration $|v_0| < c_0,\ |v_1| > c_1.$

Proof. We shall follow the same argument as in the proof of Theorem 2.3. We first consider the gas behind the reaction front. In the case of a detonation, let B' and B'' be the two points on the intersection of the Rayleigh line with $\mathcal{C}$. It follows from the proof of Lemma 4.2 that B'' lies on the Hugoniot curve with center B'. Thus, applying the argument used in the proof of Theorem 2.3 (replacing A_0 by B' and A_1 by B''), we have along Δ parametrized by $\alpha \geq 0$

$$\frac{ds}{d\alpha} > 0 \text{ at } B', \quad \frac{ds}{d\alpha} < 0 \text{ at } B''$$

and

$$c^2 < v^2 \text{ at } B', \quad c^2 > v^2 \text{ at } B''.$$

The case of a deflagration follows similarly.

In order to study the property ahead of a reaction front, let us consider the backward Crussard curve of a given burnt state $A_1 = (\tau_1, p_1)$, i.e., the locus of all possible unburnt states $A_0 = (\tau, p)$ that can be connected to A_1 by a combustion wave. These states satisfy

$$\varepsilon_0(\tau, p) - \varepsilon_1 + \frac{1}{2}(p_1 + p)(\tau - \tau_1) = 0,$$

where $\varepsilon_0(\tau, p)$ is the equation of state of the unburnt gas. Assuming as for Lemma 4.1 that the reaction is exothermic,

$$\varepsilon_0(\tau_1, p_1) > \varepsilon_1,$$

we obtain that the corresponding curve lies below the center A_1 (see Figure 4.6).

We must again exclude the part of the curve corresponding to $\frac{(p-p_1)}{(\tau-\tau_1)} \geq 0$, and there remain two branches, the upper (resp. the lower) part representing the states that can be connected to A_1 by a deflagration (resp.

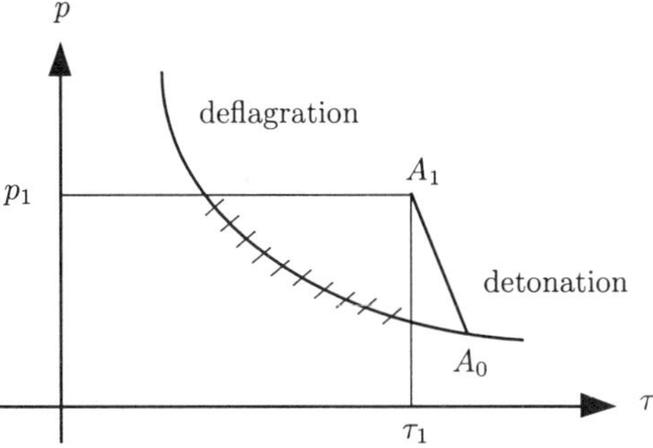

FIGURE 4.6. Backward Crussard curve of the state A_1.

detonation) since $p \geq p_1$ (resp. $p \leq p_1$). Now let A_0 be a state on the "detonation part" and $\mathcal{H}_0$ be the Hugoniot function relative to the unburnt gas with center A_0 (see (2.18))

$$\mathcal{H}_0(\tau, p) = \varepsilon_0(\tau, p) - \varepsilon_0 + \frac{1}{2}(p_0 + p)(\tau - \tau_0).$$

It follows from (2.31) that along the line A_0A_1 the entropy of the unburnt gas has exactly one critical point, which is a maximum. Thus, if $\Delta = A_0A_1$ is parametrized by $\alpha \geq 0$, $\frac{ds}{d\alpha} > 0$ at A_0, and, concluding as in the proof of Theorem 2.3 or 4.1, we deduce, since $\tau_1 < \tau_0$, that

$$|v_0| > c_0.$$

If A_0 lies on the "deflagration part", $\tau_1 < \tau_0$ and we get instead

$$|v_0| < c_0,$$

which ends the proof. □

Remark 4.2. The four possibilities enumerated in Theorem 4.1 — weak or strong deflagration or detonation — are mathematically admissible, i.e., are compatible with the conservation laws. But some of them are ruled out by physical considerations: in particular, weak detonations and strong deflagrations are not admissible, at least for the simple model that we have assumed (Williams 1985, Fickett and Davis 1979). The exclusion of weak detonations and strong deflagrations does not mean that this kind of wave does not exist at all. But these require special considerations apart from the simple model presented here and are believed to represent some particular phenomena (see Williams 1985, Section 6.1.3, 6.2.2).

Let us see how studying the characteristics can enhance these considerations. For instance, assume that we have a steady (σ constant) reaction front, moving to the right, the unburnt gas being on the right and the burnt gas on the left. Then, due to Theorem 4.1, the $C_{\pm}$ characteristics

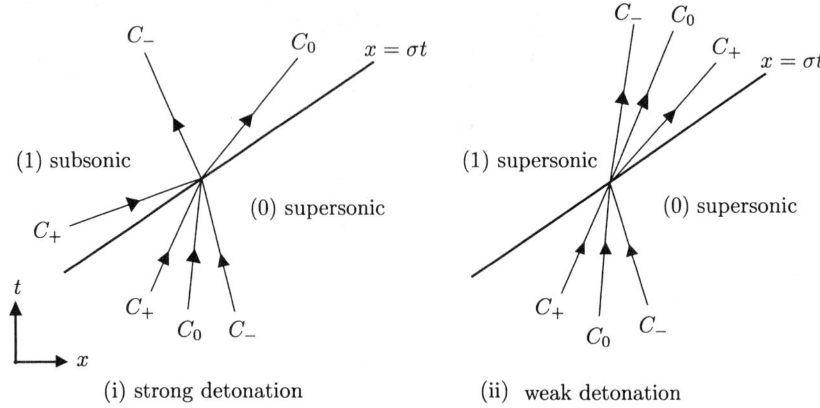

FIGURE 4.7.1. Characteristics in the case of detonations.

with slope $u \pm c$ (see Chapter I, Example 5.3) are as depicted in Figures 4.7.1 and 4.7.2.

The considerations of characteristics and boundary conditions preceding the Definition 5.1 of the Lax entropy conditions show that in case (i), which is the analog of an admissible 3-shock, the gas flow is completely determined by the Rankine–Hugoniot conditions (the inequalities (5.12) and (5.13) in Chapter I hold with $j = k-1 = 2$). In cases (ii) and (iii), there remains one degree of indeterminacy, for instance the velocity of the shock front can be chosen arbitrarily (in case (ii), $j = k = 3$, while in case (iii) $j = k = 2$). In case (iv), there are two degrees of freedom ($j = 3, k = 2$). Note in particular that the C_+ characteristics diverge from the discontinuity. □

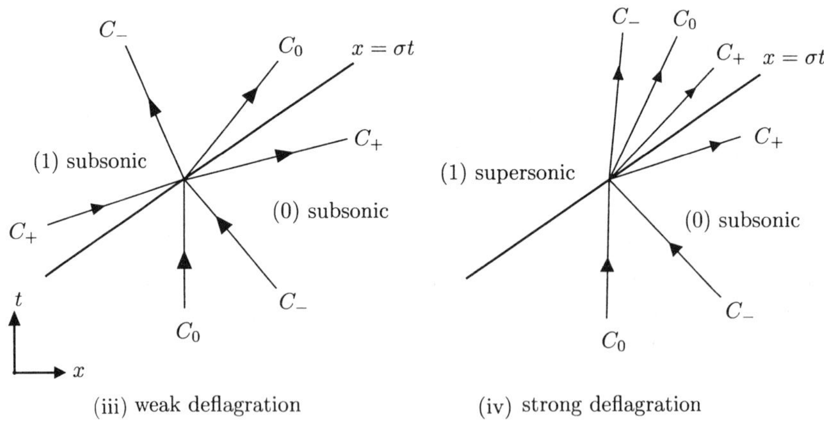

FIGURE 4.7.2. Characteristics in the case of deflagrations.

As in Section 3, it is also convenient to study the projection on the (u,p) plane of the Rankine–Hugoniot conditions. For specificity, we shall study the case where the combustion front propagates to the right: we suppose $M < 0$, and accordingly v_0 and v_1 are negative. For instance, if $u_0 = 0$ the unburnt gas is at rest and σ is > 0 (Figure 4.8).

Assume that the two parts of the Crussard curve may be parametrized by p; more precisely, assume that $\tau = \tau(p)$ with $p \geq p^0 = p_{B_0}$ for a detonation and $p \leq p_0$ for a deflagration (see Figures 4.1 and 4.3). Then, from equations (4.4) and (4.5), we get

$$u - u_0 = -\frac{(p-p_0)}{M(p)} = (p-p_0)\sqrt{(\tau-\tau_0)/(p-p_0)}.$$

In the case of a detonation, $p \geq p^0 = p_{B_0} > p_0$ and $\tau \leq \tau_0$, we have

$$u - u_0 = \sqrt{(\tau_0 - \tau(p))(p-p_0)}, \tag{4.17}$$

and u increases from u_0 to $+\infty$ with, moreover,

$$\lim_{p \to p^0_+} \frac{du}{dp} = +\infty.$$

For a deflagration, $p \leq p_0$ and $\tau > \tau_0$, we get

$$u - u_0 = \sqrt{(\tau_0 - \tau(p))(p-p_0)} \tag{4.18}$$

with

$$\lim_{p \to p_{0-}} \frac{du}{dp} = +\infty.$$

We obtain the curve depicted in Figure 4.9.

It is easily seen that there is a unique straight line passing through the point (u_0, p_0) that is tangent to the curve at both C.J. points. Indeed, differentiating

$$(u-u_0)^2 = -(p-p_0)(\tau-\tau_0), \tag{4.19}$$

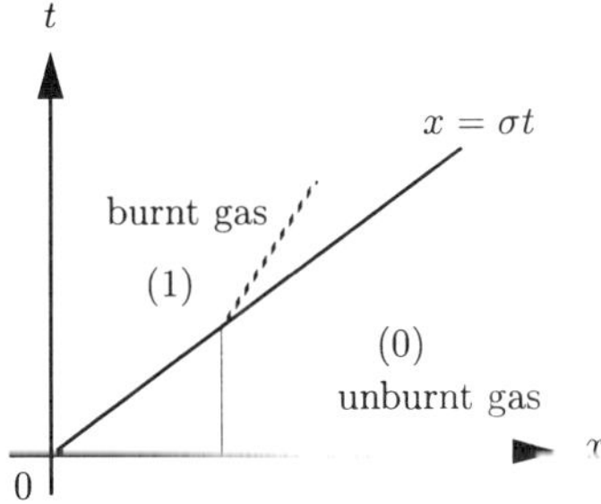

FIGURE 4.8. Combustion front.

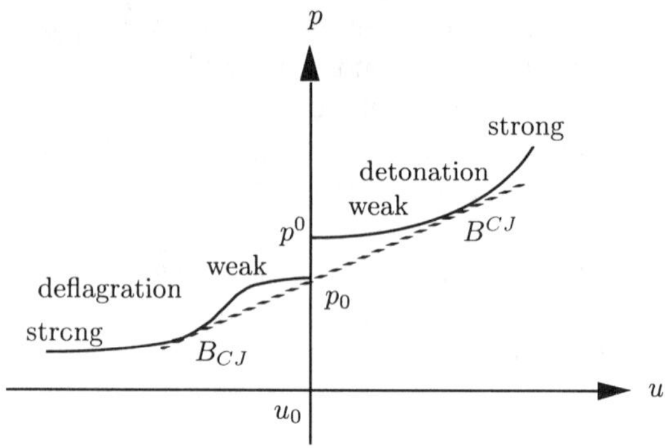

FIGURE 4.9. States that can be connected to the unburnt state (0) by a combustion wave.

we get

$$2(u-u_0)\frac{du}{dp} = -(p-p_0)\frac{d\tau}{dp} - (\tau-\tau_0).$$

Thus, a line issuing from (u_0, p_0) is tangent at the point (u, p) if

$$\frac{-(p-p_0)d\tau/dp - (\tau-\tau_0)}{2(u-u_0)} = \frac{u-u_0}{p-p_0}.$$

Together with (4.19), this yields

$$\frac{d\tau}{dp} = \frac{\tau-\tau_0}{p-p_0},$$

which characterizes the C.J. points. In particular, the value u^{CJ} at the C.J. detonation point is given by

$$u^{CJ} = u_0 + \sqrt{(\tau^{CJ}-\tau_0)(p_0-p^{CJ})},$$

where p^{CJ} and τ^{CJ} are the solutions with $p > p_0$, $\tau < \tau_0$, of

$$\text{(4.20)} \qquad \begin{cases} \varepsilon(\tau,p) - \varepsilon_0 + \dfrac{1}{2}(p_0+p)(\tau-\tau_0) = 0, \\ \dfrac{p-p_0}{\tau-\tau_0} = -\dfrac{c^2}{\tau^2}, \end{cases}$$

and the C.J. detonation speed is

$$\text{(4.21)} \qquad \sigma^{CJ} = c^{CJ} + u^{CJ}.$$

For the C.J. deflagration, we consider the solutions of (4.20) with $p < p_0$, $\tau \geq \tau_0$, with (4.18) and (4.21).

Example 4.3. Let γ denote the adiabatic exponent, defined as the negative logarithmic slope of the isentrope (for a polytropic ideal gas, this definition coincides with the definitions of Section 1.2)

$$\gamma = -\Big(\frac{\tau}{p}\Big)\frac{\partial p(\tau, s)}{\partial \tau}\Big|_s = -\Big(\frac{\partial \text{Log}\, p}{\partial \text{Log}\, \tau}\Big)\Big|_s .$$

Since the Crussard curve and the isentrope are both tangent to the Rayleigh line at point C.J., equating the slopes in the $(\text{Log}\, p, \text{Log}\, \tau)$-plane gives

$$\gamma^{CJ} = -\Big(\frac{\partial \text{Log}\, p}{\partial \text{Log}\, \tau}\Big)\Big|_{s=s^{CJ}} = \frac{1-(p_0/p^{CJ})}{(\tau_0/\tau^{CJ}-1)} .$$

If p_0 can be neglected (a case encountered for a liquid or a solid) and if the unburnt state is at rest ($u_0 = 0$), we obtain by (4.4) and (4.5)

$$\begin{aligned}
\sigma^{CJ} &= -v_0, \\
M^{CJ} &= \frac{v^{CJ}}{\tau^{CJ}} = -\frac{p^{CJ}}{u^{CJ}} = -\rho_0 \sigma^{CJ}, \\
(M^{CJ})^2 &= \frac{p^{CJ}}{\tau_0 - \tau^{CJ}} .
\end{aligned}$$

We get the simple relations

$$\begin{aligned}
\tau^{CJ} &= \frac{\tau_0 \gamma^{CJ}}{\gamma^{CJ}+1}, \quad p^{CJ} = \frac{\rho_0 (\sigma^{CJ})^2}{\gamma^{CJ}+1}, \\
\frac{v^{CJ}}{v_0} &= \frac{\gamma^{CJ}}{\gamma^{CJ}+1}, \quad u^{CJ} = \frac{\sigma^{CJ}}{\gamma^{CJ}+1}, \\
c^{CJ} &= \sigma^{CJ} - u^{CJ} = \gamma^{CJ} u^{CJ}.
\end{aligned}$$

For a constant γ - law $\varepsilon(\tau, p) = \frac{p\tau}{\gamma-1} - q$, a further computation gives

$$(\sigma^{CJ})^2 = 2q(\gamma^2 - 1)$$

for the C.J. detonation state. □

Let us study more precisely the flows involving a *detonation* process. On the basis of physical considerations (a closer study of the wave structure, to which we shall return in the next section), we shall assume that only C.J. detonations and strong detonations are admissible. Instead of weak detonations, we consider composite waves involving a Chapman–Jouguet process and a nonreacting rarefaction wave. For instance, in the (u,p)-plane, the state (u_0, p_0) can be connected to a state (u,p) with $u < u^{CJ}$ by a C.J. detonation followed by a 3-rarefaction wave (see Figure 4.10).

This is indeed possible since the fan of the rarefaction wave is bordered on the right by the line with slope $\lambda = u + c$ and we know by (4.21) that, at the C.J.detonation point,

$$\lambda = \sigma^{CJ} = c^{CJ} + u^{CJ}.$$

Thus, in Figure 4.11, the detonation part of the curve in Figure 4.9 is

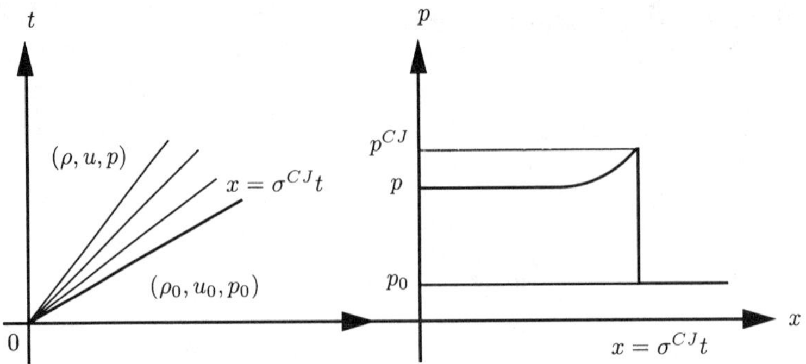

FIGURE 4.10. Composite wave: C.J. detonation and 3-rarefaction wave.

replaced for $u < u^{CJ}$ by

$$u = u^{CJ} + \Psi^{CJ}(p), \quad p < p^{CJ},$$

which is the equation of the 3-rarefaction curve passing through the C.J. detonation (see (3.10)). Note that the rarefaction curve trough B^{CJ} is an isentrope (see (3.8)) that is tangent to the detonation curve, as we have already noticed in the proof of Lemma 4.3.

The locus of burnt states that can be connected to the state (0) by a strong detonation or a C.J. detonation followed by a 3-rarefaction is then

$$u = \begin{cases} u_0 + \sqrt{(\tau(p) - \tau_0)(p_0 - p)}, & p \geq p^{CJ}, \\ u_0 + \sqrt{(\tau^{CJ} - \tau_0)(p_0 - p^{CJ})} + \Psi^{CJ}(p), & p < p^{CJ}. \end{cases} \tag{4.22}$$

The function in (4.22) is easily seen to be increasing.

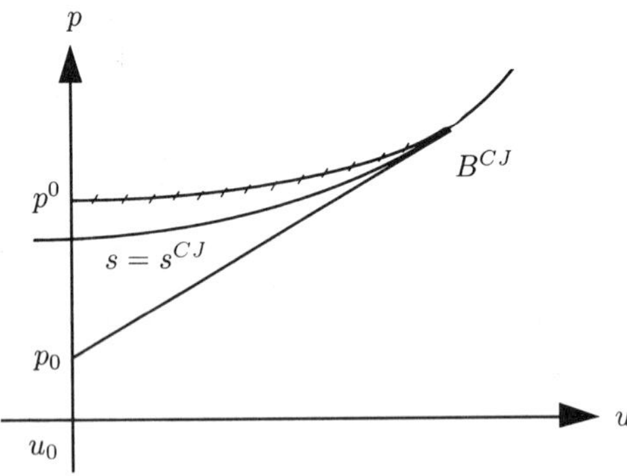

FIGURE 4.11. Isentrope through the C.J. detonation state.

This enables us to solve the Riemann problem for a reacting gas in the particular case of a flow involving a detonation. We assume that the right state is the unburnt state (0) and the left state is the burnt state (1):

$$(\rho, u, p)(x, 0) = \begin{cases} (\rho_L, u_L, p_L) = (\rho_1, u_1, p_1), & x < 0, \\ (\rho_R, u_R, p_R) = (\rho_0, u_0, p_0), & x > 0. \end{cases}$$

By analogy with the general theory of Chapter I, we look for a solution where the left state is connected to the right state by a 1-wave (shock or rarefaction propagating in the burnt gas), a 2-contact discontinuity and a strong detonation, or a compound wave involving a C.J. detonation and a 3-rarefaction wave (see Figure 4.12).

Let p^{CJ} be the pressure at the C.J. detonation point, which is entirely determined by the right state and the equation of state of the burnt gas. As in Section 3, we look at the intersection in the (u, p) plane of the curve (3.16) (the locus of states to which the left state can be connected by a 1-wave) with the curve $\mathcal{S}^d$ (4.22), which we have described above. This intersection (u^*, p^*) gives the velocity and the pressure at the two intermediate constant states. It remains to determine ρ_I and ρ_{II}, which is done following the lines of Section 3.

There are four cases according to the relative positions of $p_L = p_1$, $p_R = p_0$, and p^{CJ}. For instance, if we assume that $p_L > p^{CJ} > p_R$, a look at Figure 4.12 shows that the corresponding solution of the Riemann problem is as depicted in Figure 4.13.

Remark 4.3. We shall not investigate in the same way the case of a flow involving a deflagration. A first reason is, as noticed in Remark 4.2, that in this simple model deflagration processes have a higher degree of indeterminacy because the gas flow relative to the front is subsonic ahead of a deflagration front. A possible remedy would be to fix, for instance, the

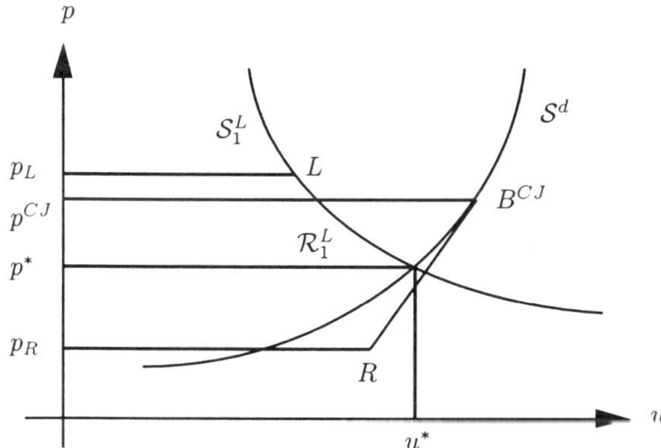

FIGURE 4.12. Solution of the Riemann problem in the (u, p)-plane.

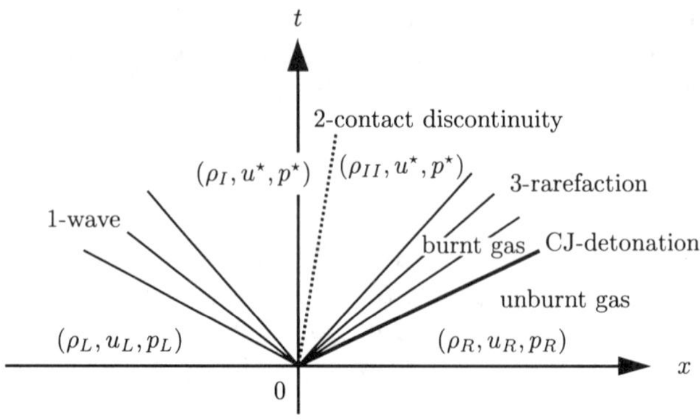

FIGURE 4.13. Solution of the Riemann problem in the (x, t)-plane.

reaction rate and then search the states that can be connected to a right state by a weak deflagration preceded by a nonreacting 3-wave (shock or rarefaction). Strong deflagrations are excluded on the same basis as weak detonation waves (see Courant and Friedrichs 1976, Section 93, Williams 1985, Section 6.1.2.3). The intersection with the curve (3.16) will then determine a possible wave pattern (see Teng, Chorin, and Liu 1982). □

Remark 4.4. In fact, weak deflagrations propagate with a definite wave speed that depends on the reaction rate, and they are nearly isobaric (see Williams 1985, Sections 6.1.2.3, 6.1.3). Note also that the exclusion of weak detonations and strong deflagrations does not mean that this kind of wave does not exist at all. But they require special considerations apart from the simple model presented here and are believed to represent some particular phenomena (see Williams 1985, Sections 6.1.3, 6.2.2). □

5 Reacting flows. The Z.N.D. model for detonations

The preceding theory is now extended to include a finite reaction rate. Thus, we do not assume anymore that the reaction takes place instantaneously. Instead, we make the assumption that a detonation process can be modeled as a nonreacting shock wave propagating in the unburnt gas initiating a chemical reaction. Hence, we suppose that the reaction is irreversible and the reaction rate is zero ahead of the shock and finite behind (the reaction is complete at the end of the reaction zone). Again, the flow is supposed to be planar and one-dimensional, and transport effects (heat conduction, viscosity) are neglected, and we consider the simplest possible chemical process $R \to P$. Recall that z is the mass fraction of the burnt gas. Then, the equations to be considered are the system (4.1) with

$$\varepsilon = \varepsilon(\rho, p, z), \quad r = r(\rho, p, z), \tag{5.1}$$

or equivalently

$$\varepsilon = \varepsilon(\tau, p, z), \quad r = r(\tau, p, z)$$

when it is more convenient. We rewrite the system in the form

$$\begin{cases} \dfrac{\partial \mathbf{\Phi}}{\partial t} + \dfrac{\partial}{\partial x}\,\mathbf{f}(\mathbf{\Phi}, z) = 0, \\ \dfrac{\partial}{\partial t}(\rho\, z) + \dfrac{\partial}{\partial x}(\rho\, z\, u) = \rho r, \end{cases} \tag{5.2}$$

where

$$\mathbf{\Phi} = \begin{pmatrix} \rho \\ \rho u \\ \rho e \end{pmatrix}, \qquad \mathbf{f}(\mathbf{\Phi}, z) = \begin{pmatrix} \rho u \\ \rho u^2 + p \\ (\rho e + p)u \end{pmatrix}.$$

We will show that there exists a family of traveling wave solutions of (5.2), i.e., solutions of the form

$$(\mathbf{\Phi}, z)(x, t) = (\hat{\mathbf{\Phi}}, \hat{z})(\xi), \quad \xi = x - \sigma t,$$

depending on the parameter σ, which is the constant velocity of the traveling wave, connecting the unburnt state $z = 0$ to the burnt state $z = 1$, and therefore satisfying

$$\begin{cases} (\hat{\mathbf{\Phi}}, \hat{z})(\xi) = (\mathbf{\Phi}_0, 0) \quad \text{for } \xi > 0, \\ (\hat{\mathbf{\Phi}}, \hat{z})(0) = (\mathbf{\Phi}_N, 0) \quad \text{at } \xi = 0, \\ \lim\limits_{\xi \to -\infty} (\hat{\mathbf{\Phi}}, \hat{z})(\xi) = (\mathbf{\Phi}_1, 1) \text{ (total reaction)}, \end{cases} \tag{5.3}$$

where $\mathbf{\Phi}_0$ is a given constant initial state. Thus, the problem is to determine the function $(\hat{\mathbf{\Phi}}, \hat{z})(\xi)$, $\xi < 0$, for a given parameter σ.

We begin by writing the usual Rankine–Hugoniot jump condition for the ordinary shock discontinuity in the unburnt gas (dropping the "ˆ")

$$\sigma[(\mathbf{\Phi}, z)] = [f(\mathbf{\Phi}, z)] \quad \text{at } \xi = 0.$$

The shock is supposed to be inert, $z = 0$ across the shock, so that

$$\sigma(\mathbf{\Phi}_N - \mathbf{\Phi}_0) = \mathbf{f}(\mathbf{\Phi}_N, 0) - \mathbf{f}(\mathbf{\Phi}_0, 0). \tag{5.4}$$

Now, assuming that the solution is smooth for $\xi < 0$, we have

$$\begin{cases} -\,\sigma\,\dfrac{d\mathbf{\Phi}}{\partial \xi} + \dfrac{d}{d\xi}\,\mathbf{f}(\mathbf{\Phi}, z) = 0, \\ -\,\sigma\,\dfrac{d}{d\xi}(\rho z) + \dfrac{d}{d\xi}(\rho z u) = \rho r. \end{cases} \tag{5.5}$$

Integrating the first equation (5.5) gives

$$\sigma(\mathbf{\Phi}(\xi) - \mathbf{\Phi}_N) = \mathbf{f}(\mathbf{\Phi}(\xi), z(\xi)) - \mathbf{f}(\mathbf{\Phi}_N, 0),$$

which together with (5.4) yields

$$\sigma(\mathbf{\Phi}(\xi) - \mathbf{\Phi}_0) = \mathbf{f}(\mathbf{\Phi}(\xi), z(\xi)) - \mathbf{f}(\mathbf{\Phi}_0, 0),$$

or equivalently

$$\begin{cases} \sigma(\rho(\xi) - \rho_0) = (\rho u)(\xi) - (\rho u)_0, \\ \sigma((\rho u)(\xi) - (\rho u)_0) = (\rho u^2 + p)(\xi) - (\rho u^2 + p)_0, \\ \sigma((\rho e)(\xi) - (\rho e)_0) = ((\rho e + p)u(\xi) - ((\rho e + p)u)_0. \end{cases}$$

We find the analog of the Rankine–Hugoniot relations. Setting

$$M = \rho_0(u_0 - \sigma) = \rho(\xi)(u(\xi) - \sigma),$$

we get as in the nonreacting case (2.10)-(2.12) or as in the Chapman–Jouguet case (4.4)-(4.6)

$$M = \frac{u(\xi) - u_0}{\tau(\xi) - \tau_0} = -\frac{p(\xi) - p_0}{u(\xi) - u_0},$$

$$M^2 = -\frac{p(\xi) - p_0}{\tau(\xi) - \tau_0}, \tag{5.6}$$

$$\varepsilon(\tau(\xi), p(\xi), z(\xi)) - \varepsilon(\tau_0, p_0, 0) + \frac{1}{2}(p(\xi) + p_0)(\tau(\xi) - \tau_0) = 0. \tag{5.7}$$

Let us next consider the second equation (5.5), which can be written in the form

$$z\Big\{\Big(-\sigma\frac{d\rho}{d\xi} + \frac{d}{d\xi}(\rho u)\Big\} + \frac{dz}{d\xi}\{\rho(-\sigma + u)\} = \rho r.$$

Since

$$\sigma\frac{d\rho}{d\xi} = \frac{d}{d\xi}(\rho u),$$

we obtain

$$\frac{dz}{d\xi} = \frac{1}{M} r(\tau(\xi),\ p(\xi), z). \tag{5.8}$$

Therefore, the problem amounts to finding the triple $(\tau(\xi), p(\xi), z(\xi))$ for $\xi < 0$ that is the solution of equations (5.6)–(5.8) with the initial condition $z(0) = 0$. We solve the equations (5.6)–(5.8) in two steps. First, considering z as a parameter, we solve (5.6) and (5.7), which gives τ and p as functions of z. Then, replacing τ and p in (5.8) by their values, we solve the ordinary differential problem

$$\begin{cases} \dfrac{dz(\xi)}{d\xi} = \dfrac{1}{M} r(\tau(z), p(z), z)(\xi), \\ z(0) = 0. \end{cases} \tag{5.9}$$

Indeed, we consider the equation of a Rayleigh line

$$M^2 = -\frac{p - p_0}{\tau - \tau_0}, \tag{5.10}$$

and we introduce for $0 \leq z \leq 1$ the family of Hugoniot curves $\mathcal{H}_z$ depending on z:

$$\varepsilon(\tau, p, z) - \varepsilon(\tau_0, p_0, 0) + \frac{1}{2}(p + p_0)(\tau - \tau_0) = 0.$$

For $z = 0$, we find the shock Hugoniot curve $\mathcal{H}_0$ with center $A_0 = (\tau_0, p_0)$, whereas for $z = 1$, $\mathcal{H}_1$ corresponds to the Crussard curve $\mathcal{C}$ with center A_0 (Figure 5.1).

Solving (5.6) and (5.7) amounts to finding, the intersection of $\mathcal{H}_z$ with the Rayleigh line Δ. Due to (5.6), for a given value of σ and thus of M, all the partial reaction states lie on Δ. By Lemma 4.2, the line intersects the Hugoniot curve $\mathcal{H}_z$ at 0 or 2 (possibly coalescing) points. At the point C_z where Δ is tangent to $\mathcal{H}_z$, the Rayleigh line is also tangent to an isentrope (see Lemma 4.3), and the flow is sonic (Lemma 4.4). The locus $\mathcal{S}$ of all such points C_z is called the *sonic locus.* The curve starts from the center A_0 on $\mathcal{H}_0$ and ends on the Crussard curve at the C.J. detonation point.

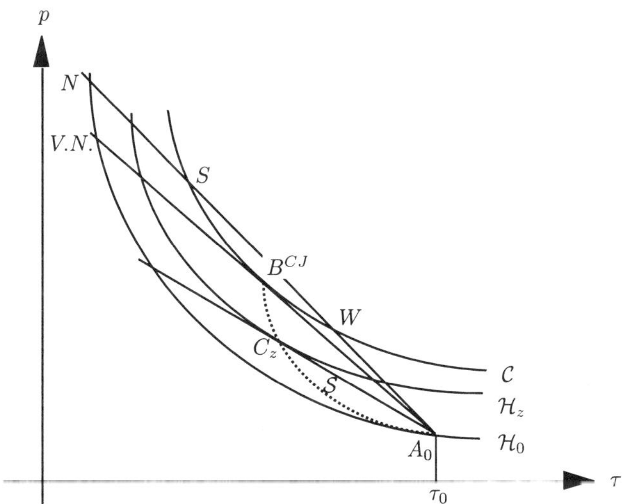

FIGURE 5.1. Hugoniot curves $\mathcal{H}_z$.

Moreover, the sonic locus $\mathcal{S}$ separates the region lying between $\mathcal{H}_0$ and $\mathcal{C}$ in the (τ, p)-plane into two zones: one (on the left of $\mathcal{S}$) where the flow is subsonic by Theorem 4.1, the other where it is supersonic (on the right) (Figure 5.1).

Provided an explicit equation of state $\varepsilon = \varepsilon(\tau, p, z)$ is known, we can obtain the function $p = p_\sigma(z)$, depending on σ, by eliminating τ from the equations (5.9) and (5.10). Each function p_σ has two branches: one subsonic, which for $z = 1$ corresponds to a strong detonation, the other supersonic, corresponding to a weak detonation. We plot in Figure 5.2 three typical curves depending on the position of σ with respect to σ^{CJ} (σ^{CJ} is indeed the value of σ for which Δ is tangent to the Crussard curve).

Now, there exist two possibilities:
(i) $\sigma \geq \sigma^{CJ}$: The Rayleigh line Δ intersects the Crussard curve $\mathcal{C}$ at two points S and W (coalescing at B^{CJ} if $\sigma = \sigma^{CJ}$). The only admissible final state is S, as follows from thermodynamical considerations (see Remark 5.1).
(ii) $\sigma < \sigma^{CJ}$: The Rayleigh line Δ does not intersect the Crussard curve $\mathcal{C}$. We see in Figure 5.2 that there may be a solution ending on the sonic locus $\mathcal{S}$ that corresponds to a maximum value of z. However, at such a point $r > 0$ (since $z < 1$) and the reaction is irreversible so that (5.8) cannot be satisfied.

For $\sigma \geq \sigma^{CJ}$ we have obtained the existence of a combustion profile of the form (5.3), traveling with speed σ. It consists of a nonreacting shock wave (represented by A_0N in the (τ, p)-plane, see Figure 5.1), followed by a chemical reaction (between N and S). The reaction is achieved in finite time, and the length of the reaction zone is finite if there exists a finite $\xi_0 > -\infty$ such that $z(\xi_0) = 1$. In this model, we have assumed that the

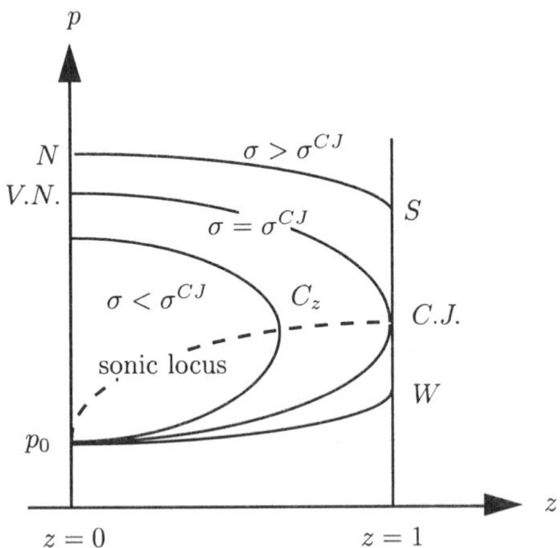

FIGURE 5.2. The curves $p = p_\sigma(z)$.

shock initiates the chemical reaction. Thus, through the shock, pressure and density are raised instantaneously from A_0 to a point N on the shock Hugoniot curve $\mathcal{H}_0$. Then, the reaction proceeds and the state is described in the (τ, p)-plane by a point on the Rayleigh line which is the intersection with the curve $\mathcal{H}_z$. If it were W (see Figure 5.1), the gas would have reached before the state S where it is in chemical ($z = 1$) and thermodynamical equilibrium ($S \in \mathcal{C}$). Since by (4.14) the entropy in W is less than in S, we cannot find an admissible transformation that leads from S to W on the Rayleigh line.

Remark 5.1. If the length of the reaction zone is finite (i.e., if there exists a finite $\xi_0 > -\infty$ such that $z(\xi_0) = 1$), a rarefaction wave may follow the reaction zone to join a final state with given pressure less than p^{CJ} as is indeed the case in the experiments. However, this supposes that the speed σ is equal to σ^{CJ} (so that the flow is sonic). This, because the speed of the head of the rarefaction wave is equal to the characteristic speed $u + c$. However at point S, the flow is subsonic and $u + c > \sigma$. In that case, the rarefaction and detonation waves will interact until the speed of the head of the rarefaction wave equals that of the detonation wave, which supposes

$$\sigma = \sigma^{CJ}.$$

In short, this justifies the famous *Chapman–Jouguet hypothesis* that only C.J. detonations are possible. The point *V.N.* on the Hugoniot curve from which the chemical reaction starts is called the *von Neumann* state. The final state is the C.J. point. The following rarefaction wave is called the *Taylor wave.* We have depicted in Figure 5.3 the corresponding pressure curve. We note that the above model excludes the possibility of weak detonations. □

These last two sections have given but an introduction to the important subject of reacting flows. The following references will convince the reader that the subject is far too wide for us to pretend to have been exhaustive.

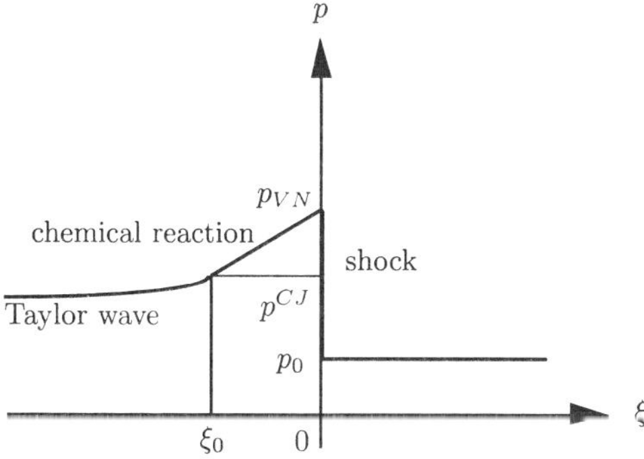

FIGURE 5.3. C.J. detonation and Taylor wave.

We have tried to develop just the simplest approaches in the spirit of the preceding sections.

Notes

For a better understanding of gas dynamics, we recommend the basic book of Courant and Friedrichs (1976) which contains a section on reacting flows, and those of Anderson (1982), Whitham (1974), and Lighthill (1978). We mention again the references already quoted in the previous chapters (Jeffrey 1976, Taniuti and Nishihara 1983, Chiang and Hsiao 1989), in particular Smoller (1994), Chapter 18, and the references therein, and Serre (1996), Chapter 4; also the important paper of Smith (1979), a very complete study of general equations of state in Menikoff and Plohr (1989), and the paper of Wagner (1987), where the equivalence of the Euler and Lagrangian equations for weak solutions is proved (on this subject, see also Dafermos 1992 and Wagner 1994). A more precise study of the rarefaction shock curves (the nonadmissible part of the shock curves) can be found in Smoller et al. (1990).

Concerning the convexity of the entropy (Section 1), see the work of Dubois (1990), and Croisille and Delorme (1991), and Croisille and Villedieu (1994) for a kinetic approach.

The solution of the Riemann problem in the case of materials with nonconvex equations of state can be found in Wendroff (1972), and Hattori (1986) for a van der Waals fluid (see Kawashima and Matsumura 1994 for the stability of shock profiles). For multi-component gas dynamics, see for instance Abgrall (1988) or Larrouturou (1990); the influence of the source term is shown in Fey et al. (1992).

Concerning combustion, we refer to the books of Courant and Friedrichs already mentioned, and those of Chorin and Marsden (1993), Fickett and Davis (1979), Williams (1985), Oran and Boris (1987), Chéret (1988), to the nice unpublished report of Thouvenin, and to the papers of Chorin (1977), Majda (1981), Teng, Chorin, and Liu (1982), Chen and Wagner (1993), Embib, Hunter, and Majda (1992), Colella, Majda, and Roytburd (1986), Bukiet (1988), Ben Artzi (1989), Tan and Zhang (1992), Gasser and Szmolyan (1993), Sheng and Tan (1994), and Liu and Ying (1995).

III

Finite difference schemes for one-dimensional systems

1 Generalities on finite difference methods for systems

Let us consider again the Cauchy problem for a general system of conservation laws

$$(1.1)\qquad \begin{cases} \dfrac{\partial \mathbf{u}}{\partial t} + \dfrac{\partial}{\partial x}\mathbf{f}(\mathbf{u}) = \mathbf{0}, & x \in \mathbb{R}, t > 0, \\ \mathbf{u}(x,0) = \mathbf{u}_0(x), \end{cases}$$

where $\mathbf{u} = (u_1, ..., u_p)^T$ is a p-vector. As usual, we assume that this system is hyperbolic, i.e., the Jacobian matrix $\mathbf{A}(\mathbf{u}) = \mathbf{f}'(\mathbf{u})$ of $\mathbf{f}(\mathbf{u})$ has p real eigenvalues ranked in increasing order,

$$a_1(\mathbf{u}) \le a_2(\mathbf{u}) \le ... \le a_p(\mathbf{u}),$$

and a complete set of eigenvectors. Moreover, we shall assume that for all $1 \le k \le p$, the kth characteristic field is either genuinely nonlinear or linearly degenerate. For the notations concerning the difference schemes, we refer to Chapter III of G.R..

Given a uniform grid with time step Δt and spatial mesh size Δx, we define an approximation $\mathbf{v}_j^n \in \mathbb{R}^p$ of $\mathbf{u}(x_j, t_n)$ at the point $(x_j = j\Delta x, t_n = n\Delta t)$ by the formula

$$(1.2a)\qquad \mathbf{v}_j^{n+1} = \mathbf{v}_j^n - \lambda\,(\mathbf{g}_{j+\frac{1}{2}}^n - \mathbf{g}_{j-\frac{1}{2}}^n),\quad j \in \mathbb{Z},\ n \ge 0$$

where $(\mathbf{v}_j^0)_{j\in\mathbb{Z}}$ is given,

$$(1.2b)\qquad \lambda = \frac{\Delta t}{\Delta x},$$

and we have

$$(1.2c)\qquad \mathbf{g}_{j+\frac{1}{2}}^n = \mathbf{g}(\mathbf{v}_{j-k+1}^n, ..., \mathbf{v}_{j+k}^n),$$

where the function $\mathbf{g} : \mathbb{R}^{p\times 2k} \to \mathbb{R}^p$ is continuous and is called the *numerical flux*. The scheme is said to be *consistent* with (1.1) if $\mathbf{g}$ satisfies

$$\mathbf{g}(\mathbf{u}, ..., \mathbf{u}) = \mathbf{f}(\mathbf{u}), \ \forall \mathbf{u} \in \mathbb{R}^p.$$

This is a $(2k+1)$-point scheme, and when $k = 1$ we have a 3-point scheme. The scheme is *essentially 3-point* if it satisfies the stronger consistency relation:

$$\mathbf{g}(\mathbf{v}_{-k+1}, .., \mathbf{v}_{-1}, \mathbf{u}, \mathbf{u}, \mathbf{v}_1, .., \mathbf{v}_k) = \mathbf{f}(\mathbf{u}), \ \forall\ \mathbf{v}_{-k+1}, ..., \mathbf{v}_k \in \mathbb{R}^p, \ \forall \mathbf{u} \in \mathbb{R}^p.$$

For the reader's convenience, we recall some basic definitions that were introduced in the scalar case in G.R., Chapter III. Scheme (1.2) can also be written in the general form

$$\mathbf{v}_j^{n+1} = \mathbf{H}(\mathbf{v}_{j-k}^n, ..., \mathbf{v}_{j+k}^n),$$

where $\mathbf{H} : \mathbb{R}^{p\times(k+1)} \to \mathbb{R}^p$ is the discrete solution operator. The converse is not true, and when a difference scheme can be written in the form (1.2a) with a numerical flux $\mathbf{g}$, it is called *conservative* , and (1.2a) is the *conservation form.* We recall that the Lax–Wendroff theorem asserts that when a conservative consistent scheme "converges" to a function $\mathbf{u}$ (in some sensible way that we do not discuss here; see Theorem 1.1 in G.R., Chapter III, Section 1.1, and also this volume, Chapter IV, Section 4.2.2), the limit $\mathbf{u}$ is a weak solution of (1.1).

We shall not recall here the classical notions of truncation error and order of accuracy (see G.R. or LeVeque 1990).

When studying the stability of the scheme, we can use the (discrete) $\mathbf{L}^p$-norms for the sequences $\mathbf{v}^n = (\mathbf{v}_j^n)$. In the scalar case, we have also introduced more specific tools: monotonicity and the "total variation diminishing" property, which are important but purely *scalar* notions. We recall the definitions just for the sake of completeness. Scheme (1.2) is *monotone* if given two sequences $v^0 = (v_j^0)$ and $w^0 = (w_j^0)$,

$$v^0 \geq w^0 \Rightarrow v^1 \geq w^1,$$

where $v \geq w$ means that $\forall j, \ v_j \geq w_j$, and $v^1 = (v_j^1)$ is the sequence obtained after one step, $v_j^1 = H(\mathbf{v}_{j-k}^0, ..., \mathbf{v}_{j+k}^0)$. Scheme (1.2) is *total variation diminishing* (T.V.D.) if

$$\forall\ v^0 = (v_j^0), \quad TV(v^1) \leq TV(v^0),$$

where $TV(v) = \sum_{j\in\mathbb{Z}} (v_{j+1} - v_j)$.

T.V.D. schemes are attractive for various reasons. A T.V.D. scheme transforms a monotone sequence, say a nondecreasing one, into a monotone (nondecreasing) sequence, and oscillations cannot occur. But while a monotone scheme is at most first-order accurate (Harten, Hyman, and Lax 1976), it is possible to derive "high order" T.V.D. schemes (at least away from sonic extrema of the solution). Moreover, there exist simple sufficient conditions that ensure that a scalar scheme is T.V.D. when it can be written in incremental or viscous form (which is always the case for all 3-point or essentially 3-point schemes). We say that scheme (1.2a) can be put in *incremental* (resp. *viscous*) form if there exist coefficients C, D (resp. Q) that are functions of $2k$ variables ($C, D, Q : \mathbb{R}^{2k} \to \mathbb{R}$) such that setting

$$\Delta v_{j+1/2} = v_{j+1} - v_j,\ C_{j+\frac{1}{2}} = C(v_{j-k+1}, ..., v_{j+k})$$

and so on, we have

$$v_j^{n+1} = v_j^n + C_{j+1/2}^n \Delta v_{j+1/2}^n - D_{j-1/2}^n \Delta v_{j-1/2}^n, \tag{1.2d}$$

respectively
(1.2*e*)

$$v_j^{n+1} = v_j^n - \frac{\lambda}{2}(f_{j+1}^n - f_{j-1}^n) + \frac{1}{2}(Q_{j+1/2}^n \Delta v_{j+1/2}^n - Q_{j-1/2}^n \Delta v_{j-1/2}^n).$$

Harten's criteria are given by the following proposition.

Proposition 1.1

Assume that the scheme (1.2) can be put in incremental (resp. viscous) form and that the incremental (resp. viscosity) coefficients satisfy for any $j \in \mathbb{Z}$ *and* $n \geq 0$

$$C_{j+1/2}^n \geq 0, \quad D_{j+1/2}^n \geq 0, \quad C_{j+1/2,k}^n + D_{j+1/2,k}^n \leq 1,$$

(resp.

$$\lambda \mid \Delta f_{j+1/2}^n / \Delta v_{j+1/2}^n \mid \leq Q_{j+1/2}^n \leq 1).$$

Then the scheme is T.V.D. □

These conditions happen to be necessary for a 3-point scheme.

Remark 1.1. These notions were introduced to mimic the properties of the continuous solution operator, which is order preserving and T.V.D. (see G.R., Chapter II, Theorem 5.2). These last properties do not hold for systems, which explains why there is no simple extension of the above notions to numerical schemes for systems. However, since these properties are easily characterized and moreover lead in general to satisfactory results, they are still considered as useful criteria for selecting relevant schemes. □

Remark 1.2. We have restricted the presentation of the difference schemes to the case of a uniform space grid (Δx constant) for simplicity only. The case of a nonuniform grid will be handled in Chapter IV, Section 4.1.2 when we deal with finite volume methods. Also, (1.2) corresponds to the use of the explicit Euler scheme for the time discretization of the "method of lines"

$$\frac{d}{dt}\mathbf{u}_j(t) + \frac{1}{\Delta x}(\mathbf{g}_{j+\frac{1}{2}}(t) - \mathbf{g}_{j-\frac{1}{2}}(t)) = \mathbf{0},$$

where $\mathbf{g}_{j+\frac{1}{2}}(t) = \mathbf{g}(\mathbf{u}_{j-k+1}(t), ...\mathbf{u}_{j+k}(t))$, and $\mathbf{u}_j(t)$ approximates $\mathbf{u}(x_j, t)$ or $\frac{1}{\Delta x}\int_{x_{j-1/2}}^{x_{j+1/2}} \mathbf{u}(x,t)dx$. One can use instead Runge–Kutta methods (see e.g. Section 7.3.4) or even implicit time schemes, but we shall not develop the corresponding theory. □

We present now the most usual ways of extending to a nonlinear hyperbolic system a finite difference scheme derived in the scalar case. Then, we shall study the L^2-stability of general linear schemes of approximation of a linear hyperbolic system. The following sections will be devoted to a deeper study of the most usual difference schemes, such as Godunov's, Roe's, Osher's, and van Leer's, followed by an introduction to the more recent kinetic schemes.

1.1 Extension of scalar schemes to systems. Some examples

Consider the usual 3-point schemes derived in the scalar case. Some of them, the Lax–Friedrichs and the Lax–Wendroff schemes, can be generalized immediately to systems.

Example 1.1. The Lax–Friedrichs scheme. This scheme is given by

$$\mathbf{v}_j^{n+1} = \frac{1}{2}(\mathbf{v}_{j+1}^n + \mathbf{v}_{j-1}^n) - \frac{\lambda}{2}(\mathbf{f}(\mathbf{v}_{j+1}^n) - \mathbf{f}(\mathbf{v}_{j-1}^n)). \tag{1.3}$$

It is associated to the numerical flux

$$\mathbf{g}^{L.F.}(\mathbf{u}, \mathbf{v}) = \frac{1}{2}(\mathbf{f}(\mathbf{u}) + \mathbf{f}(\mathbf{v})) - \frac{1}{2\lambda}(\mathbf{v} - \mathbf{u}).$$

It is first-order accurate and in the scalar case, under the so-called *C.F.L. stability condition*

$$\lambda \max_u |f'(u)| \leq 1, \tag{1.4}$$

it is monotone (see G.R., Chapter III, Examples 2.1 and 3.1). □

Example 1.2. The Lax–Wendroff scheme. This scheme can be written

$$\begin{cases} \mathbf{v}_j^{n+1} = \mathbf{v}_j^n - \dfrac{\lambda}{2}(\mathbf{f}(\mathbf{v}_{j+1}^n) - \mathbf{f}(\mathbf{v}_{j-1}^n)) \\ \qquad + \dfrac{\lambda^2}{2}\{\mathbf{A}_{j+\frac{1}{2}}^n(\mathbf{f}(\mathbf{v}_{j+1}^n) - \mathbf{f}(\mathbf{v}_j^n)) - \mathbf{A}_{j-\frac{1}{2}}^n(\mathbf{f}(\mathbf{v}_j^n) - \mathbf{f}(\mathbf{v}_{j-1}^n))\}. \end{cases} \tag{1.5}$$

Here $\mathbf{A}_{j+\frac{1}{2}}^n$ is either the Jacobian matrix of $\mathbf{f}$ evaluated at some average state, for instance

$$\mathbf{A}_{j+\frac{1}{2}}^n = \mathbf{A}(\frac{\mathbf{v}_{j+1}^n + \mathbf{v}_j^n}{2}),$$

or

$$\mathbf{A}_{j+\frac{1}{2}}^n = \mathbf{A}(\mathbf{v}_j^n, \mathbf{v}_{j+1}^n),$$

where $\mathbf{A} = \mathbf{A}(\mathbf{u}, \mathbf{v})$ is a $p \times p$ matrix satisfying

$$\begin{cases} \mathbf{f}(\mathbf{v}) - \mathbf{f}(\mathbf{u}) = \mathbf{A}(\mathbf{u}, \mathbf{v})(\mathbf{v} - \mathbf{u}), \\ \mathbf{A}(\mathbf{u}, \mathbf{u}) = \mathbf{f}(\mathbf{u}). \end{cases} \tag{1.6}$$

Such a matrix $\mathbf{A}(\mathbf{u}, \mathbf{v})$ is called a *Roe's matrix.* The construction and properties of Roe's matrices will be discussed in Section 3.

The Lax–Wendroff scheme is second-order accurate and is associated to the numerical flux

$$\mathbf{g}^{L.W.}(\mathbf{u}, \mathbf{v}) = \frac{1}{2}(\mathbf{f}(\mathbf{u}) + \mathbf{f}(\mathbf{v})) - \frac{\lambda}{2} \begin{cases} \mathbf{A}\Big(\dfrac{\mathbf{u}+\mathbf{v}}{2}\Big)(\mathbf{f}(\mathbf{v}) - \mathbf{f}(\mathbf{u})) \\ \mathbf{A}^2(\mathbf{u}, \mathbf{v})(\mathbf{v} - \mathbf{u}), \end{cases}$$

according to the choice of $\mathbf{A}^n_{j+1/2}$. In the scalar case, it is not T.V.D. since its viscosity coefficient $\lambda^2 A^2_{j+1/2}$ does not satisfy Harten's criteria.

Let us mention a variant of the above scheme, known as the two-step Richtmyer's version, that avoids the computation of the matrices $\mathbf{A}(\mathbf{u}, \mathbf{v})$ and the products $\mathbf{Af}$. This scheme reads

$$\begin{cases} \mathbf{v}^{n+\frac{1}{2}}_{j+\frac{1}{2}} = \dfrac{1}{2}(\mathbf{v}^n_{j+1} + \mathbf{v}^n_j) - \dfrac{\lambda}{2}(\mathbf{f}(\mathbf{v}^n_{j+1}) - \mathbf{f}(\mathbf{v}^n_j)), \\ \mathbf{v}^{n+1}_j = \mathbf{v}^n_j - \lambda(\mathbf{f}(\mathbf{v}^{n+\frac{1}{2}}_{j+\frac{1}{2}}) - \mathbf{f}(\mathbf{v}^{n+\frac{1}{2}}_{j-\frac{1}{2}})). \end{cases} \tag{1.7}$$

One can check easily that (1.7) is a second-order scheme and that

$$\mathbf{g}(\mathbf{u}, \mathbf{v}) = \mathbf{f}\Big(\frac{\mathbf{u}+\mathbf{v}}{2} - \frac{\lambda}{2}(\mathbf{f}(\mathbf{v}) - \mathbf{f}(\mathbf{u}))\Big)$$

corresponds to the numerical flux. □

Upwind schemes for systems are not so directly written. Another way of constructing finite difference schemes for systems consists in using a scalar scheme for the p characteristic quantities. Denote by $\mathbf{r}_1(\mathbf{u}), ..., \mathbf{r}_p(\mathbf{u})$ (respectively $\mathbf{l}_1(\mathbf{u}), ..., \mathbf{l}_p(\mathbf{u})$) a complete system of (right) eigenvectors of $\mathbf{A}(\mathbf{u})$ (resp. $\mathbf{A}^T(\mathbf{u})$), forming a dual basis of $\mathbb{R}^p$

$$\mathbf{l}_j^T \cdot \mathbf{r}_k = \delta_j^k.$$

The matrices $\mathbf{T}(\mathbf{u})$ with columns $(\mathbf{r}_1(\mathbf{u}), ..., \mathbf{r}_p(\mathbf{u}))$, and $\mathbf{T}^{-1}(\mathbf{u})$ with rows $(\mathbf{l}_1^T(\mathbf{u}), ..., \mathbf{l}_p^T(\mathbf{u}))$ satisfy

$$\mathbf{T}^{-1}(\mathbf{u})\mathbf{A}(\mathbf{u})\mathbf{T}(\mathbf{u}) = \operatorname{diag}\Big(a_i(\mathbf{u})\Big) \equiv \boldsymbol{\Lambda}(\mathbf{u}). \tag{1.8}$$

Now, let us set

$$a^+ = \max(a, 0), \;\; a^- = \min(a, 0),$$

so that

$$a = a^+ + a^-, \;\; |a| = a^+ - a^-,$$

and denote

$$\Lambda^{\pm}(\mathbf{u}) = \operatorname{diag}\left(a_k^{\pm}(\mathbf{u})\right).$$

If we now define $\mathbf{A}^+$ and $\mathbf{A}^-$ by

$$\mathbf{A}^{\pm}(\mathbf{u}) = \mathbf{T}(\mathbf{u})\Lambda^{\pm}(\mathbf{u})\mathbf{T}^{-1}(\mathbf{u}), \tag{1.9}$$

clearly we have

$$\mathbf{A}(\mathbf{u}) = \mathbf{A}^+(\mathbf{u}) + \mathbf{A}^-(\mathbf{u}), \quad |\mathbf{A}(\mathbf{u})| = \mathbf{A}^+(\mathbf{u}) - \mathbf{A}^-(\mathbf{u}).$$

Example 1.3. The upwind scheme in the linear case. Assume that $\mathbf{A}$ is constant; introducing the *characteristic* variables

$$\mathbf{w} = \mathbf{T}^{-1}\mathbf{u}, \text{ i.e., } \mathbf{w}_k = \mathbf{l}_k^T\mathbf{u},$$

we get a system of p decoupled scalar equations,

$$\frac{\partial \mathbf{w}}{\partial t} + \Lambda \frac{\partial \mathbf{w}}{\partial x} = \mathbf{0}.$$

This gives a natural way of extending the scalar upwind scheme to a (linear) system by applying it to each scalar characteristic equation. The upwind or Courant–Isaacson–Rees scheme applied to each equation gives

$$\mathbf{w}_j^{n+1} = \mathbf{w}_j^n - \frac{\lambda}{2}\Lambda(\mathbf{w}_{j+1}^n - \mathbf{w}_{j-1}^n) + \frac{\lambda}{2}|\Lambda|(\mathbf{w}_{j+1}^n - 2\mathbf{w}_j^n + \mathbf{w}_{j-1}^n),$$

and the scheme in the original conservative variables can be written

$$\mathbf{v}_j^{n+1} = \mathbf{v}_j^n - \frac{\lambda}{2}\mathbf{A}(\mathbf{v}_{j+1}^n - \mathbf{v}_{j-1}^n) + \frac{\lambda}{2}|\mathbf{A}|(\mathbf{v}_{j+1}^n - 2\mathbf{v}_j^n + \mathbf{v}_{j-1}^n). \tag{1.10a}$$

We can also write it in upwind form

$$\mathbf{v}_j^{n+1} = \mathbf{v}_j^n - \lambda\{\mathbf{A}^+(\mathbf{v}_{j+1}^n - \mathbf{v}_j^n) + \mathbf{A}^-(\mathbf{v}_j^n - \mathbf{v}_{j-1}^n)\} \tag{1.10b}$$

with numerical flux

$$\mathbf{g}^u(\mathbf{u}, \mathbf{v}) = \mathbf{A}^+\mathbf{u} + \mathbf{A}^-\mathbf{v}.$$

This scheme has various extensions in the nonlinear case such as Godunov's and Roe's scheme, which we shall detail in the next sections. □

For second-order schemes with flux limiters using the Lax–Wendroff and the above upstream scheme as building block, such as Sweby's or Davis' (see G.R., Chapter IV, Section 2), they can be extended in two ways. One way consists of replacing the ratio of consecutive increments r, for instance, by

$$r_{j+\frac{1}{2}}^{n+} = \frac{(\Delta\mathbf{v}_{j-\frac{1}{2}}^n, \Delta\mathbf{v}_{j+\frac{1}{2}}^n)}{(\Delta\mathbf{v}_{j+\frac{1}{2}}^n, \Delta\mathbf{v}_{j+\frac{1}{2}}^n)},$$

$$r_{j+\frac{1}{2}}^{n-} = \frac{(\Delta\mathbf{v}_{j-\frac{1}{2}}^n, \Delta\mathbf{v}_{j+\frac{1}{2}}^n)}{(\Delta\mathbf{v}_{j-\frac{1}{2}}^n, \Delta\mathbf{v}_{j-\frac{1}{2}}^n)},$$

where (.,.) denotes the Euclidean inner product in $\mathbb{R}^p$ (see Davis 1984b for details). Another way consists of modifying the flux componentwise (relative to the eigenvector decomposition), as we shall now explain.

Example 1.4. Schemes with flux limiters. Assume first that $\mathbf{A}$ is constant. Then, the Lax–Wendroff scheme (1.5) corresponds to the numerical flux

$$\mathbf{g}^{L.W}(\mathbf{u},\mathbf{v}) = \frac{1}{2}\mathbf{A}(\mathbf{u}+\mathbf{v}) - \frac{\lambda}{2}\mathbf{A}^2(\mathbf{v}-\mathbf{u}).$$

Since $\mathbf{A} = \mathbf{A}^+ + \mathbf{A}^-$, we can write

$$\frac{1}{2}\mathbf{A}(\mathbf{u}+\mathbf{v}) = \mathbf{A}^+\mathbf{u} + \mathbf{A}^-\mathbf{v} + \frac{1}{2}|\mathbf{A}|(\mathbf{v}-\mathbf{u})$$

and

$$\mathbf{g}^{L.W}(\mathbf{u},\mathbf{v}) = \mathbf{A}^+\mathbf{u} + \mathbf{A}^-\mathbf{v} + \frac{1}{2}(\mathbf{I}-\lambda|\mathbf{A}|)|\mathbf{A}|(\mathbf{v}-\mathbf{u}).$$

Then, we notice that we have $\mathbf{A}^+\mathbf{A}^- = \mathbf{0}$, which implies

$$(\mathbf{I}-\lambda|\mathbf{A}|)|\mathbf{A}| = (\mathbf{I}-\lambda\mathbf{A}^+)\mathbf{A}^+ - (\mathbf{I}+\lambda\mathbf{A}^-)\mathbf{A}^-.$$

Hence, the Lax–Wendroff flux can be written in terms of the upwind flux (see (1.10)),

$$\mathbf{g}^{L.W}(\mathbf{u},\mathbf{v}) = \mathbf{g}^u(\mathbf{u},\mathbf{v}) + \frac{1}{2}(\mathbf{I}-\lambda\mathbf{A}^+)\mathbf{A}^+(\mathbf{v}-\mathbf{u}) - \frac{1}{2}(\mathbf{I}+\lambda\mathbf{A}^-)\mathbf{A}^-(\mathbf{v}-\mathbf{u}).$$

Setting

$$\nu_k = \lambda a_k, \quad \sigma_k = \text{ sgn } a_k$$

and writing $\Delta\mathbf{v}_{j+\frac{1}{2}}$ in the basis of eigenvectors

$$\Delta\mathbf{v}_{j+\frac{1}{2}} = \sum_{k=1}^{p} \alpha_{k,j+\frac{1}{2}}\mathbf{r}_{k,j+\frac{1}{2}},$$

we get

$$\mathbf{g}^{L.W.}_{j+\frac{1}{2}} = \mathbf{g}^u_{j+\frac{1}{2}} + \frac{1}{2}\Big(\sum_{a_k\geq 0}(1-\lambda a_k)a_k\alpha_{k,j+\frac{1}{2}}\mathbf{r}_k - \sum_{a_k<0}(1+\lambda a_k)a_k\alpha_{k,j+\frac{1}{2}}\mathbf{r}_k\Big)$$

or

$$\mathbf{g}^{L.W.}_{j+\frac{1}{2}} = \mathbf{g}^u_{j+\frac{1}{2}} + \frac{1}{2}\sum_{k=1}^{p}(\sigma_k - \nu_k)a_k\alpha_{k,j+\frac{1}{2}}\mathbf{r}_k.$$

We then limit the flux in each characteristic direction

$$\mathbf{g}_{j+\frac{1}{2}} = \mathbf{g}^u_{j+\frac{1}{2}} + \frac{1}{2}\sum_{k=1}^{p}\varphi_{k,j+\frac{1}{2}}(\sigma_k - \nu_k)a_k\alpha_{k,j+\frac{1}{2}}\mathbf{r}_k, \tag{1.11}$$

where the limiter $\varphi_{k,j+\frac{1}{2}}$ satisfies

$$\varphi_{k,j+\frac{1}{2}} = \varphi(r_{k,j+\frac{1}{2}}), \quad r_{k,j+\frac{1}{2}} = \begin{cases} \dfrac{\alpha_{k,j-\frac{1}{2}}}{\alpha_{k,j+\frac{1}{2}}} & \text{if } \sigma_k = \operatorname{sgn} a_k > 0, \\ \dfrac{\alpha_{k,j+\frac{3}{2}}}{\alpha_{k,j+\frac{1}{2}}} & \text{if } \sigma_k = \operatorname{sgn} a_k < 0, \end{cases}$$

and φ is for instance

$$\varphi(r) = \max\{0, \min(\phi r, 1), \min(r, \phi)\}, \qquad \phi \in [1,2].$$

For $\phi = 1$, this is the usual minmod limiter since $\varphi_{j+\frac{1}{2}} r_{j+\frac{1}{2}} =$ minmod $(\Delta v_{j+\frac{1}{2}}, \Delta v_{j-\frac{1}{2}})$, where

$$\text{minmod}(a,b) = \begin{cases} s \min(|a|,|b|) \text{ if } s = \operatorname{sgn}(a) = \operatorname{sgn}(b), \\ 0 \qquad \text{otherwise.} \end{cases}$$

For $\phi = 2$ we get the so-called superbee limiter (see G.R., Chapter IV, Section 2.3).

In the general nonlinear case (following the ideas of G.R., Chapter IV, Section 2.2), we set

$$\Delta \mathbf{f}^+_{j+\frac{1}{2}} = \mathbf{f}_{j+1} - \mathbf{g}^u_{j+\frac{1}{2}}, \quad \Delta \mathbf{f}^-_{j+\frac{1}{2}} = \mathbf{g}^u_{j+\frac{1}{2}} - \mathbf{f}_j,$$

where $\mathbf{g}^u$ is the numerical flux of Roe's or Godunov's scheme, which we shall discuss in the next sections. Then, from (1.5) we obtain

$$\begin{aligned} \mathbf{g}^{L.W.}_{j+\frac{1}{2}} &= \frac{1}{2}(\mathbf{f}_{j+1} + \mathbf{f}_j) - \frac{\lambda}{2}\mathbf{A}_{j+\frac{1}{2}}(\mathbf{f}_{j+1} - \mathbf{f}_j) \\ &= \mathbf{g}^u_{j+\frac{1}{2}} + \frac{1}{2}(\Delta \mathbf{f}^+_{j+\frac{1}{2}} - \Delta \mathbf{f}^-_{j+\frac{1}{2}}) \\ &\quad - \frac{\lambda}{2}(\mathbf{A}^+_{j+\frac{1}{2}} + \mathbf{A}^-_{j+\frac{1}{2}})(\Delta \mathbf{f}^+_{j+\frac{1}{2}} + \Delta \mathbf{f}^-_{j+\frac{1}{2}}). \end{aligned}$$

If we assume

$$\mathbf{A}^+ \Delta \mathbf{f}^- = \mathbf{A}^- \Delta \mathbf{f}^+ = 0, \tag{1.12}$$

which indeed holds in the linear case, we get

$$\mathbf{g}^{L.W.}_{j+\frac{1}{2}} = \mathbf{g}^u_{j+\frac{1}{2}} + \frac{1}{2}(\mathbf{I} - \lambda \mathbf{A}^+_{j+\frac{1}{2}})\Delta \mathbf{f}^+_{j+\frac{1}{2}} - \frac{1}{2}(\mathbf{I} + \lambda \mathbf{A}^-_{j+\frac{1}{2}})\Delta \mathbf{f}^-_{j+\frac{1}{2}}.$$

Now, as previously, we decompose $\Delta \mathbf{v}_{j+\frac{1}{2}}$ on the basis $\mathbf{r}_{k,j+\frac{1}{2}}$ of eigenvectors of $\mathbf{A}_{j+\frac{1}{2}}$,

$$\begin{aligned} \mathbf{A}_{j+\frac{1}{2}} \mathbf{r}_{k,j+\frac{1}{2}} &= a_{k,j+\frac{1}{2}} \mathbf{r}_{k,j+\frac{1}{2}}, \\ \Delta \mathbf{v}_{j+\frac{1}{2}} &= \sum_{k=1}^{p} \alpha_{k,j+\frac{1}{2}} \mathbf{r}_{k,j+\frac{1}{2}} \end{aligned}$$

(where the dependence of $\mathbf{A}$, a, α, and $\mathbf{r}$ on $\mathbf{u}$ is omitted in the notations). We get

$$\mathbf{g}_{j+\frac{1}{2}}^{L.W.} = \mathbf{g}_{j+\frac{1}{2}}^{u} + \frac{1}{2} \sum_{a_k>0} (1 - \lambda a_{k,j+\frac{1}{2}}) a_{k,j+\frac{1}{2}} \alpha_{k,j+\frac{1}{2}} \mathbf{r}_{k,j+\frac{1}{2}}$$
$$- \frac{1}{2} \sum_{a_k<0} (1 + \lambda a_{k,j+\frac{1}{2}}) a_{k,j+\frac{1}{2}} \alpha_{k,j+\frac{1}{2}} \mathbf{r}_{k,j+\frac{1}{2}},$$

or

$$\mathbf{g}_{j+\frac{1}{2}}^{L.W.} = \mathbf{g}_{j+\frac{1}{2}}^{u} + \frac{1}{2} \sum_{k=1}^{p} (\sigma_{k,j+\frac{1}{2}} - \nu_{k,j+\frac{1}{2}}) a_{k,j+\frac{1}{2}} \alpha_{k,j+\frac{1}{2}} \mathbf{r}_{k,j+\frac{1}{2}}.$$

Finally, we limit the flux as in formula (1.11),

$$\mathbf{g}_{j+\frac{1}{2}} = \mathbf{g}_{j+\frac{1}{2}}^{u} + \frac{1}{2} \sum_{k=1}^{p} \varphi(r_{k,j+\frac{1}{2}}) (\sigma_{k,j+\frac{1}{2}} - \nu_{k,j+\frac{1}{2}}) a_{k,j+\frac{1}{2}} \alpha_{k,j+\frac{1}{2}} \mathbf{r}_{k,j+\frac{1}{2}},$$

where

$$r_{k,j+\frac{1}{2}} = \begin{cases} \dfrac{((\sigma-\nu)a\alpha)_{k,j-\frac{1}{2}}}{((\sigma-\nu)a\alpha)_{k,j+\frac{1}{2}}} & \text{if } \sigma_{k,j+\frac{1}{2}} > 0, \\[2ex] \dfrac{((\sigma-\nu)a\alpha)_{k,j+\frac{3}{2}}}{((\sigma-\nu)a\alpha)_{k,j+\frac{1}{2}}} & \text{if } \sigma_{k,j+\frac{1}{2}} < 0. \end{cases}$$

Note that we can use the same (Hui and Loh 1992) or a different (Swanson and Turkel 1992, Jorgenson and Turkel 1993) limiter in each characteristic variable (see also Yee 1987). □

Similarly we can easily extend the second-order schemes with the "modified flux" of Harten by modifying the flux componentwise following the ideas developed in the scalar case (see G.R., Chapter IV, Section 1).

Example 1.5. Schemes with "modified flux". As for the Lax–Wendroff scheme, we introduce an average value, more precisely a symmetric averaging operator V and the corresponding average state

$$\overline{\mathbf{v}}_{j+\frac{1}{2}} = V(\mathbf{v}_j, \mathbf{v}_{j+1}),$$

and set

$$\mathbf{r}_{k,j+\frac{1}{2}} = \mathbf{r}_k(\overline{\mathbf{v}}_{j+\frac{1}{2}}).$$

Then, we define

$$\alpha_{k,j+\frac{1}{2}} = \mathbf{l}_k^T(\overline{\mathbf{v}}_{j+\frac{1}{2}}) \Delta \mathbf{v}_{j+\frac{1}{2}}$$

so that

$$\Delta \mathbf{v}_{j+\frac{1}{2}} = \sum_{k=1}^{p} \alpha_{k,j+\frac{1}{2}} \mathbf{r}_{k,j+\frac{1}{2}}.$$

Let us set, in complete analogy with the scalar case,

$$a_{k,j+\frac{1}{2}} = a_k(\overline{\mathbf{v}}_{j+\frac{1}{2}}).$$

If using Roe's extension technique, we shall rather take for $a_{k,j+\frac{1}{2}}$ and $\mathbf{r}_{k,j+\frac{1}{2}}$ the eigenvalues and eigenvectors of an average matrix $\mathbf{A}(\mathbf{v}_j, \mathbf{v}_{j+1})$ satisfying (1.6). As we have already written, such a matrix will be constructed in Section 3. Then, the formula (1.6) of G.R., Chapter IV, Section 1, leads us to introduce the scheme

$$\mathbf{v}_j^{n+1} = \mathbf{v}_j^n - \frac{\lambda}{2}(\tilde{\mathbf{f}}_{j+1}^n - \tilde{\mathbf{f}}_{j-1}^n) + \frac{1}{2}(\tilde{\mathbf{Q}}_{j+\frac{1}{2}}^n \Delta\mathbf{v}_{j+\frac{1}{2}}^n - \tilde{\mathbf{Q}}_{j-\frac{1}{2}}^n \Delta\mathbf{v}_{j-\frac{1}{2}}^n).$$

In this formula, the modified flux is

$$\tilde{\mathbf{f}}_j = \mathbf{f}(\mathbf{v}_j) + \frac{\mathbf{h}_j}{\lambda},$$

where

$$\mathbf{h}_j = \sum_{k=1}^{p} h_{k,j}\mathbf{r}_{k,j+\frac{1}{2}},$$

and the viscosity term

$$\tilde{\mathbf{Q}}_{j+\frac{1}{2}} \Delta\mathbf{v}_{j+\frac{1}{2}} = \sum_{k=1}^{p} Q_k(\lambda a_{k,j+\frac{1}{2}} + \frac{\Delta h_{k,j+\frac{1}{2}}}{\alpha_{k,j+\frac{1}{2}}})\alpha_{k,j+\frac{1}{2}}\mathbf{r}_{k,j+\frac{1}{2}}.$$

Here, following formula (1.12) in G.R., Chapter IV, $h_{k,j}$ is defined by

$$\begin{aligned} h_{k,j} = \frac{1}{2} s_{k,j} \min \{&(Q_k(\lambda a_{k,j+\frac{1}{2}}) - \lambda^2 a_{k,j+\frac{1}{2}}^2)|\alpha_{k,j+\frac{1}{2}}|, \\ &(Q_k(\lambda a_{k,j-\frac{1}{2}}) - \lambda^2 a_{k,j-\frac{1}{2}}^2)|\alpha_{k,j-\frac{1}{2}}|\}, \end{aligned}$$

$$s_{k,j} = \begin{cases} \operatorname{sgn} \alpha_{k,j\pm\frac{1}{2}} & \text{if } \alpha_{k,j-\frac{1}{2}}\alpha_{k,j+\frac{1}{2}} > 0, \\ 0 & \text{otherwise.} \end{cases}$$

Note that the function Q_k, still satisfying

$$|x| \le Q_k(x) \le 1,$$

may depend on the characteristic field. We obtain by comparison with the Lax–Wendroff scheme an overall second-order accurate scheme. Furthermore, we can define a TV discrete norm by

$$TV(\mathbf{v}_{j+\frac{1}{2}}) = \sum_j |\Delta\mathbf{v}_{j+\frac{1}{2}}|,$$

where for instance the norm is the discrete $\mathbf{L}^1$-norm

$$|\Delta\mathbf{v}_{j+\frac{1}{2}}| = \sum_{k=1}^{p} |\alpha_{k,j+\frac{1}{2}}|.$$

Then, it is easy to show that, at least in the linear case, the resulting scheme is T.V.D. under the same C.F.L. restriction as in the scalar case (see Harten 1983 and 1984), Yee, Warming, and Harten 1985). □

Remark 1.3. Let us note that if a scheme is written in viscous form,

$$(1.13)\quad \begin{cases} \mathbf{v}_j^{n+1} = \mathbf{v}_j^n - \dfrac{\lambda}{2}(\mathbf{f}(\mathbf{v}_{j+1}^n) - \mathbf{f}(\mathbf{v}_{j-1}^n)) + \\ \qquad \dfrac{1}{2}(\mathbf{Q}_{j+\frac{1}{2}}^n(\mathbf{v}_{j+1}^n - \mathbf{v}_j^n) - \mathbf{Q}_{j-\frac{1}{2}}^n(\mathbf{v}_j^n - \mathbf{v}_{j-1}^n)), \end{cases}$$

the viscosity coefficient must then be replaced by a viscosity matrix $\mathbf{Q}$. The matrix may be diagonal, $\mathbf{Q} = \mathbf{I}$ for Lax–Friedrich's scheme or $\mathbf{Q} = \mathrm{diag}(Q_k)$ as above when working on the characteristic variables. But for the Lax–Wendroff scheme using Roe's average, the viscosity matrix is not diagonal, and we have

$$\mathbf{Q}_{j+\frac{1}{2}}^{LW} = \lambda^2 \mathbf{A}^2(\mathbf{v}_{j+1}, \mathbf{v}_j)$$

(see also Einfeldt 1988). Then, for a general scheme, if $\mathbf{Q}$ is smooth enough, second-order accuracy requires

$$\mathbf{Q}(\mathbf{u}, ..., \mathbf{u}) = \lambda^2 \mathbf{A}^2(\mathbf{u}).$$

Also, as a generalization of the results of G.R., Chapter III, Section 4 concerning the schemes satisfying an entropy condition, we can compare viscosity matrices (see Tadmor 1987 and 1988 for details). □

1.2 L^2-stability

Consider the linear hyperbolic system

$$(1.14)\quad \frac{\partial \mathbf{u}}{\partial t} + \mathbf{A}\frac{\partial \mathbf{u}}{\partial x} = \mathbf{0},$$

where $\mathbf{A}$ is a constant $p \times p$ matrix, and the following linear difference scheme (as in G.R. Chapter III, Section 1.3):

$$(1.15)\quad \mathbf{v}_j^{n+1} = \sum_{\ell=-k}^{+k} \mathbf{C}_\ell \mathbf{v}_{j+\ell}^n,$$

where the $\mathbf{C}_\ell$, $-k \leq \ell \leq k$ are $p \times p$ constant matrices that are in fact polynomials in $\lambda\mathbf{A}$. For the sake of convenience, we extend this scheme to the whole real line by setting

$$(1.16)\quad \mathbf{v}^{n+1}(x) = \sum_{\ell=-k}^{+k} \mathbf{C}_\ell \mathbf{v}^n(x + \ell\Delta x),$$

where

$$\mathbf{v}^n(x) = \mathbf{v}_\Delta(x, t_n) = \mathbf{v}_j^n,\ x_{j-\frac{1}{2}} < x < x_{j+\frac{1}{2}}.$$

Then, the scheme (1.15) is L^2-stable if there exists a constant C independent of Δx and Δt such that

$$\|\mathbf{v}^n\|_{(\mathbf{L}^2(\mathbb{R}))^p} \leq C\|\mathbf{v}^0\|_{(\mathbf{L}^2(\mathbb{R}))^p}. \tag{1.17}$$

By using the Fourier transform

$$\hat{\varphi}(\xi) = (2\pi)^{-\frac{1}{2}} \int_{\mathbb{R}} e^{-ix\xi}\varphi(x)dx,$$

(1.15) can be equivalently written

$$\hat{\mathbf{v}}^{n+1}(\xi) = \mathbf{G}^a(\xi)\hat{\mathbf{v}}^n(\xi), \tag{1.18}$$

where the matrix

$$\mathbf{G}^a(\xi) = \sum_{\ell=-k}^{+k} \mathbf{C}_\ell e^{i\ell\Delta x\xi} \tag{1.19}$$

is called the *amplification matrix* of the scheme. Iterating on (1.18) gives

$$\hat{\mathbf{v}}^n(\xi) = \mathbf{G}^a(\xi)^n\hat{\mathbf{v}}^0(\xi).$$

The L^2-stability property thus requires that the matrices $\mathbf{G}^n(\xi)$ are uniformly bounded, i.e., satisfy the "power-boundedness condition"

$$\|\mathbf{G}^a(\xi)^n\|_{\mathcal{L}(\mathbf{L}^2)} \leq C, \forall n. \tag{1.20}$$

Let $\rho(\mathbf{G})$ denote the spectral radius of a matrix $\mathbf{G}$. We have the following necessary and sufficient condition.

Proposition 1.2

Assume that $\mathbf{A}$ *is a constant matrix. The linear difference scheme (1.15) is* L^2*-stable if and only if its amplification matrix* $\mathbf{G}^a$ *defined by (1.19) satisfies*

$$\rho(\mathbf{G}^a(\xi)) \leq 1, \forall \xi \in \mathbb{R}. \tag{1.21}$$

Proof. Since by assumption the coefficient matrices $\mathbf{C}_\ell$ are polynomials in $\lambda\mathbf{A}$, the matrices $\mathbf{A}$ and $\mathbf{C}_\ell$ are all simultaneously diagonalizable by $\mathbf{T}$ (see (1.8)), and so are the $\mathbf{G}^a(\xi)^n$. The result follows easily. □

Remark 1.4. If we assume that $\mathbf{A}$ is no longer constant but depends on (x, t), a general necessary stability condition is given by

$$\rho(\mathbf{G}^a(\xi))^n \leq C, \forall n,$$

or, equivalently, by

$$\rho(\mathbf{G}^a(\xi)) \leq 1 + c\Delta t, \; 0 < \Delta t < \Delta t_0, \forall \xi. \tag{1.22}$$

This condition is known as the *von Neumann condition.* Condition (1.21), which is also sufficient in our case, is known as the strict von Neumann condition. There exists an extensive literature on the L^2-stability of difference schemes. Indeed the "power boundedness condition" in the more general context has been characterized by Kreiss 1968, see also Lax and Wendroff (1964), Richtmyer and Morton (1967), and then Tadmor (1981), LeVeque and Trefethen (1983).

Sufficient stability conditions are obtained by introducing dissipation in the difference scheme, as we shall discuss below. □

Remark 1.5. Now assume that we have a 3-point linear scheme written in viscous form,

$$\mathbf{v}_j^{n+1} = \mathbf{v}_j^n - \frac{\lambda}{2}\mathbf{A}(\mathbf{v}_{j+1}^n - \mathbf{v}_{j-1}^n) + \frac{1}{2}\mathbf{Q}(\mathbf{v}_{j+1}^n - 2\mathbf{v}_j^n + \mathbf{v}_{j-1}^n),$$

where $\mathbf{Q}$ is the constant (viscosity) matrix

$$\mathbf{Q} = \mathbf{C}_{-1} + \mathbf{C}_1 = \mathbf{I} - \lambda\mathbf{C}_0\mathbf{A}.$$

Then $\mathbf{Q}$ has the same eigenvectors as $\mathbf{A}$; thus, it is diagonalizable with real eigenvalues Q_k, and the condition for stability can be written

$$(\lambda a_k)^2 \le Q_k \le 1, \;\; k = 1, ..., p. \tag{1.23}$$

Indeed, by using the basis of eigenvectors of $\mathbf{A}$, the eigenvalues of $\mathbf{G}^a(\xi)$ are easily computed in terms of Q_k and a_k, and the proof then mimics that of G.R., Proposition 1.4, Chapter III. For the unique second order (linear) scheme, the condition becomes

$$\lambda\rho(\mathbf{A}) \le 1,$$

which is a C.F.L. like condition. □

We shall now study the dissipative and dispersive characters of a numerical scheme using Fourier analysis or discrete Fourier modes; we shall also introduce the equivalent system.

1.3 Dissipation, dispersion

Let us study first the differential system (1.14). After Fourier transforming, it becomes

$$\frac{\partial \hat{\mathbf{u}}}{\partial t} + i\xi\mathbf{A}\hat{\mathbf{u}} = 0.$$

The exact amplification matrix

$$\mathbf{G}^{ex}(\xi) = \exp(-i\xi\triangle t\mathbf{A}),$$

such that

$$\hat{\mathbf{u}}(\xi, t + \Delta t) = \mathbf{G}^{ex}(\xi)\hat{\mathbf{u}}(\xi, t)$$

obviously satisfies

$$\rho(\mathbf{G}^{ex}(\xi)) = 1.$$

One can equivalently see that (1.14) admits elementary solutions of the form

$$\mathbf{u}(x,t) = \hat{\mathbf{u}} e^{i(kx-\omega t)}$$

iff the phase velocity ω/k agrees with an eigenvalue of $\mathbf{A}$. Since these eigenvalues are real, the amplitude remains constant.

For a difference scheme, dissipation or diffusion yields the attenuation (damping) of the amplitude of the Fourier components of a discrete solution. Following Kreiss (1964), we define more precisely the dissipative character of difference schemes.

Definition 1.1

The linear difference scheme (1.15) is called dissipative *(in the sense of Kreiss) of order* $2q$ *if there exists some constant* $\delta > 0$ *such that its amplification matrix (1.19) satisfies*

$$\rho(\mathbf{G}^a(\xi)) \leq 1 - \delta|\xi|^{2q}, \quad \forall \xi \in \mathbb{R}, |\xi| \leq \pi. \tag{1.24}$$

In fact, we take for q the smallest integer such that (1.24) is satisfied.

As for the scalar case, it is easy to prove that a 3-point linear scheme is dissipative with order 4 iff it is L^2-stable (i.e., (1.22) holds) and

$$0 < Q_k < 1, \quad \forall k = 1, ..., p,$$

where Q_k are the eigenvalues of $\mathbf{Q}$, and dissipative with order 2 iff

$$(\lambda a_k)^2 < Q_k < 1.$$

Thus, a second-order accurate L^2- stable linear scheme is never dissipative with order 2 but is dissipative with order 4 iff $\mathbf{A}$ has no zero eigenvalue and

$$\rho(\lambda \mathbf{A}) < 1.$$

Note also that if a general scheme is linearized (see G.R., Chapter III, Remark 1.3) by freezing the coefficients at some given state, it will not be linearly dissipative at a point $\mathbf{u}$ where the matrix $\mathbf{A}(\mathbf{u})$ is singular. Again, for a further study of dissipative schemes, we refer to Kreiss (1964), and Richtmyer, and Morton (1967) (see also G.R., Chapter V, Section 5.1).

Dissipation or diffusion corresponds to an even-order term in the truncation error, or in the equivalent system (see G.R., Chapter III, Remark 1.2). Indeed, for a first-order scheme, the *equivalent* (or *modified*) system, which the scheme approximates with second-order accuracy, can be written

$$\frac{\partial \mathbf{u}}{\partial t} + \mathbf{A}\frac{\partial \mathbf{u}}{\partial x} = \lambda \Delta x \frac{\partial}{\partial x}(\mathbf{B}(\mathbf{u},\lambda)\frac{\partial \mathbf{u}}{\partial x}),$$

where

$$\mathbf{B}(\mathbf{u},\lambda) = \frac{1}{2\lambda^2}\Big\{\sum_j j^2 \frac{\partial \mathbf{H}}{\partial v_j}(\mathbf{u},\mathbf{u},...) - \lambda^2 \mathbf{A}^2(\mathbf{u})\Big\}.$$

In this formula $\mathbf{H} : \mathbb{R}^{p\times(k+1)} \to \mathbb{R}^p$ is the discrete solution operator. For instance, for the Lax–Friedrichs (linear) scheme, we find the diffusive system

$$\frac{\partial \mathbf{u}}{\partial t} + \mathbf{A}\frac{\partial \mathbf{u}}{\partial x} = \frac{\Delta x}{2\lambda}(1 - \lambda^2 \mathbf{A}^2)\frac{\partial^2 \mathbf{u}}{\partial x^2}.$$

For a second-order scheme, the equivalent system involves third-order differential terms (see Lerat 1981), and we can then study the dispersive character of the scheme. A scheme is *dispersive* if different Fourier components of the solution travel at different speeds. Consider first the continuous problem, i.e., a partial differential equation that stands for either the conservation law or the equivalent equation of a numerical scheme. By linearity, it is sufficient as above to consider a unique "Fourier mode"

$$\mathbf{u}(x,t) = \hat{\mathbf{u}} e^{i(kx-\omega t)},$$

where $\varphi = kx - \omega t$ is the *phase* and $\omega = \omega(k)$ is the frequency, that is a function of the *wave number* k. This function $\omega(k)$ is obtained by substituting this mode $\mathbf{u}$ in the equation, which gives the "dispersion relation" between ω and k. If this function is imaginary, it does not correspond to the description of wave phenomena, and this analysis is irrelevant. The *phase velocity* is $\frac{\omega(k)}{k}$, and the waves are called dispersive if the phase velocity is not constant but depends on k. For instance, if u is a solution of the linear scalar advection equation

$$\frac{\partial u}{\partial t} + a\frac{\partial u}{\partial x} = 0,$$

we find $\omega(k) = ak$, the phase velocity is constant, and the amplitude does not decay, as we have already seen above. For a diffusive equation

$$\frac{\partial u}{\partial t} + a\frac{\partial u}{\partial x} - b\frac{\partial^2 u}{\partial x^2} = 0,$$

we get $\omega(k) = ak - ibk^2$. The imaginary part yields a damping of the amplitude by a factor $\exp(-bk^2)$ but does not change the phase velocity, whereas if u is solution of a dispersive equation with an odd-order term,

$$\frac{\partial u}{\partial t} + a\frac{\partial u}{\partial x} + c\frac{\partial^3 u}{\partial x^3} = 0,$$

we have $\omega(k) = ak - ck^3$ and $\frac{\omega(k)}{k} = a - ck^2$. In fact, when there is a superposition of wavetrains of different wave numbers k, it is the *group velocity* defined by $\omega'(k) = \frac{d\omega(k)}{dk}$ that is the propagation velocity for the wave number k (see Whitham 1974, Chapter 11).

For example, the equivalent equation for the Lax–Wendroff method,

$$\frac{\partial u}{\partial t} + a\frac{\partial u}{\partial x} = -\frac{1}{6}a(\Delta x)^2(1 - \lambda^2 a^2)\frac{\partial^3 u}{\partial x^3},$$

is dispersive with $c = \frac{1}{6}a(\Delta x)^2(1 - \lambda^2 a^2) > 0$ if we assume $a > 0$ and a C.F.L. condition $\lambda a < 1$, and we have $\omega(k)/k = a - ck^2 < a$, i.e., a lagging phase error opposite to a leading phase error when $\omega(k)/k > a$. In general, the lower-order odd (resp. even) term gives the leading dispersive (resp. diffusive) error term. For details, we refer to Richtmyer and Morton (1967), LeVeque (1990); concerning group velocity in finite difference schemes see Trefethen (1982); for S_α^β schemes and the study of k-dispersion relative to the kth characteristic field see Lerat (1981), for some examples see Song and Tang (1993), Desideri et al. (1987), Beux et al. (1993).

We can also observe the phase error directly on the equation of the difference scheme. We can either follow the lines of the preceding section (1.16)–(1.19) or look for elementary discrete solutions of the form

$$u_j^n = \hat{u} e^{i(kj\Delta x - \omega n \Delta t)}.$$

We obtain a discrete dispersion relation $\omega = \omega(k, \Delta x, \Delta t)$ by substituting this single Fourier mode in the formula defining the scheme. On the one hand,

$$u_j^{n+1} = \hat{u} e^{i(kj\Delta x - \omega(n+1)\Delta t)} = e^{-i\omega\Delta t} u_j^n,$$

and on the other hand,

$$u_j^{n+1} = \hat{u} \sum_{\ell} C_\ell e^{i((j+\ell)k\Delta x - \omega n \Delta t)} = g^a(k) u_j^n,$$

which yields the discrete dispersion relation $e^{-i\omega\Delta t} = g^a(k)$, where g^a is the (scalar) amplification factor (1.19). The real part of $e^{-i\omega\Delta t}$ (which yields a diffusion or dissipation error) is the modulus of the amplification factor g^a and has already been considered above. The study of the imaginary part, i.e., the argument of g^a, reveals the existence of a "phase error" (or dispersion error). For instance, the amplification factor of the Lax–Wendroff scheme is

$$e^{-i\omega\Delta t} = g^a(k) = 1 - \lambda^2 a^2 (1 - \cos(k\Delta x)) + i\lambda a \sin(k\Delta x),$$

which gives, if $\omega = (a' + ib')k$, $\tan a'k\Delta t$, and after some Taylor expansions (for small $\Delta t = \lambda \Delta x$)

$$a' = a - ck^2 = a - \frac{1}{6} a(k\Delta x)^2 (1 - \lambda^2 a^2) < a$$

(within Δx^3). The discrete Fourier modes present a lagging phase error when $a > 0$ as noted previously (see Sod 1987, Chapter III, Section 3.6).

2 Godunov's method

The most natural finite difference method to solve the Cauchy problem (1.1) is the Godunov method (G.R., Chapter III, Example 2.3). Let us recall its main features.

2.1 Godunov's method for systems

With the usual assumptions of Section 1, we know that the Riemann problem

$$\text{(2.1)} \qquad \begin{cases} \dfrac{\partial \mathbf{w}}{\partial t} + \dfrac{\partial}{\partial x}\mathbf{f}(\mathbf{w}) = \mathbf{0}, & x \in \mathbb{R}, t > 0, \\ \mathbf{w}(x,0) = \begin{cases} \mathbf{u}_L, & x < 0, \\ \mathbf{u}_R, & x > 0 \end{cases} \end{cases}$$

has an entropy solution

$$\text{(2.2)} \qquad \mathbf{w}(x,t) = \mathbf{w}_R\Big(\frac{x}{t}; \mathbf{u}_L, \mathbf{u}_R\Big)$$

that consists of at most $(p+1)$ constant states separated by shock waves, rarefaction waves, or contact discontinuities, at least when $|\mathbf{u}_L - \mathbf{u}_R|$ is small enough (see Chapter I, Section 6). A solution of this form is unique.

Given an approximation $\mathbf{v}^n = (\mathbf{v}_j^n)_{j\in\mathbb{Z}}$ of $\mathbf{u}(., t_n)$ (where $\mathbf{v}_j^n$ is now a column vector of $\mathbb{R}^p$), we define the approximation $\mathbf{v}^{n+1} = (\mathbf{v}_j^{n+1})_{j\in\mathbb{Z}}$ of $\mathbf{u}(., t_{n+1})$ as follows:

(i) we extend the sequence $\mathbf{v}^n$ as a piecewise constant function $\mathbf{v}_\Delta(., t_n)$ defined by

$$\text{(2.3)} \qquad \mathbf{v}_\Delta(., t_n) = \mathbf{v}_j^n, \; x_{j-\frac{1}{2}} < x < x_{j+\frac{1}{2}},$$

(ii) we solve the Cauchy problem

$$\text{(2.4)} \qquad \begin{cases} \dfrac{\partial \mathbf{w}}{\partial t} + \dfrac{\partial}{\partial x}\mathbf{f}(\mathbf{w}) = \mathbf{0}, & x \in \mathbb{R}, t > 0, \\ \mathbf{w}(x,0) = \mathbf{v}_\Delta(., t_n), \end{cases}$$

(iii) we project ($\mathbf{L}^2$-projection) the solution $\mathbf{w}(., \Delta t)$ onto the piecewise constant functions, i.e., we set

$$\text{(2.5)} \qquad \mathbf{v}_j^{n+1} = \frac{1}{\Delta x}\int_{x_{j-\frac{1}{2}}}^{x_{j+\frac{1}{2}}} \mathbf{w}(x, \Delta t)dx.$$

Provided we assume the C.F.L. condition

$$\text{(2.6)} \qquad \lambda \max |a_k(\mathbf{v}_j^n)| \le \frac{1}{2}, \; 1 \le k \le p,$$

so that the waves issued from the points $x_{j-\frac{1}{2}}$ and $x_{j+\frac{1}{2}}$ do not interact, the solution of (2.4) is in fact obtained by solving a juxtaposition of local Riemann problems and

$$\text{(2.7)} \quad \mathbf{w}(x,t) = \mathbf{w}_R\Big(\frac{x - x_{j+\frac{1}{2}}}{\Delta t}; \mathbf{v}_j^n, \mathbf{v}_{j+1}^n\Big), \quad x_j < x < x_{j+1}, \quad j \in \mathbb{Z}.$$

In order to derive a more explicit form of the scheme, let us integrate the equation (2.4) over the rectangle $(x_{j-\frac{1}{2}}, x_{j+\frac{1}{2}}) \times (0, \Delta t)$. Since the function is piecewise smooth, we obtain

$$\int_{x_{j-\frac{1}{2}}}^{x_{j+\frac{1}{2}}} (\mathbf{w}(x,\Delta t) - \mathbf{w}(x,0))dx$$
$$+ \int_0^{\Delta t} (\mathbf{f}(\mathbf{w}(x_{j+\frac{1}{2}} - 0,t)) - \mathbf{f}(\mathbf{w}(x_{j-\frac{1}{2}} + 0,t)))dt = 0.$$

Using (2.3) and (2.5), we get

$$\Delta x(\mathbf{v}_j^{n+1} - \mathbf{v}_j^n) + \int_0^{\Delta t} (\mathbf{f}(\mathbf{w}(x_{j+\frac{1}{2}} - 0,t)) - \mathbf{f}(\mathbf{w}(x_{j-\frac{1}{2}} + 0,t)))dt = 0.$$

Hence, we have by (2.7)

$$\mathbf{v}_j^{n+1} = \mathbf{v}_j^n - \lambda\{\mathbf{f}(\mathbf{w}_R(0-;\mathbf{v}_j^n,\mathbf{v}_{j+1}^n)) - \mathbf{f}(\mathbf{w}_R(0+;\mathbf{v}_{j-1}^n,\mathbf{v}_j^n))\}.$$

Now, we have observed in the scalar case (G.R., Chapter III) that the function $\xi \to \mathbf{f}\Big(\mathbf{w}_R(\xi;\mathbf{u}_L,\mathbf{u}_R)\Big)$ is continuous at the origin because of the Rankine–Hugoniot conditions, so that Godunov's method can be written in the form

$$\mathbf{v}_j^{n+1} = \mathbf{v}_j^n - \lambda\{\mathbf{f}(\mathbf{w}_R(0;\mathbf{v}_j^n,\mathbf{v}_{j+1}^n)) - \mathbf{f}(\mathbf{w}_R(0;\mathbf{v}_{j-1}^n,\mathbf{v}_j^n))\}, \tag{2.8}$$

and its numerical flux is given by

$$\mathbf{g}(\mathbf{u},\mathbf{v}) = \mathbf{f}(\mathbf{w}_R(0;\mathbf{u},\mathbf{v})). \tag{2.9}$$

We have also seen in the scalar case that the above derivation remains valid with a relaxed C.F.L. condition, which can be written here as

$$\lambda \max |a_k(\mathbf{v}_j^n)| \le 1,\ \ 1 \le k \le p. \tag{2.10}$$

Let us notice that since

$$\begin{cases} \mathbf{w}_R(\frac{x}{t};\mathbf{u}_L,\mathbf{u}_R) = \mathbf{u}_L, & \text{if } \frac{x}{t} \le a_1(\mathbf{u}_L), \\ \mathbf{w}_R(\frac{x}{t};\mathbf{u}_L,\mathbf{u}_R) = \mathbf{u}_R, & \text{if } \frac{x}{t} \ge a_p(\mathbf{u}_R), \end{cases}$$

the numerical flux satisfies

$$\mathbf{g}(\mathbf{u},\mathbf{v}) = \begin{cases} \mathbf{f}(\mathbf{u}), & \text{if } a_p < 0, \\ \mathbf{f}(\mathbf{v}), & \text{if } a_1 > 0, \end{cases}$$

and thus coincides with the upwind scheme when all eigenvalues have the same sign.

Let us consider the *linear* case (1.14)

$$\mathbf{f}(\mathbf{u}) = \mathbf{A}\mathbf{u}$$

and $\mathbf{A}$ is a $p \times p$ matrix whose eigenvalues are real and distinct,

$$a_1 < a_2 < \dots < a_p.$$

As previously, we denote by $\mathbf{r}_1, ..., \mathbf{r}_p$ (respectively $\mathbf{l}_1, ...\mathbf{l}_p$) corresponding (right) eigenvectors of $\mathbf{A}$ (resp. $\mathbf{A}^T$) forming a dual basis of $\mathbb{R}^p$. Then (see Chapter I, Section 1), the Cauchy problem

$$(2.11)\qquad \begin{cases} \dfrac{\partial \mathbf{u}}{\partial t} + \mathbf{A}\dfrac{\partial \mathbf{u}}{\partial x} = \mathbf{0}, & x \in \mathbb{R}, t > 0, \\ \mathbf{u}(x,0) = \mathbf{u}_0(x) \end{cases}$$

can be solved explicitly. Indeed, the system can be decomposed in p decoupled scalar equations involving the characteristic variables, and we get

$$\mathbf{u}(x,t) = \sum_{k=1}^{p} \mathbf{l}_k^T \mathbf{u}_0(x - a_k t)\mathbf{r}_k.$$

Now, we consider the Riemann problem (2.1) in this *linear* case. Setting

$$(2.12)\qquad \mathbf{u}_L = \sum_{k=1}^{p} \alpha_{kL}\mathbf{r}_k, \quad \mathbf{u}_R = \sum_{k=1}^{p} \alpha_{kR}\mathbf{r}_k,$$

we obtain for $a_m < \xi < a_{m+1},\ 0 \le m \le p$, with the convention $a_0 = -\infty, a_{p+1} = +\infty$,

$$(2.13)\qquad \mathbf{w}_R(\xi; \mathbf{u}_L, \mathbf{u}_R) = \mathbf{w}_m = \sum_{k=1}^{m} \alpha_{kR}\mathbf{r}_k, + \sum_{k=m+1}^{p} \alpha_{kL}\mathbf{r}_k.$$

Hence, we have, if the index m is such that $a_m < 0 < a_{m+1}$,

$$(2.14)\qquad \mathbf{A}\mathbf{w}_R(0; \mathbf{u}_L, \mathbf{u}_R) = \sum_{k=1}^{m} a_k\alpha_{kR}\mathbf{r}_k + \sum_{k=m+1}^{p} a_k\alpha_{kL}\mathbf{r}_k.$$

Clearly (2.14) remains valid when $a_m = 0$ or $a_{m+1} = 0$.

Let us give a more compact expression for (2.14). As in (1.8), we introduce a $p \times p$ nonsingular matrix $\mathbf{T}$ such that

$$\mathbf{A} = \mathbf{T}\Lambda\mathbf{T}^{-1}, \Lambda = \text{ diag } (a_k).$$

We can suppose that the kth column of $\mathbf{T}$ is indeed the right eigenvector $\mathbf{r}_k$ while the kth row of $\mathbf{T}^{-1}$ is the left eigenvector $\mathbf{l}_k^T$. Then, we define $\mathbf{A}^+$ and $\mathbf{A}^-$ by (1.9),

$$\mathbf{A}^{\pm} = \mathbf{T}\Lambda^{\pm}\mathbf{T}^{-1}, \Lambda^{\pm} = \text{ diag } (a_k^{\pm}).$$

Now, it follows at once from (2.14) that

$$(2.15)\qquad \mathbf{A}\mathbf{w}_R(0; \mathbf{u}_L, \mathbf{u}_R) = \mathbf{A}^-\mathbf{u}_R + \mathbf{A}^+\mathbf{u}_L.$$

Hence, in the linear case, Godunov's method becomes

$$\mathbf{v}_j^{n+1} = \mathbf{v}_j^n - \lambda\{(\mathbf{A}^-\mathbf{v}_{j+1}^n + \mathbf{A}^+\mathbf{v}_j^n) - (\mathbf{A}^-\mathbf{v}_j^n + \mathbf{A}^+\mathbf{v}_{j-1}^n)\}$$

and becomes the standard upwind or C.I.R. scheme (see Example 1.3)

$$(2.16)\qquad \mathbf{v}_j^{n+1} = \mathbf{v}_j^n - \lambda\{\mathbf{A}^+(\mathbf{v}_j^n - \mathbf{v}_{j-1}^n) + \mathbf{A}^-(\mathbf{v}_{j+1}^n - \mathbf{v}_j^n)\},$$

which can also be written as in (1.10a)

$$\mathbf{v}_j^{n+1} = \mathbf{v}_j^n - \frac{\lambda}{2}\mathbf{A}(\mathbf{v}_{j+1}^n - \mathbf{v}_{j-1}^n) + \frac{\lambda}{2}|\mathbf{A}|(\mathbf{v}_{j+1}^n - 2\mathbf{v}_j^n + \mathbf{v}_{j-1}^n).$$

The numerical flux is (see (2.9)) $\mathbf{g}(\mathbf{u},\mathbf{v}) = \mathbf{A}\mathbf{w}_R(0;\mathbf{u},\mathbf{v})$ and thus (as already obtained in Example 1.3)

$$\mathbf{g}(\mathbf{u},\mathbf{v}) = \frac{1}{2}\mathbf{A}(\mathbf{u}+\mathbf{v}) - \frac{1}{2}|\mathbf{A}|(\mathbf{v}-\mathbf{u}) = \mathbf{A}^+\mathbf{u} + \mathbf{A}^-\mathbf{v}. \tag{2.17}$$

In the following, we shall need these expressions related to the linear Riemann problem:

$$\int_0^{\Delta x/2} \mathbf{w}_R\Big(\frac{x}{\Delta t};\mathbf{u}_L,\mathbf{u}_R\Big)dx,\quad \int_{-\Delta x/2}^{0} \mathbf{w}_R\Big(\frac{x}{\Delta t};\mathbf{u}_L,\mathbf{u}_R\Big)dx.$$

Lemma 2.1

When $\mathbf{f}(\mathbf{u}) = \mathbf{A}\mathbf{u}$, *we have if* $\lambda \max_k |a_k| \le \frac{1}{2}$,

(2.18*a*)

$$\frac{2}{\Delta x}\int_0^{\Delta x/2} \mathbf{w}_R\Big(\frac{x}{\Delta t};\mathbf{u}_L,\mathbf{u}_R\Big)dx = \mathbf{u}_R - \lambda\mathbf{A}(\mathbf{u}_R-\mathbf{u}_L) - \lambda|\mathbf{A}|(\mathbf{u}_R-\mathbf{u}_L),$$

(2.18*b*)

$$\frac{2}{\Delta x}\int_{-\Delta x/2}^{0} \mathbf{w}_R\Big(\frac{x}{\Delta t};\mathbf{u}_L,\mathbf{u}_R\Big)dx = \mathbf{u}_L - \lambda\mathbf{A}(\mathbf{u}_R-\mathbf{u}_L) + \lambda|\mathbf{A}|(\mathbf{u}_R-\mathbf{u}_L).$$

Proof. These expressions have already been computed in the scalar case (see G.R., Chapter III, (2.10) and (4.12)). They are obtained, respectively, by integrating (2.11) on the domains $(0, \frac{\Delta x}{2}) \times (0,\Delta t)$ and $(-\frac{\Delta x}{2}, 0) \times (0,\Delta t)$. □

In order to apply the scheme to gas dynamics, we first recall the equations.

2.2 The gas dynamics equations in a moving frame

We start from the gas dynamics equations in Eulerian coordinates (see Chapter I, Example 2.4):

$$\frac{\partial \mathbf{U}}{\partial t} + \frac{\partial}{\partial x}\mathbf{F}(\mathbf{U}) = \mathbf{0}, \tag{2.19a}$$

$$\mathbf{U} = \begin{pmatrix}\rho\\ \rho u\\ \rho e\end{pmatrix},\quad \mathbf{F}(\mathbf{U}) = \begin{pmatrix}\rho u\\ \rho u^2+p\\ (\rho e+p)u\end{pmatrix}. \tag{2.19b}$$

Let us derive the gas dynamics equations in a moving frame. Denoting by $v = v(x,t)$ the frame velocity, we consider the differential system

$$\frac{dx}{dt} = v(x,t), \tag{2.20}$$

and, for all $\xi \in \mathbb{R}$, we denote by $t \to x(\xi, t)$ the solution of (2.20) that satisfies the initial condition

$$x(0) = \xi.$$

Then (ξ, t) are the coordinates associated with the velocity field v. If we set

$$J(\xi, t) = \frac{\partial x}{\partial \xi}(\xi, t), \tag{2.21}$$

we have

$$\frac{\partial J}{\partial t}(\xi, t) = \frac{\partial}{\partial \xi}\Big(\frac{\partial x}{\partial t}(\xi, t)\Big) = \frac{\partial}{\partial \xi} v(x, t) = \frac{\partial v}{\partial x}\frac{\partial x}{\partial \xi}(\xi, t) = J\,\frac{\partial v}{\partial x}.$$

Now, given a function $\varphi = \varphi(x, t)$ expressed in Eulerian coordinates, we denote by $\overline{\varphi} = \overline{\varphi}(\xi, t)$ this function expressed in the moving frame coordinates, i.e.,

$$\overline{\varphi}(\xi, t) = \varphi(x(\xi, t), t).$$

Then, an easy computation shows that

$$\begin{aligned} \frac{\partial}{\partial t}(\overline{\varphi} J) &= J\Big(\overline{\frac{\partial \varphi}{\partial t}} + \overline{\frac{\partial}{\partial x}(\varphi v)}\Big) \\ &= J\{\overline{\frac{\partial \varphi}{\partial t}} + \overline{\frac{\partial}{\partial x}(\varphi u)} + \overline{\frac{\partial}{\partial x}\Big(\varphi(v-u)\Big)}\}. \end{aligned} \tag{2.22}$$

We obtain

$$\begin{cases} \dfrac{\partial}{\partial t}(\overline{\rho} J) + J\,\overline{\dfrac{\partial}{\partial x}\Big(\rho(u-v)\Big)} = 0, \\[2mm] \dfrac{\partial}{\partial t}(\overline{\rho u} J) + J\{\overline{\dfrac{\partial}{\partial x}\Big(\rho u(u-v)\Big)} + \overline{\dfrac{\partial p}{\partial x}}\} = 0, \\[2mm] \dfrac{\partial}{\partial t}(\overline{\rho e} J) + J\{\overline{\dfrac{\partial}{\partial x}\Big(\rho e(u-v)\Big)} + \overline{\dfrac{\partial p u}{\partial x}}\} = 0. \end{cases} \tag{2.23a}$$

Since

$$\frac{\partial}{\partial \xi} = \Big(\frac{\partial x}{\partial \xi}\Big)\frac{\partial}{\partial x} = J\,\frac{\partial}{\partial x},$$

(2.23a) gives by suppressing the bars for simplicity

$$\begin{cases} \dfrac{\partial}{\partial t}(\rho J) + \dfrac{\partial}{\partial \xi}(\rho(u-v)) = 0, \\[2mm] \dfrac{\partial}{\partial t}(\rho u J) + \dfrac{\partial}{\partial \xi}(\rho u(u-v)) + \dfrac{\partial p}{\partial \xi} = 0, \\[2mm] \dfrac{\partial}{\partial t}(\rho e J) + \dfrac{\partial}{\partial \xi}(\rho e(u-v)) + \dfrac{\partial}{\partial \xi}(p u) = 0. \end{cases} \tag{2.23b}$$

For $v = 0$ system (2.23) takes the form of the equations in Eulerian coordinates, while for $v = u$ we obtain the *Lagrangian* formulation (see Chapter I, Example 2.3)

$$(2.24) \quad \begin{cases} \dfrac{\partial}{\partial t}(\overline{\rho}J) = 0, \\ \dfrac{\partial}{\partial t}(\overline{\rho u}J) + J\overline{\dfrac{\partial p}{\partial x}} = 0, \\ \dfrac{\partial}{\partial t}(\overline{\rho e}J) + J\overline{\dfrac{\partial}{\partial x}(pu)} = 0. \end{cases} \quad or \quad \begin{cases} \dfrac{\partial}{\partial t}(\rho J) = 0, \\ \dfrac{\partial}{\partial t}(\rho u J) + \dfrac{\partial p}{\partial \xi} = 0, \\ \dfrac{\partial}{\partial t}(\rho e J) + \dfrac{\partial}{\partial \xi}(pu) = 0. \end{cases}$$

In the case of Lagrangian coordinates, we can put the equations in a more classical form (see Introduction, Example 1.4). If we introduce the specific volume and a mass variable m such that

$$dm = \rho_0 d\xi,$$

then the equations in Lagrangian coordinates can be written

$$(2.25) \quad \begin{cases} \dfrac{\partial \tau}{\partial t} - \dfrac{\partial u}{\partial m} = 0, \\ \dfrac{\partial u}{\partial t} + \dfrac{\partial p}{\partial m} = 0, \\ \dfrac{\partial e}{\partial t} + \dfrac{\partial}{\partial m}(pu) = 0, \end{cases}$$

with $p = p(\tau, \varepsilon) = p(\tau, e - \frac{u^2}{2})$. By setting

$$(2.26) \quad \mathbf{V} = \begin{pmatrix} \tau \\ u \\ e \end{pmatrix}, \qquad \mathbf{F}(\mathbf{V}) = \begin{pmatrix} -u \\ p \\ pu \end{pmatrix},$$

$\Omega = \{\mathbf{V}; \tau > 0, u \in \mathbb{R}, e - \frac{u^2}{2} > 0\}$, we can write the equations in the general form (1.1):

$$\frac{\partial \mathbf{V}}{\partial t} + \frac{\partial}{\partial m}\mathbf{F}(\mathbf{V}) = \mathbf{0}.$$

2.3 Godunov's method in Lagrangian coordinates

The gas is divided into slabs $(\xi_{j-\frac{1}{2}}, \xi_{j+\frac{1}{2}})$ with thickness $\Delta\xi_j$ that is not necessarily uniform (see Figure 2.1). As usual, the initial condition for a quantity v is given by the slab averages

$$v_j^0 = (\Delta\xi_j)^{-1} \int_{\xi_{j-1/2}}^{\xi_{j+1/2}} v(\xi, 0)d\xi.$$

In view of (2.9), we have to compute $\mathbf{F}\Big(\mathbf{w}_R(0; \mathbf{U}_j^n, \mathbf{U}_{j+1}^n)\Big)$. Remember that

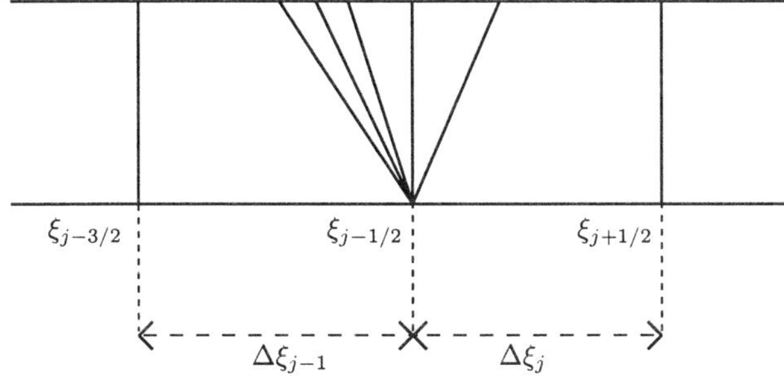

FIGURE 2.1. Grid in Lagrangian coordinates.

the eigenvalues satisfy (see Chapter I, (2.23))

$$a_1 < a_2 = 0 < a_3.$$

Let $u^n_{j+\frac{1}{2}}$ and $p^n_{j+\frac{1}{2}}$ be the values of u and p at the contact discontinuity between $\mathbf{V}^n_j$ and $\mathbf{V}^n_{j+1}$. Indeed, the second field is linearly degenerate, so that (Chapter I, Theorem 4.2) the 2-Riemann invariants, and in particular p and u, are constant. Hence

$$\mathbf{F}\Big(\mathbf{w}_R(0;\mathbf{V}^n_j,\mathbf{V}^n_{j+1})\Big) = \Big(-u^n_{j+\frac{1}{2}}, p^n_{j+\frac{1}{2}}, (pu)^n_{j+\frac{1}{2}}\Big)^T.$$

The Godunov scheme applied to the system (2.25) can be written

$$(2.27)\qquad \begin{cases} \tau_j^{n+1} = \tau_j^n + \dfrac{\Delta t}{\Delta m_j}(u^n_{j+\frac{1}{2}} - u^n_{j-\frac{1}{2}}), \\ u_j^{n+1} = u_j^n - \dfrac{\Delta t}{\Delta m_j}(p^n_{j+\frac{1}{2}} - p^n_{j-\frac{1}{2}}), \\ e_j^{n+1} = e_j^n - \dfrac{\Delta t}{\Delta m_j}((pu)^n_{j+\frac{1}{2}} - (pu)^n_{j-\frac{1}{2}}), \end{cases}$$

where the mass variable increment is defined by

$$(2.28)\qquad \Delta m_j = \rho^0_j \Delta\xi_j$$

and

$$p^n_j = p(\tau^n_j, \varepsilon^n_j), \qquad \varepsilon^n_j = e^n_j - \frac{(u^n_j)^2}{2}.$$

Let us derive another equivalent form for the Godunov scheme. We have not yet specified the movement of the grid. We use an approximation of $u = \frac{dx}{dt}$, and we set

$$(2.29)\qquad x^0_{j+\frac{1}{2}} = \xi_{j+\frac{1}{2}}.$$

Then $x^n_{j+\frac{1}{2}}$, which is the Eulerian coordinate of the interface $\xi_{j+\frac{1}{2}}$ at time t_n, is updated according to

$$x^{n+1}_{j+\frac{1}{2}} = x^n_{j+\frac{1}{2}} + \Delta t u^n_{j+\frac{1}{2}}. \tag{2.30}$$

We can check by induction that

$$\rho^n_j(x^n_{j+\frac{1}{2}} - x^n_{j-\frac{1}{2}}) = \Delta m_j. \tag{2.31}$$

First, this is true for $n = 0$ by assumption. Suppose that it holds for some n; we have

$$\begin{aligned}\Delta m_j \tau^{n+1}_j &= \Delta m_j \tau^n_j + \Delta t(u^n_{j+\frac{1}{2}} - u^n_{j-\frac{1}{2}})\\ &= (x^n_{j+\frac{1}{2}} - x^n_{j-\frac{1}{2}}) + \Delta t(u^n_{j+\frac{1}{2}} - u^n_{j-\frac{1}{2}}) = x^{n+1}_{j+\frac{1}{2}} - x^{n+1}_{j-\frac{1}{2}}.\end{aligned}$$

Hence

$$\rho^{n+1}_j(x^{n+1}_{j+\frac{1}{2}} - x^{n+1}_{j-\frac{1}{2}}) = \Delta m_j,$$

which proves the desired result.

The Godunov method in Lagrangian coordinates can thus be written

$$\begin{cases}\Delta m_j = \rho^0_j(x^0_{j+\frac{1}{2}} - x^0_{j-\frac{1}{2}}),\\ x^{n+1}_{j+\frac{1}{2}} = x^n_{j+\frac{1}{2}} + \Delta t u^n_{j+\frac{1}{2}}\end{cases} \tag{2.32a}$$

$$\begin{cases}\rho^{n+1}_j = (x^{n+1}_{j+\frac{1}{2}} - x^{n+1}_{j-\frac{1}{2}})^{-1}\Delta m_j,\\ u^{n+1}_j = u^n_j - \dfrac{\Delta t}{\Delta m_j}(p^n_{j+\frac{1}{2}} - p^n_{j-\frac{1}{2}}),\\ e^{n+1}_j = e^n_j - \dfrac{\Delta t}{\Delta m_j}((pu)^n_{j+\frac{1}{2}} - (pu)^n_{j-\frac{1}{2}}).\end{cases} \tag{2.32b}$$

Remark 2.1. The Godunov method in Lagrangian coordinates can be interpreted as a finite-volume method. Indeed, the equations (2.23b) can be written

$$\frac{\partial}{\partial t}(\varphi J) + \frac{\partial \mathbf{f}}{\partial \xi} = 0.$$

Let us integrate these equations on $(\xi_{j-\frac{1}{2}}, \xi_{j+\frac{1}{2}})$; we get

$$\frac{d}{dt}\int_{\xi_{j-\frac{1}{2}}}^{\xi_{j+\frac{1}{2}}} \varphi J d\xi + \int_{\xi_{j-\frac{1}{2}}}^{\xi_{j+\frac{1}{2}}} \frac{\partial \mathbf{f}}{\partial \xi} d\xi = 0$$

or

$$\frac{d}{dt}\int_{x_{j-\frac{1}{2}}}^{x_{j+\frac{1}{2}}} \varphi d\xi + (\mathbf{f}_{j+\frac{1}{2}} - \mathbf{f}_{j-\frac{1}{2}}) = 0,$$

where the dependence of $\mathbf{f}, \varphi$, and x on t is omitted. Assuming that φ is constant in each cell $(x_{j-\frac{1}{2}}^n, x_{j+\frac{1}{2}}^n)$, we integrate by the Euler explicit method

$$\Delta x_j^{n+1}\varphi_j^{n+1} = \Delta x_j^n \varphi_j^n - \Delta t(\mathbf{f}_{j+\frac{1}{2}}^n - \mathbf{f}_{j-\frac{1}{2}}^n).$$

This gives (in the case $v = u$), if (ρ, u, e) are constant in each cell,

$$\begin{cases} \Delta x_j^{n+1}\rho_j^{n+1} = \Delta x_j^n \rho_j^n = \dots = \Delta m_j, \\ \Delta x_j^{n+1}(\rho u)_j^{n+1} = \Delta x_j^n(\rho u)_j^n - \Delta t\{p_{j+\frac{1}{2}}^n - p_{j-\frac{1}{2}}^n\}, \\ \Delta x_j^{n+1}(\rho e)_j^{n+1} = \Delta x_j^n(\rho e)_j^n - \Delta t\{(pu)_{j+\frac{1}{2}}^n - (pu)_{j-\frac{1}{2}}^n\}, \end{cases}$$

and thus

$$\begin{cases} \Delta x_j^n \rho_j^n = \Delta m_j, \\ \Delta m_j^{n+1}u_j^{n+1} = \Delta m_j^n u_j^n - \Delta t\{p_{j+\frac{1}{2}}^n - p_{j-\frac{1}{2}}^n\}, \\ \Delta m_j^{n+1}e_j^{n+1} = \Delta m_j^n e_j^n - \Delta t\{(pu)_{j+\frac{1}{2}}^n - (pu)_{j-\frac{1}{2}}^n\}. \end{cases}$$

This is the Godunov scheme provided $(u_{j+\frac{1}{2}}^n, p_{j+\frac{1}{2}}^n)$ are defined by the solution of the Riemann problem. This finite-volume method can be easily extended to multidimensional problems and is widely used in practice (see Chapter IV, Section 4). □

2.4 Godunov's method in Eulerian coordinates (direct method)

We consider now the system (2.19), where

$$\mathbf{U} = (\rho, \rho u, \rho e)^T, \ \mathbf{F}(\mathbf{U}) = \Big(\rho u, \rho u^2 + p, (\rho e + p)u\Big)^T.$$

Let us recall that the eigenvalues of $\mathbf{F}'(\mathbf{U})$ are $a_1 = u - c < a_2 = u < a_3 = u + c$ (see Chapter I, Example 2.4). One has to solve the Riemann problem at the point $x_{j+\frac{1}{2}}$ between the states $\mathbf{U}_j^n$ and $\mathbf{U}_{j+1}^n$. The Godunov scheme can be written

$$(2.33) \quad \begin{cases} \rho_j^{n+1} = \rho_j^n - \dfrac{\Delta t}{\Delta x_j}\{(\rho u)_{j+\frac{1}{2}}^n - (\rho u)_{j-\frac{1}{2}}^n\}, \\ (\rho u)_j^{n+1} = (\rho u)_j^n - \dfrac{\Delta t}{\Delta x_j}\{(\rho u^2 + p)_{j+\frac{1}{2}}^n - (\rho u^2 + p)_{j-\frac{1}{2}}^n\}, \\ (\rho e)_j^{n+1} = (\rho e)_j^n - \dfrac{\Delta t}{\Delta x_j}\{((\rho e + p)u)_{j+\frac{1}{2}}^n - ((\rho e + p)u)_{j-\frac{1}{2}}^n\}. \end{cases}$$

The point $x_{j+\frac{1}{2}}$ no longer corresponds to the contact discontinuity, and we must test the position of the line $x = x_{j+\frac{1}{2}}$ with respect to the different

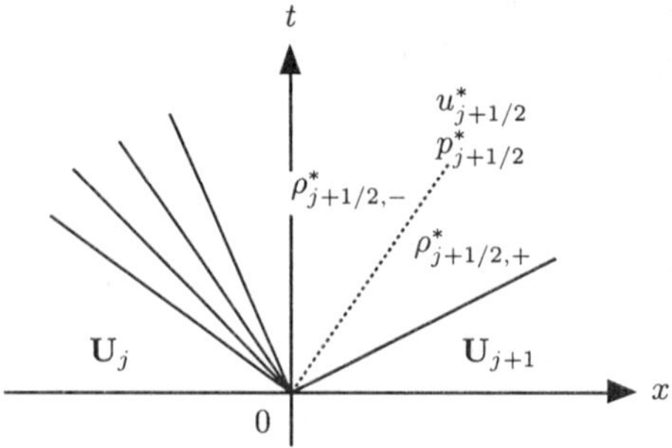

FIGURE 2.2. Example of Riemann problem.

waves. For instance, in the case of Figure 2.2 (see also Chapter II, Figures 3.1 and 3.10), the line $x = x_{j+\frac{1}{2}}$ lies between the rarefaction fan and the contact discontinuity, so that

$$(u^n_{j+\frac{1}{2}}, p^n_{j+\frac{1}{2}}) = (u^*_{j+\frac{1}{2}}, p^*_{j+\frac{1}{2}}),$$

$$\rho^n_{j+\frac{1}{2}} = \rho_{j+\frac{1}{2},-},$$

and

$$e^n_{j+\frac{1}{2}} = e_{j+\frac{1}{2},-},$$

with obvious notations (which are slightly different from those of Chapter II).

2.5 Godunov's method in Eulerian coordinates (Lagrangian step + projection)

As we have just seen, the Godunov method in a Lagrangian grid is easier to handle. However, many physical models use the Eulerian coordinates, and this justifies the present method, which couples a Lagrangian step followed by a projection back onto the Eulerian grid, which we suppose is fixed, with nodes $x_{j+\frac{1}{2}}$.

i) *The Lagrangian step*

At the beginning of each Lagrangian step, the Lagrangian and Eulerian grids coincide and $\xi_{j+\frac{1}{2}} = x_{j+\frac{1}{2}}$. The piecewise constant approximate values of (ρ, u, p) on the Lagrangian grid at time t_n are thus (ρ^n_j, u^n_j, e^n_j). We compute $u^n_{j+\frac{1}{2}}, p^n_{j+\frac{1}{2}}$ values of u and p on the contact discontinuity $x = \xi_{j+\frac{1}{2}}$. Following (2.30), we define the updated Eulerian coordinates of the Lagrangian zone (Figure 2.3) by

$$x^*_{j+\frac{1}{2}} = x_{j+\frac{1}{2}} + \Delta t u^n_{j+\frac{1}{2}}, \tag{2.34}$$

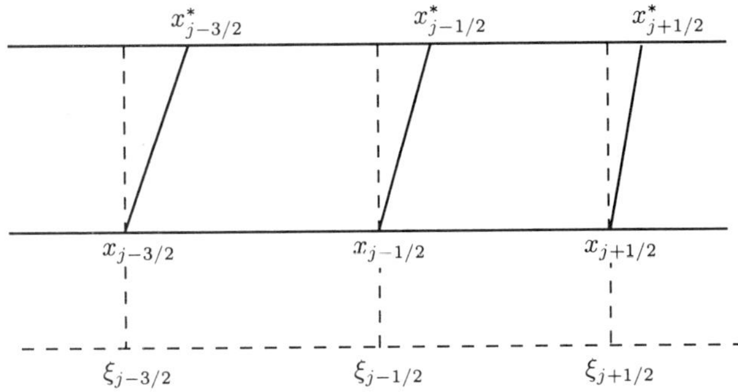

FIGURE 2.3. Updating of the Lagrangian grid.

where the dependence of $x^*_{j+\frac{1}{2}}$ on n is omitted. In other words, if $x_{j+\frac{1}{2}} = x(\xi_{j+\frac{1}{2}}, t_n)$ then $x^*_{j+\frac{1}{2}} \approx x(\xi_{j+\frac{1}{2}}, t_{n+1})$.

Now, following (2.32), we compute $\overline{\rho}_j^{n+1}, \overline{u}_j^{n+1}, \overline{e}_j^{n+1}$, which are the constant values of $\overline{\rho}^{n+1}, \overline{u}^{n+1}, \overline{e}^{n+1}$ in each cell $(x^*_{j-\frac{1}{2}}, x^*_{j+\frac{1}{2}})$:

$$(2.35)\qquad \begin{cases} \overline{\rho}_j^{n+1} = (x^*_{j+\frac{1}{2}} - x^*_{j-\frac{1}{2}})^{-1}\Delta m_j^n, \text{ where } \Delta m_j^n = \rho_j^n \Delta x_j, \\ \overline{u}_j^{n+1} = u_j^n - \dfrac{\Delta t}{\Delta m_j^n}\{p^n_{j+\frac{1}{2}} - p^n_{j-\frac{1}{2}}\}, \\ \overline{e}_j^{n+1} = e_j^n - \dfrac{\Delta t}{\Delta m_j^n}\{(pu)^n_{j+\frac{1}{2}} - (pu)^n_{j-\frac{1}{2}}\}. \end{cases}$$

ii) *The projection step*
The values $\overline{\rho}^{n+1}, \overline{u}^{n+1}, \overline{e}^{n+1}$ are projected back onto the Eulerian grid; denoting any of these quantities by v, we thus define

$$(2.36)\qquad v_j^{n+1} = (\Delta x_j)^{-1} \int_{x_{j-\frac{1}{2}}}^{x_{j+\frac{1}{2}}} \overline{v}^{n+1}(x)dx.$$

We break up the integral into three parts and compute separately the corresponding integrals

$$\int_{x_{j-\frac{1}{2}}}^{x_{j+\frac{1}{2}}} \overline{v}^{n+1}(x)dx = \int_{x_{j-\frac{1}{2}}}^{x^*_{j-\frac{1}{2}}} + \int_{x^*_{j-\frac{1}{2}}}^{x^*_{j+\frac{1}{2}}} + \int_{x^*_{j+\frac{1}{2}}}^{x_{j+\frac{1}{2}}} \overline{v}^{n+1}(x)dx.$$

On the one hand, we have

$$(2.37)\qquad \int_{x^*_{j-\frac{1}{2}}}^{x^*_{j+\frac{1}{2}}} \overline{v}^{n+1}(x)dx = \Delta x_j^* \overline{v}_j^{n+1}$$

and, on the other hand,

$$\int_{x^*_{j+\frac{1}{2}}}^{x_{j+\frac{1}{2}}} \overline{v}^{n+1}(x)dx = (x_{j+\frac{1}{2}} - x^*_{j+\frac{1}{2}}) \begin{cases} \overline{v}^{n+1}_{j+1}, \text{ if } u^n_{j+\frac{1}{2}} < 0, \\ \overline{v}^{n+1}_{j}, \text{ if } u^n_{j+\frac{1}{2}} \geq 0, \end{cases}$$

which we can write

$$\int_{x^*_{j+\frac{1}{2}}}^{x_{j+\frac{1}{2}}} \overline{v}^{n+1}(x)dx = -\Delta t u^n_{j+\frac{1}{2}} v^{n+1}_{j+\frac{1}{2}+\varepsilon(j,n)}, \tag{2.38}$$

where

$$\varepsilon(j,n) = \begin{cases} -\dfrac{1}{2}, & \text{if } u^n_{j+\frac{1}{2}} \geq 0 \\ \dfrac{1}{2}, & \text{if } u^n_{j+\frac{1}{2}} < 0. \end{cases}$$

In the same way, we have

$$\int_{x_{j-\frac{1}{2}}}^{x^*_{j-\frac{1}{2}}} \overline{v}^{n+1}(x)dx = (x^*_{j-\frac{1}{2}} - x_{j-\frac{1}{2}}) \begin{cases} \overline{v}^{n+1}_{j}, \text{ if } u^n_{j-\frac{1}{2}} < 0, \\ \overline{v}^{n+1}_{j-1}, \text{ if } u^n_{j-\frac{1}{2}} \geq 0, \end{cases}$$

so that

$$\int_{x_{j-\frac{1}{2}}}^{x^*_{j-\frac{1}{2}}} \overline{v}^{n+1}(x)dx = \Delta t u^n_{j-\frac{1}{2}} \overline{v}^{n+1}_{j-\frac{1}{2}+\varepsilon(j-1,n)}.$$

We get

$$\begin{aligned} \Delta x_j v^{n+1}_j &= \Delta x^*_j \overline{v}^{n+1}_j \\ &\quad - \Delta t (u^n_{j+\frac{1}{2}} \overline{v}^{n+1}_{j+\frac{1}{2}+\varepsilon(j,n)} - u^n_{j-\frac{1}{2}} \overline{v}^{n+1}_{j-\frac{1}{2}+\varepsilon(j-1,n)}). \end{aligned} \tag{2.39}$$

In particular, we obtain

$$\begin{aligned} \Delta m^{n+1}_j = \Delta x_j \rho^{n+1}_j &= \Delta x^*_j \, \overline{\rho}^{n+1}_j \\ &\quad - \Delta t (u^n_{j+\frac{1}{2}} \overline{\rho}^{n+1}_{j+\frac{1}{2}+\varepsilon(j,n)} - u^n_{j-\frac{1}{2}} \overline{\rho}^{n+1}_{j-\frac{1}{2}+\varepsilon(j-1,n)}), \end{aligned}$$

or

$$\Delta m^{n+1}_j = \Delta m^n_j - \Delta t (u^n_{j+\frac{1}{2}} \, \overline{\rho}^{n+1}_{j+\frac{1}{2}+\varepsilon(j,n)} - u^n_{j-\frac{1}{2}} \, \overline{\rho}^{n+1}_{j-\frac{1}{2}+\varepsilon(j-1,n)}). \tag{2.40}$$

Also, we have

$$\begin{aligned} \Delta x_j (\rho u)^{n+1}_j &= \Delta x^*_j \, \overline{\rho u}^{n+1}_j \\ &\quad - \Delta t (u^n_{j+\frac{1}{2}} \, \overline{\rho u}^{n+1}_{j+\frac{1}{2}+\varepsilon(j,n)} - u^n_{j-\frac{1}{2}} \, \overline{\rho u}^{n+1}_{j-\frac{1}{2}+\varepsilon(j-1,n)}), \end{aligned}$$

and hence

$$\begin{aligned} \Delta m^{n+1}_j u^{n+1}_j &= \Delta m^n_j \, \overline{u}^{n+1}_j \\ &\quad - \Delta t (u^n_{j+\frac{1}{2}} \, \overline{\rho u}^{n+1}_{j+\frac{1}{2}+\varepsilon(j,n)} - u^n_{j-\frac{1}{2}} \, \overline{\rho u}^{n+1}_{j-\frac{1}{2}+\varepsilon(j-1,n)}), \end{aligned}$$

and by (2.35)

(2.41)
$$\begin{cases} \Delta m_j^{n+1} u_j^{n+1} = \Delta m_j^n u_j^n - \Delta t (p_{j+\frac{1}{2}}^n - p_{j-\frac{1}{2}}^n) \\ \qquad - \Delta t \{ u_{j+\frac{1}{2}}^n \, \overline{\rho u}_{j+\frac{1}{2}+\varepsilon(j,n)}^{n+1} - u_{j-\frac{1}{2}}^n \, \overline{\rho u}_{j-\frac{1}{2}+\varepsilon(j-1,n)}^{n+1} \}. \end{cases}$$

We have similarly

(2.42)
$$\begin{cases} \Delta m_j^{n+1} e_j^{n+1} = \Delta m_j^n e_j^n - \Delta t \{ (pu)_{j+\frac{1}{2}}^n - (pu)_{j-\frac{1}{2}}^n \} \\ \qquad - \Delta t \{ u_{j+\frac{1}{2}}^n \, \overline{\rho e}_{j+\frac{1}{2}+\varepsilon(j,n)}^{n+1} - u_{j-\frac{1}{2}}^n \, \overline{\rho e}_{j-\frac{1}{2}+\varepsilon(j-1,n)}^{n+1} \}. \end{cases}$$

The scheme is thus defined by

$$(2.43) \quad \begin{cases} \rho_j^{n+1} = \rho_j^n - \dfrac{\Delta t}{\Delta x_j} \{ u_{j+\frac{1}{2}}^n \, \overline{\rho}_{j+\frac{1}{2}+\varepsilon(j,n)}^{n+1} \\ \qquad - u_{j-\frac{1}{2}}^n \, \overline{\rho}_{j-\frac{1}{2}+\varepsilon(j-1,n)}^{n+1} \}, \\ \rho_j^{n+1} u_j^{n+1} = \rho_j^n u_j^n - \dfrac{\Delta t}{\Delta x_j} (p_{j+\frac{1}{2}}^n - p_{j-\frac{1}{2}}^n) \\ \qquad - \dfrac{\Delta t}{\Delta x_j} \{ u_{j+\frac{1}{2}}^n \, \overline{\rho u}_{j+\frac{1}{2}+\varepsilon(j,n)}^{n+1} - u_{j-\frac{1}{2}}^n \, \overline{\rho u}_{j-\frac{1}{2}+\varepsilon(j-1,n)}^{n+1} \}, \\ \rho_j^{n+1} e_j^{n+1} = \rho_j^{n+1} e_j^n - \dfrac{\Delta t}{\Delta x_j} \{ (pu)_{j+\frac{1}{2}}^n - (pu)_{j-\frac{1}{2}}^n \} \\ \qquad - \dfrac{\Delta t}{\Delta x_j} \{ u_{j+\frac{1}{2}}^n \, \overline{\rho e}_{j+\frac{1}{2}+\varepsilon(j,n)}^{n+1} - u_{j-\frac{1}{2}}^n \, \overline{\rho e}_{j-\frac{1}{2}+\varepsilon(j-1,n)}^{n+1} \}, \end{cases}$$

together with (2.35). The main difference between this scheme (Lagrangian + remapping) and the direct Eulerian scheme lies in the way the convection terms are handled. It is simpler to implement but introduces some extra dissipation.

2.6 Godunov's method in a moving grid

Starting now from equations (2.23), we use the finite-volume formulation. Solving the Riemann problem for the Godunov method amounts to computing the solution on the line $\frac{x}{t} = v_{j+\frac{1}{2}}^n$. This gives $u_{j+\frac{1}{2}}^n, p_{j+\frac{1}{2}}^n, \rho_{j+\frac{1}{2}}^n, e_{j+\frac{1}{2}}^n$. As in Remark 2.1, we have to discretize an equation of the form

$$\frac{\partial}{\partial t}(\varphi J) + \frac{\partial \mathbf{f}}{\partial \xi} = \mathbf{0},$$

which by integration yields

$$\Delta x_j^{n+1} \varphi_j^{n+1} = \Delta x_j^n \varphi_j^n - \Delta t (\mathbf{f}_{j+\frac{1}{2}}^n - \mathbf{f}_{j-\frac{1}{2}}^n).$$

In the present case, we obtain

$$\begin{cases} \Delta x_j^{n+1}\rho_j^{n+1} = \Delta x_j^n \rho_j^n - \Delta t\{(\rho(u-v))_{j+\frac{1}{2}}^n - (\rho(u-v))_{j-\frac{1}{2}}^n\}, \\ \Delta x_j^{n+1}(\rho u)_j^{n+1} = \Delta x_j^n (\rho u)_j^n - \Delta t\{p_{j+\frac{1}{2}}^n - p_{j-\frac{1}{2}}^n\} \\ \qquad\qquad - \Delta t\{(\rho u(u-v))_{j+\frac{1}{2}}^n - (\rho u(u-v))_{j-\frac{1}{2}}^n\}, \\ \Delta x_j^{n+1}(\rho e)_j^{n+1} = \Delta x_j^n (\rho e)_j^n - \Delta t\{(pu)_{j+\frac{1}{2}}^n - (pu)_{j-\frac{1}{2}}^n\} \\ \qquad\qquad - \Delta t\{(\rho e(u-v))_{j+\frac{1}{2}}^n - (\rho e(u-v))_{j-\frac{1}{2}}^n\}, \end{cases}$$

which gives

$$\begin{cases} \Delta m_j^{n+1} = \Delta m_j^n - \Delta t\{(\rho(u-v))_{j+\frac{1}{2}}^n - (\rho(u-v))_{j-\frac{1}{2}}^n\}, \\ \Delta m_j^{n+1}u_j^{n+1} = \Delta m_j^n u_j^n - \Delta t\{p_{j+\frac{1}{2}}^n - p_{j-\frac{1}{2}}^n\} \\ \qquad\qquad - \Delta t\{(\rho u(u-v))_{j+\frac{1}{2}}^n - (\rho u(u-v))_{j-\frac{1}{2}}^n\}, \\ \Delta m_j^{n+1}e_j^{n+1} = \Delta m_j^n e_j^n - \Delta t\{(pu)_{j+\frac{1}{2}}^n - (pu)_{j-\frac{1}{2}}^n\} \\ \qquad\qquad - \Delta t\{(\rho e(u-v))_{j+\frac{1}{2}}^n - (\rho e(u-v))_{j-\frac{1}{2}}^n\}, \end{cases} \tag{2.44}$$

with

$$\Delta m_j^n = \rho_j^n \Delta x_j, \quad x_{j+\frac{1}{2}}^{n+1} = x_{j+\frac{1}{2}}^n + \Delta t v_{j+\frac{1}{2}}^n .$$

For $v = u$ (2.44) coincides with the Lagrangian scheme (2.32), while for $v = 0$ we find the direct Eulerian scheme (2.34). The strategy for choosing $v_{j+\frac{1}{2}}^n$ is a crucial step in the method.

Remark 2.2. We might also consider a scheme combining a Lagrangian step and a projection on a moving grid. The details are left to the reader. □

In Godunov's method, we need to solve one Riemann problem per mesh point $x_{j+\frac{1}{2}}$ at each time step, and this must be done iteratively (for the numerical techniques to obtain this solution, see for instance Colella and Glaz (1985), Loh and Hui (1990). This may be considered as expensive when the problem to solve does not present strong shock discontinuities. In the following, we shall aim to develop approximate Riemann solvers that are simpler to implement and also cheaper to use. A popular method due to Roe (1989) is based on a linearization process.

3 Roe's method

3.1 Presentation of Roe's method for a general system

Roe's scheme for solving system (1.1) numerically is based on the use of an approximate Riemann solver that is obtained by linearizing the system (2.4). We begin by introducing the following definition:

Definition 3.1

We call $\mathbf{A}(\mathbf{u}, \mathbf{v})$ *a* Roe-type linearization *if* $(\mathbf{u}, \mathbf{v}) \to \mathbf{A}(\mathbf{u}, \mathbf{v})$ *is a mapping from* $\Omega \times \Omega$ *into the set of* $p \times p$ *matrices with the following properties:*

$$\mathbf{f}(\mathbf{v}) - \mathbf{f}(\mathbf{u}) = \mathbf{A}(\mathbf{u}, \mathbf{v})(\mathbf{v} - \mathbf{u}), \tag{3.1}$$

$$\begin{cases} \textit{The } p \times p \textit{ matrix } \mathbf{A}(\mathbf{u}, \mathbf{v}) \textit{ has real eigenvalues and} \\ \textit{a corresponding set of eigenvectors that form a basis of } \mathbb{R}^p, \end{cases} \tag{3.2}$$

$$\mathbf{A}(\mathbf{u}, \mathbf{u}) = \mathbf{A}(\mathbf{u}). \tag{3.3}$$

Let us prove the existence of such a linearization when the system (1.1) admits an entropy (Harten–Lax Theorem; see Harten 1983).

Theorem 3.1

Assume that the hyperbolic system (1.1) has a strictly convex entropy U*. Then there exists at least one Roe-type linearization.*

Proof. We look for a matrix $\mathbf{A} = \mathbf{A}(\mathbf{u}, \mathbf{v})$ of the form

$$\mathbf{A} = \mathbf{S} \cdot \mathbf{P},$$

where $\mathbf{S}$ and $\mathbf{P}$ are $p \times p$ matrices and $\mathbf{P}$ is, moreover, symmetric positive definite. Then $\mathbf{A}$ is similar to the symmetric matrix $\mathbf{P}^{\frac{1}{2}}\,\mathbf{S}\mathbf{P}^{\frac{1}{2}}$, which already meets the condition (3.2). It remains to construct the matrices $\mathbf{S} = \mathbf{S}(\mathbf{u}, \mathbf{v})$ and $\mathbf{P} = \mathbf{P}(\mathbf{u}, \mathbf{v})$. The construction follows the arguments of the proof of Theorem 3.2 in the Introduction.

Let U be a strictly convex entropy. Then $U''(\mathbf{u})$ is a symmetric positive-definite matrix, and the mapping $\mathbf{u} \to U'(\mathbf{u})$ is one to one, so that we can define the change of variables (the entropy variables, see the Introduction, Section 3, Theorem 3.2)

$$\mathbf{w}^T = U'(\mathbf{u}),$$

and we set

$$\mathbf{g}(\mathbf{w}) = \mathbf{f}(\mathbf{u}(\mathbf{w})).$$

Let us show that the matrix $\mathbf{g}'(\mathbf{w})$ is symmetric. On the one hand, we know that the matrix $U''(\mathbf{u})\mathbf{f}'(\mathbf{u})$ is symmetric and therefore the matrix

$$\mathbf{f}'(\mathbf{u})U''(\mathbf{u})^{-1} = U''(\mathbf{u})^{-1}(U''(\mathbf{u})\mathbf{f}'(\mathbf{u}))U''(\mathbf{u})^{-1}$$

is also symmetric. On the other hand, we have $\mathbf{u} = \mathbf{u}(\mathbf{w})$ and

$$\mathbf{u}'(\mathbf{w})\mathbf{w}'(\mathbf{u}) = \mathbf{I}$$

so that, since $\mathbf{w}'(\mathbf{u}) = U''(\mathbf{u})$,

$$\mathbf{u}'(\mathbf{w}) = U''(\mathbf{u})^{-1}.$$

Hence, the matrix

$$\mathbf{g}'(\mathbf{w}) = \mathbf{f}'(\mathbf{u}(\mathbf{w})) \cdot \mathbf{u}'(\mathbf{w}) = \mathbf{f}'(\mathbf{u}(\mathbf{w})) \cdot U''(\mathbf{u})^{-1}$$

is symmetric.

Now, given two states $\mathbf{u}_1, \mathbf{u}_2 \in \Omega$, we set

$$\mathbf{w}_1 = \mathbf{w}(\mathbf{u}_1), \mathbf{w}_2 = \mathbf{w}(\mathbf{u}_2)$$

and write

$$\begin{aligned}\mathbf{g}(\mathbf{w}_2) - \mathbf{g}(\mathbf{w}_1) &= \int_0^1 \frac{d}{d\theta}\mathbf{g}(\theta\mathbf{w}_2 + (1-\theta)\mathbf{w}_1)d\theta \\ &= \{\int_0^1 \mathbf{g}'(\theta\mathbf{w}_2 + (1-\theta)\mathbf{w}_1)d\theta\} \cdot (\mathbf{w}_2 - \mathbf{w}_1).\end{aligned}$$

Defining

$$\mathbf{S} = \mathbf{S}(\mathbf{u}_1, \mathbf{u}_2) = \int_0^1 \mathbf{g}'(\theta\mathbf{w}_2 + (1-\theta)\mathbf{w}_1)d\theta,$$

we obtain that the matrix $\mathbf{S}$ is symmetric and satisfies

$$\mathbf{f}(\mathbf{u}_2) - \mathbf{f}(\mathbf{u}_1) = \mathbf{g}(\mathbf{w}_2) - \mathbf{g}(\mathbf{w}_1) = \mathbf{S}(\mathbf{w}_2 - \mathbf{w}_1).$$

On the other hand, we have

$$\begin{aligned}\mathbf{w}_2 - \mathbf{w}_1 &= \int_0^1 \frac{d}{d\theta}\mathbf{w}(\theta\mathbf{u}_2 + (1-\theta)\mathbf{u}_1)d\theta \\ &= \{\int_0^1 \mathbf{w}'(\theta\mathbf{u}_2 + (1-\theta)\mathbf{u}_1)d\theta\} \cdot (\mathbf{u}_2 - \mathbf{u}_1).\end{aligned}$$

We thus define

$$\mathbf{P} = \mathbf{P}(\mathbf{u}_1, \mathbf{u}_2) = \int_0^1 \mathbf{w}'(\theta\mathbf{u}_2 + (1-\theta)\mathbf{u}_1)d\theta,$$

so that

$$\mathbf{w}_2 - \mathbf{w}_1 = \mathbf{P}(\mathbf{u}_2 - \mathbf{u}_1),$$

and it follows from the symmetry of the matrix $\mathbf{w}'(\mathbf{u}) = U''(\mathbf{u})$ that the matrix $\mathbf{P}$ is symmetric. Hence, we obtain

$$\mathbf{f}(\mathbf{u}_2) - \mathbf{f}(\mathbf{u}_1) = \mathbf{SP}(\mathbf{u}_2 - \mathbf{u}_1),$$

so that the condition (3.1) holds.

Finally, we notice that

$$\mathbf{S}(\mathbf{u}, \mathbf{u}) = \int_0^1 \mathbf{g}'(\theta\mathbf{w} + (1-\theta)\mathbf{w})d\theta = \mathbf{g}'(\mathbf{w}(\mathbf{u})),$$

$$\mathbf{P}(\mathbf{u}, \mathbf{u}) = \int_0^1 \mathbf{w}'(\mathbf{u})d\theta = \mathbf{w}'(\mathbf{u}),$$

and also

$$\mathbf{A}(\mathbf{u},\mathbf{u}) = \mathbf{S}(\mathbf{u},\mathbf{u})\mathbf{P}(\mathbf{u},\mathbf{u}) = \mathbf{g}'(\mathbf{w}(\mathbf{u}))\cdot\mathbf{w}'(\mathbf{u}) = \mathbf{f}'(\mathbf{u}) = \mathbf{A}(\mathbf{u}),$$

which is the consistency condition (3.3). □

Remark 3.1. It is straightforward to check that

$$\mathbf{P}(\mathbf{u}_2,\mathbf{u}_1) = \mathbf{P}(\mathbf{u}_1,\mathbf{u}_2),\ \ \mathbf{S}(\mathbf{u}_2,\mathbf{u}_1) = \mathbf{S}(\mathbf{u}_1,\mathbf{u}_2),$$

and therefore the Roe linearization constructed in the proof of the above theorem satisfies

$$\mathbf{A}(\mathbf{u},\mathbf{v}) = \mathbf{A}(\mathbf{v},\mathbf{u}).$$

This means that we can reverse the roles of $\mathbf{u}$ and $\mathbf{v}$. □

Given a Roe-type linearization, we now construct the Roe scheme. Instead of the Riemann problem (2.1), we consider the linearized Riemann problem

$$\begin{cases} \dfrac{\partial \mathbf{w}}{\partial t} + \mathbf{A}(\mathbf{u}_L,\mathbf{u}_R)\dfrac{\partial \mathbf{w}}{\partial x} = \mathbf{0}, & x\in\mathbb{R}, t>0,\\ \mathbf{w}(x,0) = \begin{cases} \mathbf{u}_L, & x<0,\\ \mathbf{u}_R, & x>0, \end{cases} \end{cases} \tag{3.4}$$

whose solution is of the form

$$\mathbf{w}(x,t) = \tilde{\mathbf{w}}_R\Big(\frac{x}{t};\mathbf{u}_L,\mathbf{u}_R\Big) \tag{3.5}$$

and can be explicitly constructed as explained in Chapter I, Section 1. Then instead of solving exactly the Cauchy problem (2.1) as in Godunov's method, we use the approximate Riemann solver (3.5). Hence, assuming that the C.F.L.-like condition

$$\lambda\,\max_j|a_k(\mathbf{v}_j^n,\mathbf{v}_{j+1}^n)| \le \frac{1}{2},\quad 1\le k\le p, \tag{3.6}$$

holds, where $a_k(\mathbf{u},\mathbf{v})$ is the kth eigenvalue of the matrix $\mathbf{A}(\mathbf{u},\mathbf{v})$, an approximate solution $\tilde{\mathbf{w}}$ of (2.4) is given by

$$\tilde{\mathbf{w}}(x,t) = \tilde{\mathbf{w}}_R\Big(\frac{x-x_{j+1/2}}{\Delta t};\mathbf{v}_j^n,\mathbf{v}_{j+1}^n\Big),\quad x_j<x<x_{j+1},\quad j\in\mathbb{Z}. \tag{3.7}$$

We thus define $\mathbf{v}_j^{n+1}$ by

$$\mathbf{v}_j^{n+1} = \frac{1}{\Delta x}\int_{x_{j-\frac12}}^{x_{j+\frac12}} \tilde{\mathbf{w}}(x,\Delta t)dx,$$

which gives

$$\mathbf{v}_j^{n+1} = \frac{1}{\Delta x}\{\int_0^{\Delta x/2} \tilde{\mathbf{w}}_R(\frac{x}{\Delta t}; \mathbf{v}_{j-1}^n, \mathbf{v}_j^n)dx + \int_{-\Delta x/2}^{0} \tilde{\mathbf{w}}_R(\frac{x}{\Delta t}; \mathbf{v}_j^n, \mathbf{v}_{j+1}^n)dx\}.$$

Using Lemma 2.1, we are able to give a simple expression for $\mathbf{v}_j^{n+1}$. It follows from (2.18) that

$$\frac{2}{\Delta x}\int_0^{\Delta x/2} \tilde{\mathbf{w}}_R(\frac{x}{\Delta t}; \mathbf{v}_{j-1}^n, \mathbf{v}_j^n)dx = \mathbf{v}_j^n - \lambda \mathbf{A}(\mathbf{v}_{j-1}^n, \mathbf{v}_j^n)(\mathbf{v}_j^n - \mathbf{v}_{j-1}^n) - \lambda |\mathbf{A}(\mathbf{v}_{j-1}^n, \mathbf{v}_j^n)|(\mathbf{v}_j^n - \mathbf{v}_{j-1}^n)$$

and

$$\frac{2}{\Delta x}\int_{-\Delta x/2}^{0} \tilde{\mathbf{w}}_R(\frac{x}{\Delta t}; \mathbf{v}_j^n, \mathbf{v}_{j+1}^n)dx = \mathbf{v}_j^n - \lambda \mathbf{A}(\mathbf{v}_j^n, \mathbf{v}_{j+1}^n)(\mathbf{v}_{j+1}^n - \mathbf{v}_j^n) - \lambda |\mathbf{A}(\mathbf{v}_j^n, \mathbf{v}_{j+1}^n)|(\mathbf{v}_{j+1}^n - \mathbf{v}_j^n).$$

Hence, setting

$$\mathbf{A}(\mathbf{v}_j^n, \mathbf{v}_{j+1}^n) = \mathbf{A}_{j+\frac{1}{2}}^n, \tag{3.8}$$

we obtain the numerical scheme

$$\mathbf{v}_j^{n+1} = \mathbf{v}_j^n - \frac{\lambda}{2}\{(\mathbf{A}_{j+\frac{1}{2}}^n - |\mathbf{A}_{j+\frac{1}{2}}^n|)(\mathbf{v}_{j+1}^n - \mathbf{v}_j^n) + (\mathbf{A}_{j-\frac{1}{2}}^n + |\mathbf{A}_{j-\frac{1}{2}}^n|)(\mathbf{v}_j^n - \mathbf{v}_{j-1}^n)\}$$

or, equivalently,

$$\mathbf{v}_j^{n+1} = \mathbf{v}_j^n - \lambda\{(\mathbf{A}_{j+\frac{1}{2}}^n)^-(\mathbf{v}_{j+1}^n - \mathbf{v}_j^n) + (\mathbf{A}_{j-\frac{1}{2}}^n)^+(\mathbf{v}_j^n - \mathbf{v}_{j-1}^n)\}. \tag{3.9}$$

We note the similarity of (3.9) with the Godunov scheme in the linear case (2.16); the constant matrix $\mathbf{A}$ is only replaced by Roe's matrix.

By (3.1)

$$\mathbf{A}_{j+\frac{1}{2}}^n(\mathbf{v}_{j+1}^n - \mathbf{v}_j^n) = \mathbf{f}(\mathbf{v}_{j+1}^n) - \mathbf{f}(\mathbf{v}_j^n),$$

and hence the numerical scheme can also be written

$$\begin{cases} \mathbf{v}_j^{n+1} = \mathbf{v}_j^n - \frac{\lambda}{2}(\mathbf{f}(\mathbf{v}_{j+1}^n) - \mathbf{f}(\mathbf{v}_{j-1}^n)) \\ \qquad + \frac{\lambda}{2}\{|\mathbf{A}_{j+\frac{1}{2}}^n|(\mathbf{v}_{j+1}^n - \mathbf{v}_j^n) - |\mathbf{A}_{j-\frac{1}{2}}^n|(\mathbf{v}_j^n - \mathbf{v}_{j-1}^n)\}. \end{cases} \tag{3.10}$$

The corresponding numerical flux function is given by

$$\mathbf{g}(\mathbf{u}, \mathbf{v}) = \frac{1}{2}(\mathbf{f}(\mathbf{u}) + \mathbf{f}(\mathbf{v})) - \frac{1}{2}|\mathbf{A}(\mathbf{u}, \mathbf{v})|(\mathbf{v} - \mathbf{u}). \tag{3.11a}$$

Let us give a more explicit form of Roe's scheme. We denote by $\mathbf{r}_k(\mathbf{u}, \mathbf{v}), 1 \le k \le p$, the eigenvectors of the matrix $\mathbf{A}(\mathbf{u}, \mathbf{v})$ associated with the eigenvalues $a_k(\mathbf{u}, \mathbf{v})$, so that

$$\mathbf{A}(\mathbf{u}, \mathbf{v})\mathbf{r}_k(\mathbf{u}, \mathbf{v}) = a_k(\mathbf{u}, \mathbf{v})\mathbf{r}_k(\mathbf{u}, \mathbf{v}), \ 1 \le k \le p.$$

We define $\alpha_k(\mathbf{u}, \mathbf{v}), 1 \le k \le p$ by $\alpha_k(\mathbf{u}, \mathbf{v}) = \mathbf{l}_k^T(\mathbf{u}, \mathbf{v})\ (\mathbf{v} - \mathbf{u})$, or

$$\mathbf{v} - \mathbf{u} = \sum_{k=1}^{p} \alpha_k(\mathbf{u}, \mathbf{v})\mathbf{r}_k(\mathbf{u}, \mathbf{v}). \tag{3.12}$$

Then (3.11a) reads

$$\mathbf{g}(\mathbf{u}, \mathbf{v}) = \frac{1}{2}(\mathbf{f}(\mathbf{u}) + \mathbf{f}(\mathbf{v})) - \sum_{k=1}^{p}(\alpha_k|a_k|\mathbf{r}_k)(\mathbf{u}, \mathbf{v}). \tag{3.11b}$$

Similarly, we can write (3.9) in the form

$$\mathbf{v}_j^{n+1} = \mathbf{v}_j^n - \lambda \sum_{k=1}^{p}\{(\alpha_k a_k^- \mathbf{r}_k)_{j+\frac{1}{2}}^n + (\alpha_k a_k^+ \mathbf{r}_k)_{j-\frac{1}{2}}^n\}, \tag{3.13}$$

where

$$(\mathbf{r}_k)_{j+\frac{1}{2}}^n = \mathbf{r}_k(\mathbf{v}_j^n, \mathbf{v}_{j+1}^n),$$

and so on.

A well-known drawback of Roe's scheme is that it may resolve nonphysical solutions (see G.R., Chapter III, Examples 2.4 and 4.1; see also Einfeldt et al. 1991). Indeed, by (3.1) and (3.11), for a stationary discontinuity, $\mathbf{f}(\mathbf{u}) = \mathbf{f}(\mathbf{v})$ implies $\mathbf{g}(\mathbf{u}, \mathbf{v}) = \mathbf{f}(\mathbf{u})$ whether or not the entropy condition is satisfied. Various entropy corrections have been proposed (see Harten and Hyman 1983, Roe 1983, Roe and Pike 1983, Huynh 1995; Yee's formula is given in Lin 1995). If we write the scheme (3.10) in viscous form (1.13), we see that the viscosity matrix is

$$\mathbf{Q}(\mathbf{u}, \mathbf{v}) = \lambda|\mathbf{A}(\mathbf{u}, \mathbf{v})|,$$

and

$$\mathbf{Q}(\mathbf{u}, \mathbf{v})(\mathbf{v} - \mathbf{u}) = \lambda \sum_{k=1}^{p}(\alpha_k|a_k|\mathbf{r}_k)(\mathbf{u}, \mathbf{v}). \tag{3.14}$$

In the basis of eigenvectors $\mathbf{r}_k(\mathbf{u}, \mathbf{v})$, the viscosity matrix is diagonal, $\mathbf{Q} = \operatorname{diag} \lambda|a_k(\mathbf{u}, \mathbf{v})|$, and we may think that Roe's scheme is not viscous enough in the neighborhood of the sonic points. Harten (1983) has proposed replacing the function $Q(x) = \lambda|x|$ in the diagonal of $\mathbf{Q}$ by a smooth parabolic regularization Q_δ,

$$Q_\delta(x) = \begin{cases} \lambda|x|, & |x| \ge \delta, \\ \lambda(\dfrac{x^2}{2\delta} + \dfrac{\delta}{2}), & |x| < \delta, \end{cases}$$

which corresponds to adding viscosity near sonic points. Unfortunately, the actual tuning of the parameter δ is empirical and depends on the problem considered. Roe's "spreading device" (Roe 1985) does not contain an arbitrary parameter but is not so easily exhibited.

We can also extend to systems the modification of Harten and Hyman that we have presented in the scalar case (see G.R., Chapter III, Example 4.1). Indeed (see Chapter I, Section 1) the initial discontinuity in the solution of the linear Riemann problem (3.4) breaks up into p discontinuity waves that propagate with speed $a_k(\mathbf{u}_L, \mathbf{u}_R)$, $1 \le k \le p$,

$$\mathbf{w}(x,t) = \sum_{k=1}^{p} w_k(x,t)\mathbf{r}_k(\mathbf{u}, \mathbf{v}).$$

where

$$w_k(x,t) = \begin{cases} \alpha_{kL}, & \dfrac{x}{t} < a_k(\mathbf{u}_L, \mathbf{u}_R), \\ \alpha_{kR}, & \dfrac{x}{t} > a_k(\mathbf{u}_L, \mathbf{u}_R), \end{cases}$$

where α_{kL} and α_{kR} are respectively the components of $\mathbf{u}_L$ and $\mathbf{u}_R$ in the basis of eigenvectors $\mathbf{r}_k(\mathbf{u}_L, \mathbf{u}_R)$. If the kth field is genuinely nonlinear (which corresponds to the convex scalar case) and if the corresponding kth elementary wave of the exact solution of the Riemann problem (2.1) is a rarefaction wave, we introduce an intermediate state

$$w_k(x,t) = \begin{cases} \alpha_{kL}, & \dfrac{x}{t} < a_k^L, \\ \alpha_k^*, & a_k^L < \dfrac{x}{t} < a_k^R, \\ \alpha_{kR}, & \dfrac{x}{t} > a_k^R. \end{cases}$$

If a_k^L, a_k^R, and δ_k are chosen as in the scalar case,

$$a_k^L = a_k(\mathbf{u}_L, \mathbf{u}_R) - \delta_k,\ a_k^R = a_k(\mathbf{u}_L, \mathbf{u}_R) + \delta_k,$$

$$\delta_k = \sup\Big(0, a_k(\mathbf{u}_L, \mathbf{u}_R) - a_k(\mathbf{u}_L, \mathbf{u}), a_k(\mathbf{u}, \mathbf{u}_R) - a_k(\mathbf{u}_L, \mathbf{u}_R)\Big),$$

the supremum is taken over all $\mathbf{u} = \theta\mathbf{u}_L + (1-\theta)\mathbf{u}_R$ and

$$\alpha_k^* = \frac{1}{2}(\alpha_{kL} + \alpha_{kR}).$$

It amounts to modifying the viscosity componentwise, i.e., to replacing the kth diagonal element $Q_k = \lambda|a_k|$ of $\mathbf{Q}$ by

$$Q_k = \begin{cases} \lambda|a_k|, & |a_k| \ge \delta_k, \\ \lambda\delta_k, & |a_k| < \delta_k. \end{cases}$$

We mention another approach based on a nonlinear smooth modification of the flux function (by a Hermite polynomial of degree 3) at sonic points only and that gives very good results (see Dubois and Mehlman 1992). Moreover, interesting considerations can be found in Lin (1995).

Finally, among the other "entropy fix" we present the LLF (Shu and Osher 1989), which we begin by detailing in the scalar case. One defines first a *Local Lax–Friedrichs scheme*, whose numerical flux is

$$g_{j+\frac{1}{2}}^{LLF} = \frac{1}{2}(f(v_j) + f(v_{j+1}) - \sigma_{j+\frac{1}{2}} \Delta v_{j+\frac{1}{2}}),$$

where

$$\sigma_{j+\frac{1}{2}} = \max\{|f'(u)|, u \in [v_j, v_{j+1}]\}.$$

The classical Lax–Friedrichs scheme would correspond to $\sigma_{j+\frac{1}{2}} = 1/\lambda$, which is the upper bound of all the $\sigma_{j+\frac{1}{2}}$ (under CFL ≤ 1). Assuming for instance that f is convex,

$$\sigma_{j+\frac{1}{2}} = \max\{|f'(v_j)|, |f'(v_{j+1})|\},$$

it is easy to check that the scheme is monotone (under CFL $\leq 1 - \varepsilon$) by computing the partial derivatives $\frac{\partial H^{LLF}}{\partial v_i}(v_{-1}, v_0, v_1)$, $i = -1, 0, 1$ (see G.R., Chapter III, Section 3.1). Its numerical viscosity is given by

$$Q_{j+\frac{1}{2}} = \lambda \sigma_{j+\frac{1}{2}}.$$

Then one considers *Roe's scheme with LLF*: at a sonic point, one turns back from the upwind flux to the above local Lax–Friedrichs flux; thus

$$g_{j+\frac{1}{2}} = \begin{cases} f(v_j) & \text{if } f'(u) \geq 0, \quad u \text{ between } v_j, v_{j+1}, \\ f(v_{j+1}) & \text{if } f'(u) \leq 0, \quad u \text{ between } v_j, v_{j+1}, \\ g_{j+\frac{1}{2}}^{LLF} & \text{otherwise.} \end{cases}$$

For a system, we apply the procedure to each characteristic field. Using the same notations as above, $(\alpha_k, a_k, \mathbf{r}_k, \mathbf{l}_k)_{j+\frac{1}{2}}^n$, the kth component Ψ_k of the LLF-flux Ψ on the eigenbasis $\mathbf{r}_{k,j+\frac{1}{2}}$ is

$$\Psi_{k,j+\frac{1}{2}} = \frac{1}{2}(\mathbf{f}(\mathbf{v}_j) + (\mathbf{f}(\mathbf{v}_{j+1}))_k - \frac{1}{2}\sigma_{k,j+\frac{1}{2}} \alpha_{k,j+\frac{1}{2}},$$

where $(\mathbf{f}(\mathbf{v}_j))_k$ denotes the kth component of $\mathbf{f}(\mathbf{v}_j)$ in the same basis, and the coefficients σ_k are defined by

$$\sigma_{k,j+\frac{1}{2}} = \max\{a_k(\mathbf{v}_j), a_k(\mathbf{v}_{j+1})\}$$

(if all the characteristic fields are genuinely nonlinear or linearly degenerate, which holds for the gas dynamics equations). The $a_k(\mathbf{v}_j)$ denote the eigenvalues of the matrix $\mathbf{A}(\mathbf{v}_j)$, whereas the $a_{k,j+\frac{1}{2}}$ denote the eigenvalues of the matrix $\mathbf{A}_{j+\frac{1}{2}}$. Then

$$\mathbf{g}(\mathbf{v}_j, \mathbf{v}_{j+1}) = \sum_k (\Phi_k \mathbf{r}_k)_{j+\frac{1}{2}},$$

where if the signs of $a_k(\mathbf{v}_j)$ and $a_k(\mathbf{v}_{j+1})$ are identical, we take for Φ_k the kth component of the upwind flux *i.e.*,

$$\Phi_{k,j+\frac{1}{2}} = \begin{cases} (\mathbf{f}(\mathbf{v}_j))_k, & \text{if both } a_k \geq 0, \\ (\mathbf{f}(\mathbf{v}_{j+1}))_k, & \text{if both } a_k < 0. \end{cases}$$

In the "sonic case," for an index k such that the signs of $a_k(\mathbf{v}_j)$ and $a_k(\mathbf{v}_{j+1})$ are not identical, we turn to the LLF flux and take for Φ_k the kth component Ψ_k.

Remark 3.2. In Roe's scheme, the approximate Riemann solver contains a priori $p-1$ intermediate states between $\mathbf{u}_i$ and $\mathbf{u}_{i+1}$ separated by the signal velocities $a_k(\mathbf{u}_i, \mathbf{u}_{i+1})$. We must then mention another drawback of Roe's scheme. When used for the approximation of the Euler equations, it may generate in some particular cases nonphysical intermediate states, for instance with negative internal energy or density (see Einfeldt et al. 1991, who describe this scheme as nonpositively conservative). Also, the eigenvalues $a_k(\mathbf{u}_i, \mathbf{u}_{i+1}), k = 1$ or 3, could lie outside the range of values $(a_k(\mathbf{u}_i), a_k(\mathbf{u}_{i+1}))$ respectively (see Vinokur 1989, Einfeldt et al. 1991).

An improvement is suggested by the HLLE scheme (for Harten, Lax, van Leer, and Einfeldt), associated to a Riemann solver that contains the left and right initial states plus only one intermediate state. A numerical approximation to the largest and smallest signal velocities is derived, for instance

$$b^L_{i+\frac{1}{2}} = \min\Big(a_{1,i+\frac{1}{2}}, a_1(\mathbf{v}_i)\Big), \quad b^R_{i+\frac{1}{2}} = \max\Big(a_{3,i+\frac{1}{2}}, a_3(\mathbf{v}_{i+1})\Big),$$

which are lower and upper bounds for the physical signal velocities (at least in the case of rarefactions, but not for shocks because of Lax entropy inequalities). Then, the intermediate state, say $\mathbf{v}_{i+\frac{1}{2}}$, is computed in order to get consistency with the integral form of the conservation laws (see G.R., Chapter III, 4.3), which means that the integrals of the exact and approximate Riemann solver over a cell coincide, which yields

$$(b^R_{i+\frac{1}{2}} - b^L_{i+\frac{1}{2}})\mathbf{v}_{i+\frac{1}{2}} = b^R_{i+\frac{1}{2}}\mathbf{v}_{i+1} - b^L_{i+\frac{1}{2}}\mathbf{v}_i - (\mathbf{f}_{i+1} - \mathbf{f}_i).$$

If the two states $\mathbf{v}_i$ and $\mathbf{v}_{i+1}$ are separated only by a shock (1- or 3- shock), the solver is exact provided one chooses for b (b^L or b^R) the exact signal velocity (σ_1 or σ_3), but contact discontinuities are not exactly resolved. Hence, one can also think of a scheme with two intermediate states: a third approximate velocity is then needed with the role of the contact discontinuity, and the two intermediate states are still computed so as to get consistency with the integral form of the conservation laws.

Besides, its simplicity, the method applied to gas dynamics does not predict non physical states since the averaging process involved keeps the approximate solution in the set of physical states Ω, which is convex. The HLLEM version introduces some antidiffusion with the aim of improving the resolution (see Harten, Lax, and van Leer 1983, Einfeldt 1988, Einfeldt et al.1991, Davis 1988, Charrier et al. 1993). □

3.2 Application to the gas dynamics equations. I. The ideal gas case

3.2.1 Computation of Roe's matrix via parameter vectors

In order to apply Roe's method to the gas dynamics equations (2.19), it remains to construct a Roe-type linearization. We present the derivation in the original way (Roe 1981), only considering the case of an equation of state of the form

$$p = (\gamma - 1)\rho\varepsilon + c_{\mathrm{ref}}^2\,(\rho - \rho_{\mathrm{ref}}), \tag{3.15}$$

which generalizes the usual ideal gas law and is known as a stiffened equation of state of Grüneisen type (see Chapter II, Example 1.2). It is convenient to set $p_\infty = c_{\mathrm{ref}}^2\rho_{\mathrm{ref}}\,/\gamma$. However, we shall try to carry on the computations in a more general case as long as possible.

In this case, we are able to find the Roe matrix in the form

$$\mathbf{A}(\mathbf{U},\mathbf{V}) = \mathbf{A}(M(\mathbf{U},\mathbf{V})),$$

where $\mathbf{A}$ is the Jacobian matrix of $\mathbf{F}, \mathbf{A} = \mathbf{F}'$, and $M(\mathbf{U},\mathbf{V})$ is some average of the two states $\mathbf{U},\mathbf{V}$ that satisfies

$$\begin{aligned} M(\mathbf{U},\mathbf{V}) &= M(\mathbf{V},\mathbf{U}),\\ M(\mathbf{U},\mathbf{U}) &= \mathbf{U}. \end{aligned}$$

The operator M will be constructed via a change of variables $\mathbf{W} \to \mathbf{U}(\mathbf{W})$, where $\mathbf{W}$ is the *parameter vector*, such that if we set

$$\mathbf{G}(\mathbf{W}) = \mathbf{F}(\mathbf{U}(\mathbf{W})), \tag{3.16}$$

then

(3.17) $\quad\mathbf{U}(\mathbf{W})$ and $\mathbf{G}(\mathbf{W})$ are homogeneous quadratic functions of $\mathbf{W}$

(up to an additive constant vector).

Lemma 3.1

Assume that there exists a change of variables $\mathbf{W} \to \mathbf{U}(\mathbf{W})$ such that (3.17) holds. The expressions

$$\mathbf{A}(\mathbf{U}_R,\mathbf{U}_L) = \mathbf{A}(\overline{\mathbf{U}}), \overline{\mathbf{U}} = \mathbf{U}(\mathbf{W}^*), \mathbf{W}^* = \frac{1}{2}(\mathbf{W}_R + \mathbf{W}_L) \tag{3.18}$$

define a Roe-type linearization.

Proof. Assume that (3.17) holds. This yields

$$\mathbf{U}_R - \mathbf{U}_L = \mathbf{U}(\mathbf{W}_R) - \mathbf{U}(\mathbf{W}_L) = \mathbf{U}'(\mathbf{W}^*)(\mathbf{W}_R - \mathbf{W}_L)$$

where

$$\mathbf{W}^* = \frac{1}{2}(\mathbf{W}_R + \mathbf{W}_L)$$

and

$$\mathbf{F}(\mathbf{U}_R) - \mathbf{F}(\mathbf{U}_L) = \mathbf{G}(\mathbf{W}_R) - \mathbf{G}(\mathbf{W}_L) = \mathbf{G}'(\mathbf{W}^*)(\mathbf{W}_R - \mathbf{W}_L).$$

Thus, we get

$$\begin{aligned}\mathbf{F}(\mathbf{U}_R) - \mathbf{F}(\mathbf{U}_L) &= \mathbf{F}'(\mathbf{U}(\mathbf{W}^*))\mathbf{U}'(\mathbf{W}^*)(\mathbf{W}_R - \mathbf{W}_L)\\ &= \mathbf{A}(\mathbf{U}(\mathbf{W}^*))(\mathbf{U}_R - \mathbf{U}_L).\end{aligned}$$

Setting

$$\overline{\mathbf{U}} = \mathbf{U}(\mathbf{W}^*), \mathbf{W}^* = \frac{1}{2}(\mathbf{W}_R + \mathbf{W}_L),$$

we have

$$\mathbf{F}(\mathbf{U}_R) - \mathbf{F}(\mathbf{U}_L) = \mathbf{A}(\overline{\mathbf{U}})(\mathbf{U}_R - \mathbf{U}_L),$$

and

$$M(\mathbf{U}, \mathbf{V}) = \overline{\mathbf{U}} \tag{3.19}$$

is the desired "Roe-averaged state." □

In this case, Roe's scheme can be written

$$\left\{\begin{aligned}\mathbf{v}_j^{n+1} = \mathbf{v}_j^n - \lambda \sum_{k=1}^{3} \alpha_{k,j+\frac{1}{2}}^n\, a_k^-(\overline{\mathbf{v}}_{j+\frac{1}{2}}^n)\mathbf{r}_k(\overline{\mathbf{v}}_{j+\frac{1}{2}}^n)\\ + \alpha_{k,j-\frac{1}{2}}^n\, a_k^+(\overline{\mathbf{v}}_{j-\frac{1}{2}}^n)\mathbf{r}_k(\overline{\mathbf{v}}_{j-\frac{1}{2}}^n),\end{aligned}\right. \tag{3.20}$$

where again $a_1(\overline{\mathbf{U}}) = \overline{u} - \overline{c}$, $a_2(\overline{\mathbf{U}}) = \overline{u}$, $a_3(\overline{\mathbf{U}}) = \overline{u} + \overline{c}$, and by (3.18) and (3.19)

$$\overline{\mathbf{v}}_{j+\frac{1}{2}}^n = M(\mathbf{v}_j^n, \mathbf{v}_{j+1}^n);$$

moreover, the coefficients α_k are defined by (3.12), i.e.,

$$\mathbf{v}_{j+1}^n - \mathbf{v}_j^n = \Delta\mathbf{v}_{j+\frac{1}{2}}^n = \sum_{k=1}^{3} \alpha_{k,j+\frac{1}{2}}^n \mathbf{r}_k(\overline{\mathbf{v}}_{j+\frac{1}{2}}^n). \tag{3.21}$$

Let us check the property (3.17) in the case of an equation of state (3.15). We write (2.19) as

$$\mathbf{U} = \begin{pmatrix}\rho\\ q\\ E\end{pmatrix}, \quad \mathbf{F}(\mathbf{U}) = \begin{pmatrix} q\\ q^2/\rho + p\\ (E+p)q/\rho\end{pmatrix},$$

where

$$q = \rho u,\ E = \rho e = \rho(\varepsilon + \frac{1}{2}u^2),\ p = p(\rho, \varepsilon) = p\Big(\rho, \frac{E}{\rho} - \frac{q^2}{2\rho^2}\Big).$$

We introduce the total specific enthalpy

$$H = e + \frac{p}{\rho}$$

and set

$$\mathbf{W} = \begin{pmatrix} w_1 \\ w_2 \\ w_3 \end{pmatrix} = \begin{pmatrix} \rho^{\frac{1}{2}} \\ \rho^{\frac{1}{2}} u \\ \rho^{\frac{1}{2}} H \end{pmatrix}. \tag{3.22}$$

Lemma 3.2

Assume an equation of state of the form (3.15). Defining the parameter vector **W** *by (3.22), the property (3.17) is satisfied. More precisely, we have*

$$\begin{cases} \mathbf{U}(\mathbf{W}) = (w_1^2, w_1 w_2, w_1 w_3 - p)^T, \\ \mathbf{G}(\mathbf{W}) = (w_1 w_2, w_2^2 + p, w_2 w_3)^T, \end{cases} \tag{3.23}$$

where

$$p + p_\infty = \frac{c_{\text{ref}}^2}{\gamma} w_1^2 - \frac{\gamma - 1}{2\gamma} w_2^2 + \frac{\gamma - 1}{\gamma} w_1 w_3.$$

Proof. It follows immediately from (3.22) that

$$\begin{aligned} \rho &= w_1^2, \\ q &= \rho^{\frac{1}{2}} (\rho^{\frac{1}{2}} u) = w_1 w_2, \\ E &= \rho H - p = w_1 w_3 - p. \end{aligned} \tag{3.24}$$

Let us now express the pressure in terms of **W**. On the one hand,

$$\rho H = \rho^{\frac{1}{2}} (\rho^{\frac{1}{2}} H) = w_1 w_3;$$

on the other hand,

$$\begin{aligned} \rho H &= \rho\Big(e + \frac{p}{\rho}\Big) = E + p = \rho\Big(\varepsilon + \frac{u^2}{2}\Big) + p \\ &= \gamma(\gamma - 1)^{-1} p - (\gamma - 1)^{-1} c_{\text{ref}}^2 (\rho - \rho_{\text{ref}}) + \frac{\rho u^2}{2}. \end{aligned}$$

With the notation p_∞, we get

$$\begin{aligned} \rho H &= \gamma(\gamma - 1)^{-1}(p + p_\infty) - (\gamma - 1)^{-1} c_{\text{ref}}^2 \rho + \frac{\rho u^2}{2} \\ &= \gamma(\gamma - 1)^{-1}(p + p_\infty) - (\gamma - 1)^{-1} c_{\text{ref}}^2 w_1^2 + \frac{w_2^2}{2}. \end{aligned}$$

Equating the two expressions for ρH, we obtain

$$p + p_\infty = \frac{c_{\text{ref}}^2}{\gamma} w_1^2 - \frac{\gamma - 1}{2\gamma} w_2^2 + \frac{\gamma - 1}{\gamma} w_1 w_3. \tag{3.25}$$

Thus, we have

$$E = \frac{w_1 w_3}{\gamma} + \frac{\gamma - 1}{2\gamma}\, w_2^2 - \frac{c_{\text{ref}}^2}{\gamma}\, w_1^2 + p_\infty.$$

Note that for an ideal gas, $p_\infty = 0 = c_{\text{ref}}$, and the expression for E simplifies into

$$E = \frac{1}{\gamma}\, w_1 w_3 + \frac{\gamma - 1}{2\gamma}\, w_2^2.$$

Consider next $\mathbf{F}(\mathbf{U})$; we have already

$$q = w_1 w_2;$$

then

$$\begin{aligned} \frac{q^2}{\rho} + p &= \rho u^2 + p \\ &= \frac{\gamma - 1}{\gamma}\, w_1 w_3 + \frac{\gamma + 1}{2\gamma}\, w_2^2 + \frac{c_{\text{ref}}^2}{\gamma}\, w_1^2 - p_\infty \end{aligned}$$

and

$$(E + p)\,\frac{q}{\rho} = (\rho e + p)u = \rho H u = w_2 w_3,$$

which ends the proof. □

Expressions (3.23) show that $\mathbf{U}$ is indeed a homogeneous quadratic function of $\mathbf{W}$ up to an additive constant vector, so that we may apply Lemma 3.1. It remains to compute the matrix $\mathbf{A}(\overline{\mathbf{U}})$. We give first in the following lemma the expression for the sound speed; this indeed can be done for a more general equation of state (see Chapter II, Section 1.2, (1.20)).

Lemma 3.3

Assume an equation of state of the form

$$p = p(\rho, \tilde{\varepsilon}), \tag{3.26}$$

where

$$\tilde{\varepsilon} = \rho\varepsilon = E - \frac{\rho u^2}{2}.$$

Defining the specific enthalpy h by

$$h = \varepsilon + \frac{p}{\rho}$$

and setting

$$\kappa = p_{\tilde{\varepsilon}} = \frac{\partial p(\rho, \tilde{\varepsilon})}{\partial \tilde{\varepsilon}}, \quad \chi = p_\rho = \frac{\partial p(\rho, \tilde{\varepsilon})}{\partial \rho}, \tag{3.27}$$

the sound velocity c satisfies the identity

$$c^2 = \kappa h + \chi. \tag{3.28}$$

Proof. The sound speed c is defined (see Chapter II, (1.4)) by

$$c^2 = \tau^2\Big(-\frac{\partial p(\tau, s)}{\partial \tau}\Big) = \frac{\partial p(\rho, s)}{\partial \rho},$$

and hence by (3.26)

$$c^2 = \frac{\partial p(\rho, s)}{\partial \rho} = \frac{\partial p(\rho, \tilde{\varepsilon})}{\partial \tilde{\varepsilon}} \frac{\partial \tilde{\varepsilon}(\rho, s)}{\partial \rho} + \frac{\partial p(\rho, \tilde{\varepsilon})}{\partial \rho}.$$

Now, the second law of thermodynamics,

$$d\varepsilon = T ds - p d\tau,$$

implies

$$d\tilde{\varepsilon} = \rho d\varepsilon + \varepsilon d\rho = \rho T ds + \Big(\varepsilon + \frac{p}{\rho}\Big) d\rho.$$

Introducing the specific enthalpy h, we have

$$\frac{\partial \tilde{\varepsilon}(\rho, s)}{\partial \rho} = h,$$

and with shorthand notations we get

$$c^2 = p_{\tilde{\varepsilon}} h + p_\rho = \kappa h + \chi,$$

which is (3.28). □

Note that in the particular case of an equation of state of the form (3.15), we have

$$\begin{cases} \kappa = \gamma - 1, \\ \chi = c_{\text{ref}}^2, \end{cases} \tag{3.29}$$

and the expression (3.28) for c^2 gives

$$c^2 = (\gamma - 1)\Big(H - \frac{u^2}{2}\Big) + c_{\text{ref}}^2,$$

since h is related to the total specific enthalpy H by

$$H = h + \frac{u^2}{2}.$$

Lemma 3.4

Assume the hypothesis of Lemma 3.3. The Jacobian matrix $\mathbf{A}$ *is given by*

$$\mathbf{A}(\mathbf{U}) = \begin{pmatrix} 0 & 1 & 0 \\ K - u^2 & (2-\kappa)u & \kappa \\ u(K - H) & H - \kappa u^2 & (1+\kappa)u \end{pmatrix}, \tag{3.30}$$

where we have set

$$K = p_\rho + p_{\tilde{\varepsilon}} \frac{u^2}{2} = \chi + \kappa \frac{u^2}{2}, \tag{3.31}$$

and the eigenvectors may be chosen as

$$\begin{cases} \mathbf{r}_1(\mathbf{U}) = (1, u - c, H - uc)^T, \\ \mathbf{r}_2(\mathbf{U}) = (1, u, H - \dfrac{c^2}{\kappa})^T, \\ \mathbf{r}_3(\mathbf{U}) = (1, u + c, H + uc)^T. \end{cases} \tag{3.32}$$

Proof. We have already observed (Chapter I, Section 2) that we can work with a nonconservative form of the system. Here, it is convenient to use

$$\begin{cases} \dfrac{\partial \rho}{\partial t} + \dfrac{\partial q}{\partial x} = 0, \\ \dfrac{\partial q}{\partial t} + \dfrac{\partial (\rho u^2)}{\partial x} + \dfrac{\partial p}{\partial x} = 0, \\ \dfrac{\partial E}{\partial t} + u \dfrac{\partial E}{\partial x} + u \dfrac{\partial p}{\partial x} + (E + p) \dfrac{\partial u}{\partial x} = 0. \end{cases} \tag{3.33}$$

We have

$$\frac{\partial p(\rho, \tilde{\varepsilon})}{\partial x} = p_\rho \frac{\partial \rho}{\partial x} + p_{\tilde{\varepsilon}} \frac{\partial \tilde{\varepsilon}}{\partial x}.$$

First

$$\frac{\partial \tilde{\varepsilon}}{\partial x} = \frac{\partial E}{\partial x} - \frac{1}{2} \frac{\partial}{\partial x} (\rho u^2),$$

and an easy computation gives

$$\frac{\partial}{\partial x}\Big(\frac{\rho u^2}{2}\Big) = \rho \frac{\partial}{\partial x}\Big(\frac{u^2}{2}\Big) + \frac{u^2}{2} \frac{\partial \rho}{\partial x} = u \frac{\partial q}{\partial x} - \frac{u^2}{2} \frac{\partial \rho}{\partial x}.$$

Thus,

$$\frac{\partial p}{\partial x} = p_{\tilde{\varepsilon}} \frac{\partial E}{\partial x} - p_{\tilde{\varepsilon}} u \frac{\partial q}{\partial x} + \Big(p_\rho + \frac{p_{\tilde{\varepsilon}} u^2}{2}\Big) \frac{\partial \rho}{\partial x}$$

and

$$\frac{\partial (\rho u^2)}{\partial x} + \frac{\partial p}{\partial x} = \Big(p_\rho + \frac{p_{\tilde{\varepsilon}} u^2}{2} - u^2\Big) \frac{\partial \rho}{\partial x} + (2 - p_{\tilde{\varepsilon}}) u \frac{\partial q}{\partial x} + p_{\tilde{\varepsilon}} \frac{\partial E}{\partial x}.$$

Now,

$$(E + p) \frac{\partial u}{\partial x} = \rho H \frac{\partial u}{\partial x} = H\Big(\frac{\partial q}{\partial x} - u \frac{\partial \rho}{\partial x}\Big),$$

so that the last equation gives

$$\frac{\partial}{\partial x}\Big((E+p)u\Big) = u\Big(-H + p_\rho + \frac{p_{\tilde{\varepsilon}} u^2}{2}\Big) \frac{\partial \rho}{\partial x} + (H - p_{\tilde{\varepsilon}} u^2) \frac{\partial q}{\partial x} + (p_{\tilde{\varepsilon}} + 1) u \frac{\partial E}{\partial x}.$$

Defining K by (3.31) or, equivalently,

$$K = c^2 - \kappa\Big(h - \frac{u^2}{2}\Big) = c^2 - \kappa(H - u^2),$$

we get the desired expression for $\mathbf{A}$. Let us note that

$$K = \frac{\partial p(\rho, \tilde{\varepsilon})}{\partial \rho} + \frac{\partial p(\rho, \tilde{\varepsilon})}{\partial \tilde{\varepsilon}} \Big(\frac{\partial \tilde{\varepsilon}}{\partial \rho}\Big)_{|q,E} = \Big(\frac{\partial p}{\partial \rho}\Big)_{|q,E}$$

since $\tilde{\varepsilon} = E - \frac{q^2}{2\rho}$ implies $(\frac{\partial \tilde{\varepsilon}}{\partial \rho})_{|q,E} = \frac{u^2}{2}$.

Using (3.28), it is then easy to check that the eigenvectors may be chosen as in (3.32). □

In the case of equation (3.15), the Jacobian (3.30) gives (in the particular case of an ideal gas, we set $c_{\text{ref}}^2 = 0$)

(3.34)

$$\mathbf{A}(\mathbf{U}) = \begin{pmatrix} 0 & 1 & 0 \\ (\gamma - 3)\frac{u^2}{2} + c_{\text{ref}}^2 & -(\gamma - 3)u & (\gamma - 1) \\ u(-H + c_{\text{ref}}^2 + (\gamma - 1)\frac{u^2}{2}) & H - (\gamma - 1)u^2 & \gamma u \end{pmatrix}.$$

From (3.32) and (3.34), we notice that for the equation of state (3.15), we need only compute the values $\overline{u}, \overline{H}, \overline{c}$ in order to determine $\mathbf{A}(\overline{\mathbf{U}})$ and the eigenvectors $\mathbf{r}_k(\overline{\mathbf{U}})$, since χ and κ are constant. In particular, for an ideal polytropic gas, $p = (\gamma - 1)\rho\varepsilon$ implies a simplification of the last component of $\mathbf{r}_2$, which can be written $\mathbf{r}_2 = (1, u, \frac{1}{2}|u|^2)^T$.

Lemma 3.5

Assume an equation of state of the form (3.15). The velocity, total specific enthalpy, and sound speed at Roe-averaged state $\overline{\mathbf{U}}$ *are given (using the notations (3.29)) by*

(3.35a)
$$\overline{u} = \frac{\sqrt{\rho_L} u_L + \sqrt{\rho_R} u_R}{\sqrt{\rho_L} + \sqrt{\rho_R}},$$

(3.35b)
$$\overline{H} = \frac{\sqrt{\rho_L} H_L + \sqrt{\rho_R} H_R}{\sqrt{\rho_L} + \sqrt{\rho_R}},$$

(3.35c)
$$\overline{c}^2 = \kappa\Big(\overline{H} - \frac{\overline{u}^2}{2}\Big) + \chi.$$

Proof. First, since the function $\mathbf{U}(\mathbf{W})$ satisfies

$$u = \frac{q}{\rho} = \frac{w_2}{w_1}, \qquad H = \frac{w_3}{w_1},$$

we have

$$\overline{u} = \frac{w_2^*}{w_1^*} \quad \text{and} \quad \overline{H} = \frac{w_3^*}{w_1^*},$$

where $\mathbf{W}^* = (w_1^*, w_2^*, w_3^*)^T$. By (3.22) and the definition of $\mathbf{W}^*$ (see (3.18)), we get (3.35a) and (3.35b). Using (3.28) once more, we finally obtain $\overline{c}$ in terms of the velocity $\overline{u}$ and the enthalpy $\overline{H}$. □

Remark 3.3. The velocity and enthalpy of the Roe averaged state appear as a convex combination,

$$\overline{u} = \theta u_L + (1-\theta)u_R, \qquad \overline{H} = \theta H_L + (1-\theta)H_R,$$

where $\theta = \frac{\sqrt{\rho_L}}{\sqrt{\rho_L}+\sqrt{\rho_R}}$. The computations can also be carried out by taking a priori the velocity to be a linear combination of u_L and u_R (see Vinokur 1989). □

3.2.2 Determination of the coefficients in Roe's scheme

In order to complete the construction of Roe's scheme, it remains to compute according to formulas (3.20) and (3.21) the coefficients α_k of $\Delta\mathbf{U}$ in the basis $(\mathbf{r}_k(\overline{\mathbf{U}}))$ of eigenvectors of $\mathbf{A}(\overline{\mathbf{U}})$,

$$\mathbf{U}_R - \mathbf{U}_L = \Delta\mathbf{U} = \sum_{k=1}^{3} \alpha_k \mathbf{r}_k(\overline{\mathbf{U}}), \tag{3.36}$$

where $\overline{\mathbf{U}}$ is Roe's average state defined in (3.18).

It will be convenient to prove first the following simple algebraic results, which are obtained in a straightforward way. To make the formulas simpler, we use the shorthand notation a for a pair (a_L, a_R).

Lemma 3.6

Define for given pair $\rho = (\rho_L, \rho_R)$ the following averaging operators by

$$m_\rho(a) = \frac{\sqrt{\rho_L}a_L + \sqrt{\rho_R}a_R}{\sqrt{\rho_L} + \sqrt{\rho_R}} \tag{3.37}$$

and

$$\hat{m}_\rho(a) = \frac{\sqrt{\rho_R}a_L + \sqrt{\rho_L}a_R}{\sqrt{\rho_L} + \sqrt{\rho_R}}, \tag{3.38}$$

We have

$$\hat{m}_\rho(\rho) = (\rho_L\rho_R)^{\frac{1}{2}}, \tag{3.39}$$

and for any $a = (a_L, a_R), b = (b_L, b_R)$, we have

$$\Delta(ab) = \hat{m}_\rho(a)\Delta b + m_\rho(b)\Delta a, \tag{3.40}$$

$$\hat{m}_\rho(\rho a) = \hat{m}_\rho(\rho) m_\rho(a). \tag{3.41}$$

Here, we use the obvious notation

$$\Delta a = a_R - a_L. \tag{3.42}$$

In the following, we drop the dependence of m_ρ on ρ since no ambiguity arises.

Let us note that, by Lemma 3.5, Roe's averaged state satisfies

$$\overline{u} = m(u), \quad \overline{H} = m(H). \tag{3.43}$$

Lemma 3.7

Assume an equation of state of the form (3.15). The coefficients α_k in formula (3.36) are given (with the notations (3.42) and (3.39)) by

$$\begin{cases} \alpha_1 = \dfrac{\Delta p - \overline{c}\hat{m}(\rho)\Delta u}{2\overline{c}^2}, \\ \alpha_2 = \Delta\rho - \dfrac{\Delta p}{\overline{c}^2}, \\ \alpha_3 = \dfrac{\Delta p + \overline{c}\hat{m}(\rho)\Delta u}{2\overline{c}^2}. \end{cases} \tag{3.44}$$

Proof. Using the expressions (3.32) for the eigenvectors, we obtain a system of three equations,

$$\begin{cases} \Delta\rho = \alpha_1 + \alpha_2 + \alpha_3, \\ \Delta q = \alpha_1(\overline{u} - \overline{c}) + \alpha_2\overline{u} + \alpha_3(\overline{u} + \overline{c}), \\ \Delta E = \alpha_1(\overline{H} - \overline{uc}) + \alpha_2\Big(\overline{H} - \dfrac{\overline{c^2}}{\kappa}\Big) + \alpha_3(\overline{H} + \overline{uc}). \end{cases} \tag{3.45}$$

By (3.40),

$$\Delta q - \overline{u}\Delta\rho = \hat{m}(\rho)\Delta u,$$

and hence the system (3.46) is equivalently written

$$\begin{aligned} \alpha_1 + \alpha_3 &= \Delta\rho - \alpha_2, \\ \alpha_3 - \alpha_1 &= \hat{m}(\rho)\frac{\Delta u}{\overline{c}}, \\ \frac{\overline{c}^2}{\kappa}\alpha_2 &= -\Delta E + \overline{H}\Delta\rho + \overline{u}\hat{m}(\rho)\Delta u. \end{aligned}$$

The third equation gives α_2, from which we deduce α_1 and α_3. Lemma 3.6 enables us to derive a simple expression for α_2. Indeed, we have

$$\begin{aligned} -\Delta E+\overline{H}\Delta\rho + \overline{u}\Big(\Delta m - \overline{u}\Delta\rho\Big) &= -\Delta\tilde{\varepsilon} - \Delta(\frac{\rho u^2}{2}) + \overline{H}\Delta\rho \\ + \overline{u}(\Delta(\rho u) - \overline{u}\Delta\rho) &= -\Delta\tilde{\varepsilon} + \overline{H}\Delta\rho - \frac{1}{2}\overline{u}^2\Delta\rho = -\Delta\tilde{\varepsilon} + \overline{h}\Delta\rho, \end{aligned}$$

and hence we obtain by (3.15)

$$\overline{c}^2\alpha_2 = \kappa(-\Delta\tilde{\varepsilon} + \overline{h}\Delta\rho) = -\Delta p + (\chi + \kappa\overline{h})\Delta\rho$$

and, by (3.35c),

$$\alpha_2 = \Delta\rho - \frac{\Delta p}{\overline{c}^2}.$$

Then, we get

$$\alpha_1 = \frac{\Delta p - \overline{c}\widehat{m}(\rho)\Delta u}{2\overline{c}^2},$$
$$\alpha_3 = \frac{\Delta p + \overline{c}\widehat{m}(\rho)\Delta u}{2\overline{c}^2}.$$

Note that in the above proof we only use the "incremental" form of (3.15),

$$\Delta p = \chi\Delta\rho + \kappa\Delta\tilde{\varepsilon}.$$

This will be used later (see formula (3.51)). □

Gathering the previous lemmas together, we have thus proved the following theorem.

Theorem 3.2

Assume an equation of state of the form (3.15). There exists a Roe-type linearization of the form $\mathbf{A}(\mathbf{U}_R, \mathbf{U}_L) = \mathbf{A}(\overline{\mathbf{U}})$, *where the velocity, total specific enthalpy, and sound velocity at Roe-averaged state* $\overline{\mathbf{U}}$ *are given by formulas (3.35). Moreover, this is the unique linearization that is the Jacobian matrix evaluated at some average state* $\overline{\mathbf{U}}$. *Finally, Roe's scheme is defined by (3.20), where the coefficients* α_k *are given by (3.44).*

Uniqueness will follow from the following remark together with the computations in the next section, where we consider the case of a general equation of state.

Remark 3.4. The expressions (3.44) can be interpreted in terms of small fluctuations (Roe and Pike 1983, Roe 1985), which gives another algebraic way to derive the Roe averaged state. Indeed, when one computes the coefficients of $\mathbf{U}_R - \mathbf{U}_L = \Delta\mathbf{U}$, where $\mathbf{U}_L$ and $\mathbf{U}_R$ are close to some given reference state $\overline{\mathbf{U}}$, on the basis of eigenvectors $\mathbf{r}_k(\overline{\mathbf{U}})$,

$$\mathbf{U}_R - \mathbf{U}_L = \Delta\mathbf{U} = \sum_k \alpha_k \mathbf{r}_k(\overline{\mathbf{U}}),$$

one finds within $O(\Delta^2)$ the analog of (3.44)

$$\begin{cases} \alpha_1 = \dfrac{\Delta p - \overline{\rho c}\Delta u}{2\overline{c}^2}, \\[2ex] \alpha_2 = \Delta\rho - \dfrac{\Delta p}{\overline{c}^2}, \\[2ex] \alpha_3 = \dfrac{\Delta p + \overline{\rho c}\Delta u}{2\overline{c}^2}, \end{cases}$$

with the same computations as in Lemma 3.7. One also checks that with the same coefficients α_i, one has

$$\Delta\mathbf{F}(\mathbf{U}) = \mathbf{F}(\mathbf{U}_R) - \mathbf{F}(\mathbf{U}_L) = \sum_k a_k(\overline{\mathbf{U}})\alpha_k \mathbf{r}_k(\overline{\mathbf{U}}).$$

The average values $(\overline{\rho}, \overline{u}, \overline{c})$ are then determined in order that the two jump relations

$$\Delta\mathbf{U} = \sum_k \alpha_k \mathbf{r}_k(\overline{\mathbf{U}}), \qquad \Delta\mathbf{F}(\mathbf{U}) = \sum_k a_k(\overline{\mathbf{U}})\alpha_k \mathbf{r}_k(\overline{\mathbf{U}})$$

are identically satisfied whether or not the states $\mathbf{U}_L$ and $\mathbf{U}_R$ are closed. In fact, we shall detail the computations in the next section. Let us just say that though the system seems overdetermined (6 equations for 3 unknowns), it turns out that one equation is trivially satisfied, two others coincide, and one is implied, so that we are left with the only three equations corresponding to $\Delta\mathbf{F}$. We obtain in particular for $\overline{u}$ a quadratic equation; one of the roots coincides as expected with (3.35a), and the other is

$$u = \frac{\sqrt{\rho_L}u_L - \sqrt{\rho_R}u_R}{\sqrt{\rho_L} - \sqrt{\rho_R}},$$

which is not satisfactory. Then $\overline{\rho} = (\rho_L \rho_R)^{\frac{1}{2}}$, and $\overline{c}$ or $\overline{H}$ is given as above. It must be emphasized that $\overline{\rho}$ does not correspond to the density of the Roe average state (see Remark 3.5).

Note lastly that if one imposes only the jump relation

$$\Delta\mathbf{F}(\mathbf{U}) = \mathbf{A}(\overline{\mathbf{U}})\Delta\mathbf{U},$$

the first equation is trivially satisfied whatever the value of $\overline{\mathbf{U}}$, and one finds again the same expressions for $\overline{u}$ and $\overline{c}$. □

Remark 3.5. Roe's scheme applied to gas dynamics in Lagrangian coordinates is studied in Munz (1994). The relation between Roe's matrices in Lagrangian and Eulerian coordinates is studied in Gallice (1995). □

3.3 Application to the gas dynamics equations. II. The "real gas" case

For an arbitrary gas, it is usual to consider the pressure as a function of density and specific internal energy ε, $p = p(\rho, \varepsilon)$. However, in the extension of Roe's scheme to an equilibrium "real gas", it is more convenient to express the equation of state in the form (3.26) already used in Lemma 3.3,

$$p = p(\rho, \tilde{\varepsilon}).$$

For instance, for a thermally (not necessarily calorically) perfect gas, one has

$$p = \rho RT(\varepsilon).$$

We have seen, moreover, that p satisfies the identity (see Chapter II, Section 1.2, (1.21))

$$p = \rho p_\rho + \tilde{\varepsilon} p_{\tilde{\varepsilon}}, \tag{3.46}$$

which generalizes (3.15) in that p_ρ and $p_{\tilde{\varepsilon}}$ are no longer constant.

In the general case, the previous method of construction of a Roe-type linearization is no longer available since there does not exist in general a parameter vector $\mathbf{W}$ such that both $\mathbf{U}$ and $\mathbf{F}(\mathbf{U})$ are homogeneous quadratic functions of $\mathbf{W}$. For instance, if we define $\mathbf{W}$ by (3.22), only the first two components of $\mathbf{U}$ and the first and last components of $\mathbf{F}(\mathbf{U})$ satisfy the requirement. A more direct approach must be used. Nonetheless, we still look for average quantities $(\overline{u}, \overline{H}, \overline{\chi}, \overline{\kappa})$ (but not necessarily an average state $\overline{\mathbf{U}}$) such that

$$\mathbf{F}(\mathbf{U}_R) - \mathbf{F}(\mathbf{U}_L) = \overline{\mathbf{A}}(\mathbf{U}_R - \mathbf{U}_L). \tag{3.47}$$

Here $\overline{\mathbf{A}}$ is the matrix (3.30) computed for the values $(\overline{u}, \overline{H}, \overline{\chi}, \overline{\kappa})$, i.e.,

$$\overline{\mathbf{A}} = \mathbf{A}(\overline{u}, \overline{H}, \overline{\chi}, \overline{\kappa}), \tag{3.48}$$

with obvious notations since in the expression (3.30) of $\mathbf{A}(\mathbf{U})$ only the variables that appear explicitly, i.e., (u, H, χ, k) are needed. The identity (3.47) gives the system

$$\begin{cases} \Delta\rho = \Delta\rho, \\ \Delta(\rho u^2) + \Delta p = \Big(\overline{\chi} + \dfrac{1}{2}\overline{\kappa}\overline{u}^2 - \overline{u}^2\Big)\Delta\rho + (2 - \overline{\kappa})\overline{u}\Delta(\rho u) \\ \qquad\qquad + \overline{\kappa}\Delta\Big(\tilde{\varepsilon} + \dfrac{\rho u^2}{2}\Big), \\ \Delta(Hq) = \overline{u}\Big(\overline{\chi} + \dfrac{1}{2}\overline{\kappa}\overline{u}^2 - \overline{H}\Big)\Delta\rho + (\overline{H} - \overline{\kappa}\overline{u}^2)\Delta(\rho u) \\ \qquad\qquad + (1 + \overline{\kappa})\overline{u}\Delta\Big(\tilde{\varepsilon} + + \dfrac{\rho u^2}{2}\Big). \end{cases} \tag{3.49}$$

The first equation is trivial. Let us look at the second equation, which reads

$$\Delta p - (\overline{\chi}\Delta\rho + \overline{\kappa}\Delta\tilde{\varepsilon}) = \Big(\frac{\overline{\kappa}}{2} - 1\Big)\{\Delta(\rho u^2) + \overline{u}^2\Delta\rho - 2\overline{u}\Delta(\rho u)\}. \tag{3.50}$$

A natural choice for the average values $\overline{u}, \overline{\chi}$, and $\overline{\kappa}$ is such that both members in (3.50) vanish (which holds for a gas with equation (3.15)):

$$\Delta p = \overline{\chi}\Delta\rho + \overline{k}\Delta\tilde{\varepsilon}, \tag{3.51}$$

$$\Delta(\rho u^2) + \overline{u}^2\Delta\rho = 2\overline{u}\Delta(\rho u). \tag{3.52}$$

Assume that we can define $\overline{\chi}$ and $\overline{\kappa}$ such that (3.51) holds. Then, using (3.40), an easy computation shows that (3.52), which is quadratic in $\overline{u}$, is satisfied if we take for $\overline{u}$ the average obtained in the preceding section, i.e., (3.35a) or (3.43a):

$$\overline{u} = m(u), u = (u_L, u_R),$$

where the averaging operator m is defined by (3.37).

Let us now consider the third equation in (3.49). Assuming (3.51), we notice that

$$\overline{u}\overline{\chi}\Delta\rho + \overline{\kappa}\overline{u}\Delta\tilde{\varepsilon} + \overline{u}\Delta\tilde{\varepsilon} = \overline{u}\Delta p + \overline{u}\Delta\tilde{\varepsilon} = \overline{u}\Delta(p + \tilde{\varepsilon}) = \overline{u}\Delta\Big(\rho H - \rho\frac{u^2}{2}\Big).$$

There remains

$$\begin{aligned}\Delta(Hq) &= \overline{u}\Big(\overline{\kappa}\frac{\overline{u}^2}{2} - \overline{H}\Big)\Delta\rho + (\overline{H} - \overline{\kappa}\overline{u}^2)\Delta(\rho u) + (1 + \overline{\kappa})\overline{u}\Delta\Big(\rho\frac{u^2}{2}\Big) \\ &\qquad + \overline{u}\Delta\Big(\rho H - \rho\frac{u^2}{2}\Big) \\ &= \overline{u}\overline{\kappa}\Big\{\frac{\overline{u}^2}{2}\Delta\rho - \overline{u}\Delta(\rho u) + \Delta(\rho\frac{u^2}{2})\Big\} - \overline{u}\overline{H}\Delta\rho + \overline{H}\Delta(\rho u) + \overline{u}\Delta(\rho H).\end{aligned}$$

Together with (3.52), we get

$$\Delta(H\rho u) = -\overline{u}\overline{H}\Delta\rho + \overline{H}\Delta(\rho u) + \overline{u}\Delta(\rho H),$$

or using once more (3.40),

$$\Delta(H\rho u) = \overline{H}\hat{m}(\rho)\Delta u + \overline{u}\Delta(\rho H),$$

which is indeed satisfied if $\overline{H}$ is defined by (3.35b) and (3.43b), i.e.,

$$\overline{H} = m(H), H = (H_L, H_R).$$

In order to define $\overline{c}$, we also need the specific enthalpy. From

$$H = h + \frac{u^2}{2},$$

we get

$$\overline{h} = \overline{H} - \frac{\overline{u}^2}{2} = m\Big(h + \frac{u^2}{2}\Big) - \frac{1}{2}(m(u))^2$$

and thus

$$\overline{h} = m(h) + (\rho_L\rho_R)^{\frac{1}{2}}\frac{(\Delta u)^2}{(\sqrt{\rho_L} + \sqrt{\rho_R})^2}. \tag{3.53}$$

The sound speed is then defined by (3.28),

$$(\overline{c}^2) = \overline{\chi} + \overline{\kappa}\overline{h}. \tag{3.54}$$

The remaining steps for computing Roe's scheme are unchanged, only replacing in (3.45) κ, which is no longer constan, by $\overline{\kappa}$. Indeed, if (3.51) and

(3.54) hold, the computations of Lemma 3.7 are still valid, in particular

$$\alpha_2 = \Delta\rho - \frac{\Delta p}{\overline{c}^2}\,.$$

In short, (3.47) holds with the same averages for $\overline{u}$ and $\overline{H}$ as were found in the case of a polytropic ideal gas (or more generally for an equation of state (3.15)) provided we choose average values of $\overline{\kappa}$ and $\overline{\chi}$ such that (3.51) is satisfied.

Theorem 3.3

Assume an equation of state of the form (3.26). There exists a Roe-type linearization $\mathbf{A}(\mathbf{U}_R, \mathbf{U}_L) = \overline{\mathbf{A}} = \mathbf{A}(\overline{u}, \overline{H}, \overline{\chi}, \overline{\kappa})$, *where the quantities* $(\overline{u}, \overline{H})$ *are given by formulas (3.35) provided we define* $(\overline{\chi}, \overline{\kappa})$ *such that (3.51) is satisfied. Finally, Roe's scheme is defined by (3.20), where the coefficients* α_k *are given by (3.44).*

For the equation of state (3.15), χ and κ are constant so that $\overline{\chi} = c^2_{\text{ref}}$ and $\overline{\kappa} = \gamma - 1$. For an arbitrary equilibrium gas, $\overline{\kappa}$ and $\overline{\chi}$ are not uniquely determined by (3.51). Arguments for a precise definition of $\overline{\chi}$ and $\overline{\kappa}$ in terms of the thermodynamic states only are developped in Vinokur and Montagné 1990 (see also Glaister 1988). In the case of a mixture of thermally perfect nonreacting gases, and for the extension to chemical equilibrium mixtures and then to chemical and vibrational nonequilibrium mixtures, we refer to Abgrall (1989-1990), Liu and Vinokur (1989), Dubroca and Morreeuw (1992), Fernandez and Larrouturou (1988) (see also Saurel et al. 1994 for related computations).

Remark 3.6. Simple identities are obtained by defining $\overline{\rho}$ *formally* by

$$\overline{\rho} = \hat{m}(\rho) = (\rho_L \rho_R)^{\frac{1}{2}}\,.$$

Indeed, we have, for instance,

$$\begin{aligned}
\Delta(\rho u) &= \overline{\rho}\Delta u + \overline{u}\Delta\rho,\\
\Delta(\rho u^2) &= 2\overline{\rho}\;\overline{u}\Delta u = \overline{u}^2\Delta\rho,\\
\Delta(\rho H) &= \overline{\rho}\Delta H + \overline{H}\Delta\rho,\\
\Delta(H\rho u) &= \overline{H}\;\overline{\rho}\Delta u + \overline{u}\Delta(\rho H),
\end{aligned}$$

and so on. However, in the case of a perfect gas, by (3.22) $\rho = w_1^2$ implies that the density of the average state $\overline{\mathbf{U}}$ is

$$\overline{\rho} = w_1^{*2} = \Big(\frac{1}{2}(w_{1L} + w_{1R})\Big)^2 = \Big(\frac{1}{2}(\sqrt{\rho_L} + \sqrt{\rho_R})\Big)^2,$$

which is not equal to $\sqrt{\rho_L \rho_R}$. □

Remark 3.7. Let us show that if $\mathbf{U}_L$ and $\mathbf{U}_R$ are connected by a single shock, (3.43), (3.53), and (3.54) are compatible with the Rankine–Hugoniot condition, i.e., are satisfied at the shock. We first note that (3.51) implies

$$\frac{\Delta p}{\Delta \rho} = \overline{\chi} + \overline{\kappa}\,\frac{\Delta \tilde{\varepsilon}}{\Delta \rho}.$$

Thus, if we assume

$$\overline{h} = \frac{\Delta \tilde{\varepsilon}}{\Delta \rho}, \tag{3.55}$$

we have

$$(\overline{c}^2) = \overline{\chi} + \overline{\kappa}\,\overline{h} = \frac{\Delta p}{\Delta \rho}, \tag{3.56}$$

and these expressions are the discrete analogs of $h = \partial\tilde{\varepsilon}(\rho, s)/\partial\rho$ and $c^2 = \partial p(\rho, s)/\partial\rho$ (see Proof of Lemma 3.3). Let us check that (3.55) and (3.56) are satisfied when $\mathbf{U}_L$ and $\mathbf{U}_R$ are the left and right states of a shock wave. Indeed, in that case the Rankine–Hugoniot conditions derived in Chapter II, Section 2, (2.3) give

$$\Delta p = \frac{(\Delta(\rho u))^2}{\Delta \rho} - \Delta(\rho u^2).$$

An easy computation using the properties of the averaging operators m and $\hat{m}$ then implies

$$\Delta p = \hat{m}(\rho)^2\,\frac{(\Delta u)^2}{\Delta \rho}.$$

Introducing the equation of the Hugoniot curve (2.12) of Chapter II,

$$\Delta\varepsilon + \frac{1}{2}(p_L + p_R)\Delta\tau = 0, \qquad \tau = \frac{1}{\rho},$$

and writing

$$\frac{1}{2}(p_L + p_R) = p_L + \frac{\Delta p}{2} = p_R - \frac{\Delta p}{2},$$

we get for a shock connecting $\mathbf{U}_L$ and $\mathbf{U}_R$

$$\frac{\Delta \tilde{\varepsilon}}{\Delta \rho} = \frac{h_L\rho_L + h_R\rho_R}{\rho_L + \rho_R} = \mathcal{M}(h), \tag{3.57}$$

where we define the average $\mathcal{M}$ by

$$\mathcal{M}(a) = \frac{a_L\rho_L + a_R\rho_R}{\rho_L + \rho_R}, \quad a = (a_L, a_R). \tag{3.58}$$

We can check that, for a shock, the expression for $\overline{h}$ defined by (3.53) is also equal to $\mathcal{M}(h)$. In short,

$$\overline{h} = \frac{\Delta \tilde{\varepsilon}}{\Delta \rho} = \mathcal{M}(h),$$

$$\overline{c}^2 = \frac{\Delta p}{\Delta \rho} = \overline{\chi} + \overline{\kappa}\,\overline{h}.$$

Note that if (3.56) holds, we have in (3.44) $\alpha_2 = 0$. Also

$$\alpha_1 = \frac{1}{2}\left(\Delta\rho - \hat{m}(\rho)\frac{\Delta u}{\overline{c}}\right) = \frac{\Delta p - \hat{m}(\rho)\overline{c}\Delta u}{2\overline{c}^2},$$

$$\alpha_3 = \frac{1}{2}\left(\Delta\rho + \hat{m}(\rho)\frac{\Delta u}{\overline{c}}\right) = \frac{\Delta p + \hat{m}(\rho)\overline{c}\Delta u}{2\overline{c}^2}.$$

These are the discrete forms of the characteristic variables (see Chapter I, Section 2, Remark 2.1). □

3.4 A Roe-type linearization based on shock curve decomposition

In Roe's method, as we have already observed, we solve a linear Riemann problem, and the initial discontinuity breaks up into p discontinuity waves that propagate with speed $a_k(\mathbf{u}_L, \mathbf{u}_R), 1 \le k \le p$. We can now think of a new Roe-type linearization where the wave speeds (i.e., the eigenvalues) are chosen to satisfy more closely the properties of the exact ones. This new Riemann solver is based on shock curve decomposition. Indeed, assume at first that when solving exactly the Riemann problem, $\mathbf{U}_L$ and $\mathbf{U}_R$ are connected by discontinuity waves only (1-shock, 2-contact discontinuity, and 3-shock). Since u and p are constant across the 2-wave, we determine the intermediate states u^* and p^* as the intersection of the shock curves in the (u, p)-plane (see Chapter II, Section 3). An easy computation using the Rankine–Hugoniot conditions shows that the speed of the 1-shock can be given uniquely in terms of $\mathbf{U}_L$ and u^* and p^* by

$$\sigma_1 = u_L + \frac{(p^* - p_L)}{\rho_L(u^* - u_L)}.$$

Similarly, the speed of the 3-shock is equal to

$$\sigma_3 = u_R + \frac{(p^* - p_R)}{\rho_R(u^* - u_R)}.$$

Now, in the general case we replace the eventual rarefaction waves by (non-admissible) shock waves. This is achieved by considering the whole shock curve in the (u, p)-plane and not only the admissible part, replacing the rarefaction curve by the nonadmissible part of the shock curve (remember that the two curves are osculatory). In the proof of Theorem 3.1, Chapter II, the values u^* and p^* are thus obtained as the intersection of the shock curves. They determine in turn the wave speeds σ_1 and σ_3 and the values ρ_L^*, ρ_R^* of the density (which we had denoted ρ_I and ρ_{II} in Chapter II, Section 3), i.e., the whole states $\mathbf{U}_L^*$ and $\mathbf{U}_R^*$ on each side of the contact discontinuity.

Let us rewrite this method in terms of a Roe-type linearization. Define $\mathbf{A}(\mathbf{U}_L, \mathbf{U}_R)$ as the matrix whose eigenvalues $\sigma_i, i = 1, 2, 3$ are the above-computed wave speeds

$$(3.59)\qquad \sigma_1 = u_L + \frac{(p^* - p_L)}{\rho_L(u^* - u_L)}, \sigma_2 = u^*, \sigma_3 = u_R + \frac{(p^* - p_R)}{\rho_R(u^* - u_R)},$$

(we have omitted the dependence in $\mathbf{U}_L, \mathbf{U}_R$) and eigenvectors

$$(3.60)\qquad \begin{cases} \mathbf{r}_1 = \left(1, u_L + \dfrac{(p^* - p_L)}{\rho_L(u^* - u_L)}, H_L + u^* \dfrac{(p^* - p_L)}{\rho_L(u^* - u_L)}\right)^T, \\ \mathbf{r}_2 = \left(1, u^*, \dfrac{(\rho\varepsilon)_R^* - (\rho\varepsilon)_L^*}{\rho_R^* - \rho_L^*} + \dfrac{(u^*)^2}{2}\right)^T, \\ \mathbf{r}_3 = \left(1, u_R + \dfrac{(p^* - p_R)}{\rho_R(u^* - u_R)}, H_R + u^* \dfrac{(p^* - p_R)}{\rho_R(u^* - u_R)}\right)^T. \end{cases}$$

The eigenvectors have been defined in such a way that they satisfy the following identity.

Lemma 3.8

The vectors $\mathbf{r}_i(\mathbf{U}_L, \mathbf{U}_R), i = 1, 2, 3$ *defined by (3.60) satisfy*

$$\begin{aligned} \mathbf{U}_L^* - \mathbf{U}_L &= (\rho_L^* - \rho_L)\mathbf{r}_1(\mathbf{U}_L, \mathbf{U}_R), \\ \mathbf{U}_R^* - \mathbf{U}_L^* &= (\rho_R^* - \rho_L^*)\mathbf{r}_2(\mathbf{U}_L, \mathbf{U}_R), \\ \mathbf{U}_R - \mathbf{U}_R^* &= (\rho_R - \rho_R^*)\mathbf{r}_3(\mathbf{U}_L, \mathbf{U}_R). \end{aligned}$$

Proof. The formula for $\mathbf{r}_1$ results from the fact that since $\mathbf{U}_L^*$ and $\mathbf{U}_L$ are connected by a 1-shock (which may be nonadmissible), the Rankine–Hugoniot condition (Chapter II, (2.10)–(2.12)) gives

$$\frac{p^* - p_L}{u^* - u_L} = -M = \frac{(u^* - u_L)\rho_L^*\rho_L}{(\rho_L^* - \rho_L)}.$$

Hence

$$\begin{aligned} (\rho u)_L^* - (\rho u)_L = \rho_L^* u^* - (\rho u)_L &= (\rho_L^* - \rho_L)\left\{u_L + \frac{(u^* - u_L)\rho_L^*}{(\rho_L^* - \rho_L)}\right\} \\ &= (\rho_L^* - \rho_L)\left\{u_L + \frac{(p^* - p_L)}{\rho_L(u^* - u_L)}\right\}, \end{aligned}$$

which gives the second component of $\mathbf{r}_1(\mathbf{U}_L, \mathbf{U}_R)$. We study similarly the last component of $\mathbf{r}_1$, which is

$$\begin{aligned} (\rho e)_L^* - (\rho e)_L &= (\rho H)_L^* - (\rho H)_L - (p^* - p_L) \\ &= (\rho_L^* - \rho_L)\left\{H_L + \rho_L^* \frac{(H_L^* - H_L)}{(\rho_L^* - \rho_L)} - \frac{(p^* - p_L)}{(\rho_L^* - \rho_L)}\right\}. \end{aligned}$$

Writing

$$H = \varepsilon + \frac{u^2}{2} + \frac{p}{\rho}$$

and using once more the Rankine–Hugoniot condition (Chapter II, Section 2, (2.12))

$$\varepsilon_L^* - \varepsilon_L + (\tau_L^* - \tau_L)\frac{(p^* + p_L)}{2} = 0,$$

we check easily that

$$\rho_L^*(H_L^* - H_L) - (p^* - p_L) = u^* \rho_L^*(u^* - u_L),$$

which implies as expected

$$\frac{\rho_L^*(H_L^* - H_L)}{(\rho_L^* - \rho_L)} - \frac{(p^* - p_L)}{(\rho_L^* - \rho_L)} = u^* \frac{(p^* - p_L)}{\rho_L(u^* - u_L)}.$$

The computations for $\mathbf{r}_3$ are identical. The formula for $\mathbf{r}_2$ is obvious; it is obtained by writing

$$\begin{aligned}(\rho e)_R^* - (\rho e)_L^* &= (\rho\varepsilon)_R^* - (\rho\varepsilon)_L^* + (\rho_R^* - \rho_L^*)\frac{(u^*)^2}{2} \\ &= (\rho_R^* - \rho_L^*)\Big\{\frac{(\rho\varepsilon)_R^* - (\rho\varepsilon)_L^*}{\rho_R^* - \rho_L^*} + \frac{(u^*)^2}{2}\Big\},\end{aligned}$$

which ends the calculations. □

Let us note that the formulas for $\mathbf{r}_1$ and $\mathbf{r}_3$ give the discrete analog of (3.32). More precisely, we can prove the following result.

Lemma 3.9

The matrix $\mathbf{A}(\mathbf{U}_L, \mathbf{U}_R)$ *defined by (3.59) and (3.60) satisfies*

$$\begin{aligned}&\mathbf{F}(\mathbf{U}_R) - \mathbf{F}(\mathbf{U}_L) = \mathbf{A}(\mathbf{U}_L, \mathbf{U}_R)(\mathbf{U}_R - \mathbf{U}_L), \\ &\mathbf{A}(\mathbf{U}_L, \mathbf{U}_R) \to \mathbf{A}(\mathbf{U}_L) = \mathbf{A}(\mathbf{U}) \text{ as } \mathbf{U}_R \to \mathbf{U}_L = \mathbf{U}.\end{aligned}$$

Proof. First, we have by the definition of σ_i and by the Rankine–Hugoniot conditions

$$\begin{aligned}\mathbf{F}(\mathbf{U}_L^*) - \mathbf{F}(\mathbf{U}_L) &= \sigma_1(\mathbf{U}_L^* - \mathbf{U}_L), \\ \mathbf{F}(\mathbf{U}_R^*) - \mathbf{F}(\mathbf{U}_L^*) &= \sigma_2(\mathbf{U}_R^* - \mathbf{U}_L^*), \\ \mathbf{F}(\mathbf{U}_R) - \mathbf{F}(\mathbf{U}_R^*) &= \sigma_3(\mathbf{U}_R - \mathbf{U}_R^*).\end{aligned}$$

Together with Lemma 3.8, this implies,

$$\begin{aligned}\mathbf{F}(\mathbf{U}_R) - \mathbf{F}(\mathbf{U}_L) &= \sigma_1(\rho_L^* - \rho_L)\mathbf{r}_1 + u^*(\rho_R^* - \rho_L^*)\mathbf{r}_2 + \sigma_3(\rho_R - \rho_R^*)\mathbf{r}_3 \\ &= \mathbf{A}(\mathbf{U}_L, \mathbf{U}_R)\{(\rho_L^* - \rho_L)\mathbf{r}_1 + (\rho_R^* - \rho_L^*)\mathbf{r}_2 + (\rho_R - \rho_R^*)\mathbf{r}_3\} \\ &= \mathbf{A}(\mathbf{U}_L, \mathbf{U}_R)(\mathbf{U}_R - \mathbf{U}_L).\end{aligned}$$

Now, let us check that

$$\frac{(p^* - p_L)}{\rho_L(u^* - u_L)} \to -c \text{ as } \mathbf{U}_R \to \mathbf{U}_L = \mathbf{U}.$$

We write as above the Rankine–Hugoniot condition for the 1-shock connecting $\mathbf{U}_L^*$ and $\mathbf{U}_L$,

$$\frac{p^* - p_L}{u^* - u_L} = -M,$$

and

$$M^2 = \frac{p^* - p_L}{\tau^* - \tau_L}.$$

Since

$$\frac{p^* - p_L}{\tau^* - \tau_L} \to -\frac{\partial p}{\partial \tau} = \frac{c^2}{\tau^2}$$

(along the Hugoniot curve with center $A_L = (\tau_L, p_L)$, $s' = 0$ at A_L), we obtain

$$\sigma_1 \to u - c = a_1, \;\; \mathbf{r}_1(\mathbf{U}_L, \mathbf{U}_R) \to \mathbf{r}_1(\mathbf{U})$$

given by (3.32). Similarly,

$$\sigma_3 \to u + c = a_3, \;\; \mathbf{r}_3(\mathbf{U}_L, \mathbf{U}_R) \to \mathbf{r}_3(\mathbf{U}).$$

Then, since $p^* = p(\rho_L^*, \rho\varepsilon_L^*) = p(\rho_R^*, \rho\varepsilon_R^*)$, we have

$$\frac{(\rho\varepsilon)_R^* - (\rho\varepsilon)_L^*}{\rho_R^* - \rho_L^*} = \frac{(p(\rho_R^*, \tilde{\varepsilon}_L^*) - p(\rho_L^*, \tilde{\varepsilon}_L^*))/(\rho_R^* - \rho_L^*)}{(p(\rho_R^*, \tilde{\varepsilon}_L^*) - p(\rho_R^*, \tilde{\varepsilon}_R^*))/((\rho\varepsilon)_R^* - (\rho\varepsilon)_L^*)}$$
$$\to -\frac{p_\rho}{p_{\tilde{\varepsilon}}} = -\frac{\chi}{\kappa}.$$

We get by (3.28)

$$\frac{(\rho\varepsilon)_R^* - (\rho\varepsilon)_L^*}{(\rho_R^* - \rho_L^*)} + \frac{(u^*)^2}{2} \to -\frac{\chi}{\kappa} + \frac{u^2}{2}$$
$$= h - \frac{c^2}{\kappa} + \frac{u^2}{2} = H - \frac{c^2}{\kappa}$$

and

$$\sigma_2 \to u = a_2, \;\; \mathbf{r}_2(\mathbf{U}_L, \mathbf{U}_R) \to \mathbf{r}_2(\mathbf{U}),$$

which ends the proof. □

Note that for a thermally perfect gas, $\chi = 0$ and

$$(\rho\varepsilon)_R^* = (\rho\varepsilon)_L^* = \frac{p^*}{\kappa} = \frac{p^*}{\gamma - 1}.$$

The extension of this construction to a multicomponent fluid made up of perfect and real gases is developed in Mehlmann (1991). In this situation, which is of practical importance, one can show that the associated scheme respects the positivity of the mass fractions and the local proportion of atoms; see also Colella and Glaz (1985).

3.5 Another Roe-type linearization

Using the nonconservative form of system (2.1), Toumi (1992) has proposed a generalized Roe-type linearization associated to a path Φ. Indeed, let Φ be a sufficiently smooth function,

$$\Phi : [0,1] \times \mathbb{R}^p \times \mathbb{R}^p \to \mathbb{R}^p \text{ with } \Phi(0;\mathbf{u},\mathbf{v}) = \mathbf{u}, \Phi(1;\mathbf{u},\mathbf{v}) = \mathbf{v}.$$

One can define a Roe-type linearization depending on the path Φ, i.e., a matrix $\mathbf{A}(\mathbf{u},\mathbf{v})_\Phi$ that together with properties (3.2) and (3.3) satisfies

$$\mathbf{A}(\mathbf{u},\mathbf{v})_\Phi(\mathbf{u}-\mathbf{v}) = \int_0^1 \mathbf{A}(\Phi(s;\mathbf{u},\mathbf{v}))\frac{\partial \Phi}{\partial s}(s;\mathbf{u},\mathbf{v})ds. \tag{3.61}$$

In the present case where $\mathbf{A} = \mathbf{f}'$, the integral on the right-hand side does not depend on the path Φ connecting $\mathbf{u}$ and $\mathbf{v}$,

$$\begin{aligned}\int_0^1 \mathbf{f}'(\Phi(s;\mathbf{u},\mathbf{v}))\frac{\partial \Phi}{\partial s}(s;\mathbf{u},\mathbf{v})ds &= \mathbf{f}(\Phi(1;\mathbf{u},\mathbf{v})) - \mathbf{f}(\Phi(0;\mathbf{u},\mathbf{v}))\\ &= \mathbf{f}(\mathbf{v}) - \mathbf{f}(\mathbf{u}).\end{aligned}$$

However, the matrix $\mathbf{A}(\mathbf{u},\mathbf{v})_\Phi$ depends on the path. Setting $\mathbf{u}(\theta) = \mathbf{u} + \theta(\mathbf{v}-\mathbf{u})$, we have

$$\mathbf{A}(\mathbf{u},\mathbf{v})_\Phi = \int_0^1\int_0^1 \mathbf{A}(\Phi(s;\mathbf{u},\mathbf{v}))\frac{\partial^2 \Phi}{\partial s \partial \mathbf{u}}(s;\mathbf{u},\mathbf{u}(\theta))d\theta ds. \tag{3.62}$$

If the "canonical" straight line is chosen, we obtain

$$\mathbf{A}(\mathbf{u},\mathbf{v})_\Phi = \int_0^1 \mathbf{A}(\mathbf{u}+s(\mathbf{v}-\mathbf{u}))ds.$$

Let us check that the Roe matrix obtained via the parameter vectors in the case of a perfect gas corresponds to the straight path in the parameter-vector variables.

Lemma 3.10

Assume the hypotheses of Lemma 3.2. The matrix (3.62) associated to the following path,

$$\Phi = \Phi_W(s;\mathbf{U}_L,\mathbf{U}_R) = \mathbf{U}\Big(\mathbf{W}_L + s(\mathbf{W}_R - \mathbf{W}_L)\Big), \tag{3.63}$$

coincides with the Roe-type linearization (3.18).

Proof. The result follows directly from the property that $\mathbf{U}(\mathbf{W})$ and $\mathbf{G}(\mathbf{W}) = \mathbf{F}(\mathbf{U}(\mathbf{W}))$ are homogeneous quadratic functions of $\mathbf{W}$, together with the fact that integrating a linear function of $\mathbf{W}$ along the path Φ gives

$$\int_0^1 (\mathbf{W}_L + s(\mathbf{W}_R - \mathbf{W}_L))ds = \frac{1}{2}(\mathbf{W}_R + \mathbf{W}_L) = \mathbf{W}^*. \tag{3.64}$$

We have, on the one hand,

$$\int_0^1 \mathbf{A}(\Phi(s;\mathbf{U}_L,\mathbf{U}_R))\frac{\partial \Phi}{\partial s}(s;\mathbf{U}_L,\mathbf{U}_R)ds$$
$$= \int_0^1 \mathbf{A}(\mathbf{U}(\mathbf{W}_L + s(\mathbf{W}_R - \mathbf{W}_L)))$$
$$\mathbf{U}'(\mathbf{W}_L + s(\mathbf{W}_R - \mathbf{W}_L)) \cdot (\mathbf{W}_R - \mathbf{W}_L)ds.$$

On the other hand,

$$\mathbf{U}_R - \mathbf{U}_L = \mathbf{U}(\mathbf{W}_R) - \mathbf{U}(\mathbf{W}_L)$$
$$= \int_0^1 \frac{d}{ds}\mathbf{U}(\mathbf{W}_L + s(\mathbf{W}_R - \mathbf{W}_L))ds$$
$$= \int_0^1 \mathbf{U}'(\mathbf{W}_L + s(\mathbf{W}_R - \mathbf{W}_L))ds \cdot (\mathbf{W}_R - \mathbf{W}_L).$$

Thus, we get
(3.65)

$$\mathbf{A}(\mathbf{U}_L,\mathbf{U}_R)_{\Phi_W} = (\int_0^1 \mathbf{A}(\mathbf{U}(\mathbf{W}_L + s(\mathbf{W}_R - \mathbf{W}_L)))$$
$$\mathbf{U}'(\mathbf{W}_L + s(\mathbf{W}_R - \mathbf{W}_L))ds) \cdot \Big(\int_0^1 \mathbf{U}'(\mathbf{W}_L + s(\mathbf{W}_R - \mathbf{W}_L))ds\Big)^{-1}.$$

Now we compute successively each integral in (3.65); this requires expressions for $\mathbf{U}'(\mathbf{W})$ and $\mathbf{G}'(\mathbf{W}) = \mathbf{A}(\mathbf{U}(\mathbf{W}))\mathbf{U}'(\mathbf{W})$. From (3.23), the Jacobian matrix $\mathbf{U}'$ for a perfect gas gives

$$\mathbf{U}'(\mathbf{W}) = \begin{pmatrix} 2w_1 & 0 & 0 \\ w_2 & w_1 & 0 \\ w_3/\gamma & w_2(\gamma-1)/\gamma & w_1/\gamma \end{pmatrix}, \tag{3.66}$$

which is linear in the w_i; by (3.64), it yields

$$\int_0^1 \mathbf{U}'(\mathbf{W}_L + s(\mathbf{W}_R - \mathbf{W}_L))ds = \mathbf{U}'(\mathbf{W}^*).$$

Now, by Lemma 3.2,

$$\mathbf{G}'(\mathbf{W}) = \begin{pmatrix} w_2 & w_1 & 0 \\ w_3(\gamma-1)/\gamma & w_2(\gamma+1)/\gamma & w_1(\gamma-1)/\gamma \\ 0 & w_3 & w_2 \end{pmatrix},$$

which is also linear in the w_i's; thus, using (3.64) again,

$$\int_0^1 \mathbf{G}'(\mathbf{W}_L + s(\mathbf{W}_R - \mathbf{W}_L))ds = \mathbf{G}'(\mathbf{W}^*). \tag{3.67}$$

It is then easy to check that

$$\mathbf{A}_{\Phi_W}(\mathbf{U}_L,\mathbf{U}_R) = \mathbf{A}(\mathbf{U}_L,\mathbf{U}_R) = \mathbf{A}(\overline{\mathbf{U}}),$$

where the matrix $\mathbf{A}(\mathbf{U}_L, \mathbf{U}_R)$ was computed above in (3.20)–(3.24) via the parameter vectors, since by (3.18)

$$\mathbf{A}(\overline{\mathbf{U}}) = \mathbf{G}'(\mathbf{W}^*)\mathbf{U}'(\mathbf{W}^*)^{-1},$$

which thus coincides with (3.65). □

These considerations lead us to define a generalized Roe matrix for a real gas by formula (3.65) associated to the same straight path (3.63) in the parameter-vector variables (3.22), i.e.,

$$\begin{aligned} \mathbf{A}_{\Phi_W}(\mathbf{U}_L, \mathbf{U}_R) = \int_0^1 & \mathbf{G}'(\mathbf{W}_L + s(\mathbf{W}_R - \mathbf{W}_L))ds \\ & (\int_0^1 \mathbf{U}'(\mathbf{W}_L + s(\mathbf{W}_R - \mathbf{W}_L))ds)^{-1}. \end{aligned} \tag{3.68}$$

Let us just sketch the ideas for the computation of $\mathbf{A}_{\Phi_W}$. As we have already observed, in (3.23) Lemma 3.2 the last component $w_1w_3 - p$ of $\mathbf{U}$ is no longer a quadratic function of $\mathbf{W}$, and we must keep the expression of $\mathbf{U}$ in the form

$$\mathbf{U}(\mathbf{W}) = (w_1^2, w_1w_2, w_1w_3 - p)^T.$$

Thus, in the expression (3.66) of $\mathbf{U}'(\mathbf{W})$, the last line is replaced by

$$w_3 - p_{w_1}, \quad -p_{w_2}, \quad w_1 - p_{w_3}.$$

Similarly, $\mathbf{G}(\mathbf{W}) = \mathbf{F}(\mathbf{U}(\mathbf{W})) = (w_1w_2, p + w_2^2, w_2w_2)^T$, and in $\mathbf{G}'(\mathbf{W})$ the second line is replaced by

$$p_{w_1}, \quad p_{w_2} + 2w_2, \quad p_{w_3}.$$

In order to compute the integrals in (3.68), we proceed as follows:
(i) We set

$$\tilde{p}_{w_i} = \int_0^1 p_{w_i}(\mathbf{W}_L + s(\mathbf{W}_R - \mathbf{W}_L))ds,$$

which gives immediately

$$\int_0^1 \mathbf{U}'(\mathbf{W}_L + s(\mathbf{W}_R - \mathbf{W}_L))ds = \begin{pmatrix} 2w_1^* & 0 & 0 \\ w_2^* & w_1^* & 0 \\ w_3^* - \tilde{p}_{w_1} & -\tilde{p}_{w_2} & w_1^* - \tilde{p}_{w_3} \end{pmatrix},$$

and

$$\int_0^1 \mathbf{G}'(\mathbf{W}_L + s(\mathbf{W}_R - \mathbf{W}_L))ds = \begin{pmatrix} w_2^* & w_1^* & 0 \\ \tilde{p}_{w_1} & \tilde{p}_{w_2} + 2w_2^* & \tilde{p}_{w_3} \\ 0 & w_3^* & w_2^* \end{pmatrix}.$$

Then formula (3.68) gives the matrix $\mathbf{A}_{\Phi_W}(\mathbf{U}_L, \mathbf{U}_R)$.
(ii) In order to express this matrix in a more practical for, we introduce

the partial derivatives

$$\frac{\partial p}{\partial \rho}|_{q,E} = P_\rho, \quad \frac{\partial p}{\partial q}|_{\rho,E} = P_q, \quad \frac{\partial p}{\partial E}|_{\rho,q} = P_E.$$

In fact $P_\rho = K$ (3.31), as we observed at the end of the proof of Lemma 3.4, and we rewrite the matrix $\mathbf{A}$ given by (3.30) in terms of P_ρ, P_q, and P_E

$$\mathbf{A}(\mathbf{U}) = \begin{pmatrix} 0 & 1 & 0 \\ P_\rho - u^2 & 2u + P_q & P_E \\ u(P_\rho - H) & H + uP_q & (1 + P_E)u \end{pmatrix}. \tag{3.69}$$

Lemma 3.11

The matrix $\mathbf{A}_{\Phi_W}(\mathbf{U}_L, \mathbf{U}_R)$ defined by (3.68) is the Jacobian matrix (3.69) $\mathbf{A}(\overline{u}, \overline{H}, \tilde{P}_\rho, \tilde{P}_q, \tilde{P}_E)$ evaluated at the Roe's averaged states $\overline{u}, \overline{H}$ given by (3.29) and the averaged values of the partial derivatives.

Proof. First we express the partial derivatives P_ρ, P_q, and P_E in terms of the p_{w_i}; we get the relations

$$\begin{aligned} P_\rho = K = p_\rho + p_{\tilde{\varepsilon}} \frac{u^2}{2} &= \chi + \kappa \frac{u^2}{2} \\ &= \frac{w_1 p_{w_1} - w_2 p_{w_2} - w_3 p_{w_3}}{2w_1(w_1 - p_{w_3})}, \end{aligned} \tag{3.70a}$$

$$P_q = -\kappa u = \frac{p_{w_2}}{(w_1 - p_{w_3})} = -uP_E, \tag{3.70b}$$

$$P_E = \kappa = \frac{p_{w_3}}{(w_1 - p_{w_3})}. \tag{3.70c}$$

Since p is a function of only two thermodynamic variables, we note that the derivatives are not independent and that the equality $P_q = -uP_E$ implies in turn

$$p_{w_2} = -p_{w_3} \frac{w_2}{w_1}. \tag{3.71}$$

Next, we set

$$\begin{cases} \tilde{P}_\rho = \dfrac{w_1^* \tilde{p}_{w_1} - w_2^* \tilde{p}_{w_2} - w_3^* \tilde{p}_{w_3}}{2w_1^*(w_1^* - \tilde{p}_{w_3})}, \\[2ex] \tilde{P}_q = \dfrac{\tilde{p}_{w_2}}{w_1^* - \tilde{p}_{w_3}}, \\[2ex] \tilde{P}_E = \dfrac{\tilde{p}_{w_3}}{w_1^* - \tilde{p}_{w_3}}. \end{cases} \tag{3.72}$$

These expressions are those obtained by replacing w_i by w_i^* and P_ρ by the corresponding expression $\tilde{P}_\rho$, where p_{w_i} is replaced by $\tilde{p}_{w_i}$.

By analogy with the identity obtained for an ideal gas,

$$\begin{aligned}\mathbf{A}(\overline{\mathbf{U}}) &= \mathbf{G}'(\mathbf{W}^*)\mathbf{U}'(\mathbf{W}^*)^{-1} \\ &= \int_0^1 \mathbf{G}'(\mathbf{W}_L + s(\mathbf{W}_R - \mathbf{W}_L))ds\Big(\int_0^1 \mathbf{U}'(\mathbf{W}_L + s(\mathbf{W}_R - \mathbf{W}_L))ds\Big)^{-1},\end{aligned}$$

we check that we obtain for the matrix $\mathbf{A}_{\Phi_W}(\mathbf{u}, \mathbf{v})$ the expression

$$\mathbf{A}(\mathbf{u}, \mathbf{v})_{\Phi_W} = \begin{pmatrix} 0 & 1 & 0 \\ \tilde{P}_\rho - \overline{u}^2 & 2\overline{u} + \tilde{P}_q & \tilde{P}_E \\ \overline{u}(\tilde{P}_\rho - \overline{H}) & \overline{H} + \overline{u}\tilde{P}_q & (1 + \tilde{P}_E)\overline{u} \end{pmatrix}, \tag{3.73}$$

which is the Jacobian matrix (3.69) evaluated at $\overline{u}, \overline{H}$, and the averaged values $\tilde{P}_{\rho,m,E}$. □

Remark 3.8. We can define $\tilde{\chi}$ and $\tilde{\kappa}$ from formulas (3.70) and (3.72),

$$\begin{aligned}\tilde{\kappa} &= \tilde{P}_E, \\ \tilde{\chi} &= \tilde{P}_\rho + \tilde{P}_q \frac{w_2^*}{w_1^*}.\end{aligned}$$

We might ask whether condition (3.51) of the preceding section is satisfied. First, since

$$\begin{aligned}\Delta p &= \int_0^1 \frac{d}{ds}(p(\mathbf{W}_L + s(\mathbf{W}_R - \mathbf{W}_L)))ds \\ &= \int_0^1 \{p'(\mathbf{W}_L + s(\mathbf{W}_R - \mathbf{W}_L))\} \cdot (\mathbf{W}_R - \mathbf{W}_L)ds,\end{aligned}$$

we have

$$\Delta p = \sum \tilde{p}_{w_i} \Delta w_i.$$

Similarly,

$$\begin{aligned}\tilde{P}_\rho \Delta\rho + \tilde{P}_q \Delta q + \tilde{P}_E \Delta(\rho H) =& \tilde{P}_\rho(2w_1^*\Delta w_1) + \tilde{P}_q(w_2^*\Delta w_1 + w_1^*\Delta w_2) \\ &+ \tilde{P}_E(w_3^*\Delta w_1 + w_1^*\Delta w_3).\end{aligned}$$

Hence, by (3.70) and (3.72),

$$\begin{aligned}&\tilde{P}_\rho \Delta\rho + \tilde{P}_q \Delta q + \tilde{P}_E \Delta(\rho H) \\ &\quad = (\tilde{p}_{w_1}\Delta w_1 + \tilde{p}_{w_2}\Delta w_2 + \tilde{p}_{w_3}\Delta w_3)\frac{w_1^*}{(w_1^* - \tilde{p}_{w_3})} \\ &\quad = (1 + \tilde{P}_E)\Delta p,\end{aligned}$$

which in turn yields

$$\Delta p = \tilde{P}_\rho \Delta\rho + \tilde{P}_q \Delta q + \tilde{P}_E \Delta E.$$

Then, in general for a "real gas,"

$$\Delta p \neq \tilde{\chi}\Delta\rho + \tilde{\kappa}\Delta(\rho\varepsilon).$$

This results from the fact that (3.71) does not hold for the mean values

$$\tilde{p}_{w_2} \neq -\tilde{p}_{w_3}\, \frac{w_2^*}{w_1^*}\,.$$

For more details we refer to Toumi 1992 (see also Glaister 1994, Coquel and Liou 1994). □

Remark 3.9. The above construction of $\mathbf{A}(\mathbf{u}, \mathbf{v})_\Phi$ is linked to the definition of a nonconservative product $(\mathbf{A}(\mathbf{U})\frac{\partial \mathbf{U}}{\partial x})_\Phi$ associated to a path (see Dal Maso, LeFloch, and Murat 1995). This was introduced to study systems in nonconservation form (see Chapter I, Section 5, Remark 5.3). The "canonical" straight line corresponds to the nonconservative product of Volpert. □

4 The Osher scheme

4.1 The scalar case

The Engquist–Osher scheme in the scalar case (see G.R., Chapter III, Example 2.5) is given by the formula

$$v_j^{n+1} = v_j - \frac{\lambda}{2}(f(v_{j+1} - f(v_{j-1})) + \frac{\lambda}{2}(\int_{v_j}^{v_{j+1}} |a(\xi)|d\xi - \int_{v_{j-1}}^{v_j} |a(\xi)|d\xi)$$

(the index n has been omitted on the right-hand side).

In the *strictly convex* case ($f'' > 0$), we can give a simple geometric interpretation. Define

$$f^+(u) = f\Big(\max(u, \overline{u})\Big),$$
$$f^-(u) = f\Big(\min(u, \overline{u})\Big),$$

where $\overline{u}$ is the only stagnation (or sonic) point, i.e., $a(\overline{u}) = f'(\overline{u}) = 0$. Then, we have

$$f(v) = f^+(v) + f^-(v) - f(\overline{u}),$$
$$|f'(v)| = f^{+\prime}(v) + f^{-\prime}(v).$$

Hence, an easy computation shows that $g^{E.O.}$ is given by

$$g^{E.O.}(u, v) = f^+(u) + f^-(v) - f(\overline{u})$$

or

$$g^{E.O.}(u, v) = f^+(u) + f^-(v)$$

since g is defined up to an additive constant. In domains where the sign of f' is constant, it reduces to the standard first-order upwind scheme.

Let us see that the Osher scheme can be interpreted as a Godunov-type scheme (see G.R., Chapter III, Section 4), i.e.,

$$g^{E.O.}(u_L, u_R) = f(w(0; u_L, u_R)), \tag{4.1}$$

where the approximate Riemann solver w is defined as follows (we still assume $f'' > 0$), setting $\xi = \frac{x}{t}$:

(4.2)

$$w(\xi; u_L, u_R) = \begin{cases} u_L, & \xi \le a(u_L), \\ a^{-1}(\xi), & a(\min(u_L, u_R)) \le \xi \le a(\max(u_L, u_R)), \\ u_R, & \xi \ge a(u_R). \end{cases}$$

For $u_L > u_R$, the (approximate) Riemann solver replaces the shock wave solution of the Riemann problem by a multivalued function that may be viewed as a nonadmissible rarefaction wave (a "compression" wave). We have thus three branches $w^i(\xi) = w^i(\xi; u_L, u_R)$:
— the constant state $w^1(\xi) = u_L$,
— the rarefaction wave $w^2(\xi) = a^{-1}(\xi)$,
— the constant state $w^3(\xi) = u_R$.

Example 4.1. In the case of Burgers' equation $a^{-1}(\xi) = \xi$, we get the function drawn in Figure 4.1 (see van Leer (1984)). □

Some care must be taken in defining the associated numerical flux. Let us show that we can still define $g^{E.O.}(u_L, u_R)$ by the formula (4.1) provided that we set

$$g^{E.O.}(u_L, u_R) = \sum_i (-1)^{i-1} f(w^i(0; u_L, u_R)), \tag{4.3}$$

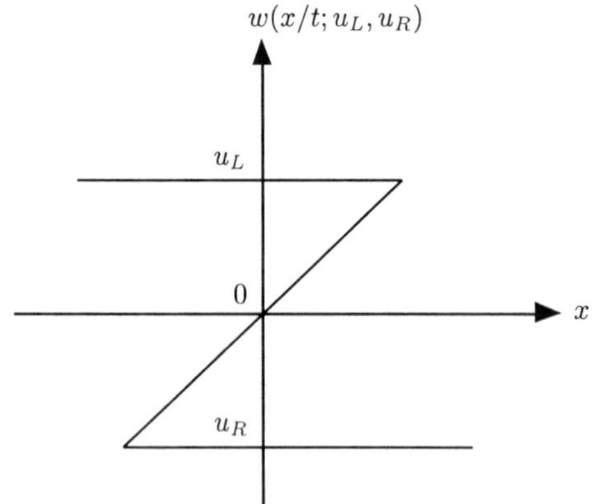

FIGURE 4.1. Compression wave for Burgers' equation.

where the sum is taken over all branches present at 0.
First, if $a(u_L) \geq 0 \geq a(u_R)$, (4.2) is multivalued at point 0, and (4.3) gives

$$g^{E.O.}(u_L, u_R) = f(w^1(0)) - f(w^2(0)) + f(w^3(0)).$$

Since $w^2(0) = \overline{u}$, we find in that case

$$g^{E.O.}(u, v) = f(u) - f(\overline{u}) + f(v).$$

Now, if $a(u_L) \leq 0 \leq a(u_R)$,

$$g^{E.O.}(u_L, u_R) = f(w(0; u_L, u_R)) = f(w^2(0)) = f(\overline{u}).$$

If 0 does not belong to the interval $(a(u_L), a(u_R))$, we find

$$g^{E.O.}(u_L, u_R) = f(w(0; u_L, u_R)) = \begin{cases} f(w^1(0)) \text{ if } a(u_L) \text{ and } a(u_R) > 0, \\ f(w^3(0)) \text{ if } a(u_L) \text{ and } a(u_R) < 0. \end{cases}$$

Since the flux is defined up to an additive constant, the formula is valid in all cases.

4.2 The Osher scheme for a system

For a system, the numerical flux is defined by

$$\text{(4.4a)} \qquad \mathbf{g}^{E.O.}(\mathbf{u}_L, \mathbf{u}_R) = \frac{1}{2}(\mathbf{f}(\mathbf{u}_L) + \mathbf{f}(\mathbf{u}_R)) - \frac{1}{2}\int_\Gamma |\mathbf{A}(\mathbf{w})| d\mathbf{w}$$

or, equivalently,

$$\text{(4.4b)}, \ \mathbf{g}^{E.O.}(\mathbf{u}_L, \mathbf{u}_R) = \mathbf{f}(\mathbf{u}_L) + \int_\Gamma \mathbf{A}^-(\mathbf{w}) d\mathbf{w} = \mathbf{f}(\mathbf{u}_R) - \int_\Gamma \mathbf{A}^+(\mathbf{w}) d\mathbf{w},$$

for a suitable path of integration Γ connecting $\mathbf{u}_L$ and $\mathbf{u}_R$ (in the state space). The idea consists of choosing the path Γ in a "natural" way using the integral curves of the (right) eigenvectors of the matrix $\mathbf{A}$. This means that we connect $\mathbf{u}_L$ to $\mathbf{u}_R$ by a sequence of intermediate states $\mathbf{u}_k$ as in the solution of the Riemann problem (Chapter I, Theorem 6.1), except that we replace the k-shock waves by multivalued k-rarefaction waves.

More precisely, let us recall (see Chapter I, Section 3) that, on the one hand, when the kth field is *genuinely nonlinear* (or equivalently in the scalar case if f is strictly convex) if $a_k(\mathbf{u}_L) < a_k(\mathbf{u}_r)$, and if $\mathbf{u}_R$ belongs to the integral curve of the vector field $\mathbf{r}_k$, then $\mathbf{u}_L$ and $\mathbf{u}_R$ can be connected by a k-rarefaction wave

$$\text{(4.5)} \qquad \begin{cases} \mathbf{v}'(\xi) = \mathbf{r}_k(\mathbf{v}(\xi)), \\ \mathbf{v}(\lambda_k(\mathbf{u}_L)) = \mathbf{u}_L, \end{cases}$$

($a_k(\mathbf{u}_L) < a_k(\mathbf{u}_R)$ since a_k is increasing along the curve). If $a_k(\mathbf{u}_R) < a_k(\mathbf{u}_L)$ and if $\mathbf{u}_R$ belongs to the k-shock curve, then the states $\mathbf{u}_L$ and $\mathbf{u}_R$ are connected by a k-shock.

Instead of the admissible part of the k-shock curve, we shall use the "non-admissible" part of the integral curve of the vector $\mathbf{r}_k$, which is osculatory (it starts from $\mathbf{u}_L$ in the opposite direction to the rarefaction curve). This amounts to connecting $\mathbf{u}_L$ and $\mathbf{u}_R$ by a multivalued rarefaction (Figure 4.2). So we define for $a_k(\mathbf{u}_R) < a_k(\mathbf{u}_L)$ a "compression" wave that is a multivalued solution by

$$\mathbf{w}(\xi; u, v) = \begin{cases} \mathbf{u}_L, & \xi \leq a_k(\mathbf{u}_L), \\ \mathbf{v}(\xi), & a_k(\mathbf{u}_R) \leq \xi \leq a_k(\mathbf{u}_L), \\ \mathbf{u}_R, & \xi \geq a_k(\mathbf{u}_R), \end{cases}$$

where $\mathbf{v}$ is the solution of (4.5) and $\xi = x/t$.

Again, for $a_k(\mathbf{u}_R) \leq \xi \leq a_k(\mathbf{u}_L)$, we have three states:

— the constant state $\mathbf{w}^1(\xi) = \mathbf{u}_L$,

— the rarefaction wave $\mathbf{w}^2(\xi) = \mathbf{v}(\xi)$,

— the constant state $\mathbf{w}^3(\xi) = \mathbf{u}_R$.

On the other hand if the kth field is *linearly degenerate*, we have seen that the set of states $\mathbf{v}$ that can be connected to a given state $\mathbf{u}$ by a k-contact discontinuity is also an integral curve of the vector field $\mathbf{r}_k$.

We want now to define an approximate Riemann solver. We shall use the following family of k-waves: rarefactions (multivalued if necessary) if the kth field is genuinely nonlinear, and contact discontinuities if the kth field is linearly degenerate. By an analog of Theorem 6.1 in Chapter I, we know that for sufficiently close initial states $\mathbf{u}_L$, $\mathbf{u}_R$, the Riemann problem has a multivalued solution consisting of at most $(p+1)$ constant states $\mathbf{u}_k$, such that $\mathbf{u}_{k+1}$ is connected to $\mathbf{u}_k$ by a k-wave of the above family and such a solution is unique. The proof follows exactly that of Theorem 6.1 (see Dubois 1989). This Riemann solver is exact when the solution of the Riemann problem has no shocks.

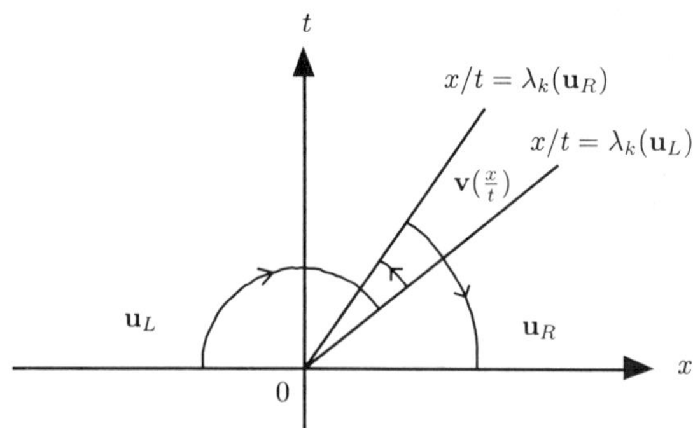

FIGURE 4.2. Compression wave in the (x, t)-plane.

If we denote by θ_k the strength of the kth wave (corresponding to ε_k in Chapter I, Theorem 6.1, but now θ_k can be ≥ 0 or ≤ 0), we can compute easily the integral in (4.4). The path Γ is the union of $\Gamma_k : \Gamma = \bigcup \Gamma_k$, each Γ_k being a portion of an integral curve of $\mathbf{r}_k$. We have thus

$$\int_\Gamma \mathbf{A}^-(\mathbf{w})d\mathbf{w} = \sum_k \int_0^{\theta_k} a_k^-(\mathbf{w}_k(\xi))\mathbf{r}_k(\mathbf{w}_k(\xi))d\xi,$$

$$\int_\Gamma \mathbf{A}^+(\mathbf{w})d\mathbf{w} = \sum_k \int_0^{\theta_k} a_k^+(\mathbf{w}_k(\xi))\mathbf{r}_k(\mathbf{w}_k(\xi))d\xi,$$

where $\mathbf{w}_k$ is solution of (4.5),

$$\mathbf{w}_k'(\xi) = \mathbf{r}_k(\mathbf{w}_k(\xi)), \text{ for } 0 \leq \xi \leq \theta_k \text{ or } 0 \geq \xi \geq \theta_k, k = 1, ..., p,$$

and

$$\begin{cases} \mathbf{w}_{k+1}(0) = \mathbf{w}_k(\theta_k) = \mathbf{u}_k, \ k = 1, ..., p-1, \\ \mathbf{w}_1(0) = \mathbf{u}_L, \\ \mathbf{w}_p(\theta_p) = \mathbf{u}_R. \end{cases}$$

On the one hand, if the kth field is linearly degenerate, we know that a_k is a k-Riemann invariant (see also Chapter I, Theorem 4.2) and is constant along the integral curve of $\mathbf{r}_k$. Hence

$$\int_0^{\theta_k} a_k^+(\mathbf{w}_k(\xi))\mathbf{r}_k(\mathbf{w}_k(\xi))d\xi = \begin{cases} \mathbf{f}(\mathbf{w}_{k+1}(0)) - \mathbf{f}(\mathbf{w}_k(0)) & \text{if } a_k(\mathbf{w}_k) > 0, \\ 0 & \text{if } a_k(\mathbf{w}_k) \leq 0. \end{cases}$$

On the other hand, if the kth field is genuinely nonlinear, $\xi \rightarrow a_k(\mathbf{w}_k(\xi))$ is strictly monotone. Let us denote by $\overline{s}_k$ the "sonic" point (if it exists) such that

$$a_k(\mathbf{w}_k(\overline{s}_k)) = 0,$$

and the corresponding "sonic" state by

$$\overline{\mathbf{u}}_k = \mathbf{w}_k(\overline{s}_k).$$

Then, according to the respective signs of $a_k(\mathbf{u}_k)$, we obtain

$$\int_0^{\theta_k} a_k^+(\mathbf{w}_k(\xi))\mathbf{r}_k(\mathbf{w}_k(\xi))d\xi = \begin{cases} \mathbf{f}(\mathbf{w}_{k+1}(0)) - \mathbf{f}(\mathbf{w}_k(0)) \\ \quad \text{if } a_k(\mathbf{w}_{k+1}(0)) > 0, a_k(\mathbf{w}_k(0)) > 0, \\ \mathbf{f}(\mathbf{w}_{k+1}(0)) - \mathbf{f}(\overline{\mathbf{u}}_k) \\ \quad \text{if } a_k(\mathbf{w}_{k+1}(0)) > 0, a_k(\mathbf{w}_k(0)) \leq 0, \\ \mathbf{f}(\overline{\mathbf{u}}_k) - \mathbf{f}(\mathbf{w}_k(0)) \\ \quad \text{if } a_k(\mathbf{w}_{k+1}(0)) \leq 0, a_k(\mathbf{w}_k(0)) > 0, \\ 0 \\ \quad \text{if } a_k(\mathbf{w}_{k+1}(0)) \leq 0, a_k(\mathbf{w}_k(0)) \leq 0. \end{cases}$$

It remains to derive the numerical flux. As in the scalar case, we sum up the different possibilities with the following formula, which can indeed be interpreted as a Godunov-type flux (see Dubois 1989)

$$\mathbf{g}^{E.O.}(\mathbf{u}_L, \mathbf{u}_R) = \sum_{k=0}^{p} \varepsilon_k \mathbf{f}(\mathbf{u}_k) + \sum_{k=1}^{p} \overline{\varepsilon}_k \mathbf{f}(\overline{\mathbf{u}}_k), \tag{4.6}$$

where the ε_k and $\overline{\varepsilon}_k$ are defined by

$$\varepsilon_k = \begin{cases} 1 & \text{if } a_k(\mathbf{u}_k) \leq 0 < a_{k+1}(\mathbf{u}_k), \\ 0 & \text{elsewhere,} \end{cases}$$

$$\overline{\varepsilon}_k = \begin{cases} 1 & \text{if } a_k(\mathbf{u}_{k-1}) \leq 0 < a_k(\mathbf{u}_k), \\ -1 & \text{if } a_k(\mathbf{u}_k) \leq 0 < a_k(\mathbf{u}_{k-1}), \\ 0 & \text{elsewhere} \end{cases}$$

(with the convention $a_0 = -\infty, a_{n+1} = +\infty$). Thus, a rarefaction wave has a positive sign, whereas a "compression" wave has a negative sign. Note that this method can be related to that developed in Section 3.4 where only shock waves were considered, whereas here only simple waves and contact discontinuities are involved.

Remark 4.1. We might also have used the reverse order ($\mathbf{u}_{k+1}$ is connected to $\mathbf{u}_k$ by a $(p-k)$-wave). This has been proposed by Osher and Solomon (1982). The ordering of the path may be significant (see Hoff 1985 for the problem of invariance or Roberts 1990 for that of slowly moving shocks).□

Remark 4.2. The Osher scheme and the Godunov and Roe scheme as well are "flux difference splitting" methods, i.e., there flux can be written

$$\mathbf{g}(\mathbf{u}, \mathbf{v}) = \frac{1}{2}(\mathbf{f}(\mathbf{u}) + \mathbf{f}(\mathbf{v})) - \frac{1}{2}(\Delta\mathbf{f}^+(\mathbf{u}, \mathbf{v}) - \Delta\mathbf{f}^-(\mathbf{u}, \mathbf{v})),$$

where

$$\Delta\mathbf{f}(\mathbf{u}, \mathbf{v}) = \mathbf{f}(\mathbf{v}) - \mathbf{f}(\mathbf{u}) = \Delta\mathbf{f}^+(\mathbf{u}, \mathbf{v}) + \Delta\mathbf{f}^-(\mathbf{u}, \mathbf{v}),$$

and $\Delta\mathbf{f}^+$ (resp. $\Delta\mathbf{f}^-$) corresponds more or less to right (resp. left) running waves, i.e., positive (resp. negative) signal speeds.

For the Osher scheme,

$$\Delta\mathbf{f}^{\pm}(\mathbf{u}, \mathbf{v}) = \int_{\Gamma} \mathbf{A}^{\pm}(\mathbf{w})d\mathbf{w} = \sum_k \int_0^{\theta_k} a_k^{\pm}(\mathbf{w}_k(\xi))\mathbf{r}_k(\mathbf{w}(\xi))d\xi,$$

while for the Godunov flux (2.9),

$$\Delta\mathbf{f}^+(\mathbf{u}, \mathbf{v}) = \mathbf{f}(\mathbf{v}) - \mathbf{f}(\mathbf{w}_R(0, \mathbf{u}, \mathbf{v})),$$
$$\Delta\mathbf{f}^-(\mathbf{u}, \mathbf{v}) = -(\mathbf{f}(\mathbf{v}) - \mathbf{f}(\mathbf{w}_R(0, \mathbf{u}, \mathbf{v})),$$

and for Roe's flux (3.13),

$$\Delta\mathbf{f}^{\pm}(\mathbf{u},\mathbf{v}) = \mathbf{A}^{\pm}(\mathbf{u},\mathbf{v})(\mathbf{v}-\mathbf{u}) = \sum_k \alpha_k a_k^{\pm}\mathbf{r}_k(\mathbf{u},\mathbf{v}).$$

Note that in the Osher or Roe scheme, the upwinding is achieved via a field by field decomposition. □

We shall now focus on the gas dynamics equations (this application was first treated in Osher and Solomon 1988).

4.3 Application to the gas dynamics system

Osher's scheme applied to the Euler equations gives the result that $\mathbf{U}_L$ is connected to $\mathbf{U}_R$ via two intermediate states as follows (see Chapter II, Section 3):
— $\mathbf{U}_L$ is connected to $\mathbf{U}_1$ by a 1-(multivalued) rarefaction wave,
— $\mathbf{U}_1$ to $\mathbf{U}_2$ by a 2-contact discontinuity,
— $\mathbf{U}_2$ to $\mathbf{U}_R$ by a 3-(multivalued) rarefaction wave.
As in the classical Riemann problem, we begin by determining the states $\mathbf{U}_1$ and $\mathbf{U}_2$ on each side of the contact discontinuity. We use the k-Riemann invariants (see Chapter I, Example 3.3), which are constant on the k-wave since it is an integral curve of $\mathbf{r}_k$. We deduce first that $\mathbf{U}_1$ and $\mathbf{U}_2$ have the same velocity $u_1 = u_2 = u^*$ and pressure $p_1 = p_2 = p^*$. We find u^* and p^* by using the 1- and 3-Riemann invariants $u \pm \ell$ and s: since $s_1^* = s_L$ and $s_2^* = s_R$, we obtain

$$\begin{aligned} u^* + \ell\Big(\rho(p^*, s_L), s_L\Big) &= u_L + \ell(\rho_L, s_L), \\ u^* - \ell\Big(\rho(p^*, s_R), s_R\Big) &= u_R - \ell(\rho_R, s_R). \end{aligned}$$

The densities ρ_1 and ρ_2 are then computed by using the equation $\rho = \rho(p, s)$.

Once $\mathbf{U}_1$ and $\mathbf{U}_2$ are computed, we need to specify the states that can be present in the 1- and 3-rarefaction waves at $\frac{x}{t} = 0$, i.e., the "sonic" states $\overline{\mathbf{U}}_1 = (\overline{\rho}_1, \overline{u}_1, s_L)$ and $\overline{\mathbf{U}}_3 = (\overline{\rho}_3, \overline{u}_3, s_R)$. These satisfy

$$\begin{aligned} a_1(\overline{\mathbf{U}}_1) &= \overline{u}_1 - c(\overline{u}_1, s_L) = 0, \\ \overline{u}_1 + \ell(\overline{\rho}_1, s_L) &= u_L + \ell(\rho_L, s_L), \\ a_3(\overline{\mathbf{U}}_3) &= \overline{u}_3 + c(\overline{u}_3, s_R) = 0, \\ \overline{u}_1 - \ell(\overline{\rho}_1, s_R) &= u_R - \ell(\rho_R, s_R). \end{aligned}$$

Formula (4.6) becomes

$$\begin{aligned} \mathbf{g}^{E.O.}(\mathbf{u}_L, \mathbf{u}_R) = {} & \varepsilon_L \mathbf{f}(\mathbf{U}_L) + \overline{\varepsilon}_1 \mathbf{f}(\overline{\mathbf{U}}_1) + \varepsilon_1 \mathbf{f}(\mathbf{U}_1) \\ & + \varepsilon_2 \mathbf{f}(\mathbf{U}_2) + \overline{\varepsilon}_3 \mathbf{f}(\mathbf{U}_3) + \varepsilon_R \mathbf{f}(\mathbf{U}_R), \end{aligned}$$

where

$$\varepsilon_L = \begin{cases} 1 & \text{if } a_1(\mathbf{U}_L) = u_L - c_L > 0, \\ 0 & \text{if } a_1(\mathbf{U}_L) < 0, \end{cases}$$

$$\overline{\varepsilon}_1 = \begin{cases} 1 & \text{if } a_1(\mathbf{U}_L) \le 0 < a_1(\mathbf{U}_1), \\ -1 & \text{if } a_1(\mathbf{U}_1) \le 0 < a_1(\mathbf{U}_L), \\ 0 & \text{elsewhere}, \end{cases}$$

$$\varepsilon_1 = \begin{cases} 1 & \text{if } a_1(\mathbf{U}_1) < 0 < u^*, \\ 0 & \text{if } a_1(\mathbf{U}_1) \ge 0 \text{ or } u^* \le 0, \end{cases}$$

$$\varepsilon_2 = \begin{cases} 1 & \text{if } u^* \le 0 < a_3(\mathbf{U}_2) = u^* + c_2, \\ 0 & \text{if } a_3(\mathbf{U}_2) \le 0 \text{ or } u^* > 0, \end{cases}$$

$$\overline{\varepsilon}_3 = \begin{cases} 1 & \text{if } a_3(\mathbf{U}_2) \le 0 < a_3(\mathbf{U}_R), \\ -1 & \text{if } a_3(\mathbf{U}_R) \le 0 < a_3(\mathbf{U}_2), \\ 0 & \text{elsewhere}, \end{cases}$$

$$\varepsilon_R = \begin{cases} 1 & \text{if } a_3(\mathbf{U}_R) \le 0, \\ 0 & \text{if } a_3(\mathbf{U}_R) > 0. \end{cases}$$

For a polytropic perfect gas (see Chapter II, Section 1.2), we know that

$$c^2 = \gamma \frac{p}{\rho}, \quad \ell = \frac{2c}{(\gamma - 1)}.$$

Moreover, the pressure as the function $p = p(c, s)$ satisfies

$$p^{\frac{(\gamma-1)}{2\gamma}} = \sqrt{\gamma}\,\frac{c}{s},$$

which implies

$$\frac{c_1}{s_L} = \frac{c_2}{s_R}.$$

Hence, the equations for $\mathbf{U}_1$ and $\mathbf{U}_2$ are easily solved:

$$c_1 = \Big(1 + \frac{s_R}{s_L}\Big)^{-1} \Big\{c_L + c_R + \frac{(\gamma - 1)}{2}(u_L - u_R)\Big\},$$

$$c_2 = \Big(1 + \frac{s_L}{s_R}\Big)^{-1} \Big\{c_L + c_R + \frac{(\gamma - 1)}{2}(u_L - u_R)\Big\},$$

$$u^* = \Big(1 + \frac{s_L}{s_R}\Big)^{-1} \Big\{u_L + \frac{2c_L}{(\gamma - 1)}\Big\} + \Big(1 + \frac{s_R}{s_L}\Big)^{-1} \Big\{u_R - \frac{2c_R}{(\gamma - 1)}\Big\}.$$

Also, one finds

$$\overline{u}_1 = \frac{(\gamma-1)}{(\gamma+1)}\left\{u_L + \frac{2c_L}{(\gamma-1)}\right\}, \quad \overline{u}_3 = \frac{(\gamma-1)}{(\gamma+1)}\left\{u_R - \frac{2c_R}{(\gamma-1)}\right\}.$$

Besides the smoothness of the flux, Osher's scheme does not need entropy correction and gives a good resolution of contact discontinuities (see Coquel and Liou 1994), which makes it a good candidate for the extension to Navier–Stokes equations in spite of its relative complexity. (It provides moreover a consistent treatment of boundary conditions Koren 1989). It is also superior on the slowly moving shock problem and the carbuncle phenomenon (Lin 1995). For the extension to chemical and vibrational equilibrium or nonequilibrium gas flows, we refer to Dubois (1989), Abgrall and Montagné (1989), Abgrall, Fezoui, and Talandier (1992). For the extension of Engquist–Osher's scheme to systems that are not strictly hyperbolic, see Bell, Colella, and Trangenstein (1989).

5 Flux vector splitting methods

5.1 General formulation

The upwind difference scheme is easy to implement in the nonlinear case when the eigenvalues of $\mathbf{A}(\mathbf{u})$ are all of one sign. When they are of mixed sign, we want to give a direct generalization of the upwind difference scheme (1.10) obtained in the linear case $\mathbf{f}(\mathbf{u}) = \mathbf{A}\mathbf{u}$, which corresponds to the numerical flux function (2.17),

$$\mathbf{g}(\mathbf{u},\mathbf{v}) = \mathbf{A}^+\mathbf{u} + \mathbf{A}^-\mathbf{v}.$$

In fact, we look for a flux splitting of the form

$$\mathbf{f}(\mathbf{u}) = \mathbf{f}^+(\mathbf{u}) + \mathbf{f}^-(\mathbf{u}), \tag{5.1}$$

and we set

$$\mathbf{g}(\mathbf{u},\mathbf{v}) = \mathbf{f}^+(\mathbf{u}) + \mathbf{f}^-(\mathbf{v}). \tag{5.2}$$

We shall require that the Jacobian matrix $\mathbf{A}_+(\mathbf{u}) \equiv \mathbf{f}^{+'}(\mathbf{u})$ (resp. $\mathbf{A}_-(\mathbf{u}) \equiv \mathbf{f}^{-'}(\mathbf{u})$) has positive (resp. negative) eigenvalues.

Consider first the *scalar* case. If the function f is strictly convex, we have already found such a decomposition when deriving the Engquist–Osher scheme (see Section 4.1). Let $\overline{u} \in \mathbb{R}$ be the only sonic point, $a(\overline{u}) = f'(\overline{u}) = 0$. We can set

$$f^+(v) = f(\max(v,\overline{u})), f^-(v) = f(\min(v,\overline{u}))$$

so that

$$f(v) = f^+(v) + f^-(v) - f(\overline{u}).$$

and then (up to an additive constant)

$$g(u, v) = f^+(v) + f^-(v).$$

Let us consider now the case of a system whose flux function is a *homogeneous* function of degree one, i.e.,

$$\mathbf{f}(\mu\mathbf{u}) = \mu\mathbf{f}(\mathbf{u}), \forall\mu \in \mathbb{R},$$

which is indeed the case for ideal gas dynamics. In that case, Euler's identity for homogeneous functions gives

$$\mathbf{f}(\mathbf{u}) = \mathbf{A}(\mathbf{u})\mathbf{u},$$

so that following Steger and Warming we define

$$\mathbf{f}^+(\mathbf{u}) = \mathbf{A}^+(\mathbf{u})\mathbf{u},\ \mathbf{f}^-(\mathbf{u}) = \mathbf{A}^-(\mathbf{u})\mathbf{u}, \tag{5.3}$$

where $\mathbf{A}^{\pm}$ is defined by (1.9). We obtain the numerical flux

$$\mathbf{g}(\mathbf{u},\mathbf{v}) = \mathbf{f}^+(\mathbf{u}) + \mathbf{f}^-(\mathbf{v}) = \frac{1}{2}(\mathbf{f}(\mathbf{u}) + \mathbf{f}(\mathbf{v})) - \frac{1}{2}(|\mathbf{A}(\mathbf{v})|\mathbf{v} - |\mathbf{A}(\mathbf{u})|\mathbf{u}).$$

Remark 5.1. Let us see why the present flux-vector splitting scheme is relatively viscous near stagnation points. The "viscosity" or dissipation term $\mathbf{D}(\mathbf{u},\mathbf{v})$ such that

$$\mathbf{g}(\mathbf{u},\mathbf{v}) = \frac{1}{2}(\mathbf{f}(\mathbf{u}) + \mathbf{f}(\mathbf{v})) - \frac{1}{2}\mathbf{D}(\mathbf{u},\mathbf{v})$$

is given by

$$\mathbf{D}(\mathbf{u},\mathbf{v}) = (\mathbf{f}^+(\mathbf{v}) - \mathbf{f}^-(\mathbf{v})) - (\mathbf{f}^+(\mathbf{u}) - \mathbf{f}^-(\mathbf{u})),$$

which in the particular case of Steger and Warming's scheme gives

$$\mathbf{D}(\mathbf{u},\mathbf{v}) = |\mathbf{A}(\mathbf{v})|\mathbf{v} - |\mathbf{A}(\mathbf{u})|\mathbf{u}.$$

Now $\mathbf{f}(\mathbf{v}) = \mathbf{f}(\mathbf{u})$ does not imply $\mathbf{f}^+(\mathbf{v}) - \mathbf{f}^-(\mathbf{v}) = \mathbf{f}^+(\mathbf{u}) - \mathbf{f}^-(\mathbf{u})$, roughly because $v = -u$ does not imply $|v|v = |u|u$! For nearby values of u and $v, v = -u$ implies that u and v are near 0. It follows that a flux-vector splitting scheme does not give a good resolution near "sonic" points such that $a_k = 0$, since there may exist nearby states $\mathbf{u}, \mathbf{v}$ such that $\operatorname{sgn} a_k(\mathbf{u}) \neq \operatorname{sgn} a_k(\mathbf{v})$. This happens for stationary contact discontinuities in the case of Euler equations (see Coquel and Liou 1992).

Also, after rearrangement, we can write Steger and Warming's viscosity term as

$$\mathbf{D}(\mathbf{u},\mathbf{v}) = \frac{1}{2}(|\mathbf{A}(\mathbf{u})| + |\mathbf{A}(\mathbf{v})|)(\mathbf{v} - \mathbf{u}) + \frac{1}{2}(|\mathbf{A}(\mathbf{v})| - |\mathbf{A}(\mathbf{u})|)(\mathbf{u} + \mathbf{v}).$$

A scheme satisfying

$$\mathbf{D}(\mathbf{u},\mathbf{v}) = |\mathbf{A}\Big(\frac{\mathbf{u}+\mathbf{v}}{2}\Big)|(\mathbf{v} - \mathbf{u}) + \mathbf{o}(|\mathbf{u} - \mathbf{v}|)$$

is called an "upstream scheme" by Harten, Lax, and van Leer (1983). Thus, the above scheme is of upstream form except near "sonic" points. □

If we set

$$\mathbf{u} = \sum_{k=1}^{p} \alpha_k(\mathbf{u})\mathbf{r}_k(\mathbf{u}),$$

where $(\mathbf{r}_k)$ is a basis of eigenvectors of $\mathbf{A}$ and the $\alpha_k(\mathbf{u})$ are the "characteristic" variables, we have

$$\mathbf{f}(\mathbf{u}) = \sum_{k=1}^{p} a_k(\mathbf{u})\alpha_k(\mathbf{u})\mathbf{r}_k(\mathbf{u})$$

and

$$\mathbf{f}^{\pm}(\mathbf{u}) = \sum_{k=1}^{p} a_k^{\pm}(\mathbf{u})\alpha_k(\mathbf{u})\mathbf{r}_k(\mathbf{u}). \tag{5.4}$$

Remark 5.2. Let us note that, for any $\mu \in \mathbb{R}$,

$$\mathbf{A}(\mu\mathbf{u}) = \mathbf{A}(\mathbf{u}),$$

and hence by (1.9)

$$\mathbf{A}^{\pm}(\mu\mathbf{u}) = \mathbf{A}^{\pm}(\mathbf{u}) \text{ and } \mathbf{f}^{\pm}(\mu\mathbf{u}) = \mu\mathbf{f}^{\pm}(\mathbf{u}).$$

By Euler's identity, this implies

$$\mathbf{f}^{\pm}(\mathbf{u}) = \mathbf{f}^{\pm'}(\mathbf{u})\mathbf{u} \equiv \mathbf{A}_{\pm}(\mathbf{u})\mathbf{u}.$$

However, $\mathbf{A}_{+}(\mathbf{u}) = \mathbf{f}^{+'}(\mathbf{u})$ is not equal to $\mathbf{A}^{+}(\mathbf{u})$ (defined by (1.9)), though $\mathbf{A}_{+}(\mathbf{u})\mathbf{u} = \mathbf{A}^{+}(\mathbf{u})\mathbf{u}$ (see Lerat 1982 for details). Also, the notations may be misleading since $\mathbf{A}_{+}(\mathbf{u})$ and $\mathbf{A}_{-}(\mathbf{u})$ do not necessarily have positive or negative eigenvalues, and one will have to check their sign. □

5.2 *Application to the gas dynamics equations. I. Steger and Warming's approach*

Let us check the flux homogeneity property in the case of the gas dynamics equations.

Lemma 5.1

We assume that the equation of state satisfies

$$p(\mu\rho, \varepsilon) = \mu p(\rho, \varepsilon).$$

Then, the flux function is homogeneous of degree 1.

Proof. We have from (2.19)

$$\mathbf{U} = \begin{pmatrix} \rho \\ q \\ E \end{pmatrix}, \quad \mathbf{F}(\mathbf{U}) = \begin{pmatrix} q \\ q^2/\rho + p \\ (E+p)m/\rho \end{pmatrix}, \quad p = p\Big(\rho, \frac{E}{\rho} - \frac{q^2}{2\rho^2}\Big).$$

Setting

$$p(\mathbf{U}) = p(\rho, \varepsilon) = p\Big(\rho, \frac{E}{\rho} - \frac{q^2}{2\rho^2}\Big),$$

we note that by assumption

$$p(\mu\mathbf{U}) = p(\mu\rho, \varepsilon) = \mu p(\rho, \varepsilon).$$

Hence

$$\mathbf{F}(\mu\mathbf{U}) = \begin{pmatrix} \mu q \\ \mu q^2/\rho + p(\mu\rho, \varepsilon) \\ (\mu E + p(\mu\rho, \varepsilon))q/\rho \end{pmatrix} = \mu \begin{pmatrix} q \\ q^2/\rho + p(\rho, \varepsilon) \\ (E + p(\rho, \varepsilon))q/\rho \end{pmatrix},$$

and

$$\mathbf{F}(\mu\mathbf{U}) = \mu\mathbf{F}(\mathbf{U}),$$

which proves the homogeneity of $\mathbf{F}$. □

Remark 5.3. Lemma 5.1 applies in the case of a polytropic ideal gas since

$$p = (\gamma - 1)\rho\varepsilon.$$

More generally, it also applies for an equation of state of a thermally (not necessarily calorically) perfect gas,

$$p = \rho T(\varepsilon). \tag{5.5}$$

For such an equation, p satisfies (3.46) (see Chapter II, Section 1.2, (1.21))

$$p = \rho p_\rho + \tilde{\varepsilon} p_{\tilde{\varepsilon}}.$$

Now, an easy computation using the expression (3.30) of $\mathbf{A}(\mathbf{U})$ found in the preceding section shows that

$$\mathbf{F}(\mathbf{U}) = \mathbf{A}(\mathbf{U})\mathbf{U} + (p - \rho p_\rho - \tilde{\varepsilon} p_{\tilde{\varepsilon}})(0, 1, u)^T.$$

Hence, the homogeneity property

$$\mathbf{F}(\mathbf{U}) = \mathbf{A}(\mathbf{U})\mathbf{U}$$

is equivalent to requiring that p satisfies (3.46).

Note that the flux Jacobian remains homogeneous for non-equilibrium mixtures of thermally perfect gases (see Grossman and Cinnella 1990). □

Let us express the split fluxes, given by the formulas (5.4),

$$\mathbf{U} = \sum_{k=1}^{3} \alpha_k(\mathbf{U})\mathbf{r}_k(\mathbf{U}),$$

and

$$\mathbf{F}^{\pm}(\mathbf{U}) = \sum_{k=1}^{3} a_k^{\pm}(\mathbf{U})\alpha_k(\mathbf{U})\mathbf{r}_k(\mathbf{U}),$$

where again we have $a_1 = u - c, a_2 = u, a_3 = u + c$, and the vectors $\mathbf{r}_k$ are given by (3.32).

We first consider the case of a thermally perfect gas where the flux is a homogeneous function. Solving the system for the (α_k) as in Lemma 3.7, we get for a thermally perfect gas (5.5)

$$\alpha_1 = \alpha_3 = \frac{\rho}{2\gamma}, \quad \alpha_2 = p\frac{(\gamma-1)}{c^2} = \rho\frac{(\gamma-1)}{\gamma}. \tag{5.6}$$

Setting

$$\mathbf{F}_k(\mathbf{U}) = a_k(\mathbf{U})\alpha_k(\mathbf{U})\mathbf{r}_k(\mathbf{U}), \quad k = 1, 2, 3,$$

we have

$$\mathbf{F}(\mathbf{U}) = \mathbf{F}_1(\mathbf{U}) + \mathbf{F}_2(\mathbf{U}) + \mathbf{F}_3(\mathbf{U}).$$

From (3.32), the $\mathbf{F}_k(\mathbf{U})$ are given by

(5.7a)

$$\mathbf{F}_1(\mathbf{U}) = a_1\frac{\rho}{2\gamma}\begin{pmatrix} 1 \\ u - c \\ H - uc \end{pmatrix}, \quad \mathbf{F}_2(\mathbf{U}) = a_2\frac{\rho(\gamma-1)}{\gamma}\begin{pmatrix} 1 \\ u \\ H - \frac{c^2}{(\gamma-1)} \end{pmatrix},$$

$$\mathbf{F}_3(\mathbf{U}) = a_3\frac{\rho}{2\gamma}\begin{pmatrix} 1 \\ u + c \\ H + uc \end{pmatrix}. \tag{5.7b}$$

(Note that $H - \frac{c^2}{(\gamma-1)} = \frac{u^2}{2}$ for a polytropic ideal gas.)

We can make precise the expressions of $\mathbf{F}^{\pm}$ in terms of the $\mathbf{F}_k$ since we are in the following situation: each eigenvalue has a single zero that we denote by $\overline{u}_i$ $(\overline{u}_1 = c, \overline{u}_2 = 0, \overline{u}_3 = -c)$. We obtain

$$\begin{cases} u \geq c & \Rightarrow a_k \geq 0 \text{ and } \mathbf{F}^{+}(\mathbf{U}) = \mathbf{F}(\mathbf{U}), \mathbf{F}^{-}(\mathbf{U}) = \mathbf{0}, \\ u \leq -c & \Rightarrow a_k \leq 0 \text{ and } \mathbf{F}^{-}(\mathbf{U}) = \mathbf{F}(\mathbf{U}), \mathbf{F}^{+}(\mathbf{U}) = \mathbf{0}. \end{cases} \tag{5.8a}$$

Thus, in supersonic regions, which are by definition those where the Mach number $|M| = |\frac{u}{c}|$ satisfies$|M| \geq 1$, we recover the upwind scheme as expected.

Now, for $-c < u < c$,

$$\begin{cases} -c < u < 0 \Rightarrow a_3 > 0 > a_2 > a_1 \\ \quad \text{and} \quad \mathbf{F}^{+}(\mathbf{U}) = \mathbf{F}_3(\mathbf{U}), \mathbf{F}^{-}(\mathbf{U}) = \mathbf{F}_1(\mathbf{U}) + \mathbf{F}_2(\mathbf{U}), \\ 0 < u < c \quad \Rightarrow a_3 > a_2 > 0 > a_1 \\ \quad \text{and} \quad \mathbf{F}^{+}(\mathbf{U}) = \mathbf{F}_2(\mathbf{U}) + \mathbf{F}_3(\mathbf{U}), \mathbf{F}^{-}(\mathbf{U}) = \mathbf{F}_1(\mathbf{U}). \end{cases} \tag{5.8b}$$

The eigenvalues of $\mathbf{F}^{+'}$ and $\mathbf{F}^{-'}$ have the correct sign (i.e., resp. ≥ 0 and ≤ 0) for a perfect gas, but this is not straightforward to prove (see Lerat 1983, who proves the result for $1 < \gamma < \frac{5}{3}$ and Vinokur and Montagné 1990 who plot their numerical values). Note that the eigenvalues of $\mathbf{A}_+(\mathbf{U}) = \mathbf{F}^{+'}(\mathbf{U})$ are not continuous in general at a zero $\overline{u}_i$.

We turn to an arbitrary equilibrium gas for which the homogeneity property is not satisfied. There no longer exists a natural flux splitting. Nonetheless, following the approach of Sanders and Prendergast (see Vinokur and Montagné 1990), we can extend the above decomposition, i.e., write $\mathbf{F}(\mathbf{U})$ in the form

$$\mathbf{F}(\mathbf{U}) = \mathbf{F}_1(\mathbf{U}) + \mathbf{F}_2(\mathbf{U}) + \mathbf{F}_3(\mathbf{U}),$$

and then define $\mathbf{F}^{\pm}(\mathbf{U})$ as in (5.8). The idea is to look for convective fluxes of the form

$$\mathbf{F}_k(\mathbf{U}) = a_k(\mathbf{U})\mathbf{U}_k$$

with the constraints

$$(5.9)\qquad \begin{cases} \displaystyle\sum_{k=1}^{3} \mathbf{U}_k = \mathbf{U} = \rho\Big(1, u, \varepsilon + \frac{u^2}{2}\Big)^T, \\ \displaystyle\sum_{k=1}^{3} \mathbf{F}_k(\mathbf{U}) = \mathbf{F}(\mathbf{U}) = \rho\Big(u, u^2 + \frac{p}{\rho}, \Big(e + \frac{p}{\rho}\Big)u\Big)^T. \end{cases}$$

Now, we set

$$\mathbf{U}_k = \rho_k\Big(1, a_k, \varepsilon_k + \frac{a_k^2}{2}\Big)^T, \quad 1 \leq k \leq 3,$$

which means that the velocity in $\mathbf{U}_k$ is equal to a_k, as was indeed the case in (5.7). The six unknowns ρ_k, ε_k are determined from the equations (5.9), which represent only five equations since one of them is trivial (the variable ρu is present in both $\mathbf{U}$ and $\mathbf{F}$). We easily obtain as in (5.6)

$$\rho_1 = \rho_3 = \alpha_1 = \alpha_3 = \frac{\rho}{2\gamma}, \quad \rho_2 = \alpha_2 = \frac{\rho(\gamma - 1)}{\gamma},$$

where γ is defined by

$$(5.10)\qquad \gamma = \rho c^2/p.$$

Also,

$$\varepsilon_1 = \varepsilon_3, \; \sum_{k=1}^{3} \varepsilon_k \rho_k = \rho\Big(\varepsilon - \frac{c^2}{2\gamma}\Big).$$

There remains one degree of freedom, which in Vinokur and Montagné (1990) is taken in the form of a nondimensional parameter β such that

$$\varepsilon_2 = \varepsilon - (1-\beta)\,\frac{c^2}{\gamma(\gamma-1)}\,.$$

Since by (5.10)

$$H - \frac{c^2}{(\gamma-1)} = H - \frac{p}{\rho} - \frac{c^2}{\gamma(\gamma-1)} = \frac{u^2}{2} + \varepsilon - \frac{c^2}{\gamma(\gamma-1)}\,,$$

the expression has been determined so that the case $\beta = 0$ corresponds to a thermally perfect gas. It is then proved by arguing on the eigenvalues of $\mathbf{A}_+$ that a convenient choice is indeed

$$\varepsilon_2 = \varepsilon - \frac{c^2}{\gamma(\gamma-1)}\,,$$

thus leading to

$$\varepsilon_1 = \varepsilon_3 = \varepsilon + \frac{c^2(2-\gamma)}{\gamma}$$

and to the same expressions for the fluxes $\mathbf{F}_k$ as above.

In short, we can still take the same expressions (5.8) for $\mathbf{F}^{\pm}$ in order to define a generalized Steger and Warming flux vector splitting.

5.3 Application to the gas dynamics equations. II. Van Leer's approach

As we have already observed, Steger and Warming's decomposition is not continuous in general at a sonic point (see Liang and Chan 1989). Another splitting of the fluxes in the range $|u| < c$ has been derived by van Leer under the requirement that the split fluxes and their Jacobians be continuous. Besides the usual conditions ("consistency" $\mathbf{F}^+ + \mathbf{F}^- = \mathbf{F}$, and the correct sign for the eigenvalues of $\mathbf{F}^{\pm'}$), further requirements are:

(i) a symmetry property,

$$\mathbf{F}^+(u) = \pm\mathbf{F}^-(u) \text{ if } \mathbf{F}(u) = \pm\mathbf{F}(-u)$$

(all other quantities but u constant).

(ii) $\mathbf{F}^{\pm}$ is a polynomial in u (of lowest possible degree).

(iii) $\mathbf{F}^{\pm'}$ has a vanishing eigenvalue for $|M| < 1$.

Note that (i) is linked to the fact that $a_1(u) = u - c = -a_3(-u)$ and is satisfied by the splitting of the preceding section. For what concerns condition (ii), in the system of gas dynamics, each component of $\mathbf{F}$ is a polynomial in u of degree ≤ 3. Condition (iii) is imposed in the hope that the eigenvalues of $\mathbf{F}^{\pm'}$ will have the correct sign (since the one that is most likely to have the wrong sign is "forced" to vanish).

We can seek cubic polynomials in u for the components of $\mathbf{F}^{\pm}$ too. For $|M| > 1, \mathbf{F}^{+}$ or $\mathbf{F}^{-} = 0$, and the continuity requirement on $\mathbf{F}^{+'}$ requires that the polynomials include a factor $(u \pm c)^2$. Thus, for each component $F_i^{\pm}$ of $\mathbf{F}^{\pm}$ we postulate in the range $|u| < c$

$$F_i^{\pm} = (u \pm c)^2(au + b), \quad i = 1, 2, 3,$$

where the coefficients of the remaining first-order factor are obtained by using the symmetry property (i), and the "consistency" $F_i = F_i^{+} + F_i^{-}$. (The notation of the components F_i is to be distinguished from the fluxes $\mathbf{F}_i$ of the preceding section.) For the splitting of the first two components of $\mathbf{F}$ (mass flux $F_1 = q = \rho u$ and momentum flux $F_2 = qu + p$), using the identity

$$q = \rho u = \frac{\rho}{4c}\{(u + c)^2 - (u - c)^2\},$$

we easily obtain

$$F_1^{\pm} = \pm\frac{\rho}{4c}(u \pm c)^2, \tag{5.11}$$

and by (5.10) ($p = \frac{\rho c^2}{\gamma}$),

$$F_2^{\pm} = F_1^{\pm}\,\frac{\{(\gamma - 1)u \pm 2c\}}{\gamma}. \tag{5.12}$$

In fact, the splitting of the third component (energy flux $F_3 = (E + p)u = \rho u^3/2 + pu + \rho\varepsilon u$) obtained in this way would differ from that derived by van Leer for an ideal gas in that (iii) is not satisfied. If γ is constant in (5.10), $\varepsilon = \frac{c^2}{\gamma(\gamma-1)}$, and the third component of $\mathbf{F}$ can be written

$$F_3 = q\Big\{\frac{u^2}{2} + \frac{c^2}{\gamma(\gamma - 1)}\Big\}.$$

Then, it is easy to check that a convenient splitting is

$$\begin{aligned} F_3^{\pm} &= F_1^{\pm}\,\frac{((\gamma - 1)u \pm 2c)^2}{2(\gamma^2 - 1)} \\ &= \frac{\gamma^2}{2(\gamma^2 - 1)}\,\frac{F_2^{\pm 2}}{F_1^{\pm}}. \end{aligned} \tag{5.13}$$

When γ is not constant, we express the flux in the form

$$F_3^{\pm} = F_1^{\pm}\Big\{\frac{((\gamma - 1)u \pm 2c)^2}{2(\gamma^2 - 1)} + \Big(\varepsilon - \frac{c^2}{\gamma(\gamma - 1)}\Big) + \beta(u \mp c)^2\Big\}.$$

Let us note the following simple facts:

— For a polytropic ideal gas, choosing $\beta = 0$ reduces to van Leer's splitting. The sign of the two corresponding nonzero eigenvalues of $\mathbf{F}^{\pm'}$ is studied in van Leer (1982), and Vinokur and Montagné (1990). In particular, we get

the right sign in the range $1 \leq \gamma \leq 3$.
— For $\beta = \frac{1}{(\gamma+1)}$, we get

$$F_3^{\pm} = F_1^{\pm} H,$$

or with more explicit notation,

$$(Eu + pu)^{\pm} = q^{\pm} H,$$

which means that the total enthalpy is preserved (see also Liou, van Leer, and Shuen 1990).
— For nonconstant γ, condition (iii) cannot be identically satisfied for any choice of β. For a thermally perfect gas, it can be shown that the choice $\beta = 0$ leads to a negative eigenvalue in the whole range $|M| < 1$, but since it is very small this is not a major drawback, and the split energy flux for nonconstant γ will indeed be taken as in the ideal gas case,

$$F_3^{\pm} = F_1^{\pm} \Big\{ \frac{((\gamma-1)u \pm 2c)^2}{2(\gamma^2-1)} + \varepsilon - \frac{c^2}{\gamma(\gamma-1)} \Big\}.$$

In short, in the range $|\frac{u}{c}| \geq 1$, we keep the upwind scheme, and the decomposition is given by formulas (5.8a), while in the range $|\frac{u}{c}| < 1$, we define a splitting

$$\mathbf{F} = \mathbf{F}^{+} + \mathbf{F}^{-}, \quad \mathbf{F}^{\pm} = (F_i^{\pm})$$

by the formulas (5.11)–(5.13). This splitting satisfies the requirements (i)–(iii). Moreover, for an ideal gas, the Jacobian matrix $\mathbf{F}^{+'}$ (resp. $\mathbf{F}^{-'}$) has positive (resp. negative) eigenvalues.

For other studies of the real gas case, we refer to Larrouturou and Fezoui (1989), Grossman and Cinella (1990), Shuen, Liou, and van Leer (1990), Müller (1988), and Montagné (1986).

We shall study in Section 7 below kinetic schemes that are also flux vector splitting methods.

Other papers related to flux splitting are Liou and Steffen (1993), Radespiel and Kroll (1995), Coquel and Liou (1994), Chen and Le Floch (1995) and for the relation with parabolized schemes Chang and Merkle (1989).

6 Van Leer's second-order method

Van Leer's method generalizes Godunov's method to obtain a second-order scheme. Following the ideas of G.R., Chapter IV, Section 3, we present it for a space grid that is not necessarily uniform. Let us recall the three main steps of the method:
(i) A "reconstruction step," which consists in constructing a piecewise linear

function $\tilde{\mathbf{v}}$ from given cell-averages $\mathbf{v}_j^n$,

(6.1) $$\tilde{\mathbf{v}}^n(x) = \mathbf{v}_j^n + (x - x_j)\frac{\mathbf{S}_j^n}{\Delta x_j}, \quad x_{j-\frac{1}{2}} < x < x_{j+\frac{1}{2}},$$

where $\Delta x_j = x_{j+\frac{1}{2}} - x_{j-\frac{1}{2}}$; the slopes $\mathbf{S}_j^n$ are to be detailed later on.
(ii) An "evolution step," which involves either an exact or an approximate Riemann solver. One solves

(6.2) $$\begin{cases} \dfrac{\partial \mathbf{w}}{\partial t} + \dfrac{\partial \mathbf{f}(\mathbf{w})}{\partial x} = \mathbf{0}, & 0 \le t \le \Delta t, \\ \mathbf{w}(x,0) = \tilde{\mathbf{v}}^n(x), \end{cases}$$

which gives $\mathbf{w}(.,\Delta t)$.
(iii) A "cell-averaging step," which gives $\mathbf{v}_j^{n+1}$. As in (2.5), we project (in the sense of $\mathbf{L}^2$) the solution $\mathbf{w}(.,\Delta t)$ onto the piecewise constant functions

(6.3) $$\mathbf{v}_j^{n+1} = \frac{1}{\Delta x_j}\int_{x_{j-\frac{1}{2}}}^{x_{j+\frac{1}{2}}} \mathbf{w}(x,\Delta t)dx.$$

6.1 Van Leer's method for systems

Provided we assume some convenient C.F.L. condition so that the waves issued from the points $x_{j-\frac{1}{2}}$ and $x_{j+\frac{1}{2}}$ do not interact, the solution of (6.2) is in fact obtained by juxtaposition of local generalized Riemann problems. In order to derive a more explicit form of the scheme, we integrate the equation (6.2) over a cell $(x_{j-\frac{1}{2}}, x_{j+\frac{1}{2}}) \times (0,\Delta t)$,

$$\int_0^{\Delta t}\int_{x_{j-\frac{1}{2}}}^{x_{j+\frac{1}{2}}} \left(\frac{\partial \mathbf{w}}{\partial t} + \frac{\partial \mathbf{f}(\mathbf{w})}{\partial x}\right)dx = \mathbf{0}.$$

We obtain

$$\int_{x_{j-\frac{1}{2}}}^{x_{j+\frac{1}{2}}} (\mathbf{w}(x,\Delta t) - \mathbf{w}(x,0))dx$$
$$+ \int_0^{\Delta t} (\mathbf{f}(\mathbf{w}(x_{j+\frac{1}{2}},t)) - \mathbf{f}(\mathbf{w}(x_{j-\frac{1}{2}},t)))dt = \mathbf{0}$$

since the flux is continuous, and then by (6.3)

$$\Delta x_j(\mathbf{v}_j^{n+1} - \mathbf{v}_j^n) + \int_0^{\Delta t} \{\mathbf{f}(\mathbf{w}(x_{j+\frac{1}{2}},t)) - \mathbf{f}(\mathbf{w}(x_{j-\frac{1}{2}},t))\}dt = \mathbf{0}.$$

We are left with the evaluation of the numerical flux

(6.4) $$\mathbf{g}_{j+\frac{1}{2}}^n = \frac{1}{\Delta t}\int_0^{\Delta t} \mathbf{f}(\mathbf{w}(x_{j+\frac{1}{2}},t))dt.$$

To begin with, using the midpoint rule, we write

$$\frac{1}{\Delta t}\int_0^{\Delta t} \mathbf{f}(\mathbf{w}(x_{j+\frac12}, t))dt = \mathbf{f}\Big(\mathbf{w}\Big(x_{j+\frac12}, \frac{\Delta t}{2}\Big)\Big) + O(\Delta t^2).$$

As in the scalar case (G.R., Chapter IV, Sections 3.2 and 3.3), we now give two ways for approximating $\mathbf{f}(\mathbf{w}(x_{j+\frac12}, \frac{\Delta t}{2}))$.
1) One can use the asymptotic expansion of the solution of the G.R.P. (6.2) associated to the piecewise linear data (6.1),

$$\mathbf{w}(x,t) = \mathbf{u}^0\Big(\frac{x - x_{j+\frac12}}{t}\Big) + t\mathbf{u}^1\Big(\frac{x - x_{j+\frac12}}{t}\Big) + O(t^2),$$

where the derivation of $\mathbf{u}^0$ and $\mathbf{u}^1$ will be detailed in the next section (in particular, $\mathbf{u}^0$ is solution of a classical Riemann problem). It implies
(6.5)

$$\mathbf{w}(x_{j+\frac12}, t) = \mathbf{u}^0(0) + t\mathbf{u}^1(0) + O(t^2) = \mathbf{w}^0(x_{j+\frac12}) + t\mathbf{w}^1(x_{j+\frac12}) + O(t^2),$$

and the scheme reads

$$\mathbf{v}_j^{n+1} = \mathbf{v}_j^n + \frac{\Delta t}{\Delta x_j}\{\mathbf{g}_{j+\frac12}^n - \mathbf{g}_{j-\frac12}^n$$

where the numerical flux is given by $\mathbf{g}_{j+\frac12}^n = \mathbf{f}(\mathbf{u}^0(0) + \frac{\Delta t}{2}\mathbf{u}^1(0))$, i.e.,

$$\mathbf{g}_{j+\frac12}^n = \mathbf{f}\Big(\mathbf{w}^0(x_{j+\frac12}) + \frac{\Delta t}{2}\mathbf{w}^1(x_{j+\frac12})\Big). \tag{6.6}$$

The argument may be double-valued, but the flux is single-valued. In fact, we can take an approximation, for example

$$\mathbf{f}\Big(\mathbf{u}^0(0) + \frac{\Delta t}{2}\mathbf{u}^1(0)\Big) \approx \mathbf{f}\Big(\mathbf{u}^0(0)\big) + \frac{\Delta t}{2}\mathbf{f}'(\mathbf{u}^0(0)\Big)\mathbf{u}^1(0),$$

or any other more convenient approximation. In the next section, for example, we shall use for the third component pu of $\mathbf{f}$

$$(pu)^0 + t(pu)^1 = p^0u^0 + t(p^0u^1 + p^1u^0) \approx (p^0 + tp^1)(u^0 + tu^1).$$

2) One uses a predictor-corrector scheme. Following an idea of Hancock, we define the updated values $\mathbf{v}_{j+\frac12,\pm}^{n+\frac12}$ at time $t_n + \frac{\Delta t}{2}$ by

$$\begin{cases} \mathbf{v}_{j+\frac12,-}^{n+\frac12} = \mathbf{v}_{j+\frac12,-}^n - \dfrac{\Delta t}{2\Delta x_j}(\mathbf{f}(\mathbf{v}_{j+\frac12,-}^n) - \mathbf{f}(\mathbf{v}_{j-\frac12,+}^n)), \\ \mathbf{v}_{j+\frac12,+}^{n+\frac12} = \mathbf{v}_{j+\frac12,+}^n - \dfrac{\Delta t}{2\Delta x_j}(\mathbf{f}(\mathbf{v}_{j+\frac32,-}^n) - \mathbf{f}(\mathbf{v}_{j+\frac12,+}^n)), \end{cases}$$

where

$$\begin{cases} \mathbf{v}_{j+\frac12,-}^n = \tilde{\mathbf{v}}(x_{j+\frac12} - 0) - \mathbf{v}_j^n + \dfrac{\mathbf{S}_j^n}{2}, \\ \mathbf{v}_{j+\frac12,+}^n = \tilde{\mathbf{v}}(x_{j+\frac12} + 0) = \mathbf{v}_{j+1}^n - \dfrac{\mathbf{S}_{j+1}^n}{2}. \end{cases}$$

Then, we solve the Riemann problem at the point $x_{j+\frac{1}{2}}$ with piecewise constant initial data $\mathbf{v}^{n+\frac{1}{2}}_{j+\frac{1}{2},\pm}$

$$\begin{cases} \dfrac{\partial \mathbf{u}}{\partial t} + \dfrac{\partial}{\partial x}\mathbf{f}(\mathbf{u}) = \mathbf{0}, \\ \mathbf{u}(x,0) = \begin{cases} \mathbf{v}^{n+\frac{1}{2}}_{j+\frac{1}{2},-}, & x < x_{j+\frac{1}{2}}, \\ \mathbf{v}^{n+\frac{1}{2}}_{j+\frac{1}{2},+}, & x > x_{j+\frac{1}{2}}, \end{cases} \end{cases}$$

whose solution is noted as usual $\mathbf{w}_R\Big(\frac{x-x_{j+\frac{1}{2}}}{t}; \mathbf{v}^{n+\frac{1}{2}}_{j+\frac{1}{2},-}, \mathbf{v}^{n+\frac{1}{2}}_{j+\frac{1}{2},+}\Big)$. We replace

$$\mathbf{w}\Big(x_{j+\frac{1}{2}}, \frac{\Delta t}{2}\Big)$$

by

$$\mathbf{w}_R(0; \mathbf{v}^{n+\frac{1}{2}}_{j+\frac{1}{2},-}, \mathbf{v}^{n+\frac{1}{2}}_{j+\frac{1}{2},+}),$$

and thus we take

$$\mathbf{g}^n_{j+\frac{1}{2}} = \mathbf{f}\Big(\mathbf{w}_R(0; \mathbf{v}^{n+\frac{1}{2}}_{j+\frac{1}{2},-}, \mathbf{v}^{n+\frac{1}{2}}_{j+\frac{1}{2},+})\Big). \tag{6.7}$$

Remark 6.1. The updated values $\mathbf{v}^{n+\frac{1}{2}}_{j+\frac{1}{2},\pm}$ can be interpreted as follows:

$$\mathbf{v}^{n+\frac{1}{2}}_{j+\frac{1}{2},\pm} \approx \mathbf{w}\Big(x_{j+\frac{1}{2}} \pm 0, \frac{\Delta t}{2}\Big),$$

and $\mathbf{w}(x_{j+\frac{1}{2}} \pm 0, \frac{\Delta t}{2})$ is evaluated by

$$\begin{cases} \mathbf{w}(x_{j+\frac{1}{2}} \pm 0, \dfrac{\Delta t}{2}) \approx \mathbf{w}(x_{j+\frac{1}{2}} \pm 0, 0) + \dfrac{\Delta t}{2}\dfrac{\partial \mathbf{u}}{\partial t} \\ \qquad\qquad = \mathbf{w}(x_{j+\frac{1}{2}} \pm 0, 0) - \dfrac{\Delta t}{2}\dfrac{\partial \mathbf{f}(\mathbf{u})}{\partial x}. \end{cases}$$

Using a finite difference approximation for $\frac{\partial \mathbf{f}(\mathbf{u})}{\partial x}$ yields

$$\mathbf{w}\Big(x_{j+\frac{1}{2}} - 0, \frac{\Delta t}{2}\Big) \approx \mathbf{v}^n_{j+\frac{1}{2},-} - \frac{\Delta t}{2\Delta x_j}\Big(\mathbf{f}(\mathbf{v}^n_{j+\frac{1}{2},-}) - \mathbf{f}(\mathbf{v}^n_{j-\frac{1}{2},+})\Big),$$

$$\mathbf{w}\Big(x_{j+\frac{1}{2}} + 0, \frac{\Delta t}{2}\Big) \approx \mathbf{v}^n_{j+\frac{1}{2},+} - \frac{\Delta t}{2\Delta x_j}\Big(\mathbf{f}(\mathbf{v}^n_{j+\frac{3}{2},-}) - \mathbf{f}(\mathbf{v}^n_{j-\frac{1}{2},+})\Big).$$

The last step corresponds to applying Godunov's scheme to these updated values.

The reconstruction step consists then in defining new slopes $\mathbf{S}^{n+1}_j$. We also proceed by prediction-correction. We can use, for instance, central differencing to define the predictor

$$\hat{\mathbf{S}}_j^{n+1} = \Delta x_j \frac{\mathbf{v}_{j+1}^{n+1} - \mathbf{v}_{j-1}^{n+1}}{x_{j+1} - x_{j-1}}. \tag{6.8}$$

Otherwise, if we follow the first approach 1), $\mathbf{S}_j^{n+1}$ can be predicted by

$$\hat{\mathbf{S}}_j^{n+1} = \mathbf{w}(x_{j+\frac{1}{2}} - 0, \Delta t) - \mathbf{w}(x_{j-\frac{1}{2}} + 0, \Delta t), \tag{6.9}$$

where $\mathbf{w}(x_{j+\frac{1}{2}} \pm 0, \Delta t)$ is computed from the first terms of the expansion (6.5),

$$\mathbf{u}(0\pm, \Delta t) \approx \mathbf{u}^0(0\pm) + \Delta t \mathbf{u}^1(0\pm) = \mathbf{w}^0(x_{j+\frac{1}{2}} \pm 0) + \Delta t \mathbf{w}^1(x_{j+\frac{1}{2}} \pm 0),$$

or even more simply by retaining the first-order term of the expansion (which corresponds to the solution of a classical Riemann problem),

$$\hat{\mathbf{S}}_j^{n+1} = \mathbf{w}^0(x_{j+\frac{1}{2}} - 0, \Delta t) - \mathbf{w}^0(x_{j-\frac{1}{2}} + 0, \Delta t).$$

In the second approach,

$$\hat{\mathbf{S}}_j^{n+1} = v_{j+\frac{1}{2},+}^{n+\frac{1}{2}} - v_{j+\frac{1}{2},-}^{n+\frac{1}{2}}.$$

In the correction steps, we limit the slope componentwise following formula (3.11) in G.R., Chapter IV, which ensures that $\tilde{\mathbf{v}}^{n+1}(x_{j\pm\frac{1}{2}})$ does not take values outside the range spanned by the neighboring mesh averages. Omitting the time superscript for each component $\delta v_{k,j}$ of $\mathbf{S}_j$, $k = 1, ..., p$, we impose

$$\delta v_{k,j} = \begin{cases} s \min\{2|\Delta v_{k,j-\frac{1}{2}}|, |\hat{\delta} v_{k,j}|, 2|\Delta v_{k,j+\frac{1}{2}}|\} \\ \text{if } s = \operatorname{sgn} \Delta v_{k,j-\frac{1}{2}} = \operatorname{sgn} \Delta v_{k,j+\frac{1}{2}} = \operatorname{sgn} \hat{\delta} v_{k,j}, \\ 0 \qquad \text{otherwise.} \end{cases} \tag{6.10}$$

There are other possibilities for limiting the slope that are not so easy to interpret. However, this correction is very important in practice.

6.2 Solution of the generalized Riemann problem

The first method (6.6) involves a generalized Riemann problem (G.R.P.)

$$\begin{cases} \dfrac{\partial \mathbf{u}}{\partial t} + \dfrac{\partial}{\partial x} \mathbf{f}(\mathbf{u}) = \mathbf{0}, \\ \mathbf{u}(x, 0) = \begin{cases} \mathbf{u}_L(x), & x < 0, \\ \mathbf{u}_R(x), & x > 0. \end{cases} \end{cases} \tag{6.11}$$

The solution is not self-similar, since the discontinuity curves are no longer straight lines, but it still consists of a sequence of shocks, rarefactions, and

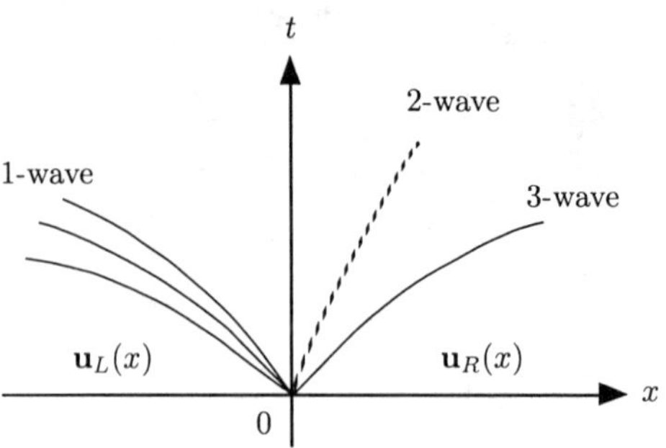

FIGURE 6.1. Solution of the G.R.P.

contact discontinuities (see Figure 6.1).

Following the arguments of G.R, Chapter IV, Section 3.3, we can use an asymptotic expansion near the origin in order to derive an approximate solution, which is obtained as a perturbation of the (classical) Riemann problem

$$\begin{cases} \dfrac{\partial \mathbf{u}^0}{\partial t} + \dfrac{\partial}{\partial x}\mathbf{f}(\mathbf{u}^0) = \mathbf{0}, \\ \mathbf{u}^0(x,0) = \begin{cases} \mathbf{u}_L(0), & x < 0, \\ \mathbf{u}_R(0), & x > 0. \end{cases} \end{cases} \tag{6.12}$$

Indeed, we have

$$\mathbf{u}(x,t) = \mathbf{u}^0\Big(\frac{x}{t}\Big) + t\mathbf{u}^1\Big(\frac{x}{t}\Big) + O(t^2), \tag{6.13}$$

where $\mathbf{u}^0(\xi) = \mathbf{w}_R\Big(\xi; \mathbf{u}_L(0), \mathbf{u}_R(0)\Big)$ is the solution of (6.12), and the computation of $\mathbf{u}^1$ is less straightforward. The derivation of $\mathbf{u}^0$ follows essentially the lines of the scalar case. We shall sketch the main ideas of the derivation of $\mathbf{u}^1$. We determine $\mathbf{u}^1$ in the regions of smoothness of $\mathbf{u}$ and its jumps across the transition curves that separate these regions (see Figure 6.1) as well as the conditions at infinity. First setting $\xi = \frac{x}{t}$, we obtain easily that $\mathbf{u}^1$ satisfies, in each smoothness region,

$$-\xi\frac{d\mathbf{u}^1}{d\xi} + \frac{d}{d\xi}\Big(\mathbf{A}(\mathbf{u}^0)\mathbf{u}^1\Big) + \mathbf{u}^1 = \mathbf{0}. \tag{6.14}$$

In order to determine the jump conditions satisfied by $\mathbf{u}^1$, let us consider a discontinuity curve $\sum$ with equation

$$\sigma(t) = \sigma_0 t + \frac{\sigma_1 t^2}{2} + O(t^3)$$

separating two smoothness regions of $\mathbf{u}$. On the one hand, if $\mathbf{u}^0$ is continuous across $\sum$,

$$[\mathbf{u}^1] + \sigma_1 [\frac{d\mathbf{u}^0}{d\xi}] = \mathbf{0}. \tag{6.15}$$

On the other hand, if $\mathbf{u}^0$ is discontinuous,

$$[(\mathbf{A}(\mathbf{u}^0) - \sigma_0)\mathbf{u}^1] = \sigma_1 [\mathbf{u}^0]. \tag{6.16}$$

The conditions as $\xi \to \pm\infty$ are derived as in the scalar case,

$$\lim_{\xi \to -\infty} \frac{d\mathbf{u}^1}{d\xi} = \frac{d\mathbf{u}_L}{d\xi}(0), \quad \lim_{\xi \to +\infty} \frac{d\mathbf{u}^1}{d\xi} = \frac{d\mathbf{u}_R}{d\xi}(0). \tag{6.17}$$

Proposition 6.1

If $|\mathbf{u}_R(0) - \mathbf{u}_L(0)|$ is small enough, there exists a unique function $\mathbf{u}^1$ that satisfies (6.14) in the domains of smoothness of $\mathbf{u}$ together with the conditions (6.15)–(6.17).

Proof. Let us write $\mathbf{u}^1$ in the basis $(\mathbf{r}_k(\mathbf{u}^0))$,

$$\mathbf{u}^1(\xi) = \sum_{k=1}^{p} \alpha_k(\xi) \mathbf{r}_k(\mathbf{u}^0). \tag{6.18}$$

We study successively the regions where $\mathbf{u}^0$ is constant, those where $\mathbf{u}^0$ is smooth, and then the discontinuities of $\mathbf{u}^0$.
(i) Outside the rarefaction waves and the discontinuity curves, $\mathbf{u}^0$ is constant and the system (6.14) becomes

$$\mathbf{u}^1 + \Big(\mathbf{A}(\mathbf{u}^0) - \xi\Big) \frac{d\mathbf{u}^1}{d\xi} = \mathbf{0}.$$

Thus, we get

$$\alpha_k + (\lambda_k(\mathbf{u}^0) - \xi)\alpha_k' = 0,$$

and we can solve each scalar equation as in the scalar case,

$$\alpha_k = a_k(\xi - \lambda_k(\mathbf{u}^0)), \tag{6.19}$$

where a_k is a constant. We obtain that $\mathbf{u}^1$ is an *affine* function of ξ in each region of smoothness where $\mathbf{u}^0$ is constant.
(ii) If $\mathbf{u}^0$ contains a k-rarefaction wave, we have

$$\begin{cases} \dfrac{d}{d\xi}\mathbf{u}^0(\xi) = \mathbf{r}_k(\mathbf{u}^0(\xi)), \\ \xi = \lambda_k(\mathbf{u}^0(\xi)), \end{cases}$$

and (6.14) becomes

$$\Big(\mathbf{A}(\mathbf{u}^0) - \xi\Big) \frac{d\mathbf{u}^1}{d\xi} + \Big(\frac{d}{d\xi}(\mathbf{A}(\mathbf{u}^0))\Big)\mathbf{u}^1 + \mathbf{u}^1 = \mathbf{0}. \tag{6.20}$$

We obtain again a linear differential system in the α_i (which can be solved explicitly in the case of the gas dynamics equations, as we shall see below). In particular, for $i = k$ the first term in (6.20) cancels, and we get a pure algebraic equation,

$$\alpha_k + \sum_j \beta_{jk}(\mathbf{u}^0)\alpha_j = 0, \tag{6.21}$$

where the $\beta_{j\ell}$ are defined by

$$\frac{d}{d\xi}\Big(\mathbf{A}(\mathbf{u}^0)\Big)\mathbf{r}_j(\mathbf{u}^0) = \sum_\ell \beta_{j\ell}(\mathbf{u}^0)\mathbf{r}_\ell(\mathbf{u}^0).$$

(It is easy to prove that $\beta_{kk} = 1$.)
(iii) Now, assume that $\mathbf{u}^0$ is continuous at a point $\xi = \sigma_0$ but $\frac{d\mathbf{u}^0}{d\xi}$ is discontinuous: this corresponds to the boundary of a rarefaction wave. By (6.15), we get for a k-rarefaction

$$\mathbf{u}^1(\sigma_0+) - \mathbf{u}^1(\sigma_0-) = \pm\sigma_1\mathbf{r}_k(\mathbf{u}^0(\sigma_0)), \tag{6.22a}$$

where the sign $+$ (resp. $-$) holds if the constant state is located on the right-hand side (resp. left) of the rarefaction wave. This shows that α_k is discontinuous at σ_0 but

$$\alpha_i, \quad k \neq i, \text{ is continuous at } \xi = \sigma_0. \tag{6.22b}$$

(iv) Then, if $\mathbf{u}^0$ is discontinuous at a point $\xi = \sigma_0$, which corresponds to a shock wave or a contact discontinuity, (6.16) gives a system of p equations in the $(2p+1)$ variables $\alpha_k(\sigma_0\pm)$ and σ_1.
(v) Finally, following the arguments of the scalar case, the conditions at $\pm\infty$ (6.17) give for $-\xi$ large enough

$$\mathbf{u}^1(\xi) = (\xi - \mathbf{A}(\mathbf{u}_L(0))\frac{d\mathbf{u}_L}{d\xi}(0), \tag{6.23a}$$

and for ξ large enough

$$\mathbf{u}^1(\xi) = (\xi - \mathbf{A}(\mathbf{u}_R(0))\frac{d\mathbf{u}_R}{d\xi}(0). \tag{6.23b}$$

We then check that (6.16) and (6.18)–(6.23) determine $\mathbf{u}^1$ in a unique way (see Bourgeade, Le Floch, and Raviart 1989). □

Remark 6.2. The theory above applies to a system with a source term

$$\frac{\partial \mathbf{u}}{\partial t} + \frac{\partial}{\partial x}\mathbf{F}(x, t, \mathbf{u}) = \mathbf{G}(x, t, \mathbf{u});$$

we refer again to Bourgeade, Le Floch, and Raviart (1989). □

6.3 The G.R.P. for the gas dynamics equations in Lagrangian coordinates

We start from equations (2.25):

$$\begin{cases} \dfrac{\partial \tau}{\partial t} - \dfrac{\partial u}{\partial m} = 0, \\ \dfrac{\partial u}{\partial t} + \dfrac{\partial p}{\partial m} = 0, \\ \dfrac{\partial e}{\partial t} + \dfrac{\partial}{\partial m}(pu) = 0, \end{cases}$$

with $p = p(\tau, \varepsilon) = p(\tau, e - \frac{u^2}{2})$, which can be written in the general form

$$\frac{\partial \mathbf{U}}{\partial t} + \frac{\partial}{\partial m}\mathbf{F}(\mathbf{U}) = \mathbf{0}, \text{ where}$$
$$\mathbf{U} = (\tau, u, e)^T, \quad \mathbf{F}(\mathbf{U}) = (-u, p, pu)^T.$$

We shall just sketch the resolution of the G.R.P. The solution of the classical Riemann problem is detailed in Chapter II, Section 3, and we may suppose that $u^0, p^0, \tau_\pm^0$ and $s_\pm^0$ are determined explicitly. We must then compute $\mathbf{u}^1$. We solve equations (6.17)–(6.21), i.e., we determine the coefficients α_k of $\mathbf{u}^1(\xi)$ in the basis $\big(\mathbf{r}_k(\mathbf{u}^0)\big)$. In fact, we use the nonconservative variables $\mathbf{v} = (\tau, u, s)^T$ (Chapter I, Example 2.3) with $p = p(\tau, s)$ since the formulas are easier to handle. Denoting for ease of notation

$$g = g(\mathbf{v}) = \sqrt{-\frac{\partial p}{\partial \tau}} = \frac{c}{\tau}, \tag{6.24a}$$

where $p = p(\tau, s)$ and

$$q = -\frac{\left(\frac{\partial p}{\partial \tau}\right)}{\left(\frac{\partial p}{\partial s}\right)}, \tag{6.24b}$$

the system (2.25) can be written in nonconservative form

$$\frac{\partial \mathbf{v}}{\partial t} + \mathbf{B}(\mathbf{v})\frac{\partial \mathbf{v}}{\partial m} = \mathbf{0},$$

where the matrix $\mathbf{B}$ is given by

$$\mathbf{B} = \begin{pmatrix} 0 & -1 & 0 \\ -g^2 & 0 & g^2/q \\ 0 & 0 & 0 \end{pmatrix}.$$

The contact discontinuity in Lagrangian coordinates corresponds to $\xi = \lambda_2(\mathbf{v}) = 0$, while the other eigenvalues are

$$\lambda_1(\mathbf{v}) = -g, \ \lambda_3(\mathbf{v}) = g,$$

and the eigenvectors can be chosen as

$$\mathbf{r}_1(\mathbf{v}) = (1, g, 0)^T, \; \mathbf{r}_2(\mathbf{v}) = (1, 0, q)^T, \; \mathbf{r}_3(\mathbf{v}) = (1, -g, 0)^T.$$

We shall assume for simplicity an equation of state of the form (3.15),

$$p = (\gamma - 1)\frac{\varepsilon}{\tau} + c_{\text{ref}}^2 \frac{(\tau_{\text{ref}} - \tau)}{\tau_{\text{ref}}\tau}.$$

Then, by (3.28), we compute

$$\frac{\partial p(\tau, s)}{\partial \tau} = -\frac{c^2}{\tau^2} = -\gamma(p + p_\infty)/\tau,$$

where p_∞ is defined as previously by $p_\infty = c_{\text{ref}}^2 \rho_{\text{ref}}/\gamma$. Thus $(p + p_\infty)\tau^\gamma$ is a function of s only and we can take

$$p + p_\infty = \frac{s}{\tau^\gamma}.$$

Then, we have

$$g^2 = \gamma \frac{s}{\tau^{\gamma+1}}, \quad q = \gamma \frac{s}{\tau}.$$

Given $\mathbf{v}^0 = (\tau_0, u_0, s_0)^T$ a solution of the classical Riemann problem, the coefficients $\alpha_k(\mathbf{v})$ of $\mathbf{v} = (\tau, u, s)^T$ in the basis $\big(\mathbf{r}_k(\mathbf{v}^0)\big)$ defined by

$$\mathbf{v} = \sum_{k=1}^{3} \alpha_k(\mathbf{v})\mathbf{r}_k(\mathbf{v}^0) \tag{6.25a}$$

satisfy

$$\begin{cases} \tau = \alpha_1 + \alpha_2 + \alpha_3, \\ u = (\alpha_1 - \alpha_3)g^0, \\ s = q^0 \alpha_2, \end{cases} \tag{6.25b}$$

or, equivalently,

$$\begin{cases} \alpha_1 = \dfrac{1}{2}\left(\dfrac{u}{g^0} + \tau - \dfrac{s}{q^0}\right), \\ \alpha_2 = \dfrac{s}{q^0}, \\ \alpha_3 = \dfrac{1}{2}\left(\dfrac{-u}{g^0} + \tau - \dfrac{s}{q^0}\right). \end{cases} \tag{6.25c}$$

Following the ideas of the preceding section, we determine the coefficients $\alpha_k(\xi)$ of $\mathbf{v}^1$ in each zone.

(i) Definition of the coefficients $a_{K,i}$; computation of $a_{0,k}$ and $a_{III,k}$. In each region K where $\mathbf{v}^0$ is constant, we know from (6.18) that the α_k are affine and more precisely that there exist constants $a_{K,i}, i = 1, 2, 3$, such that

$$\begin{cases} \alpha_1(\xi) = a_{K,1}(\xi + g^0), \\ \alpha_2(\xi) = a_{K,2}\xi, \\ \alpha_3(\xi) = a_{K,3}(\xi - g^0). \end{cases} \tag{6.26}$$

Assume for specificity that the (classical) Riemann problem corresponds to case 3 of Figure 3.8, Chapter II (1-rarefaction, 3-shock). In Lagrangian coordinates, we have a centered rarefaction wave propagating to the left and a shock traveling to the right. The zone $K = 0$ corresponds to the initial condition $\mathbf{v}_L^0 = \mathbf{v}_L(0)$ and is bordered on the right by the line $\xi = -g_L^0 = -g(\mathbf{v}_L^0)$; zone I lies between the tail of the rarefaction wave $\xi = -g(\mathbf{v}_I^0) = -g_I^0$ and the contact discontinuity $x = 0$, from which in turn starts zone II, which ends at $\xi = \sigma_3^0$ (the shock wave), and zone III corresponds to the right state $\mathbf{v}_R^0 = \mathbf{v}_R(0)$ (see Figure 6.2). Now, defining the coefficients $\alpha_{L,k}$ and $\alpha_{R,k}$ by

$$\mathbf{v}_L^1 = \frac{d\mathbf{v}_L}{dx}(0) = \sum_{k=1}^{3} \alpha_{L,k}\mathbf{r}_k(\mathbf{v}_L^0),\ \mathbf{v}_L^0 = \mathbf{v}_L(0), \tag{6.27a}$$

$$\mathbf{v}_R^1 = \frac{d\mathbf{v}_R}{dx}(0) = \sum_{k=1}^{3} \alpha_{R,k}\mathbf{r}_k(\mathbf{v}_R^0),\ \mathbf{v}_R^0 = \mathbf{v}_R(0), \tag{6.27b}$$

and using the conditions (6.17) at $\xi = \pm\infty$, we have

$$a_{0,k} = \alpha_{L,k}, \quad a_{III,k} = \alpha_{R,k}, \quad k = 1, 2, 3, \tag{6.28}$$

which determines completely the coefficients α_k of $\mathbf{v}^1$ in zone 0 and in zone III. The other coefficients will be determined below.

(ii) Determination of $\alpha_2(\xi)$ in the rarefaction; computation of $a_{I,2}$. Between zones 0 and I (i.e., in the 1-rarefaction that is bordered by the lines $\xi = -g_L^0$

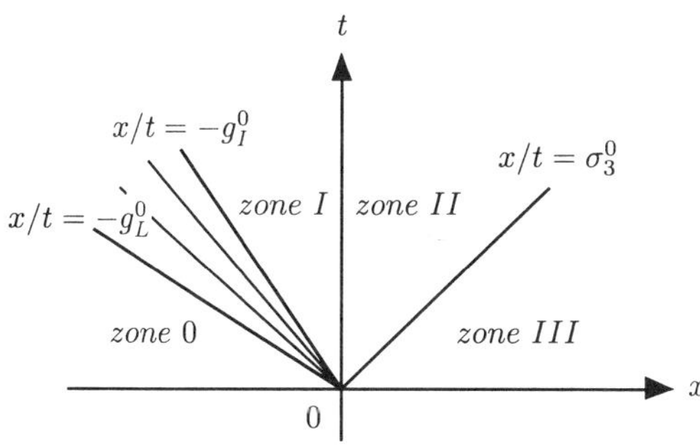

FIGURE 6.2. The four zones in the Riemann problem.

and $\xi = -g_I^0$) we solve the system (6.19), which consists of one algebraic equation giving $\alpha_1(\xi)$ in terms of $\alpha_2(\xi)$ and $\alpha_3(\xi)$ and a system of two linear differential equations for $\alpha_2(\xi)$ and $\alpha_3(\xi)$. For an equation of state of the form (3.15), and taking into account the jump conditions (6.21), we can solve explicitly the differential system. We use the fact that for a rarefaction

$$\frac{d}{d\xi}\, s^0(\xi) = 0,\ g^0(\xi) = -\lambda_1(\mathbf{v}(\xi)) = \xi.$$

Then, we note that by (6.22b), $\alpha_2(\xi)$ and $\alpha_3(\xi)$ are continuous across the line $\xi = -g_L^0$; hence the coefficients $\alpha_{L,k} = a_{0,k}$ appear, and for $-g_L^0 < \xi < -g_I^0$, we get

$$(6.29)\quad \begin{cases} \alpha_1(\xi) = \dfrac{\alpha_2(\xi)}{2(\gamma+1)} - \dfrac{\alpha_3(\xi)}{2}, \\ \alpha_2(\xi) = -\big(\dfrac{-\xi}{g_L^0}\big)^{\frac{(\gamma-1)}{(\gamma+1)}} g_L^0 \alpha_{L,2}, \\ \alpha_3(\xi) = \dfrac{g_L^0}{3\gamma-1}\Big(\dfrac{-g_L^0}{\xi}\Big)^{\frac{1}{2}}\Big\{-1 + \Big(\dfrac{-\xi}{g_L^0}\Big)^{\frac{3\gamma-1}{2(\gamma+1)}}\Big\}\alpha_{L,2} \\ \qquad\qquad - 2g_L^0\Big(\dfrac{-g_L^0}{\xi}\Big)^{\frac{1}{2}}\alpha_{L,3}. \end{cases}$$

$\alpha_2(\xi)$ and $\alpha_3(\xi)$ are also continuous across the tail of the 1-rarefaction $\xi = -g_I^0$, i.e.,

$$\alpha_j(-g_I^0-) = \alpha_j(-g_I^0+), \quad j = 2, 3.$$

In particular, α_2 is continuous at $\xi = -g_I^0$. Hence, equating the corresponding expressions (6.26) and (6.29), we get

$$-a_{I,2}g_I^0 = \alpha_2(-g_I^0) = -\Big(\frac{g_I^0}{g_L^0}\Big)^{\frac{(\gamma-1)}{(\gamma+1)}} g_L^0\alpha_{L,2},$$

or

$$(6.30)\quad a_{I,2} = \Big(\frac{g_I^0}{g_L^0}\Big)^{\frac{-2}{(\gamma+1)}}\alpha_{L,2} = \Big(\frac{q_L^0}{q_I^0}\Big)\alpha_{L,2}.$$

(iii) Definition of $p^1(\xi)$; continuity of $p^1(\xi)$ at the contact discontinuity. Now, we give an easy though important result concerning the contact discontinuity: defining $p^1(\xi)$ by

$$(6.31)\quad p^1 = \Big(\frac{\partial p}{\partial \tau}\Big)^0 \tau^1 + \Big(\frac{\partial p}{\partial s}\Big)^0 s^1,$$

we have

Lemma 6.1

The functions u^1 and p^1 are continuous at $\xi = 0$.

Proof. Recall that the contact discontinuity is characterized by

$$[u] = [p] = 0.$$

Using

$$u = u^0 + tu^1 + O(t^2), p = p^0 + tp^1 + O(t^2)$$

and identifying the terms of order 0, 1 yields

$$[u^0] = [p^0] = 0$$

and

$$[u^1] = [p^1] = 0,$$

which gives the result. □

We shall set

$$u_*^0 = u^0(0), \quad p_*^0 = p^0(0),$$
$$u_*^1 = u^1(0), \quad p_*^1 = p^1(0).$$

(iv) The jump conditions. For the 3-shock separating zones II and III, in order to specify the jump conditions, we introduce again the functions $\tau = h_a(p) = h_a(\tau_a, u_a; p)$ and $\Phi_a(p) = \Phi(\tau_a, u_a; p)$ of (3.2) and (3.4), Chapter II, Section 3. If the state $\mathbf{u} = (\tau, u, p)$ is connected to the state $\mathbf{u}_a = (\tau_a, u_a, p_a)$, we write

$$\tau_a = \tau_a^0 + t\tau_a^1 + ..., \qquad u_a = u_a^0 + tu_a^1 + ...$$
$$\tau^0 + t\tau^1 + ... = h(\tau_a, u_a; p^0 + tp^1 + ...)$$

and

$$u^0 + tu^1 + ... = u_a^0 + tu_a^1 + ... + \Phi(\tau_a, u_a; p^0 + tp^1 + ...).$$

This gives

$$u^1 = u_a^1 + \Big\{\Big(\frac{\partial \Phi}{\partial \tau_a}\Big)^0 \tau_a^1 + \Big(\frac{\partial \Phi}{\partial p_a}\Big)^0 p_a^1 + \Big(\frac{\partial \Phi}{\partial p}\Big)^0 p^1\Big\}, \tag{6.32a}$$

where the derivatives of Φ are computed at (τ_a^0, u_a^0, p^0) and a similar formula for τ^1

$$\tau^1 = \Big(\frac{\partial h}{\partial \tau_a}\Big)^0 \tau_a^1 + \Big(\frac{\partial h}{\partial p_a}\Big)^0 p_a^1 + \Big(\frac{\partial h}{\partial p}\Big)^0 p^1. \tag{6.32b}$$

These formulas can be made explicit for the equation of state (3.15). We have

$$\begin{cases} \dfrac{\partial \Phi}{\partial \tau_a}(p) = \dfrac{u - u_a}{2\tau_a}, \\[2ex] \dfrac{\partial \Phi}{\partial p_a}(p) = -W_a(p) - \dfrac{\mu^2}{2}\, \dfrac{(u - u_a)}{p + p_\infty + \mu^2(p_a + p_\infty)}, \\[2ex] \dfrac{\partial \Phi}{\partial p}(p) = W_a(p) - \dfrac{(u - u_a)}{2(p + p_\infty + \mu^2(p_a + p_\infty))}, \end{cases} \tag{6.33}$$

where $\mu^2 = \frac{(\gamma-1)}{(\gamma+1)}$ and $W_a(p)^2 = \frac{(1-\mu^2)\tau_a}{p+p_\infty+\mu^2(p_a+p_\infty)}$. We shall apply this formula below with $\mathbf{u} = \mathbf{u}(\sigma_3^0-)$ and $\mathbf{u}_a = \mathbf{u}(\sigma_3^0+)$.
(v) Computation of u_*^1, p_*^1. These values of u^1 and p^1 at $\xi = 0$ will in turn be used to determine the coefficients $a_{K,i}$, $K = I, II$; they will also be needed in the following section. In the case of the classical Riemann problem, u_*^0, p_*^0 are determined by the intersection of two curves. Similarly, u_*^1, p_*^1 are found as the solution of an algebraic system of two equations. Each equation is obtained by working in the zones I and II, respectively. We give first a technical result.

Lemma 6.2

Let $\xi \in$ zone K, $\mathbf{v}^1(\xi) = \left(\tau^1(\xi), u^1(\xi), s^1(\xi)\right)^T = \sum \alpha_k \mathbf{r}_k(\mathbf{v}^0)$ and $p^1(\xi)$ be defined by (6.31). We have

$$\begin{aligned}\tau^1(\xi) &= (a_{K,1} + a_{K,2} + a_{K,3})\xi + g^0(a_{K,1} - a_{K,3}),\\ u^1(\xi) &= g^0\{\xi(a_{K,1} - a_{K,3}) + (a_{K,1} + a_{K,3})\},\\ p^1(\xi) &= -(g^0)^2(a_{K,1} + a_{K,3})\xi,\end{aligned}$$

where the coefficients $a_{K,i}$ are defined in (6.26).

Proof. For any $\mathbf{v}^1 = (\tau^1, u^1, s^1)^T = \sum \alpha_k \mathbf{r}_k(\mathbf{v}^0)$, we have by (6.25)

$$\alpha_1 + \alpha_2 + \alpha_3 = \tau^1.$$

If, moreover, the α_k are given by (6.26), we get

$$\tau^1(\xi) = (a_{K,1} + a_{K,2} + a_{K,3})\xi + g^0(a_{K,1} - a_{K,3}).$$

Next, by (6.25b), (6.26),

$$\alpha_1 - \alpha_3 = \frac{u^1}{g^0},$$

and hence

$$u^1(\xi) = g^0\{\xi(a_{K,1} - a_{K,3}) + (a_{K,1} + a_{K,3})\}.$$

Then

$$\alpha_1 + \alpha_3 = \tau^1 - \frac{s^1}{q^0}.$$

By (6.24a) ($q = -\frac{p_\tau}{p_s}$, $g^2 = -p_\tau$) and (6.31), we have the identity

$$\tau^1 - \frac{s^1}{q^0} = -(g^0)^{-2}\left\{\left(\frac{\partial p}{\partial \tau}\right)^0 \tau^1 + \left(\frac{\partial p}{\partial s}\right)^0 s^1\right\} = -\frac{p^1}{(g^0)^2}. \tag{6.34}$$

Thus

$$\alpha_1 + \alpha_3 = -\frac{p^1}{(g^0)^2}$$

and

$$p^1(\xi) = -(g^0)^2(a_{K,1} + a_{K,3})\xi,$$

which ends the proof. □

In order to obtain the system of equations defining (u_*^1, p_*^1), we consider the zones I and II. There the functions α_k are affine and the values at the boundary $\xi = 0-$ (resp $\xi = 0+$) are written in terms of the values at the other boundary of the zone $\xi = -g_I^0$ (resp. $\xi = (\sigma_3^0-)$). More precisely, in zone I, by definition (6.25c) of α_3, setting $\mathbf{v}^1(0-) = (\tau_I^1, u_*^1, s_I^1)$, we get

$$\alpha_3(0-) = \frac{1}{2}\Big(-\frac{u_*^1}{g_I^0} + \tau_I^1 - \frac{s_I^1}{q_I^0}\Big).$$

We have, by (6.34),

$$\tau_I^1 - \frac{s_I^1}{q_I^0} = -\frac{p_*^1}{(g_I^0)^2}.$$

Hence

$$\alpha_3(0-) = -\frac{1}{2}\Big(\frac{u_*^1}{g_I^0} + \frac{p_*^1}{(g_I^0)^2}\Big).$$

Then, on the one hand, we know by (6.26) that since we are in zone I,

$$\alpha_3(0-) = -a_{I,3}g_I^0$$

and

$$\alpha_3(-g_I^0+) = -2a_{I,3}g_I^0.$$

On the other hand, due to (6.29) and the continuity of α_3, we have another expression for $\alpha_3(-g_I^0+) = \alpha_3(-g_I^0-)$, which enables us to determine the coefficients $a_{I,3}$ and leads thus to a first equation in u_*^1, p_*^1. We have

$$u_*^1 + \frac{p_*^1}{g_I^0} = C_L, \tag{6.35}$$

where

$$C_L = (g_L^0 g_I^0)^{\frac{1}{2}}\Big\{-u_L^1 + \frac{p_L^1}{g_L^0} + \Big(\frac{p_L^1}{g_L^0} + g_L^0\tau_L^1\Big)\frac{\big(\frac{g_I^0}{g_L^0}\big)^{\frac{(3\gamma-1)}{2(\gamma+1)}} - 1}{(3\gamma-1)}\Big\}.$$

In zone II, we use the jump relations (6.32) and compute separately $u^1(\sigma_3^0\pm)$. First, using Lemma 6.2 in zone III for $\mathbf{v}^1(\sigma_3^0+)$ gives

(6.36)
$$\begin{cases} \tau^1(\sigma_3^0+) = \sigma_3^0(\alpha_{R,1} + \alpha_{R,2} + \alpha_{R,3}) + g_R^0(\alpha_{R,1} - \alpha_{R,3}) = \sigma_3^0\tau_R^1 + u_R^1, \\ u^1(\sigma_3^0+) = g_R^0\{\sigma_3^0(\alpha_{R,1} - \alpha_{R,3}) + g_R^0(\alpha_{R,1} + \alpha_{R,3})\} = \sigma_3^0 u_R^1 - p_R^1, \\ p^1(\alpha_3^0+) = -\sigma_3^0(g_R^0)^2(\alpha_{R,1} + \alpha_{R,3}) = \sigma_3^0 p_R^1. \end{cases}$$

Similarly, in zone II, for $\mathbf{v}^1(\sigma_3^0-)$ and $\mathbf{v}^1(0+)$ we have four equations in $u^1(\sigma_3^0-), p^1(\sigma_3^0-), u_*^1, p_*^1$. Eliminating the constants $a_{II,1}$ and $a_{II,3}$, we get

$$\text{(6.37)} \qquad \begin{cases} u^1(\sigma_3^0-) = u_*^1 - \sigma_3^0 \dfrac{p_*^1}{(g_{II}^0)^2}, \\ p^1(\sigma_3^0-) = -\sigma_3^0 u_*^1 + p_*^1. \end{cases}$$

Substituting (6.36) and (6.37) in the jump relation (6.32) yields the second equation for u_*^1 and p_*^1,

$$\text{(6.38)} \qquad \begin{aligned} &\Big\{1 - \sigma_3^0\Big(\frac{\partial \Phi_R}{\partial p}\Big)_{II}^0\Big\}u_*^1 + \Big\{-\frac{\sigma_3^0}{(g_{II}^0)^2} + \Big(\frac{\partial \Phi_R}{\partial p}\Big)_{II}^0\Big\}p_*^1 \\ &= -\Big(\frac{\partial \Phi_R}{\partial \tau_a}\Big)_{II}^0(\sigma_3^0\tau_R^1 + u_R^1) - \Big(\frac{\partial \Phi_R}{\partial p_a}\Big)_{II}^0\{\sigma_3^0 p_R^1 - (g_R^0)^2 u_R^1\}, \end{aligned}$$

where the derivatives of Φ, computed at (τ_R^0, p_R^0, p_*^0), are explicitly given for the equation of state (3.15) by (6.33). The system (6.35), (6.38) is always numerically solvable and gives u_*^1 and p_*^1.

(vi) Computation of $a_{I,k}$ and $a_{II,k}$. The coefficients $a_{I,k}$ and $a_{II,k}$, $k = 1, 3$, are given in terms of u_*^1 and p_*^1. We have

$$\text{(6.39)} \qquad a_{K,1} = \frac{\big(u_*^1 - \frac{p_*^1}{g_K^0}\big)}{(g_K^0)^2}, \qquad a_{K,3} = \frac{\big(u_*^1 + \frac{p_*^1}{g_K^0}\big)}{(g_K^0)^2}, \quad \text{for } K = I, II.$$

As a corollary, using (6.25b), (6.26), and (6.39), we get explicitly $\tau^1(0\pm)$:

$$\tau_I^1 = \tau^1(0-) = (a_{K,1} - a_{K,3})g_I^0 = \frac{u_*^1}{g_I^0},$$

$$\tau_{II}^1 = \tau^1(0+) = (a_{K,1} - a_{K,3})g_{II}^0 = \frac{u_*^1}{g_{II^0}},$$

so that $\mathbf{v}^1(0\pm)$ is now known at the contact discontinuity.

There remains to compute the coefficient $a_{II,2}$. For instance, for a 3-shock, from (6.36) we get

$$\begin{aligned} \tau^1(\sigma_3^0+) &= \sigma_3^0\tau_R^1 + u_R^1, \\ p^1(\sigma_3^0+) &= \sigma_3^0 p_R^1 - (g_R^0)^2 u_R^1, \end{aligned}$$

and from (6.37)

$$p^1(\sigma_3^0-) = -\sigma_3^0 u_*^1 + p_*^1.$$

Together with (6.32), this yields

(6.40)

$$\begin{cases} a_{II,2} = -\,u_*^1\dfrac{\sigma_3^0 + g_{II}^0}{\sigma_3^0(g_{II}^0)^2} + \Big(\dfrac{\partial h}{\partial p}\Big)_{II}^0\Big\{-u_*^1 + \dfrac{p_*^1}{\sigma_3^0}\Big\} \\ \qquad + \Big(\dfrac{\partial h}{\partial \tau_a}\Big)_R^0\Big\{\tau_R^1 + \dfrac{u_R^1}{\sigma_3^0}\Big\} + \Big(\dfrac{\partial h}{\partial p_a}\Big)_R^0\Big\{p_R^1 - (g_{II}^0)^2\dfrac{u_R^1}{\sigma_3^0}\Big\}. \end{cases}$$

We have thus computed all the coefficients given by (6.28), (6.30), (6.39) and (6.40) in the particular case of a 1-rarefaction and a 3-shock. The other cases are computed in a similar way.

6.4 Use of the G.R.P. in van Leer's method

According to (6.5), we have to determine u^0, p^0, u^1 and p^1 at the contact discontinuity. As in Section 2.3, the values of u and p at the contact discontinuity between $\mathbf{U}_j^n, \mathbf{U}_{j+1}^n$ are denoted by $u_{j+\frac{1}{2}}^n$, $p_{j+\frac{1}{2}}^n$; we shall also denote by $(\frac{du}{dt})_{j+\frac{1}{2}}^n$ and $(\frac{dp}{dt})_{j+\frac{1}{2}}^n$ the corresponding values of u^1 and p^1. When other values φ for which φ^0 and φ^1 are not continuous are needed, we denote the corresponding values by $\varphi_{j+\frac{1}{2}\pm}^n$ and $\frac{d}{dt}\varphi_{j+\frac{1}{2}\pm}^n$. Then since by (6.6)

$$\mathbf{g}_{j+\frac{1}{2}}^n = \mathbf{f}(\mathbf{w}^0(0) + \frac{\Delta t}{2}\mathbf{w}^1(0)),$$

we first set

$$(6.41)\qquad \begin{cases} u_{j+\frac{1}{2}}^{n+\frac{1}{2}} = u_{j+\frac{1}{2}}^n + \dfrac{\Delta t}{2}(\dfrac{du}{dt})_{j+\frac{1}{2}}^n, \\ p_{j+\frac{1}{2}}^{n+\frac{1}{2}} = p_{j+\frac{1}{2}}^n + \dfrac{\Delta t}{2}(\dfrac{dp}{dt})_{j+\frac{1}{2}}^n, \\ (pu)_{j+\frac{1}{2}}^{n+\frac{1}{2}} = p_{j+\frac{1}{2}}^{n+\frac{1}{2}} u_{j+\frac{1}{2}}^{n+\frac{1}{2}}, \end{cases}$$

where in the last equation (6.41) we have approximated $(pu)^0 + t(pu)^1$ by

$$(pu)^0 + t(pu)^1 \approx p^0u^0 + t(p^0u^1 + p^1u) = (p^0 + tp^1)(u^0 + tu^1).$$

Then

$$(6.42)\qquad \begin{cases} \tau_j^{n+1} = \tau_j^n + \dfrac{\Delta t}{\Delta m_j}(u_{j+\frac{1}{2}}^{n+\frac{1}{2}} - u_{j-\frac{1}{2}}^{n+\frac{1}{2}}), \\ u_j^{n+1} = u_j^n - \dfrac{\Delta t}{\Delta m_j}(p_{j+\frac{1}{2}}^{n+\frac{1}{2}} - p_{j-\frac{1}{2}}^{n+\frac{1}{2}}), \\ e_j^{n+1} = e_j^n + \dfrac{\Delta t}{\Delta m_j}\Big((pu)_{j+\frac{1}{2}}^{n+\frac{1}{2}} - (pu)_{j-\frac{1}{2}}^{n+\frac{1}{2}}\Big), \end{cases}$$

where as in (2.28)

$$(6.43)\qquad \Delta m_j = \rho_j^0 \Delta\xi_j$$

and

$$p_j^n = p(\tau_j^n, \varepsilon_j^n),\ \ \varepsilon_j^n = e_j^n - \frac{(u_j^n)^2}{2}.$$

Let us derive another equivalent form for van Leer's scheme. We have not yet specified the motion of the grid. We solve numerically the equation

$$\frac{dx}{dt} = u.$$

Setting

(6.44) $$x^0_{j+\frac{1}{2}} = \xi_{j+\frac{1}{2}},$$

then $x^n_{j+\frac{1}{2}}$, which is the Eulerian coordinate of the interface $\xi_{j+\frac{1}{2}}$ at time t_n, is updated according to

(6.45) $$x^{n+1}_{j+\frac{1}{2}} = x^n_{j+\frac{1}{2}} + \Delta t\; u^{n+\frac{1}{2}}_{j+\frac{1}{2}}.$$

We can check by induction that

(6.46) $$\rho^n_j (x^n_{j+\frac{1}{2}} - x^n_{j-\frac{1}{2}}) = \Delta m_j.$$

First, this is true for $n = 0$ by assumption. Suppose that it holds for some n; we have

$$\begin{aligned} \Delta m_j \tau^{n+1}_j &= \Delta m_j \tau^n_j + \Delta t (u^{n+\frac{1}{2}}_{j+\frac{1}{2}} - u^{n+\frac{1}{2}}_{j-\frac{1}{2}}) \\ &= (x^n_{j+\frac{1}{2}} - x^n_{j-\frac{1}{2}}) + \Delta t (u^{n+\frac{1}{2}}_{j+\frac{1}{2}} - u^{n+\frac{1}{2}}_{j-\frac{1}{2}}) = x^{n+1}_{j+\frac{1}{2}} - x^{n+1}_{j-\frac{1}{2}}. \end{aligned}$$

Hence

(6.47) $$\rho^{n+1}_j (x^{n+1}_{j+\frac{1}{2}} - x^{n+1}_{j-\frac{1}{2}}) = \Delta m_j,$$

which proves the desired result.

The method in Lagrangian coordinates can thus be written

(6.48) $$\begin{cases} \Delta m_j = \rho^0_j (x^0_{j+\frac{1}{2}} - x^0_{j-\frac{1}{2}}), \\ x^{n+1}_{j+\frac{1}{2}} = x^n_{j+\frac{1}{2}} + \Delta t\; u^{n+\frac{1}{2}}_{j+\frac{1}{2}}, \\ \rho^{n+1}_j = (x^{n+1}_{j+\frac{1}{2}} - x^{n+1}_{j-\frac{1}{2}})^{-1} \Delta m_j, \\ u^{n+1}_j = u^n_j - \dfrac{\Delta t}{\Delta m_j} (p^{n+\frac{1}{2}}_{j+\frac{1}{2}} - p^{n+\frac{1}{2}}_{j-\frac{1}{2}}), \\ e^{n+1}_j = e^n_j - \dfrac{\Delta t}{\Delta m_j} \Big((pu)^{n+\frac{1}{2}}_{j+\frac{1}{2}} - (pu)^{n+\frac{1}{2}}_{j-\frac{1}{2}} \Big). \end{cases}$$

Remark 6.3. As in Remark 2.1, van Leer's scheme can be interpreted as a finite-volume method. Indeed, each of the equations (2.25) can be written

$$\frac{\partial}{\partial t} (\varphi J) + J \frac{\partial f}{\partial x} = 0.$$

By integrating this equation on $(\xi_{j-\frac{1}{2}}, \xi_{j+\frac{1}{2}})$, we get

$$\frac{d}{dt} \int_{\xi_{j-\frac{1}{2}}}^{\xi_{j+\frac{1}{2}}} \varphi J d\xi + \int_{\xi_{j-\frac{1}{2}}}^{\xi_{j+\frac{1}{2}}} \frac{\partial f}{\partial x} J d\xi = 0$$

or

$$\frac{d}{dt}\int_{x_{j-\frac{1}{2}}}^{x_{j+\frac{1}{2}}} \varphi dx + (f_{j+\frac{1}{2}} - f_{j-\frac{1}{2}}) = 0$$

(the dependence of f, φ, and x on t is omitted). We use a midpoint rule for the time integration

$$\Delta x_j^{n+1}\varphi_j^{n+1} = \Delta x_j^n \varphi_j^n - \Delta t(f_{j+\frac{1}{2}}^{n+\frac{1}{2}} - f_{j-\frac{1}{2}}^{n+\frac{1}{2}}),$$

where φ_j is the average value over the cell.

This gives for $\varphi = \rho, \rho u, \rho e$,

$$\begin{cases} \Delta x_j^{n+1}\rho_j^{n+1} = \Delta x_j^n \rho_j^n = \ldots = \Delta m_j, \\ \Delta x_j^{n+1}(\rho u)_j^{n+1} = \Delta x_j^n (\rho u)_j^n - \Delta t(p_{j+\frac{1}{2}}^{n+\frac{1}{2}} - p_{j-\frac{1}{2}}^{n+\frac{1}{2}}), \\ \Delta x_j^{n+1}(\rho e)_j^{n+1} = \Delta x_j^n (\rho e)_j^n - \Delta t\{(pu)_{j+\frac{1}{2}}^{n+\frac{1}{2}} - (pu)_{j-\frac{1}{2}}^{n+\frac{1}{2}}\}, \end{cases}$$

and thus

$$\begin{cases} \Delta x_j^n \rho_j^n = \Delta m_j, \\ \Delta m_j^{n+1} u_j^{n+1} = \Delta m_j^n u_j^n - \Delta t(p_{j+\frac{1}{2}}^{n+\frac{1}{2}} - p_{j-\frac{1}{2}}^{n+\frac{1}{2}}), \\ \Delta m_j^{n+1} e_j^{n+1} = \Delta m_j^n e_j^n - \Delta t\{(pu)_{j+\frac{1}{2}}^{n+\frac{1}{2}} - (pu)_{j-\frac{1}{2}}^{n+\frac{1}{2}}\}. \end{cases}$$

This is the above scheme, provided $u_{j+\frac{1}{2}}^{n+\frac{1}{2}}, p_{j+\frac{1}{2}}^{n+\frac{1}{2}}$ are defined by the expansion of the solution of the generalized Riemann problem. □

For the reconstruction step, we need to define the slopes $\mathbf{S} = (\delta\rho, \delta u, \delta e)$. By (6.8)

$$\begin{aligned} \hat{\mathbf{S}}_j^{n+1} &= \mathbf{w}(x_{j+\frac{1}{2}} - 0, \Delta t) - \mathbf{w}(x_{j-\frac{1}{2}} + 0, \Delta t) \\ &= \mathbf{v}_{j+\frac{1}{2}-}^{n+1} - \mathbf{v}_{j-\frac{1}{2}+}^{n+1}. \end{aligned}$$

Thus

$$\begin{aligned} (\delta\tau)_j^{n+1} &= \tau_{j+\frac{1}{2}-}^{n+1} - \tau_{j-\frac{1}{2}+}^{n+1}, \\ (\delta u)_j^{n+1} &= u_{j+\frac{1}{2}}^{n+1} - u_{j-\frac{1}{2}}^{n+1}, \\ (\delta e)_j^{n+1} &= e_{j+\frac{1}{2}-}^{n+1} - e_{j-\frac{1}{2}+}^{n+1}, \end{aligned}$$

where

$$\begin{aligned} \tau_{j+\frac{1}{2}\pm}^{n+1} &= \tau_{j+\frac{1}{2}\pm}^n + \Delta t\Big(\frac{d\tau}{dt}\Big)_{j+\frac{1}{2}\pm}^n, \\ u_{j+\frac{1}{2}\pm}^{n+1} &= u_{j+\frac{1}{2}\pm}^n + \Delta t\Big(\frac{du}{dt}\Big)_{j+\frac{1}{2}}^n, \\ e_{j+\frac{1}{2}\pm}^{n+1} &= e_{j+\frac{1}{2}\pm}^n + \Delta t\Big(\frac{de}{dt}\Big)_{j+\frac{1}{2}\pm}^n, \end{aligned}$$

and this is followed by a correction procedure. The slope is taken to be equal to zero if the slab-average reaches an extremum or varies in a way opposite to the variation of the average values and is limited so that the values at the end of the mesh (or less restrictively the average value) lie between the neighboring slab-averages (see G.R., Chapter IV, Section 3 for details).

Remark 6.4. In fact, it is more convenient to work with ρ and, moreover, τ does not appear in the formulas. Thus, we can suppose that ρ is an affine function and then correct $\delta\rho$ instead of $\delta\tau$. We can also think of limiting the pressure. For a polytropic ideal gas law

$$p = p(\tau, \varepsilon) = (\gamma - 1)\frac{\varepsilon}{\tau}, \quad \varepsilon = e - \frac{u^2}{2},$$

we can write

$$\frac{\delta p}{p} = \frac{\delta\varepsilon}{\varepsilon} - \frac{\delta\tau}{\tau}.$$

Limiting the physical rather than the conservative variables may reduce the oscillations. □

As we have done for Godunov's method, we can now derive a numerical method in Eulerian coordinates that couples a Lagrangian step and then a remapping onto the fixed Eulerian grid, with nodes $x_{j+\frac{1}{2}}$. Let us note that in (6.1) $\mathbf{v}_j^n$ may indeed be viewed as the cell-average of $\tilde{\mathbf{v}}^n(x)$,

$$\mathbf{v}_j^n = \frac{1}{\Delta x_j}\int_{x_{j-\frac{1}{2}}}^{x_{j+\frac{1}{2}}} \tilde{\mathbf{v}}^n(x)dx = \frac{1}{2}\Big(\tilde{\mathbf{v}}^n(x_{j+\frac{1}{2}}) + \tilde{\mathbf{v}}^n(x_{j-\frac{1}{2}})\Big),$$

while $\mathbf{S}_j^n$ is related to the first "moment"

$$\mathbf{S}_j^n = \frac{12}{(\Delta x_j)^2}\int_{x_{j-\frac{1}{2}}}^{x_{j+\frac{1}{2}}} \tilde{\mathbf{v}}^n(x)(x - x_j)dx.$$

i) The Lagrangian step.
At the beginning of each Lagrangian step, the Lagrangian and Eulerian grids coincide and $\xi_{j+\frac{1}{2}} = x_{j+\frac{1}{2}}$ (see Figure 2.3). Following (2.34), we define the updated Eulerian coordinates of the Lagrangian zone by

$$x^*_{j+\frac{1}{2}} = x_{j+\frac{1}{2}} + \Delta t u^n_{j+\frac{1}{2}},$$

where the dependence of $x^*_{j+\frac{1}{2}}$ on n is omitted. In other words, if $x_{j+\frac{1}{2}} = x(\xi_{j+\frac{1}{2}}, t_n)$, then $x^*_{j+\frac{1}{2}} \approx x(\xi_{j+\frac{1}{2}}, t_{n+1})$. We obtain by formulas (6.48) the quantities $\rho_*^{n+1}, u_*^{n+1}, e_*^{n+1}$, which are piecewise affine functions in each interval. In fact, we shall assume that ρ_*^{n+1} is an affine function of x in each interval $(x^*_{j-\frac{1}{2}}, x^*_{j+\frac{1}{2}})$ (this avoids a costly operation during the remapping procedure, see step (iii) and Remark 6.5), while u_*^{n+1}, e_*^{n+1} are affine functions of the mass variable m. The quantities $\rho_*^{n+1}, u_*^{n+1}, e_*^{n+1}$ are not necessarily linear in the interval $(x_{j-\frac{1}{2}}, x_{j+\frac{1}{2}})$ of the Eulerian grid, which, at the next Lagrangian step, coincides with the Lagrangian grid (see for instance Figure 6.3). Therefore, we must "remap," i.e., replace ρ_*^{n+1}, u_*^{n+1}, e_*^{n+1} by linear functions $\tilde{\rho}^{n+1}$, $\tilde{u}^{n+1}$, $\tilde{e}^{n+1}$, which share the same cell-average and first moment (w.r.t. x for ρ, w.r.t. m for u and e). This is the object of the projection step. We shall first detail the updating of the mass coordinates.

ii) Definition of the updated mass coordinates $m_{j+\frac{1}{2}}^{n+1}$.

Using the density versus the space coordinates for the definition of $m_{j+\frac{1}{2}}^{n+1}$,

$$\rho_*^{n+1}(x) = \rho_j^{n+1} + (x - x_j^*)\frac{(\delta\rho)_j^{n+1}}{\Delta x_j^*}, x_{j-\frac{1}{2}}^* < x < x_{j+\frac{1}{2}}^*, \tag{6.49}$$

the mass $m_{j+\frac{1}{2}}^{n+1} - m_{j+\frac{1}{2}}^{n}$ convected across $x_{j+\frac{1}{2}}$ is

$$m_{j+\frac{1}{2}}^{n} - m_{j+\frac{1}{2}}^{n+1} = \int_{x_{j+\frac{1}{2}}}^{x_{j+\frac{1}{2}}^*} \rho_*^{n+1}(x)dx.$$

Some easy computation gives

$$m_{j+\frac{1}{2}}^{n} - m_{j+\frac{1}{2}}^{n+1} = (x_{j+\frac{1}{2}}^* - x_{j+\frac{1}{2}})\Big\{\rho_j^{n+1} + \frac{(\delta\rho)_j^{n+1}}{2}\Big(1 - \frac{x_{j+\frac{1}{2}}^* - x_{j+\frac{1}{2}}}{\Delta x_j^*}\Big)\Big\},$$

which yields a simple equation for $m_{j+\frac{1}{2}}^{n+1}$.

Remark 6.5. The mass coordinates $m_{j+\frac{1}{2}}^{n+1}$ at time t_{n+1} of the Eulerian zone boundaries could be computed using $\tau_*^{n+1}(m)$ in the (original) Lagrangian zone. Consider for instance the usual case of Figure 6.3, which corresponds also to Figure 2.3 ($u_{j-\frac{1}{2}}^n > 0, u_{j+\frac{1}{2}}^n > 0$; the Lagrangian zone has moved to the right). We can define $m_{j+\frac{1}{2}}^{n+1}$ from the following equality obtained

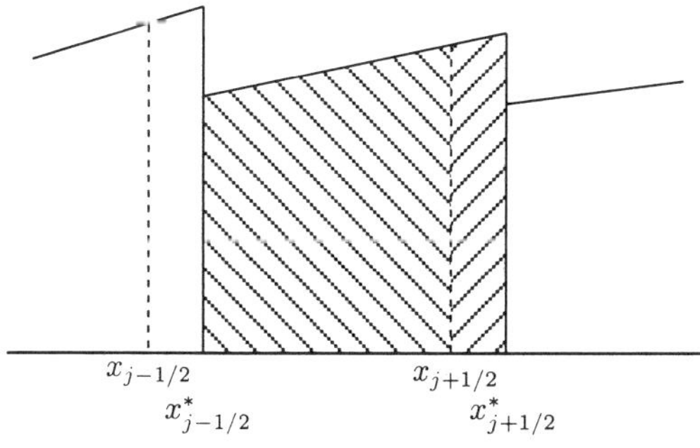

FIGURE 6.3. Necessity of "remapping."

with the specific volume

$$(6.50)\qquad \tau_*^{n+1}(m) = \tau_j^{n+1} + (m - m_j^n)\,\frac{(\delta\tau)_j^{n+1}}{\Delta m_j^n}\,,\, m_{j-\frac12}^n < m < m_{j+\frac12}^n\,,$$

defined on the (original) Lagrangian zone:

$$\begin{aligned}\int_{j+\frac12,n}^{j+\frac12,n+1} \tau_*^{n+1}(m)dm &= \int \Big\{\tau_j^{n+1} + (m - m_j^n)\,\frac{(\delta\tau)_j^{n+1}}{\Delta m_j^n}\Big\}dm \\ &= x_{j+\frac12}^* - x_{j+\frac12}\,,\end{aligned}$$

where the integral lies over $(m_{j+\frac12}^n, m_{j+\frac12}^{n+1})$ and

$$m_j^n = \frac{(m_{j+\frac12}^n + m_{j-\frac12}^n)}{2} = m_{j+\frac12}^n - \frac{\Delta m_j^n}{2}\,.$$

Using the obvious identities

$$(b-m)^2 - (a-m)^2 = (b-a)(a+b-2m)$$

and

$$m_{j+\frac12}^n + m_{j+\frac12}^{n+1} - 2\Big(m_{j+\frac12}^n - \frac{\Delta m_j^n}{2}\Big) = \Delta m_j^n + m_{j+\frac12}^{n+1} - m_{j+\frac12}^n\,,$$

we compute easily

$$\begin{aligned}&\int \Big\{\tau_j^{n+1} + (m - m_j^n)\,\frac{(\delta\tau)_j^{n+1}}{\Delta m_j^n}\Big\}dm \\ &\qquad = (m_{j+\frac12}^{n+1} - m_{j+\frac12}^n)\Big\{\tau_j^{n+1} + \frac{(\delta\tau)_j^{n+1}}{2}\Big(1 - \frac{m_{j+\frac12}^{n+1} - m_{j+\frac12}^n}{\Delta m_j^n}\Big)\Big\},\end{aligned}$$

which gives a quadratic equation for the mass $m_{j+\frac12}^{n+1} - m_{j+\frac12}^n$ convected across $x_{j+\frac12}$.

Let us check that it is equivalent to use, in the original Lagrangian zone, the density $\rho(x)$ given by (6.48),

$$\rho_*^{n+1}(x) = \rho_j^{n+1} + (x - x_j^*)\,\frac{(\delta\rho)_j^{n+1}}{\Delta x_j^*}\,,\, x_{j-\frac12}^* < x < x_{j+\frac12}^*\,,$$

instead of the specific volume $\tau(m)$ given by (6.50),

$$\tau_*^{n+1}(m) = \tau_j^{n+1} + (m - m_j)\,\frac{(\delta\tau)_j^{n+1}}{\Delta m_j}\,,\, m_{j-\frac12} < m < m_{j+\frac12}$$

(we have dropped in $m_{j+\frac12} = m_{j+\frac12}^n$ the superscript corresponding to the time t_n). Indeed, consider first the averaged values; since in the pure Lagrangian method, $x_{j+\frac12}^{n+1}$ corresponds to what we have denoted by $x_{j+\frac12}^*$, we have by (6.48)

$$\rho_j^{n+1} = (\Delta x_j^*)^{-1} \int_{x^*_{j-\frac{1}{2}}}^{x^*_{j+\frac{1}{2}}} \rho_*^{n+1}(x)dx = (\Delta x_j^*)^{-1}\Delta m_j = \frac{1}{\tau_j^{n+1}}, \tag{6.51a}$$

where

$$\tau_j^{n+1} = (\Delta m_j)^{-1} \int_{m_{j-\frac{1}{2}}}^{m_{j+\frac{1}{2}}} \tau_*^{n+1}(m)dm$$

is the average value of $\tau_*^{n+1}(m)$. Then

$$\begin{aligned}(\delta\rho)_j^{n+1} &= \frac{12}{(\Delta x_j^*)^2} \int_{x^*_{j-\frac{1}{2}}}^{x^*_{j+\frac{1}{2}}} \rho_*^{n+1}(x)(x - x_j^*)dx \\ &= \frac{12}{(\Delta x_j^*)^2} \int_{m_{j-\frac{1}{2}}}^{m_{j+\frac{1}{2}}} (x - x_j^*)d(m - m_j).\end{aligned}$$

Integrating by parts yields

$$\begin{aligned}(\delta\rho)_j^{n+1} &= -\frac{12}{(\Delta x_j^*)^2} \int_{m_{j-\frac{1}{2}}}^{m_{j+\frac{1}{2}}} (m - m_j)\Big(\frac{dx}{dm}\Big)dm \\ &= -\frac{12}{(\Delta x_j^*)^2} \int_{x^*_{j-\frac{1}{2}}}^{x^*_{j+\frac{1}{2}}} (m - m_j)dx \\ &= -\frac{12}{(\Delta x_j^*)^2} \int_{m_{j-\frac{1}{2}}}^{m_{j+\frac{1}{2}}} (m - m_j)\tau_*^{n+1}(m)dm \\ &= -\frac{12}{(\Delta x_j^*)^2} \frac{(\Delta m_j)^2}{12} \delta\tau_j^{n+1},\end{aligned}$$

so that

$$(\delta\rho)_j^{n+1} = -\frac{\delta\tau_j^{n+1}}{(\tau_j^{n+1})^2}, \tag{6.51b}$$

since $\delta\tau_j^{n+1}$ is the first moment of $\tau_*^{n+1} = \tau_j^{n+1} + (m - m_j)(\delta\tau)_j^{n+1}/\Delta m_j$. Formulas (6.51) prove indeed that we can convert τ into ρ and vice versa.□

iii) The projection step.

The values $\rho_*^{n+1}, u_*^{n+1}, e_*^{n+1}$ are projected back on the Eulerian grid. We define a piecewise linear density on the Eulerian grid $(x_{j-\frac{1}{2}}, x_{j+\frac{1}{2}})$ that shares the same average ρ_j^{n+1} and slope $(\hat{\delta}\rho)_j^{n+1}$ as $\rho_*^{n+1}(x)$ (the hat corresponds to the fact that we have not yet limited the slope). We compute

$$\rho_j^{n+1} = (\Delta x_j)^{-1} \int_{x_{j-\frac{1}{2}}}^{x_{j+\frac{1}{2}}} \rho_*^{n+1}(x)dx$$

by taking into account the fact that the function is piecewise linear on $(x^*_{j-\frac{1}{2}}, x^*_{j+\frac{1}{2}})$ (in the case of Figure 6.3 for instance, the function is linear on $(x^*_{j-\frac{1}{2}}, x^*_{j+\frac{1}{2}})$, but we might have three pieces if $u^n_{j-\frac{1}{2}} > 0$, $u^n_{j+\frac{1}{2}} < 0$). We have already noted that the slope is given by the first "moment," i.e., we have the following obvious equality:

$$\int_{x_{j-\frac{1}{2}}}^{x_{j+\frac{1}{2}}} (x - x_j)(ax + b)dx = a\,\frac{(\Delta x_j)^3}{12},$$

from which we deduce that $(\hat{\delta}\rho)^{n+1}_j$ is given by

$$(\hat{\delta}\rho)^{n+1}_j = \frac{12}{\Delta x_j^2} \int_{x_{j-\frac{1}{2}}}^{x_{j+\frac{1}{2}}} \rho^{n+1}_*(x)(x - x_j)dx.$$

Now, having defined an updated mass coordinate $m^{n+1}_{j+\frac{1}{2}}$, we can project u and e with respect to the mass variable. We get

$$u^{n+1}_j = (\Delta m^{n+1}_j)^{-1} \int_{m_{j-\frac{1}{2}}}^{m_{j+\frac{1}{2}}} u^{n+1}_*(m)dm,$$

where $m_{j\pm\frac{1}{2}}$ stands for $m^{n+1}_{j\pm\frac{1}{2}}$, and

$$(\hat{\delta}u)^{n+1}_j = \frac{12}{(\Delta m^{n+1}_j)^2} \int_{m_{j-\frac{1}{2}}}^{m_{j+\frac{1}{2}}} u^{n+1}_*(m)(m - m^{n+1}_j)dm,$$

where

$$m^{n+1}_j = \frac{1}{2}(m^{n+1}_{j+\frac{1}{2}} + m^{n+1}_{j-\frac{1}{2}}),$$

with similar formulas for e.

For practical use, the costly computation of the first moment for defining the slope can be replaced by interpolation and monotonicity correction, which we now detail.

(iv) Monotonicity correction.

When correcting the slopes, we can take advantage of the fact that the projection is done onto the Eulerian grid, and use a milder kind of limiting than (6.10) (see van Leer 1979). Hence, we take for limiting the slope $\hat{\delta}v_k$ of a quantity v_k ($v_k = \rho, u$, or e),

$$(6.52)\qquad \delta v_{k,j} = \begin{cases} s \min\left\{ \dfrac{2|\Delta v_{k,j-\frac{1}{2}}|}{\sigma_{j+\frac{1}{2}}}, |\hat{\delta}v_{k,j}|, \dfrac{2|\Delta v_{k,j+\frac{1}{2}}|}{(1-\sigma_{j+\frac{1}{2}})} \right\} \\ \qquad \text{if } s = \operatorname{sgn}\Delta v_{k,j-\frac{1}{2}} = \operatorname{sgn}\Delta v_{k,j+\frac{1}{2}} = \operatorname{sgn}\hat{\delta}v_{k,j}, \\ 0 \quad \text{otherwise}, \end{cases}$$

where

$$\sigma_{j+\frac{1}{2}} = \frac{m^n_{j+\frac{1}{2}} - m^{n+1}_{j+\frac{1}{2}}}{\Delta m^n_j}$$

is the mass fraction of the Lagrangian slab $(m^n_{j-\frac{1}{2}}, m^n_{j+\frac{1}{2}})$ that has crossed the Eulerian zone boundary $x_{j+\frac{1}{2}}$. The limitation is to be used on the Lagrangian quantities once we have determined $m^{n+1}_{j+\frac{1}{2}}$, i.e., between steps (ii) and (iii).

The functions $\tilde{\rho}^{n+1}, \tilde{u}^{n+1}, \tilde{e}^{n+1}$ are thus reconstructed, and $\tilde{\rho}^{n+1}$ can be converted into $\tilde{\tau}^{n+1}$ for the next Lagrangian step. The new Lagrangian zones can now be defined, coinciding again with the Eulerian grid. Other references for schemes using G.R.P. are Ben Artzi and Falcovitz (1989), Ben Artzi and Birman (1990), and in the case of reactive flows Falcovitz and Ben Artzi (1992–93) and the references therein.

7 Kinetic schemes for the Euler equations

The kinetic schemes we introduce in this section will eventually be written in the general form of finite difference schemes. It seems, however, essential for a better understanding of their properties to have first a quick look at the underlying kinetic theory, skipping most difficulties since it is out of the scope of this book to present a complete treatment of this theory. We refer to Cercignani (1988), Cercignani et al. (1994), Lions (1990), Perthame (1996) for a rigorous approach.

7.1 The Boltzmann equation

First, we consider for simplicity a rarefied monatomic perfect gas and assume in this section that the dimension is $d = 3$ (however, we keep the notation d since in dimension $d = 1$ or $d = 2$ most formulas remain valid). The Boltzmann equation

$$\frac{\partial f}{\partial t} + \mathbf{v} \cdot \operatorname{grad}_{\mathbf{x}} f = Q(f, f) \tag{7.1}$$

describes the time evolution of the one-particle distribution $f(t, \mathbf{x}, \mathbf{v})$ for a gas in the phase space $(\mathbf{x}, \mathbf{v}) \in \mathbb{R}^{2d}$, where $\mathbf{x}$ is the position vector and $\mathbf{v}$ the molecular velocity. One can say that f is the expected mass density in the phase space, and thus

$$\rho = \rho(\mathbf{x}, t) = \int_{\mathbb{R}^d} f(t, \mathbf{x}, \mathbf{v}) d\mathbf{v}$$

will represent the mass per unit volume, i.e., the density in the physical space.

The equation (7.1) can be justified by considering the limit (Boltzmann–Grad limit) as $N \to \infty$ of a system of N interacting particles: the gas is made of many identical molecules considered as hard spheres with diameter a (Na^2 is fixed, $a \to 0$) and we assume elastic, binary collisions and no external force fields. We have $f = NmP = MP$ where P represents the probability density that a random molecule has velocity $\mathbf{v} \pm d\mathbf{v}$ and position $\mathbf{x} \pm d\mathbf{x}$, N is the number of molecules, m the mass of molecule, and M the total mass. Since P is a probability density, $\int_{\mathbb{R}^{2d}} P d\mathbf{x} d\mathbf{v} = 1$ and

$$\int_{\mathbb{R}^{2d}} f d\mathbf{x} d\mathbf{v} = M = Nm.$$

We shall not give the precise expression for the collision operator $Q(f, f)$; it is a quadratic integral operator, which acts on the velocity dependence of f and is symmetric ($Q(f,g) = Q(g,f)$). We mention the main property in the following theorem.

Theorem 7.1

The states of thermodynamic equilibrium characterized by $Q(f, f) = 0$ *are obtained for the Maxwellian distributions*

$$f(\mathbf{v}) = A \exp\left(-\beta|\mathbf{v} - \mathbf{u}|^2\right), \tag{7.2}$$

where $A \in \mathbb{R}^+, \mathbf{u} \in \mathbb{R}^d, \beta \in \mathbb{R}^+$ *are arbitrary parameters.*

Proof. The proof of this result relies on the following important facts:
(i) The microscopic collisional invariants, i.e., the functions $\varphi = \varphi(\mathbf{v})$ satisfying

$$\int_{\mathbb{R}^d} Q(f,g)\varphi(\mathbf{v}) d\mathbf{v} = 0, \forall f, g \geq 0,$$

are exactly the functions $\varphi(\mathbf{v}) = a + \mathbf{b} \cdot \mathbf{v} + c|\mathbf{v}|^2$ (where a, c are constants and $\mathbf{b}$ is a constant vector). The elementary invariants are the components of the vector

$$\mathbf{K}(\mathbf{v}) = (1, \mathbf{v}, |\mathbf{v}|^2)^T \in \mathbb{R}^{d+2}. \tag{7.3}$$

Thus, for $g = f \geq 0$, we obtain

$$\int_{\mathbb{R}^d} Q(f,f)\mathbf{K}(\mathbf{v}) d\mathbf{v} = \mathbf{0}, \tag{7.4a}$$

and for f satisfying (7.1),

$$\frac{\partial}{\partial t} \int_{\mathbb{R}^d} \mathbf{K}(\mathbf{v}) f(t, \mathbf{x}, \mathbf{v}) d\mathbf{v} + \int_{\mathbb{R}^d} \mathbf{v} \cdot \operatorname{grad}_{\mathbf{x}} f(t, \mathbf{x}, \mathbf{v}) \mathbf{K}(\mathbf{v}) d\mathbf{v} = \mathbf{0}. \tag{7.4b}$$

(ii) Whatever the distribution function $f \geq 0$, the following Boltzmann inequality holds:

$$\int_{\mathbb{R}^d} \log f Q(f,f) d\mathbf{v} \leq 0, \tag{7.5}$$

where equality holds if and only if log f is an invariant (the "if" part is obvious),

$$f(\mathbf{v}) = \exp\,(a + \mathbf{b}\cdot\mathbf{v} + c|\mathbf{v}|^2)$$

or, with an appropriate choice of constants, f is a Maxwellian,

$$f(\mathbf{v}) = A\,\exp\,\Big(-\beta(\mathbf{v}-\mathbf{u})^2\Big),$$

which gives the result. □

As a consequence of (7.5), we find that introducing the function

$$h(r) = r\,\log\,r, \tag{7.6}$$

we have

$$\frac{\partial}{\partial t}\int_{\mathbb{R}^d} h(f)d\mathbf{v} + \int_{\mathbb{R}^d}\mathbf{v}\cdot\mathrm{grad}_{\mathbf{x}}\,h(f)d\mathbf{v} \le 0, \tag{7.7}$$

where equality holds iff f is a Maxwellian (part of the H-theorem). Indeed, we write

$$\frac{\partial}{\partial t}\int_{\mathbb{R}^d} h(f)d\mathbf{v} + \int_{\mathbb{R}^d}\mathbf{v}\cdot\mathrm{grad}_{\mathbf{x}}\,h(f)d\mathbf{v} = \int_{\mathbb{R}^d} h'(f)\Big(\frac{\partial f}{\partial t} + \mathbf{v}\cdot\mathrm{grad}\,f\Big)d\mathbf{v}$$
$$= \int_{\mathbb{R}^d} h'(f)Q(f,f)d\mathbf{v} = \int_{\mathbb{R}^d}(\log\,f+1)Q(f,f)d\mathbf{v} = \int_{\mathbb{R}^d}\log\,fQ(f,f)d\mathbf{v}$$

since 1 is a collision invariant. Now, by (7.5) the last integral is ≤ 0 and vanishes iff f is a Maxwellian.

The function $h(r) = r\,\log\,r$, which is the microscopic or kinetic entropy, is strictly convex. If we define

$$H(f) = \int_{\mathbb{R}^d} h(f)d\mathbf{v} = \int_{\mathbb{R}^d} f\,\log\,fd\mathbf{v} \tag{7.8a}$$

(H is closely related to the thermodynamic entropy as we shall see below), and

$$\boldsymbol{\Psi}(f) = \Big(\Psi_i(f)\Big), \tag{7.8b}$$

where

$$\Psi_i(f) = \int_{\mathbb{R}^d} v_i h(f)d\mathbf{v} = \int_{\mathbb{R}^d} v_i f\,\log\,fd\mathbf{v},$$

then (7.7) yields

$$\frac{\partial}{\partial t}H(f) + \mathrm{div}_{\mathbf{x}}\,\boldsymbol{\Psi}(f) \le 0. \tag{7.8c}$$

Remark 7.1. If we integrate over $\mathbb{R}^d$, or for more realistic situations over a region $\mathcal{R}$ filled by gas, and if we assume moreover that $\mathcal{R}$ is bounded by nonporous solid walls and at no point of the boundary of $\mathcal{R}$ is heat flowing from the gas, we can prove that (for any solution f of (7.1)) the quantity

$$\mathcal{H} = \int_{\mathcal{R}} H(f) d\mathbf{x} = \int_{\mathbb{R}^d \times \mathcal{R}} h(f) d\mathbf{v} d\mathbf{x}$$

decreases, i.e.,

$$\frac{d\mathcal{H}}{dt} \leq 0$$

and $\mathcal{H}$ is constant if and only if f is a Maxwellian.

The Boltzmann equation describes the evolution ("relaxation") towards a state of minimum for $\mathcal{H}$. Under the above assumption, the final state ($t \to +\infty$) is a steady state and thus a Maxwellian. In particular, the distribution function in an equilibrium state (i.e., a steady state with no energy exchange with the surroundings) is a Maxwellian. □

Given a distribution function f, let us define the following macroscopic quantities (indeed the moments of f):

$$\mathbf{U}(\mathbf{x}, t) = \begin{pmatrix} \rho \\ \rho \mathbf{u} \\ \rho e \end{pmatrix} (\mathbf{x}, t) = \int_{\mathbb{R}^d} f(t, \mathbf{x}, \mathbf{v}) \begin{pmatrix} 1 \\ \mathbf{v} \\ |\mathbf{v}|^2/2 \end{pmatrix} d\mathbf{v}. \tag{7.9}$$

Because of the probabilistic meaning of f, ρ represents the mass per unit volume and $\mathbf{u}$ represents the velocity. The specific total energy e satisfies

$$\rho e = \int_{\mathbb{R}^d} f(t, \mathbf{x}, \mathbf{v}) \frac{|\mathbf{v}|^2}{2} d\mathbf{v} = \int_{\mathbb{R}^d} f(t, \mathbf{x}, \mathbf{v}) \frac{|\mathbf{v} - \mathbf{u}|^2}{2} d\mathbf{v} + \rho \frac{|\mathbf{u}|^2}{2}.$$

Thus

$$\rho \varepsilon = \int_{\mathbb{R}^d} f(t, \mathbf{x}, \mathbf{v}) \frac{|\mathbf{v} - \mathbf{u}|^2}{2} d\mathbf{v}$$

represents the internal energy (per unit volume). The velocity

$$\mathbf{C} = \mathbf{v} - \mathbf{u}$$

is called the peculiar velocity. We also define the stress tensor $\boldsymbol{\pi} = (\pi_{ij})_{1 \leq i,j \leq 3}$

$$\pi_{ij} = \int_{\mathbb{R}^d} f(t, \mathbf{x}, \mathbf{v}) C_i C_j d\mathbf{v}, \quad 1 \leq i, j \leq 3.$$

We have

$$\sum_{i=1}^{3} \pi_{ii} = \int_{\mathbb{R}^d} f(t, \mathbf{x}, \mathbf{v}) |\mathbf{C}|^2 d\mathbf{v} = 2\rho\varepsilon,$$

and it is convenient to identify $\frac{1}{3}\left(\sum_{i=1}^{3} \pi_{ii}\right) = \frac{2}{3}\rho\varepsilon$ with the gas pressure p; thus

$$p = \frac{2\rho\varepsilon}{3} = \frac{1}{d} \int_{\mathbb{R}^d} f(t, \mathbf{x}, \mathbf{v}) |\mathbf{u} - \mathbf{v}|^2 d\mathbf{v}. \tag{7.10}$$

Indeed, for a monatomic perfect gas, ε is a function of temperature only; we get $\frac{p}{\rho}$ = constant if temperature = constant. This property is characteristic of a perfect gas, for which Boyle's law, $p = \rho RT$, holds (R is the Boltzmann constant of the gas). With the above identification, we are led to define the kinetic temperature T (it is convenient to note the vectors of $\mathbb{R}^{d+2}$ in transposed form, such as K in (7.3), for which the notation is easily distinguishable from the temperature) by

$$T = \frac{2\varepsilon}{3R},$$

and we obtain for a monatomic perfect gas (in dimension $d = 3$)

$$\varepsilon = \frac{3RT}{2}. \tag{7.11}$$

Remark 7.2. Note that with the notations of the preceding chapters (Introduction, Example 3.3 or Chapter II, Section 1.2), $p = (\gamma - 1)\rho\varepsilon$ and here $p = 2\rho\varepsilon/3$; thus $\gamma = \frac{5}{3}$, and the specific heat at constant volume is $C_v = \frac{R}{(\gamma-1)} = 3R/2$. Also, after some computations the relation defining the thermodynamic entropy $TdS = d\varepsilon + pd\tau$, which yields $S - S_0 = C_v \log \frac{\varepsilon}{\rho^{\gamma-1}}$, gives

$$S = R \log\left(\frac{T^{\frac{3}{2}}}{\rho}\right) + S_0 \tag{7.12}$$

in dimension $d = 3$ for a monatomic gas. □

Last we introduce the heat flow vector $\mathbf{Q} = (Q_i)$

$$Q_i = \frac{1}{2}\int_{\mathbb{R}^d} f(t, \mathbf{x}, \mathbf{v})C_i|\mathbf{C}|^2 d\mathbf{v}, 1 \leq i \leq 3.$$

Then, from (7.1) and (7.4), we deduce the following result.

Proposition 7.1

Assume that f is a solution of (7.1). Then, the vector $\mathbf{U}$ *defined by (7.9) satisfies the system*

$$\begin{cases} \dfrac{\partial\rho}{\partial t} + \displaystyle\sum_{j=1}^{d} \frac{\partial}{\partial x_j}(\rho u_j) = 0, \\ \dfrac{\partial\rho u_i}{\partial t} + \displaystyle\sum_{j=1}^{d} \frac{\partial}{\partial x_j}(\rho u_i u_j + \pi_{ij}) = 0, \ 1 \leq i \leq d, \\ \dfrac{\partial\rho e}{\partial t} + \displaystyle\sum_{j=1}^{d} \frac{\partial}{\partial x_j}\{\rho u_j e + \sum_{k=1}^{d} \pi_{jk}u_k + Q_j\} = 0. \end{cases}$$

We shall detail a similar computation a little later (in the proof of Proposition 7.2). Thus, in general, the equations satisfied by the moments contain a stress tensor (π_{ij}) and a heat flux vector $\mathbf{Q}$.

Assume now that f is a Maxwellian. Substituting (7.2) in (7.9), we find that the parameter $\mathbf{u}$ in (7.2) is indeed the velocity defined by (7.9), while the other coefficients satisfy (in the case $d = 3$)

$$\beta = \frac{3}{4\varepsilon} = \frac{1}{2RT}, \qquad A = \Big(\frac{4\pi\varepsilon}{3}\Big)^{-\frac{3}{2}} \rho = (2\pi RT)^{-\frac{3}{2}} \rho,$$

and thus

$$f(\mathbf{v}) = (2\pi RT)^{-\frac{3}{2}} \rho \exp\Big(-\frac{|\mathbf{v}-\mathbf{u}|^2}{2RT}\Big). \tag{7.13}$$

Moreover, $\pi_{jk} = p\delta_{jk}$ and $Q_j = 0$ from which we deduce the following corollary.

Corollary 7.1

Assume that f is a Maxwellian (7.13). The associated vector $\mathbf{U}$ satisfies the Euler equations

$$\begin{cases} \dfrac{\partial \rho}{\partial t} + \displaystyle\sum_{j=1}^{d} \frac{\partial}{\partial x_j}(\rho u_j) = 0, \\[2ex] \dfrac{\partial \rho u_i}{\partial t} + \displaystyle\sum_{j=1}^{d} \frac{\partial}{\partial x_j}(\rho u_i u_j) + \frac{\partial p}{\partial x_i} = 0, \ 1 \le i \le d, \\[2ex] \dfrac{\partial \rho e}{\partial t} + \displaystyle\sum_{j=1}^{d} \frac{\partial}{\partial x_j}\Big((\rho e + p)u_j\Big) = 0, \end{cases} \tag{7.14}$$

where, for a monatomic perfect gas in dimension $d = 3$,

$$p = \rho RT = \frac{2\rho\varepsilon}{3}.$$

Defining the function $H = H(\mathbf{U})$ (i.e., $H(M(\mathbf{v};\rho,\mathbf{u},T))$ by (7.8) for the Maxwellian distribution $f = M$), we have

$$\begin{aligned} H &= \int_{\mathbb{R}^d} M \log(M) d\mathbf{v} \\ &= \log\big(\rho(2\pi RT)^{-\frac{3}{2}}\big) \int_{\mathbb{R}^d} M d\mathbf{v} - \frac{1}{2RT}\int M|\mathbf{v}-\mathbf{u}|^2 d\mathbf{v} \\ &= \rho \log\big(\rho(2\pi RT)^{-\frac{3}{2}}\big) - \frac{\rho\varepsilon}{RT} \\ &= \rho\Big\{\log\big(\rho T^{-\frac{3}{2}}\big) - \frac{3}{2}\log(2\pi R) - \frac{3}{2}\Big\} \\ &= \rho \log\big(\rho T^{-\frac{3}{2}}\big) - \frac{3}{2}\rho\{\log(2\pi R) + 1\}, \end{aligned}$$

and thus

$$H(\mathbf{U}) = \rho\{\log\big(\rho T^{-\frac{3}{2}}\big) + C\}. \tag{7.15}$$

Lemma 7.1

The function $H(\mathbf{U})$ *is a strictly convex function of* $\mathbf{U}$.

Proof. Since H is a function of (ρ, ε) only, following the results of Chapter II, Section 1, it is enough to prove that

$$\mathrm{tr}H'' = \frac{\partial^2 H}{\partial \rho^2} + \frac{\partial^2 H}{\partial \varepsilon^2} \geq 0,$$

$$\det H'' = \frac{\partial^2 H}{\partial \rho^2}\,\frac{\partial^2 H}{\partial \varepsilon^2} - \Big(\frac{\partial^2 H}{\partial \rho \partial \varepsilon}\Big)^2 \geq 0,$$

where H'' is the 2×2 Hessian matrix.

By the chain rule, we compute

$$\mathrm{tr}H'' = \int_{\mathbb{R}^d} h'(M)\Big\{\frac{\partial^2 M}{\partial \rho^2} + \frac{\partial^2 M}{\partial \varepsilon^2}\Big\}d\mathbf{v} + \int_{\mathbb{R}^d} h''(M)\Big(\Big(\frac{\partial M}{\partial \rho}\Big)^2 + \Big(\frac{\partial M}{\partial \varepsilon}\Big)^2\Big)d\mathbf{v}.$$

Since h is convex, the second integral is nonnegative, and it is easy to check that the first integral vanishes. Indeed, we can write $h'(M)$ in the form

$$h'(M) = a(\rho, \varepsilon, \mathbf{u}) + b(\rho, \varepsilon, \mathbf{u})|\mathbf{v}|^2$$

and

$$\int_{\mathbb{R}^d} h'(M)\Big\{\frac{\partial^2 M}{\partial \rho^2} + \frac{\partial^2 M}{\partial \varepsilon^2}\Big\}d\mathbf{v} = a\Big\{\frac{\partial^2}{\partial \rho^2} + \frac{\partial^2}{\partial \varepsilon^2}\Big\}\int_{\mathbb{R}^d} M d\mathbf{v} + b\Big\{\frac{\partial^2}{\partial \rho^2} + \frac{\partial^2}{\partial \varepsilon^2}\Big\}\int_{\mathbb{R}^d} M|\mathbf{v}|^2 d\mathbf{v} = 0.$$

This yields

$$\mathrm{tr}H'' \geq 0.$$

With a similar argument, we get

$$\det H'' = \Big(\int_{\mathbb{R}^d} h''(M)\Big(\frac{\partial M}{\partial \rho}\Big)^2 d\mathbf{v}\Big)\Big(\int_{\mathbb{R}^d} h''(M)\Big(\frac{\partial M}{\partial \varepsilon}\Big)^2 d\mathbf{v}\Big) - \Big(\int_{\mathbb{R}^d} h''(M)\Big(\frac{\partial M}{\partial \rho}\,\frac{\partial M}{\partial \varepsilon}\Big)d\mathbf{v}\Big)^2 \geq 0$$

by the Cauchy–Schwarz inequality. □

Remark 7.3. We can give a more fundamental interpretation of the convexity of H (Brenier 1992). We can write

$$H(\mathbf{U}) = \text{ Min } \Big\{\int_{\mathbb{R}^d} h(f)dv;\, f(v) \geq 0,\, \int_{\mathbb{R}^d} f(v)\mathbf{K}(v)dv = \mathbf{U}\Big\},$$

where $\mathbf{K}, \mathbf{U}$ are given by (7.3) and (7.9), and the minimum is reached for $f = M$. Since we minimize a convex functional under linear constraints, the minimum $H(\mathbf{U})$ is a convex function of the constraints. (See also Croisille and Delorme 1991). □

The above expression (7.15) for H shows that it is related to the macroscopic entropy S given by (7.12),

$$\frac{H}{\rho} - C = \log\,(\rho T^{-\frac{3}{2}}) = -\frac{(S - S_0)}{R},$$

or

$$\frac{H}{\rho} = -\frac{s}{R} + C'.$$

Also, the function $\Psi = \Psi(\mathbf{U})$ (see (7.8)) satisfies

$$\begin{aligned}\Psi &= \int_{\mathbb{R}^d} M \log\,(M)\mathbf{v}d\mathbf{v} \\ &= \log\left(\rho(2\pi RT)^{-\frac{3}{2}}\right)\int_{\mathbb{R}^d} M\mathbf{v}d\mathbf{v} - \frac{1}{2RT}\int M\mathbf{v}|\mathbf{v}-\mathbf{u}|^2 d\mathbf{v} \\ &= \rho\mathbf{u}\log\rho(2\pi RT)^{-\frac{3}{2}} - \mathbf{u}\Big(\frac{\rho\varepsilon}{RT}\Big),\end{aligned}$$

and thus

$$\Psi = \mathbf{u}H.$$

Now, from (7.5), (7.7), and the first equation (7.14), we obtain the entropy equality

$$\frac{\partial}{\partial t}H(\mathbf{U}) + \operatorname{div}_{\mathbf{x}}\,\mathbf{u}H(\mathbf{U}) = 0,$$

which is satisfied (for smooth $\mathbf{U}$) since M is a Maxwellian. Together with Lemma 7.1, this proves that $-\rho S$ is convex and is indeed an entropy function (in the sense of Lax, see Introduction, Section 3 or Chapter I, Section 5), with entropy flux $-\rho\mathbf{u}S$. It is known that the entropies may be written $\rho G(\frac{\rho^{\gamma-1}}{\varepsilon})$ or $\rho G(\frac{\rho^{\gamma-1}}{T})$ (here we have $\gamma = \frac{5}{3}$) for some function G.

When the gas is not assumed to be monatomic, we must increase the dimension of the phase space. Let us present a more general model where the distribution function f depends not only on $\mathbf{x}$ and $\mathbf{v}$ but also on an internal energy (or temperature) microscopic variable $\theta \in \mathbb{R}+$, with density $n(\theta)$. Therefore, we have $f = f(t, \mathbf{x}, \mathbf{v}, \theta) = f(\mathbf{v}, \theta)$ (with shorthand notations) and

$$(7.16a)\qquad (\rho, \rho\mathbf{u}, \rho e)^T = \int_{\mathbb{R}^d\times\mathbb{R}+} f(\mathbf{v},\theta)\Big(1, \mathbf{v}, \frac{|\mathbf{v}|^2}{2} + \theta\Big)^T n(\theta)d\theta d\mathbf{v},$$

where $e = \varepsilon + \frac{|\mathbf{u}|^2}{2}$,

$$(7.16b)\qquad \rho\varepsilon = \int_{\mathbb{R}^d\times\mathbb{R}+} f(\mathbf{v},\theta)\Big\{\frac{|\mathbf{v}-\mathbf{u}|^2}{2} + \theta\Big\}n(\theta)d\theta d\mathbf{v} = \rho(\varepsilon_k + \varepsilon_i),$$

and

$$(7.16c)\qquad \begin{cases} \rho\varepsilon_k = \displaystyle\int_{\mathbb{R}^d\times\mathbb{R}+} f(\mathbf{v},\theta)\,\frac{|\mathbf{v}-\mathbf{u}|^2}{2}\,n(\theta)d\theta d\mathbf{v}, \\ \rho\varepsilon_i = \displaystyle\int_{\mathbb{R}^d\times\mathbb{R}+} f(\mathbf{v},\theta)\theta n(\theta)d\theta d\mathbf{v}. \end{cases}$$

We define the temperature by

$$(7.17)\qquad T = \frac{2\varepsilon_k}{dR}.$$

Now, for simplicity, we consider an internal energy ε_i, which has the form

$$(7.18a)\qquad \varepsilon_i = \delta RT,$$

where the coefficient $\delta > 0$ satisfies

$$(7.18b)\qquad \frac{1}{\gamma-1} = \delta + \frac{d}{2} \iff \delta = \frac{2-d(\gamma-1)}{2(\gamma-1)}.$$

Thus, from (7.16b),

$$\varepsilon = \varepsilon_i + \varepsilon_k = RT\Big(\delta + \frac{d}{2}\Big) = \frac{RT}{(\gamma-1)}.$$

This particular form of ε_i is obtained for a density of energy $n(\theta) = \theta^{\delta-1}$, and

$$\rho\varepsilon_i = \int_{\mathbb{R}^d\times\mathbb{R}+} f(\mathbf{v},\theta)\theta^\delta d\theta d\mathbf{v}.$$

The Maxwellian equilibrium function satisfying

$$(\rho,\rho\mathbf{u},\rho e)^T = \int_{\mathbb{R}^d\times\mathbb{R}+} M(\mathbf{v},\theta)\Big(1,\mathbf{v},\frac{|\mathbf{v}|^2}{2}+\theta\Big)^T n(\theta)d\theta d\mathbf{v}$$

is then given by

$$(7.19)\qquad M(\mathbf{v},\theta) = (2\pi RT)^{-\frac{d}{2}}\,\rho\exp\Big(-\frac{|\mathbf{v}-\mathbf{u}|^2}{2RT}\Big)\,\frac{1}{\theta_0}\exp\Big(-\frac{\theta}{RT}\Big),$$

where

$$\theta_0 = \int_{\mathbb{R}_+} \exp\Big(-\frac{\theta}{RT}\Big)n(\theta)d\theta.$$

The pressure is given by (see (7.10))

$$(7.20a)\qquad \begin{aligned} p &= \frac{1}{d}\sum_{i=1}^{d}\int_{\mathbb{R}^d\times\mathbb{R}+} M(\mathbf{v},\theta)(v_i-u_i)^2 n(\theta)d\theta d\mathbf{v}, \\ &= \frac{1}{d}\int_{\mathbb{R}^d\times\mathbb{R}+} M(\mathbf{v},\theta)|\mathbf{v}-\mathbf{u}|^2 n(\theta)d\theta d\mathbf{v} = \frac{2\rho\varepsilon_k}{d}, \end{aligned}$$

i.e.,

(7.20*b*) $$p = \rho RT = (\gamma - 1)\rho\varepsilon.$$

We note that in (7.19) log M depends on $\mathbf{v}$ and θ through $\frac{|\mathbf{v}|^2}{2} + \theta$. The elementary invariants in (7.3) are now $(1, \mathbf{v}, \frac{|\mathbf{v}|^2}{2} + \theta)$, and the kinetic entropy is again the function $h(r) = r \log r$.

Remark 7.4. For an ideal monatomic gas in dimension $d = 3$ (resp. $d = 1$) we have $\gamma = \frac{5}{3}$, (resp. $\gamma = 3$), $RT = \frac{2\varepsilon}{3}$ (resp. $RT = 2\varepsilon$), and by (7.18b), $\delta = 0$. Thus by (7.18a) the ε_i vanish, which means that we drop in that case the dependence on θ.

Otherwise, $\gamma = \frac{\delta+d+2}{\delta+d} < \frac{5}{3}$ (in $d = 3$) if $\delta > 0$. □

7.2 The B.G.K. model

We present now a simplified collision term that is at the source of the B.G.K. model (from Bhatnagar-Gross-Krook). The B.G.K. model describes the evolution of f through the equation

(7.21) $$\frac{\partial f}{\partial t} + \mathbf{v} \cdot \text{ grad } f = \frac{M(\mathbf{v}) - f}{\nu},$$

where $M(\mathbf{v}; \rho, \mathbf{u}, T) = M(\mathbf{v})$ is the (local) Maxwellian (see (7.11)) given by

(7.22) $$M(\mathbf{v}) = \Big(2\pi RT(\mathbf{x}, t)\Big)^{-\frac{d}{2}} \rho(\mathbf{x}, t) \exp\Big(-\frac{|\mathbf{v} - \mathbf{u}|^2}{2RT}(\mathbf{x}, t)\Big)$$

or by (7.19) if we consider the dependence on the internal energy θ. In (7.21), the collision term

$$J(f) = \frac{1}{\nu}(M(\mathbf{v}) - f)$$

is constructed in order to satisfy the following properties that were proved for Q,

(7.23) $$\int_{\mathbb{R}^d} J(f)\mathbf{K}(\mathbf{v})d\mathbf{v} = 0, \quad \forall f \geq 0,$$

where $\mathbf{K}(\mathbf{v})$ is the vector of collision invariants (7.3), and

(7.24) $$\int_{\mathbb{R}^d} \log f J(f) d\mathbf{v} \leq 0,$$

with equality if and only if f is a Maxwellian M. The constant ν is a small parameter meant to tend to 0 ($\frac{1}{\nu}$ is the collision frequency, and can also be viewed as the relaxation time). Thus, from (7.24), we get

$$\int_{\mathbb{R}^d} M(\mathbf{v})\mathbf{K}(\mathbf{v})d\mathbf{v} = \int_{\mathbb{R}^d} f(\mathbf{v})\mathbf{K}(\mathbf{v})d\mathbf{v},$$

and the parameters ρ, $\mathbf{u}$, $E = \rho(\varepsilon + \frac{|\mathbf{v}|^2}{2})$ in the Maxwellian are indeed the moments of f, given by (7.9),

$$\mathbf{U}(\mathbf{x}, t) = \begin{pmatrix} \rho \\ \rho\mathbf{u} \\ \rho e \end{pmatrix}(\mathbf{x}, t) = \int_{\mathbb{R}^d} f(t, \mathbf{x}, \mathbf{v}) \begin{pmatrix} 1 \\ \mathbf{v} \\ |\mathbf{v}|^2/2 \end{pmatrix} d\mathbf{v},$$

or by (7.16) if M is defined by (7.19). As ν tends to 0, f is expected to tend to the Maxwellian distribution $M(\mathbf{v}; \mathbf{U})$ whose moments satisfy the Euler equations (7.14), with the equation of state (7.10) or (7.11): $p = \rho RT = 2\rho\varepsilon/3$, $\varepsilon = \frac{3RT}{2}$. (For more precise existence and stability results, we refer to Perthame 1989).

As we have noticed in Remark 7.1, the entropy $\mathcal{H}(f) = \int_{\mathbb{R}^d} H(f)d\mathbf{x} = \int_{\mathbb{R}^{2d}} h(f)d\mathbf{v}d\mathbf{x}$, where $h(f) = f \log f$, decreases as $t \to +\infty$ towards a minimum, which corresponds to a Maxwellian distribution. Thus M realizes the minimum of $\mathcal{H}(f)$ under the constraints $\int_{\mathbb{R}^d} f(\mathbf{v})(1, \mathbf{v}, \frac{|\mathbf{v}|^2}{2})^T d\mathbf{v} = (\rho, \mathbf{0}, \rho\varepsilon)^T$.

Now it is interesting, in view of the numerical applications, to consider other entropy functions and their associated "equilibrium" functions. Given a "kinetic entropy" h, we can build a B.G.K. model, i.e., find an equilibrium function N that realizes the minimum of the entropy $\int h$, and consider the equation (7.21) associated with N.

For this, given a function $h(f, \theta)$, a "kinetic entropy", which is a strictly convex function of f (satisfying $h(0, \theta) = 0$ and say $h'(0, \theta) = 0$ or some more stringent conditions that we do not discuss), we consider the problem of minimization,

$$H(\rho, \varepsilon) \equiv \min \Big\{ \int_{\mathbb{R}^d \times \mathbb{R}+} h\Big(f(\mathbf{v}, \theta), \theta\Big) n(\theta) d\theta d\mathbf{v} \Big\}, \tag{7.25}$$

where the minimum is taken over all $f \geq 0$ satisfying the constraints

$$\int_{\mathbb{R}^d \times \mathbb{R}+} f(\mathbf{v}, \theta)\Big(1, \mathbf{v}, \frac{|\mathbf{v}|^2}{2} + \theta\Big)^T n(\theta) d\theta d\mathbf{v} = (\rho, \mathbf{0}, \rho\varepsilon)^T. \tag{7.26}$$

Lemma 7.2

Problem (7.25), (7.26) admits a unique solution N.

Proof. We just sketch the proof. It can be shown that H is nonincreasing in ε and has a unique minimum obtained for a function $f = N(\mathbf{v}, \theta; \rho, \varepsilon)$ such that

$$h'(N) = a - b\Big(\frac{|\mathbf{v}|^2}{2} + \theta\Big) \Big(\text{or } \Big[a - b\Big(\frac{|\mathbf{v}|^2}{2} + \theta\Big)\Big]_+ \Big),$$

where $h' = \frac{\partial h}{\partial f}$ and the constants a and b are such that N satisfies the constraints (7.26). Indeed, provided the constants a and b are found, we have, since h' is strictly convex,

$$h(f) \geq h(N) + h'(N)(f - N),$$

which yields

$$\int_{\mathbb{R}^d \times \mathbb{R}+} h(f)n(\theta)d\theta d\mathbf{v} \geq \int_{\mathbb{R}^d \times \mathbb{R}+} h(N)n(\theta)d\theta d\mathbf{v}$$
$$+ \int_{\mathbb{R}^d \times \mathbb{R}+} \left(a - b\left(\frac{|\mathbf{v}|^2}{2} + \theta\right)\right)(f - N)n(\theta)d\theta d\mathbf{v},$$

and the last integral vanishes. For details, we refer to Perthame (1990) and Coron and Perthame (1991). □

Let us observe that by construction,

$$\int_{\mathbb{R}^d \times \mathbb{R}+} N(\mathbf{v} - \mathbf{u})\left(1, \mathbf{v}, \frac{|\mathbf{v}|^2}{2} + \theta\right)^T n(\theta)d\theta d\mathbf{v} = \left(\rho, \rho\mathbf{u}, \rho\varepsilon + \rho\frac{|\mathbf{u}|^2}{2}\right)^T.$$

We then consider the B.G.K. model, equation (7.21), associated with this function, $N(\mathbf{v}, \theta; \rho, \varepsilon)$,

$$\frac{\partial f}{\partial t} + \mathbf{v} \cdot \operatorname{grad}_{\mathbf{x}} f = \frac{N(\mathbf{v} - \mathbf{u}) - f}{\nu}. \tag{7.27}$$

Setting

$$\left(\rho, \rho\mathbf{u}, \rho\varepsilon + \rho\frac{|\mathbf{u}|^2}{2}\right)^T = \int_{\mathbb{R}^d \times \mathbb{R}+} f(\mathbf{v}, \theta)\left(1, \mathbf{v}, \frac{|\mathbf{v}|^2}{2} + \theta\right)^T n(\theta)d\theta d\mathbf{v}, \tag{7.28}$$

we have

$$\int_{\mathbb{R}^d \times \mathbb{R}+} \left(f - N(\mathbf{v} - \mathbf{u})\right)(1, \mathbf{v}, \frac{|\mathbf{v}|^2}{2} + \theta)^T n(\theta)d\theta d\mathbf{v} = \mathbf{0}.$$

It can be proved that (7.27) admits h as a kinetic entropy, i.e., we have formally

$$\frac{d}{dt} \int_{\mathbb{R}^d \times \mathbb{R}^d \times \mathbb{R}+} h(f)n(\theta)d\theta d\mathbf{v}d\mathbf{x} \leq 0$$

for any solution f of (7.27).

Proposition 7.2

Let N be the unique solution of problem (7.25), (7.26). As ν tends to 0, the quantities $\rho, \rho\mathbf{u}, \rho\varepsilon$ defined by (7.28) satisfy (formally) the Euler system (7.14) with the equation of state

$$p(\rho, \varepsilon) = \frac{1}{d} \int_{\mathbb{R}^d \times \mathbb{R}+} N(\mathbf{v}, \theta; \rho, \varepsilon)|\mathbf{v} - \mathbf{u}|^2 n(\theta)d\theta d\mathbf{v}.$$

Moreover, the system admits H defined by (7.25) as a convex entropy, i.e.,

$$\frac{\partial H}{\partial t} + \mathit{div}\,(H\mathbf{u}) = 0$$

for any smooth solution of the Euler equations (7.14).

By "formally" we mean that we assume that as $\nu \to 0, f \to N$ in a convenient sense, which enables us to pass to the limit in the integrals!

Proof. By multiplying the equation (7.27) by $(1, \mathbf{v}, \frac{|\mathbf{v}|^2}{2} + \theta)$ and integrating w.r.t. $\mathbf{v}$ and θ, we obtain the first equation

$$\frac{\partial}{\partial t}\int_{\mathbb{R}^d\times\mathbb{R}+} f d\mathbf{v} d\theta + \sum_{j=1}^{d}\frac{\partial}{\partial x_j}\int_{\mathbb{R}^d\times\mathbb{R}+} v_j f n(\theta) d\theta d\mathbf{v} = 0,$$

and the remaining two equations are

$$\begin{cases} \dfrac{\partial}{\partial t}\displaystyle\int_{\mathbb{R}^d\times\mathbb{R}+} f\mathbf{v} d\mathbf{v} d\theta + \sum_{j=1}^{d}\frac{\partial}{\partial x_j}\int_{\mathbb{R}^d\times\mathbb{R}+} v_j f\mathbf{v} n(\theta) d\theta d\mathbf{v} = \mathbf{0}, \\ \dfrac{\partial}{\partial t}\displaystyle\int_{\mathbb{R}^d\times\mathbb{R}+} f\Big(\frac{|\mathbf{v}|^2}{2} + \theta\Big) d\mathbf{v} n(\theta) d\theta \\ \qquad + \displaystyle\sum_{j=1}^{d}\frac{\partial}{\partial x_j}\int_{\mathbb{R}^d\times\mathbb{R}+} v_j f\Big(\frac{|\mathbf{v}|^2}{2} + \theta\Big) n(\theta) d\theta d\mathbf{v} = 0. \end{cases}$$

We want to prove that the system obtained by letting $\nu \to 0$ is the Euler system. In fact, the first equation is exactly

$$\frac{\partial \rho}{\partial t} + \operatorname{div}(\rho\mathbf{u}) = 0.$$

We now consider the other equations; using

$$\int_{\mathbb{R}^d\times\mathbb{R}+} (v_i - u_i) f n(\theta) d\theta d\mathbf{v} = 0,$$

they can be written,

$$\frac{\partial}{\partial t}(\rho u_i) + \sum_{j=1}^{d}\frac{\partial}{\partial x_j}\{\rho u_j u_i + \int_{\mathbb{R}^d\times\mathbb{R}+} (v_j - u_j)(v_i - u_i) f n(\theta) d\theta d\mathbf{v}\} = 0.$$

$1 \le i \le d$, and, with $E = \rho\varepsilon + \rho\frac{|\mathbf{u}|^2}{2}$,

$$\frac{\partial E}{\partial t} + \sum_{j=1}^{d}\frac{\partial}{\partial x_j}\Big\{E u_j + \int_{\mathbb{R}^d\times\mathbb{R}+} (v_j - u_j)\Big(\frac{|\mathbf{v}|^2}{2} + \theta\Big) f n(\theta) d\theta d\mathbf{v}\Big\} = 0$$

or

$$\frac{\partial E}{\partial t} + \operatorname{div}(E\mathbf{u}) + \sum_{j=1}^{d}\frac{\partial}{\partial x_j}\int_{\mathbb{R}^d\times\mathbb{R}+} (v_j - u_j)\{\sum_{i=1}^{d} u_i(v_i - u_i) + \frac{|\mathbf{v} - \mathbf{u}|^2}{2} + \theta\} f n(\theta) d\theta d\mathbf{v} = 0.$$

Now, we have assumed that as $\nu \to 0, f(\mathbf{v}, \theta) \to N(\mathbf{v} - \mathbf{u}, \theta)$; using, moreover, the fact that by symmetry

$$\int_{\mathbb{R}^d\times\mathbb{R}+} v_i v_j N(\mathbf{v}, \theta) d\mathbf{v} d\theta = 0, i \neq j \text{ (corresponding to } \pi_{ij}),$$

and

$$\int_{\mathbb{R}^d\times\mathbb{R}+} v_i\Big(\frac{|\mathbf{v}|^2}{2}+\theta\Big)N(\mathbf{v},\theta)n(\theta)d\theta d\mathbf{v}=0, \forall i, \text{ (corresponding to } Q_i),$$

we obtain for $1\le i\le d$,

$$\frac{\partial}{\partial t}(\rho u_i)+\text{ div }(\rho\mathbf{u}u_i)+\sum_{j=1}^{d}\frac{\partial}{\partial x_j}\int_{\mathbb{R}^d\times\mathbb{R}+}N(\mathbf{v},\theta)v_j^2 n(\theta)d\theta d\mathbf{v}=0$$

and

$$\frac{\partial E}{\partial t}+\text{ div }(E\mathbf{u})+\sum_{j=1}^{d}\frac{\partial}{\partial x_j}u_j\int_{\mathbb{R}^d\times\mathbb{R}+}v_j^2 N(\mathbf{v},\theta)n(\theta)d\theta d\mathbf{v}=0.$$

We finally obtain, as expected, the Euler system

$$\begin{cases}\dfrac{\partial\rho}{\partial t}+\text{div }(\rho\mathbf{u})=0,\\[2mm] \dfrac{\partial}{\partial t}(\rho u_i)+\text{div }(\rho\mathbf{u}u_i)+\dfrac{\partial p}{\partial x_i}=0,\quad 1\le i\le d,\\[2mm] \dfrac{\partial}{\partial t}\Big(\rho\Big(\varepsilon+\dfrac{|\mathbf{u}|^2}{2}\Big)\Big)+\text{ div }\Big(\rho\mathbf{u}\Big(\varepsilon+\dfrac{|\mathbf{u}|^2}{2}\Big)+p\mathbf{u}\Big)=0\end{cases}$$

with

$$p=\frac{2}{d}\int_{\mathbb{R}^d\times\mathbb{R}+}N(\mathbf{v},\theta)\frac{|\mathbf{v}|^2}{2}n(\theta)d\theta d\mathbf{v}=\int_{\mathbb{R}^d\times\mathbb{R}+}N(\mathbf{v},\theta)v_j^2 n(\theta)d\theta d\mathbf{v}.$$

Moreover, it can be proved that the function $H(\rho,\varepsilon)$ defined in (7.25), which is thus given by

$$H(\rho,\varepsilon)=\int_{\mathbb{R}^d\times\mathbb{R}+}h\Big(N(\mathbf{v},\theta),\theta\Big)n(\theta)d\theta d\mathbf{v},$$

is a convex function of (ρ,ε) (the proof follows the lines of Lemma 7.1) and

$$\frac{\partial H}{\partial t}+\text{ div }(H\mathbf{u})=0$$

for any smooth solution of the Euler equation (7.14), which means that H is a convex entropy for the Euler system. □

Example 7.1. If we choose the functions

$$h_n(f)=\frac{f^{n+1}}{(n+1)},$$

the associated M_n, which satisfy

$$h_n'(M_n)=[a_n-b_n(\frac{|\mathbf{v}|^2}{2}+\theta)]_+,$$

are thus the functions

$$M_n(\mathbf{v}, \theta) = (a_n)^{\frac{1}{n}} \left[1 - \frac{b_n}{a_n}\left(\frac{|\mathbf{v}|^2}{2} + \theta\right)\right]_+^{\frac{1}{n}}$$

for some constants a_n, b_n such that (7.26) holds. As $n \to +\infty$,

$$M_n(\mathbf{v}, \theta) \to a 1_{\{\frac{|\mathbf{v}|^2}{2} + \theta \leq b\}},$$

where the constants a and b are again given by the constraints (7.26, and 1_X denotes the usual characteristic function of a set X. □

7.3 *The kinetic scheme*

In this section, we restrict ourselves to the one-dimensional case $d = 1$. The Maxwellian distribution is then, by (7.19), $M(v - u, \theta; \rho, \varepsilon)$ if we set

$$M(v, \theta; \rho, \varepsilon) = \frac{\rho}{\theta_0}(2\pi RT)^{-\frac{1}{2}} \exp\left\{-\frac{v^2}{2RT} - \frac{\theta}{RT}\right\}. \tag{7.29}$$

The temperature T is a convenient variable, which is related to ε in the simple case we are considering by

$$RT = (\gamma - 1)\varepsilon$$

and

$$\varepsilon_i = \delta RT,$$

where δ is defined by (7.18). The constant R is often incorporated in T, in that case $T = (\gamma - 1)\varepsilon$ and $p = \rho T$.

7.3.1 The time discretization

A numerical scheme in *time* is obtained by the "splitting" of the transport and collision parts of the equation. Roughly, it consists of two steps (transport + relaxation); the first step solves

$$\frac{\partial f}{\partial t} + v \cdot \operatorname{grad} f = 0 \tag{7.30}$$

(collisionless molecule transport) and the second step solves

$$J(f) = 0$$

(relaxation to thermodynamic equilibrium, whose solution is a local Maxwellian). More precisely, given initial conditions $\rho_0 = \rho(x, 0)$, $u_0 = u(x, 0)$, and, $\varepsilon_0 = \varepsilon(x, 0)$, we define an "equilibrium function"

$$f^0(x, v, \theta) = M(v - u_0, \theta; \rho_0, \varepsilon_0),$$

where M is the exact Maxwellian (7.29) or a function that is more convenient for numerical purposes, as we shall detail below. For the first step, we solve the linear transport equation (7.30),

$$\begin{cases} \dfrac{\partial f}{\partial t} + v \cdot \dfrac{\partial f}{\partial x} = 0, & t \in]0, \Delta t], x \in \mathbb{R}, \\ f(x, 0; v, \theta) = f^0(x, v, \theta), & x \in \mathbb{R} \end{cases}$$

and then define the updated quantities

$$\begin{pmatrix} \rho \\ \rho u \\ \rho e \end{pmatrix} (x, t) = \int_{\mathbb{R}} f(x, t; v, \theta) \begin{pmatrix} 1 \\ v \\ \frac{v^2}{2} + \theta \end{pmatrix} n(\theta) d\theta dv,$$

where

$$f(x, t; v, \theta) = f^0(x - vt, v, \theta).$$

For the next step, we start with $\rho_1 = \rho(x, \Delta t), u_1 = u(x, \Delta t), \varepsilon_1 = \varepsilon(x, \Delta t)$ and follow the same procedure.

In fact, it is more convenient to drop the dependence on θ by integrating first w.r.t. θ and, moreover, we get a more general formalism: we introduce instead of f^0 the functions f_0, g_0 defined by

$$\text{(7.31)} \qquad \begin{cases} f_0(x, v) = \displaystyle\int_{\mathbb{R}+} M(v - u_0, \theta; \rho_0, \varepsilon_0) n(\theta) d\theta, \\ g_0(x, v) = \displaystyle\int_{\mathbb{R}+} M(v - u_0, \theta; \rho_0, \varepsilon_0) \theta n(\theta) d\theta. \end{cases}$$

We see from (7.16) and (7.18) that

$$\int_{\mathbb{R}} g_0(x, v) dv = \int_{\mathbb{R} \times \mathbb{R}+} M(v - u_0, \theta; \rho_0, \varepsilon_0) \theta n(\theta) d\theta dv = \rho_0 \varepsilon_i = \delta \rho_0 R T_0,$$

and thus

$$\text{(7.32)} \qquad \int_{\mathbb{R}} g_0(x, v) dv = \delta R T_0 \int_{\mathbb{R}} f_0(x, v) dv.$$

Also, we have

$$\text{(7.33)} \qquad \begin{pmatrix} \rho_0 \\ \rho_0 u_0 \\ \rho_0 e_0 \end{pmatrix} (x) = \int_{\mathbb{R}} \begin{pmatrix} f_0 \\ v f_0 \\ \frac{v^2}{2} f_0 + g_0 \end{pmatrix} (x, v) dv.$$

Next, we solve

$$\text{(7.34a)} \qquad \begin{cases} \dfrac{\partial f}{\partial t} + v \cdot \dfrac{\partial f}{\partial x} = 0, & t \in]0, \Delta t], x \in \mathbb{R}, \\ f(x, 0; v) = f_0(x, v), \end{cases}$$

$$\text{(7.34b)} \qquad \begin{cases} \dfrac{\partial g}{\partial t} + v \cdot \dfrac{\partial g}{\partial x} = 0, & t \in]0, \Delta t], x \in \mathbb{R}, \\ g(x, 0; v) = g_0(x, v), \end{cases}$$

and then for $t \leq \Delta t$,

$$\mathbf{U}(x,t) = \begin{pmatrix} \rho \\ \rho u \\ \rho e \end{pmatrix}(x,t) = \int_{\mathbb{R}} \begin{pmatrix} f \\ vf \\ (\frac{v^2}{2})f + g \end{pmatrix}(x,t;v)dv. \tag{7.35}$$

For a general step, if f_n and g_n are known, we solve

$$\begin{cases} \dfrac{\partial f}{\partial t} + v \cdot \dfrac{\partial f}{\partial x} = 0, \quad t \in]t_n, t_{n+1}], \quad x \in \mathbb{R}, \\ f(x, t_n + 0; v) = f_n(x,v) \end{cases} \tag{7.36a}$$

and

$$\begin{cases} \dfrac{\partial g}{\partial t} + v \cdot \dfrac{\partial g}{\partial x} = 0, \quad t \in]t_n, t_{n+1}], \quad x \in \mathbb{R}, \\ g(x, t_n + 0; v) = g_n(x,v), \end{cases} \tag{7.36b}$$

which gives

$$f(x,t;v) = f_n\Big(x - v(t-t_n), v\Big), t \in]t_n, t_{n+1}],$$

$$g(x,t;v) = g_n\Big(x - v(t-t_n), v\Big).$$

We define $\rho_{n+1}, (\rho u)_{n+1}, (\rho e)_{n+1}$ by

$$\begin{pmatrix} \rho_{n+1} \\ (\rho u)_{n+1} \\ (\rho e)_{n+1} \end{pmatrix}(x) = \int_{\mathbb{R}} \begin{pmatrix} f_n \\ vf_n \\ \frac{v^2}{2} f_n + g_n \end{pmatrix}(x - v\Delta t, v)dv, \tag{7.37}$$

and we associate to these quantities the equilibrium function $M(v - u_{n+1}, \theta; \rho_{n+1}, \varepsilon_{n+1})$. Then, we define

$$f_{n+1}(x,v) = \int_{\mathbb{R}+} M(v - u_{n+1}, \theta; \rho_{n+1}, \varepsilon_{n+1})d\theta,$$

$$g_{n+1}(x,v) = \int_{\mathbb{R}+} M(v - u_{n+1}, \theta; \rho_{n+1}, \varepsilon_{n+1})\theta d\theta.$$

This says that in the second (collision) phase, mass, momentum, and energy are conserved.

Lemma 7.3

The quantities $\rho_{n+1}, (\rho u)_{n+1}, (\rho e)_{n+1}$ defined by (7.34) and (7.36) are first-order approximations (in time) of the solution of the Euler equation (7.14).

Proof. The result means that if $(\rho_n, (\rho u)_n, (\rho e)_n) = (\rho, \rho u, \rho e)(x, t_n)$ is the exact solution of the Euler equation at time t_n, the quantities defined above are first-order approximations of $(\rho, \rho u, \rho e)(x, t_{n+1})$. Let us first note that we can define the quantities $(\rho, \rho u, \rho e)$ not only at time t_{n+1} but at any time, for instance

$$(\rho, \rho u)^T(x,t) = \int_{\mathbb{R}} f(x,v,t)(1,v)^T dv,$$

where f is the solution of (7.36). It is then easy to see that ρ is indeed the solution of (7.14). (The first equation in (7.14) is solved exactly.) The argument follows the proof of Proposition 7.2. Indeed, let us integrate the first equation (7.36) w.r.t. v; we get

$$\begin{aligned} 0 &= \int_{\mathbb{R}} \Big(\frac{\partial f}{\partial t} + v \cdot \frac{\partial f}{\partial x} \Big) dv = \frac{\partial}{\partial t} \int_{\mathbb{R}} f(x,v,t)dv + \frac{\partial}{\partial x} \int_{\mathbb{R}} v f(x,v,t)dv \\ &= \frac{\partial \rho}{\partial t} + \frac{\partial (\rho u)}{\partial x}. \end{aligned}$$

Thus, if the initial conditions are exact, so is ρ at any time t. The other quantities are only first-order approximations. Multiplying (7.36) by v and integrating w.r.t. v, we obtain

$$\begin{aligned} 0 &= \int_{\mathbb{R}} \Big(\frac{\partial f}{\partial t} + v \cdot \frac{\partial f}{\partial x} \Big) v dv \\ &= \frac{\partial}{\partial t} \int_{\mathbb{R}} f(x,v,t) v dv + \frac{\partial}{\partial x} \int_{\mathbb{R}} v^2 f(x,v,t)dv \\ &= \frac{\partial}{\partial t}(\rho u) + \frac{\partial}{\partial x} 2\Big(\rho e - \int_{\mathbb{R}} g(x,v,t)dv\Big). \end{aligned}$$

As in (7.32), we have exactly at time t_n

$$\int_{\mathbb{R}} g(x,v,t_n)dv = \rho_n R T_n \delta,$$

and hence

$$2\Big(\rho e - \int_{\mathbb{R}} g(x,v,t_n)dv\Big) = (\rho u^2 + p)(t_n),$$

which is only satisfied at the order $O(\Delta t)$ for $t \in]t_n, t_{n+1}]$. For instance, we write for simplicity the first step on $[0, \Delta t]$, using a Taylor expansion, the definition (7.35), and equations (7.34), (7.32), and (7.18):

$$\begin{aligned} \rho u(x,t) &= \rho u(x,0) + \Delta t \frac{\partial}{\partial t}(\rho u)(x,0) + O(\Delta t^2) \\ &= \rho u(x,0) + \Delta t \frac{\partial}{\partial t} \int_{\mathbb{R}} f(x,v,t) v dv + O(\Delta t^2) \\ &= \rho u(x,0) + \Delta t \int_{\mathbb{R}} v \frac{\partial}{\partial t} f(x,v,t)dv + O(\Delta t^2) \\ &= \rho u(x,0) - \Delta t \int_{\mathbb{R}} v^2 \frac{\partial}{\partial x} f(x,v,t)dv + O(\Delta t^2), \end{aligned}$$

where the derivatives are taken at time $t = 0$. From which we obtain the result since

$$\begin{aligned}\rho u(x,t) &= \rho u(x,0) - 2\Delta t \frac{\partial}{\partial x}\Big(\rho_0 e_0 - \int_{\mathbb{R}} g_0(x,v)dv\Big) + O(\Delta t^2)\\ &= \rho u(x,0) - 2\Delta t \frac{\partial}{\partial x}(\rho_0 e_0 - RT_0\delta)\int_{\mathbb{R}} f_0(x,v)dv + O(\Delta t^2)\\ &= \rho u(x,0) - 2\Delta t \frac{\partial}{\partial x}\rho_0(e_0 - RT_0\delta) + O(\Delta t^2)\\ &= \rho u(x,0) - \Delta t \frac{\partial}{\partial x}(\rho u^2 + p)(x,0) + O(\Delta t^2),\end{aligned}$$

gives the exact solution ρu of the Euler equations within $O(\Delta t^2)$. The proof for ρe is similar. □

For numerical purposes, it is interesting to replace $\int_{\mathbb{R}+} Mn(\theta)d\theta$ by an expression that is easier to handle. In practice, Perthame (1990) has considered a numerical scheme associated with a function χ satisfying the following properties,

$$\text{(7.38)}\qquad \begin{cases} \chi \geq 0,\\ \displaystyle\int_{\mathbb{R}} \chi(w)(1,w^2)dw = (1,1),\\ \chi(-w) = \chi(w)\Big(\Rightarrow \displaystyle\int_{\mathbb{R}} \chi(w)wdw = 0\Big). \end{cases}$$

We then introduce the functions

$$\text{(7.39)}\qquad \begin{cases} f_0(x,v) = \rho_0(x)\Big(T_0(x)\Big)^{\frac{1}{2}}\chi\Big(\dfrac{v-u_0(x)}{T_0^{\frac{1}{2}}(x)}\Big),\\ g_0(x,v) = \delta T_0(x) f_0(x,v) = \delta\rho_0(x)\Big(T_0(x)\Big)^{\frac{1}{2}}\chi\Big(\dfrac{v-u_0(x)}{T_0^{\frac{1}{2}}(x)}\Big), \end{cases}$$

where δ is defined by (7.18b). In fact, we can take for f and g two functions χ and ζ. This will be detailed below in Section 7.3.3.

Remark 7.5. We observe that the function χ is constructed in order to take the place of

$$(2\pi)^{-\frac{1}{2}}\exp\Big(-\frac{v^2}{2}\Big) = T_0^{\frac{1}{2}}\rho_0^{-1}\int_{\mathbb{R}+} M(v;\rho_0,\varepsilon_0,v,\theta)n(\theta)d\theta,$$

where we have incorporated the constant R in the temperature. We have used the identity defining θ_0,

$$\int_{\mathbb{R}+}\exp\Big(-\frac{\theta}{T}\Big)n(\theta)d\theta = \theta_0,$$

which implies

$$\int_{\mathbb{R}+} M(v;\rho_0,\varepsilon_0,v,\theta)d\theta = \frac{\rho_0}{(2\pi T_0)^{\frac{1}{2}}}\ \exp\,(-\frac{v^2}{2T}).$$

Indeed, on the one hand,

$$(2\pi)^{-\frac{1}{2}}\int_{\mathbb{R}} \exp\,\Big(-\frac{w^2}{2}\Big)(1,w,w^2)^T dw = (1,0,1)^T$$

and, on the other hand, the last two properties of χ imply

$$\int_{\mathbb{R}} \chi(w)(1,w,w^2)^T dw = (1,0,1)^T.$$

With this in mind, we see that the definition of f_0 and g_0 mimics (7.31), (7.32). □

It is easy to check that (7.33) still holds, i.e.,

$$(\rho_0,\rho_0 u_0)^T = \int_{\mathbb{R}} f_0(x,v)(1,v)^T dv,$$

$$\rho_0 e_0 = \rho_0\Big(\varepsilon_0 + \frac{u_0^2}{2}\Big) = \int_{\mathbb{R}}\Big\{\frac{|v|^2}{2} f_0 + g_0\Big\}(x,v)dv.$$

We then solve the linear transport equation (7.34) with f_0, g_0 given by (7.39), and define the updated quantities $\mathbf{U}(x,t) = (\rho,\rho u,\rho e)(x,t)$ by (7.35). As in Lemma 7.3, we can prove the following lemma.

Lemma 7.4

Assume that f and g are solutions of (7.34), where f_0, g_0 are given by (7.39). Then, the quantities $(\rho,\rho u,\rho e)(x,t)$ defined for $0 \le t \le \Delta t$ by (7.35) are first-order approximations (in time) of the solution of the Euler equation (7.14).

The description of a general time step follows the same lines.

7.3.2 The space discretization

For the *space discretization*, we follow the ideas of Godunov's scheme: we assume that the functions $\rho_0, \rho_0 u_0, E_0$ are piecewise constant over the cells $(x_{j-\frac{1}{2}}, x_{j+\frac{1}{2}})$,

$$\mathbf{U}(x,0) = \mathbf{U}_0(x) = \mathbf{U}_j^0 = (\rho_j^0,\rho u_j^0,\rho e_j^0), \quad x\in(x_{j-\frac{1}{2}},x_{j+\frac{1}{2}}).$$

As above, we solve (7.34) and obtain $\mathbf{U}(x,t)$ by (7.35). Lastly, we project on the grid

$$\mathbf{U}_j^1 = \frac{1}{\Delta x}\int_{x_{j-\frac{1}{2}}}^{x_{j+\frac{1}{2}}} \mathbf{U}(x,\Delta t)dx. \tag{7.40}$$

Thus, the scheme results in three steps: if $\mathbf{U}_n(x) = (\mathbf{U}_j^n) = (\rho_j^n,\rho u_j^n,\rho e_j^n)$ constant over the cell $(x_{j-\frac{1}{2}},x_{j+\frac{1}{2}})$ is known, we compute $\mathbf{U}_{n+1}(x) =$

$(\mathbf{U}_j^{n+1})$ by:
(i) reconstruction of f_n, g_n by (7.39):

$$\text{(7.41}a\text{)} \qquad \begin{cases} f_n(x,v) = \rho_n (T_n)^{-\frac{1}{2}} \chi\Big(\dfrac{v-u_n}{T_n^{\frac{1}{2}}}\Big), \\ g_n(x,v) = \delta T_n f_n(x,v) = \delta \rho_n T_n^{\frac{1}{2}} \chi\Big(\dfrac{v-u_n}{T_n^{\frac{1}{2}}}\Big). \end{cases}$$

(ii) evolution by (7.34),

$$\text{(7.41}b_1\text{)} \qquad \begin{cases} \dfrac{\partial f}{\partial t} + v \cdot \dfrac{\partial f}{\partial x} = 0, \quad t \in]t_n, t_{n+1}], \quad x \in \mathbb{R}, \\ f(x, t_n; v) = f_n(x,v), \end{cases}$$

$$\text{(7.41}b_2\text{)} \qquad \begin{cases} \dfrac{\partial g}{\partial t} + v \cdot \dfrac{\partial g}{\partial x} = 0, \quad t \in]t_n, t_{n+1}], \quad x \in \mathbb{R}, \\ g(x, t_n; v) = g_n(x,v), \end{cases}$$

(iii) projection: defining $\mathbf{U}(., t_{n+1})$ by (7.35), where $f, g(., t_{n+1})$ are solutions of (7.41b), set

$$\text{(7.41}c\text{)} \qquad \mathbf{U}_j^{n+1} = \frac{1}{\Delta x} \int_{x_{j-\frac{1}{2}}}^{x_{j+\frac{1}{2}}} \mathbf{U}(x, t_{n+1}) dx.$$

We assume from now on that χ is compactly supported,

$$\text{(7.42)} \qquad \chi(w) = 0, \text{ for } |w| > w_M.$$

As was the case for Godunov's or van Leer's scheme, we can derive a very simple formula for $\mathbf{U}_j^n$.

Proposition 7.3

Let χ satisfy (7.38), (7.42). Under the C.F.L. condition

$$\text{(7.43)} \qquad \lambda \max \{|u_j^n| + w_M (T_j^n)^{\frac{1}{2}}\} \le 1, \quad \lambda = \frac{\Delta t}{\Delta x},$$

the difference scheme (7.41) can be written

$$\text{(7.44}a\text{)} \qquad \mathbf{U}_j^{n+1} = \mathbf{U}_j^n - \lambda\{\mathbf{F}(\mathbf{U}_j^n, \mathbf{U}_{j+1}^n) - \mathbf{F}(\mathbf{U}_{j-1}^n, \mathbf{U}_j^n)\},$$

where the numerical flux $\mathbf{F}(\mathbf{U}, \mathbf{V})$ is given by

$$\text{(7.44}b\text{)} \qquad \mathbf{F}(\mathbf{U}, \mathbf{V}) = \mathbf{F}^+(\mathbf{U}) + \mathbf{F}^-(\mathbf{V})$$

and
(7.44*c*)

$$\mathbf{F}^+(\mathbf{U}) = \int_{w \ge -\frac{u}{\sqrt{T}}} \begin{pmatrix} w\sqrt{T} + u \\ (w\sqrt{T} + u)^2 \\ \frac{1}{2}(w\sqrt{T}+u)^3 + \delta T (w\sqrt{T}+u) \end{pmatrix} \rho \chi(w) dw,$$

(7.44d)

$$\mathbf{F}^-(\mathbf{U}) = \int_{w \leq -\frac{u}{\sqrt{T}}} \begin{pmatrix} w\sqrt{T} + u \\ (w\sqrt{T} + u)^2 \\ \frac{1}{2}(w\sqrt{T} + u)^3 + \delta T(w\sqrt{T} + u) \end{pmatrix} \rho\chi(w)dw.$$

Proof. Let us integrate (7.34) w.r.t. (v, x, t) over $\mathbb{R} \times (x_{j-\frac{1}{2}}, x_{j+\frac{1}{2}}) \times (0, \Delta t)$:

$$\begin{aligned} 0 &= \int_{\mathbb{R}\times(x_{j-\frac{1}{2}}, x_{j+\frac{1}{2}})\times(0,\Delta t)} \Big(\frac{\partial f}{\partial t} + v \cdot \frac{\partial f}{\partial x}\Big) dv\, dx\, dt \\ &= \int_{\mathbb{R}\times(x_{j-\frac{1}{2}}, x_{j+\frac{1}{2}})} \{f(x, v, \Delta t) - f(x, v, 0)\} dv dx \\ &\qquad + \int_{\mathbb{R}\times(0,\Delta t)} v\{f(x_{j+\frac{1}{2}}, v, t) - f(x_{j-\frac{1}{2}}, v, t)\} dv dt. \end{aligned}$$

By (7.35) and (7.40), the first integral on the right-hand side is equal to

$$\int_{(x_{j-\frac{1}{2}}, x_{j+\frac{1}{2}})} \{\rho(x, \Delta t) - \rho(x, 0)\} dx = \Delta x(\rho_j^1 - \rho_j^0).$$

The second integral is

$$\begin{aligned} &\int_{\mathbb{R}\times(0,\Delta t)} v\{f_0(x_{j+\frac{1}{2}} - vt, v) - f_0(x_{j-\frac{1}{2}} - vt, v)\} dv dt \\ &\qquad = \int_{\mathbb{R}} \Big\{ \int_{(0,\Delta t)} v\{f_0(x_{j+\frac{1}{2}} - vt, v) - f_0(x_{j-\frac{1}{2}} - vt, v)\} dv \Big\} dt, \end{aligned}$$

with

$$f_0(x, v) = \rho_0 \Big(T_0(x)\Big)^{-\frac{1}{2}} \chi\Big(\frac{v - u_0(x)}{T_0^{\frac{1}{2}}(x)}\Big).$$

Setting

$$w = \frac{v - u_0}{\sqrt{T_0}},$$

the condition $|w| \leq w_M$, where $[-w_M, w_M]$ is the support of χ, means $|\frac{(v-u_0)}{\sqrt{T_0}}| \leq w_M$ and implies that

$$|v| \leq |u_0| + w_M \sqrt{T_0}.$$

Now, by the C.F.L. condition (7.43),

$$\Delta t(|u_0| + \sqrt{T}_0 w_M) \leq \Delta x,$$

the integrals are limited to the v such that $|v| \leq \frac{\Delta x}{\Delta t}$. This yields that the quantity $x_{j+\frac{1}{2}} - vt$ remains for $t \in (0, \Delta t)$, and $v > 0$, in the cell $(x_{j-\frac{1}{2}}, x_{j+\frac{1}{2}})$, where $\mathbf{U}_0(x)$ is constant and given by $\mathbf{U}_j^0 = (\rho_j^0, \rho u_j^0, \rho e_j^0)$

(resp. for $v < 0$ in the cell $(x_{j+\frac{1}{2}}, x_{j+\frac{3}{2}})$, where $\mathbf{U}_0(x) = \mathbf{U}_{j+1}^0 = (\rho_{j+1}^0, \rho u_{j+1}^0, \rho e_{j+1}^0)$). We can thus write

$$\begin{aligned}
&\int_{\mathbb{R}} \{\int_{(0,\Delta t)} v f_0(x_{j+\frac{1}{2}} - vt, v)dt\}dv \\
&= \int_{v>0} \{\int_{(0,\Delta t)} v f_0(x_{j+\frac{1}{2}} - vt, v)dt\}dv \\
&\qquad\qquad + \int_{v\leq 0} \{\int_{(0,\Delta t)} v f_0(x_{j+\frac{1}{2}} - vt, v)dt\}dv \\
&= \Delta t\{\int_{v>0} \rho_j^0 (T_j^0)^{-\frac{1}{2}} \chi\Big(\frac{v-u_j^0}{\sqrt{T_j^0}}\Big)dv \\
&\qquad\qquad + \int_{v\leq 0} \rho_{j+1}^0 (T_{j+1}^0)^{-\frac{1}{2}} \chi\Big(\frac{v-u_{j+1}^0}{\sqrt{T_j^0}}\Big)dv\}.
\end{aligned}$$

After a change of variable $w = \frac{(v-u)}{\sqrt{T}}$, we get by (7.44c) the first component of $\mathbf{F}^+(\mathbf{U}_j^0) + \mathbf{F}^-(\mathbf{U}_{j+1}^0)$. Similarly,

$$\begin{aligned}
&\int_{\mathbb{R}} \{\int_{(0,\Delta t)} v f_0(x_{j-\frac{1}{2}} - vt, v)dt\}dv \\
&= \int_{v>0} \{\int_{(0,\Delta t)} v f_0(x_{j-\frac{1}{2}} - vt, v)dt\}dv \\
&\qquad\qquad + \int_{v\leq 0} \{\int_{(0,\Delta t)} v f_0(x_{j-\frac{1}{2}} - vt, v)dt\}dv \\
&= \Delta t\{\int_{v>0} \rho_{j-1}^0 (T_{j-1}^0)^{-\frac{1}{2}} \chi\Big(\frac{v-u_{j-1}^0}{\sqrt{T_j^0}}\Big)dv \\
&\qquad\qquad + \int_{v\leq 0} \rho_j^0 (T_j^0)^{-\frac{1}{2}} \chi\Big(\frac{v-u_j^0}{\sqrt{T_j^0}}\Big)dv\},
\end{aligned}$$

which gives the first component of $\mathbf{F}^+(\mathbf{U}_{j-1}^0) + \mathbf{F}^-(\mathbf{U}_j^0)$.

The other components are computed in the same way. This yields

$$\Delta x(\mathbf{U}_j^1 - \mathbf{U}_j^0) + \Delta t\{\mathbf{F}^+(\mathbf{U}_j^0) + \mathbf{F}^-(\mathbf{U}_{j+1}^0) - \mathbf{F}^+(\mathbf{U}_{j-1}^0) - \mathbf{F}^-(\mathbf{U}_j^0)\} = 0,$$

and proves that the scheme is indeed given by (7.44).

The formulas can be written in the more compact form

$$\mathbf{F}^\pm(\mathbf{U}) = \rho \int_{\mathbb{R}} v_\pm \Big(1, v, \frac{|v|^2}{2} + \delta T\Big)^T \chi\Big(\frac{v-u}{\sqrt{T}}\Big)\frac{dv}{\sqrt{T}}, \tag{7.45}$$

which gives a flux-vector splitting scheme whose numerical flux is clearly consistent with the flux of the Euler equations. □

An important feature of this scheme is its positivity-preserving property.

Theorem 7.2

Assume the hypotheses of Proposition 7.3. Then, the difference scheme (7.44) satisfies

$$\rho_j^0 \geq 0, T_j^0 \geq 0, \quad \forall j \quad \Rightarrow \quad \rho_j^n, T_j^n \geq 0, \quad \forall j, \forall n > 0, \tag{7.46}$$

and is L^1-stable.

Proof. Assume

$$\rho_j^0 \geq 0, T_j^0 \geq 0, \forall j,$$

and consider the first step. The functions f_0 and g_0 in (7.39) are thus non-negative, and f, g remain such after the transport by (7.34). Now consider the moments; by (7.41c)

$$\rho_j^1 = \frac{1}{\Delta x} \int_{\mathbb{R}\times(x_{j-\frac{1}{2}}, x_{j+\frac{1}{2}})} f_0(x - v\Delta t, v) dv dx \geq 0,$$

and

$$E_j^1 = \frac{1}{\Delta x} \int_{\mathbb{R}\times(x_{j-\frac{1}{2}}, x_{j+\frac{1}{2}})} \Big\{ \frac{v^2}{2} f_0(x - v\Delta t, v) + g_0(x - v\Delta t, v) \Big\} dv dx.$$

We can write

$$\begin{aligned} \Delta x E_j^1 \geq \int_{\mathbb{R}\times(x_{j-\frac{1}{2}}, x_{j+\frac{1}{2}})} & \frac{1}{2} \{ (u_j^1)^2 + (v - u_j^1)^2 \\ & + 2u_j^1 (v - u_j^1) \} f_0(x - v\Delta t, v) dv dx \\ \geq (u_j^1)^2 \frac{\rho_j^1}{2} + u_j^1 & \int_{\mathbb{R}\times(x_{j-\frac{1}{2}}, x_{j+\frac{1}{2}})} (v - u_j^1) f_0(x - v\Delta t, v) dv dx, \end{aligned}$$

and the last integral vanishes. Since

$$E_j^1 = \rho_j^1 \Big\{ \varepsilon_j^1 + \frac{(u_j^1)^2}{2} \Big\},$$

this yields

$$RT_j^1 = (\gamma - 1)\varepsilon_j^1 \geq 0.$$

Consider next the L^1-stability. Since the quantities are nonnegative and since the scheme (7.44) is obviously conservative, we have

$$||\rho_1||_{\mathbf{L}^1(\mathbb{R})} = \sum_{j\in\mathbb{Z}} \rho_j^1 = \sum_{j\in\mathbb{Z}} \rho_j^0 = ||\rho_0||_{\mathbf{L}^1(\mathbb{R})},$$

and similarly

$$\sum_{j\in\mathbb{Z}} E_j^1 = ||E_0||_{\mathbf{L}^1(\mathbb{R})}.$$

Then

$$\sum_{j\in\mathbb{Z}} \rho_j^1|u_j^1| = ||\rho_1 u_1||_{\mathbf{L}^1(\mathbb{R})}$$

$$\leq (||\rho_1||_{\mathbf{L}^1(\mathbb{R})}||\rho_1(u_1)^2||_{\mathbf{L}^1(\mathbb{R})})^{\frac{1}{2}} \leq (2||\rho_0||_{\mathbf{L}^1(\mathbb{R})}||E_1||_{\mathbf{L}^1(\mathbb{R})})^{\frac{1}{2}}$$

$$\leq ||\rho_0||_{\mathbf{L}^1(\mathbb{R})} + ||E_0||_{\mathbf{L}^1(\mathbb{R})},$$

which ends the proof. □

Example 7.2. Let us take for χ the function

$$\chi(w) = \frac{1}{2\sqrt{3}}1_{\{|w|\leq\sqrt{3}\}},$$

which is the unique step function satisfying (7.38). The integrals (7.44c) are then very easily computed, and in the case $\delta = 0, \gamma = 3$, we recover van Leer's flux-splitting formulas (Section 5.3). Indeed, from (7.45), we get

$$\mathbf{F}^+(\mathbf{U}) = \Big(\frac{\rho}{2\sqrt{3T}}\Big)\int_0^{u+\sqrt{3T}}\Big(v, v^2, \frac{v^3}{2}\Big)^T dv$$

$$= \Big(\frac{\rho}{2\sqrt{3T}}\Big)(u+\sqrt{3T})^2\Big(\frac{1}{2}, \frac{(u+\sqrt{3T})}{3}, \frac{(u+\sqrt{3T})^2}{8}\Big)^T.$$

Now, from the identity

$$p = \frac{\rho c^2}{\gamma},$$

we obtain

$$c^2 = \gamma T = 3T,$$

so that formulas (5.11)–(5.13) of Section 5, giving the components of van Leer's split flux, yield

$$F_1+ = \frac{\rho}{4c}(u+c) = \frac{\rho}{4\sqrt{3T}}(u+\sqrt{3T})^2,$$

$$F_2+ = \frac{2}{3}(u+c)F_1+ = \frac{\rho}{6\sqrt{3T}}(u+\sqrt{3T})^3,$$

$$F_3+ = \frac{1}{4}(u+c)^2F_1+ = \frac{\rho}{4\sqrt{3T}}(u+\sqrt{3T})^4,$$

which coincide with the components of $\mathbf{F}^+(\mathbf{U})$. We have similar formulas for F_i-.

We know that the corresponding B.G.K. model is the limit of entropic models. Indeed, with the notation of Example 7.1, we have $b = 3T/2$ and $a = \rho/2\sqrt{3T}$, and the function $f_0(x,v) = a1_{\{|v-u|^2\leq 2bT\}}$ satisfies the constraints (7.26) since

$$\frac{\rho}{2\sqrt{3T}}\int_{\mathbb{R}} 1_{\{|v|^2\leq 3T\}}\Big(1, \frac{|v|^2}{2}\Big)^T dv = (\rho, \frac{\rho T}{2})^T = (\rho, \rho\varepsilon)^T$$

(see Remark 7.3; the case $\delta = 0, \gamma = 3$ gives $T = 2\varepsilon$).

In the general case (γ and $\delta \neq 0$ satisfying (7.18b)) we would like the functions f_0, g_0, associated with this χ by (7.39), to be the integrals w.r.t. θ of an "equilibrium" function N, i.e.,

$$f_0(x,v) = \frac{\rho}{\sqrt{T}}\chi\Big(\frac{v-u}{\sqrt{T}}\Big) = \int_{\mathbb{R}+} N(v-u,\theta)n(\theta)d\theta$$

and

$$g_0(x,v) = \delta T f_0(x,v) = \delta\rho\sqrt{T}\chi\Big(\frac{v-u}{\sqrt{T}}\Big) = \int_{\mathbb{R}+} N(v,\theta)\theta n(\theta)d\theta$$

(see also Remark 7.5). For instance, the above identities hold with the function

$$N(v,\theta) = \frac{\rho}{2\theta_0\sqrt{3T}} 1_{\{\frac{|v|^2}{2} \leq \frac{3T}{2}\}} \exp\Big(-\frac{\theta}{T}\Big)$$

(we can also replace $\frac{1}{\theta_0}\exp(-\frac{\theta}{T})$ by a characteristic function $c1_{\{\theta\leq d\}}$ such that $\int_{\mathbb{R}+} 1_{\{\theta\leq d\}}n(\theta)d\theta = 1$, $\int_{\mathbb{R}+} 1_{\{\theta\leq d\}}\theta n(\theta)d\theta = \delta T$. However, it is not clear that an entropy inequality exists in that case, and this motivates another choice for χ, which we shall now present. □

7.3.3 A maximum principle on the entropy

In order to satisfy the entropy inequality and a maximum principle on an entropy function, we are led to modify the above technique and for f and g deal with two functions χ and ζ rather than with two kinetic variables v and θ. More precisely, one chooses (δ is defined by (7.8) with $d = 1$)

$$\chi(w) = a\Big(1 - \frac{w^2}{b}\Big)_+^{\delta}, \tag{7.47a}$$

$$\zeta(w) = \tilde{a}\Big(1 - \frac{w^2}{b}\Big)_+^{\delta+1} = \tilde{b}\Big(\chi(w)\Big)^{\frac{(\gamma+1)}{(3-\gamma)}}, \tag{7.47b}$$

and set

$$f(v) = f(v;\rho,u,T) = \frac{\rho}{\sqrt{T}}\chi\Big(\frac{v-u}{\sqrt{T}}\Big), \tag{7.48a}$$

$$g(v) = g(v;\rho,u,T) = \rho\sqrt{T}\zeta\Big(\frac{v-u}{\sqrt{T}}\Big), \tag{7.48b}$$

where the constants a, b are given by the constraints (7.38) and $\tilde{a}, \tilde{b}$ by

$$\int_{\mathbb{R}} \zeta(w)dw = \delta.$$

Thus

$$\int_{\mathbb{R}} f(1,v)^T dv = (\rho, 0) \tag{7.49}$$

and

$$\int_{\mathbb{R}} g dv = \rho T \delta,$$

so that, as previously,

$$\int_{\mathbb{R}} \Big\{ \frac{|v|^2}{2} f + g \Big\} dv = \rho \frac{(T + u^2)}{2} + \rho T \delta = \frac{\rho T}{(\gamma - 1)} + \frac{\rho u^2}{2} = \rho e.$$

The associated numerical split flux is naturally

$$(7.50a) \quad \begin{cases} \mathbf{F}^{\pm}(\mathbf{U}) = \rho \int_{\mathbb{R}} v_{\pm} \Big\{ \Big(1, v, \frac{|v|^2}{2}\Big)^T \chi\Big(\frac{v-u}{\sqrt{T}}\Big) \\ \qquad\qquad + (0, 0, T)^T \zeta\Big(\frac{v-u}{\sqrt{T}}\Big)\Big\} \frac{dv}{\sqrt{T}} \end{cases}$$

or

$$(7.50b) \qquad \mathbf{F}^{\pm}(\mathbf{U}) = \int_{\mathbb{R}} v_{\pm} \Big(f, vf, \frac{|v|^2}{2} f + g\Big)^T dv$$

which is the analog of (7.45).

We choose for an entropy function (in the sense of Lax),

$$(7.51) \qquad S = \rho T^{\frac{-1}{(\gamma-1)}};$$

since $p = \rho T$, and $\log \frac{p}{\rho^\gamma} = \log \big(\frac{T}{\rho^{(\gamma-1)}}\big)$, we check easily that S is indeed of the form $-\varphi(\log(\frac{p}{\rho^\gamma}))$ where φ is a nonpositive, nondecreasing concave function, which implies (see Harten 1983, Tadmor 1986) that $U = \rho S$ satisfies $\frac{\partial(\rho S)}{\partial t} + \frac{\partial(\rho u S)}{\partial x} \leq 0$. Tadmor has proven a maximum principle for the entropy

$$S(x, t + \Delta t) \leq \max \{S(y,t); |y - x| \leq \Delta t ||u||_{\mathbf{L}^\infty}\}$$

and the analogous property for numerical schemes such as Lax–Friedrichs' and Godunov's. Here, we can derive a discrete entropy inequality and a maximum principle on the above entropy S for the kinetic scheme (Khobalatte and Perthame 1994).

Theorem 7.3

Consider the difference scheme (7.44a,b)

$$\mathbf{U}_j^{n+1} = \mathbf{U}_j^n - \lambda\{\mathbf{F}(\mathbf{U}_j^n, \mathbf{U}_{j+1}^n) - \mathbf{F}(\mathbf{U}_{j-1}^n, \mathbf{U}_j^n)\}$$

with

$$\mathbf{F}(\mathbf{U}, \mathbf{V}) = \mathbf{F}^+(\mathbf{U}) + \mathbf{F}^-(\mathbf{V}),$$

where $\mathbf{F}^{\pm}$ *are defined by (7.50). Under the C.F.L. condition* $\lambda \max\big(|u_i^n| + \sqrt{T_i^n / b}\big) \leq 1$, *the scheme satisfies the property* (7.46),

$$\rho_j^0 \geq 0, T_j^0 \geq 0, \forall j \;\Rightarrow\; \rho_j^n \geq 0, T_j^n \geq 0, \quad \forall j, \forall n > 0,$$

together with the entropy inequality

$$(\rho\pi_v)_i^{n+1} - (\rho\pi_v)_i^n + \lambda\{G_{i+\frac{1}{2}}^n - G_{i-\frac{1}{2}}^n\} \leq 0, \quad \forall v > 0, \tag{7.52}$$

and a maximum principle on the entropy (7.51)

$$S_i^{n+1} = \frac{\rho_i^{n+1}}{(T_i^{n+1})^{\frac{1}{(\gamma-1)}}} \leq \max\,(S_{i-1}^n, S_i^n, S_{i+1}^n). \tag{7.53}$$

In (7.52), π_v is a one-parameter family of degenerated entropy functions,

$$\pi_v = \begin{cases} 0 & \text{if } \dfrac{\rho^{\gamma-1}}{T} < v, \\ 1 & \text{if } \dfrac{\rho^{\gamma-1}}{T} = v, \\ +\infty & \text{otherwise.} \end{cases} \tag{7.54}$$

Note that $\rho\pi_v$ is obtained as the limit as p tends to $+\infty$ of the convex entropies

$$\rho\Big(\frac{S^{\gamma-1}}{v}\Big)^p = \rho\Big(\frac{\rho^{\gamma-1}}{Tv)}\Big)^p.$$

The numerical entropy flux G associated with $\rho\pi_v$,

$$G_{i+\frac{1}{2}}^n = G(\mathbf{U}_j^n, \mathbf{U}_{j+1}^n),$$

can also be split

$$G(\mathbf{U}, \mathbf{V}) = G^+(\mathbf{U}) + G^-(\mathbf{V}),$$

with $G^\pm$ such that

$$G^\pm(\mathbf{U}) = F_\rho^\pm(\mathbf{U})\pi_v(\mathbf{U}),$$

where $F_\rho^\pm(\mathbf{U})$ is the numerical mass flux (first component of $\mathbf{F}^+(\mathbf{U})$). Thus, G is consistent:

$$G(\mathbf{U}, \mathbf{U}) = \rho u \pi_v.$$

Proof of Theorem 7.3. Let us first set

$$f_i^n(v) = f(v; \rho_i^n, u_i^n, T_i^n), \quad g_i^n(v) = g(v; \rho_i^n, u_i^n, T_i^n)$$

and

$$\begin{aligned} \overline{f}_i(v) &= f_i^n(v) - \lambda\{v_+ f_i^n(v) - v_- f_{i+1}^n(v) - v_+ f_{i-1}^n(v) + v_- f_i^n(v)\} \\ &= f_i^n(v) - \lambda\{v_+\Big(f_i^n(v) - f_{i-1}^n(v)\Big) - v_-\Big(f_{i+1}^n(v) - f_i^n(v)\Big)\}, \\ \overline{g}_i(v) &= g_i^n(v) - \lambda\{v_+ g_i^n(v) - v_- g_{i+1}^n(v) - v_+ g_{i-1}^n(v) + v_- g_i^n(v)\}, \end{aligned} \tag{7.55}$$

so that

$$\mathbf{U}_i^{n+1} = \int_{\mathbb{R}} \Big(\overline{f}_i(v), v\overline{f}_i(v), \frac{|v|^2}{2}\overline{f}_i(v) + \overline{g}_i(v)\Big)^T dv,$$

$$\mathbf{F}^{\pm}(\mathbf{U}_i^n) = \pm \int_{\mathbb{R}} v_{\pm}\Big(f_i^n, vf_i^n(v), \frac{|v|^2}{2} f_i^n + g_i^n\Big)^T dv.$$

The first property (7.46) follows as in Theorem 7.2.

Let us sketch the proof of the inequality (7.53). We consider the function

$$h = h(f,g) = (f^{\gamma+1}g^{\gamma-3})^{\frac{1}{(\gamma-1)}}. \tag{7.56}$$

The function h is a convex function of f and g, and the exponents have been chosen in order that the power of $(1-w^2/b)_+$ vanishes. Thus, we get

$$h_i^n = (S_i^n)^2 \tilde{b}^{\frac{(\gamma-3)}{(\gamma-1)}} 1_{\{|v-u_i^n|^2 \le bT\}} = c(S_i^n)^2 1_{\{|v-u_i^n|^2\le bT\}},$$

where $c = \tilde{b}^{\frac{(\gamma-3)}{(\gamma-1)}}$. We obtain by (7.55)

$$\overline{h}_i = h(\overline{f}_i, \overline{g}_i) \le h_{i-1}^n(\lambda v_+) + h_i^n(1-\lambda v_+ - \lambda v_-) + h_{i+1}^n(\lambda v_-), \tag{7.57}$$

which yields

$$\overline{h}_i \le c \max\{S_{i-1}^n, S_i^n, S_{i+1}^n\}^2. \tag{7.58}$$

We then consider the problem of minimization,

$$j(\rho,\Sigma) = \min \int_{\mathbb{R}} \{\frac{|v|^2}{2} f + g\} dv, \tag{7.59}$$

where the minimum is taken over all functions $f \ge 0, g \ge 0$ satisfying the constraints (7.49) and such that $h(f,g) \le \Sigma$ (for given Σ and δ), and set $j(\rho,\Sigma) = \frac{\rho\tau}{(\gamma-1)}$, which defines τ in terms of j.

Lemma 7.5

Problem (7.59) admits the following solution: the minimum of j is obtained for the functions $f(v;\rho,0,\tau) = (\frac{\rho}{\sqrt{\tau}})\chi(\frac{v}{\sqrt{\tau}})$, $g(v;\rho,0,\tau) = \rho\sqrt{\tau}\zeta(\frac{v}{\sqrt{\tau}})$ where χ and ζ are defined by (7.47) and $j(\rho,\Sigma) = \frac{\rho}{\gamma-1}(\frac{\rho}{\sqrt{\Sigma/c}})^{\gamma-1}$.

We apply the results of Lemma 7.5 with

$$\Sigma = c \max(S_{i-1}^n, S_i^n, S_{i+1}^n)^2.$$

Since by (7.58), $\overline{h}_i = h(\overline{f}_i, \overline{g}_i) \le \Sigma$, $\overline{f}_i(v+u_i^n)$, and $\overline{g}_i(v+u_i^n)$ satisfy the constraints, we can write

$$\begin{aligned}\frac{\rho_i^{n+1}T_i^{n+1}}{(\gamma-1)} &= \int_{\mathbb{R}}\Big\{\frac{|v|^2}{2}\overline{f}_i(v+u_i^{n+1}) + \overline{g}_i(v+u_i^{n+1})\Big\}dv \\ &\ge j(\rho_i^{n+1},\Sigma) = \frac{\rho_i^{n+1}}{(\gamma-1)}\Big(\frac{\rho_i^{n+1}}{\sqrt{\Sigma/c}}\Big)^{\gamma-1},\end{aligned}$$

and we get by (7.51)

$$S_i^{n+1} \leq (\frac{\Sigma}{c})^{\frac{1}{2}} = \max\,(S_{i-1}^n, S_i^n, S_{i+1}^n)$$

as expected.

The proof of (7.52) relies on a similar argument: one considers the function

$$k(f,g) = f\Big(\frac{h(f,g)}{v^2}\Big)^p = f\Big(\frac{(f^{\gamma+1}g^{\gamma-3})^{\frac{1}{(\gamma-1)}}}{v^2}\Big)^p,$$

for which an inequality of the form (7.57) is derived, and uses another minimization problem (analogous to (7.25)),

$$H(\rho,T) = \min\,\{\int_{\mathbb{R}} f\Big(h(f,g)\Big)^p dv\},$$

where the minimum is taken over all $f, g \geq 0$ satisfying the constraints (7.49) and

$$\int_{\mathbb{R}}\Big\{f(v)\frac{|v|^2}{2} + g(v)\Big\}dv = \frac{\rho T}{(\gamma-1)}.$$

The minimum is obtained for some functions f_p and g_p,

$$f_p = \frac{\rho a_p}{\sqrt{T}}\Big(1 - \frac{w^2}{b_p T}\Big)_+^{\delta+\frac{1}{2p}},$$

$$g_p = c_p T F_p\Big(1 - \frac{w^2}{b_p T}\Big)_+ = c_p \rho a_p \sqrt{T}\Big(1 - \frac{w^2}{b_p T}\Big)_+^{\delta+1+\frac{1}{2p}},$$

where the constants a_p, b_p, c_p are determined by the constraints. Then

$$\int_{\mathbb{R}} k(f_p, g_p)dv \leq \int_{\mathbb{R}} \overline{k}_i dv \leq \int_{\mathbb{R}} \{\lambda v_+ h_{i-1}^n + (1-\lambda v_+ - \lambda v_-)h_i^n + \lambda v_- h_{i+1}^n\}dv,$$

and (7.52) follows by letting p tend to $+\infty$. We skip the details of the proof. □

Remark 7.6. We are able to prove an entropy inequality for these functions f_p and g_p with entropy functions that are not degenerate, but no maximum principle on the entropy; this is obtained for the limit functions as p goes to $+\infty$. This should be compared with Example 7.1. This remark and the results of this section are taken from Khobalatte-Perthame (1994). □

Proof of Lemma 7.5. We just sketch the proof, which follows from classical arguments of the calculus of variations after writing the Euler–Lagrange equations of the minimization problem with constraints. The minimum is achieved for $f(v;\rho,0,\tau) = (\frac{\rho}{\sqrt{\tau}})\chi(\frac{v}{\sqrt{\tau}})$, $g(v;\rho,0,\tau) = (\rho\sqrt{\tau})\zeta(\frac{v}{\sqrt{\tau}})$, defined by (7.47). Let us just note that, as we have already observed, the exponents in formulas (7.47) are such that

$$\delta(\gamma+1) + (\gamma-3)(\delta+1) = 0,$$

which implies that

$$h(f,g) = (f^{\gamma+1}g^{\gamma-3})^{\frac{1}{(\gamma-1)}} = c\Big(\frac{\rho}{T^{\frac{1}{(\gamma-1)}}}\Big)^2 1_{\{|v|^2\le bT\}}.$$

The constraint inequality becomes an equality for the functions f and g, and we obtain the identity

$$c\Big(\frac{\rho}{T^{\frac{1}{(\gamma-1)}}}\Big)^2 = \Sigma, \text{ or } T = \Big(\frac{\rho}{(\frac{\Sigma}{c})^{\frac{1}{2}}}\Big)^{(\gamma-1)},$$

which in turn gives j. □

7.3.4 Second-order-accurate schemes

The extension of these schemes to obtain second-order accuracy can be done in two ways. The first uses a Chapman–Enskog-type analysis for the time discretization and takes initial conditions f_1, g_1, whose Taylor expansions are higher-order approximations of the exact solutions (see the proof of Lemma 7.3) (we refer to Deshpande 1986, Perthame 1990–1992). Then, following the MUSCL approach, one takes piecewise linear functions for $\mathbf{U}(x,0) = \mathbf{U}_0(x)$.

The other method is simpler and can be extended easily to 2-dimensional systems: one divides the cell $(x_{j-\frac{1}{2}}, x_{j+\frac{1}{2}})$ into two parts and takes functions that are constant on a half cell: $\mathbf{U}_i^-$ on $(x_{j-\frac{1}{2}}, x_j)$ and $\mathbf{U}_i^+$ on $(x_j, x_{j+\frac{1}{2}})$. In one dimension, it is equivalent to introducing slopes; we can develop a MUSCL-type scheme, and it is easy to prove second-order accuracy. Indeed, we define

$$\text{(7.60)} \qquad \begin{cases} \rho_i^\pm = \rho_i \pm D\rho_i, \\ u_i^\pm = \overline{u}_i \pm Du_i, \\ s_i^\pm = \overline{s}_i \pm Ds_i, \end{cases}$$

where $\overline{u}_i, \overline{s}_i$ are second order close to u_i, s_i and will be chosen to guarantee conservation of momentum and energy. The limited variables are $(\rho, u, s = \frac{\rho^{(\gamma-1)}}{T} = \frac{\rho^\gamma}{p})$; u and s are not the conservative variables but are chosen for the limitation procedure with the aim of getting a maximum principle on the entropy. The increments or (loosely speaking) the slopes are obtained starting from a centered prediction of the slope that is slightly corrected, for instance

$$\text{(7.61)} \qquad \begin{cases} D\rho_i = \operatorname{sgn}(\rho_{i+1}-\rho_{i-1}) \min\Big(\frac{|\rho_{i+1}-\rho_{i-1}|}{4}, \rho_i\Big), \\ Du_i = \operatorname{sgn}(u_{i+1}-u_{i-1}) \min\Big(\frac{|u_{i+1}-u_{i-1}|}{4}, \sqrt{T_i/\gamma-1}\Big), \\ Ds_i = \operatorname{sgn}(s_{i+1}-s_{i-1}) \min\Big(\frac{|s_{i+1}-s_{i-1}|}{4}, \frac{s_i}{4}\Big). \end{cases}$$

The values $\overline{u}_i$ and $\overline{s}_i$ are second-order modifications of u_i, s_i computed in order that $\mathbf{U}_i^{\pm} = \left(\rho_i^{\pm}, \rho_i^{\pm} u_i^{\pm}, \rho_i^{\pm} e_i^{\pm} = \rho_i^{\pm} \frac{u_i^{\pm 2}}{2} + \frac{\rho_i^{\pm\gamma}}{s_i^{\pm}(\gamma-1)}\right)$ satisfy

$$\mathbf{U}_i = \frac{1}{2}(\mathbf{U}_i^+ + \mathbf{U}_i^-).$$

Thus

$$(7.62) \qquad \begin{cases} \rho_i^+ u_i^+ + \rho_i^- u_i^- = 2\rho_i u_i, \\ \dfrac{\rho_i^+ u_i^{+2}}{2} + \dfrac{\rho_i^- u_i^{-2}}{2} + \dfrac{\dfrac{(\rho_i^+)^\gamma}{s_i^+} + \dfrac{(\rho_i^-)^\gamma}{s_i^-}}{(\gamma - 1)} = 2E_i. \end{cases}$$

Substituting the expressions (7.60) for $\rho_i^{\pm}, u_i^{\pm}$, the first equation (7.62) yields

$$\overline{u}_j = u_j - Du_j \frac{D\rho_j}{\rho_j},$$

and the second equation gives in turn $\overline{s}_i$ as the positive root of a quadratic polynomial.

Thus, starting from $\mathbf{U}_i = (\rho_i, \rho_i u_i, \rho_i e_i)$, we have defined $\mathbf{U}_i^{\pm}$, which we use to derive the numerical flux. In fact, if we use a particular (second-order) Runge–Kutta 2-stage scheme in time, one can prove that the resulting scheme is second order in space and time and preserves the positivity of ρ and T. Thus, we consider the scheme

$$\begin{cases} \mathbf{U}_i^{n,1} = \mathbf{U}_i^n - \lambda\{\mathbf{F}_{i+\frac{1}{2}}^n - \mathbf{F}_{i-\frac{1}{2}}^n\}, \\ \mathbf{U}_i^{n,2} = \mathbf{U}_i^{n,1} - \lambda\{\mathbf{F}_{i+\frac{1}{2}}^{n,1} - \mathbf{F}_{i-\frac{1}{2}}^{n,1}\}, \\ \mathbf{U}_i^{n+1} = \dfrac{1}{2}(\mathbf{U}_i^n + \mathbf{U}_i^{n,2}), \end{cases}$$

where we have set

$$\begin{aligned} \mathbf{F}_{i+\frac{1}{2}} &= \mathbf{F}^+(\mathbf{U}_i^+) + \mathbf{F}^-(\mathbf{U}_{i+1}^-), \\ \mathbf{F}_{i+\frac{1}{2}}^1 &= \mathbf{F}^+(\mathbf{U}_i^{1+}) + \mathbf{F}^-(\mathbf{U}_{i+1}^{1-}), \end{aligned}$$

and $\mathbf{F}^{\pm}$ are defined in (7.50) and $\mathbf{U}_i^{1\pm}$ are computed from $\mathbf{U}_i^1$ through the formulas (7.60)–(7.62). With a further limitation on the only slope Δs,

$$|\Delta s_i| \leq \max\{s_{i-1}, s_i, s_{i+1}\} - s_i,$$

(and no limiter on ρ and u) one obtains, moreover, a maximum principle on $s_{i\pm\frac{1}{2}}$ up to second order terms, which yields a very satisfactory damping of the oscillations (see Khobalatte and Perthame 1994, and Perthame and Qiu 1994).

Remark 7.7. A first drawback concerning the use of kinetic schemes, which was already mentioned in Section 5, Remark 5.1 for all flux-splitting schemes, is the poor resolution of contact discontinuities and slip surfaces that are heavily smeared. Thus, the application to viscous external flows, where a precise capture of boundary layers is important, requires some modifications (see De Vuyst 1995).

Another limitation seems to be the lack of generality since, as we have presented them, they rely heavily on the underlying kinetic theory. However, they have been extended to more general equations of states (Coron and Perthame 1991), nonequilibrium flow (De Vuyst 1992, Villedieu 1994), multiphase (Zelmanse 1995), MHD (Khanfir 1995). □

Notes

In this chapter we have chosen to detail some of the finite difference schemes that are the most often mentioned and used in the applications to computational fluid dynamics, as well as some interesting variants or extensions of these methods. But we do not pretend to have an exhaustive survey of the field and we precise that more references will be given in the next chapter, in particular concerning applications other than gas dynamics. Apart from the kinetic schemes, they were already presented in the scalar case (and we refer to G.R., Chapter III and IV), but we could not develop all of the previously mentionned ones such as S^{α}_{β}-schemes (Lerat 1981, Peyret and Taylor 1983, Hirsch 1990), Godunov-type schemes (see references in Remark 3.2), F.C.T. (Kunhardt and Wu 1987, Salari and Steinberg 1994 and the references therein), Quasi-monotone, PPM (Colella and Woodward 1984), ENO (Shu and Osher 1989, Yang 1990), staggered mesh (Nessyahu and Tadmor 1990, Sanders and Weiser 1992, Huynh 1995). Neither do we introduce different methods such as Local Extremum Diminishing (Kim and Jameson 1995), compact schemes (Cockburn and Shu 1994), random choice methods (Glimm 1965, Chorin 1977), nor study the extension to implicit (Lerat 1990, Yee 1987, Yee, Warming, and Harten 1985, Yee, Klopfer, and Montagné 1990, Mulder and van Leer 1985, Liang and Chan 1989, Blunt and Rubin 1992, Collins et al. 1995), front or shock-tracking methods (Chern et al. 1986, Mao 1992, Charrier and Tessieras 1986, Henshaw 1987, Davis 1992, LeVeque and Shyue 1995, and the references therein), nonconservative schemes (Raviart and Sainsaulieu 1992, Karni 1992, Harabetian and Pego 1993, Kumbaro 1992, Colombeau et al. 1989, Adamczewski et al. 1990, Berger and Colombeau 1995), approximation of nonstrictly hyperbolic systems (Freistühler and Pitman 1992, Tveito and Winther 1995) or systems of mixed type (Stewart and Wendroff 1984, Shu 1992). Except in Sections 2 and 6, we have mainly considered the application to the equations of gas dynamics in Eulerian coordinates; for a Lagrangian approach, see Munz (1994), for detonation waves and reactive flows LeVeque and Shyue (1995), Hilditch and Colella (1995). We did not consider systems with source terms (Roe 1986, Leveque and Yee 1990, Sweby 1989, Mulder and van Leer 1985,

Glaister 1993, Mehlman 1991, Chalabi 1992, Griffiths et al. 1992, Bermudez and Vasquez 1994, Klingenstein 1994, Greenberg and Leroux 1996, Fey et al. 1992, Schroll et al. 1994) or relaxation (S. Jin 1995, Jin and Xin 1995, Bereux and Sainsaulieu 1995), viscosity (Harabetian 1992). For boundary conditions, a brief survey will be given in Chapter V. We did not take up the study of theoretical aspects of schemes such as nonlinear stability of discrete shocks (Liu and Xin 1993 and the references therein), or convergence and error estimates; in fact, concerning systems, few results are available (Tveito and Winther 1993, Chen and Liu 1993, Chen and Le Floch 1995, Isaacson and Temple 1995). We did not either get on to the effective implementation (Ajmani et al. 1994 for instance).

We mention other main references for the kinetic theory: Perthame and Pulvirenti (1993) and the references therein for the B.G.K. model; Lions, Perthame, and Tadmor (1994) for a kinetic formulation in the scalar case, and for p-systems Bardos (1987), Whitham (1974) Section 6.3. For (kinetic) schemes, see Deshpande (1986), Pullin (1980), Kaniel (1988), Harten, Lax, and van Leer (1985), Mandal and Deshpande (1988, 1994)(see also Coron and Perthame 1991, Reitz 1981, Macrossan 1989, Prendergast and Xu 1993, Yang and Huang 1995, Xu et al. 1995, Croisille 1990, De Vuyst 1994, Estivalezes and Villedieu 1996).

Several recent papers provide a comparative study of some shock capturing schemes (Sod 1978, Montagné, Yee, and Vinokur 1987, Chargy et al. 1990, Roberts 1990, Yang and Przekwas 1992, Quirk 1992, Menne et al. 1995, Hannapel et al. 1995, Lin 1995; and Rider 1994 for methods in Lagrangian coordinates).

We have tried to present the underlying theory, computations and results as simply as possible and to mention the main contributors to the subject. We shall not list all of the references here, especially since it is impossible to quote all the authors who have contributed to recent developments in the field. We just quote some important contributions that have not yet been given: first of all the very complete text of Hirsch 1990; also many interesting papers such as Colella (1985), Hall (1990), LeVeque (1988), Toro (1989).

IV

The case of multidimensional systems

1 Generalities on multidimensional hyperbolic systems

1.1 Definitions

We shall mainly consider in this section a two-dimensional $p \times p$ hyperbolic system

$$\frac{\partial \mathbf{u}}{\partial t} + \frac{\partial}{\partial x}\mathbf{f}(\mathbf{u}) + \frac{\partial}{\partial y}\mathbf{g}(\mathbf{u}) = \mathbf{0}. \tag{1.1}$$

We shall not give much detail on the theoretical aspects of the problem, which are not fully understood and are more complex than in the scalar case $p = 1$ or in the one-dimensional case. Let us just recall the definition of hyperbolicity (see the Introduction, Section 1) for a system in dimension d, which can be written

$$\frac{\partial \mathbf{u}}{\partial t} + \sum_{j=1}^{d} \frac{\partial}{\partial x_j}\mathbf{f}_j(\mathbf{u}) = \mathbf{0}, \tag{1.2a}$$

where $\mathbf{u} = (u_1, ..., u_p)^T \in \Omega \subset \mathbb{R}^p$; $\mathbf{f}_j(\mathbf{u}) = (f_{1j}(\mathbf{u}), ..., f_{pj}(\mathbf{u}))^T \in \mathbb{R}^p$, $j = 1, ..., d$; and $\mathbf{x} \in \mathbb{R}^d, t \in \mathbb{R}+$. We shall also use the more compact form

$$\frac{\partial \mathbf{u}}{\partial t} + \operatorname{div} \mathbb{F}(\mathbf{u}) = \mathbf{0}, \quad \mathbb{F} = (\mathbf{f}_1, ..., \mathbf{f}_d)^T. \tag{1.2b}$$

We define for $\mathbf{u} = (u_1, ...u_p)^T \in \Omega$ and $\boldsymbol{\omega} = (\omega_1, ...\omega_d)^T \in \mathbb{R}^d$ the $p \times p$ matrix $\mathbf{A}(\mathbf{u}, \boldsymbol{\omega})$ "in the direction $\boldsymbol{\omega}$" by

$$\mathbf{A}(\mathbf{u}, \boldsymbol{\omega}) = \sum_{j=1}^{d} \mathbf{f}_j'(\mathbf{u})\omega_j, \tag{1.3}$$

where $\mathbf{f}_j'(u) = (\partial f_{ij}/\partial u_k)_{1 \leq i,k \leq p}$ is the Jacobian matrix of $\mathbf{f}_j(\mathbf{u})$. With shorthand notation, we shall also write

$$\mathbf{A}(\mathbf{u}, \boldsymbol{\omega}) = \mathbb{F}'(\mathbf{u}) \cdot \boldsymbol{\omega}.$$

Definition 1.1

The system (1.2) is called hyperbolic *if for any* $\mathbf{u} \in \Omega$ *and any direction* $\boldsymbol{\omega} \in \mathbb{R}^d$, $|\boldsymbol{\omega}| = 1$, *the matrix* $\mathbf{A}(\mathbf{u}, \boldsymbol{\omega})$ *defined by (1.3) has p real eigenvalues*

$$\lambda_1(\mathbf{u}, \boldsymbol{\omega}) \le \lambda_2(\mathbf{u}, \boldsymbol{\omega}) \le \dots \le \lambda_p(\mathbf{u}, \boldsymbol{\omega}),$$

with a complete family of (right) eigenvectors $\mathbf{r}_k(\mathbf{u}, \boldsymbol{\omega})$,

$$\mathbf{A}(\mathbf{u}, \boldsymbol{\omega})\mathbf{r}_k(\mathbf{u}, \boldsymbol{\omega}) = \lambda_k(\mathbf{u}, \boldsymbol{\omega})\mathbf{r}_k(\mathbf{u}, \boldsymbol{\omega}).$$

Moreover, the system is *strictly hyperbolic* at a state $\mathbf{u}$ if the eigenvalues are distinct.

The notions of *genuine nonlinearity* ($\forall \mathbf{u}$, $D_{\mathbf{u}}\lambda_k(\mathbf{u}, \boldsymbol{\omega}) \cdot \mathbf{r}_k(\mathbf{u}, \boldsymbol{\omega}) \neq 0$) and *linear degeneracy* ($\forall \mathbf{u}$, $D_{\mathbf{u}}\lambda_k(\mathbf{u}, \boldsymbol{\omega}) \cdot \mathbf{r}_k(\mathbf{u}, \boldsymbol{\omega}) = 0$) can then be defined for the kth field $\lambda_k(\mathbf{u}, \boldsymbol{\omega})$ in the direction $\boldsymbol{\omega}$, as well as the *k-Riemann invariants* (i.e., a function $w : \mathbb{R}^p \to \mathbb{R}$ such that $Dw(\mathbf{u}) \cdot \mathbf{r}_k(\mathbf{u}, \boldsymbol{\omega}) = 0$). In particular, if the kth field is linearly degenerate, $\lambda_k(\cdot, \boldsymbol{\omega})$ is a k-Riemann invariant.

We shall also denote as usual the eigenvectors of $\mathbf{A}(\mathbf{u}, \boldsymbol{\omega})^T$ by $\mathbf{l}_k(\mathbf{u}, \boldsymbol{\omega})$ and normalize the vectors as in Chapter I, (2.33), (2.34) (in particular, $D_{\mathbf{u}}\lambda_k(\mathbf{u}, \boldsymbol{\omega}) \cdot \mathbf{r}_k(\mathbf{u}, \boldsymbol{\omega}) = 1$ for a genuinely nonlinear field).

The system is symmetrizable if there exists a matrix $\mathbf{S}(\mathbf{u})$, smoothly varying with $\mathbf{u}$, such that the matrices $\mathbf{S}(\mathbf{u})\mathbf{A}_j(\mathbf{u})$ where $\mathbf{A}_j(\mathbf{u}) = \mathbf{f}_j'(u)$ are symmetric, $j = 1, \dots, d$. This property holds if we assume that the system admits a strictly convex entropy (Godunov and Mock's theorem, see Introduction, Theorem 3.1; see Harten 1983 for Euler equations, Godunov 1986 for further examples).

Example 1.1. Ideal gas isentropic dynamics. Let us consider the system in dimension $d = 2$

$$\begin{cases} \dfrac{\partial \rho}{\partial t} + \dfrac{\partial}{\partial x}(\rho u) + \dfrac{\partial}{\partial y}(\rho v) = 0, \\[2mm] \dfrac{\partial}{\partial t}(\rho u) + \dfrac{\partial}{\partial x}(\rho u^2 + p) + \dfrac{\partial}{\partial y}(\rho u v) = 0, \\[2mm] \dfrac{\partial}{\partial t}(\rho v) + \dfrac{\partial}{\partial x}(\rho u v) + \dfrac{\partial}{\partial y}(\rho v^2 + p) = 0, \end{cases} \tag{1.4}$$

with $p = p(\rho) > 0$ (p is the pressure). Setting $\mathbf{U} = (\rho, u, v)^T$, the system (1.4) is of the form (1.1). We find the eigenvalues by working with the nonconservative variables $\mathbf{V} = (p, u, v)^T$,

$$\begin{cases} \dfrac{\partial p}{\partial t} + u\dfrac{\partial p}{\partial x} + v\dfrac{\partial p}{\partial y} + \rho p'\Big(\dfrac{\partial u}{\partial x} + \dfrac{\partial v}{\partial y}\Big) = 0, \\[2mm] \rho\Big(\dfrac{\partial u}{\partial t} + u\dfrac{\partial u}{\partial x} + v\dfrac{\partial u}{\partial y}\Big) + \dfrac{\partial p}{\partial x} = 0, \\[2mm] \rho\Big(\dfrac{\partial v}{\partial t} + u\dfrac{\partial v}{\partial x} + v\dfrac{\partial v}{\partial y}\Big) + \dfrac{\partial p}{\partial y} = 0. \end{cases}$$

The matrix $\mathbf{A}(\mathbf{V}, \boldsymbol{\omega})$ is given by

$$\mathbf{A}(\mathbf{V}, \boldsymbol{\omega}) = \begin{pmatrix} \mathbf{u} \cdot \boldsymbol{\omega} & \rho p' \omega_1 & \rho p' \omega_2 \\ \omega_1 / \rho & \mathbf{u} \cdot \boldsymbol{\omega} & 0 \\ \omega_2 / \rho & 0 & \mathbf{u} \cdot \boldsymbol{\omega} \end{pmatrix},$$

where we have set $\mathbf{u} = (u, v)^T$, and

$$\mathbf{u} \cdot \boldsymbol{\omega} = u\omega_1 + v\omega_2$$

denotes the scalar product of the two vectors (which we shall also write $\mathbf{u}^T\boldsymbol{\omega}$). Assuming $p'(\rho) > 0$, the matrix $\mathbf{A}(\mathbf{V}, \boldsymbol{\omega})$ has 3 real eigenvalues, which are given by

$$\lambda_0 = \mathbf{u} \cdot \boldsymbol{\omega}, \quad \lambda_\pm = \mathbf{u} \cdot \boldsymbol{\omega} \pm \sqrt{p'(\rho)}|\boldsymbol{\omega}| = \mathbf{u} \cdot \boldsymbol{\omega} \pm \sqrt{p'(\rho)}$$

for $|\boldsymbol{\omega}| = 1$, and is thus strictly hyperbolic. We shall set

$$c^2 = p'(\rho),$$

so that $\lambda_\pm = \mathbf{u} \cdot \boldsymbol{\omega} \pm c$. Assuming more specifically $p(\rho) = A\rho^\gamma, \gamma > 1$, we note that $c^2 = p'(\rho) = \frac{\gamma p}{\rho}$. The corresponding eigenvectors can then be chosen as

$$\mathbf{r}_0(\mathbf{V}, \boldsymbol{\omega}) = (0, -\omega_2, \omega_1)^T,$$

which is independent of $\mathbf{u}$ and yields a linearly degenerate field since $(-\omega_2, \omega_1)$ is orthogonal to $\boldsymbol{\omega}$, and

$$\mathbf{r}_\pm(\mathbf{V}, \boldsymbol{\omega}) = (c\rho, \pm\omega_1, \pm\omega_2)^T.$$

The associated fields are genuinely nonlinear provided $\rho p'' + 2p' \neq 0$ since

$$D_{\mathbf{V}}\lambda_\pm(\mathbf{V}, \boldsymbol{\omega}) \cdot \mathbf{r}_\pm(\mathbf{V}, \boldsymbol{\omega}) = \pm \frac{(\rho p'' + 2p')}{2p'}$$

(which holds for a polytropic gas, in which case the expression is > 0, see (2.30) in Chapter I). □

We shall be especially interested in the one-dimensional "projected equations"

$$\frac{\partial \mathbf{u}}{\partial t} + \frac{\partial}{\partial \zeta}(\mathbb{F} \cdot \boldsymbol{\omega})(\mathbf{u}) = \mathbf{0}, \tag{1.5a}$$

where

$$\zeta = \mathbf{x} \cdot \boldsymbol{\omega} = \sum_{j=1}^{d} x_j \omega_j, \quad \mathbb{F} \cdot \boldsymbol{\omega} = \sum_{j=1}^{d} \mathbf{f}_j(\mathbf{u})\omega_j. \tag{1.5b}$$

Note that ζ is the variable in the direction of $\boldsymbol{\omega}$, which appears for instance when we solve a Riemann problem where the initial data are given on each side of the set $\{\zeta = 0\}$ (line in the two-dimensional case, plane in dimension

$d = 3$). The system (1.5a) is obtained when we look for a solution $\mathbf{u}$ of (1.2) that depends only on $\mathbf{x} \cdot \boldsymbol{\omega}$,

$$\mathbf{u}(\mathbf{x}, t) = \mathbf{v}(\mathbf{x} \cdot \boldsymbol{\omega}, t), \tag{1.5c}$$

where $\mathbf{v}(\zeta, t) : \mathbb{R} \times \mathbb{R}_+ \to \mathbb{R}^p$. We find that the Jacobian is

$$(\mathbb{F} \cdot \boldsymbol{\omega})'(\mathbf{v}) = \mathbf{A}(\mathbf{v}, \boldsymbol{\omega}),$$

and the eigenvalues are again the $\lambda_k(\cdot, \boldsymbol{\omega})$. The system (1.2) is thus hyperbolic iff the one-dimensional projected equations (1.5) are hyperbolic for any $\boldsymbol{\omega} \in \mathbb{R}^d$.

1.2 *Characteristics*

We first give the following definition (where $\varphi : \mathbb{R}^d \times [0, \infty[\to \mathbb{R}$).

Definition 1.2

A (hyper)surface Σ in $\mathbb{R}^d \times [0, \infty[$ with equation $\varphi(\mathbf{x}, t) = 0$ is characteristic *for the system (1.2) at a point $(\mathbf{x}_0, t_0)$ if the matrix*

$$\frac{\partial \varphi}{\partial t} \mathbf{I} + \sum_{j=1}^{d} \mathbf{f}_j'(\mathbf{u}) \frac{\partial \varphi}{\partial x_j}$$

is singular at this point.

In dimension $d = 1$, Σ is a curve; in dimension $d = 2$, Σ is a surface. We shall say that Σ is characteristic if Σ is characteristic at each point. Setting

$$n_t = \frac{\partial \varphi}{\partial t}, \quad \boldsymbol{\nu} = \left(\frac{\partial \varphi}{\partial x_j}\right)^T \in \mathbb{R}^d, \quad \mathbf{n} = (n_t, \boldsymbol{\nu}) \in \mathbb{R}^{d+1},$$

we know that $\mathbf{n}$ is a normal vector to Σ, and we can suppose without loss of generality that the vector $\boldsymbol{\nu}$ is a unit vector in $\mathbb{R}^d$. If Σ is characteristic, we obtain the following lemma (where $\lambda_k(\mathbf{u}, \boldsymbol{\nu})$ denotes the kth eigenvalue of the matrix $\mathbf{A}(\mathbf{u}, \boldsymbol{\nu})$ defined by (1.3)).

Lemma 1.1

Consider a characteristic surface Σ for system (1.2) and let $\mathbf{n} = (n_t, \boldsymbol{\nu})$, where $\boldsymbol{\nu}$ is a unit vector in $\mathbb{R}^d$, denote an outward normal vector to Σ. There exists locally an index k, $1 \le k \le p$, such that

$$n_t = -\lambda_k(\mathbf{u}, \boldsymbol{\nu}).$$

Proof. By definition, if Σ is characteristic at a point $(\mathbf{x}, t)$, the matrix

$$M(\mathbf{u}, \mathbf{n}) = n_t \mathbf{I} + \mathbb{F}'(\mathbf{u}) \cdot \boldsymbol{\nu} = n_t \mathbf{I} + \mathbf{A}(\mathbf{u}, \boldsymbol{\nu}) \tag{1.6}$$

is not invertible, which implies that $-n_t$ is an eigenvalue of $\mathbf{A}(\mathbf{u}, \boldsymbol{\nu})$,

$$n_t = -\lambda_k(\mathbf{u}, \boldsymbol{\nu}), \tag{1.7}$$

and

$$M(\mathbf{u}, \mathbf{n}) = \mathbf{A}(\mathbf{u}, \boldsymbol{\nu}) - \lambda_k(\mathbf{u}, \boldsymbol{\nu})\mathbf{I}.$$

By a continuity argument, the index k, which a priori depends on the point $(\mathbf{x}, t)$, will be the same for all points in a neighborhood. □

Of course, in the one-dimensional case, Definition 1.2 coincides with the usual ones given in the Introduction or in Chapter I: a curve $x = \varphi(t)$ is characteristic if $\frac{\partial \varphi}{\partial t} - a(u(x,t)) = 0$ in the scalar case (resp. $\frac{\partial \varphi}{\partial t} - \lambda_k(\mathbf{u}(\mathbf{x}, t)) = 0$ for a one-dimensional system).

Remark 1.1. For an initial boundary value problem in dimension $d = 2$, one often says that the boundary of a domain $\mathcal{O}$ in $\mathbb{R}^2$ is "noncharacteristic" if $\mathbf{A}(\mathbf{u}, \boldsymbol{\nu}) = \mathbf{f}'(\mathbf{u})\nu_1 + \mathbf{g}'(\mathbf{u})\nu_2$ is invertible, where $\boldsymbol{\nu}$ is the normal to $\partial\mathcal{O}$. Let $\varphi(x, y) = 0$ denote the equation of $\partial\mathcal{O}$, and consider the cylindrical domain Q in $\mathbb{R}^2 \times [0, \infty[$ built on $\mathcal{O}$. Then $\partial\mathcal{O}$ is noncharacteristic iff the lateral surface Γ of the cylinder Q is noncharacteristic in the sense of Definition 1.2. Indeed, this results from Lemma 1.1 since the normal to Γ is $(0, \frac{\partial \varphi}{\partial x}, \frac{\partial \varphi}{\partial y})^T = (0, \boldsymbol{\nu})$. □

A *characteristic equation* is obtained as follows: one makes an appropriate linear combination of the equations of the system (1.2),

$$\boldsymbol{\alpha} \cdot \Big\{ \frac{\partial \mathbf{u}}{\partial t} + \sum_j \frac{\partial}{\partial x_j} \mathbf{f}_j(\mathbf{u}) \Big\} = 0, \quad \boldsymbol{\alpha} = (\alpha_i)^T \in \mathbb{R}^p,$$

or

$$\sum_{i=1}^{p} \alpha_i \Big\{ \frac{\partial u_i}{\partial t} + \sum_{j,k=1}^{d} \frac{\partial f_{ij}}{\partial u_k} \frac{\partial u_k}{\partial x_j} \Big\} = 0,$$

in order that the resulting equation contains only derivatives of each u_k in directions lying on a d-dimensional surface, instead of $d + 1$ as in (1.2) (for $d \geq 2$, we cannot in general impose that the directional derivatives have the same direction). We write equivalently that they are normal to the normal vector to a surface Σ; if the equation of Σ is $\varphi(\mathbf{x}, t) = 0$, its normal vector is, as we have just seen, $(n_t = \frac{\partial \varphi}{\partial t}, \boldsymbol{\nu} = (\frac{\partial \varphi}{\partial x_j})) \in \mathbb{R} \times \mathbb{R}^d$. The directional derivative for each u_k is $(\alpha_k, \Sigma_{ij} \alpha_i \frac{\partial f_{ij}}{\partial u_k} \mathbf{e}_j) \in \mathbb{R} \times \mathbb{R}^d$ ($\mathbf{e}_j$ is the standard basis of $\mathbb{R}^d$), and thus we impose

$$\alpha_k \frac{\partial \varphi}{\partial t} + \sum_{i,j} \alpha_i \frac{\partial f_{ij}}{\partial u_k} \frac{\partial \varphi}{\partial x_j} = 0, \quad k = 1, ..., p.$$

This is a $p \times p$ linear system in the α_i, that has a nontrivial solution. This supposes that the determinant vanishes, i.e.,

$$\det \Big(\frac{\partial \varphi}{\partial t} \mathbf{I} + \sum_j \mathbf{f}_j'(\mathbf{u}) \frac{\partial \varphi}{\partial x_j} \Big) = 0,$$

which is the definition of a characteristic surface. Therefore, for a characteristic equation, at any point $(\mathbf{x}, t)$, the directions in which all u_k $(1 \leq k \leq p)$ are differentiated lie in the tangent space to a surface Σ that is characteristic at this point (see Holt 1977 for examples).

The main property is that singularities in the solution propagate along characteristic surfaces. Indeed, writing that a solution $\mathbf{u}$, which is assumed to be continuous across Σ, has discontinuous derivatives leads to exactly the same system as above (Whitham 1974, Section 5.9, Courant and Friedrichs 1976, Chapter II, no. 32).

If we can find p independent equations, the system is in characteristic form. In the multidimensional case, it is not obvious that a simplification may occur if the number of different directions involved for each characteristic equation is too important. However, in dimension $d = 2$, setting for $\boldsymbol{\omega} = (\omega_1, \omega_2)^T \in \mathbb{R}^2$,

$$\boldsymbol{\omega}^{\perp} = (-\omega_2, \omega_1),$$

which is the unit vector directly orthogonal to $\boldsymbol{\omega}$, we obtain the following result.

Lemma 1.2

Consider a characteristic surface Σ and let $\mathbf{n} = (n_t, \boldsymbol{\nu})$, where $n_t = -\lambda_k(\mathbf{u}, \boldsymbol{\nu})$ and $\boldsymbol{\nu}$ is a unit vector in $\mathbb{R}^2$, denote an outward normal vector to Σ. A smooth solution $\mathbf{u}$ of (1.1) satisfies

$$\mathbf{l}_k^T(\mathbf{u}, \boldsymbol{\nu}) \frac{d\mathbf{u}}{ds_k} + (\mathbf{l}_k^T(\mathbf{u}, \boldsymbol{\nu}) \mathbf{A}(\mathbf{u}, \boldsymbol{\nu}^{\perp}))(\boldsymbol{\nu}^{\perp} \cdot \mathbf{grad}\, \mathbf{u}) = 0, \tag{1.8}$$

where $\mathbf{l}_k(\mathbf{u}, \boldsymbol{\omega})$ denote the eigenvectors of $\mathbf{A}(\mathbf{u}, \boldsymbol{\omega})^T$ and

$$\frac{d\mathbf{u}}{ds_k} = \frac{\partial \mathbf{u}}{\partial t} + \lambda_k(\mathbf{u}, \boldsymbol{\nu}) \Big(\nu_1 \frac{\partial \mathbf{u}}{\partial x} + \nu_2 \frac{\partial \mathbf{u}}{\partial y} \Big).$$

Proof. Note first that by (1.7), the vector $(n_t, \boldsymbol{\nu})$, which is normal to Σ, is orthogonal to $(1, \lambda_k(\mathbf{u}, \boldsymbol{\nu})\boldsymbol{\nu})$ and $(0, \boldsymbol{\nu}^{\perp})$; thus, each equation in (1.8) contains only derivatives in directions that lie in the plane tangent to Σ; they are particular characteristic equations.

Now, assuming that $\mathbf{u}$ is a smooth solution, we have

$$\frac{\partial \mathbf{u}}{\partial t} + \frac{\partial \mathbf{f}(\mathbf{u})}{\partial x} + \frac{\partial \mathbf{g}(\mathbf{u})}{\partial y} = \frac{\partial \mathbf{u}}{\partial t} + \mathbf{f}'(\mathbf{u}) \frac{\partial \mathbf{u}}{\partial x} + \mathbf{g}'(\mathbf{u}) \frac{\partial \mathbf{u}}{\partial y}.$$

For any unit vector $\boldsymbol{\omega} \in \mathbb{R}^2$ and any $\mathbf{w} \in \mathbb{R}^2$, since $\mathbf{w} = (\boldsymbol{\omega} \cdot \mathbf{w})\boldsymbol{\omega} + (\boldsymbol{\omega}^{\perp} \cdot \mathbf{w})\boldsymbol{\omega}^{\perp}$, we have the obvious relation

$$\mathbf{w} \cdot \mathbf{u} = (\boldsymbol{\omega} \cdot \mathbf{w})(\boldsymbol{\omega} \cdot \mathbf{u}) + (\boldsymbol{\omega}^{\perp} \cdot \mathbf{w})(\boldsymbol{\omega}^{\perp} \cdot \mathbf{u}), \tag{1.9}$$

which yields, setting $\mathbb{F}(\mathbf{u}) = (\mathbf{f}(\mathbf{u}), \mathbf{g}(\mathbf{u}))^T$ and $\mathbf{grad}\, \mathbf{u} = \left(\frac{\partial \mathbf{u}}{\partial x}, \frac{\partial \mathbf{u}}{\partial y} \right)^T$,

$$\mathbb{F}'(\mathbf{u}) \cdot \mathbf{grad}\, \mathbf{u} = (\mathbb{F}'(\mathbf{u}) \cdot \boldsymbol{\omega})(\mathbf{grad}\, \mathbf{u} \cdot \boldsymbol{\omega}) + (\mathbb{F}'(\mathbf{u}) \cdot \boldsymbol{\omega}^{\perp})(\mathbf{grad}\, \mathbf{u} \cdot \boldsymbol{\omega}^{\perp}).$$

Thus, smooth solutions satisfy for any unit vector $\boldsymbol{\omega} \in \mathbb{R}^2$

$$\frac{\partial \mathbf{u}}{\partial t} + (\mathbb{F}'(\mathbf{u}) \cdot \boldsymbol{\omega})(\mathbf{grad}\,\mathbf{u} \cdot \boldsymbol{\omega}) + (\mathbb{F}'(\mathbf{u}) \cdot \boldsymbol{\omega}^{\perp})(\mathbf{grad}\,\mathbf{u} \cdot \boldsymbol{\omega}^{\perp}) = \mathbf{0},$$

and in particular for $\boldsymbol{\omega} = \boldsymbol{\nu} = (\frac{\partial \varphi}{\partial x}, \frac{\partial \varphi}{\partial y})^T$,

$$\frac{\partial \mathbf{u}}{\partial t} + \mathbf{A}(\mathbf{u}, \boldsymbol{\nu})(\mathbf{grad}\,\mathbf{u} \cdot \boldsymbol{\nu}) + \mathbf{A}(\mathbf{u}, \boldsymbol{\nu}^{\perp})(\mathbf{grad}\,\mathbf{u} \cdot \boldsymbol{\nu}^{\perp}) = \mathbf{0}. \tag{1.10}$$

Now, (1.10) implies

$$\begin{aligned}
&\mathbf{l}_k(\mathbf{u}, \boldsymbol{\nu}) \cdot \Big\{ \frac{\partial \mathbf{u}}{\partial t} + \operatorname{div} \mathbb{F}(\mathbf{u}) \Big\} \\
&= \mathbf{l}_k^T(\mathbf{u}, \boldsymbol{\nu}) \Big\{ \frac{\partial \mathbf{u}}{\partial t} + \lambda_k(\mathbf{u}, \boldsymbol{\nu}) \boldsymbol{\nu} \cdot \mathbf{grad}\,\mathbf{u} + \mathbf{A}(\mathbf{u}, \boldsymbol{\nu}^{\perp})(\boldsymbol{\nu}^{\perp} \cdot \mathbf{grad}\,\mathbf{u}) \Big\} = 0.
\end{aligned}$$

Consider the differentiation along the integral curves $C_k(\boldsymbol{\nu})$ of the differential system

$$\frac{d\mathbf{x}}{dt} = \lambda_k(\mathbf{u}, \boldsymbol{\nu}) \boldsymbol{\nu}. \tag{1.11}$$

These curves are the integral curves of the vector field $\lambda_k(\mathbf{u}, \boldsymbol{\nu})\boldsymbol{\nu}$, which is tangent to Σ, and they lie on Σ. Indeed, we can assume that the curve is parametrized by t, and we have then

$$\frac{dx}{dt} = \lambda_k(\mathbf{u}, \boldsymbol{\nu}) \nu_1, \qquad \frac{dy}{dt} = \lambda_k(\mathbf{u}, \boldsymbol{\nu}) \nu_2,$$

and hence

$$\frac{d}{dt} \varphi(t, x(t), y(t)) = \frac{\partial \varphi}{\partial t} + \lambda_k(\mathbf{u}, \boldsymbol{\nu}) \Big\{ \nu_1 \frac{\partial \varphi}{\partial x} + \nu_2 \frac{\partial \varphi}{\partial y} \Big\} = 0.$$

The differentiation operator along $C_k(\boldsymbol{\nu})$ is given by

$$\frac{d}{ds_k} = \frac{\partial}{\partial t} + \lambda_k(\mathbf{u}, \boldsymbol{\nu})(\boldsymbol{\nu} \cdot \nabla), \tag{1.12}$$

i.e.,

$$\frac{d\mathbf{u}}{ds_k} = \frac{\partial \mathbf{u}}{\partial t} + \lambda_k(\mathbf{u}, \boldsymbol{\nu}) \Big(\frac{\partial \mathbf{u}}{\partial x} \nu_1 + \frac{\partial \mathbf{u}}{\partial y} \nu_2 \Big),$$

which gives the desired result. □

Remark 1.2. In the scalar case ($p = 1$), for a smooth solution u of (1.2), we can introduce the characteristic curves as in the one-dimensional case (Introduction, Section 2); they are the integral curves $t \to \mathbf{x}(t)$ of the ordinary differential system

$$\frac{d\mathbf{x}}{dt} = \mathbf{a}(u(\mathbf{x}, t)), \quad \mathbf{a} = \mathbb{F}' = (f_1', \ldots, f_d')^T,$$

i.e.,

$$\frac{dx_j}{dt} = a_j(u(\mathbf{x},t)), \quad j = 1, ..., d, \; a_j = f_j'.$$

We check easily that

$$\frac{d}{dt}(u(\mathbf{x}(t),t)) = 0,$$

so that again u is constant along characteristics which are thus straight lines. We can then follow the method of characteristics and prove the existence of a smooth solution for t small enough; if there exists $\mathbf{y}$ such that $\mathrm{div}_{\mathbf{x}}\mathbf{a}(u_0(\mathbf{y})) < 0$, the critical time is

$$t^* = \left\{- \min_{y\in\mathbb{R}^d} \mathrm{div}\, \mathbf{a}(u_0(\mathbf{y}))\right\}^{-1}.$$

Otherwise, there exists a smooth global solution (for details, we refer to Majda 1984, Section 3.1). □

Remark 1.3. Notice that the definition (1.11) of $C_k(\boldsymbol{\nu})$ and (1.12) can be extended in any dimension d. In the particular case where we look for a smooth solution of the form (1.5c)

$$\mathbf{u}(\mathbf{x},t) = \mathbf{v}(\mathbf{x}\cdot\boldsymbol{\omega},t),$$

where $\boldsymbol{\omega}$ is a fixed unit vector in $\mathbb{R}^d$ and $\mathbf{v}(\zeta,t) : \mathbb{R}\times\mathbb{R}_+ \to \mathbb{R}^p$, we have seen that $\mathbf{v}$ is a solution of a system (1.5a) in dimension $d = 1$,

$$\frac{\partial\mathbf{v}}{\partial t} + \frac{\partial}{\partial\zeta}(\mathbb{F}\cdot\boldsymbol{\omega})(\mathbf{v}) = \frac{\partial\mathbf{v}}{\partial t} + \mathbf{A}(\mathbf{v}\cdot\boldsymbol{\omega})\frac{\partial\mathbf{v}}{\partial\zeta} = \mathbf{0}.$$

One can write this system in characteristic form following the arguments of Chapter I, Section 5. Setting for fixed $\boldsymbol{\omega}$

$$\frac{d\mathbf{v}}{ds_k} = \frac{\partial}{\partial t} + \lambda_k(\mathbf{v},\boldsymbol{\omega})\frac{\partial\mathbf{v}}{\partial\zeta},$$

which is the differentiation along the characteristic curve

$$\frac{d\zeta}{dt} = \lambda_k(\mathbf{v}(\zeta,t),\boldsymbol{\omega}), \tag{1.13}$$

we get

$$\mathbf{l}_k^T(\mathbf{v},\boldsymbol{\omega})\left\{\frac{\partial\mathbf{v}}{\partial t} + \mathbf{A}(\mathbf{v}\cdot\boldsymbol{\omega})\frac{\partial\mathbf{v}}{\partial\zeta}\right\} = \mathbf{l}_k^T(\mathbf{v},\boldsymbol{\omega})\frac{d\mathbf{v}}{ds_k} = 0.$$

Note that $\frac{d\mathbf{v}}{ds_k}$ does coincide with $\frac{d\mathbf{u}}{ds_k}$ defined by (1.12),

$$\begin{aligned}\frac{d\mathbf{u}}{ds_k} &= \frac{\partial\mathbf{u}}{\partial t} + \lambda_k(\mathbf{u},\boldsymbol{\omega})\boldsymbol{\omega}\cdot\mathbf{grad}\,\mathbf{u}\\ &= \frac{\partial\mathbf{v}}{\partial t} + \lambda_k(\mathbf{u},\boldsymbol{\omega})|\boldsymbol{\omega}|\frac{\partial\mathbf{v}}{\partial\zeta} = \frac{d\mathbf{v}}{ds_k},\end{aligned}$$

which justifies the use of the same notation. Also, (1.13) yields

$$\frac{d}{dt}(\mathbf{x}\cdot\boldsymbol{\omega}) = \lambda_k(\mathbf{u},\boldsymbol{\omega}),$$

which shows that the projection of $C_k(\boldsymbol{\omega})$ in the direction $\boldsymbol{\omega}$ (which lies in the (ζ, t)-plane $\boldsymbol{\omega}\mathbb{R}\times\mathbb{R}_+$ of the $(\mathbf{x}, t)$-space $\mathbb{R}^d\times\mathbb{R}_+$) coincides with the usual characteristic of the one-dimensional projected equation (1.5). □

1.3 Simple plane-waves

A plane-centered rarefaction wave is a "self-similar" continuous (piecewise C^1) solution

$$\mathbf{u}(\mathbf{x},t) = \mathbf{v}\Big(\frac{\mathbf{x}\cdot\boldsymbol{\omega}}{t}\Big), \tag{1.14}$$

where $\mathbf{v}:\mathbb{R}\to\mathbb{R}^p$ is a curve in $\mathbb{R}^p$, and $\boldsymbol{\omega}$ is a given constant vector in $\mathbb{R}^d$. We shall set

$$\xi = \frac{\mathbf{x}\cdot\boldsymbol{\omega}}{t}.$$

For any fixed $\bar{\xi}$, $\mathbf{u}$ is constant on the (hyper)planes $\mathbf{x}\cdot\boldsymbol{\omega} = \bar{\xi}t$.

As in the one-dimensional case (see Chapter I, Section 3.1, Theorem 3.1), we find that $\mathbf{v}$ must satisfy

$$(\mathbf{A}(\mathbf{v}(\xi),\boldsymbol{\omega}) - \xi\mathbf{I})\mathbf{v}'(\xi) = \mathbf{0},$$

so that either $\mathbf{v}$ is constant or $\mathbf{v}'$ is an eigenvector of $\mathbf{A}(\mathbf{v},\boldsymbol{\omega})$ associated with ξ,

$$\begin{cases} \mathbf{v}'(\xi) = \mathbf{r}_k(\mathbf{v}(\xi),\boldsymbol{\omega}), \\ \xi = \lambda_k(\mathbf{v}(\xi),\boldsymbol{\omega}). \end{cases}$$

This can be solved iff the kth field is genuinely nonlinear in the direction $\boldsymbol{\omega}$. If this is the case, we can connect two states $\mathbf{u}_L$ and $\mathbf{u}_R$ by a k-plane-centered rarefaction wave if $\mathbf{u}_R$ lies on the integral curve $\mathcal{R}_k(\mathbf{u}_L,\boldsymbol{\omega})$ of $\mathbf{r}_k(\cdot,\boldsymbol{\omega})$,

$$\begin{cases} \mathbf{v}'(\xi) = \mathbf{r}_k(\mathbf{v}(\xi),\boldsymbol{\omega}), \quad \lambda_k(\mathbf{u}_L,\boldsymbol{\omega}) < \xi < \lambda_k(\mathbf{u}_L,\boldsymbol{\omega}) + \varepsilon, \\ \mathbf{v}(\lambda_k(\mathbf{u}_L,\boldsymbol{\omega})) = \mathbf{u}_L, \end{cases} \tag{1.15}$$

which exists for ε small enough and such that $\lambda_k(\mathbf{v}(\xi),\boldsymbol{\omega})$ increases from $\lambda_k(\mathbf{u}_L,\boldsymbol{\omega})$ to $\lambda_k(\mathbf{u}_R,\boldsymbol{\omega})$.

When $\boldsymbol{\omega}$ varies, the k-rarefaction curves $\mathcal{R}_k(\mathbf{u}_L,\boldsymbol{\omega})$ form a (rarefaction) cone $\mathcal{R}_k(\mathbf{u}_L)$.

Example 1.1. (revisited) The rarefaction curve $\mathcal{R}_\pm(\mathbf{u}_L,\boldsymbol{\omega})$ is defined by

$$\mathbf{V}'(\xi) = \mathbf{r}_\pm(\mathbf{V}(\xi),\boldsymbol{\omega}) = \Big(p', \pm\omega_1\frac{\sqrt{p'(\rho)}}{\rho}, \pm\omega_2\frac{\sqrt{p'(\rho)}}{\rho}\Big).$$

We can parametrize the curve by ρ and, after elementary computations (similar to those of the p-system in Chapter I, Section 7), we find

$$u = u(\rho) = u_L \pm \omega_1 \int_{\rho_L}^{\rho} \sqrt{p'(r)}\,\frac{dr}{r},$$

$$v = v(\rho) = v_L \pm \omega_2 \int_{\rho_L}^{\rho} \sqrt{p'(r)}\,\frac{dr}{r},$$

and the rarefaction cone $\mathcal{R}_{\pm}(\mathbf{u}_L)$ is given in both cases by

$$(u-u_L)^2 + (v-v_L)^2 = \Big(\int_{\rho_L}^{\rho} \sqrt{p'(r)}\,\frac{dr}{r}\Big)^2.$$

Under the assumption made for genuine nonlinearity, $p' > 0$ and $\rho p'' + 2p' > 0$, we can discriminate between the fields. Indeed, the hypothesis $\rho p'' + 2p' > 0$ implies that λ_+ (resp. λ_-) increases (resp. decreases) along $\mathcal{R}_+(\mathbf{u}_L)$ (resp. $\mathcal{R}_-(\mathbf{u}_L)$) since

$$\frac{d\lambda_\pm}{d\rho} = D_{\mathbf{v}}\lambda_\pm \cdot \mathbf{r}_\pm = \pm\Big(\frac{p''}{2\sqrt{p'}} + \frac{\sqrt{p'}}{\rho}\Big) = \pm\frac{(\rho p'' + 2p')}{2\rho\sqrt{p'}}.$$

Given $\mathbf{u}_R \in \mathcal{R}_\pm(\mathbf{u}_L)$, it can be connected to $\mathbf{u}_L$ by a plane rarefaction + wave (resp. − wave) if $\rho_R > \rho_L$ (resp. $\rho_R < \rho_L$). □

As in the one-dimensional case (Chapter I, Section 3.2), we can look more generally for continuous piecewise C^1 solutions of the form

$$\mathbf{u}(\mathbf{x},t) = \mathbf{v}(\varphi(\mathbf{x}\cdot\boldsymbol{\omega}, t)), \tag{1.16}$$

where $\varphi(\zeta,t) : \mathbb{R}\times\mathbb{R} \to \mathbb{R}$, $\mathbf{v}(\varphi) : \mathbb{R} \to \mathbb{R}^p$, and $\boldsymbol{\omega}$ is a *given* constant vector in $\mathbb{R}^d$. Such a function is called a *simple plane-wave* solution: it is constant at any time t along the hyperplane perpendicular to the direction $\boldsymbol{\omega}$ and takes its values on a curve in $\mathbb{R}^p$. Substituting (1.16) into (1.2) gives

$$\mathbf{v}'\frac{\partial\varphi}{\partial t} + \mathbf{A}(\mathbf{v},\boldsymbol{\omega})\mathbf{v}'\frac{\partial\varphi}{\partial\zeta} = \mathbf{0}. \tag{1.17}$$

Following the arguments of Chapter I, Section 3.2, we solve (1.17) in two steps. First, we take for $\mathbf{v}(\varphi) : \mathbb{R} \to \mathbb{R}^p$ an integral curve of the nonlinear ordinary differential equation

$$\begin{cases} \mathbf{v}'(\varphi) = \mathbf{r}_k(\mathbf{v}(\varphi),\boldsymbol{\omega}), \\ \mathbf{v}(0) = \mathbf{v}_0. \end{cases} \tag{1.18}$$

Such a solution exists at least locally. Then, given $\mathbf{v} = \mathbf{v}(\cdot,\boldsymbol{\omega})$, φ satisfies the first-order quasi-linear equation

$$\begin{cases} \dfrac{\partial\varphi}{\partial t} + \lambda_k(\mathbf{v},\boldsymbol{\omega})\dfrac{\partial\varphi}{\partial\zeta}, & \zeta\in\mathbb{R}, t>0, \\ \varphi(\xi,0) = \varphi_0(\xi). \end{cases}$$

As in Chapter I, formula (3.20), we obtain that φ is constant along the characteristics given by (1.13),

$$\frac{d\zeta}{dt} = \lambda_k(\mathbf{v}(\varphi(\zeta,t)),\boldsymbol{\omega}),$$

which are thus straight lines in the (ζ,t)-plane, and

$$\varphi(\zeta,t) = \varphi_0(\zeta - \lambda_k(\mathbf{v}(\varphi(\zeta,t)),\boldsymbol{\omega})t)$$

so long as φ remains smooth. In the variables $(\mathbf{x},t)$, the above characteristics give (hyper)planes

$$\mathbf{x}\cdot\boldsymbol{\omega} = \lambda_k(\mathbf{u},\boldsymbol{\omega})t + C$$

along which $\mathbf{u}$ is constant. Note that these (hyper)planes are obviously characteristic in the sense of Definition 1.2. Finally, we have the implicit formula

$$\mathbf{u}(\mathbf{x},t) = \mathbf{v}(\varphi_0(\mathbf{x}\cdot\boldsymbol{\omega} - \lambda_k(\mathbf{u},\boldsymbol{\omega})t)).$$

We can study some particular simple plane-waves. On the one hand, assuming that the kth field is genuinely nonlinear in the direction $\boldsymbol{\omega}$, and normalizing the vector $\mathbf{r}_k(\mathbf{v},\boldsymbol{\omega})$ so that

$$D_{\mathbf{v}}\lambda_k(\mathbf{v},\boldsymbol{\omega})\cdot\mathbf{r}_k(\mathbf{v},\boldsymbol{\omega}) = 1,$$

we see that the equation in φ resumes in Burgers' equation (see Example 3.4, Chapter I). We recover as particular simple plane-waves the rarefaction waves studied above, corresponding to $\varphi(\zeta,t) = \frac{\zeta}{t}$. In the one-dimensional case $d = 1$, for a k-rarefaction wave, the characteristics of the kth field form a fan of straight lines in the (x,t)-plane (see Chapter I, Example 3.4). In the multidimensional case, for a plane rarefaction wave (1.14) connecting $\mathbf{u}_L$ and $\mathbf{u}_R$, the characteristics (1.18) form a converging pencil of (hyper)planes in the $(\mathbf{x},t)$-space passing through $t = 0$, $\mathbf{x}\cdot\boldsymbol{\omega} = 0$ (more generally $t = t_0$, $(\mathbf{x}-\mathbf{x}_0)\cdot\boldsymbol{\omega} = 0$).

On the other hand, assuming that the kth field is linearly degenerate in the direction $\boldsymbol{\omega}$, we see that $\lambda_k(\mathbf{v},\boldsymbol{\omega})$ is constant on a k-simple plane-wave since the variation of λ_k on the curve where $\mathbf{v}$ takes its values is given by

$$D_{\mathbf{v}}\lambda_k\cdot\mathbf{v}' = D_{\mathbf{v}}\lambda_k(\mathbf{v},\boldsymbol{\omega})\cdot\mathbf{r}_k(\mathbf{v},\boldsymbol{\omega}) = 0.$$

Thus, the functions

$$\mathbf{u}(\mathbf{x},t) = \mathbf{v}(\varphi_0(\mathbf{x}\cdot\boldsymbol{\omega} - \overline{\lambda}_k t)),\quad \overline{\lambda}_k = \lambda_k(\mathbf{u},\boldsymbol{\omega})$$

are k-simple plane-waves, and the characteristics (1.18) of the kth field form a pencil of parallel (hyper)planes

$$\mathbf{x}\cdot\boldsymbol{\omega} - \overline{\lambda}_k t = C.$$

Example 1.1. (revisited). The integral curve of $\mathbf{r}_0(\mathbf{u},\boldsymbol{\omega}) = (0,-\omega_2,\omega_1)^T = (0,\boldsymbol{\omega}^\perp)$ corresponding to the linearly degenerate field $\lambda_0(\mathbf{u},\boldsymbol{\omega}) = \mathbf{u}\cdot\boldsymbol{\omega}$

issued from $\mathbf{u}_L$ is a straight line (intersection of the planes $p = p_L$, or equivalently $\rho = \rho_L$ and $(u - u_L)\omega_1 + (v - v_L)\omega_2 = 0$) on which λ_0 is obviously constant ($\lambda_0(\mathbf{u}, \boldsymbol{\omega}) = \mathbf{u}_L \cdot \boldsymbol{\omega}$). The set spanned by these curves as $\boldsymbol{\omega}$ varies is the plane $\rho = \text{const.} = \rho_L$. □

Remark 1.4. The definition and existence of general rarefaction waves that are not plane simple-waves can be found in Alinhac (1989). □

Now we can also look for some particular *discontinuous* solutions.

1.4 Shock waves

Let us recall that a weak solution satisfies the Rankine–Hugoniot condition along the (hyper)surfaces of discontinuity Σ (see the Introduction, (2.8))

$$n_t[\mathbf{u}] + \sum_{j=1}^{d} [\mathbf{f}_j(\mathbf{u})] n_{x_j} = \mathbf{0}, \tag{1.19}$$

where $\mathbf{n} = (n_t, \mathbf{n_x})$ is the normal vector to Σ. We can define in particular *plane* shock-wave solutions

$$\mathbf{u}(\mathbf{x}, t) = \begin{cases} \mathbf{u}_L, & \mathbf{x} \cdot \boldsymbol{\omega} - \sigma t < 0, \\ \mathbf{u}_R, & \mathbf{x} \cdot \boldsymbol{\omega} - \sigma t > 0, \end{cases} \tag{1.20a}$$

which satisfy the Rankine–Hugoniot condition across the (hyper)plane $\Sigma = \{(\mathbf{x}, t) \in \mathbb{R}^d \times \mathbb{R}_+ / \mathbf{x} \cdot \boldsymbol{\omega} - \sigma t = 0\}$,

$$-\sigma(\mathbf{u}_R - \mathbf{u}_L) + \sum_{j=1}^{d} (\mathbf{f}_j(\mathbf{u}_R) - \mathbf{f}_j(\mathbf{u}_L))\omega_j = \mathbf{0}, \tag{1.20b}$$

or

$$-\sigma[\mathbf{u}] + [\mathbb{F}(\mathbf{u})] \cdot \boldsymbol{\omega} = \mathbf{0},$$

where $\mathbf{u}_L, \mathbf{u}_R$ are *constant* states. Assuming, for instance that the front propagates along the x_d-axis, i.e., choosing $\boldsymbol{\omega} = (0, 0..., 1)^T$, we get

$$\mathbf{u}(\mathbf{x}, t) = \begin{cases} \mathbf{u}_L, & x_d < \sigma t, \\ \mathbf{u}_R, & x_d > \sigma t, \end{cases}$$

$\Sigma = \{(\mathbf{x}, t) / x_d = \sigma t\}$, and $\mathbf{u}$ satisfies a one-dimensional jump condition

$$\sigma(\mathbf{u}_R - \mathbf{u}_L) + (\mathbf{f}_d(\mathbf{u}_R) - \mathbf{f}_d(\mathbf{u}_L)) = \mathbf{0}.$$

The theory of these shock waves is developed in Chapter I. Thus, for given $\boldsymbol{\omega}$ and $\mathbf{u}_L$, assuming that the system is strictly hyperbolic, the Rankine–Hugoniot set of $\mathbf{u}_L$ contains locally the k-shock curves $\mathcal{S}_k(\mathbf{u}_L, \boldsymbol{\omega})$ associated to a (simple) eigenvalue $\lambda_k(\mathbf{u}, \boldsymbol{\omega})$ (see Chapter I, Theorem 4.1). When $\boldsymbol{\omega}$ varies, the curves $\mathcal{S}_k(\mathbf{u}_L, \boldsymbol{\omega})$ form a "cone." The curve $\mathcal{S}_k(\mathbf{u}_L, \boldsymbol{\omega})$ is tangent to $\mathbf{r}_k(\mathbf{u}_L, \boldsymbol{\omega})$ at $\mathbf{u}_L$ and can be parametrized by

$$(1.21a) \qquad \mathbf{u}(\varepsilon, \boldsymbol{\omega}) = \mathbf{u}_L + \varepsilon\, \mathbf{r}_k(\mathbf{u}_L, \boldsymbol{\omega}) + O(\varepsilon^2),$$

$$(1.21b) \qquad \sigma = \lambda_k(\mathbf{u}_L, \boldsymbol{\omega}) + \frac{\varepsilon}{2} D_{\mathbf{u}}\lambda_k(\mathbf{u}_L, \boldsymbol{\omega}) \cdot \mathbf{r}_k(\mathbf{u}_L, \boldsymbol{\omega}) + O(\varepsilon^2).$$

Now, assume first that the kth field is genuinely nonlinear, i.e.,

$$D\lambda_k(\mathbf{u}, \boldsymbol{\omega}) \cdot \mathbf{r}_k(\mathbf{u}, \boldsymbol{\omega}) \neq 0.$$

Since $(n_t, \mathbf{n_x}) = (-\sigma, \boldsymbol{\omega})$ is normal to Σ, it follows from (1.21b) that the eigenvalues $\lambda_k(\mathbf{u}, \boldsymbol{\omega}) - \sigma$ of the matrix (1.6),

$$M(\mathbf{u}, \mathbf{n}) = \frac{\partial \varphi}{\partial t}\mathbf{I} + \sum_{j=1}^{d} \mathbf{A}_j(\mathbf{u}) \frac{\partial \varphi}{\partial x_j} = -\sigma \mathbf{I} + \mathbf{A}(\mathbf{u}, \boldsymbol{\omega}),$$

are $\neq 0$ (i.e., $M(\mathbf{u}, \mathbf{n})$ is invertible) for $\mathbf{u} = \mathbf{u}_L$ and $\mathbf{u}_R$, which means that Σ is not characteristic.

If the kth field is linearly degenerate, i.e.,

$$D\lambda_k(\mathbf{u}, \boldsymbol{\omega}) \cdot \mathbf{r}_k(\mathbf{u}, \boldsymbol{\omega}) = 0,$$

then $M(\mathbf{u}, \mathbf{n})$ is singular and,

$$\sigma = \lambda_k(\mathbf{u}_L, \boldsymbol{\omega}) = \lambda_k(\mathbf{u}_R, \boldsymbol{\omega}) = \overline{\lambda}_k,$$

and so the (hyper)plane $\mathbf{x} \cdot \boldsymbol{\omega} - \sigma t$ is characteristic. Thus, if $\mathbf{u}_R \in \mathcal{S}_k(\mathbf{u}_L, \boldsymbol{\omega})$, which is now an integral curve of $\mathbf{r}_k(\mathbf{u}, \boldsymbol{\omega})$, we have a k-contact discontinuity

$$\mathbf{u}(\mathbf{x}, t) = \begin{cases} \mathbf{u}_L, & \mathbf{x} \cdot \boldsymbol{\omega} - \overline{\lambda}_k t < 0, \\ \mathbf{u}_R, & \mathbf{x} \cdot \boldsymbol{\omega} - \overline{\lambda}_k t > 0. \end{cases}$$

For a planar shock connecting two states $\mathbf{u}_-$ and $\mathbf{u}_+$, we can impose the Lax entropy conditions. These conditions were obtained in the one-dimensional case (see Chapter I, Section 5) by considering the number of characteristics impinging on Σ or emerging from Σ considered as a boundary. This number corresponds to the number of positive and negative eigenvalues of the matrix $M(\mathbf{u}_\pm, \mathbf{n})$ defined by (1.6). For instance, if $p = d = 2$, a 1-planar shock connecting $\mathbf{u}_-$ and $\mathbf{u}_+$ satisfies the Lax entropy conditions if

$$\lambda_1(\mathbf{u}_+, \boldsymbol{\omega}) < \sigma < \lambda_1(\mathbf{u}_-, \boldsymbol{\omega}), \quad \sigma < \lambda_2(\mathbf{u}_+, \boldsymbol{\omega}),$$

and a 2-shock if

$$\lambda_2(\mathbf{u}_+, \boldsymbol{\omega}) < \sigma < \lambda_2(\mathbf{u}_-, \boldsymbol{\omega}), \quad \lambda_1(\mathbf{u}_-, \boldsymbol{\omega}) < \sigma.$$

In particular, if $\boldsymbol{\omega} = (0, 1)^T$, the $\lambda_i(\mathbf{u}, \boldsymbol{\omega})$, $i = 1, 2$, are in this case the eigenvalues of $\mathbf{A}(\mathbf{u}, \boldsymbol{\omega}) = \mathbf{f}_2'(\mathbf{u}) = \mathbf{g}'(\mathbf{u})$.

Remark 1.5. One can prove more generally the existence of multidimensional shock fronts (at least for sufficiently short times t) that are not plane shock waves. These are piecewise smooth solutions $\mathbf{u}(\mathbf{x},t) = \mathbf{u}_+(\mathbf{x},t)$, $\mathbf{u}_-(\mathbf{x},t)$ on each side of a hypersurface Σ in the $(\mathbf{x},t)$-space, satisfying (1.19) across Σ. For instance, in the 2-dimensional case, if we assume that, at least for short times, $\Sigma = \{(x,y,t), x = \sigma(y,t)\}$, with $\sigma(y,0) = 0$, then $\mathbf{u}(x,y,t) = \mathbf{u}_-(x,y,t)$ for $x < \sigma(y,t)$ and $\mathbf{u}(x,y,t) = \mathbf{u}_+(x,y,t)$ for $x > \sigma(y,t)$, and $\mathbf{u}_+$ and $\mathbf{u}_-$ are linked by the Rankine–Hugoniot condition across Σ.

The construction of a shock front, i.e., the existence of $\mathbf{u}_+$ satisfying (1.2) and (1.19), appears as a free boundary value problem since the surface Σ is not known in advance in the nonlinear case. If the equation of Σ is given by $\varphi(\mathbf{x},t) = 0$, and the normal to Σ by $(n_t, \mathbf{n_x}) = (\frac{\partial\varphi}{\partial t}, \frac{\partial\varphi}{\partial x_1}, \dots, \frac{\partial\varphi}{\partial x_d})^T$, we shall require first that Σ is not characteristic, which means, as we have already seen, that the matrix (1.6),

$$M(\mathbf{u},\mathbf{n}) = \frac{\partial\varphi}{\partial t}\mathbf{I} + \sum_{j=1}^{d} \mathbf{A}_j(\mathbf{u})\frac{\partial\varphi}{\partial x_j} = n_t\mathbf{I} + \mathbf{A}(\mathbf{u},\mathbf{n_x}),$$

is invertible for $\mathbf{u} = \mathbf{u}_+$ and $\mathbf{u}_-$. Following the example of planar shock waves, we can also impose a number of boundary conditions for more general shock fronts. However, they are no longer sufficient (for $d > 1$) to ensure that the problem is well-posed. Hence, stability conditions are needed (see Majda 1984).

The shock front solutions are constructed as progressing waves emanating from a "shock front" with initial data $\mathbf{u}_0(\mathbf{x}) = \mathbf{u}_{0+}(\mathbf{x}), \mathbf{u}_{0-}(\mathbf{x})$ given on each side of a surface Σ_0. The stability of these shock fronts is studied by perturbation of plane shock waves (see Majda 1984 for the case of strong shocks, Métivier 1989 for weak shocks). For the thorough study of a solution presenting two shock waves in the case $p = d = 2$, and of their interactions see Métivier (1986). □

2 The gas dynamics equations in two space dimensions

The Euler system of gas dynamics in two dimensions is given by

$$(2.1)\qquad \begin{cases} \dfrac{\partial\rho}{\partial t} + \dfrac{\partial}{\partial x}(\rho u) + \dfrac{\partial}{\partial y}(\rho v) = 0, \\ \dfrac{\partial}{\partial t}(\rho u) + \dfrac{\partial}{\partial x}(\rho u^2 + p) + \dfrac{\partial}{\partial y}(\rho uv) = 0, \\ \dfrac{\partial}{\partial t}(\rho v) + \dfrac{\partial}{\partial x}(\rho uv) + \dfrac{\partial}{\partial y}(\rho v^2 + p) = 0, \\ \dfrac{\partial}{\partial t}(\rho e) + \dfrac{\partial}{\partial x}((\rho e + p)u) + \dfrac{\partial}{\partial y}((\rho e + p)v) = 0 \end{cases}$$

(see Example 1.3 in the Introduction). The pressure p is related to ε and ρ by an equation of state of the form

$$p = p(\rho, \varepsilon), \quad e = \varepsilon + \frac{1}{2}(u^2 + v^2).$$

For a polytropic ideal gas, we have $p = (\gamma - 1)\rho\varepsilon$, where $\gamma > 1$ is a constant. When the flow is isentropic, the system (2.1) reduces to the system (1.4) of Example 1.1. Setting

$$\text{(2.2a)} \quad \mathbf{U} = \begin{pmatrix} \rho \\ \rho u \\ \rho v \\ \rho e \end{pmatrix}, \ \mathbf{f}(\mathbf{U}) = \begin{pmatrix} \rho u \\ \rho u^2 + p \\ \rho u v \\ (\rho e + p) u \end{pmatrix}, \ \mathbf{g}(\mathbf{U}) = \begin{pmatrix} \rho v \\ \rho u v \\ \rho v^2 + p \\ (\rho e + p) v \end{pmatrix},$$

the system (2.1) can be written in the general form (1.1),

$$\text{(2.2b)} \qquad \frac{\partial \mathbf{U}}{\partial t} + \frac{\partial}{\partial x}\mathbf{f}(\mathbf{U}) + \frac{\partial}{\partial y}\mathbf{g}(\mathbf{U}) = \mathbf{0}.$$

Note that the Jacobian matrices of $\mathbf{f}$ and $\mathbf{g}$ cannot be diagonalized simultaneously (but any linear combination of both can be diagonalized).

2.1 Entropy and entropy variables

Let us recall that the system is endowed with an entropy

$$\text{(2.3a)} \qquad \Phi(\mathbf{U}) = -\rho s,$$

with associated entropy flux

$$\text{(2.3b)} \qquad \Psi(\mathbf{U}) = -\rho s u \quad (\text{i.e., } \Psi_i = -\rho s u_i).$$

In these expressions, we consider the (thermodynamic) specific entropy s as a function $s = s(\tau, \varepsilon)$, which is strictly convex, and we have seen that $-\rho s(\frac{1}{\rho}, \frac{E}{\rho} - \frac{|\rho \mathbf{u}|^2}{2\rho^2})$ is then a strictly convex function of the conservative variables $\mathbf{U} = (\rho, \rho\mathbf{u}, E)^T$ (see the Introduction, Example 3.3, and Chapter II, Section 1). The system (2.1) is thus symmetrizable, and the entropy variables that symmetrize the system (see Theorem 3.2, Introduction) are given by the following lemma.

Lemma 2.1

The entropy variables

$$\mathbf{V}^T = \Phi'(\mathbf{U})$$

for the Euler system (2.2) are given by

$$\text{(2.4)} \qquad \mathbf{V}^T = T^{-1}\Big(h - Ts - \frac{|\mathbf{u}|^2}{2}, \mathbf{u}, -1\Big),$$

where h is the specific enthalpy, $h = \varepsilon + \frac{p}{\rho}$.

Proof. Indeed, in order to compute

$$\mathbf{V}^T = \Phi'(\mathbf{U}) = -\Big(\frac{\partial(\rho s)}{\partial \rho}, \frac{\partial(\rho s)}{\partial(\rho u)}, \frac{\partial(\rho s)}{\partial(\rho v)}, \frac{\partial(\rho s)}{\partial E}\Big),$$

where ρs is considered as a function of the conservative variables $\mathbf{U}$,

$$\rho s(\mathbf{U}) = \rho s\Big(\frac{1}{\rho}, \frac{E}{\rho} - \frac{|\rho \mathbf{u}|^2}{2\rho^2}\Big),$$

we start from the relation

$$Tds = d\varepsilon + pd\tau,$$

which gives $\frac{\partial s(\tau,\varepsilon)}{\partial \varepsilon} = \frac{1}{T}$ and $\frac{\partial s(\tau,\varepsilon)}{\partial \tau} = \frac{p}{T}$. We then compute the partial derivatives using the chain rule; for instance,

$$\begin{aligned}\frac{\partial(\rho s(\mathbf{U}))}{\partial \rho} &= s + \rho\frac{\partial s(\tau,\varepsilon)}{\partial \tau}\Big(-\frac{1}{\rho^2}\Big) + \frac{\partial s(\tau,\varepsilon)}{\partial \varepsilon}\Big(-\frac{E}{\rho^2} + \frac{|\mathbf{u}|^2}{\rho^3}\Big)\\ &= s + T^{-1}\Big(-\frac{p}{\rho} - \frac{E}{\rho} + |\mathbf{u}|^2\Big) = T^{-1}\Big(-h + Ts + \frac{|\mathbf{u}|^2}{2}\Big),\end{aligned}$$

where the specific entropy $h = \frac{p}{\rho} + \varepsilon$ was introduced in Chapter III, Section 3, and so on. This gives (2.4). □

Notice that setting

$$\mu = h - Ts = \varepsilon + p\tau - Ts,$$

we can also write (2.4) as

$$\mathbf{V}^T = T^{-1}\Big(\mu - \frac{|\mathbf{u}|^2}{2}, \mathbf{u}, -1\Big).$$

We have, moreover, the relation

$$Td(\rho s) = d(\rho\varepsilon) - \mu d\rho,$$

i.e., $\mu = -\frac{\partial(\rho s)(\rho,\rho\varepsilon)}{\partial \rho}$ (μ is the Gibbs potential). For a polytropic ideal gas, $T = \frac{\varepsilon}{C_v}$, $\gamma = \frac{C_p}{C_v}$, $s = C_v \log \varepsilon\tau^{(\gamma-1)}$ up to an additive constant, $h = C_p T$, and $\mu = T(C_p - s)$.

For the sake of completeness, let us give the expression of the conjugate (or polar) function $\Phi^\star$ (resp. $\Psi^\star$) of the entropy Φ (resp. Ψ), defined by

$$\Phi^\star(\mathbf{V}) = \mathbf{V}^T\mathbf{U}(\mathbf{V}) - \Phi(\mathbf{U}(\mathbf{V})) \tag{2.5a}$$

and

$$\Psi^\star(\mathbf{V}) = \mathbf{V}^T\mathbb{F}(\mathbf{U}(\mathbf{V})) - \Psi(\mathbf{U}(\mathbf{V})), \tag{2.5b}$$

i.e.,

$$\Psi = (\Psi_i), \quad \Psi_i^\star = \mathbf{V}^T\mathbf{f}_i - \Psi_i$$

($\Phi^\star$ and $\Psi^\star$ were defined in the proof of Theorem 3.2, Introduction). The functions $\Phi^\star$ and $\Psi^\star$ satisfy

$$D_{\mathbf{V}}\Phi^\star = \mathbf{V}^T\mathbf{U}'(\mathbf{V}) + \mathbf{U}(\mathbf{V}) - \Phi'\mathbf{U}'(\mathbf{V}) = \mathbf{U}(\mathbf{V})$$

and

$$D_{\mathbf{V}}\Psi^\star = \mathbf{V}^T\mathbb{F}'(\mathbf{U}(\mathbf{V}))\mathbf{U}'(\mathbf{V}) + \mathbf{F}(\mathbf{U}(\mathbf{V})) - \Psi'(\mathbf{U}(\mathbf{V}))\mathbf{U}'(\mathbf{V}) = \mathbf{F}(\mathbf{U}(\mathbf{V}))$$

since $\Psi' = \Phi'\mathbf{F}'$.

In the case of the gas dynamics equations, the expressions for Φ^* and Ψ^* are particularly simple.

Lemma 2.2

The functions $\Phi^\star$ and $\Psi^\star$ defined by (2.4) and (2.5) are given in the case of the gas dynamics equations by

$$\Phi^\star = \frac{p}{T}, \quad \Psi^\star = \frac{p\mathbf{u}}{T}.$$

Proof. The computations are very simple. First, we have by definition

$$\begin{aligned}\Phi^\star &= T^{-1}\Big(\mu - \frac{|\mathbf{u}|^2}{2}, \mathbf{u}, -1\Big)(\rho, \rho\mathbf{u}, E)^T + \rho s \\ &= T^{-1}\Big\{\rho\Big(\varepsilon + \frac{p}{\rho} - Ts\Big) - \rho\frac{|\mathbf{u}|^2}{2} + \rho|\mathbf{u}|^2 - E\Big\} + \rho s = \frac{p}{T}.\end{aligned}$$

Similarly, we find

$$\Psi_1^\star = T^{-1}\Big(\mu - \frac{|\mathbf{u}|^2}{2}, u, v, -1\Big)(\rho u, \rho u^2 + p, \rho uv, Eu + pu)^T + \rho su = u\frac{p}{T}$$

and

$$\Psi_2^\star = v\frac{p}{T}.$$

We refer to Bourdel et al. (1989), Mazet and Bourdel (1995), and Croisille (1990) for more general results and utilization of these formulas. □

2.2 Invariance of the Euler equations

Let us now study the invariance of the Euler equations through a Galilean transformation. We consider a frame $\mathcal{R}'$ moving with uniform speed $\mathbf{v}$ w.r.t. the reference frame $\mathcal{R}$ of $\mathbb{R}^2 \times \mathbb{R}_+$; thus, the new independent variables in $\mathcal{R}'$ are

$$\mathbf{x}' = \mathbf{x} - \mathbf{v}t, \quad t' = t.$$

Looking at the dependent variables in the frame $\mathcal{R}'$, the density ρ, pressure p, and internal energy ε are invariant,

$$\rho' = \rho, \quad \varepsilon' = \varepsilon, \quad p' = p,$$

while the velocity $\mathbf{u}$ is transformed into

$$\mathbf{u}' = \mathbf{u} - \mathbf{v}.$$

The expression "Galilean invariance" means that setting

$$\mathbf{U}' = (\rho', \rho'\mathbf{u}', E'),$$

$\mathbf{U}'$ satisfies the Euler equations

$$\frac{\partial \mathbf{U}'}{\partial t'} + \frac{\partial \mathbf{f}(\mathbf{U}')}{\partial x'} + \frac{\partial \mathbf{g}(\mathbf{U}')}{\partial y'} = \mathbf{0}$$

with the same functions $\mathbf{f}$ and $\mathbf{g}$ as in (2.2), i.e.,

$$\mathbf{f}(\mathbf{U}') = (\rho' u', \rho' u'^2 + p', \rho' u' v', (\rho' e' + p')u')^T,$$

with the analog for $\mathbf{g}(\mathbf{U}')$.

Remark 2.1. Assume a general pressure law

$$p = p(\mathbf{U}) = p'(\mathbf{U}').$$

Then, Galilean invariance requires in particular that the functions p and p' coincide. In fact, this presupposes a pressure law of the form $p = p(\rho, \rho\varepsilon)$. Indeed, $p = p'$ implies, setting $\mathbf{m} = \rho\mathbf{u}$, $E = \rho\varepsilon + \frac{\rho|\mathbf{u}|^2}{2}$,

$$p(\rho, \mathbf{m}, E) = p\Big(\rho, \mathbf{m} - \rho\mathbf{v}, E + \rho\frac{1}{2}|\mathbf{v}|^2 - \rho\mathbf{v}^T\mathbf{u}\Big);$$

differentiating this identity w.r.t. $\mathbf{v}$ gives

$$0 = -\rho\frac{\partial p}{\partial m_i} + \rho(v_i - u_i)\frac{\partial p}{\partial E}, \quad 1 \le i \le 2$$

(for ease of notation, we have set $\mathbf{u} = (u_i)$), and at $\mathbf{v} = \mathbf{0}$ it yields

$$-\rho\frac{\partial p}{\partial m_i} + m_i\frac{\partial p}{\partial E} = 0.$$

Integrating this differential system, we obtain that p is a function of the form

$$p(\rho, \mathbf{m}, E) = p\Big(\rho, E - \frac{|\mathbf{m}|^2}{2\rho}\Big) = p(\rho, \rho\varepsilon),$$

which gives the result. □

Lemma 2.3

Assuming an equation of state of the form

$$p = p(\rho, \rho\varepsilon), \tag{2.6}$$

the Euler equations (2.1) are invariant under Galilean transformations.

Proof. The invariance of the equations is easily established using the rules

$$\begin{aligned} \text{grad}_{\mathbf{x}'} &= \text{grad}_{\mathbf{x}}, \\ \frac{\partial}{\partial t'} &= \frac{\partial}{\partial t} + \mathbf{v} \cdot \text{grad}_{\mathbf{x}} = \frac{\partial}{\partial t} + u \frac{\partial}{\partial x} + v \frac{\partial}{\partial y}. \end{aligned}$$

Let us check, for instance, the first equation:

$$\begin{aligned} \frac{\partial \rho}{\partial t'} + \text{div}_{\mathbf{x}'} \rho \mathbf{u}' &= \frac{\partial \rho}{\partial t} + \mathbf{v} \cdot \text{grad}_{\mathbf{x}} \rho + \text{div}_{\mathbf{x}} \rho (\mathbf{u} - \mathbf{v}) \\ &= \frac{\partial \rho}{\partial t} + \mathbf{v} \cdot \text{grad}_{\mathbf{x}} \rho + \text{div}_{\mathbf{x}} \rho \mathbf{u} - \mathbf{v} \cdot \text{grad}_{\mathbf{x}} \rho \\ &= \frac{\partial \rho}{\partial t} + \text{div}_{\mathbf{x}} \rho \mathbf{u} = 0. \end{aligned}$$

The other computations are similar. □

Remark 2.2. We have chosen to write the equation of state in the form (2.6) rather than in the usual form $p = p(\rho, \varepsilon)$ because of Remark 2.1; moreover, as we have already seen, it leads to simpler algebraic computations (see (3.26), Chapter III, Section 3.2). □

We come now to the rotational invariance of the Euler equations which is used extensively in the numerical schemes. Let us first specify some notations. A rotation of angle θ in the (x, y)-plane corresponds to the matrix

$$R = \begin{pmatrix} \cos\theta & -\sin\theta \\ \sin\theta & \cos\theta \end{pmatrix}. \tag{2.7}$$

Setting

$$\mathbf{X} = (x, y)^T = R\tilde{\mathbf{X}}, \quad \tilde{\mathbf{X}} = (\zeta, \tau)^T = (\mathbf{X} \cdot \mathbf{n}, \mathbf{X} \cdot \mathbf{n}^\perp)^T,$$

and

$$\mathbf{n} = R\,\mathbf{e}_1 = (\cos\theta, \sin\theta)^T, \quad \mathbf{n}^\perp = R\mathbf{e}_2 = (-\sin\theta, \cos\theta)^T, \tag{2.8}$$

we see geometrically that $(\mathbf{n}, \mathbf{n}^\perp)$ are the new basis vectors obtained from $(\mathbf{e}_1, \mathbf{e}_2)$ by the rotation R. If M is a point with coordinates (x, y) (in the basis $(\mathbf{e}_1, \mathbf{e}_2)$), then (ζ, τ) are the coordinates of M in the basis $(\mathbf{n}, \mathbf{n}^\perp)$.

Similarly, we can write

$$\mathbf{u} = u\mathbf{e}_1 + v\mathbf{e}_2 = (\mathbf{u} \cdot \mathbf{n})\mathbf{n} + (\mathbf{u} \cdot \mathbf{n}^\perp)\mathbf{n}^\perp,$$

and the vector

$$\tilde{\mathbf{u}} = (u_n, u_\tau)^T = (\mathbf{u} \cdot \mathbf{n}, \mathbf{u} \cdot \mathbf{n}^\perp)^T \tag{2.9}$$

gives the coefficients of the velocity vector in the basis $(\mathbf{n}, \mathbf{n}^\perp)$, which we can translate precisely by

$$\tilde{\mathbf{u}} = R^{-1}\mathbf{u} \circ R, \quad \text{i.e., } \tilde{\mathbf{u}}(\tilde{\mathbf{X}}) = \tilde{\mathbf{u}}(R^{-1}\mathbf{X}) = R^{-1}\mathbf{u}(\mathbf{X}) = R^{-1}\mathbf{u}(R\tilde{\mathbf{X}}),$$

or algebraically (equality of matrices),

$$(u_n, u_\tau)^T = \tilde{\mathbf{u}} = R^{-1}\mathbf{u} = R^{-1}(u, v)^T.$$

Now, since ρ, ε, and e are not changed, we define

$$\tilde{\mathbf{U}}(\zeta, \tau, t) = (\rho, \rho\tilde{\mathbf{u}}, \rho e)^T,$$

and the invariance means that $\tilde{\mathbf{U}}$ is a solution of the Euler equations, i.e., we have

$$\frac{\partial \tilde{\mathbf{U}}}{\partial t} + \frac{\partial \mathbf{f}(\tilde{\mathbf{U}})}{\partial \zeta} + \frac{\partial \mathbf{g}(\tilde{\mathbf{U}})}{\partial \tau} = \mathbf{0}$$

with the same functions $\mathbf{f}$ and $\mathbf{g}$ as in (2.2),

$$\begin{aligned} \mathbf{f}(\tilde{\mathbf{U}}) &= (\rho u_n, \rho(u_n)^2 + p, \quad \rho u_n u_\tau, (\rho e + p)u_n))^T, \\ \mathbf{g}(\tilde{\mathbf{U}}) &= (\rho u_\tau, \rho u_n u_\tau, \rho(u_\tau)^2 + p, (\rho e + p)u_\tau))^T. \end{aligned}$$

Lemma 2.4

The Euler equations (2.1) are invariant under rotation.

Proof. We consider a rotation with angle θ in the (x, y)-plane, corresponding to the matrix (2.7), and we use the notations (2.8), (2.9). The computations can be put in a rather compact form using matrix notations. For the reader's convenience, we present a detailed proof. Since

$$\tilde{\mathbf{X}} = (\zeta, \tau)^T = R^{-1}\mathbf{X},$$

setting

$$\nabla = \left(\frac{\partial}{\partial x}, \frac{\partial}{\partial y}\right)^T, \quad \tilde{\nabla} = \left(\frac{\partial}{\partial \zeta}, \frac{\partial}{\partial \tau}\right)^T,$$

we have

$$\tilde{\nabla} = R^{-1}\nabla, \tag{2.10a}$$

which implies

$$\tilde{\nabla} \cdot \tilde{\mathbf{u}} = \nabla \cdot \mathbf{u}, \tag{2.10b}$$

i.e.,

$$\frac{\partial u}{\partial x} + \frac{\partial v}{\partial y} = \operatorname{div}_{\mathbf{X}}\mathbf{u} = \nabla \cdot \mathbf{u} = R\tilde{\nabla} \cdot R\tilde{\mathbf{u}} = \tilde{\nabla} \cdot \tilde{\mathbf{u}} = \operatorname{div}_{\tilde{\mathbf{X}}}\tilde{\mathbf{u}} = \frac{\partial u_n}{\partial \zeta} + \frac{\partial u_\tau}{\partial \tau}$$

(the dot "$\cdot$" means the scalar product of vectors in $\mathbb{R}^2$, which is invariant under rotations, *i.e.* $R\mathbf{a} \cdot R\mathbf{b} = \mathbf{a} \cdot \mathbf{b}$). This gives the invariance of the first equation of (2.1) and also of the last since $|\mathbf{u}| = |\tilde{\mathbf{u}}|$ and $\nabla \cdot p\mathbf{u} = \tilde{\nabla} \cdot p\tilde{\mathbf{u}}$.

Consider now the second and third equations of the system (2.1), which we can write

$$\frac{\partial}{\partial t}(\rho\mathbf{u})(\mathbf{X}, t) + \boldsymbol{\nabla} \cdot (\rho u\mathbf{u} + p\mathbf{e}_1, \rho v\mathbf{u} + p\mathbf{e}_2)^T,$$

where we use the notation

$$\nabla \cdot (\mathbf{v}, \mathbf{w})^T = \frac{\partial \mathbf{v}}{\partial x} + \frac{\partial \mathbf{w}}{\partial y}. \tag{2.11}$$

Multiplying on the left by R^{-1}, we get

$$\frac{\partial}{\partial t}(\rho\tilde{\mathbf{u}})(\tilde{\mathbf{X}}, t) + R^{-1}\nabla \cdot (\rho u\mathbf{u} + p\mathbf{e}_1, \rho v\mathbf{u} + p\mathbf{e}_2)^T.$$

Now, using the obvious identities

$$R^{-1}\nabla \cdot (\mathbf{v}, \mathbf{w})^T = R^{-1}\Big(\frac{\partial \mathbf{v}}{\partial x} + \frac{\partial \mathbf{w}}{\partial y}\Big)$$
$$= \Big(\frac{\partial R^{-1}v}{\partial x} + \frac{\partial R^{-1}\mathbf{w}}{\partial y}\Big) = \nabla \cdot (R^{-1}\mathbf{v}, R^{-1}\mathbf{w})^T$$

and

$$\nabla \cdot (\mathbf{v} + \mathbf{v}', \mathbf{w} + \mathbf{w}')^T = \frac{\partial(\mathbf{v} + \mathbf{v}')}{\partial x} + \frac{\partial(\mathbf{w} + \mathbf{w}')}{\partial y}$$
$$= \nabla \cdot (\mathbf{v}, \mathbf{w})^T + \nabla \cdot (\mathbf{v}', \mathbf{w}')^T,$$

we get

$$R^{-1}\nabla\cdot(\rho u\mathbf{u} + p\mathbf{e}_1, \rho v\mathbf{u} + p\mathbf{e}_2)^T$$
$$= \nabla \cdot (R^{-1}(\rho u\mathbf{u} + p\mathbf{e}_1), R^{-1}(\rho v\mathbf{u} + p\mathbf{e}_2))^T$$
$$= \nabla \cdot (\rho u\, R^{-1}\mathbf{u}, \rho v R^{-1}\mathbf{u})^T + R^{-1}\nabla \cdot (p\mathbf{e}_1, p\mathbf{e}_2))^T.$$

Using again the invariance of the scalar product in $\mathbb{R}^2$ under rotation, we have

$$\nabla \cdot (\rho u\tilde{\mathbf{u}}, \rho v\tilde{\mathbf{u}})^T = \tilde{\nabla} \cdot (\rho u_n\tilde{\mathbf{u}}, \rho u_\tau\tilde{\mathbf{u}})^T,$$

which yields

$$R^{-1}\nabla \cdot (\rho u\mathbf{u} + p\mathbf{e}_1, \rho v\mathbf{u} + p\mathbf{e}_2)^T = \tilde{\nabla} \cdot (\rho u_n\tilde{\mathbf{u}}, \rho u_\tau\tilde{\mathbf{u}}) + \tilde{\nabla} \cdot (p\mathbf{e}_1, p\mathbf{e}_2)^T$$
$$= \tilde{\nabla} \cdot (\rho u_n\tilde{\mathbf{u}} + p\mathbf{e}_1, \rho u_\tau\tilde{\mathbf{u}} + p\mathbf{e}_2)$$

and proves the invariance.

Let us define the transformation R^{-1} on vectors of $\mathbb{R}^4 = \mathbb{R} \times \mathbb{R}^2 \times \mathbb{R}$ by

$$R^{-1}(a, \mathbf{b}, c) = (a, R^{-1}\mathbf{b}, c). \tag{2.12}$$

Thus

$$R^{-1}\mathbf{U} = (\rho, \rho\tilde{\mathbf{u}}, \rho e)^T = \tilde{\mathbf{U}}.$$

To prove the rotational invariance of the Euler equations, we have checked that

$$R^{-1}\nabla \cdot \mathbb{F}(\mathbf{U}) = \tilde{\nabla} \cdot \mathbb{F}(\tilde{\mathbf{U}}), \quad \text{where } \mathbb{F} = (\mathbf{f}, \mathbf{g})^T$$

with the notations (2.10), (2.11). □

Remark 2.3. We can choose to expand the above identities. Let us check for instance that

$$\frac{\partial \rho}{\partial t} + \frac{\partial(\rho u_n)}{\partial \zeta} + \frac{\partial(\rho u_\tau)}{\partial \tau} = 0.$$

Since $\mathbf{X}' = R^{-1}\mathbf{X}$, we have

$$\frac{\partial}{\partial \zeta} = \cos\,\theta \frac{\partial}{\partial x} + \sin\,\theta \frac{\partial}{\partial y}, \qquad \frac{\partial}{\partial \tau} = -\sin\,\theta \frac{\partial}{\partial x} + \cos\,\theta \frac{\partial}{\partial y},$$

and

$$\begin{aligned}\frac{\partial(\rho u_n)}{\partial \zeta} + \frac{\partial(\rho u_\tau)}{\partial \tau} &= \Big(\cos\,\theta \frac{\partial}{\partial x} + \sin\,\theta \frac{\partial}{\partial y}\Big)(\rho(u \cos\,\theta + v \sin\,\theta)) \\ &\quad + \Big(-\sin\,\theta \frac{\partial}{\partial x} + \cos\,\theta \frac{\partial}{\partial y}\Big)(\rho(-u \sin\,\theta + v \cos\,\theta)) \\ &= \frac{\partial(\rho u)}{\partial x} + \frac{\partial(\rho v)}{\partial y}.\end{aligned}$$

Consider now the second equation:

$$\begin{aligned}&\frac{\partial}{\partial t}(\rho u_n) + \frac{\partial}{\partial \zeta}(\rho(u_n)^2 + p) + \frac{\partial}{\partial \tau}(\rho u_n u_\tau) = \frac{\partial}{\partial t}(\rho(u \cos\,\theta + v \sin\,\theta)) \\ &+ \Big(\cos\,\theta \frac{\partial}{\partial x} + \sin\,\theta \frac{\partial}{\partial y}\Big)(\rho(u_n)^2 + p) - \Big(\sin\,\theta \frac{\partial}{\partial x} - \cos\,\theta \frac{\partial}{\partial y}\Big)(\rho u_n u_\tau) \\ &= \frac{\partial}{\partial t}(\rho(u \cos\,\theta + v \sin\,\theta)) + \frac{\partial}{\partial x} \rho u_n(u_n \cos\,\theta - u_\tau \sin\,\theta) \\ &\quad + \frac{\partial}{\partial y} \rho u_n(u_n \sin\,\theta + u_\tau \cos\,\theta) + \Big(\cos\,\theta \frac{\partial p}{\partial x} + \sin\,\theta \frac{\partial p}{\partial y}\Big) \\ &= \frac{\partial}{\partial t}(\rho(u \cos\,\theta + v \sin\,\theta)) + \frac{\partial}{\partial x} \rho(u \cos\,\theta + v \sin\,\theta)u \\ &\quad + \frac{\partial}{\partial y} \rho(u \cos\,\theta + v \sin\,\theta)v + \Big(\cos\,\theta \frac{\partial p}{\partial x} + \sin\,\theta \frac{\partial p}{\partial y}\Big).\end{aligned}$$

We obtain a combination of the second and third equations in (2.2),

$$\begin{aligned}&\cos\,\theta \Big\{\frac{\partial}{\partial t}(\rho u) + \frac{\partial}{\partial x}(\rho u^2 + p) + \frac{\partial}{\partial y}(\rho u v)\Big\} \\ &\quad + \sin\,\theta \Big\{\frac{\partial}{\partial t}(\rho v) + \frac{\partial}{\partial x}(\rho u v) + \frac{\partial}{\partial y}(\rho v^2 + p)\Big\} = 0.\end{aligned}$$

The other computations are similar.

We can also use the notation of the projected equations introduced in (1.5) and (1.9), i.e., in dimension $d = 2$,

$$\begin{aligned}\frac{\partial}{\partial \zeta} &= \cos\,\theta \frac{\partial}{\partial x} + \sin\,\theta \frac{\partial}{\partial y} = \mathbf{n} \cdot \nabla, \\ \frac{\partial}{\partial \tau} &= -\sin\,\theta \frac{\partial}{\partial x} + \cos\,\theta \frac{\partial}{\partial y} = \mathbf{n}^\perp \cdot \nabla.\end{aligned}$$

As we have already seen in (1.9), we can write

$$\nabla \cdot \mathbf{u} = (\mathbf{n} \cdot \nabla)(\mathbf{n} \cdot \mathbf{u}) + (\mathbf{n}^{\perp} \cdot \nabla)(\mathbf{n}^{\perp} \cdot \mathbf{u}) = \tilde{\nabla} \cdot \tilde{\mathbf{u}},$$

which gives (2.10)b. Also, for $\mathbb{F} = (\mathbf{f}, \mathbf{g})^T$,

$$\begin{aligned} \operatorname{div}_{\mathbf{X}} \cdot \mathbb{F}(\mathbf{U}(\mathbf{X}, t)) &= \boldsymbol{\nabla} \cdot \mathbb{F} = (\mathbf{n} \cdot \boldsymbol{\nabla})(\mathbf{n} \cdot \mathbb{F}) + (\mathbf{n}^{\perp} \cdot \boldsymbol{\nabla})(\mathbf{n}^{\perp} \cdot \mathbb{F}) \\ &= \tilde{\boldsymbol{\nabla}}(\mathbf{n} \cdot \mathbb{F}^{\mathbf{T}}, \mathbf{n}^{\perp} \cdot \mathbb{F}^{\mathbf{T}})^{\mathbf{T}}, \end{aligned}$$

where

$$\mathbf{n} \cdot \mathbb{F} = \cos\,\theta \mathbf{f} + \sin\,\theta \mathbf{g}, \quad \mathbf{n}^{\perp} \cdot \mathbb{F} = -\sin\,\theta \mathbf{f} + \cos\,\theta \mathbf{g}.$$

Thus, for any system $\frac{\partial}{\partial t}\mathbf{U} + \frac{\partial}{\partial x}\mathbf{f}(\mathbf{U}) + \frac{\partial}{\partial y}\mathbf{g}(\mathbf{U}) = \mathbf{0}$, we have

$$\frac{\partial \mathbf{U}}{\partial t} + \frac{\partial(\mathbf{n} \cdot \mathbb{F})}{\partial \zeta} + \frac{\partial(\mathbf{n}^{\perp} \cdot \mathbb{F})}{\partial \tau} = \mathbf{0}.$$

We now focus on the corresponding system for the Euler equations. Consider the second and third equations, which we can write

$$\frac{\partial}{\partial t}(\rho \mathbf{u}) + \frac{\partial}{\partial \zeta}(\mathbf{n} \cdot \mathbf{F}_{23}) + \frac{\partial}{\partial \tau}(\mathbf{n}^{\perp} \cdot \mathbf{F}_{23}) = \mathbf{0}$$

with $\mathbf{F}_{23} = (\mathbf{f}_{23}, \mathbf{g}_{23}),\ \mathbf{f}_{23} = (f_2, f_3)$, i.e.,

$$\begin{aligned} \mathbf{n} \cdot \mathbf{F}_{23} &= \mathbf{n} \cdot (\rho u \mathbf{u} + p \mathbf{e}_1, \rho v \mathbf{u} + p \mathbf{e}_2)^T \\ &= \cos\,\theta(\rho u \mathbf{u} + p \mathbf{e}_1) + \sin\,\theta(\rho v \mathbf{u} + p \mathbf{e}_2), \\ \mathbf{n}^{\perp} \cdot \mathbf{F}_{23} &= \mathbf{n}^{\perp}(\rho u \mathbf{u} + p \mathbf{e}_1, \rho v \mathbf{u} + p \mathbf{e}_2)^T \\ &= -\sin\,\theta(\rho u \mathbf{u} + p \mathbf{e}_1) + \cos\,\theta(\rho v \mathbf{u} + p \mathbf{e}_2). \end{aligned}$$

In order to obtain the system in the dependent variables $\tilde{\mathbf{U}}$, we take the scalar product of this system of two equations with $\mathbf{n}$ and $\mathbf{n}^{\perp}$ (which corresponds to applying R^{-1}),

$$\text{(2.13}a) \quad \mathbf{n} \cdot \left\{ \frac{\partial}{\partial t}(\rho \mathbf{u})(\mathbf{X}, t) + \tilde{\boldsymbol{\nabla}} \cdot (\mathbf{n} \cdot \mathbf{F}_{23}, \mathbf{F}_{23} \cdot \mathbf{n}^{\perp})^T \right\} = 0,$$

$$\text{(2.13}b) \quad \mathbf{n}^{\perp} \cdot \left\{ \frac{\partial}{\partial t}(\rho \mathbf{u})(\mathbf{X}, t) + \tilde{\boldsymbol{\nabla}} \cdot (\mathbf{n} \cdot \mathbf{F}_{23}, \mathbf{F}_{23} \cdot \mathbf{n}^{\perp})^T \right\} = 0.$$

We obtain for the equation (2.13a)

$$\begin{aligned} \mathbf{n} \cdot \left\{ \frac{\partial}{\partial t}(\rho \mathbf{u}) + \tilde{\boldsymbol{\nabla}} \cdot (\mathbf{n} \cdot \mathbf{F}_{23}, \mathbf{n}^{\perp} \cdot \mathbf{F}_{23})^T \right\} &= \frac{\partial}{\partial t}(\rho \mathbf{u} \cdot \mathbf{n}) \\ &+ \frac{\partial}{\partial \zeta}\mathbf{n} \cdot (\mathbf{n} \cdot \mathbf{F}_{23}) + \frac{\partial}{\partial \tau}\mathbf{n} \cdot (\mathbf{n}^{\perp} \cdot \mathbf{F}_{23}) \end{aligned}$$

and

$$\begin{aligned} \mathbf{n} \cdot (\mathbf{F}_{23} \cdot \mathbf{n}) &= \mathbf{n} \cdot \{\cos\,\theta(\rho u \mathbf{u} + p \mathbf{e}_1) + \sin\,\theta(\rho v \mathbf{u} + p \mathbf{e}_2)\} \\ &= \cos\,\theta(\rho u \mathbf{u} \cdot \mathbf{n} + p \mathbf{e}_1 \cdot \mathbf{n}) + \sin\,\theta(\rho v \mathbf{u} \cdot \mathbf{n} + p \mathbf{e}_2 \cdot \mathbf{n}), \end{aligned}$$

which is the scalar product with $\mathbf{n}$ of the vector $\rho\mathbf{u}(\mathbf{u}\cdot\mathbf{n})+p\mathbf{n}$ with component

$$(\rho u\mathbf{u}\cdot\mathbf{n}+p\mathbf{e}_1\cdot\mathbf{n},\rho v\mathbf{u}\cdot\mathbf{n}+p\mathbf{e}_2\cdot\mathbf{n}),$$

and this scalar product gives

$$(\rho\mathbf{u}(\mathbf{u}\cdot\mathbf{n})+p\mathbf{n})\cdot\mathbf{n}=\rho u_n^2+p=f_2(\tilde{\mathbf{U}}).$$

Similarly,

$$\begin{aligned}\mathbf{n}\cdot(\mathbf{F}_{23}\cdot\mathbf{n}^{\perp})&=\mathbf{n}\cdot\{-\sin\,\theta(\rho u\mathbf{u}+p\mathbf{e}_1)+\cos\,\theta(\rho v\mathbf{u}+p\mathbf{e}_2)\}\\&=(-\sin\,\theta(\rho u\mathbf{u}\cdot\mathbf{n}+p\mathbf{e}_1\cdot\mathbf{n}+\cos\,\theta(\rho v\mathbf{u}\cdot\mathbf{n}+p\mathbf{e}_2\cdot\mathbf{n})\end{aligned}$$

is the scalar product with $\mathbf{n}^{\perp}$ of the vector $\rho\mathbf{u}(\mathbf{u}\cdot\mathbf{n})+p\mathbf{n}$ and

$$(\rho\mathbf{u}(\mathbf{u}\cdot\mathbf{n})+p\mathbf{n})\cdot\mathbf{n}^{\perp}=\rho u_n u_\tau=g_2(\tilde{\mathbf{U}}).$$

Hence, we have

$$\frac{\partial}{\partial\zeta}\mathbf{n}\cdot(\mathbf{n}\cdot\mathbf{F}_{23})+\frac{\partial}{\partial\tau}\mathbf{n}\cdot(\mathbf{n}^{\perp}\cdot\mathbf{F}_{23})=\frac{\partial}{\partial\zeta}(\rho u_n^2)+\frac{\partial}{\partial\tau}(\rho u_\tau u_n).$$

The computations for the equation (2.13b) are similar:

$$\begin{aligned}\mathbf{n}^{\perp}\cdot\Big\{\frac{\partial}{\partial t}(\rho\mathbf{u})(\mathbf{X},t)+\tilde{\boldsymbol{\nabla}}\cdot(\mathbf{n}\cdot\mathbf{F}_{23},\mathbf{F}_{23}\cdot\mathbf{n}^{\perp})^T\Big\}\\=\frac{\partial}{\partial t}(\rho u_\tau)+\frac{\partial}{\partial\zeta}(\rho u_n u_\tau)+\frac{\partial}{\partial\tau}(\rho u_\tau^2+p).\end{aligned}$$

We can check easily that, using (2.12), we have also proven

$$R^{-1}\mathbf{n}\cdot\mathbb{F}(\mathbf{U})=\mathbf{f}(\tilde{\mathbf{U}}),\ R^{-1}\mathbf{n}^{\perp}\cdot\mathbb{F}(\mathbf{U})=\mathbf{g}(\tilde{\mathbf{U}}),\tag{2.14a}$$

which implies in particular by differentiation

$$\mathbf{A}(\mathbf{U},\mathbf{n})=R\mathbf{A}(\tilde{\mathbf{U}})R^{-1},\ \text{where }\mathbf{A}(\cdot)=\mathbf{f}'(\cdot),\tag{2.14b}$$

a formula that is used in numerical schemes. □

2.3 Eigenvalues

Let us now study the hyperbolicity of the system (2.2). As in Chapter I, Example 2.4, we use the nonconservative variables (ρ,u,v,s) to compute the eigenvalues of the matrix $\mathbf{A}(\mathbf{U},\boldsymbol{\omega})$. From

$$Tds=d\varepsilon-\frac{p}{\rho^2}d\rho,$$

we get

$$\begin{aligned}T\Big(\frac{\partial s}{\partial t}+u\frac{\partial s}{\partial x}+v\frac{\partial s}{\partial y}\Big)&=\frac{\partial\varepsilon}{\partial t}+u\frac{\partial\varepsilon}{\partial x}+v\frac{\partial\varepsilon}{\partial y}\\&-\Big(\frac{p}{\rho^2}\Big)\Big(\frac{\partial\rho}{\partial t}+u\frac{\partial\rho}{\partial x}+v\frac{\partial\rho}{\partial y}\Big)=0,\end{aligned}$$

so that the system can be written in nonconservation form

$$
(2.15)\qquad \begin{cases}
\dfrac{\partial \rho}{\partial t} + u\dfrac{\partial \rho}{\partial x} + \rho\dfrac{\partial u}{\partial x} + v\dfrac{\partial \rho}{\partial y} + \rho\dfrac{\partial v}{\partial y} = 0,\\[2mm]
\dfrac{\partial u}{\partial t} + u\dfrac{\partial u}{\partial x} + v\dfrac{\partial u}{\partial y} + \left(\dfrac{1}{\rho}\right)\dfrac{\partial p}{\partial x} = 0,\\[2mm]
\dfrac{\partial v}{\partial t} + u\dfrac{\partial v}{\partial x} + v\dfrac{\partial v}{\partial y} + \left(\dfrac{1}{\rho}\right)\dfrac{\partial p}{\partial y} = 0,\\[2mm]
\dfrac{\partial s}{\partial t} + u\dfrac{\partial s}{\partial x} + v\dfrac{\partial s}{\partial y} = 0.
\end{cases}
$$

We take $p = p(\rho, s)$ and define the local speed of sound c by

$$c^2 = \frac{\partial p}{\partial \rho}(\rho, s) = p_\rho.$$

For instance, for a polytropic ideal gas, $p = A(s)\rho^\gamma$, $c^2 = \gamma\frac{p}{\rho}$. Denoting the velocity field by

$$\mathbf{u} = (u, v)^T,$$

the Mach number is defined by

$$M = \frac{|\mathbf{u}|}{c}.$$

Replacing in (2.15) $\frac{\partial p}{\partial x}$ (resp. $\frac{\partial p}{\partial y}$) by $p_\rho\frac{\partial \rho}{\partial x} + p_s\frac{\partial s}{\partial x}$ (resp. $p_\rho\frac{\partial \rho}{\partial y} + p_s\frac{\partial s}{\partial y}$), we compute the matrix $\mathbf{A}(\mathbf{U}, \boldsymbol{\omega})$,

$$\mathbf{A}(\mathbf{U}, \boldsymbol{\omega}) = \begin{pmatrix} \mathbf{u}\cdot\boldsymbol{\omega} & \rho\omega_1 & \rho\omega_2 & 0\\ p_\rho\omega_1/\rho & \mathbf{u}\cdot\boldsymbol{\omega} & 0 & p_s\omega_1/\rho\\ p_\rho\omega_2/\rho & 0 & \mathbf{u}\cdot\boldsymbol{\omega} & p_s\omega_2/\rho\\ 0 & 0 & 0 & \mathbf{u}\cdot\boldsymbol{\omega} \end{pmatrix},$$

and the eigenvalues are easily found to be $\mathbf{u}\cdot\boldsymbol{\omega}$ and $\mathbf{u}\cdot\boldsymbol{\omega} \pm c|\boldsymbol{\omega}|$. Thus, we have two simple eigenvalues, which for $|\boldsymbol{\omega}| = 1$ are $\lambda_1 = \mathbf{u}\cdot\boldsymbol{\omega} - c$ and $\lambda_4 = \mathbf{u}\cdot\boldsymbol{\omega} + c$, and give genuinely nonlinear fields ("acoustic" or "pressure" waves), and $\lambda_2 = \lambda_3 = \mathbf{u}\cdot\boldsymbol{\omega}$ (of multiplicity 2) associated to the entropy waves and "vorticity" (or shear) waves, which are linearly degenerate. We can choose

$$(2.16a)\qquad \mathbf{r}_1 = \begin{pmatrix} -\rho/c\\ \omega_1\\ \omega_2\\ 0 \end{pmatrix}, \quad \mathbf{r}_4 = \begin{pmatrix} \rho/c\\ \omega_1\\ \omega_2\\ 0 \end{pmatrix},$$

while the eigenvectors associated to the eigenvalue $\mathbf{u}\cdot\boldsymbol{\omega}$ may be taken as

$$(2.16b)\qquad \mathbf{r}_2 = \begin{pmatrix} p_s\\ 0\\ 0\\ -p_\rho \end{pmatrix}, \quad \mathbf{r}_3 = \begin{pmatrix} 0\\ -\omega_2\\ \omega_1\\ 0 \end{pmatrix}.$$

The vector $\mathbf{r}_2(\mathbf{U}, \boldsymbol{\omega})$ does not depend on $\boldsymbol{\omega}$, and $\mathbf{r}_3(\mathbf{U}, \boldsymbol{\omega}) = (0, \boldsymbol{\omega}^\perp, 0)$ does not depend on $\mathbf{U}$; its form explains the term "shear wave" which we have used above (see Section 2.5 below; see also Roe 1983 and Roe 1986a, and Rumsey et al. 1993 for the use of shear waves in numerical schemes).

Remark 2.4. One can prove more generally that the multiple eigenvalues of a hyperbolic system give linearly degenerate fields, and therefore genuinely nonlinear fields are simple (see Chapter I, Remark 6.1). □

The 2,3-Riemann invariants in the direction $\boldsymbol{\omega}$ are $\mathbf{u} \cdot \boldsymbol{\omega}$ and p (since we have a system of four equations, and the multiplicity of the eigenvalue is 2, we can find only $4 - 2 = 2$ Riemann invariants whose gradients $Dw(\mathbf{U})$ satisfying $Dw(\mathbf{U}) \cdot \mathbf{r}_i(\mathbf{U}, \boldsymbol{\omega}) = 0$, $i = 2, 3$, are independent; see Chapter I, Section 3.2). The three 1- (resp. 4-) Riemann invariants are $u + \ell\omega_1, v + \ell\omega_2, s$ (resp. $u - \ell\omega_1, v - \ell\omega_2, s$), where $\ell(\rho, s)$ is the function defined (in Chapter I, formula (3.18)) by

$$\frac{\partial \ell}{\partial \rho} = \frac{c}{\rho}.$$

We could as well take $\mathbf{u} \cdot \boldsymbol{\omega} \pm \ell$, $\mathbf{u} \cdot \boldsymbol{\omega}^\perp$, s.

Remark 2.5. Let us see why, as is well known, the simple eigenvalues $\mathbf{u}\cdot\boldsymbol{\omega}\pm c$ are associated to the sound (or acoustic) waves (see Courant and Friedrichs 1976, Section 11, Chorin and Marsden 1978, Section 3.1, J.D.Anderson 1982, Section 7.5, Whitham 1974, Section 6.6). Indeed, consider a small, smooth (therefore isentropic as we shall see in (2.20) below) perturbation

$$\rho = \rho_0 + \delta\rho_1,$$

where ρ_0 is constant, and take for simplicity $\mathbf{u}_0 = 0$, so that $\mathbf{u} = \delta\mathbf{u}_1$ is small. Since $ds = 0$, we can write

$$dp = \frac{\partial p}{\partial \rho}(\rho, s)d\rho = c^2 d\rho,$$

which yields

$$\frac{\partial p}{\partial x} = c^2 \frac{\partial \rho}{\partial x}, \quad \frac{\partial p}{\partial y} = c^2 \frac{\partial \rho}{\partial y}.$$

Substituting these expressions in the equations (2.15), and neglecting terms of order higher than one in δ, we obtain

$$\begin{cases} \dfrac{\partial \rho_1}{\partial t} + \rho_0\Big(\dfrac{\partial u_1}{\partial x} + \dfrac{\partial v_1}{\partial y}\Big) = 0, \\ \dfrac{\partial u_1}{\partial t} + \Big(\dfrac{1}{\rho_0}\Big)\dfrac{\partial p}{\partial \rho}(\rho, s)\dfrac{\partial \rho_1}{\partial x} = 0, \\ \dfrac{\partial v_1}{\partial t} + \Big(\dfrac{1}{\rho_0}\Big)\dfrac{\partial p}{\partial \rho}(\rho, s)\dfrac{\partial \rho_1}{\partial y} = 0. \end{cases}$$

Now, also expanding $\frac{\partial p}{\partial \rho}(\rho, s) = c^2$ about ρ_0,

$$\frac{\partial p}{\partial \rho}(\rho, s) = c_0^2 + O(\delta),$$

we differentiate the first equation (resp. the second and third) w.r.t. t, (resp. w.r.t. x and y) and obtain that the disturbance ρ_1 satisfies the wave equation associated to the velocity c_0,

$$\frac{\partial^2(\rho_1)}{\partial t^2} = c_0^2\Big(\frac{\partial^2 \rho_1}{\partial x^2} + \frac{\partial^2 \rho_1}{\partial y^2}\Big) = c_0^2 \Delta \rho_1.$$

In the one-dimensional case, the general solution is a function of the form $f(x + c_0 t) + g(x - c_0 t)$, where f and g are arbitrary, i.e., a superposition of two waves traveling with constant speed $\pm c_0$; small disturbances propagate with speed c_0. □

Remark 2.6. We might also have used the primitive variables (ρ, u, v, p). From

$$dp = c^2 d\rho + \frac{\partial p}{\partial s} ds,$$

we get from the first and last equations (2.1),

$$\begin{aligned}\frac{\partial p}{\partial t} + u\frac{\partial p}{\partial x} + v\frac{\partial p}{\partial y} &= c^2\frac{\partial \rho}{\partial t} + \frac{\partial p}{\partial s}\frac{\partial s}{\partial t}\\ &\quad + u\Big\{c^2\frac{\partial \rho}{\partial x} + \frac{\partial p}{\partial s}\frac{\partial s}{\partial x}\Big\} + v\Big\{c^2\frac{\partial \rho}{\partial y} + \frac{\partial p}{\partial s}\frac{\partial s}{\partial y}\Big\}\\ &= c^2\Big\{\frac{\partial \rho}{\partial t} + u\frac{\partial \rho}{\partial x} + v\frac{\partial \rho}{\partial y}\Big\}.\end{aligned}$$

Setting $\operatorname{grad} p = \nabla p = (\frac{\partial p}{\partial x}, \frac{\partial p}{\partial y})^T$, this yields

$$\frac{\partial p}{\partial t} + \mathbf{u}\cdot\operatorname{grad} p - c^2\Big\{\frac{\partial \rho}{\partial t} + \mathbf{u}\cdot\operatorname{grad}\rho\Big\} = 0. \tag{2.17}$$

Using the primitive variables (u, v, p), the system can be written in the form

$$\Big\{\frac{\partial \rho}{\partial t} + \mathbf{u}\cdot\operatorname{grad}\rho\Big\} - c^{-2}\Big\{\frac{\partial p}{\partial t} + \mathbf{u}\cdot\operatorname{grad} p\Big\} = 0, \tag{2.18a}$$

$$\begin{cases}\dfrac{\partial u}{\partial t} + u\dfrac{\partial u}{\partial x} + v\dfrac{\partial u}{\partial y} + \Big(\dfrac{1}{\rho}\Big)\dfrac{\partial p}{\partial x} = 0,\\[2mm] \dfrac{\partial v}{\partial t} + u\dfrac{\partial v}{\partial x} + v\dfrac{\partial v}{\partial y} + \Big(\dfrac{1}{\rho}\Big)\dfrac{\partial p}{\partial y} = 0,\\[2mm] \dfrac{\partial p}{\partial t} + u\dfrac{\partial p}{\partial x} + v\dfrac{\partial p}{\partial y} + \rho c^2\Big(\dfrac{\partial u}{\partial x} + \dfrac{\partial v}{\partial y}\Big) = 0.\end{cases} \tag{2.18b}$$

We consider the nonconservative system (2.18b) of the last three equations in $\mathbf{V} = (u, v, p)$. The density is supposed to be determined from p through the equation (2.18a), which holds along the particles' paths, as we shall see below. The corresponding matrix $\mathbf{A}(\mathbf{V},\boldsymbol{\omega}) = \mathbf{A}_1(\mathbf{V})\omega_1 + \mathbf{A}_2(\mathbf{V})\omega_2$ is then

$$\begin{pmatrix} \mathbf{u}\cdot\boldsymbol{\omega} & 0 & \omega_1/\rho \\ 0 & \mathbf{u}\cdot\boldsymbol{\omega} & \omega_2/\rho \\ \rho c^2\omega_1 & \rho c^2\omega_2 & \mathbf{u}\cdot\boldsymbol{\omega} \end{pmatrix},$$

whose (simple) eigenvalues are naturally $\mathbf{u}\cdot\boldsymbol{\omega}$ and $\mathbf{u}\cdot\boldsymbol{\omega} \pm c$. The complete matrix associated to system (2.18)b completed by the equation

$$\frac{\partial \rho}{\partial t} + u\frac{\partial \rho}{\partial x} + v\frac{\partial \rho}{\partial y} + \rho\frac{\partial u}{\partial x} + \rho\frac{\partial v}{\partial y} = 0$$

is

$$\begin{pmatrix} \mathbf{u}\cdot\boldsymbol{\omega} & \rho\omega_1 & \rho\omega_2 & 0 \\ 0 & \mathbf{u}\cdot\boldsymbol{\omega} & 0 & \omega_1/\rho \\ 0 & 0 & \mathbf{u}\cdot\boldsymbol{\omega} & \omega_2/\rho \\ 0 & \rho c^2\omega_1 & \rho c^2\omega_2 & \mathbf{u}\cdot\boldsymbol{\omega} \end{pmatrix},$$

and the eigenvectors are $\mathbf{r}_1 = (1, -c\frac{\omega_1}{\rho}, -c\frac{\omega_2}{\rho}, c^2)^T$, $\mathbf{r}_2 = (1,0,0,0)^T$, $\mathbf{r}_3 = (1, -\omega_2, \omega_1, 0)^T$, $\mathbf{r}_4 = (1, c\frac{\omega_1}{\rho}, c\frac{\omega_2}{\rho}, c^2)^T$. □

We shall need the expression for the eigenvectors in conservative variables in Section 4.3.3 (for Roe's scheme); setting $H = \frac{(E+p)}{\rho}$, which is the total enthalpy, we have

$$\mathbf{r}_1(\mathbf{U},\boldsymbol{\omega}) = (1, u - c\omega_1, v - c\omega_2, H - \mathbf{u}\cdot\boldsymbol{\omega}\, c)^T,$$
$$\mathbf{r}_4(\mathbf{U},\boldsymbol{\omega}) = (1, u + c\omega_1, v + c\omega_2, H + \mathbf{u}\cdot\boldsymbol{\omega}\, c)^T,$$
$$\mathbf{r}_2(\mathbf{U},\boldsymbol{\omega}) = \Big(1, u, v, \frac{|\mathbf{u}|^2}{2}\Big)^T, \mathbf{r}_3(\mathbf{U},\boldsymbol{\omega}) = (0, -\omega_2, \omega_1, \mathbf{u}\cdot\boldsymbol{\omega}^{\perp})^T.$$

By rotational invariance, these values can also be obtained from the formula (2.14b), $\mathbf{A}(\mathbf{U},\boldsymbol{\omega}) = R\mathbf{A}(\tilde{\mathbf{U}})R^{-1}$, where R is a rotation such that $\boldsymbol{\omega} = R\mathbf{e}_1$, so that we need only compute the eigenvectors of $\mathbf{f}' = \mathbf{A}$. Note that the vector $\mathbf{r}_2(\mathbf{U},\boldsymbol{\omega})$ is an eigenvector of both $\mathbf{f}'$ and $\mathbf{g}'$.

2.4 Characteristics

Let us introduce now the particle path, or trajectory. It is an integral curve of the velocity field $\mathbf{u} = (u,v)^T$, i.e., a curve $t \to \mathbf{x}(t) = (x(t), y(t))$, parametrized by t, such that

$$\frac{d}{dt}\mathbf{x}(t) = \mathbf{u}(\mathbf{x}(t), t), \tag{2.19a}$$

i.e.,

$$\frac{dx}{dt} = u, \quad \frac{dy}{dt} = v.$$

One sometimes uses instead of "particle path" the term *streamline*; a streamline is a function $s \to \mathbf{x}(s) = (x(s), y(s))$, which is an integral curve at fixed time t,

$$\frac{d}{ds}\mathbf{x}(s) = \mathbf{u}(\mathbf{x}(s), t).$$

A streamline coincides with a particle path for a stationary flow.

The particle derivative is the differential along the particle path,

$$\frac{D}{Dt} = \frac{\partial}{\partial t} + \mathbf{u} \cdot \nabla = \frac{\partial}{\partial t} + u\frac{\partial}{\partial x} + v\frac{\partial}{\partial y}. \tag{2.19b}$$

In fact this corresponds to (1.12) with $\boldsymbol{\nu} = \frac{\mathbf{u}}{|\mathbf{u}|}$ and $\lambda_k(\mathbf{u}, \boldsymbol{\nu}) = \mathbf{u} \cdot \boldsymbol{\nu} = |\mathbf{u}|$, i.e., $\lambda_k(\mathbf{u}, \boldsymbol{\nu})\boldsymbol{\nu} = \mathbf{u}$, and the trajectories are the corresponding curves C_k (1.11) ($k = 2$ or 3).

Note that with the notation $\frac{D}{Dt}$, the last equation (2.15) can be written

$$\frac{Ds}{Dt} = 0, \tag{2.20}$$

which means that the entropy is constant along the particle paths (in the smooth parts of the flow). Also, the equation (2.17) can be written equivalently

$$\frac{Dp}{Dt} - c^2\frac{D\rho}{Dt} = 0. \tag{2.21}$$

Recall that a characteristic surface $\varphi(x, y, t) = 0$ is such that the matrix $\frac{\partial \varphi}{\partial t}\mathbf{I} + \mathbf{f}'(\mathbf{U})\frac{\partial \varphi}{\partial x} + \mathbf{g}'(\mathbf{U})\frac{\partial \varphi}{\partial y}$ is singular, which, by Lemma 1.1, implies that $-\frac{\partial \varphi}{\partial t}$ is an eigenvalue of $\mathbf{A}(\mathbf{u}, \boldsymbol{\nu})$; for instance, the characteristic surfaces associated to the eigenvalue λ_4 satisfy

$$-\frac{\partial \varphi}{\partial t} = \mathbf{u} \cdot \boldsymbol{\nu} + c,$$

where $\boldsymbol{\nu} = (\frac{\partial \varphi}{\partial x}, \frac{\partial \varphi}{\partial y})$, which gives the condition

$$\frac{\partial \varphi}{\partial t} + u\frac{\partial \varphi}{\partial x} + v\frac{\partial \varphi}{\partial y} + c = 0.$$

Lemma 2.5

The envelope of all the characteristic surfaces through a point (x_0, y_0, t_0) consists of the streamline through the point and a conoid.

Proof. Consider first the envelope of the tangent planes to the characteristic surfaces through the point $M_0 = (t_0, x_0, y_0)$ (it corresponds to a linearization about the state $\mathbf{U}(M_0)$). The equation of the tangent plane to a surface $\varphi(t, x, y) = 0$ at the point M_0 is

$$\frac{\partial \varphi}{\partial t}(M_0)(t - t_0) + \frac{\partial \varphi}{\partial x}(M_0)(x - x_0) + \frac{\partial \varphi}{\partial y}(M_0)(y - y_0) = 0.$$

If the characteristic surface is associated, say, to the eigenvalue λ_4, we add the condition

$$\frac{\partial\varphi}{\partial t}(M_0) + u_0\frac{\partial\varphi}{\partial x}(M_0) + v_0\frac{\partial\varphi}{\partial y}(M_0) + c(M_0) = 0,$$

together with the normalization

$$\Big(\frac{\partial\varphi}{\partial x}(M_0)\Big)^2 + \Big(\frac{\partial\varphi}{\partial y}(M_0)\Big)^2 = 1.$$

We can set

$$\frac{\partial\varphi}{\partial x}(M_0) = \cos\,\beta,\ \frac{\partial\varphi}{\partial y}(M_0) = \sin\,\beta,$$

and we get

$$\frac{\partial\varphi}{\partial t}(M_0) + u_0\cos\,\beta + v_0\sin\,\beta + c_0 = 0.$$

Hence, the tangent plane satisfies

$$-(u_0\cos\,\beta + v_0\sin\,\beta + c_0)(t-t_0) + \cos\,\beta(x-x_0) + \sin\,\beta(y-y_0) = 0$$

or

$$\cos\,\beta(x-x_0-u_0(t-t_0)) + \sin\,\beta(y-y_0-v_0(t-t_0)) = c_0(t-t_0). \tag{2.22}$$

The envelope of this family of planes is derived by differentiating (2.22) w.r.t. β,

$$-\sin\,\beta(x-x_0-u_0(t-t_0)) + \cos\,\beta(y-y_0-v_0(t-t_0)) = 0, \tag{2.23}$$

and we obtain from (2.22), (2.23)

$$\begin{cases} x-x_0-u_0(t-t_0) = \cos\,\beta\,c_0(t-t_0), \\ y-y_0-v_0(t-t_0) = \sin\,\beta\,c_0(t-t_0); \end{cases} \tag{2.24}$$

and finally, eliminating β between the two equations (2.24), we get the (sonic) cone through the point $M_0 = (t_0, x_0, y_0)$,

$$(x-x_0-u_0(t-t_0))^2 + (y-y_0-v_0(t-t_0))^2 = (c_0(t-t_0))^2. \tag{2.25}$$

The intersection of the cone (2.25) with the tangent plane (2.22) is precisely the line (2.24), which is called a bicharacteristic (see Holt 1977, for instance).

If we consider the envelope of the characteristic surfaces, we write similarly the identity

$$\frac{\partial\varphi}{\partial t}dt + \frac{\partial\varphi}{\partial x}dx + \frac{\partial\varphi}{\partial y}dy = 0,$$

and then the condition that the surfaces be characteristic can be written

$$\frac{\partial\varphi}{\partial t} + u\frac{\partial\varphi}{\partial x} + v\frac{\partial\varphi}{\partial y} + c = 0,$$

and the normalization identity

$$\nu_1 = \frac{\partial \varphi}{\partial x} = \cos\,\beta, \quad \nu_2 = \frac{\partial \varphi}{\partial y} = \sin\,\beta.$$

We have

$$\cos\,\beta\Big(\frac{dx}{dt} - u\Big) + \sin\,\beta\Big(\frac{dy}{dt} - v\Big) = c.$$

We differentiate w.r.t. β,

$$-\sin\,\beta\Big(\frac{dx}{dt} - u\Big) + \cos\,\beta\Big(\frac{dy}{dt} - v\Big) = 0$$

and obtain

$$\begin{cases} \dfrac{dx}{dt} - u = c\cos\,\beta, \\ \dfrac{dy}{dt} - v = c\sin\,\beta. \end{cases} \tag{2.26}$$

Finally,

$$\Big(\frac{dx}{dt} - u\Big)^2 + \Big(\frac{dy}{dt} - v\Big)^2 = c^2. \tag{2.27}$$

The characteristic surface touches the envelope at the line of tangency (2.26), which is the integral curve of the vector field $\mathbf{u} + c\boldsymbol{\nu}$ and is usually called a bicharacteristic; we shall call it B_k. If Σ is a characteristic surface, both curves $B_4(\frac{d\mathbf{x}}{dt} = \mathbf{u} + c\boldsymbol{\nu})$ and C_4 (defined by (1.11) i.e., $\frac{d\mathbf{x}}{dt} = (\mathbf{u} \cdot \boldsymbol{\nu} + c)\boldsymbol{\nu}$) lie on Σ.

We have similar results with the other eigenvalues. Notice that the cone associated to λ_1, obtained by changing c to $-c$, coincides with (2.27). Now, if we take $c = 0$, corresponding to the double eigenvalue, the corresponding conoid (2.27) degenerates into the streamline. □

We have already obtained the characteristic equations (2.20), (2.21), which hold along the streamlines. The other characteristic equations are obtained following exactly the computations of Example 5.1, Chapter I.

The equation (2.21) together with (2.15a) gives

$$\frac{\partial p}{\partial t} + u\frac{\partial p}{\partial x} + v\frac{\partial p}{\partial y} + \rho c^2\Big\{\frac{\partial u}{\partial x} + \frac{\partial v}{\partial y}\Big\} = 0$$

or, with $\operatorname{div}\mathbf{u} = \nabla \cdot \mathbf{u} = \frac{\partial u}{\partial x} + \frac{\partial v}{\partial y}$,

$$\frac{\partial p}{\partial t} + \mathbf{u} \cdot \operatorname{grad} p + \rho c^2 \operatorname{div}\mathbf{u} = 0. \tag{2.28}$$

The equations (2.15b) and (2.15c) can be written with $\operatorname{grad} = \nabla = (\frac{\partial}{\partial x}, \frac{\partial}{\partial y})^T$,

$$\frac{\partial \mathbf{u}}{\partial t} + (\mathbf{u} \cdot \operatorname{grad})\mathbf{u} + \Big(\frac{1}{\rho}\Big)\operatorname{grad} p = \mathbf{0} \tag{2.29}$$

or, with the particle derivative $\frac{D}{Dt}$ defined in (2.19),

$$\frac{D\mathbf{u}}{Dt} + \left(\frac{1}{\rho}\right)\operatorname{grad} p = \mathbf{0}.$$

We can now take the scalar product of (2.29) by $\boldsymbol{\mu}$, where $\boldsymbol{\mu}$ is any unit vector, to obtain the characteristic equation

$$\boldsymbol{\mu} \cdot \left\{ \frac{D\mathbf{u}}{Dt} + \left(\frac{1}{\rho}\right)\operatorname{grad} p \right\} = 0 \tag{2.30}$$

(the differentiation takes place in a plane tangent to the particle path and parallel to $\boldsymbol{\mu}$).

If we multiply (2.30), written for some unit vector $\boldsymbol{\omega}$, by $\pm\rho c$ and add it to (2.28), we get

$$\frac{\partial p}{\partial t} + \mathbf{u} \cdot \operatorname{grad} p + \rho c^2 \operatorname{div} \mathbf{u} \pm \rho c \boldsymbol{\omega} \frac{D\mathbf{u}}{Dt} \pm c\boldsymbol{\omega} \operatorname{grad} p = 0,$$

which yields

$$\left(\frac{D}{Dt} \pm c\boldsymbol{\omega} \cdot \operatorname{grad}\right)p \pm \rho c\left(\boldsymbol{\omega} \cdot \frac{D}{Dt} \pm c \operatorname{div}\right)\mathbf{u} = 0,$$

or

$$\left(\frac{D}{Dt} \pm c\boldsymbol{\omega} \cdot \operatorname{grad}\right)p \pm \rho c\boldsymbol{\omega} \cdot \left(\frac{D}{Dt} \pm c\boldsymbol{\omega} \operatorname{div}\right)\mathbf{u} = 0. \tag{2.31}$$

The characteristic equations are thus

$$\begin{cases} \dfrac{Dp}{Dt} - c^2 \dfrac{D\rho}{Dt} = 0 \ \left(\text{can be equivalently replaced by } \dfrac{Ds}{Dt} = 0\right), \\ \boldsymbol{\mu} \cdot \left(\dfrac{D\mathbf{u}}{Dt} + \left(\dfrac{1}{\rho}\right)\operatorname{grad} p\right) = 0, \\ \left(\dfrac{D}{Dt} + c\boldsymbol{\omega} \cdot \operatorname{grad}\right)p + \rho c\boldsymbol{\omega} \cdot \left(\dfrac{D}{Dt} + c\boldsymbol{\omega}\operatorname{div}\right)\mathbf{u} = 0, \\ \left(\dfrac{D}{Dt} - c\boldsymbol{\omega} \cdot \operatorname{grad}\right)p + \rho c\boldsymbol{\omega} \cdot \left(\dfrac{D}{Dt} - c\boldsymbol{\omega}\operatorname{div}\right)\mathbf{u} = 0. \end{cases} \tag{2.32}$$

It is easy to check that in (2.31) the differential operators inside each bracket all act in one plane. Indeed, we can write equation (2.31) using the decomposition $\mathbf{u} = (\mathbf{u} \cdot \boldsymbol{\omega})\boldsymbol{\omega} + (\mathbf{u} \cdot \boldsymbol{\omega}^\perp)\boldsymbol{\omega}^\perp$ and $\operatorname{div} \mathbf{u} = (\boldsymbol{\omega} \cdot \operatorname{grad})(\mathbf{u} \cdot \boldsymbol{\omega}) + (\boldsymbol{\omega}^\perp \cdot \operatorname{grad})(\mathbf{u} \cdot \boldsymbol{\omega}^\perp)$ in the form

$$\begin{aligned} &\left\{\frac{\partial}{\partial t} + (\mathbf{u} + c\boldsymbol{\omega}) \cdot \operatorname{grad}\right\}p + \rho c\boldsymbol{\omega} \cdot \left\{\frac{\partial}{\partial t} + \mathbf{u} \cdot \operatorname{grad} + c\boldsymbol{\omega} \operatorname{div}\right\}\mathbf{u} \\ &= \left\{\frac{\partial}{\partial t} + (\mathbf{u} + c\boldsymbol{\omega}) \cdot \operatorname{grad}\right\}p + \rho c\left\{\frac{\partial}{\partial t} + (\mathbf{u} + c\boldsymbol{\omega}) \cdot \operatorname{grad}\right\}(\mathbf{u} \cdot \boldsymbol{\omega}) \\ &\quad + \rho c^2(\boldsymbol{\omega}^\perp \cdot \operatorname{grad})(\mathbf{u} \cdot \boldsymbol{\omega}^\perp) = 0. \end{aligned}$$

The operator $\{\frac{\partial}{\partial t} + (\mathbf{u} + c\boldsymbol{\omega}) \cdot \operatorname{grad}\}$ acts along an integral curve of the vector $\mathbf{u} + c\boldsymbol{\omega}$, i.e.,

$$\frac{dx}{dt} = u + c\omega_1, \quad \frac{dy}{dt} = u + c\omega_2,$$

which is a bicharacteristic B_4 (see (2.26)). Thus in (2.31), the directions of differentiation lie in a plane passing through this bicharacteristic (tangent to the sonic cone) and through $(0, \boldsymbol{\omega}^\perp)$ (i.e., the intersection with the plane $t = 0$ is orthogonal to $\boldsymbol{\omega}$).

Remark 2.7. We have derived some characteristic equations using the particular expression of the eigenvalues in terms of the velocity field. We might also consider the form mentioned in Lemma 1.2 with the notation (1.12),

$$\frac{d}{ds_4} = \frac{\partial}{\partial t} + (\mathbf{u} \cdot \boldsymbol{\omega} + c)\boldsymbol{\omega} \cdot \operatorname{grad}.$$

We have for instance

$$\begin{aligned}\frac{\partial}{\partial t} + (\mathbf{u} + c\boldsymbol{\omega}) \cdot \operatorname{grad} &= \frac{\partial}{\partial t} + (\mathbf{u} \cdot \boldsymbol{\omega} + c)\boldsymbol{\omega} \cdot \operatorname{grad} + (\mathbf{u} \cdot \boldsymbol{\omega}^\perp)\boldsymbol{\omega}^\perp \cdot \operatorname{grad} \\ &= \frac{d}{ds_4} + (\mathbf{u} \cdot \boldsymbol{\omega}^\perp)\boldsymbol{\omega}^\perp \cdot \operatorname{grad},\end{aligned}$$

and we get

$$\begin{aligned}&\frac{\partial p}{\partial t} + (\mathbf{u} + c\boldsymbol{\omega}) \cdot \operatorname{grad} p + \rho c\Big\{\frac{\partial}{\partial t} + (\mathbf{u} + c\boldsymbol{\omega}) \cdot \operatorname{grad}\Big\}(\mathbf{u} \cdot \boldsymbol{\omega}) \\ &\quad + \rho c^2(\boldsymbol{\omega}^\perp \cdot \operatorname{grad})(\mathbf{u} \cdot \boldsymbol{\omega}^\perp) \\ &= \frac{dp}{ds_4} + \rho c \frac{d}{ds_4}(\mathbf{u} \cdot \boldsymbol{\omega}) + (\mathbf{u} \cdot \boldsymbol{\omega}^\perp)\boldsymbol{\omega}^\perp \cdot \operatorname{grad} p \\ &\quad + \rho c(\mathbf{u} \cdot \boldsymbol{\omega}^\perp)\boldsymbol{\omega}^\perp \cdot \operatorname{grad}(\mathbf{u} \cdot \boldsymbol{\omega}) + \rho c^2(\boldsymbol{\omega}^\perp \cdot \operatorname{grad})(\mathbf{u} \cdot \boldsymbol{\omega}^\perp) = 0.\end{aligned}$$

Thus, as we have already observed, the equation contains only derivatives in the directions $(1, \lambda_k(\mathbf{u}, \boldsymbol{\omega})\boldsymbol{\omega})$ and $(0, \boldsymbol{\omega}^\perp)$. □

2.5 Plane wave solutions. Self-similar solutions

2.5.1 Simple plane waves and contact discontinuities

For what concerns simple plane wave solutions (see Chapter I, Section 3.2)

$$\mathbf{U}(\mathbf{x}, t) = \mathbf{V}(\varphi(\mathbf{x} \cdot \boldsymbol{\omega}, t)),$$

we are led to consider the integral curves (1.18) of the vector fields $\mathbf{r}_k$ given by (2.16) in the system of variables $\mathbf{U} = (\rho, u, v, s)^T$.

For $k = 1$ and $k = 4$, using the Riemann invariants given in Section 2.3 above, we obtain that a state $\mathbf{U}$ belongs to the set $\mathcal{R}_k(\mathbf{U}_L, \boldsymbol{\omega})$ if

$$\left\{\begin{aligned} u + \ell(\rho, s)\omega_1 &= u_L + \ell(\rho_L, s_L)\omega_1, \\ v + \ell(\rho, s)\omega_2 &= v_L + \ell(\rho_L, s_L)\omega_2, \\ s &= s_L, \end{aligned}\right. \tag{2.33}$$

and in particular that $\mathbf{u} \cdot \boldsymbol{\omega}^{\perp}$ is continuous,

$$\mathbf{u} \cdot \boldsymbol{\omega}^{\perp} = \mathbf{u}_L \cdot \boldsymbol{\omega}^{\perp}.$$

For instance, for a perfect gas, $\ell = \frac{2c}{(\gamma-1)}$, (see Example 3.2, Chapter II). When $\boldsymbol{\omega}$ varies in (2.33), we obtain the set

$$s = s_L, \quad (u - u_L)^2 + (v - v_L)^2 = (\ell(\rho_L, s_L) - \ell(\rho, s_L))^2.$$

For $k = 2$ or 3, $\lambda_2 = \lambda_3 = \mathbf{u} \cdot \boldsymbol{\omega}$; the system is not strictly hyperbolic and we must extend slightly the results of Sections 1.3, 1.4. If $\mathbf{r}$ belongs to the eigenspace spanned by $\mathbf{r}_2$ and $\mathbf{r}_3$, for a curve

$$\frac{d\mathbf{U}}{d\varphi} = \mathbf{r}(\mathbf{U})$$

we have the relations

$$\frac{dp}{d\varphi} = p_\rho \frac{d\rho}{d\varphi} + p_s \frac{ds}{d\varphi} = 0$$

and

$$\omega_1 \frac{du}{d\varphi} + \omega_2 \frac{dv}{d\varphi} = 0$$

(these relations hold obviously for $\mathbf{r}_2$ and $\mathbf{r}_3$). Hence, the set of states $\mathbf{U}$ that can be connected to a given state $\mathbf{U}_L$ by a plane contact discontinuity corresponding to the eigenvalue $\mathbf{u} \cdot \boldsymbol{\omega}$ is a two-dimensional manifold in the (ρ, u, v, s)-space

$$\left\{\begin{aligned} & p = p_L, \\ & \omega_1(u - u_L) + \omega_2(v - v_L) = 0, \end{aligned}\right. \tag{2.34}$$

which results from the fact that p and $\mathbf{u}\cdot\boldsymbol{\omega}$ are Riemann invariants. When $\boldsymbol{\omega}$ varies, these manifolds span the set $p = p_L$. In particular, (2.34) says that normal velocity components are continuous, that there is a discontinuity of the tangential velocity component only, which is characteristic of a shear wave (for a pure "contact " or "entropy" wave

$$\frac{d\mathbf{U}}{d\varphi} = \mathbf{r}_2(\mathbf{U}, \boldsymbol{\omega}) = (p_s, 0, 0, -p_\rho)^T,$$

it is obvious that $\frac{d\mathbf{u}}{d\varphi} = 0$, and hence there is no discontinuity in the velocity). The plane of discontinuity in the (x, y, t)-space is:

$$\mathbf{x} \cdot \boldsymbol{\omega} = (\mathbf{u} \cdot \boldsymbol{\omega})t = (\mathbf{u}_L \cdot \boldsymbol{\omega})t \text{ (i.e., } \sigma = \lambda_2 = \mathbf{u} \cdot \boldsymbol{\omega} = \mathbf{u}_L \cdot \boldsymbol{\omega}).$$

2.5.2 Plane shock waves

The computations for a 1- or 4-plane shock follow exactly those for the one-dimensional case. One writes the Rankine–Hugoniot conditions across the plane $\mathbf{x}\cdot\boldsymbol{\omega} = \sigma t$ as follows:

$$(2.35)\qquad \begin{cases} \sigma[\rho] = \boldsymbol{\omega}\cdot[\rho\mathbf{u}], \\ \sigma[\rho u] = \omega_1[\rho u^2 + p] + \omega_2[\rho uv], \\ \sigma[\rho v] = \omega_1[\rho uv] + \omega_2[\rho v^2 + p], \\ \sigma[\rho e] = \boldsymbol{\omega}\cdot[(\rho e + p)\mathbf{u}] = \omega_1[(\rho e + p)u] + \omega_2[(\rho e + p)v]. \end{cases}$$

The analog of the velocity relative to the discontinuity (see Chapter II, formula (2.6)) is now

$$v = \mathbf{u}\cdot\boldsymbol{\omega} - \sigma,$$

so that the first equation (2.35) is equivalent to

$$\rho v = \rho_L v_L,$$

and we shall set

$$\mathcal{M} = \rho v = \rho_L v_L.$$

Note that, as in the one-dimensional case, $\mathcal{M} = 0$ corresponds to a contact discontinuity, which we have just studied (see (2.34)). In the same way, multiplying the second (resp. the third) equation by ω_1 (resp. ω_2) and adding, which corresponds to taking the scalar product by $\boldsymbol{\omega}$ of the system of two equations

$$\sigma[\rho\mathbf{u}] = (\boldsymbol{\omega}\cdot[\rho u\mathbf{u} + p\mathbf{e}_1], \boldsymbol{\omega}\cdot[\rho v\mathbf{u} + p\mathbf{e}_2])^T,$$

yields, together with the first equation,

$$\rho v^2 + p = \rho_L v_L^2 + p_L.$$

And if we take the scalar product by $\boldsymbol{\omega}^\perp = (-\omega_2, \omega_1)^\perp$, we get

$$\rho v\mathbf{u}\cdot\boldsymbol{\omega}^\perp = \rho_L v_L \mathbf{u}_L\cdot\boldsymbol{\omega}^\perp.$$

Thus, if $\mathcal{M} \neq 0$,

$$\mathbf{u}\cdot\boldsymbol{\omega}^\perp = \mathbf{u}_L\cdot\boldsymbol{\omega}^\perp,$$

i.e., there is no change in the tangential velocity component (normal to $\boldsymbol{\omega}$), which implies that

$$(2.36)\qquad \mathbf{u} - \mathbf{u}_L = ((\mathbf{u} - \mathbf{u}_L)\cdot\boldsymbol{\omega})\boldsymbol{\omega}$$

is collinear to $\boldsymbol{\omega}$. Also, we check that the last equation gives

$$\Big\{\rho\Big(\varepsilon + \frac{v^2}{2}\Big) + p\Big\}v = \Big\{\rho_L\Big(\varepsilon_L + \frac{v_L^2}{2}\Big) + p_L\Big\}v_L,$$

and we have the exact analog of (2.7) in Chapter II.

Then, we have

$$\mathcal{M} = \frac{(\mathbf{u} - \mathbf{u}_L) \cdot \boldsymbol{\omega}}{(\tau - \tau_L)} \tag{2.37}$$

and

$$\mathcal{M}v + p = \rho v^2 + p = \rho_L v_L^2 + p_L = \mathcal{M}v_L + p_L,$$

which gives

$$\mathcal{M} = -\frac{(p - p_L)}{(v - v_L)} = -\frac{(p - p_L)}{(\mathbf{u} - \mathbf{u}_L) \cdot \boldsymbol{\omega}}$$

and

$$\mathcal{M}^2 = -\frac{(p - p_L)}{(\tau - \tau_L)}. \tag{2.38}$$

Finally, we also obtain the equation of the Hugoniot curve (see Chapter II, (2.18)),

$$\varepsilon - \varepsilon_L + \frac{1}{2}(p + p_L)(\tau - \tau_L) = 0. \tag{2.39}$$

Assuming that the Hugoniot curve can be parametrized by p, i.e., can be represented by an equation of the form

$$\tau = h_L(p),$$

and using (2.36)–(2.38), we proceed as in the one-dimensional case to obtain the shock curves in the $(\mathbf{u}, p)$-space

$$\begin{aligned} \mathbf{u} - \mathbf{u}_L &= ((\mathbf{u} - \mathbf{u}_L) \cdot \boldsymbol{\omega})\boldsymbol{\omega} = \pm(-(p - p_L)(\tau - \tau_L))^{1/2}\boldsymbol{\omega} \\ &= \pm\Big(\frac{(p - p_L)(\rho - \rho_L)}{\rho\rho_L}\Big)^{1/2}\boldsymbol{\omega}. \end{aligned} \tag{2.40}$$

The speed σ is given by (2.35),

$$\sigma = \boldsymbol{\omega} \cdot \frac{(\rho\mathbf{u} - \rho_L\mathbf{u}_L)}{(\rho - \rho_L)} = \boldsymbol{\omega} \cdot \mathbf{u}_L + \rho\boldsymbol{\omega} \cdot \frac{(\mathbf{u} - \mathbf{u}_L)}{(\rho - \rho_L)},$$

$$\sigma = \boldsymbol{\omega} \cdot \mathbf{u}_L + \tau_L\boldsymbol{\omega} \cdot \frac{(\mathbf{u} - \mathbf{u}_L)}{(\tau_L - \tau)},$$

and by (2.37), (2.38) we have

$$\begin{aligned} \sigma &= \boldsymbol{\omega} \cdot \mathbf{u}_L \pm \tau_L\Big(-\frac{(p - p_L)}{(\tau - \tau_L)}\Big)^{1/2} \\ &= \boldsymbol{\omega} \cdot \mathbf{u}_L \pm \Big(\frac{\rho(p - p_L)}{\rho_L(\rho - \rho_L)}\Big)^{1/2}. \end{aligned}$$

The sign "$-$" (resp. "$+$") corresponds to an admissible shock for the first field $\lambda_1 = \mathbf{u} \cdot \boldsymbol{\omega} - c = \mathbf{u} \cdot \boldsymbol{\omega} - (\frac{\partial p}{\partial \rho}(\rho, s))^{1/2}$ (resp. the fourth field $\lambda_4 =$

$\mathbf{u} \cdot \boldsymbol{\omega} + c = \mathbf{u} \cdot \boldsymbol{\omega} + (\frac{\partial p}{\partial \rho}(\rho, s))^{1/2}$, and the "Lax entropy conditions"

$$\lambda_1(\mathbf{u}_R, \boldsymbol{\omega}) < \sigma < \lambda_1(\mathbf{u}_L, \boldsymbol{\omega}), \quad \sigma < \lambda_2(\mathbf{u}_R, \boldsymbol{\omega})$$

(resp.

$$\lambda_4(\mathbf{u}_R, \boldsymbol{\omega}) < \sigma < \lambda_4(\mathbf{u}_L, \boldsymbol{\omega}), \quad \sigma > \lambda_3(\mathbf{u}_L, \boldsymbol{\omega}))$$

hold.

When $\boldsymbol{\omega}$ varies, we see from (2.40) that the set spanned by the shock curves is

$$|\mathbf{u}-\mathbf{u}_L|^2 = (u-u_L)^2+(v-v_L)^2 = -(p-p_L)(\tau-\tau_L) = \frac{(p-p_L)(\rho-\rho_L)}{\rho\rho_L}.$$

2.5.3 Plane Riemann problem

In the case of the gas dynamics equations that are invariant under rotation, it is important to consider solutions $\mathbf{U}(\mathbf{x}, t) = \mathbf{V}(\mathbf{x} \cdot \boldsymbol{\omega}, t)$ in the direction of, say, $\boldsymbol{\omega} = (1, 0)^T$. Since then, $\mathbf{U} = \mathbf{U}(x, t)$ and $\mathbf{A}(\mathbf{U}, \boldsymbol{\omega}) = \mathbf{f}'(\mathbf{U})$, we are led to the system

$$(2.41) \qquad \begin{cases} \dfrac{\partial \rho}{\partial t} + \dfrac{\partial}{\partial x}(\rho u) = 0, \\ \dfrac{\partial}{\partial t}(\rho u) + \dfrac{\partial}{\partial x}(\rho u^2 + p) = 0, \\ \dfrac{\partial}{\partial t}(\rho v) + \dfrac{\partial}{\partial x}(\rho u v) = 0, \\ \dfrac{\partial}{\partial t}(\rho e) + \dfrac{\partial}{\partial x}((\rho e + p)u) = 0. \end{cases}$$

System (2.41) is a one-dimensional system that is hyperbolic but not strictly hyperbolic. However, it is easy to prove that the result of Chapter I, Theorem 6.1, concerning the solution of the Riemann problem can be extended in this particular case (see Remark 6.1, Chapter I). The solution has the same structure as that of the system, which is described in Chapter II, Section 3, Remark 3.5. It consists of at most four constant states separated by a 1-wave (shock or rarefaction), a 2- (or 3-) contact discontinuity, and 4-waves (shock or rarefaction), as illustrated in Figure 2.1. Across a 1- and 4-waves, v is continuous as results from (2.33) and (2.36) (with $\boldsymbol{\omega} = (1, 0)^T$), while across the 2-wave p and u are continuous, as results from (2.34); in primitive variables, $\mathbf{U}_{L,R} = (\rho, u, v, p)^T_{L,R}$, $\mathbf{U}_I = (\rho_I, u^*, v_L, p^*)^T$, $\mathbf{U}_{II} = (\rho_{II}, u^*, v_R, p^*)^T$.

Remark 2.8. The Riemann problem for (1.1) is the initial value problem with piecewise constant initial data in each of the four quadrants, or more generally in sectors meeting at the origin. The problem is invariant under the transformation $(x, y, t) \to (cx, cy, ct)$, and the solution is self-similar as in (2.42), $\mathbf{u}(x, y, t) = \mathbf{v}(\frac{x}{t}, \frac{y}{t})$. The construction of explicit solutions has been investigated in the scalar case by Wagner (1983), Lindquist (1986),

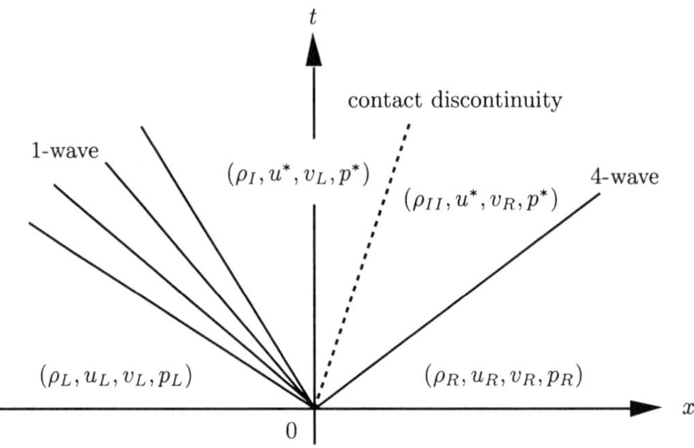

FIGURE 2.1. Solution of the Riemann problem.

Klingenberg and Osher (1988) essentially when $g = f$, and by Chang and Hsio (1989), and Zhang and Zheng (1989) in general. The case of linear hyperbolic systems of two equations has been investigated by Gilquin et al. (1992), and Abgrall (1993). In the case of isentropic or polytropic ideal gas dynamics, the Riemann problem is studied in Zhang and Zheng (1990), Schulz and Rinne (1993), see also Glimm et al. (1985). □

2.5.4 Characteristic equations

Introducing the differentiation along the particle paths,

$$\frac{D}{Dt} = \frac{\partial}{\partial t} + u\frac{\partial}{\partial x},$$

system (2.41) written in the variables ρ, u, s reads

$$\begin{aligned} &\frac{D\rho}{Dt} + \rho\frac{\partial u}{\partial x} = 0, \\ &\rho\frac{Du}{Dt} + \frac{\partial p}{\partial x} = 0, \\ &\frac{Dv}{Dt} = 0, \\ &\frac{Ds}{Dt} = 0. \end{aligned}$$

The last two equations $\frac{Ds}{Dt} = 0$, $\frac{Dv}{Dt} = 0$, are characteristic, corresponding to the double eigenvalue $\mathbf{u} \cdot \boldsymbol{\omega} = u$ (since $\boldsymbol{\omega} = (1,0)^T$); the characteristic equations that correspond to the simple eigenvalues $\mathbf{u} \cdot \boldsymbol{\omega} \pm c = u \pm c$ are

$$\frac{\partial p}{\partial t} + (u \pm c)\frac{\partial p}{\partial x} \pm \rho c\left\{\frac{\partial u}{\partial t} + (u \pm c)\frac{\partial u}{\partial x}\right\} = 0;$$

the differential operator $\frac{\partial}{\partial t} + (u \pm c)\frac{\partial}{\partial x}$ acts along the characteristic $\frac{dx}{dt} = u \pm c$.

2.5.5 Self-similar solutions

If we are interested in a self-similar (or pseudostationary) solution

$$\mathbf{U}(x, y, t) = \mathbf{V}\Big(\frac{x}{t}, \frac{y}{t}\Big) \tag{2.42}$$

of system (2.1) (see Remark 2.8 above), we set $\Xi = (\xi, \zeta)$, where $\xi = \frac{x}{t}$, $\zeta = \frac{y}{t}$, and

$$\begin{aligned}\frac{\partial}{\partial t} &= -\frac{x}{t^2}\frac{\partial}{\partial \xi} - \frac{y}{t^2}\frac{\partial}{\partial \zeta} = -\frac{1}{t}\Big(\xi\frac{\partial}{\partial \xi} + \zeta\frac{\partial}{\partial \zeta}\Big),\\ \frac{\partial}{\partial x} &= \frac{1}{t}\frac{\partial}{\partial \xi}, \qquad \frac{\partial}{\partial y} = \frac{1}{t}\frac{\partial}{\partial \zeta}.\end{aligned}$$

The equation of conservation of mass in (2.1) becomes

$$-\xi\frac{\partial \rho}{\partial \xi} - \zeta\frac{\partial \rho}{\partial \zeta} + \frac{\partial}{\partial \xi}(\rho u) + \frac{\partial}{\partial \zeta}(\rho v) = 0,$$

keeping the same notations for the function $\mathbf{U}$ and $\mathbf{V}$, or

$$\frac{\partial(\rho(u-\xi))}{\partial \xi} + \frac{\partial(\rho(v-\zeta))}{\partial \zeta} = -2\rho.$$

Hence, we are led to set

$$\tilde{\mathbf{u}} = \mathbf{u} - \Xi = (u - \xi, v - \zeta).$$

The other equations are computed in the same way, and system (2.1) becomes

$$\begin{aligned}&\frac{\partial(\rho\tilde{u})}{\partial \xi} + \frac{\partial(\rho\tilde{v})}{\partial \zeta} = -2\rho,\\ &\frac{\partial}{\partial \xi}(\rho\tilde{u}^2 + p) + \frac{\partial}{\partial \zeta}(\rho\tilde{u}\tilde{v}) = -3\rho\tilde{u},\\ &\frac{\partial}{\partial \xi}(\rho\tilde{u}\tilde{v}) + \frac{\partial}{\partial \zeta}(\rho\tilde{v}^2 + p) = -3\rho\tilde{v},\\ &\frac{\partial}{\partial \xi}((\rho\tilde{e} + p)\tilde{u}) + \frac{\partial}{\partial \zeta}((\rho\tilde{e} + p)\tilde{v}) = -\rho|\tilde{\mathbf{u}}|^2 - 2\rho\Big(\tilde{e} + \frac{p}{\rho}\Big),\end{aligned}$$

where $\tilde{e} = e + \frac{(\tilde{u}^2+\tilde{v}^2)}{2} = e + \frac{|\tilde{\mathbf{u}}|^2}{2}$. Thus, the system satisfied by a self-similar solution is the steady Euler system in the variable $\tilde{\mathbf{U}} = (\rho, \rho\tilde{u}, \rho\tilde{v}, \rho\tilde{e})$ with a source term

$$\frac{\partial \mathbf{f}(\tilde{\mathbf{U}})}{\partial \xi} + \frac{\partial \mathbf{g}(\tilde{\mathbf{U}})}{\partial \zeta} = \tilde{\mathbf{S}},$$

where $\mathbf{f}, \mathbf{g}$ are defined by (2.2a). The overtaking of two shocks in steady flow (above system without source term) is studied in Chang and Hsiao (1989), see also Marshall and Plohr (1992), Glaz (1991).

Remark 2.9. Let us have a look at the equation of two-dimensional steady flow that we have just encountered, i.e.,

$$\frac{\partial \mathbf{f}(\mathbf{U})}{\partial x} + \frac{\partial \mathbf{g}(\mathbf{U})}{\partial y} = \mathbf{0}, \tag{2.43}$$

where $\mathbf{f}, \mathbf{g}$ are defined by (2.2a). If we seek a smooth, simple wave solution $\mathbf{U}(\mathbf{x}) = \mathbf{V}(\mathbf{x} \cdot \boldsymbol{\omega})$ constant along the characteristic lines $\mathbf{x} \cdot \boldsymbol{\omega} = \text{const.}$, we find that

$$(\omega_1 \mathbf{f}'(\mathbf{V}) + \omega_2 \mathbf{g}'(\mathbf{V}))\mathbf{V}' = \mathbf{A}(\mathbf{V}, \boldsymbol{\omega})\mathbf{V}' = \mathbf{0},$$

and $\mathbf{A}(\mathbf{V}, \boldsymbol{\omega})$ is not invertible. We have already computed

$$\det \mathbf{A}(\mathbf{V}, \boldsymbol{\omega}) - \lambda \mathbf{I}) = (\lambda - \mathbf{u} \cdot \boldsymbol{\omega})^2 \{(\mathbf{u} \cdot \boldsymbol{\omega} - \lambda)^2 - c^2 |\boldsymbol{\omega}|^2).$$

Thus, the roots $\boldsymbol{\omega}$ of

$$\det \mathbf{A}(\mathbf{V}, \boldsymbol{\omega}) = (\mathbf{u} \cdot \boldsymbol{\omega})^2 (\mathbf{u} \cdot \boldsymbol{\omega}^2 - c^2 |\boldsymbol{\omega}|^2) = 0$$

are firstly those of

$$\mathbf{u} \cdot \boldsymbol{\omega} = 0,$$

which always exist and correspond to the streamlines (since the direction of the characteristics $\mathbf{x} \cdot \boldsymbol{\omega} = \text{const.}$ is then $\mathbf{u}$), and secondly those of the quadratic polynomial

$$(\mathbf{u} \cdot \boldsymbol{\omega})^2 - c^2 |\boldsymbol{\omega}|^2 = 0.$$

Considered as a polynomial in $\frac{\omega_1}{\omega_2}$ (slope of the characteristic), it has real roots iff

$$|\mathbf{u}|^2 \geq c^2,$$

i.e., in supersonic regions where the system (2.41) is thus hyperbolic. There, we get

$$\frac{\omega_1}{\omega_2} = \frac{(uv \pm c(u^2 + v^2 - c^2)^{1/2})}{(u^2 - c^2)}.$$

Since

$$\mathbf{u} \cdot \frac{\boldsymbol{\omega}}{|\boldsymbol{\omega}|} = |\mathbf{u}| \cos(\mathbf{u}, \boldsymbol{\omega}) = \pm c,$$

the characteristics $\mathbf{x} \cdot \boldsymbol{\omega} = c$ (which are orthogonal to $\boldsymbol{\omega}$) make with the streamlines an angle $\pm\mu$ such that $\sin \mu = \frac{c}{|\mathbf{u}|} = \frac{1}{M}$, where M is the Mach number, and are often called Mach lines.

Using the characteristic equations, we can transform the system into characteristic form (following the arguments of Section 1.2, which can easily be extended). For details concerning this "hodograph" transformation, and more generally steady supersonic plane-flows, we refer to Courant and Friedrichs (1976), J.D. Anderson (1982), Whitham (1974), Menikoff (1988). A physical example of 2-D stationary simple waves is given by the flow around a bend (Prandtl–Meyer expansion waves), which is studied in the above references (respectively in Section 111, Section 4.13, Section 6.17).

The flow may also be subsonic ($M = \frac{|\mathbf{u}|}{c} < 1$) so that the sound wave characteristics become complex and the system is elliptic. The flow is called transonic if it involves mixed subsonic–supersonic regions. The steady equations (2.43) are of mixed type, while the time-dependent problem is always hyperbolic. □

3 Multidimensional finite difference schemes

3.1 Direct approach

3.1.1 Difference schemes in conservation form

Many schemes can be extended to two dimensions in the finite difference setting. We consider a two-dimensional uniform Cartesian spatial grid $\Delta = \Delta x \times \Delta y$ with space increments Δx and Δy, and we set

$$\lambda_x = \frac{\Delta t}{\Delta x}, \quad \lambda_y = \frac{\Delta t}{\Delta y}. \tag{3.1}$$

Let $\mathbf{u}_{j,k}^n$ denote an approximation of the solution at the grid point ($x_j = j\Delta x$, $y_k = k\Delta y$, $t_n = n\Delta t$), and define the sequences $\mathbf{u}^n$ by $\mathbf{u}^n = (\mathbf{u}_{j,k}^n)$, $j, k \in \mathbb{Z}, n \in \mathbb{N}$. For approximating the system

$$\frac{\partial \mathbf{u}}{\partial t} + \frac{\partial}{\partial x}\mathbf{f}(\mathbf{u}) + \frac{\partial}{\partial y}\mathbf{g}(\mathbf{u}) = 0, \tag{3.2}$$

we can use a finite difference scheme of the form

$$\mathbf{u}^{n+1} = \mathbf{H}_\Delta(\mathbf{u}^n), \tag{3.3a}$$

i.e.,

$$\mathbf{u}_{j,k}^{n+1} = \mathbf{H}(\mathbf{u}_{j-J,k-K}^n, \ldots \mathbf{u}_{j+J,k+K}^n), \tag{3.3b}$$

where $\mathbf{H} : \mathbb{R}^{(2J+1)\times(2K+1)\times p} \to \mathbb{R}^p$ and $\mathbf{u}^0 = (\mathbf{u}_{j,k}^0)$ are given, for instance,

$$\mathbf{u}_{j,k}^0 = \frac{1}{\Delta x \Delta y}\int_{\Omega_{j,k}} \mathbf{u}_0(\boldsymbol{\xi})d\boldsymbol{\xi},$$

$$\Omega_{j,k} = (x_{j-1/2}, x_{j+1/2}) \times (y_{k-1/2}, y_{k+1/2}).$$

The scheme is in *conservation form* if there exist continuous functions $\mathbf{F} : \mathbb{R}^{2J\times(2K+1)\times p} \to \mathbb{R}^p$ and $\mathbf{G} : \mathbb{R}^{(2J+1)\times 2K\times p} \to \mathbb{R}^p$ such that

$$\mathbf{u}_{j,k}^{n+1} = \mathbf{u}_{jk}^n - \lambda_x(\mathbf{F}_{j+1/2,k}^n - \mathbf{F}_{j-1/2,k}^n) - \lambda_y(\mathbf{G}_{j,k+1/2}^n - \mathbf{G}_{j,k-1/2}^n), \tag{3.4}$$

where

$$\mathbf{F}^n_{j+1/2,k} = \mathbf{F}(\mathbf{u}^n_{j-J+1,k-K}, ...\mathbf{u}^n_{j+J,k+K}),$$
$$\mathbf{G}^n_{j,k+1/2} = \mathbf{G}(\mathbf{u}^n_{j-J,k-K+1}, ...\mathbf{u}^n_{j+J,k+K}).$$

For instance, if $J = K = 1$ (9-point schemes), each component of $\mathbf{F}$ and $\mathbf{G}$ is a function of six variables,

$$\mathbf{F}_{j+1/2,k} = \mathbf{F}(u_{j,k-1}, u_{j,k}, u_{j,k+1}, v_{j+1,k-1}, v_{j+1,k}, v_{j+1,k+1}),$$
$$\mathbf{G}_{j,k+1/2} = \mathbf{G}(u_{j-1,k}, u_{j,k}, u_{j+1,k}, v_{j-1,k+1}, v_{j,k+1}, v_{j+1,k+1}).$$

Let us observe that the conservation form implies: if $\mathbf{u}_{j-J,k} = ... = \mathbf{u}_{j+J,k} = \mathbf{u}_k$, $\forall -K \le k \le K$, then $\mathbf{F}_{j-1/2,k} = \mathbf{F}_{j+1/2,k}$; similarly, if $\mathbf{u}_{j,k-K} = ... = \mathbf{u}_{j,k+K} = \mathbf{u}_j$, $\forall -J \le j \le J$, then $\mathbf{G}_{j,k+1/2} = \mathbf{G}_{j,k-1/2}$.

We shall assume, moreover, that the numerical fluxes $\mathbf{F}$ and $\mathbf{G}$ are consistent with $\mathbf{f}$ and $\mathbf{g}$, respectively, i.e.,

$$\mathbf{F}(\mathbf{u}, \mathbf{u}, ..., \mathbf{u}) = \mathbf{f}(\mathbf{u}), \quad \mathbf{G}(\mathbf{u}, \mathbf{u}, ..., \mathbf{u}) = \mathbf{g}(\mathbf{u}).$$

One can study the order of accuracy of a difference scheme exactly as in the one-dimensional case (G.R., Chapter III, Section 1.2) by expanding a smooth solution in Taylor series and deriving the equivalent system (see Lerat 1981, Jeng and Chen 1992, Billet 1989). For instance, a 9-point conservative scheme is second-order accurate if the fluxes $\mathbf{F}(\mathbf{u}_{-1}, \mathbf{u}_0, \mathbf{u}_1, \mathbf{v}_{-1}, \mathbf{v}_0, \mathbf{v}_1)$ and $\mathbf{G}(\mathbf{u}_{-1}, \mathbf{u}_0, \mathbf{u}_1, \mathbf{v}_{-1}, \mathbf{v}_0, \mathbf{v}_1)$ satisfy

$$(3.5a) \qquad \begin{cases} \sum_j \Big(\dfrac{\partial \mathbf{F}}{\partial u_j} + \dfrac{\partial \mathbf{F}}{\partial v_j}\Big)(\mathbf{u}, \mathbf{u}, ..., \mathbf{u}) = \mathbf{A}(\mathbf{u}), \\ \sum_j \Big(\dfrac{\partial \mathbf{G}}{\partial u_j} + \dfrac{\partial \mathbf{G}}{\partial v_j}\Big)(\mathbf{u}, \mathbf{u}, ..., \mathbf{u}) = \mathbf{B}(\mathbf{u}), \end{cases}$$

which comes from the consistency (order one), and

$$(3.5b) \qquad \begin{cases} \sum_j \Big(\dfrac{\partial \mathbf{F}}{\partial u_j} - \dfrac{\partial \mathbf{F}}{\partial v_j}\Big)(\mathbf{u}, ..., \mathbf{u}) = \lambda_x \mathbf{A}^2(\mathbf{u}), \\ \sum_j \Big(\dfrac{\partial \mathbf{G}}{\partial u_j} - \dfrac{\partial \mathbf{G}}{\partial v_j}\Big)(\mathbf{u}, ..., \mathbf{u}) = \lambda_y \mathbf{B}^2(\mathbf{u}), \end{cases}$$

$$(3.5c) \qquad \begin{cases} \sum_j j\Big(\dfrac{\partial \mathbf{F}}{\partial u_j} + \dfrac{\partial \mathbf{F}}{\partial v_j}\Big)(\mathbf{u}, ..., \mathbf{u}) = -\dfrac{\lambda_y}{2} \mathbf{A}(\mathbf{u})\mathbf{B}(\mathbf{u}), \\ \sum_j j\Big(\dfrac{\partial \mathbf{G}}{\partial u_j} + \dfrac{\partial \mathbf{G}}{\partial v_j}\Big)(\mathbf{u}, ..., \mathbf{u}) = -\dfrac{\lambda_x}{2} \mathbf{B}(\mathbf{u})\mathbf{A}(\mathbf{u}). \end{cases}$$

For details concerning the study of 9-point linear schemes ($J = K = 3$), we refer to Lerat (1981).

Example 3.1. The Lax–Wendroff scheme. It is derived, as in the one-dimensional case, from a Taylor expansion of a (smooth) solution and corresponds to the numerical fluxes

$$\begin{aligned}\mathbf{F}(\mathbf{u}_{j,k-1},\dots,\mathbf{u}_{j+1,k+1}) &= \frac{1}{2}(\mathbf{f}(\mathbf{u}_{j,k})+\mathbf{f}(\mathbf{u}_{j,k}))\\ &- \frac{\lambda_x}{2}\mathbf{A}_{j+1/2,k}(\mathbf{f}(\mathbf{u}_{j+1,k})-\mathbf{f}(\mathbf{u}_{j,k}))\\ &- \frac{\lambda_y}{8}\Big\{\mathbf{A}(\mathbf{u}_{j,k})(\mathbf{g}(\mathbf{u}_{j,k+1})-\mathbf{g}(\mathbf{u}_{j,k-1}))\\ &\qquad + \mathbf{A}(\mathbf{u}_{j+1,k})(\mathbf{g}(\mathbf{u}_{j+1,k+1})-\mathbf{g}(\mathbf{u}_{j+1,k-1}))\Big\},\end{aligned}$$

where for instance $\mathbf{A}_{j+1/2,k} = \mathbf{A}(\frac{\mathbf{u}_{j,k}+\mathbf{u}_{j+1,k}}{2})$, with an analogous formula for $\mathbf{G}$ obtained by exchanging λ_x and λ_y, $\mathbf{A}$ and $\mathbf{B}$, $\mathbf{f}$ and $\mathbf{g}$, j and k. Thus, one notices the presence of "crossed derivatives" terms. The scheme is second-order accurate, but in the linear case it is not the only one since the six linear relations (3.5), which are required for second-order accuracy, do not determine the nine coefficients uniquely as was the case in dimension one G.R., Chapter III, Section 1.3; see Lax and Wendroff (1964). □

If we restrict ourselves to the *scalar* case ($p = 1$),

$$\frac{\partial u}{\partial t} + \frac{\partial}{\partial x}f(u) + \frac{\partial}{\partial y}g(u) = 0, \tag{3.6a}$$

then recall that Theorem 5.2 in Chapter II gives for $u_0 \in \mathbf{L}^\infty(\mathbb{R}^2)$ the existence and uniqueness of the entropy solution u of the scalar conservation law (3.6a) satisfying

$$u(\mathbf{x},0) = u_0(\mathbf{x}). \tag{3.6b}$$

The consistency of the numerical fluxes ensures that, in the scalar case, when the sequence of approximate solutions converges in some sensible way, the limit is indeed a weak solution, i.e. satisfies the Rankine–Hugoniot jump condition (Lax–Wendroff Theorem, see G.R., Chapter III, Theorem 1.1).

Still, in the scalar case, the scheme (3.3) is *monotone* if H is a nondecreasing function of each of its arguments.

In many cases, we use much simpler formulas where $\mathbf{F}_{j+1/2,k} = \mathbf{F}(\mathbf{u}_{j-J+1,k},\dots,u_{j+J,k})$ (resp. $\mathbf{G}_{j,k+1/2} = \mathbf{G}(\mathbf{u}_{j,k-K+1},\dots,u_{j,k+K})$) depends only on the $2J$ values $\mathbf{u}_{p,k}$, $j-J+1 \le p \le j+J$, (resp. on the $2K$ values $\mathbf{u}_{j,q}$, $k-K+1 \le q \le k+K$). This occurs naturally if we start from one-dimensional numerical fluxes $\mathbf{F}: \mathbb{R}^{2J\times p} \to \mathbb{R}^p$ and $\mathbf{G}: \mathbb{R}^{2K\times p} \to \mathbb{R}^p$ consistent respectively with $\mathbf{f}$ and $\mathbf{g}$. We can also take a combination of one-dimensional numerical fluxes consistent respectively with $2\mathbf{f}$ and $2\mathbf{g}$, i.e., construct

$$\mathbf{u}_{j,k}^{n+1} = \mathbf{u}_{j,k}^n + \frac{\lambda_x}{2}(\mathbf{F}_{j+1/2,k}^n - \mathbf{F}_{j-1/2,k}^n) + \frac{\lambda_y}{2}(\mathbf{G}_{j,k+1/2}^n - \mathbf{G}_{j,k-1/2}^n).$$

This scheme corresponds to the discretization of the system (3.2) written in the form

$$\frac{1}{2}\left\{\frac{\partial \mathbf{u}}{\partial t}+\frac{\partial}{\partial x}2\mathbf{f}(\mathbf{u})\right\}+\frac{1}{2}\left\{\frac{\partial \mathbf{u}}{\partial t}+\frac{\partial}{\partial y}2\mathbf{g}(\mathbf{u})\right\}=\mathbf{0}.$$

If F and G are the numerical fluxes of monotone schemes, the resulting scheme is monotone in the scalar case.

Example 3.2. The Lax–Friedrichs scheme. We take for F and G the one-dimensional Lax–Friedrichs numerical flux (see G.R., Chapter III, Example 2.1), consistent with $2f$ (i.e., we replace f in the formula by $2f$). The above construction gives

$$\begin{aligned}
u_{j,k}^{n+1} &= u_{j,k}^n + \frac{u_{j+1,k}^n - 2u_{j,k}^n + u_{j-1,k}^n}{4} - \frac{\lambda_x}{2}\{f(u_{j+1,k}^n) - f(u_{j-1,k}^n)\} \\
&\quad + \frac{u_{j,k+1}^n - 2u_{j,k}^n + u_{j,k-1}^n}{4} - \frac{\lambda_y}{2}\{g(u_{j,k+1}^n - g(u_{j,k-1}^n))\} \\
&= \frac{u_{j+1,k}^n + u_{j-1,k}^n + u_{j,k+1}^n + u_{j,k-1}^n}{4} - \frac{\lambda_x}{2}\{f(u_{j+1,k}^n) - f(u_{j-1,k}^n\} \\
&\quad - \frac{\lambda_y}{2}\{g(u_{j,k+1}^n) - g(u_{j,k-1}^n)\}.
\end{aligned}$$

This scheme is monotone if $\lambda_x \max|f'| \leq \frac{1}{2}$ and $\lambda_y \max|g'| \leq \frac{1}{2}$. This has to be compared with the one-dimensional C.F.L. condition $\lambda \max|f'| \leq 1$.

If we had just added the usual Lax–Friedrichs fluxes in each direction, we would have obtained the following formula:

$$\begin{aligned}
u_{j,k}^{n+1} &= u_{j,k}^n + \frac{u_{j+1,k}^n - 2u_{j,k}^n + u_{j-1,k}^n}{2} - \frac{\lambda_x}{2}\{f(u_{j+1,k}^n) - f(u_{j-1,k}^n)\} \\
&\quad + \frac{u_{j,k+1}^n - 2u_{j,k}^n + u_{j,k-1}^n}{2} - \frac{\lambda_y}{2}\{g(u_{j,k+1}^n - g(u_{j,k-1}^n))\} \\
&= \frac{1}{2}(u_{j+1,k}^n + u_{j-1,k}^n + u_{j,k+1}^n + u_{j,k-1}^n - 2u_{j,k}^n) \\
&\quad - \frac{\lambda_x}{2}\{f(u_{j,k+1}^n) - f(u_{j,k-1}^n)\} - \frac{\lambda_y}{2}\{g(u_{j,k+1}^n) - g(u_{j,k-1}^n)\},
\end{aligned}$$

in which the dependence on $u_{j,k}^n$ is effective. □

Example 3.3. "5-point" schemes. For $J = K = 1$ (a 9-point scheme), the simpler formulas $F_{j+1/2,k}^n = F(u_{j-J+1,k}^n, \ldots, u_{j+J,k})$ (resp. $G_{j,k+1/2}^n = G(u_{j,k-K+1}^n, \ldots, u_{j,k+K}^n)$) give that F and G are functions of only two variables,

$$F_{j+1/2,k} = F(u_{j,k}, u_{j+1,k}), \quad G_{j,k+1/2} = G(u_{j,k}, u_{j,k+1}),$$

and $u_{j,k}^{n+1}$ depends on only five values. However, we cannot get second-order accuracy for which at least six points are necessary. Indeed, relations (3.5a),

(3.5b) yield, as expected,

$$\frac{\partial F}{\partial u_0} = \frac{a}{2}(1+\lambda_x a), \quad \frac{\partial F}{\partial v_0} = \frac{a}{2}(1-\lambda_x a),$$
$$\frac{\partial G}{\partial u_0} = \frac{b}{2}(1+\lambda_y b), \quad \frac{\partial G}{\partial v_0} = \frac{b}{2}(1-\lambda_y b),$$

but $\frac{\partial F}{\partial u_{\pm 1}} = \frac{\partial F}{\partial v_{\pm 1}} = \frac{\partial G}{\partial u_{\pm 1}} = \frac{\partial G}{\partial v_{\pm 1}} = 0$ is clearly incompatible with (3.5c), which means that "crossed" terms must be involved. □

Example 3.4. The two-step Lax–Wendroff scheme. This last drawback can be avoided by using the two-step version of the Lax–Wendroff scheme proposed by Richtmyer (G.R., Chapter III, (2.19); see Richtmyer and Morton 1967, Section 13.4). This two-step scheme is given by:

$$\begin{aligned} u_{j,k}^{n+1/2} &= \frac{1}{4}(u_{j+1,k}^n + u_{j-1,k}^n + u_{j,k+1}^n + u_{j,k-1}^n) \\ &\quad - \frac{\lambda_x}{2}(f(u_{j+1,k}^n) - f(u_{j-1,k}^n)) - \frac{\lambda_y}{2}(g(u_{j,k+1}^n) - g(u_{j,k-1}^n)) \\ u_{j,k}^{n+1} &= u_{j,k}^n - \lambda_x(f(u_{j+1,k}^{n+1/2}) - f(u_{j-1,k}^{n+1/2})) - \lambda_y(g(u_{j,k+1}^{n+1/2}) - g(u_{j,k-1}^{n+1/2})). \end{aligned}$$

The two steps involve staggered meshes, and the resulting scheme is in fact 9-point (see Figure 3.1). □

In this frame of finite difference schemes constructed from one-dimensional schemes, the quasi-monotone schemes (Cockburn 1990), schemes with flux limiter can be extended to the two-dimensional case (Spekreijse 1987, Venkatakrishnan 1995). Also, a fully multidimensional (one step) extension of the F.C.T. scheme has been derived (Zalesak 1979). Similarly, Colella has derived upwind methods (Colella 1990, Saltzman 1994, Pember et al. 1995) and ENO schemes can be extended to two- and three-dimensional flows (Harten 1986, Shu et al. 1992, Casper and Atkins 1993).

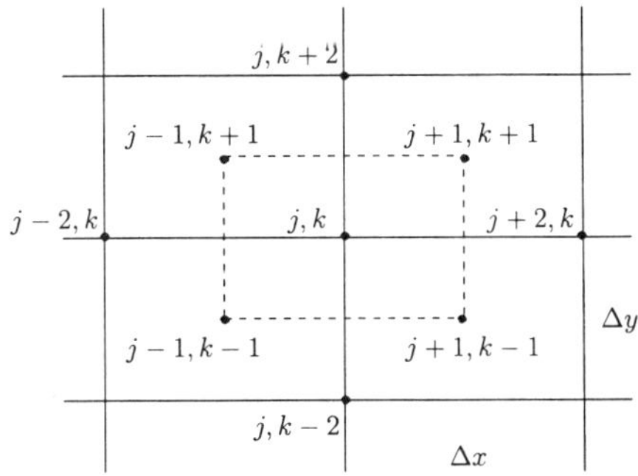

FIGURE 3.1. The two-step Lax–Wendroff scheme.

3.1.2 L^2-stability

One can first study the L^2-stability in the linear case

$$\frac{\partial \mathbf{u}}{\partial t} + \mathbf{A}\frac{\partial \mathbf{u}}{\partial x} + \mathbf{B}\frac{\partial \mathbf{u}}{\partial y} = \mathbf{0}, \tag{3.7}$$

where $\mathbf{A}$ and $\mathbf{B}$ are constant matrices. Following the ideas of G.R., Chapter III, Section 1.3 and the present Chapter III, Section 1.2, we consider a linear scheme that can be written

$$\mathbf{u}_{j,k}^{n+1} = \sum_{\substack{\ell=-J,\ldots J\\ m=-K,\ldots,K}} \mathbf{C}_{\ell,m}\mathbf{u}_{j+\ell,k+m}^{n},$$

where the matrices $\mathbf{C}$ are polynomials in $\lambda_x\mathbf{A}$ and $\lambda_y\mathbf{B}$. By extending the scheme to the whole space $\mathbb{R}^2$,

$$\mathbf{u}^{n+1}(x,y) = \sum_{\ell,m} \mathbf{C}_{\ell,m}\mathbf{u}^{n}(x+\ell\Delta x, y+m\Delta y),$$

and using the Fourier transform

$$\hat{\varphi}(\xi,\eta) = (2\pi)^{-1}\int_{\mathbb{R}^2} e^{i(x\xi+y\eta)}\varphi(x,y)dx\,dy, \quad (\xi,\eta)\in\mathbb{R}^2,$$

we get

$$\hat{\mathbf{u}}^{n+1}(\xi,\eta) = \mathbf{G}^a(\xi,\eta)\hat{\mathbf{u}}^{n}(\xi,\eta),$$

where the amplification matrix is defined by

$$\mathbf{G}^a(\xi,\eta) = \sum_{\ell,m}\mathbf{C}_{\ell,m}e^{i(\ell\xi\Delta+m\eta\Delta y)}.$$

A simple, necessary condition for L^2-stability, known as the von Neumann condition, is that the spectral radius of $\mathbf{G}^a$ be less than 1. A necessary and sufficient condition is that the powers $(\mathbf{G}^a)^n(\xi,\eta)$ of the amplification matrix be bounded uniformly in (ξ,η) and n. This condition was proven by Kreiss to be equivalent to the so-called resolvent condition or to the Hermitian norm condition (see Chapter III, Remark 1.2).

Note that $\mathbf{G}^a$ depends on (ξ,η) but also on $(\lambda_x\mathbf{A},\lambda_y\mathbf{B})$, which means that besides the dissipation linked to the modulus of the eigenvalues of $\mathbf{G}^a$ and the "phase error" induced by the imaginary part, there appears "numerical anisotropy" due to the dependence of $\mathbf{G}^a$ on the advection direction. A thorough study of $\mathbf{G}^a$ in the general case is difficult, and one often assumes particular values (such as square mesh) in order to carry out the computations (see Desideri et al. 1987 for an example).

Following the same approach as in Chapter III, Section 1.3, we see that by Fourier transform the linear system (3.7) gives

$$\frac{\partial \hat{\mathbf{u}}}{\partial t} + i(\xi\mathbf{A}+\eta\mathbf{B})\hat{\mathbf{u}} = \mathbf{0},$$

and the exact amplification matrix is

$$\mathbf{G}^{ex}(\xi,\eta) = \exp(-i(\xi\mathbf{A}+\eta\mathbf{B})\Delta t).$$

If the system is hyperbolic, the matrix $\xi\mathbf{A}+\eta\mathbf{B}$ has real eigenvalues, is diagonalizable, and

$$\rho(\mathbf{G}^{ex}(\xi,\eta)) = 1.$$

Equivalently, if we look for an elementary solution of the form

$$\mathbf{u}(\mathbf{x},t) = \hat{\mathbf{u}}e^{i(\mathbf{k}\cdot\mathbf{x}-\omega t)},$$

where $\mathbf{k} = (\xi,\eta)^T$ is the vector wave number and ω the frequency, ω must be an eigenvalue of $\xi\mathbf{A}+\eta\mathbf{B}$ and is therefore real, so that the amplitude remains constant.

In the scalar case

$$\frac{\partial u}{\partial t} + a\frac{\partial u}{\partial x} + b\frac{\partial u}{\partial y} = 0,$$

substituting an elementary wave

$$u(\mathbf{x},t) = \hat{u}e^{i(\mathbf{k}\cdot\mathbf{x}-\omega t)},$$

we get the dispersion relation (see Chapter III, Section 1.3) $\omega(\mathbf{k}) = \mathbf{c}\cdot\mathbf{k}$, where $\mathbf{c} = (a,b)^T$ is the advection vector. The phase surfaces $\mathbf{k}\cdot\mathbf{x}-\omega t =$ constant are parallel planes that propagate in the direction $\mathbf{k}$ with normal velocity $\frac{\omega}{|\mathbf{k}|}$, and the phase velocity is $\frac{\omega\mathbf{k}}{|\mathbf{k}|^2}$. Note that the group velocity is grad $\omega(\mathbf{k}) = \mathbf{c}$, which is constant and represents the advection direction, whereas the phase velocity may point in any direction (Whitham 1974 Chapter 11, Higdon 1996).

Looking for discrete Fourier mode solutions of the scalar numerical scheme

$$u_{j,k}^n = \hat{u}\, e^{i(\xi j\Delta x+\eta k\Delta y-\omega n\Delta t)}$$

leads to the discrete dispersion relation

$$e^{i\omega\Delta t} = g^a(\xi,\eta),$$

where g^a is the amplification factor. Writing $\tan(\omega\Delta t) = -\arg(g^a(\xi,\eta))$ yields by implicit derivation the discrete group velocity $(\frac{\partial\omega}{\partial\xi}, \frac{\partial\omega}{\partial\eta})^T$. Again, the error in group velocity yields an error not only in the speed (dispersion) but also in the direction (anisotropy) (see Trefethen 1982).

Example 3.5. The upwind scheme The natural upwind approximation of the scalar linear equation

$$\frac{\partial u}{\partial t} + a\frac{\partial u}{\partial x} + b\frac{\partial u}{\partial y} = 0, \quad a>0,\ b>0,$$

whose exact solutions satisfy $u(x,y,t) = u(x-at, y-bt)$, would be

$$u_{j,k}^{n+1} = u_{j,k}^n - \lambda_x a(u_{j,k}^n - u_{j-1,k}^n) - \lambda_y b(u_{j,k}^n - u_{j,k-1}^n).$$

It involves only three points: j, k; $j-1, k$; and $j, k-1$. Setting $\nu_x = a\lambda_x$, $\nu_y = b\lambda_y$, we write

$$u_{j,k}^{n+1} = u_{j,k}^n - \nu_x(u_{j,k}^n - u_{j-1,k}^n) - \nu_y(u_{j,k}^n - u_{j,k-1}^n);$$

the coefficient of amplification is

$$\begin{aligned} g^a(\xi,\eta) = 1 - \nu_x(1-\cos\xi\Delta x) - \nu_y(1-\cos\eta\Delta y) \\ - i(\nu_x \sin\xi\Delta x + \nu_y \sin\eta\Delta y). \end{aligned}$$

For $\xi\Delta x = \eta\Delta y = \pi$, we find the necessary condition

$$|1 - 2(\nu_x + \nu_y)| \leq 1$$

or, equivalently,

$$0 \leq \nu_x + \nu_y = \Delta t\Big(\frac{a}{\Delta x}\Big) + \Big(\frac{b}{\Delta y}\Big) \leq 1.$$

This C.F.L. condition seems too restrictive (for a given mesh, it is most restrictive if the convection $\mathbf{c} = (a,b)^T$ is parallel to $(\Delta x, \Delta y)^T$).

Instead, consider the split scheme (see Section 3.2 below)

$$\begin{aligned} u_{j,k}^{n,1} &= u_{j,k}^n - \lambda_x a(u_{j,k}^n - u_{j-1,k}^n) = (1-\nu_x)u_{j,k}^n + \nu_x u_{j-1,k}^n, \\ u_{j,k}^{n+1} &= u_{j,k}^{n,1} - \lambda_y b(u_{j,k}^{n,1} - u_{j,k-1}^{n,1}) = (1-\nu_y)u_{j,k}^{n,1} + \nu_y u_{j,k-1}^{n,1}, \end{aligned}$$

which differs from the above scheme by a second-order cross term

$$\begin{aligned} u_{j,k}^{n+1} = u_{j,k}^n &- \nu_x(u_{j,k}^n - u_{j-1,k}^n) - \nu_y(u_{j,k}^n - u_{j,k-1}^n) \\ &+ \nu_x\nu_y\{(u_{j,k}^n - u_{j-1,k}^n) - (u_{j,k-1}^n - u_{j-1,k-1}^n)\}. \end{aligned}$$

The amplification factor is the product

$$\begin{aligned} g^a(\xi,\eta) = \ & (1-\nu_x(1-\cos\xi\Delta x) - i\nu_x \sin\xi\Delta x) \\ & (1-\nu_y(1-\cos\eta\Delta y) - i\nu_y \sin\eta\Delta y)\ , \end{aligned}$$

and the split scheme is stable under the C.F.L. condition

$$0 \leq \nu_x \leq 1, \quad 0 \leq \nu_y \leq 1,$$

which appears to be less restrictive. □

Together with the preceding examples, this shows that there are several ways of extending a one-dimensional scheme to two dimensions. For a study of optimum linear schemes for advection, see Roe and Sidilkover (1992).

3.1.3. Total variation

Let us now define the following "$\mathbf{L}^1(\Delta)$ norm" of a scalar sequence $u = (u_{j,k})$,

$$\|u\|_{\mathbf{L}^1(\Delta)} = \Delta x \Delta y \sum_{j,k} |u_{j,k}|,$$

and the total variation of u by

$$TV(u) = \sum_{j,k} \{\Delta y |u_{j+1,k} - u_{j,k}| + \Delta x |u_{j,k+1} - u_{j,k}|\}.$$

If we introduce the notation

$$TV_{x,1}(u) = \Delta y \sum_{j,k} |u_{j+1,k} - u_{j,k}|, \tag{3.8a}$$

we have

$$TV_{x,1}(u) = \Delta y \sum_k \Big\{ \sum_j |u_{j+1,k} - u_{j,k}| \Big\} = \Delta y \sum_k TV_x u_{\cdot,k}.$$

$TV_{x,1}(u)$ is the one-dimensional (in y) $\mathbf{L}^1$-norm of the sequence $k \to TV_x u_{\cdot,k}$, where $TV_x u_{\cdot,k}$ is the total variation of the sequence $j \to u_{j,k}$. Similarly, setting

$$TV_{y,1}(u) = \Delta x \sum_{j,k} |u_{j,k+1} - u_{j,k}|, \tag{3.8b}$$

then

$$TV(u) = TV_{x,1}(u) + TV_{y,1}(u). \tag{3.8c}$$

It is the discrete norm associated with the continuous B.V. norm (see G.R., Chapter II, Section 1). Assuming that the numerical fluxes F and G are Lipschitz continuous, we can prove that a scheme (3.4) that is monotone converges to the unique entropy solution. More precisely, we define

$$\begin{aligned} &\Omega_{jk} = (x_{j-1/2}, x_{j+1/2}) \times (y_{k-1/2}, y_{k+1/2}), \\ &u_\Delta(\mathbf{x}, t) = u_{j,k}^n, \mathbf{x} \in \Omega_{jk}, \quad t_n < t \le t_{n+1}, \end{aligned} \tag{3.9}$$

and

$$u_{j,k}^0 = \frac{1}{\Delta x\, \Delta y} \int_{\Omega_{jk}} u_0(\mathbf{x}) d\mathbf{x},$$

and assume, moreover, that $u_0 \in BV(\mathbb{R}^2)$. One can then prove the following result.

Theorem 3.1

The sequence u_Δ associated to a monotone scheme (3.4) converges by (3.9) in $\mathbf{L}^\infty(0, T; \mathbf{L}^1_{\text{loc}}(\mathbb{R}^2))$ to the unique entropy solution of the scalar Cauchy problem (3.6) as $\Delta x \to 0$, $\Delta y \to 0$, $\Delta t = \lambda_x \Delta x = \lambda_y \Delta y$, with λ_x, λ_y kept constant.

Proof. The convergence of monotone conservative schemes is proved exactly as in the one-dimensional case (see G.R., Chapter III, Theorem 3.4, and Crandall and Majda 1980, Leroux 1979, Sanders 1983). In particular, it can be proved, using Crandall–Tartar's Lemma (see G.R., Chapter II, Lemma

5.2 and Chapter III, Theorem 3.2) that monotone schemes are contractions in $\mathbf{L}^1$,

$$\|H_\Delta(u) - H_\Delta(v)\|_{\mathbf{L}^1(\Delta)} \leq \|u - v\|_{\mathbf{L}^1(\Delta)}.$$

They are also L^∞-stable and T.V.D. □

Remark 3.1. A generalization of this result, the proof of which does not involve B.V. estimates but instead uses measure-valued solutions and DiPerna's uniqueness result, can be found in Szepessy (1991), Coquel and Le Floch (1993). We shall discuss this result in Remark 3.3 and in Section 4.2.3.

For similar results concerning the approximation of conservation laws with source terms, we refer to Chalabi (1992). □

We state now the limiting result concerning the maximum order of a T.V.D. scheme.

Proposition 3.1

A scheme (3.3) that is T.V.D. w.r.t. the norm (3.7) is at most first-order accurate.

Proof. Goodman and LeVeque (1988) have proven that, given a 2-dimensional T.V.D. scheme, there exists a set of "one space dimensional" data such that the restriction of the scheme to these data gives a monotone one dimensional scheme; we refer to Goodman and LeVeque for details. □

Remark 3.2. In fact, in several space dimensions, an estimate of the total variation of either the exact solution of a hyperbolic system (see Rauch 1986) or its approximate solution often fails. For all these reasons, the notion of T.V.D. or T.V.B. (total variation bounded) schemes is not as well adapted as in the one-dimensional scalar case (see also Lemma 3.4 below). □

We can still define an incremental form. Setting

$$(3.10)\qquad \begin{cases} \Delta u_{j+1/2,k} = \Delta_x u_{j,k} = u_{j+1,k} - u_{j,k}, \\ \Delta u_{j,k+1/2} = \Delta_y u_{j,k} = u_{j,k+1} - u_{j,k}, \end{cases}$$

we say that scheme (3.3) can be put in incremental form if there exist coefficients C_x, D_x, C_y, D_y such that

$$(3.11)\qquad \begin{cases} u_{j,k}^{n+1} = u_{j,k}^n + C_{x,j+1/2,k}^n \Delta u_{j+1/2,k}^n - D_{x,j-1/2,k}^n \Delta u_{j-1/2,k}^n \\ \qquad\qquad + C_{y,j,k+1/2}^n \Delta u_{j,k+1/2}^n - D_{y,j,k-1/2}^n \Delta u_{j,k-1/2}^n. \end{cases}$$

This form can be useful for proving monotonicity (see Spekreijse 1987) and L^∞ estimates.

Lemma 3.1

Assume that the scheme (3.3) can be put in incremental form (3.11) and that the incremental coefficients satisfy for any j, k, n

$$C^n_{x,j+1/2,k} \geq 0, \quad D^n_{x,j-1/2,k} \geq 0, \quad C^n_{y,j,k+1/2} \geq 0, \quad D^n_{y,j,k-1/2} \geq 0,$$
$$C^n_{x,j+1/2,k} + D^n_{x,j-1/2,k} + C^n_{y,j,k+1/2} + D^n_{y,j,k-1/2} \leq 1.$$

Then, the scheme is L^∞-stable.

Proof. By assumption, we can write $u^{n+1}_{j,k}$ as a convex combination of the values $u^n_{j,k}$, $u^n_{j\pm1,k}$, $u^n_{j,k\pm1}$. □

Remark 3.3. The result holds if the incremental form differs from (3.11) by a term $E^n_{j,k}$ such that $n|E^n_{j,k}|$ can be bounded. It has been applied for higher-order schemes built from monotone or E-schemes following a corrected antidiffusive flux approach (see G.R. Chapter IV, Section 1): the one-dimensional numerical E-flux $F^n_{j+1/2,k}$ (resp. $G^n_{j,k+1/2}$) is corrected by an antidiffusive flux $h^n_{j+1/2,k}$ (resp. $k^n_{j,k+1/2}$)

$$\tilde{F}^n_{j+1/2,k} = \frac{F^n_{j+1/2,k} + h^n_{j+1/2,k}}{\lambda_x}, \quad \tilde{G}^n_{j,k+1/2} = \frac{G^n_{j,k+1/2} + k^n_{j,k+1/2}}{\lambda_y}.$$

This L^∞-estimate is used by Coquel and Le Floch (1993) to prove the convergence of this type of schemes. As we shall detail in Section 4.2.3, the L^∞-estimate ensures that a Young measure ν can be constructed from the family (u_h). However, since L^∞-stability is not sufficient to ensure the convergence of the nonlinear terms, some other estimates are needed (which are linked to the local entropy production of the scheme). We just sketch the ideas of the proof. The E-schemes satisfy, for any convex entropy U and the associated numerical entropy fluxes ψ_x and ψ_y, a discrete entropy inequality (see G.R., Chapter III, Theorem 4.3)

$$\begin{aligned} U(u^{n+1}_{j,k}) \leq U(u^n_{j,k}) &- \lambda_x(\psi^n_{x,j+1/2,k} - \psi^n_{x,j-1/2,k}) \\ &- \lambda_y(\psi^n_{y,j,k+1/2} - \psi^n_{y,j,k-1/2}). \end{aligned}$$

Following some ideas of the proof of Theorem 1.1 in G.R., Chapter IV, the authors show for the modified scheme an estimate of the form

$$\begin{aligned} U(u^{n+1}_{j,k}) \leq U(u^n_{j,k}) &- \lambda_x(\psi^n_{x,j+1/2,k} - \psi^n_{x,j-1/2,k}) \\ &- \lambda_y(\psi^n_{y,j,k+1/2} - \psi^n_{y,j,k-1/2}) + R^n_{j,k}. \end{aligned}$$

The term $R^n_{j,k}$ is shown to tend to zero. More precisely, it is evaluated by obtaining a sharp evaluation of the entropy production generated by the whole scheme for the particular entropy $U(u) = \frac{u^2}{2}$. Note that the proof is based on a decomposition of the scheme in a form analogous to (3.6), i.e., as a convex combination of the two one-dimensional schemes

$$\begin{aligned} u_{j,k}^{n+1} &= \frac{1}{2}\{u_{j,k}^n - 2\lambda_x(F_{j+1/2,k}^n - F_{j-1/2,k}^n)\} \\ &+ \frac{1}{2}\{u_{j,k}^n - 2\lambda_y(G_{j,k+1/2}^n - G_{j,k-1/2}^n)\}. \end{aligned}$$

The sharp estimate of the entropy production leads to a weak uniform B.V. estimate

$$\begin{aligned} \Delta t \sum_{j\le T/\Delta t} TV(u^n) &= \Delta t \sum_n \sum_{j,k} \{\Delta y|u_{j+1,k} - u_{j,k}| + \Delta x|u_{j,k+1} - u_{j,k}\} \\ &\le C\,\Delta t^{-2/3}; \end{aligned}$$

it does not imply a compactness property but is enough to pass to the limit, which proves that the Young measure ν associated with u_h is indeed a measure-valued solution (see Definition 4.1 below). Together with DiPerna's uniqueness result (see Remark 3.1) the discrete entropy inequalities imply that ν is the unique entropy weak L^∞-solution. □

3.2 Dimensional splitting

However, still assuming a Cartesian grid, another popular way of constructing multidimensional schemes starting from one-dimensional schemes is to use a splitting method. The most practical calculations use a dimensional splitting or alternating direction technique (see Beam and Warming 1976, Woodward and Colella 1984, Yee, Warming, and Harten 1985, Daru and Lerat 1983, van Leer and Mulder 1983, Le Gruyer and Le Roux 1983, Glaister 1988, Osher and Solomon 1982, Clarke et al. 1993). Note that some attemps have been made to use operator splitting in some particular cases (Dukowicz and Dvinsky 1992, Baraille et al. 1992, Buffard and Hérard 1993, Buffard 1993, Liou and Steffen 1993, Chalabi and Vila 1992).

Let us first present the general time-stepping technique, without specifying the space discretization or the chosen decomposition. Suppose that we have discretized in some way the terms $\frac{\partial}{\partial x} f(u)$ and $\frac{\partial}{\partial y} g(u)$, and consider the method of lines. We shall even restrict ourselves to the simple case of a linear differential system

$$\frac{du}{dt} + (A + B)u = 0, \tag{3.12}$$

where A and B are constant matrices. The simplest procedure consists of two steps. Assume that we have an approximation u^n of u at time t_n. In order to compute u^{n+1}, in the first step one solves the equations

$$\begin{cases} \dfrac{du}{dt} + Au = 0, \quad t \in (t_n, t_{n+1}), \\ u(t_n) = u^n, \end{cases} \tag{3.13a}$$

which give $u^{n,1} = u(t_{n+1}) = e^{-\Delta t A} u^n$.

In the second step, one solves

$$\text{(3.13}b\text{)} \qquad \begin{cases} \dfrac{du}{dt} + Bu = 0, \\ u(t_n) = u^{n,1}, \end{cases}$$

and then one takes

$$\text{(3.14)} \qquad u^{n+1} = u(t_{n+1}) = e^{-\Delta t B} u^{n,1} = e^{-\Delta t B} e^{-\Delta t A} u^n.$$

Thus, we have

$$u^{n+1} = (e^{-\Delta t B} e^{-\Delta t A})^n u^0.$$

This leads to a first-order-in-time method, as we shall see later. A more subtle time stepping introduces three steps: one solves (3.13a) during a half timestep, then (3.13b) on a timestep, and last (3.13a) on a half timestep. Hence, we first solve

$$\text{(3.15}a\text{)} \qquad \begin{cases} \dfrac{du}{dt} + Au = 0, \quad t \in (t_n, t_{n+1/2}), \\ u(t_n) = u^n, \end{cases}$$

which gives

$$u^{n,1} = u(t_{n+1/2}) = e^{-\Delta t A/2} u^n.$$

In the second step, one solves

$$\text{(3.15}b\text{)} \qquad \begin{cases} \dfrac{dv}{dt} + Bv = 0, \quad t \in (t_n, t_{n+1}), \\ v(t_n) = u^{n,1}, \end{cases}$$

which gives

$$u^{n,2} = v(t_{n+1}) = e^{-\Delta t B} u^{n,1}.$$

Finally,

$$\text{(3.15}c\text{)} \qquad \begin{cases} \dfrac{dw}{dt} + Aw = 0, \quad t \in (t_{n+1/2}, t_{n+1}), \\ w(t_{n+1/2}) = u^{n,2}, \end{cases}$$

which yields

$$\text{(3.16)} \qquad u^{n+1} = w(t_{n+1}) = e^{-\Delta t A/2} u^{n,2} = e^{-\Delta t A/2} e^{-\Delta t B} e^{-\Delta t A/2} u^n.$$

Lemma 3.2

The scheme (3.13), (3.14) yields a first-order accurate method if the matrices A and B do not commute, whereas (3.15), (3.16) is second-order accurate.

Proof. We use the Taylor expansion of the exponential function

$$\begin{aligned}
&e^{-\Delta t A/2} e^{-\Delta t B} e^{-\Delta t A/2} \\
&= \Big(I - \frac{\Delta t}{2} A + \frac{\Delta t^2}{8} A^2 + \dots\Big)\Big(I - \Delta t B + \frac{\Delta t^2}{2} B^2 + \dots\Big) \\
&\qquad \Big(I - \frac{\Delta t}{2} A + \frac{\Delta t^2}{8} A^2 + \dots\Big) \\
&= I - \Delta t(A + B) + \frac{\Delta t^2}{2} (A^2 + AB + BA + B^2) + \dots \\
&= e^{-\Delta t(A+B)} + 0(\Delta t^3),
\end{aligned}$$

which proves that the resulting 3-timestep scheme is second-order accurate (this is Strang's result). If, however, the matrices A and B do not commute,

$$\begin{aligned}
e^{-\Delta t B} e^{-\Delta t A} &= \Big(I - \frac{\Delta t}{2} B + \frac{\Delta t^2}{2} B^2 + \dots\Big)\Big(I - \Delta t A + \frac{\Delta t^2}{2} A^2 + \dots\Big) \\
&= I - \Delta t(A + B) + \frac{\Delta t^2}{2} (A^2 + 2BA + B^2) + \dots \\
&= e^{-\Delta t(A+B)} + 0(\Delta t^2),
\end{aligned}$$

and the two-step scheme is only first-order accurate. □

Remark 3.4. The convergence of scheme (3.13) is a very simple example of the application of the Trotter formula, which holds for generators of unbounded operators A, B in a Banach space,

$$\exp\{t(A + B)\} = \lim_{n\to\infty} \Big\{\exp\Big(\frac{tA}{n}\Big)\exp\Big(\frac{tB}{n}\Big)\Big\}^n.$$

(For a detailed result and proof, see Chorin and al. 1978). □

Let us now study the application for a two-dimensional conservation law (3.6)

$$\begin{aligned}
&\frac{\partial u}{\partial t} + \frac{\partial}{\partial x} f(u) + \frac{\partial}{\partial y} g(u) = 0, \\
&u(x, y, 0) = u_0(x, y).
\end{aligned}$$

Using the notations of G.R., Chapter II, Section 5, we can write the unique entropy solution as $u(\cdot, t) = S(t)u_0$, where $S(t)$ is the solution operator. We consider the one-dimensional conservation laws

$$\frac{\partial u}{\partial t} + \frac{\partial}{\partial x} f(u) = 0$$

and

$$\frac{\partial u}{\partial t} + \frac{\partial}{\partial y} g(u) = 0.$$

The associated solution operators $S_x(t)$ and $S_y(t)$ take the place of the above exponential function. Thus, the fractional step (dimensional splitting) method approximates $u(x, y, n\Delta t) = S(n\Delta t)u_0(x, y)$ by $(S_y(\Delta t) S_x(\Delta t))^n u_0(x, y)$. (The accuracy of this semi-discrete splitting is studied in Teng 1994).

In order to define an alternate-direction, fully discrete scheme, we introduce finite difference one-dimensional schemes,

$$u_j^{n+1} = H_x(u_{j-J}^n, ..., u_{j+J}^n), \tag{3.17a}$$

$$u_j^{n+1} = H_y(u_{j-K}^n, ..., u_{j+K}^n), \tag{3.17b}$$

and we set

$$\begin{cases} u_{j,k}^{n,1} = H_x(u_{j-J,k}^n, ..., u_{j+J,k}^n), \\ u_{j,k}^{n+1} = H_y(u_{j,k-K}^{n,1}, ..., u_{j,k-K}^{n,1}). \end{cases} \tag{3.18}$$

Lemma 3.3

Assume that the difference schemes (3.17) are in conservation form and consistent with f and g, respectively. Then, the resulting scheme (3.18) can be put in conservation form and is consistent. Moreover, in the scalar case, if the schemes are monotone, the scheme (3.18) is also monotone.

Proof. By assumption on scheme (3.17a), we can write $u_{j,k}^{n,1}$ in the form

$$u_{j,k}^{n,1} = u_{j,k}^n - \lambda_x\{F_{j+1/2,k}^n - F_{j-1/2,k}^n)\},$$

and similarly,

$$u_{j,k}^{n+1} = u_{j,k}^{n,1} - \lambda_y\{G_{j,k+1/2}^{n,1} - G_{j,k-1/2}^{n,1})\},$$

where

$$G_{j,k+1/2}^{n,1} = G(u_{j,k-K+1}^{n,1}, ..., u_{j,k+K}^{n,1}).$$

Thus, (3.18) yields

$$u_{j,k}^{n+1} = u_{j,k}^n - \lambda_x\{F_{j+1/2,k}^n - F_{j-1/2,k}^n)\} - \lambda_y\{G_{j,k+1/2}^{n,1} - G_{j,k-1/2}^{n,1}).$$

Substituting the expressions for $u_{j,k}^{n,1}$ in G, we check that we can put the scheme in conservation form. Indeed, we have

$$G_{j,k+1/2}^{n,1} = \tilde{G}(u_{j-J,k-K+1}^n, ..., u_{j+J,k+K}^n) = \tilde{G}_{j,k+1/2}^n$$

for some function $\tilde{G}$. It is also consistent: taking

$$u_{j-J,k-K+1}^n = ... = u_{j+J,k+K}^n - u,$$

we have by the consistency of F with f,

$$u_{j,k-K+1}^{n,1} = ... = u_{j,k+K}^{n,1} = u,$$

which implies by the consistency of G with g

$$G(u, ..., u) = g(u).$$

The fact that the resulting scheme (3.18) is monotone if both schemes (3.17) are monotone is easily proven. Since H_x and H_y are nondecreasing functions of each of their arguments,

$$H_\Delta(u^n_{j-J,k-K}, ... u^n_{j+J,k+K}) = H_y(H_x(...), ..., H_x(...))$$

is also a nondecreasing function. □

Is it possible to prove an analogous property for a scheme split from T.V.D. schemes? In order to study this question, we assume that the schemes (3.17) are T.V.D. and can be written in incremental form,

$$\begin{aligned} (3.19a) \qquad u_j^{n+1} &= H_x(u^n_{j-J}, ..., u^n_{j+J}) \\ &= u_j^n + C^n_{x,j+1/2}\Delta u^n_{j+1/2} - D^n_{x,j-1/2}\Delta u^n_{j-1/2} \\ (3.19b) \qquad u_j^{n+1} &= H_y(u^n_{j-K}, ..., u^n_{j+K}) \\ &= u_j^n + C^n_{y,j+1/2}\Delta u^n_{j+1/2} - D^n_{y,j-1/2}\Delta u^n_{j-1/2}, \end{aligned}$$

where the coefficients C, D are defined by

$$C^n_{x,j+1/2} = C_x(u^n_{j-J+1}, ... u^n_{j+J})$$

for some function $C_x : \mathbb{R}^{2J} \to \mathbb{R}$, and so on. We can define an alternate-direction scheme by (3.18). Let us consider the simple linear constant coefficient case

$$\frac{\partial u}{\partial t} + a\frac{\partial u}{\partial x} + b\frac{\partial u}{\partial y} = 0,$$

and assume moreover that the coefficients $C^n_{x,j+1/2}$, $D^n_{x,j+1/2}$, $C^n_{y,j+1/2}$, $D^n_{y,j+1/2}$ do not depend on j, k, n and satisfy the T.V.D. property (see G.R., Chapter III, (3.24)), i.e.,

$$(3.20a) \quad C^n_{x,j+1/2} = C_x,\ C^n_{y,j+1/2} = C_y,\ D^n_{x,j+1/2} = D_x,\ D^n_{y,j+1/2} = D_y$$

with

$$(3.20b) \qquad C \geq 0, \quad D \geq 0, \quad C + D = Q \leq 1.$$

This occurs if schemes (3.17a) and (3.17b) coincide and are, for instance, 3-point conservative consistent schemes (see Chapter III, Section 1, and G.R., Chapter III, formula (1.38)). Even in this simple case, we cannot prove that the resulting scheme is T.V.D., only that the total variation is bounded. More precisely, we obtain then the following estimate.

Lemma 3.4

Assume that the difference schemes (3.17) are T.V.D. and that (3.20) holds. Then, scheme (3.18) satisfies

$$TV(u^{n+1}) \leq TV(u^n) + (\Delta x + \Delta y)\sum_{i,j} |\Delta_x \Delta_y u^n_{i,j}|.$$

Proof. Let us write the schemes (3.17) in incremental form,

$$u_{j,k}^{n,1} = H_x(u_{j-J,k}^n, ..., u_{j+J,k}^n) = u_j^n + C_x \Delta_x u_{j,k}^n - D_x \Delta_x u_{j-1,k}^n$$

where we have used the notations (3.10), and

$$u_{j,k}^{n+1} = H_y(u_{j,k-K}^{n,1}, ..., u_{j,k+K}^{n,1}) = u_{j,k}^{n,1} + C_y \Delta_y u_{j,k}^{n,1} - D_y \Delta_y u_{j,k-1}^{n,1}.$$

It is easy to prove that the operators Δ_x and Δ_y commute. Indeed, by (3.20), we have

$$\begin{aligned} \Delta_x \Delta_y u_{j,k} &= \Delta_x(u_{j,k+1} - u_{j,k}) = u_{j+1,k+1} - u_{j,k+1} - u_{j+1,k} + u_{j,k} \\ &= \Delta_y(u_{j+1,k} - u_{j,k}) = \Delta_y \Delta_x u_{j,k}. \end{aligned}$$

By assumption (3.20), this yields that the operators H_x, H_y, Δ_x, and Δ_y also commute; for instance, the two expressions

$$\begin{aligned} \Delta_y H_x u_{j,k} &= \Delta_y(u_{j,k} + C_x \Delta_x u_{j,k} - D_x \Delta_x u_{j-1,k}) \\ &= \Delta_y u_{j,k} + C_x \Delta_y \Delta_x u_{j,k} - D_x \Delta_y \Delta_x u_{j-1,k} \end{aligned} \tag{3.21a}$$

and

$$H_x \Delta_y u_{j,k} = \Delta_y u_{j,k} + C_x \Delta_x \Delta_y u_{j,k} - D_x \Delta_x \Delta_y u_{j-1,k} \tag{3.21b}$$

coincide. Then, by (3.8),

$$\begin{aligned} TV(u_{j,k}) &= \Delta y \sum_{j,k} |\Delta_x u_{j,k}| + \Delta x \sum_{j,k} |\Delta_y u_{j,k}| \\ &= \Delta y \sum_k TV_x(u_{\cdot,k}) + \Delta x \sum_j TV_y(u_{j,\cdot}), \end{aligned}$$

where TV_x or TV_y denotes the (one-dimensional) TV norm (see Chapter III, Section 1). Thus, since the operators commute, we can write

$$TV(H_y H_x u_{j,k}) = \Delta y \sum_k TV_x(H_x H_y u)_{\cdot,k} + \Delta x \sum_j TV_y(H_y H_x u)_{j,\cdot},$$

and if the one-dimensional schemes are T.V.D.,

$$\begin{aligned} TV(H_y H_x u_{j,k}) &\leq \Delta y \sum_k TV_x(H_y u)_{\cdot,k} + \Delta x \sum_j TV_y(H_x u)_{j,\cdot} \\ &= \Delta y \sum_{j,k} |\Delta_x H_y u_{j,k}| + \Delta x \sum_{j,k} |\Delta_y H_x u_{j,k}|. \end{aligned}$$

By (3.21) and (3.20b), we obtain, setting $Q = C + D$

$$\begin{aligned} &\sum_{j,k} |\Delta_y H_x u_{j,k}| \\ &\quad \leq \sum_{j,k} |\Delta_y u_{j,k}| + C_x \sum_{j,k} |\Delta_x \Delta_y u_{j,k}| + D_x \sum_{j,k} |\Delta_x \Delta_y u_{j-1,k}| \\ &\quad \leq \sum_{j,k} |\Delta_y u_{j,k}| + Q_x \sum_{j,k} |\Delta_x \Delta_y u_{j,k}|. \end{aligned}$$

Thus

$$TV(H_y H_x u_{j,k}) \leq TV(u_{j,k}) + (Q_x \Delta x + Q_y \Delta y) \sum_{j,k} |\Delta_x \Delta_y u_{j,k}|,$$

which gives the result

$$TV(u^{n+1}) \leq TV(u^n) + (\Delta x + \Delta y) \sum_{j,k} |\Delta_x \Delta_y u^n_{j,k}|.$$

Now, using again the fact that both schemes are T.V.D. and that the operators commute, we can write

$$\sum_{j,k} |\Delta_x \Delta_y u^n_{j,k}| = \sum_{j,k} \Delta_x \Delta_y H_y H_x u^{n-1}_{j,k}| = \sum_{j,k} |\Delta_x H_x \Delta_y H_y u^{n-1}_{j,k}|$$
$$\leq \sum_{j,k} |\Delta_x \Delta_y H_y u^{n-1}_{j,k}| = \sum_{j,k} |\Delta_y H_y \Delta_x u^{n-1}_{j,k}| \leq \sum_{j,k} |\Delta_y \Delta_x u^{n-1}_{j,k}|.$$

We obtain, then, that for $n \leq \frac{T}{\Delta t}$,

$$TV(u^n) \leq TV(u^0) + T\Big(\frac{Q_x}{\lambda_x} + \frac{Q_y}{\lambda_y}\Big) \sum_{j,k} |\Delta_x \Delta_y u^0_{j,k}|,$$

and thus a scheme that is T.V.B. □

For instance, if we take the same 3-point upwind scheme in each direction,

$$C_x + D_x = Q_x = \lambda_x |a|, \quad C_y + D_y = Q_y = \lambda_y |b|,$$

we get

$$TV(u^n) \leq TV(u^0) + T(|a| + |b|) \sum_{j,k} |\Delta_x \Delta_y u^0_{j,k}|.$$

Remark 3.5. In fact, following the ideas of Lemma 3.2, when using an alternating direction scheme, one can solve (3.13a) during a half timestep $\frac{\Delta t}{2}$, then (3.13b) during a time step Δt, and finally one solves (3.13a) during a half timestep $\frac{\Delta t}{2}$. It is then necessary to preserve the global order of discretization to use a second-order (in time) method such as a Runge–Kutta method or an implicit multistep method (see Beam and Warming 1976) instead of the backward Euler method. Anyway, the drawbacks of such alternating direction techniques are obvious since the grid directions play an overdetermined role. Stability is studied in Serre (1992). □

4 Finite-volume methods

Actually, the most usual extensions include a finite-element or a finite-volume formulation on structured or unstructured meshes. In structured (Cartesian or curvilinear) meshes, there exist locally two axes, the center of the cell admits a natural parametrization by (i, j), and each cell is surrounded by a fixed number of neighboring cells. This allows A.D.I. techniques and easy implementation on vector computers. Unstructured meshes, with the use of an automatic mesh generator and adaptive grid refinement, offer a greater flexibility when dealing with complex geometries and limit grid orientation effects. So let us consider now the general case of an unstructured mesh, which means that we do not consider a rectangular grid Δ as in the previous section but a "triangulation" $\mathcal{T}_h$ of the computational domain $\mathcal{O}$ by triangles or quadrilaterals.

4.1 Definition of the finite-volume method

4.1.1 General principles

In a *finite-volume* method, the computational domain $\mathcal{O}$ is composed of cells, or control volumes, Ω_i with center $\mathbf{c}_i$, which are either the elements of the triangulation or constructed from these elements (see Examples 4.1 and 4.2). On Ω_i, $\mathbf{u}(\cdot, t)$ is approximated by a constant $\mathbf{u}_i(t)$, which should be considered as an approximation of the mean value of $\mathbf{u}$ over the cell Ω_i rather than of the value at point $\mathbf{c}_i$,

$$\mathbf{u}_i(t) \cong \frac{1}{|\Omega_i|} \int_{\Omega_i} \mathbf{u}(\mathbf{x}, t) d\mathbf{x},$$

where $|\Omega_i|$ denotes the area of Ω_i. The differential system defining $\mathbf{u}_i(t)$ is obtained as follows. First, integrating the system (3.2) over Ω_i,

$$\int_{\Omega_i} \Big(\frac{\partial \mathbf{u}}{\partial t} + \operatorname{div} \mathbb{F}(\mathbf{u}) \Big) d\mathbf{u} = \mathbf{0}, \quad \mathbb{F} = (\mathbf{f}, \mathbf{g})^{\mathbf{T}},$$

yields

$$\frac{\partial}{\partial t} \Big(\int_{\Omega_i} \mathbf{u}(\mathbf{x}, t) d\mathbf{x} \Big) + \int_{\partial \Omega_i} \mathbb{F}(\mathbf{u}(\cdot, t)) \cdot \mathbf{n}_i \, d\sigma = \mathbf{0}, \tag{4.1}$$

where $\partial\Omega_i$ is the boundary of Ω_i and $\mathbf{n}_i$ the outward unit normal vector to Ω_i (and the dot "$\cdot$" stands for the product $\mathbb{F} \cdot \mathbf{n} = \cos\theta \, \mathbf{f} + \sin\theta \, \mathbf{g}$ if $\mathbf{n} = (\cos\theta, \sin\theta)^T$). The first term in (4.1) is naturally approximated by

$$\frac{\partial}{\partial t} \Big(\int_{\Omega_i} \mathbf{u}(\mathbf{x}, t) d\mathbf{x} \Big) \cong |\Omega_i| \frac{\partial \mathbf{u}_i(t)}{\partial t}.$$

Since the approximation is not continuous across $\partial\Omega_i$, we have to discretize the "residual" $\int_{\partial\Omega_i} \mathbb{F}(u(\cdot, t_n)) \cdot \mathbf{n}_i d\sigma$, which represents the flux across the boundary of the cell at time t_n. It can be written

$$\int_{\partial\Omega_i} \mathbb{F}(\mathbf{u}) \cdot \mathbf{n}_i \, d\sigma = \sum_{e \subset \partial\Omega_i, e = \Gamma_{ij}} \int_{\Gamma_{ij}} \mathbb{F}(\mathbf{u}) \cdot \mathbf{n}_i \, d\sigma,$$

where the sum is taken over all the edges e of the cell, and $\partial\Omega_i = \cup\Gamma_{ij}$, where $\Gamma_{ij} = \Omega_i \cap \Omega_j$ is the face separating Ω_i and Ω_j. Note that the control cells are usually assumed to satisfy the properties of a finite-element triangulation: the Ω_i's are nonoverlapping sets and, if e is a given edge of $\partial\Omega_i$, there exists a unique Ω_j such that $e = \Omega_i \cap \Omega_j$.

The problem is then to define the numerical fluxes approximating $\int_{\Gamma_{ij}} \mathbb{F}(\mathbf{u}) \cdot \mathbf{n}_i \, d\sigma$, using only the values $\mathbf{u}_i(t)$. In fact, we shall only detail here the case of internal fluxes and not the fluxes at the boundary of $\mathcal{O}$ (for instance, solid wall or inflow–outflow conditions will be considered later on in Chapter V). The usual way consists in introducing a function $\mathbf{\Phi}$ such that for $e = \Gamma_{ij} \subset \partial\Omega_i$

$$\int_{\Gamma_{ij}} \mathbb{F}(\mathbf{u}) \cdot \mathbf{n}_i \, d\sigma \cong |e| \mathbf{\Phi}(\mathbf{u}_i, \mathbf{u}_j, \mathbf{n}_e), \tag{4.2}$$

where $\mathbf{n}_e$ denotes the unit normal to e pointing in the direction of Ω_j (thus outward to Ω_i), and $|e|$ the length of e. Though the notation for the numerical flux is not completely satisfying, it means that we have assumed that the numerical flux depends "only" on the values on each side of the edge and on the normal direction to the edge (it depends also on the continuous flux $\mathbb{F}$). This yields a "method of lines"

$$|\Omega_i| \frac{\partial \mathbf{u}_i(t)}{\partial t} + \sum_{e \subset \partial\Omega_i, e=\Gamma_{ij}} |e| \Phi(\mathbf{u}_i, \mathbf{u}_j, \mathbf{n}_e) = \mathbf{0}.$$

Eventually, we approximate this ordinary differential system by the explicit Euler scheme to obtain the formula

$$|\Omega_i|(\mathbf{u}_i^{n+1} - \mathbf{u}_i^n) + \Delta t \Big\{ \sum_{e \subset \partial\Omega_i, e=\Gamma_{ij}} |e| \Phi(\mathbf{u}_i^n, \mathbf{u}_j^n, \mathbf{n}_e) \Big\} = \mathbf{0}, \tag{4.3}$$

where $\mathbf{u}_i^n \cong \mathbf{u}_i(t_n)$, and where $\mathbf{u}_i^0$ is given.

4.1.2 Properties

In the general case, the numerical fluxes $\mathbf{\Phi}$ are assumed to be locally Lipschitz continuous and must satisfy some conditions:
Conservation:

$$\mathbf{\Phi}(\mathbf{u}_i, \mathbf{u}_j, \mathbf{n}) = -\mathbf{\Phi}(\mathbf{u}_j, \mathbf{u}_i, -\mathbf{n}). \tag{4.4}$$

This property (which is directly inherited from the continuous flux (4.2)) means that, in the absence of source term, the approximate flux at the boundary separating Ω_i and Ω_j is the *same* as the flux at the boundary separating Ω_j and Ω_i (since $-\mathbf{n}_e$ is the unit normal to e pointing in the direction of Ω_i), i.e., there is only one exchange term per edge e separating Ω_i and Ω_j. In the one-dimensional case it reduces to the fact that we can write the numerical flux at the boundary $x_{j+1/2}$ of the cells $(x_{j-1/2}, x_{j+1/2})$ and $(x_{j+1/2}, x_{j+3/2})$ in the form $g_{j+1/2}$. As previously, conservation ensures that when the scheme converges (in a strong way), the limit satisfies the

Rankine–Hugoniot condition and is thus a weak solution of the conservation law (see Proposition 4.1 below).

Consistency:

$$\mathbf{\Phi}(\mathbf{u}, \mathbf{u}, \mathbf{n}) = \mathbb{F}(\mathbf{u}) \cdot \mathbf{n}. \tag{4.5}$$

Again, this is natural from (4.2).

Usually, the flux $\mathbf{\Phi}(\mathbf{u}_i^n, \mathbf{u}_j^n, \mathbf{n})$ is defined by solving exactly or approximately a one-dimensional Riemann problem, in the direction $\mathbf{n}$ normal to the edge $e = \Gamma_{ij}$, associated to the (continuous) flux $\mathbb{F}(\mathbf{u}) \cdot \mathbf{n}$. More precisely, we define as in Section 1.1 new variables ζ (normal) and τ (tangential) by

$$\zeta = \cos\,\theta\, x + \sin\,\theta\, y = \mathbf{X} \cdot \mathbf{n}, \quad \tau = -\sin\,\theta\, x + \cos\,\theta\, y = \mathbf{X} \cdot \mathbf{n}^{\perp},$$

where θ is the angle of the normal to an edge with the x-axis, i.e., $\mathbf{n} = (\cos\,\theta, \sin\,\theta)$, $\mathbf{n}^{\perp} = (-\sin\,\theta, \cos\,\theta)$ is directly orthogonal to $\mathbf{n}$, and $\mathbf{X} = (x, y)$. The system (3.2) is transformed into

$$\frac{\partial \mathbf{v}}{\partial t} + \frac{\partial}{\partial \zeta}(\mathbb{F} \cdot \mathbf{n})(\mathbf{v}) + \frac{\partial}{\partial \tau}(\mathbb{F} \cdot \mathbf{n}^{\perp})(\mathbf{v}) = \mathbf{0},$$

where

$$\mathbf{v}(\zeta, \tau, t) = \mathbf{u}(x(\zeta, \tau), y(\zeta, \tau), t)$$

and

$$\mathbb{F} \cdot \mathbf{n} = \cos\,\theta\, \mathbf{f} + \sin\,\theta\, \mathbf{g}, \quad \mathbb{F} \cdot \mathbf{n}^{\perp} = \sin\,\theta\, \mathbf{f} + \cos\,\theta\, \mathbf{g},$$

i.e.,

$$R^{-1}\mathbb{F} = (\mathbb{F} \cdot \mathbf{n}, \mathbb{F} \cdot \mathbf{n}^{\perp})^T$$

if R is the rotation with angle θ in $\mathbb{R}^2$.

Now, if $\mathbf{u}$ is a given constant on each side of the line $\zeta = 0$, the associated Cauchy problem reduces to solving the one-dimensional Riemann projected problem in the direction $\mathbf{n}$,

$$\frac{\partial \mathbf{v}}{\partial t} + \frac{\partial}{\partial \zeta}(\mathbb{F} \cdot \mathbf{n})(\mathbf{v}) = \mathbf{0},$$

$$\mathbf{v}(\zeta, 0) = \begin{cases} \mathbf{u}_i, & \zeta < 0, \\ \mathbf{u}_j, & \zeta > 0, \end{cases}$$

and the flux through the (extended) edge is $\mathbb{F} \cdot \mathbf{n}(\mathbf{w}_R(0; \mathbf{u}_i, \mathbf{u}_j))$.

Thus, we shall take more generally a one-dimensional numerical flux $\varphi(\mathbf{u}, \mathbf{v})$ associated to a 3-point difference scheme. We introduce in the notations the dependence on the continuous flux (the conservative scheme with numerical flux $\varphi(\mathbf{u}, \mathbf{v}) = \varphi(\mathbf{f}; \mathbf{u}, \mathbf{v})$ approximates the system $\frac{\partial \mathbf{u}}{\partial t} + \frac{\partial}{\partial x}\mathbf{f}(\mathbf{u}) = \mathbf{0}$, and φ is consistent with $\mathbf{f}$). We define $\mathbf{\Phi}$ by

$$\mathbf{\Phi}(\mathbf{u}_i, \mathbf{u}_j, \mathbf{n}) = \varphi(\mathbb{F} \cdot \mathbf{n}; \mathbf{u}_i, \mathbf{u}_j). \tag{4.6}$$

The finite-volume method is said to be *monotone* if this underlying flux is that of a 3-point monotone scheme (see G.R., Chapter III, Section 3). This gives a "first-order" method.

Remark 4.1. Some care must be taken in the notations. For instance, if we choose as above the Godunov flux

$$\varphi(\mathbf{f};\mathbf{u},\mathbf{v}) = \mathbf{f}(\mathbf{w}_r(0;\mathbf{u},\mathbf{v}))$$

and then define

$$\mathbf{\Phi}(\mathbf{u}_i,\mathbf{u}_j,\mathbf{n}) = \mathbb{F}\cdot\mathbf{n}(\mathbf{w}_R(0;\mathbf{u}_i,\mathbf{u}_j)) = R\,\mathbf{f}(\mathbf{w}_R(0;\tilde{\mathbf{u}}_i,\tilde{\mathbf{u}}_j),$$

the conservation property is not satisfied since in general

$$\mathbf{f}(\mathbf{w}_R(0;\mathbf{u},\mathbf{v})) \neq \mathbf{f}(\mathbf{w}_R(0;\mathbf{v},\mathbf{u})).$$

We must in fact introduce $\mathbf{f}$ in the notations $\mathbf{w}_R(0;\mathbf{v},\mathbf{u}) = \mathbf{w}_R(\mathbf{f})(0;\mathbf{v},\mathbf{u})$ for the solution of the Riemann problem associated to $\mathbf{f}$ and the Riemann data $\mathbf{u}$ and $\mathbf{v}$. Indeed, when we change $\mathbf{n}$ to $-\mathbf{n}$, it corresponds after the rotation R to a change in the orientation of the x-axis, or equivalently to changing $\mathbf{f}$ to $-\mathbf{f}$. Now, we observe that

$$\mathbf{w}_R(-\mathbf{f})(0;\mathbf{v},\mathbf{u}) = \mathbf{w}_R(\mathbf{f})(0;\mathbf{u},\mathbf{v}),$$

and thus

$$\begin{aligned}\mathbf{\Phi}(\mathbf{u}_j,\mathbf{u}_i,-\mathbf{n}) &= -\mathbb{F}\cdot\mathbf{n}(\mathbf{w}_R(-\mathbb{F}\cdot\mathbf{n})(0;\mathbf{u}_j,\mathbf{u}_i))\\ &= -\mathbb{F}\cdot\mathbf{n}(\mathbf{w}_R(\mathbb{F}\cdot\mathbf{n})(0;\mathbf{u}_i,\mathbf{u}_j)) = -\mathbf{\Phi}(\mathbf{u_i},\mathbf{u_j},\mathbf{n}),\end{aligned}$$

so that the flux is indeed conservative. □

Remark 4.2. The work that is going on today concerning the study of the solution of the two-dimensional Riemann problem (see Remark 2.7 in this chapter) may lead to the derivation of truly bidimensional schemes in the spirit of Godunov's scheme. □

If we want to apply the scheme to the gas dynamics equations, we can, moreover, require that $\mathbf{\Phi}$ be invariant under rotation. Setting for a rotation R in $\mathbb{R}^2$, as in (2.12),

$$R(a,\mathbf{b},c)^T = (a,R\mathbf{b},c)^T,$$

so that R denotes either the 2×2 matrix (2.7) or the 4×4 matrix

$$R = \begin{pmatrix} 1 & 0 & 0 & 0\\ 0 & \cos\theta & -\sin\theta & 0\\ 0 & \sin\theta & \cos\theta & 0\\ 0 & 0 & 0 & 1\end{pmatrix},$$

depending on whether it acts on $\mathbb{R}^2$ or $\mathbb{R}\times\mathbb{R}^2\times\mathbb{R}$, we have the obvious definition of
rotational invariance:

$$\mathbf{\Phi}(\mathbf{U}_i,\mathbf{U}_j,\mathbf{n}) = R\mathbf{\Phi}(R^{-1}\mathbf{U}_i,R^{-1}\mathbf{U}_j,R^{-1}\mathbf{n}), \tag{4.7}$$

for any rotation R in $\mathbb{R}^2$.

Recall that (see Section 2.2, Lemma 2.4 and Remark 2.3) if $\mathbf{U} = (\rho, \rho\mathbf{u}, \rho e)^T$,

$$\tilde{\mathbf{U}} = R^{-1}\mathbf{U} = (\rho, \rho\tilde{\mathbf{u}}, \rho e)^T, \text{ where } \tilde{\mathbf{u}} = (u_n, u_\tau)^T,$$

and if $\mathbb{F} = (\mathbf{f}, \mathbf{g})^T$

$$\begin{aligned}
\mathbb{F} \cdot \mathbf{n} &= \cos\theta\, \mathbf{f} + \sin\theta\, \mathbf{g} \\
&= (\rho u_n, \rho u_n u + p \cos\theta, \rho\, u_n v + p \sin\theta, (\rho e + p) u_n)^T, \\
\mathbb{F} \cdot \mathbf{n}^\perp &= -\sin\theta\, \mathbf{f} + \cos\theta\, \mathbf{g} \\
&= (\rho u_\tau, \rho u_\tau u - p \sin\theta, \rho\, u_\tau v + p \cos\theta, (\rho e + p) u_\tau)^T.
\end{aligned}$$

Let $\mathbf{V}$ be defined as above by

$$\mathbf{V}(\zeta, \tau) = \mathbf{U}(x(\zeta, \tau), y(\zeta(\tau)),$$

so that

$$R^{-1}\mathbf{U}(x, y) = R^{-1}\mathbf{V}(\zeta, \tau) = \tilde{\mathbf{U}}(\zeta, \tau).$$

Letting R^{-1} act on $\frac{\partial \mathbf{V}}{\partial t} + \frac{\partial}{\partial \zeta}\mathbb{F}(\mathbf{V}) \cdot \mathbf{n} + \frac{\partial}{\partial \tau}\mathbb{F}(\mathbf{V}) \cdot \mathbf{n}^\perp = \mathbf{0}$ (in fact, it acts only on the two components in the middle), we have seen that

$$\begin{aligned}
R^{-1}&\left\{ \frac{\partial \mathbf{V}}{\partial t} + \frac{\partial}{\partial \zeta}(\mathbb{F}(\mathbf{V}) \cdot \mathbf{n}) + \frac{\partial}{\partial \tau}(\mathbb{F}(\mathbf{V}) \cdot \mathbf{n}^\perp) \right\} \\
&= \frac{\partial \tilde{\mathbf{U}}}{\partial t} + \frac{\partial}{\partial \zeta}\mathbf{f}(\tilde{\mathbf{U}}) + \frac{\partial}{\partial \tau}\mathbf{g}(\tilde{\mathbf{U}}) = \mathbf{0},
\end{aligned}$$

and the invariance of the Euler equations comes from the identities (2.14) (see Remark 2.3)

$$R^{-1}(\mathbb{F} \cdot \mathbf{n})(\mathbf{U})) = \mathbf{f}(\tilde{\mathbf{U}}), \quad R^{-1}(\mathbb{F} \cdot \mathbf{n}^\perp(\mathbf{U})) = \mathbf{g}(\tilde{\mathbf{U}}).$$

Considering as above a rotation R with angle θ such that $R^{-1}\mathbf{n} = \mathbf{e}_1$, where $\mathbf{e}_1$ is the first basis vector $\mathbf{e}_1 = (1, 0)$, we have for a rotational invariant flux satisfying (4.7)

$$\mathbf{\Phi}(\mathbf{U}_i, \mathbf{U}_j, \mathbf{n}) = R\mathbf{\Phi}(R^{-1}\mathbf{U}_i, R^{-1}\mathbf{U}_j, \mathbf{e}_1) = R\mathbf{\Phi}(\tilde{\mathbf{U}}_i, \tilde{\mathbf{U}}_j, \mathbf{e}_1).$$

If we denote by $(\Phi_1, \Phi_2, \Phi_3, \Phi_4)$ the four components of $\mathbf{\Phi}(\mathbf{U}_i, \mathbf{U}_j, \mathbf{n})$ and by $(\tilde{\Phi}_1, \tilde{\Phi}_2, \tilde{\Phi}_3, \tilde{\Phi}_4)$ those of $\mathbf{\Phi}(\tilde{\mathbf{U}}_i, \tilde{\mathbf{U}}_j, \mathbf{e}_1)$, rotational invariance supposes $\Phi_1 = \tilde{\Phi}_1$, $\Phi_4 = \tilde{\Phi}_4$, while Φ_2, Φ_3 are deduced from $\tilde{\Phi}_2$, $\tilde{\Phi}_3$ by a rotation with angle θ,

$$(\Phi_2, \Phi_3)^T = R(\tilde{\Phi}_2, \tilde{\Phi}_3)^T = (\cos\tilde{\Phi}_2 + \sin\tilde{\Phi}_3, -\sin\theta\Phi_2 + \cos\theta\Phi_3)^T.$$

Now, if we take as above in (4.6) a flux $\mathbf{\Phi}$ of the form

$$\mathbf{\Phi}(\mathbf{U}_i, \mathbf{U}_j, \mathbf{n}) = \boldsymbol{\varphi}(\mathbb{F} \cdot \mathbf{n}; \mathbf{U}_i, \mathbf{U}_j),$$

since

$$\mathbf{\Phi}(\tilde{\mathbf{U}}_i, \tilde{\mathbf{U}}_j, \mathbf{e}_1) = \boldsymbol{\varphi}(\mathbf{f}; \tilde{\mathbf{U}}_i, \tilde{\mathbf{U}}_j),$$

rotational invariance is equivalent to

$$\varphi(\tilde{\mathbf{U}}_i, \tilde{\mathbf{U}}_j, \mathbf{e}_1) = R\,\varphi(\mathbf{f}; \tilde{\mathbf{U}}_i, \tilde{\mathbf{U}}_j). \tag{4.8}$$

If we write the expressions of the usual 3-point fluxes, we can see that the last equality is satisfied by many usual schemes. Indeed, from (2.14),

$$\mathbb{F} \cdot \mathbf{n}(\mathbf{U}) = R\,\mathbf{f}(\tilde{\mathbf{U}}). \tag{4.9a}$$

We get by differentiation

$$\mathbf{A}(\mathbf{U}, \mathbf{n}) = (\mathbb{F} \cdot \mathbf{n})'(\mathbf{U}) = R\mathbf{f}'(\tilde{\mathbf{U}})R^{-1} = R\mathbf{A}(\tilde{\mathbf{U}})R^{-1}, \tag{4.9b}$$

so that for any $\mathbf{V} = R\,\tilde{\mathbf{V}} \in \mathbb{R}^4$

$$\mathbf{A}(\mathbf{U}, \mathbf{n})\mathbf{V} = R\mathbf{A}(\tilde{\mathbf{U}})\tilde{\mathbf{V}}. \tag{4.9c}$$

We shall consider Roe's matrix in Section 4.3.1 below and see that

$$\mathbf{A}_n(\mathbf{U}_L, \mathbf{U}_R)\mathbf{V} = R\mathbf{A}_n(\tilde{\mathbf{U}}_L, \tilde{\mathbf{U}}_R)\tilde{\mathbf{V}}. \tag{4.9d}$$

We shall also obtain a similar result for Osher's and Steger and Warming's fluxes. Substituting the identities (4.9) in the formula of a difference scheme proves indeed that (4.8) holds for most numerical fluxes. Hence, the flux can be equivalently defined by

$$\mathbf{\Phi}(\mathbf{U}_i, \mathbf{U}_j, \mathbf{n}) = R\,\varphi(\mathbf{f}; \tilde{\mathbf{U}}_i, \tilde{\mathbf{U}}_j),$$

which yields much simpler computations (see, for instance Mulder 1992, Selmin and Quartapelle 1993). Note that the rotational invariance of the Euler equations is also used in more sophisticated schemes derived in order to prevent grid alignment problems. Indeed, in formulas (4.2) and (4.3), the interface normal is chosen as the direction for wave propagation. If the waves are not aligned with the grid, they may be misrepresented (see van Leer 1992, Roe 1985; also Coirier and Powell 1995, Noelle 1994). Many attempts have been made at adapting a computational grid or minimizing grid orientation effects (Bourgeat and Koebbe 1992), for instance by taking into account the directions in which information is propagated and using a "rotated Riemann solver" (Davis 1984, Levy et al. 1993, Rumsey et al. 1993, LeVeque and Walder 1991, Fey 1995), or including "tangential wave propagation" as well as normal wave propagation (LeVeque 1988). We shall proceed with the derivation of a "truly multidimensional" solver in Section 4.3.2.

4.1.3 Examples

Let us now give some examples of "finite-volume" methods.

Example 4.1. "Cell center" scheme.

Starting from a triangulation $\mathcal{T}_h = \cup\, T_i$ of $\mathcal{O} \subset \mathbb{R}^2$ (which we assume is regular enough), one defines a control cell as a triangle $\Omega_i = T_i$, and the

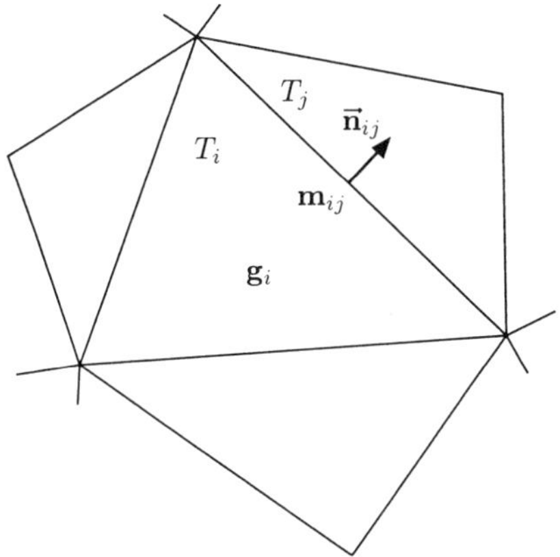

FIGURE 4.1. "Cell center."

center of the cell is the centroid $\mathbf{g}_i$ of the triangle T_i (see Figure 4.1). Approximating the solution of (3.2) by a function that is piecewise constant on each triangle, $\mathbf{w}_h(t)_{|T_i} = \mathbf{w}_i(t)$ and integrating the equation over T_i gives

$$|T_i| \frac{\partial \mathbf{w}_i}{\partial t} + \int_{\partial T_i} (\mathbf{f}(\mathbf{w}_i)n_{ix} + \mathbf{g}(\mathbf{w}_i)n_{iy})d\sigma = \mathbf{0}, \quad \forall i,$$

and the integral on the boundary is then discretized via a numerical flux function as we have already explained. One often interprets this "finite-volume method" as a "finite-element method", which is sometimes called "cell centered"; indeed, since

$$\mathbf{u}_i \cong \frac{1}{|T_i|} \int_{T_i} \mathbf{u}(\mathbf{x})d\mathbf{x} = \mathbf{u}(\mathbf{g}_i, t) + O(h^2),$$

one can say that the values $\mathbf{u}_i$ are associated to the centroid $\mathbf{g}_i$. If we introduce the space $\mathbf{W}_h = \{\mathbf{w}_h \in \mathbf{L}^2(\mathcal{O})^p; \forall T_i \in \mathcal{T}_h, \exists \mathbf{w}_i \in \mathbb{R}, \mathbf{w}_{h|T_i} = \mathbf{w}_i\}$, we can associate a variational problem: find $\mathbf{w}_h : (0,T) \to \mathbf{W}_h$, $\mathbf{w}_h(t)_{|T_i} = \mathbf{w}_i(t)$, such that

$$\int_\Omega \Big\{ \frac{\partial \mathbf{w}_h}{\partial t} + \frac{\partial}{\partial x}\mathbf{f}(\mathbf{w}_h) + \frac{\partial}{\partial y}\mathbf{g}(\mathbf{w}_h) \Big\} \cdot \varphi_h dx\, dy = 0, \quad \forall \varphi_h \in \mathbf{W}_h.$$

However, the above formula is not correct since the functions of $\mathbf{W}_h$ are discontinuous. Instead, writing for any $\varphi_h \in \mathbf{W}_h$

$$\varphi_h = \sum \varphi_h(g_i) 1_{T_i},$$

we obtain

$$\int_{T_i} \Big\{ \frac{\partial \mathbf{w}_i(t)}{\partial t} + \frac{\partial}{\partial x}\mathbf{f}(\mathbf{w}_i(t)) + \frac{\partial}{\partial y}\mathbf{g}(\mathbf{w}_i(t)) \Big\} dx\, dy = \mathbf{0}, \quad \forall i,$$

and using Green's formula we find the finite-volume method (for the relation with the streamline diffusion finite-element method, see Hansbo 1994).

One can also consider unstructured quadrilaterals or a "triangulation" made of mixed triangular–quadrilateral elements. □

Example 4.2. "Cell vertex" scheme.

In many situations, the quantities are defined at the vertices $\mathbf{a}_i$ of the triangulation rather than at the barycenter $\mathbf{g}_i$ of each triangle. Thus, starting from a triangulation $\mathcal{T}_h$ of $\mathcal{O} \subset \mathbb{R}^2$ (which again is regular enough), one defines the centers as the vertices $\mathbf{a}_i$ of the triangles (the nodes of the triangulation), and the control cell $\Omega_i = C_i$ associated to $\mathbf{a}_i$ as the "dual cell" of the node $\mathbf{a}_i$: the polygonal boundary of C_i is obtained by joining, for each triangle T having the vertex $\mathbf{a}_i$ in common, the midpoint of each triangle edge issued from $\mathbf{a}_i$ to the triangle barycenter $\mathbf{g}$. The boundary is thus composed of medians (Vijayasundaram 1986, Stoufflet 1983, Fezoui 1985, Angrand and Lafon 1993). Note that $\mathbf{a}_i$ is not necessarily the centroid of C_i. The cell C_i consists of the union of quadrilateral regions of type $R_i : C_i = \cup R_i^T$ (where R_i^T is the region R_i belonging to the triangle T) for all triangles T having the vertex $\mathbf{a}_i$ in common (see Figure 4.2). The boundary between two neighboring cells C_i and C_j consists of segments $e_{i,j}$ and $\varepsilon_{i,j}$ crossing at triangle edge midpoints $\mathbf{m}_{ij}$. When integrating the system over C_i, on R_i Green's formula gives, in particular, a flux across the segment e_{ij}.

Some variants are found where the boundary of C_i is obtained by joining the barycenters of two neighboring triangles (in Perthame and Qiu 1994 for instance). We can also consider the perpendicular bisectors of the edges

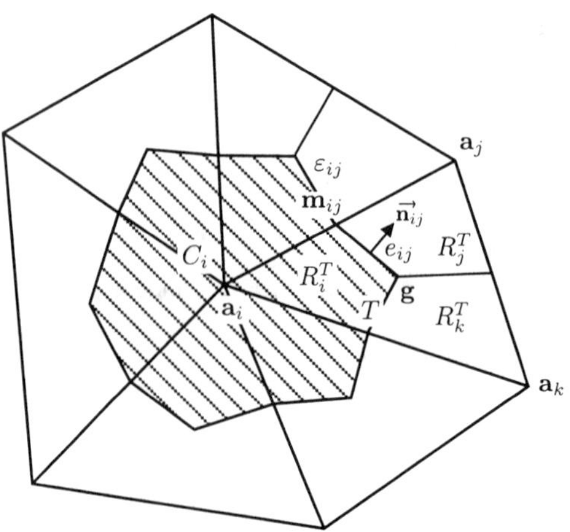

FIGURE 4.2. "Cell vertex."

of the triangles, which gives then for C_i the Voronoï cell associated to the Delaunay triangulation (see Mavripilis 1990, W.K. Anderson 1994). □

Remark 4.3. When starting from quadrilateral elements, with vertices characterized by (i, j), the dual cell is obtained by taking the centers of the neighboring cells $(i \pm \frac{1}{2}, j \pm \frac{1}{2})$ for the vertices of the dual cell, and the boundary of the dual cell is obtained by joining them to the middle ($i \pm \frac{1}{2}$ and $i, j \pm \frac{1}{2}$) of the edges (see Figure 4.3). It is then easy to compute by interpolation the gradients at the centers (i, j) of the dual meshes. Note that when approximating a system, the dual mesh is sometimes used for approximating one of the components of the unknown vector (typically for compressible Navier–Stokes equations the pressure can be evaluated at the nodes i, j while the velocity is evaluated at the centers; it occurs also in codes for two-phase flows in reservoir simulation). In structured meshes, a shifted mesh may be used for the computation of gradients (see Peyret and Taylor 1983, Koren 1988). □

Example 4.3. Weighted finite volume.

In this example, we start again from a triangulation of $\mathbb{R}^2$, $\mathcal{T}_h = \cup T$, composed of triangles, and we introduce the finite element space of type (1) $\mathbf{X}_h$ consisting of piecewise linear ($\mathbf{P}_1$) continuous functions characterized by their values at the nodes $\mathbf{a}_i$, $i \in I$. A basis of $\mathbf{X}_h$ consists of the shape functions $\psi_j \in \mathbf{X}_h$ such that (using the Kronecker symbol)

$$\psi_j(a_i) = \delta_i^j, \quad i, j \in I$$

(for simplicity, we consider the scalar case). The support of ψ_i, which we denote by $\operatorname{supp}(\psi_i)$, consists of the union of the triangles T having the vertex $\mathbf{a}_i$ in common. The $\mathbf{L}^2$-projection πg of a function $g \in \mathbf{L}^2(\mathcal{O})$ on

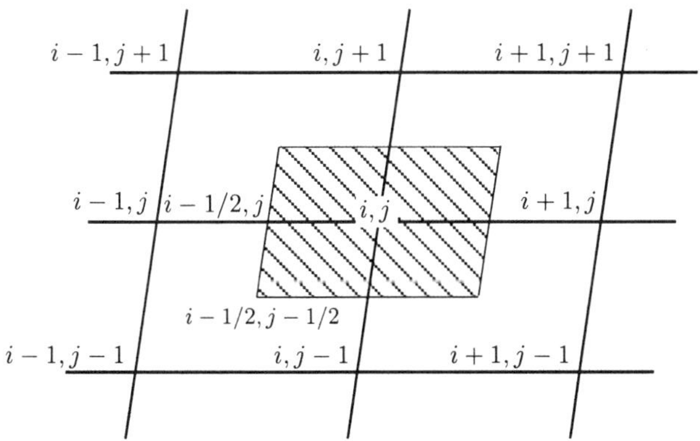

FIGURE 4.3. "Dual cell" in the structured case.

$\mathbf{X}_h$ is defined by

$$\int \pi g \, \varphi \, d\mathbf{x} = \int g \, \varphi \, d\mathbf{x}, \quad \forall \varphi \in \mathbf{X}_h,$$

and using a $\mathbf{P}_1$ quadrature formula (which corresponds to the principle of "mass-lumping", i.e., to diagonalizing the "mass matrix" $(\int \varphi_i \varphi_j \, d\mathbf{x})$),

$$\int f \, d\mathbf{x} \cong \int \Big(\sum_{i \in I} f(a_i) \psi_i \Big) d\mathbf{x} = \sum_{i \in I} f(a_i) \int \psi_i \, d\mathbf{x}$$

gives "the value of g" at point a_i,

$$\pi g(a_i) = \frac{(\int g \, \psi_i \, d\mathbf{x})}{(\int \psi_i \, d\mathbf{x})}, \quad i \in I.$$

We associate a fictitious cell Ω_i to a node $\mathbf{a}_i$ as follows. We define Ω_i as the volume (area) of $\text{supp}(\psi_i)$ weighted by ψ_i, i.e., for any $g : \mathbb{R}^2 \to \mathbb{R}$,

$$\int_{\Omega_i} g \, d\mathbf{x} = \int_{\text{supp}(\psi_i)} g \psi_i \, d\mathbf{x}.$$

Let us compute the "area" of Ω_i,

$$|\Omega_i| = \int_{\mathbb{R}^2} \psi_i \, d\mathbf{x}.$$

Since

$$\sum_i \psi_i(\mathbf{x}) = 1, \quad \forall \mathbf{x} \in \Omega,$$

we have

$$\int_{T \cap \Omega_i} d\mathbf{x} = \frac{1}{3} |T|,$$

and

$$|\Omega_i| = \frac{1}{3} \sum_{T / a_i \in T} |T|$$

is exactly the area of the dual cell C_i of Example 4.2. Formula (4.2) becomes

$$|\Omega_i| (u_i^{n+1} - u_i^n) - \Delta t \int_{\mathbb{R}^2} \mathbb{F}(u) \cdot \text{grad} \, \psi_i \, d\mathbf{x} = 0.$$

Let us give an example of a situation where these weighted finite volumes are used. In problems from petroleum reservoir simulation, the continuous flux has a prescribed direction $\mathbf{v}$,

$$\mathbb{F}(u) = \mathbf{v} f(u), \quad f : \mathbb{R} \to \mathbb{R},$$

with $\mathbf{v} = \text{grad} \, p \in \mathbb{R}^2$. Let us replace p by its interpolate,

$$p \cong \pi p = \sum_j p(a_j) \psi_j.$$

Since Σ_j grad $\psi_j(x) = 0$, we can write

$$\text{grad}\, p \cong \sum_j (p(a_j i) - p(a_i))\text{grad}\, \psi_j,$$

and we obtain the scheme

$$|\Omega_i|(u_i^{n+1} - u_i^n) - \Delta t \sum_j (p(a_j) - p(a_i)) f(u_{ij}^n) \int_{\mathbb{R}^2} \text{grad}\, \psi_i \cdot \text{grad}\, \psi_j \, d\mathbf{x} = 0,$$

where the exchange term $f(u_{ij}^n)$ can be defined in an upwind manner

$$f(u_{ij}^n) = \begin{cases} f(u_i^n) & \text{if } p(a_i) = p_i > p_j = p(a_j), \\ f(u_i^n) & \text{if } p(a_i) = p_i < p_j = p(a_j). \end{cases}$$

For details we refer to Eymard and Gallouët (1983). □

Example 4.4. Rectangular mesh.

If we start from a rectangular mesh $\Omega = \cup\, \Omega_{j,k}$, where the rectangle $\Omega_{j,k} = (x_{j-1/2}, x_{j+1/2}) \times (y_{k-1/2}, y_{k+1/2})$ has center $\mathbf{a}_{jk} = (x_j, y_k)$, we can find the classical finite difference schemes. Setting

$$\Delta x_j = x_{j+1/2} - x_{j-1/2}, \quad \Delta y_k = y_{k+1/2} - y_{k-1/2},$$

we have $|\Omega_{j,k}| = \Delta x_j \Delta y_k$, and formula (4.3) becomes

$$\mathbf{u}_{j,k}^{n+1} = \mathbf{u}_{j,k}^n - \frac{\Delta t}{\Delta x_j}(\mathbf{F}_{j+1/2,k}^n - \mathbf{F}_{j-1/2,k}^n) - \frac{\Delta t}{\Delta y_k}(\mathbf{G}_{j,k+1/2}^n - \mathbf{G}_{j,k-1/2}^n),$$

where $\mathbf{F}_{j+1/2,k}^n$ denotes the flux at the edge $x = x_{j+1/2}$ (with length Δy_k) between $\Omega_{j,k}$ and $\Omega_{j+1,k}$, and so on. Defining

$$\mathbf{u}_{j,k}^0 = \frac{1}{\Delta x_j \Delta y_k} \int_{\Omega_{j,k}} \mathbf{u}_0(\mathbf{x}) d\mathbf{x}, \ \lambda_{x_j} = \frac{\Delta t}{\Delta x_j} \ \lambda_{y_k} = \frac{\Delta t}{\Delta y_k},$$

we get exactly scheme (3.3) with variable mesh size,

$$\mathbf{u}_{j,k}^{n+1} = \mathbf{u}_{j,k}^n - \lambda_{x_j}(\mathbf{F}_{j+1/2,k}^n - \mathbf{F}_{j-1/2,k}^n) - \lambda_{y_k}(\mathbf{G}_{j,k+1/2}^n - \mathbf{G}_{j,k-1/2}^n).$$

We can take for $\mathbf{F}$ and $\mathbf{G}$ the numerical flux associated to any 3-point scheme. □

4.2 General results

Most stability and convergence results obtained in the one-dimensional case extend to a Cartesian grid (see Crandall and Majda 1980, Sanders 1983, Osher and Sanders 1983). Hence, we shall mainly consider in this section the case of an unstructured grid, and particularly the case of cell-centered triangles (Example 4.1). Let us first have a closer look at the order of accuracy.

4.2.1 Order

Let u be a smooth (scalar) solution of (3.6), and $\Phi(u, v, n)$ a C^1 (scalar) function of its arguments u and v; the truncation error is obtained by substituting u in the formula (4.3) (divided by $|K|$). Indeed, setting

$$u_K(t) = \frac{1}{|K|} \int_K u(\mathbf{x}, t) d\mathbf{x},$$

the truncation error in K is defined by

$$\begin{aligned} \varepsilon_K(t) =& u_K(t + \Delta t) - u_K(t) \\ &+ \frac{\Delta t}{|K|} \sum_{e \subset \partial K, e = K \cap K'} |e| \Phi(u_K(t), u_{K'}(t), \mathbf{n}_e). \end{aligned} \tag{4.10}$$

We can write

$$\begin{aligned} u_K(t + \Delta t) - u_K(t) &= \frac{1}{|K|} \int_K (u(\mathbf{x}(t + \Delta t) - u(\mathbf{x}, t)) d\mathbf{x} \\ &= \frac{1}{|K|} \int_K \frac{\partial u}{\partial t}(\mathbf{x}, t + s) d\mathbf{x}\, ds. \end{aligned}$$

Now, taking into account the fact that u is solution of (3.5), we can replace $\frac{\partial u}{\partial t}$ by $-(\frac{\partial}{\partial x} f(u) + \frac{\partial}{\partial y} g(u))$ and use the divergence theorem

$$\begin{aligned} u_K(t + \Delta t) - u_K(t) &= -\frac{1}{|K|} \int_K \int_0^{\Delta t} \operatorname{div}(\mathbb{F}(u)) d\mathbf{x}\, ds \\ &= -\frac{1}{|K|} \int_0^{\Delta t} \int_{\partial K} \mathbb{F}(u) \cdot \mathbf{n}\, d\sigma\, ds, \end{aligned}$$

which yields

$$\begin{aligned} \varepsilon_K(t) = -\frac{1}{|K|} \sum_e \Big\{ \int\!\!\int_0^{\Delta t} \mathbb{F}(u(\sigma, t + s)) \cdot \mathbf{n}_e d\sigma\, ds \\ + \Delta t\, |e| \Phi(u_K(t), u_{K'}(t), \mathbf{n}_e) \Big\}. \end{aligned}$$

Using quadrature formulas, we write

$$\int_0^{\Delta t} \mathbb{F}(u(\sigma, t + s)) \cdot \mathbf{n}_e\, ds = \Delta t \{ \mathbb{F}(u(\sigma, t)) \cdot \mathbf{n}_e + \tilde{\varepsilon}(\sigma, t) \}, \quad \forall \sigma \in e, \tag{4.11}$$

$$\int_e \mathbb{F}(u(\sigma, t)) \cdot \mathbf{n}_e d\sigma = |e| \{ \mathbb{F}(u(\mathbf{m}_e, t)) \cdot \mathbf{n}_e + O(h^2) \},$$

$$u_K(t) = \frac{1}{|K|} \int_K u(\mathbf{x}, t) d\mathbf{x} = u(\mathbf{g}, t) + O(h^2),$$

where $\mathbf{g}$ is the centroid of K, $\mathbf{m}_e$ is the midpoint of the edge e of K, and h the diameter of K. In fact, some care must be taken in the evaluation of the $O(\Delta, t)$ term $\tilde{\varepsilon}$ in (4.11). Since the sum comes from the integral $\int_K \int_0^{\Delta t}$ div

$(\mathbb{F}(u))d\mathbf{x}\,ds$, it is in fact $O(\Delta t)O(h^2)$. Indeed, by Taylor series expansion, we can write

$$\int_e \tilde{\varepsilon}(\sigma, t)d\sigma = |e|\Big\{\frac{\Delta t}{2}\left(\mathbb{F}'(u(\mathbf{g}, t)) \cdot \mathbf{n}_e\right)\frac{\partial u}{\partial t}(\mathbf{g}, t) + O(h) + O(\Delta t)\Big\}.$$

Since $\sum_{e \in K} n_e = 0$, summing on all the edges e of K, we get

$$\sum_{e \subset K} \int_e \tilde{\varepsilon}(\sigma, t)d\sigma = |e|\Delta t\{O(h) + O(\Delta t)\}.$$

Now, setting

$$u_e = u(\mathbf{m}_e, t),$$

we have

$$\begin{aligned}\Phi(u_K(t), u_{K'}(t), \mathbf{n}_e) = \Phi(u_e, u_e, \mathbf{n}_e) &+ \frac{\partial \Phi}{\partial u}(u_e, u_e, \mathbf{n}_e)(u_K(t) - u_e) \\ &+ \frac{\partial \Phi}{\partial v}(u_e, u_e, \mathbf{n}_e)(u_{K'}(t) - u_e) + O(h^2)\end{aligned}$$

and

$$u_K(t) - u_e = u_K(t) - u(\mathbf{m}_e, t) = \operatorname{grad} u \cdot (\mathbf{g} - \mathbf{m}_e) + O(h^2).$$

Due to the consistency of Φ, (4.10) becomes

$$\begin{aligned}\varepsilon_K(t) = \frac{\Delta t}{|K|}\sum_{e \subset \partial K} |e|\Big\{&\frac{\partial \Phi}{\partial u}(u_e, u_e, \mathbf{n}_e)(u_K(t) - u_e) \\ &+ \frac{\partial \Phi}{\partial v}(u_e, u_e, \mathbf{n}_e)(u_{K'}(t) - u_e) + O(h^2) + O(\Delta t)\Big\},\end{aligned}$$

and hence

$$(4.12)\quad \left\{\begin{aligned}\varepsilon_K(t) = \frac{\Delta t}{|K|}\operatorname{grad} u \cdot \sum_{e \subset \partial K} |e|\Big\{&\frac{\partial \Phi}{\partial u}(u_e, u_e, \mathbf{n}_e)(\mathbf{g} - \mathbf{m}_e) \\ &+ \frac{\partial \Phi}{\partial v}(u_e, u_e, \mathbf{n}_e)(\mathbf{g}' - \mathbf{m}_e) + O(h^2) + \Delta t O(h)\Big\}.\end{aligned}\right.$$

Lemma 4.1

Assume that we are in the situation of Example 4.1, with, moreover, a uniform triangulation by equilateral triangles T with side h. If the numerical flux Φ satisfies

$$\sum_{e \subset \partial T}\Big\{\frac{\partial \Phi}{\partial u}(u, u, \mathbf{n}_e) - \frac{\partial \Phi}{\partial v}(u, u, \mathbf{n}_e)\Big\}\mathbf{n}_e = 0, \quad \forall u \in \mathbb{R},$$

or

$$\sum_{e \subset \partial t}\Big\{\frac{\partial \Phi}{\partial u}(u, u, \mathbf{n}_e) + \frac{\partial \Phi}{\partial u}(u, u, -\mathbf{n}_e)\Big\}\mathbf{n}_e = 0, \quad \forall u \in \mathbb{R},$$

the scheme (4.3) is first-order accurate.

Proof. Starting from formula (4.12) now with $K = T$, we see that some terms simplify since

$$\mathbf{m}_e - g = |\mathbf{m}_e - \mathbf{g}|\mathbf{n}_e = -(\mathbf{m}_e - \mathbf{g}'), \quad |K| = |T| = 3|\mathbf{m}_e - \mathbf{g}|\,\frac{|e|}{2}.$$

Thus

$$\varepsilon_T = \frac{2\Delta t}{3}\operatorname{grad} u \cdot \sum_{e\subset\partial T}\Big\{-\frac{\partial\Phi}{\partial u}(u_e, u_e, \mathbf{n}_e) + \frac{\partial\Phi}{\partial v}(u_e, u_e, \mathbf{n}_e))\mathbf{n}_e + O(\Delta t) + O(h)\Big\}.$$

We can write the same formula with the fluxes evaluated at the centroid: setting

$$u = u(\mathbf{g}, t),$$

we have

$$\varepsilon_K = \frac{2\Delta t}{3}\operatorname{grad} u \cdot \sum_{e\subset\partial T}\Big\{-\frac{\partial\Phi}{\partial u}(u, u, \mathbf{n}_e) + \frac{\partial\Phi}{\partial v}(u, u, \mathbf{n}_e))\mathbf{n}_e + O(\Delta t) + O(h)\Big\}.$$

Differentiating the conservation relation (4.4), we have

$$\frac{\partial\Phi}{\partial v}(u, u, \mathbf{n}) - \frac{\partial\Phi}{\partial u}(u, u, -\mathbf{n}) = 0$$

and

$$\frac{\partial\Phi}{\partial u}(u, u, \mathbf{n}) - \frac{\partial\Phi}{\partial v}(u, u, \mathbf{n}) = \frac{\partial\Phi}{\partial u}(u, u, \mathbf{n}) + \frac{\partial\Phi}{\partial u}(u, u, -\mathbf{n}),$$

so that we can also write $\varepsilon_K(t)$ in the following form,

$$\varepsilon_K(t) = -\frac{2\Delta t}{3}\operatorname{grad} u \cdot \sum_{e\subset\partial T}\Big\{\frac{\partial\boldsymbol{\Phi}}{\partial u}(u, u, \mathbf{n}_e) + \frac{\partial\boldsymbol{\Phi}}{\partial u}(u, u, -\mathbf{n}_e)\mathbf{n}_e + O(\Delta t) + O(h)\Big\}.$$

Assuming that

$$\sum\Big\{\frac{\partial\Phi}{\partial u}(u, u, \mathbf{n}_e) - \frac{\partial\Phi}{\partial v}(u, u, \mathbf{n}_e)\Big\}\mathbf{n}_e = 0$$

or that

$$\sum\Big\{\frac{\partial\Phi}{\partial u}(u, u, \mathbf{n}_e) - \frac{\partial\Phi}{\partial v}(u, u, -\mathbf{n}_e)\Big\}\mathbf{n}_e = 0,$$

the first-order term in $\varepsilon_K(t)$ vanishes, and we obtain a first-order accurate method

$$\varepsilon_K(t) = O(\Delta t) + O(h),$$

i.e., $\varepsilon_K(t) = O(h)$ under a C.F.L. condition. □

The conditions of the lemma are satisfied by the usual schemes. Indeed, in the formula for the numerical flux Φ, the terms that do not depend on F will be taken out of the "Σ" so that we can use the identity $\Sigma_e \mathbf{n}_e = 0$, while those that do depend on F (and thus on $\mathbf{n}_e$) are frequently symmetric in u and v or linear in $\mathbf{n}$. For instance, the Lax–Friedrichs scheme gives

$$\Phi(u, v, \mathbf{n}) = \mathbb{F}(u) \cdot \mathbf{n} + \mathbb{F}(v) \cdot \mathbf{n} - \frac{(v-u)}{2\lambda},$$

and hence

$$\begin{aligned} \frac{\partial \Phi}{\partial u}(u, u, \mathbf{n}) &= \mathbb{F}'(u) \cdot \mathbf{n} + \frac{1}{2\lambda}, \\ \frac{\partial \Phi}{\partial v}(u, u, \mathbf{n}) &= \mathbb{F}'(u) \cdot \mathbf{n} - \frac{1}{2\lambda}, \end{aligned}$$

and

$$\sum \Big(\frac{\partial \Phi}{\partial u}(u(\mathbf{m}, t), u(\mathbf{m}, t), \mathbf{n}_e) - \frac{\partial \Phi}{\partial v}(u(\mathbf{m}, t), u(\mathbf{m}, t), \mathbf{n}_e) \Big) \mathbf{n}_e = 0.$$

In other cases, such as the upwind scheme (Godunov's),

$$\Phi(u, v, \mathbf{n}) = \begin{cases} \mathbb{F}(u) \cdot \mathbf{n} & \text{if } \mathbb{F}'(u) \cdot \mathbf{n} > 0, \\ \mathbb{F}(u) \cdot \mathbf{n} & \text{if } \mathbb{F}'(u) \cdot \mathbf{n} < 0. \end{cases}$$

We can see that away from "sonic" points,

$$\frac{\partial \Phi}{\partial u}(u, u, \mathbf{n}_e) + \frac{\partial \Phi}{\partial u}(u, u, -\mathbf{n}_e) = 0.$$

When the grid is not uniform, the scheme is not first-order accurate. Note that even in the one-dimensional case, when the length Δx_i of the mesh is not uniform, it is easy to see that conservation and consistency do not imply that the truncation error tends to zero, unless the ratio (sup Δx_i)/(inf Δx_i) $\to 1$ as sup $\Delta x_i \to 0$ (and a second-order scheme gives a first-order scheme on an irregular grid (see Pike 1987; see also Jeng and Chen 1992, Turkel 1985, Wendroff and White 1988, Durlofsky et al. 1993 for numerical experiments).

In any case, the conservation property implies somehow a cancellation of the errors in adjacent cells, and we shall prove the analog of the Lax–Wendroff Theorem. It says that in the scalar case ($p = 1$), when the scheme "converges" in some sense that will be specified below, the limit is a weak solution to the conservation law. Remember that such a solution exists (see G.R., Chapter II, Section 5).

4.2.2 The Lax–Wendroff Theorem

For the proof of the analog of the Lax–Wendroff theorem in the case of an unstructured mesh, we need the following classical approximation result.

Lemma 4.2

Let $\mathcal{T}_h = \cup K$ be a countable family of triangulations of a bounded set $\mathcal{O} \subset \mathbb{R}^2$, and define for $u \in \mathbf{L}^1(\mathcal{O})$

$$u_K = \frac{1}{|K|} \int_K u(\mathbf{x}) d\mathbf{x}$$

and

$$\pi_h u = \sum_k u_K 1_K.$$

Then $\pi_h u \to u$ in $\mathbf{L}^1(\mathcal{O})$ as $h \to 0$, where $h = \sup|K|, K \in \mathcal{T}_h$.

Proof. Let us first consider the case of a continuous function $\varphi \in \mathbf{C}^0(\mathcal{O})$. Then

$$\pi_h \varphi \to \varphi \quad \text{a.e. as } h \to 0$$

(more precisely, outside the set of measure zero composed of the union of the boundaries of the sets $K \in \mathcal{T}_h$ for the countable family $\mathcal{T}_h$), and

$$|\pi_h \varphi| \leq \|\varphi\|_{\mathbf{L}^\infty(\mathcal{O})}.$$

Thus $\pi_h \varphi \to \varphi$ in $\mathbf{L}^1(\mathcal{O})$.

Now, for $u \in \mathbf{L}^1(\mathcal{O})$,

$$\begin{aligned} \|\pi_h u\|_{\mathbf{L}^1(\mathcal{O})} &= \int_{\mathcal{O}} |\pi_h u| d\mathbf{x} = \int_{\mathcal{O}} |\sum_K u_K 1_K| d\mathbf{x} \\ &\leq \sum_K |u_K|\, |K| = \sum_K \int_K |u(\mathbf{x})| d\mathbf{x} = \|u\|_{\mathbf{L}^1(\mathcal{O})}, \end{aligned}$$

and for any $\varphi \in \mathbf{C}^0(\mathcal{O})$,,

$$\|\pi_h u - u\|_{\mathbf{L}^1(\mathcal{O})} \leq \|\pi_h u - \pi_h \varphi\|_{\mathbf{L}^1(\mathcal{O})} + \|\pi_h \varphi - \varphi\|_{\mathbf{L}^1(\mathcal{O})} + \|u - \varphi\|_{\mathbf{L}^1(\mathcal{O})}.$$

Given ε, we choose $\varphi \in \mathbf{C}^0(\mathcal{O})$ such that $\|u - \varphi\|_{\mathbf{L}^1(\mathcal{O})} \leq \varepsilon$. Then for h small enough, we have $\|\pi_h \varphi - \varphi\|_{\mathbf{L}^1(\mathcal{O})} \leq \varepsilon$, and the result follows. □

Let us now introduce basis functions on each triangle, which will enable us to reconstruct conveniently an affine function from its values at the vertices.

Lemma 4.3

Let us consider a triangle T with vertices $\mathbf{a}_i$, $1 \leq i \leq 3$. Let e_i denote the edge of T opposite to $\mathbf{a}_i$, with outside unit normal $\mathbf{n}_i$. There exists an affine function $\mathbf{p}_i : T \to \mathbb{R}^2$, for $i = 1, 2, 3$, such that

$$\text{(4.13a)} \quad \mathbf{p}_i(\mathbf{x}) \cdot \mathbf{n}_i = 1, \quad \forall \mathbf{x} \in e_i; \quad \mathbf{p}_i(\mathbf{x}) \cdot \mathbf{n}_j = 0, \quad \forall \mathbf{x} \in e_j, \quad j \neq i.$$

Moreover, the $\mathbf{p}_i$ *satisfy*

(4.13*b*)
$$\sum_{k=1}^{3} \mathbf{n}_k \mathbf{p}_k(\mathbf{x})^T = \mathbf{I}, \quad \forall \mathbf{x} \in T$$

and

(4.13*c*)
$$\operatorname{div} \mathbf{p}_i = \frac{|e_i|}{|T|}, \quad i = 1, 2, 3.$$

Proof. It is easy to construct the $\mathbf{p}_i$ satisfying (4.13a) on a reference triangle with vertices $\mathbf{a}_1 = \mathbf{x}_1 = (0,0)^T$, $\mathbf{a}_2 = \mathbf{x}_2 = (0,1)^T$, $\mathbf{a}_3 = \mathbf{x}_3 = (1,0)^T$. We take

$$\mathbf{p}_1(\mathbf{x}) = \sqrt{2}\mathbf{x}, \quad \mathbf{p}_2(\mathbf{x}) = \mathbf{x} - (0,1)^T, \quad \mathbf{p}_3(\mathbf{x}) = \mathbf{x} - (1,0)^T.$$

On any triangle T, we get for the function associated to the edge e_i opposite to the vertex $\mathbf{a}_i$

$$\mathbf{p}_i(\mathbf{x}) = (\mathbf{x} - \mathbf{a}_i)\frac{|ei|}{2|T|},$$

where the constants are chosen to satisfy (4.13c).

Let us check the property (4.13b); it means that for all $\mathbf{v} \in \mathbb{R}^2$

$$\sum_k (\mathbf{p}_k(\mathbf{x}) \cdot \mathbf{v})\mathbf{n}_k = \mathbf{v}.$$

Let us introduce the notation

$$\mathbf{B}(\mathbf{x}) = \sum_k \mathbf{n}_k \mathbf{p}_k(\mathbf{x})^T.$$

For any $\mathbf{x} \in T$, $\mathbf{B}(x)$ is a 2×2 matrix, each component of which is an affine function of $\mathbf{x}$. On any edge e_j, we have

$$\mathbf{B}(\mathbf{x})\mathbf{n}_j = \sum_k (\mathbf{p}_k(\mathbf{x}) \cdot \mathbf{n}_j)\mathbf{n}_k = \sum_k \delta_{j,k}\mathbf{n}_k = \mathbf{n}_j, \quad \mathbf{x} \in e_j.$$

Thus, at the vertex $\mathbf{a}_k = e_i \cap e_j (i \neq k, j \neq k)$,

$$\mathbf{B}(\mathbf{a}_k)\mathbf{n}_i = \mathbf{n}_i \quad \text{and} \quad \mathbf{B}(\mathbf{a}_k)\mathbf{n}_j = \mathbf{n}_j,$$

which proves that $\mathbf{B}(\mathbf{a}_k) = \mathbf{I}$ since $(\mathbf{n}_i, \mathbf{n}_j)$ form a basis of $\mathbb{R}^2$, for $k = 1, 2, 3$. Since the coefficients are affine functions, we obtain

$$(\mathbf{B}(\mathbf{a}_i) = \mathbf{I} \quad \text{and} \quad \mathbf{B}(\mathbf{a}_j) = \mathbf{I}) \Longrightarrow \mathbf{B}(\mathbf{x}) = \mathbf{I} \quad \text{on } e_k,$$

which yields

$$\mathbf{B}(\mathbf{x}) = \mathbf{I} \quad \text{on } e_k \Longrightarrow \mathbf{B}(\mathbf{x}) = \mathbf{I} \quad \text{on } T.$$

Note that the above property is directly linked to the fact that

(4.14)
$$\sum_{i=1}^{3} |e_i|\mathbf{n}_i = \mathbf{0},$$

which we shall use again. □

Let us define the piecewise constant function associated to the finite-volume scheme (4.3),

(4.15) $$v_\delta(\mathbf{x}, t) = u_i^n, \quad \mathbf{x} \in \Omega_i, \quad t \in [n\Delta t, (n+1)\Delta t)[, \quad \delta = (h, \Delta t),$$

and set

(4.16) $$u_i^0 = \frac{1}{|\Omega_i|} \int_{\Omega_i} u_0(\mathbf{x}) d\mathbf{x}.$$

We assume for simplicity that we are in the situation of Example 4.1, i.e., we are given a triangulation $\mathcal{T}_h = \cup T_i$ of $\Omega \subset \mathbb{R}^2$, and $\Omega_i = T_i$. We introduce on the one hand regularity assumptions on the mesh: we assume that there exist positive constants a, b such that

(4.17a) $$ah^2 \leq |T_i| \leq bh^2, \quad \forall T_i \in \mathcal{T}_h$$

and

(4.17b) $$ah \leq |e| \leq bh, \quad \forall e \in \mathcal{E} = \{e \subset \partial T_i, T_i \in \mathcal{T}_h\}.$$

On the other hand, we assume a C.F.L.-like stability condition

(4.17c) $$\Delta t \leq Ch.$$

Let us now prove the analog of the Lax–Wendroff theorem. The result justifies the use of conservative numerical fluxes, and the proof presents some interesting arguments, though the assumptions of strong convergence are unrealistic. Indeed, at the present time, no general finite-volume scheme on triangles has yet been proven to converge strongly (only weakly in L^∞) (see Remark 4.5 below).

Proposition 4.1

Let (4.3) be a conservative finite-volume scheme with Lipschitz continuous numerical flux Φ consistent with equation (3.6a), and let u_i^0 be given by (4.16). Assume that there exists a sequence δ_k which tends to 0 such that

(4.18) $\|v_{\delta_k}\|_{\mathbf{L}^\infty(\mathbb{R}^2 \times (0,+\infty))} \leq C,$

(4.19) v_{δ_k} *converges in* $\mathbf{L}^1_{\text{loc}}(\mathbb{R}^2 \times (0,+\infty))$ *and a.e. to a function* v.

Then v *is a weak solution of (3.6).*

Proof. Let $\varphi \in \mathbf{C}_0^1(\mathbb{R}^2 \times \mathbb{R}+)$ be a test function with compact support, and set

$$\varphi_i^n = \frac{1}{|T_i|} \int_{T_i} \varphi(\mathbf{x}, t_n) d\mathbf{x}.$$

By multiplying the equation (4.3) by φ_i^n and summing over i (such that $T_i \in \mathcal{T}_h$) and $n \geq 0$, we obtain

$$\Delta t \sum_{i,n\geq 0} \int_{T_i} \Big\{\varphi(\mathbf{x}, t_n)\, \frac{(u_i^{n+1} - u_i^n)}{\Delta t}\Big\} d\mathbf{x}$$
$$+ \Delta t \sum_{i,n\geq 0} \varphi_i^n \sum_{e=\Gamma_{ij}\in\partial T_i} |e|\Phi(u_i^n, u_j^n, \mathbf{n}_e)\} = 0.$$

The summation reduces, in fact, to the indices i, n such that $T_i \times (0, t_n) \cap \operatorname{supp}\varphi \neq \phi$. A summation by parts gives for the first term

$$\Delta t \sum_{i,n\geq 0} \int_{T_i} \Big\{\varphi(\mathbf{x}, t_n)\, \frac{(u_i^{n+1} - u_i^n)}{\Delta t}\Big\} d\mathbf{x}$$
$$= \Delta t \sum_{i,n\geq 1} \int_{T_i} u_i^n \Big\{ \frac{\varphi(\mathbf{x}, t_{n-1}) - \varphi(\mathbf{x}, t_n)\}}{\Delta t} \Big\} d\mathbf{x} - \sum_i \int_{T_i} \varphi(\mathbf{x}, 0) u_i^0 \, d\mathbf{x}.$$

Consider the last term,

$$\sum_i \int_{T_i} \varphi(\mathbf{x}, 0) u_i^0 \, d\mathbf{x} = \int_{\mathbb{R}^2} \varphi(\mathbf{x}, 0) \sum u_i^0 1_{T_i}(\mathbf{x}) d\mathbf{x}.$$

By (4.16)

$$u_i^0 = \frac{1}{|T_i|} \int_{T_i} u_0(\mathbf{x}) d\mathbf{x},$$

so that

$$\sum_i u_i^0 1_{T_i}(\mathbf{x}) = v_\delta(\mathbf{x}, 0) = \pi_h u_0,$$

and $\pi_h u_0$ converges to u_0 in $\mathbf{L}^1$ as $h \to 0$ by Lemma 4.2. Hence, we have

$$\sum_i \int_{T_i} \varphi(\mathbf{x}, 0) u_i^0 d\mathbf{x} \to \int_{\mathbb{R}^2} \varphi(\mathbf{x}, 0) u_0(\mathbf{x}) d : bfx \text{ as } h \to 0. \tag{4.20}$$

Now, setting for $(\mathbf{x}, t) \in \mathbb{R}^2 \times \mathbb{R}+$

$$\varphi_\Delta(\mathbf{x}, t) = \varphi(\mathbf{x}, t_n), \quad \text{if } t_n \leq t < t_{n+1},\ n \geq 0,$$

we have

$$v_\delta(\mathbf{x}, t)\{\varphi_\Delta(\mathbf{x}, t) - \varphi(\mathbf{x}, t - \Delta t)\}$$
$$= u_i^n\{\varphi(\mathbf{x}, t_n) - \varphi(\mathbf{x}, t_{n-1}\}, \quad \mathbf{x} \in T_i,\ t_n < t < t_{n+1},$$

and

$$\Delta t \sum_{i,n\geq 1} \int_{T_i} u_i^n \Big\{ \frac{\varphi(\mathbf{x}, t_n) - \varphi(\mathbf{x}, t_{n-1})\}}{\Delta t} \Big\} d\mathbf{x}$$
$$= \int_{\mathbb{R}^2} \int_{\Delta t}^{+\infty} v_\delta(\mathbf{x}, t) \Big\{ \frac{\varphi_\Delta(\mathbf{x}, t) - \varphi_\Delta(\mathbf{x}, t - \Delta t)}{\Delta t} \Big\} d\mathbf{x}\, dt.$$

The term $1_{[\Delta t,+\infty)}(t)\,\frac{\varphi_\Delta(\mathbf{x},t)-\varphi_\Delta(x,t-\Delta t)}{\Delta t}$ converges to $\frac{\partial\varphi}{\partial t}(x,t)$ as $\Delta t\to 0$ and, by assumption, v_δ is bounded. It yields

$$\int_{\mathbb{R}^2\times(0,+\infty)} v_\delta(\mathbf{x},t)\{1_{[\Delta t,+\infty)}(t)\,\frac{\varphi_\Delta(\mathbf{x},t)-\varphi_\Delta(\mathbf{x},t-\Delta t)}{\Delta t} - \frac{\partial\varphi}{\partial t}(\mathbf{x},t)\}d\mathbf{x}\,dt \to 0$$

as $\Delta t\to 0$. Now, since v_δ converges in $\mathbf{L}^1_{\mathrm{loc}}$ towards v,

$$\int_{\mathbb{R}^2\times(0,+\infty)} v_\delta(\mathbf{x},t)\,\frac{\partial\varphi}{\partial t}(\mathbf{x},t)d\mathbf{x}\,dt \to \int_{\mathbb{R}^2\times(0,+\infty)} v(\mathbf{x},t)\,\frac{\partial\varphi}{\partial t}(\mathbf{x},t)d\mathbf{x}\,dt \quad \text{as } \delta\to 0. \tag{4.21}$$

We have thus proven that

$$\int_{\mathbb{R}^2}\int_{\Delta t}^{+\infty} v_\delta(\mathbf{x},t)\Big\{\frac{\varphi_\Delta(\mathbf{x},t)-\varphi_\Delta(\mathbf{x},t-\Delta t)\}}{\Delta t}\Big\}d\mathbf{x}\,dt \to \int_{\mathbb{R}^2\times(0,+\infty)} v(\mathbf{x},t)\,\frac{\partial\varphi}{\partial t}(\mathbf{x},t)\}d\mathbf{x}\,dt \quad \text{as } \delta\to 0.$$

Let us next study the term

$$\Delta t\sum_{i,n\geq 0}\Big\{\frac{1}{|T_i|}\int_{T_i}\varphi(\mathbf{x},t_n)\sum_{e=\Gamma_{ij}\subset\partial T_i}|e|\Phi(u_i^n,u_j^n,\mathbf{n}_e)\Big\}=0.$$

First, we introduce the function $\mathbf{p}_e(\mathbf{x})$ associated to an edge e by Lemma 4.3 (with slightly different notations). Then we define

$$\tilde{\mathbf{F}}^n(\mathbf{x}) = \sum_{e=\Gamma_{ij}\subset\partial T_i}\mathbf{p}_e(\mathbf{x})\Phi(u_i^n,u_j^n,\mathbf{n}_e), \quad \mathbf{x}\in T_i$$

(the dependence of $\mathbf{p}_e(\mathbf{x})$ on T_i does not appear in the notation but is effective). Thus, by (4.13),

$$\operatorname{div}\tilde{\mathbf{F}}^n(\mathbf{x}) = \sum_{e=\Gamma_{ij}\subset\partial T_i}\frac{|e|}{|\Omega_i|}\,\Phi(u_i^n,u_j^n,\mathbf{n}_e), \quad \mathbf{x}\in T_i$$

and

$$\begin{aligned}&\sum_i\int_{T_i}\varphi(\mathbf{x},t_n)\Big\{\sum_e\frac{|e|}{|T_i|}\,\Phi(u_i^n,u_j^n,\mathbf{n}_e)\Big\}d\mathbf{x}\\ &=\sum_i\Big\{\int_{T_i}\varphi(\mathbf{x},t_n)\operatorname{div}\tilde{\mathbf{F}}^n(\mathbf{x})d\mathbf{x}\Big\}\\ &=-\sum_i\Big\{\int_{\Omega_i}\operatorname{grad}\varphi(\mathbf{x},t_n)\tilde{\mathbf{F}}^n(\mathbf{x})d\mathbf{x}\Big\}+\sum_i\sum_e\int_e\varphi(\mathbf{x},t_n)\tilde{\mathbf{F}}^n(\mathbf{x})\cdot\mathbf{n}_e\,d\mathbf{x}.\end{aligned}$$

For each edge $e = \Gamma_{ij} = T_i \cap T_j = \Gamma_{ji}$, we get two terms: one coming from the integral on T_i and by (4.13),

$$\tilde{\mathbf{F}}^n(\mathbf{x})_{|\Gamma_{ij}} \cdot \mathbf{n}_e = \Phi(u_i^n, u_j^n, \mathbf{n}_e),$$

the other from the integral on T_j, with the opposite normal,

$$\tilde{\mathbf{F}}^n(\mathbf{x})_{|\Gamma_{ij}} \cdot (-\mathbf{n}_e) = \Phi(u_j^n, u_i^n, -\mathbf{n}_e).$$

Thanks to the conservativity (4.4), these two terms cancel,

$$\Phi(u_j^n, u_i^n, -\mathbf{n}_e) = -\Phi(u_i^n, u_j^n, \mathbf{n}_e),$$

and

$$\sum_i \sum_e \int_e \varphi(\mathbf{x}, t_n) \tilde{\mathbf{F}}^n(\mathbf{x}) \cdot \mathbf{n}_e d\mathbf{x} = 0.$$

We are left with the term

$$\sum_i \int_{T_i} \operatorname{grad} \varphi(\mathbf{x}, t_n) \cdot \tilde{\mathbf{F}}^n(\mathbf{x}) d\mathbf{x}.$$

Using now the consistency property (4.5), we write

$$\Phi(u_i^n, u_j^n, \mathbf{n}_e) = \Phi(u_i^n, u_j^n, \mathbf{n}_e) - \Phi(u_i^n, u_i^n, \mathbf{n}_e) + \mathbb{F}(u_i^n) \cdot \mathbf{n}_e,$$

which leads us to define

$$\mathbf{F}^n(\mathbf{x}) = \sum_{e \in \partial\Omega_i} \mathbf{p}_e(\mathbf{x}) \mathbf{n}_e \cdot \mathbb{F}(u_i^n) \; \mathbf{F}^n(\mathbf{x}) = \mathbb{F}(u_i^n), \quad \mathbf{x} \in T_i,$$

and, summing over $n \geq 0$, to consider the limit of the term

$$\Delta t \sum_{n,i} \int_{T_i} \operatorname{grad} \varphi(\mathbf{x}, t_n) \cdot \mathbf{F}^n(\mathbf{x}) d\mathbf{x} = \Delta t \sum_n \int_{\mathbb{R}^2} \operatorname{grad} \varphi(\mathbf{x}, t_n) \cdot \mathbf{F}^n(\mathbf{x}) d\mathbf{x}.$$

We see easily that as $\delta \to 0$,

$$\begin{aligned} \Delta t \sum_n \int_{\mathbb{R}^2} & \operatorname{grad} \varphi(\mathbf{x}, t_n) \cdot \mathbf{F}^n(\mathbf{x}) dx \\ & \to \int_{\mathbb{R}^2 \times (0,+\infty)} \operatorname{grad} \varphi(\mathbf{x}, t) \cdot \mathbb{F}(v(\mathbf{x}, t)) d\mathbf{x} \, dt. \end{aligned} \tag{4.22}$$

There remains to study the term R corresponding to $\Phi(u_i^n, u_j^n, \mathbf{n}_e) - \Phi(u_i^n, u_i^n, \mathbf{n}_e)$,

$$\begin{aligned} \frac{R}{\Delta t} &= \sum_{n,i} \int_{T_i} \operatorname{grad} \varphi(\mathbf{x}, t_n) \cdot \{\mathbf{F}^n(\mathbf{x}) - \tilde{\mathbf{F}}^n(\mathbf{x})\} dx \\ &= \sum_{n,i} \int_{T_i} \operatorname{grad} \varphi(\mathbf{x}, t_n) \cdot \sum_{e=\Gamma_{ij}} \mathbf{p}_e(\Phi(u_i^n, u_j^n, \mathbf{n}_e) - \Phi(u_i^n, u_i^n, \mathbf{n}_e)) d\mathbf{x}. \end{aligned}$$

The regularity assumptions (4.17) imply that the $\mathbf{p}_e$ are uniformly bounded, and hence

$$R \leq C\Delta t \sum_{n,i} \int_{T_i} |\text{grad}\, \varphi(\mathbf{x}, t_n| \sum_{j/\Gamma_{ij}\subset\partial T_i} |\Phi(u_i^n, u_j^n, \mathbf{n}_e) - \Phi(u_i^n, u_i^n, \mathbf{n}_e)| d\mathbf{x}.$$

Since Φ is locally Lipschitz continuous, it remains only to estimate the right-hand side of

$$\begin{aligned} R &\leq C\Delta t \sum_{n,i} \int_{T_i} |\text{grad}\, \varphi(\mathbf{x}, t_n)| \sum_{j/\Gamma_{ij}\subset\partial T_i} |u_i^n - u_j^n| d\mathbf{x} \\ &\leq C\Delta t \sum_{n,i} \|\text{grad}\, \varphi\|_{\mathbf{L}^\infty} |\Omega_i| \sum_{j/\Gamma_{ij}\subset\partial T_i} |u_i^n - u_j^n|, \end{aligned}$$

where the summation is limited to those i such that $T_i \cap$ supp $\varphi \neq \phi$. Now, recall that

$$u_i^n = \frac{1}{|T_i|\Delta t} \int_{T_i\times(t_n,t_{n+1})} v_\delta(\mathbf{x}, t) d\mathbf{x}\, dt.$$

Then, setting

$$v_i^n = \frac{1}{|T_i|\Delta t} \int_{T_i\times(t_n,t_{n+1})} v(\mathbf{x}, t) d\mathbf{x}\, dt$$

and writing

$$|u_i^n - u_j^n| \leq |u_i^n - v_i^n| + |v_i^n - v_j^n| + |u_j^n - v_j^n|,$$

we get first, since v_δ converges in $\mathbf{L}^1_{\text{loc}}$ to v,

$$\Delta t \sum_{i,n} |T_i|\ |u_i^n - v_i^n| = \int_{\text{supp}\,(\varphi)} |v_\delta - v| d\mathbf{x}\, dt \to 0 \text{ as } \delta \to 0.$$

For the term $|u_j^n - v_j^n|$, we can conclude in an analogous way since, thanks to the assumption of nondegeneracy of the mesh, we can write

$$\sum_i \sum_{j/\Gamma_{ij}\subset\partial T_i} |T_i|\ |u_j^n - v_j^n| \leq 3C \sum_j |T_j|\ |u_j^n - v_j^n|.$$

There remains to prove that the term

$$\Delta t \sum_{i,n} |T_i| \sum_{j/\Gamma_{ij}\subset\partial T_i} |v_i^n - v_j^n|$$

converges towards 0. Given ε, by the density of $\mathbf{C}^1$ in $\mathbf{L}^1$ we can find a continuously differentiable function ψ with support in supp (φ), such that

$$\|u - \psi\|_{\mathbf{L}^1(\text{supp}\,(\varphi))} \leq \varepsilon.$$

Setting

$$\psi_i^n = \frac{1}{\Delta t |T_i|} \int_{T_i\times(t_n,t_{n+1})} \psi(\mathbf{x}) d\mathbf{x},$$

by the Lipschitz continuity of ψ, we have

$$\Delta t \sum_i |T_i| \sum_{j/\Gamma_{ij}\in\partial T_i} |\psi_i^n - \psi_j^n| \to 0, \text{ as } \delta \to 0.$$

Using again the assumption of nondegeneracy, we find

$$\Delta t \sum_{i,n} |T_i| \sum_{j/\Gamma_{ij}\subset\partial T_i} |\psi_j^n - v_j^n| \le 3C\|v - \psi\|_{\mathbf{L}^1(\mathrm{supp}\,(\varphi))} \le \varepsilon.$$

Finally, combining (4.20)–(4.22), we conclude that the limit v satisfies

$$\int_{\mathbb{R}^2\times[0,+\infty)} \Big\{v(x,t)\,\frac{\partial\varphi}{\partial t}(\mathbf{x},t) + \mathbb{F}(v(x,t))\cdot \operatorname{grad}\varphi(\mathbf{x},t)\Big\}d\mathbf{x}\,dt + \int_{\mathbb{R}^2} u_0(\mathbf{x})\varphi(\mathbf{x},0)d\mathbf{x} = 0.$$

Hence v is a weak solution of (3.6). □

We now specify some of the main stability results that are available.

4.2.3 Stability

We can prove in the scalar case L^∞-stability, which is the first necessary step for the above-mentioned convergence results.

Proposition 4.2

Assume that in scheme (4.3) the numerical flux Φ is associated by (4.6) to a 3-point one-dimensional monotone numerical flux φ (which is Lipschitz continuous). Then, the scheme (4.3) is L^∞-stable and satisfies

$$\sum_i |\Omega_i|\,|u_i^{n+1}| \le \sum_i |\Omega_i|\,|u_i^n|$$

under the C.F.L.-like condition

$$\lambda C(\varphi) \le 1,$$

where $C(\varphi)$ is the Lipschitz constant of φ, and $\lambda = \Delta t \sup_i(\frac{|\partial\Omega_i|}{|\Omega_i|})$.

In the formula for λ, $|\partial\Omega_i|$ denotes the perimeter of Ω_i; hence for a uniform triangulation by triangles with side h, we have $\lambda = 4\sqrt{3}\frac{\Delta t}{h}$. In the one-dimensional case, it corresponds to a C.F.L. of 1/2.

Proof. We can write, again using (4.14),

$$\begin{aligned} u_i^{n+1} &= u_i^n - \frac{\Delta t}{|\Omega_i|}\Big\{\sum_{j/e=\Gamma_{ij}\subset\partial\Omega_i} |e|\Phi(u_i^n, u_j^n, \mathbf{n}_e)\Big\} \\ &= u_i^n - \frac{\Delta t}{|\Omega_i|}\Big\{\sum_{j/e=\Gamma_{ij}\subset\partial\Omega_i} |e|(\Phi(u_i^n, u_j^n, \mathbf{n}_e) - \mathbb{F}\cdot\mathbf{n}(u_i^n))\Big\} \end{aligned}$$

and for a consistent numerical flux

$$u_i^{n+1} = u_i^n - \frac{\Delta t}{|\Omega_i|}\{\sum |e|(\Phi(u_i^n, u_j^n, \mathbf{n}_e) - \Phi(u_i^n, u_i^n, \mathbf{n}_e))\}.$$

Multiplying out and dividing each term of the sum by the factor $(u_j^n - u_i^n)$, we get

$$u_i^{n+1} = u_i^n\Big\{1 - \frac{\Delta t}{|\Omega_i|}\sum |\Gamma_{ij}| \frac{\Phi(u_i^n, u_i^n, \mathbf{n}_{ij}) - \Phi(u_i^n, u_j^n, \mathbf{n}_{ij})}{u_j^n - u_i^n}\Big\}$$
$$+ \frac{\Delta t}{|\Omega_i|}\Big\{\sum |\Gamma_{ij}| u_j^n \frac{\Phi(u_i^n, u_i^n, \mathbf{n}_{ij}) - \Phi(u_i^n, u_j^n, \mathbf{n}_{ij})}{u_j^n - u_i^n}\Big\},$$

where we have denoted $\mathbf{n}_e = \mathbf{n}_{ij}$ for $e = \Gamma_{ij}$. This corresponds, in fact, to expressing the one-dimensional 3-point flux $\varphi(\mathbb{F}\cdot\mathbf{n}; u, v)$ in terms of the incremental coefficients

$$\frac{C(u,v)}{\lambda} = \frac{\varphi(\mathbb{F}\cdot\mathbf{n}; u, u) - \varphi(\mathbb{F}\cdot\mathbf{n}; u, v)}{v - u},$$
$$\frac{D(u,v)}{\lambda} = \frac{\varphi(\mathbb{F}\cdot\mathbf{n}; u, u) - \varphi(\mathbb{F}\cdot\mathbf{n}; u, v)}{v - u}.$$

We then follow the lines of G.R., Chapter III, Section 3.2 (Theorem 3.6), just replacing the 1-D coefficient $\lambda = \frac{\Delta t}{\Delta x}$ by $\lambda_i = \frac{\Delta t |\partial\Omega_i|}{|\Omega_i|}$ ($|\partial\Omega_i|$ denotes the perimeter of Ω_i). The condition in the 1-D case,

$$0 \le C_{j+1/2} \le \frac{1}{2}, \quad 0 \le D_{j-1/2} \le \frac{1}{2},$$

which yields

$$C_{j+1/2} + D_{j-1/2} \le 1,$$

is now replaced by

$$0 \le C_{i,j} = \frac{\Delta t |\Gamma_{ij}|}{|\Omega_i|} \frac{\Phi(u_i^n, u_i^n, \mathbf{n}_{ij}) - \Phi(u_i^n, u_j^n, \mathbf{n}_{ij})}{u_j^n - u_i^n} \tag{4.23a}$$

and

$$\frac{C_{i,j}|\partial\Omega_i|}{|\Gamma_{ij}|} \le 1. \tag{4.23b}$$

The coefficient $C_{i,j}$ is indeed positive for a monotone scheme since

$$C_{i,j} = \frac{\lambda_i |\Gamma_{ij}|}{|\partial\Omega_i|} C(u_i^n, u_j^n)$$

and $C(u,v) \le 0$ ($v \to \varphi(\cdot, v)$ is nonincreasing); (4.23a) also holds for other schemes such as E-schemes.

Inequality (4.23b), which is equivalent to

$$\lambda_i C(u_i^n, u_j^n) \le 1,$$

holds if

$$\lambda_i \frac{\Phi(u_i^n, u_i^n, \mathbf{n}_{ij}) - \Phi(u_i^n, u_j^n, \mathbf{n}_{ij})}{u_j^n - u_i^n} \leq 1,$$

which appears as a C.F.L.-like condition (linked to the modulus of continuity of φ). This yields

$$\sum_{j, e=\Gamma_{ij} \subset \partial\Omega_i} C_{i,j} = \frac{\lambda_i}{|\partial\Omega_i|} \sum_j |\Gamma_{ij}| C(u_i^n, u_j^n) \leq 1$$

and proves in turn that u_i^{n+1} is a convex combination of u_i^n and the neighboring values u_j^n,

$$u_i^{n+1} = u_i^n(1 - \sum_j C_{i,j}) + \sum_j C_{i,j} u_j^n = u_i^n + \sum_j C_{i,j}(u_j^n - u_i^n).$$

It implies first the local maximum principle

$$\min(u_i^n, \min_{e \subset \partial\Omega_i} u_j^n) \leq u_i^{n+1} \leq \max(u_i^n, \max_{e \subset \partial\Omega_i} u_j^n)$$

and also

$$|u_i^{n+1}| \leq (1 - \sum_j C_{i,j})\, |u_i^n| + \sum_j C_{i,j}\, |u_j^n|.$$

Then, multiplying by $|\Omega_i|$ and summing over a finite set, for instance corresponding to the cells intersecting a given ball B_R, which we denote by $i \in J(R)$, we write

$$\sum_{i \in J(R)} |\Omega_i|\, |u_i^{n+1}| \leq \sum_{i \in J(R)} |\Omega_i| \{(1 - \sum_j C_{i,j}) |u_i^n| + \sum_j C_{i,j} |u_j^n|\}. \tag{4.24}$$

We now observe that in the right-hand side of (4.24) we can reorder the terms and sum over the indices "j", provided we take care of the "boundary" elements Ω_j which do not intersect B_R but then intersect B_{R+h} (if h is the maximum of the diameters of the Ω_i). Since by the conservation relation (4.4)

$$\begin{aligned}
|\Omega_i| C_{i,j} &= \Delta t |\Gamma_{ij}| \frac{\mathbb{F} \cdot \mathbf{n}_{ij}(u_i^n) - \Phi(u_i^n, u_j^n, \mathbf{n}_{ij})}{u_j^n - u_i^n}, \\
|\Omega_j| C_{j,i} &= \Delta t |\Gamma_{ji}| \frac{-\mathbb{F} \cdot \mathbf{n}_{ij}(u_j^n) - \Phi(u_j^n, u_i^n, -\mathbf{n}_{ij})}{u_i^n - u_j^n} \\
&= \Delta t |\Gamma_{ij}| \frac{\mathbb{F} \cdot \mathbf{n}_{ij}(u_j^n) - \Phi(u_i^n, u_j^n, \mathbf{n}_{ij})}{u_j^n - u_i^n},
\end{aligned}$$

we get for an "interior" element Ω_j that in (4.24), $|u_j^n|$ is multiplied by

$$\sum_{i \text{ neighbors of } j} |\Omega_i|\, C_{i,j} + |\Omega_j| (1 - \sum_{i \text{ neighbors of } j} C_{j,i}) = |\Omega_j|.$$

Indeed, for an index i such that Ω_j and Ω_i are neighbors,

$$|\Omega_i| C_{i,j} - |\Omega_j| C_{j,i} = \Delta t |\Gamma_{ij}| \frac{\mathbb{F} \cdot \mathbf{n}_{ij}(u_j^n) - \mathbb{F} \cdot \mathbf{n}_{ij}(u_i^n)}{u_j^n - u_i^n},$$

so that by (4.14), summing up over i, the sum vanishes. Finally, we obtain as expected

$$\sum_{i \in J(R)} |\Omega_i| \, |u_i^{n+1}| \leq \sum_{i \in J(R+h)} |\Omega_i| \, |u_i^n|,$$

and then let $R \to \infty$. □

Remark 4.4. Note that we could have taken any convex function of u in place of $|u|$ (in particular $\eta(u) = |u|^p$), which yields

$$\sum_{i \in J(R)} |\Omega_i| \eta(u_i^{n+1}) \leq \sum_{i \in J(R+h)} |\Omega_i| \, \eta(u_i^n)$$

and provides other estimates. □

For what concerns B.V.-stability, we have already observed that assumption (4.19) is unrealistic, and no (strong) B.V.-estimate is available in general. For the reader's convenience, we just state that the "natural" total variation of the sequence $u = (u_i)$ is

$$TV(u) = \|u_\Delta\|_{BV(\mathbb{R}^2)},$$

where u_Δ is the piecewise constant function associated to u. An easy computation gives

$$TV(u) = \frac{1}{2} \sum_{ij} |\Gamma_{ij}| \, |u_i - u_j|,$$

which coincide, with (3.7) in the case of rectangles.

Such B.V.-estimates, which hold in the one-dimensional case or on a rectangular 2-D mesh, would provide relative compactness of the sequence (v_{δ_k}) in $\mathbf{L}^1_{\mathrm{loc}}$ and imply the convergence of v_{Δ_k} (or a subsequence) towards u and of $\mathbb{F}(v_{\Delta_k})$ towards $\mathbb{F}(u)$.

Let us sketch the main ideas that are involved in the currently known convergence results initiated by Szepessy for a multidimensional scalar conservation law. The L^∞-bound (4.18) is indeed satisfied by many schemes (Proposition 4.2 above) and yields convergence of (v_{δ_k}) for the weak star topology in $\mathbf{L}^\infty(\mathbb{R}^2 \times (0, +\infty))$ to some u. Instead of (strong) B.V.-estimates, "weak" B.V.-estimates are obtained, but they cannot allow this relative compactness, and it is well known that one cannot conclude that $\mathbb{F}(v_{\delta_k})$ tends to $\mathbb{F}(u)$.

Hence, one requires more powerful tools namely Young's measure (Tartar 1983, Ball 1989) and entropy-measure-valued solution (DiPerna 1985). A *Young measure* on $\mathbb{R}^d \times (0, +\infty)$ is a parametrized family $(\mathbf{x}, t) \in \mathbb{R}^d \times$

$(0, +\infty) \to \nu_{\mathbf{x},t}$ of probability measures on $\mathbb{R}$. For a continuous function $g \in \mathbf{C}^0(\mathbb{R})$,

$$< \nu_{\mathbf{x},t}, g(\cdot) > = \int_{\mathbb{R}} g(\lambda) d\nu_{\mathbf{x},t}(\lambda) = \mu_g(\mathbf{x}, t)$$

denotes the pairing between the probability $\nu_{\mathbf{x},t}$ and the function g; for $g(\lambda) = \lambda$, we write $< \nu_{\mathbf{x},t}, \lambda > = \int_{\mathbb{R}} \lambda d\nu_{x,t}(\lambda)$. In the particular case where $\nu_{\mathbf{x},t}$ is a Dirac measure δ_u,

$$< \delta_u, g(\cdot) > = \int_{\mathbb{R}} g(\lambda)\delta_u(\lambda) = g(u).$$

In general, $(\mathbf{x}, t) \in \mathbb{R}^d \times (0, \infty) \to \mu_g(\mathbf{x}, t)$ is measurable, and we can assume for simplicity that it is in $\mathbf{L}^\infty(\mathbb{R}^d \times (0, +\infty))$.

Definition 4.1

A measure-valued (mv) solution of (3.6) is a measurable map from $\mathbb{R}^d \times (0, +\infty)$ *to the space* Prob($\mathbb{R}$) *of nonnegative Radon measure with unit mass, such that for all* $\varphi \in \mathbf{C}_0^1(\mathbb{R}^d \times (0, +\infty))$

$$\int \Big\{ < \nu_{\mathbf{x},t}, \lambda > \frac{\partial \varphi}{\partial t} + \sum_j < \nu_{\mathbf{x},t}, f_j(\lambda) > \frac{\partial \varphi}{\partial x_j} \Big\} d\mathbf{x}\, dt = 0.$$

An entropy-measure-valued solution satisfies, moreover, a weak entropy inequality (see DiPerna 1985, Szepessy 1989). The uniqueness theorem of DiPerna concerning entropy mv-solutions states that if at time zero an mv-entropy solution is a Dirac measure $\delta_{u_0}(x)$, then, for time $t > 0$, it remains a Dirac measure $\delta_{u(\mathbf{x},t)}$, where u is the unique entropy weak solution given in Chapter II. This result has been generalized by Szepessy (1989).

Thanks to DiPerna's result (DiPerna 1985, Theorem 2.1), the uniform bound (4.18) ensures that a Young measure $(\mathbf{x}, t) \in \mathbb{R}^d \times (0, +\infty) \to \nu_{\mathbf{x},t}$ can be constructed that represents all $w^\star$-composite limits of $(v_{\delta k})$ for the weak star topology in $\mathbf{L}^\infty(\mathbb{R}^2 \times (0, +\infty))$. This means that for any $g \in \mathbf{C}^0(\mathbb{R})$

$$g(v_{\delta_k}(\mathbf{x}, t)) \to \mu_g(\mathbf{x}, t) = < \nu_{\mathbf{x},t}, g(\cdot) > \ (\mathbf{L}^\infty \text{ weak star limit}) \text{ as } \delta_k \to 0$$

i.e.,

$$\int g(v_{\delta_k})\varphi(\mathbf{x}, t) d\mathbf{x}\, dt \to \int < \nu_{\mathbf{x},t}, g(\cdot) > \varphi(\mathbf{x}, t) d\mathbf{x}\, dt.$$

We can thus give a meaning to "lim $\mathbb{F}(v_{\delta_k})$", which is defined by setting "lim $\mathbb{F}(v_{\delta_k})$" $= \mu_{\mathbb{F}}(\mathbf{x}, t) = < \nu_{\mathbf{x},t}, \mathbb{F}(\cdot) >$ and yields the existence of a measure-valued solution. The aim is to prove that $\nu_{\mathbf{x},t}$ is the Dirac measure $\delta_{u(\mathbf{x},t)}$

$$< \delta_{u(\mathbf{x},t)}, g(\cdot) > = \int_{\mathbb{R}} g(\lambda)\delta_u(\lambda) = g(u(x, t)).$$

The "weak" B.V.-estimates provide the extra information that enables one to show that the limit u of (v_{δ_k}) satisfies the weak entropy inequality; then, by Szepessy's uniqueness theorem, u is the unique entropy solution and (v_{δ_k}) converges strongly in $\mathbf{L}^p, 1 \leq p < +\infty$.

Remark 4.5. At the present time, in Champier, Gallouët and Herbin (1993), estimate (4.19), weak B.V.-stability and convergence are proved for an upstream finite volume scheme in the case $\mathbb{F}(u) = \mathbf{v}f(u)$, with div $\mathbf{v} = 0$ (see Example 4.3). Also convergence results and error estimates (following Kuznetsov's theorem, see G.R., the Appendix to Chapter III) for finite-volume monotone or E-shemes have been obtained by Cockburn et al. (1994) and Benharbit et al. (1993). Both approaches use in particular the relation (4.14), which yields

$$\sum_{e \subset \partial\Omega_i} |a| \mathbb{F}(u) \cdot \mathbf{n}_e = \mathbb{F}(\mathbf{u}) \cdot \sum_{e \subset \partial\Omega_i} |e| \, \mathbf{n}_e = 0$$

and induces the convex decomposition

$$u_i^{n+1} = \sum_{e \subset \partial\Omega_i} \frac{|e|}{|\partial\Omega_i|} u_{i,e}^{n+1},$$

where

$$u_{i,e}^{n+1} = u_i^n - \Delta t \frac{|\partial\Omega_i|}{|\Omega_i|} \{\Phi(u_i^n, u_j^n, \mathbf{n}_e) - \mathbb{F}(u_i^n) \cdot \mathbf{n}_e\}$$

or, equivalently,

$$u_{i,e}^{n+1} = u_i^n - \Delta t \frac{|\partial\Omega_i|}{|\Omega_i|} \{\Phi(u_i^n, u_j^n, \mathbf{n}_e) - \Phi(u_i^n, u_i^n, \mathbf{n}_e\}$$

due to the consistency of Φ. These last expressions are now purely one-dimensional, and one can prove for them a discrete maximum principle in the spirit of Proposition 4.2, and discrete entropy inequalities that lead rather simply to L^1-stability. Weak B.V. estimates are obtained by studying more sharply the local entropy production. We refer to the above-mentioned papers for details; see also Szepessy (1991), Perthame (1994). Perthame et al. (1994) prove the strong convergence in $\mathbf{L}^2$ of the so-called N-scheme (Roe and Sidilkover 1992) on general triangulations for a linear constant coefficient equation. □

We turn now to the description of some of the usual schemes in their two-dimensional finite volume extensions, together with the application to gas dynamics.

4.3 Usual schemes

Let us start with Roe's scheme, which is actually one of the most popular schemes. We first give some detailed computations of the straightforward finite-volume extension of the one-dimensional scheme, and then we shall present a recent attempt to derive a truly multidimensional scheme, which follows the same approach for finding an appropriate linearization.

4.3.1 Roe's scheme

Following the ideas of the previous sections, we want to detail the standard two-dimensional Roe's scheme (see Chapter III, Section 3.1). We first linearize the system (1.1) and then integrate the linearized system above the control cells. Let us set

$$\mathbf{A}_\omega(\mathbf{U}) = \mathbf{A}(\mathbf{U}, \boldsymbol{\omega}),$$

where $\mathbf{A}(\mathbf{U}, \boldsymbol{\omega})$ is defined by (1.3). We construct a Roe linearization in each direction $\boldsymbol{\omega}$, i.e., a matrix $A_\omega(\mathbf{U}_L, \mathbf{U}_R)$ such that

$$\begin{cases} (i) & \mathbf{A}_\omega(\mathbf{U}, \mathbf{U}) = \mathbf{A}_\omega(\mathbf{U}) = \mathbf{A}(\mathbf{U}, \boldsymbol{\omega}), \\ (ii) & \mathbf{A}_\omega(\mathbf{U}, \mathbf{V})(\mathbf{U} - \mathbf{V}) = \mathbb{F} \cdot \boldsymbol{\omega}(\mathbf{U}) - \mathbb{F} \cdot \boldsymbol{\omega}(\mathbf{V}), \\ (iii) & \mathbf{A}_\omega(\mathbf{U}, \mathbf{V}) \text{ has real eigenvalues and a complete set} \\ & \qquad \text{of eigenvectors.} \end{cases}$$

If "parameter vectors" $\mathbf{W}$ (see Chapter III, Section 3) are available, we can define

$$\mathbf{A}_\omega(\mathbf{U}_L, \mathbf{U}_R) = \mathbf{A}(\overline{\mathbf{U}}, \boldsymbol{\omega}),$$

where $\overline{\mathbf{U}}$ is the Roe's averaged state $\overline{\mathbf{U}} = \mathbf{U}(\mathbf{W}^*)$, $\mathbf{W}^* = \frac{(\mathbf{W}_L + \mathbf{W}_R)}{2}$.

When integrating the linearized system over C_i, Green's formula gives, in particular, a flux across the edge $e = \Gamma_{ij}$. We integrate the linearized system with matrix $A_{\mathbf{n}_{ij}}(\mathbf{U}_i, \mathbf{U}_j)$ in the direction $\mathbf{n}_{ij}$ normal to Γ_{ij} from data taken at each side of the edge. Then Roe's scheme is associated to the following flux (see Chapter III (3.11), and Lemma 4.1):

$$\begin{aligned} \boldsymbol{\Phi}(\mathbf{U}_i, \mathbf{U}_j, \mathbf{n}) &= \frac{1}{2}\{\mathbb{F} \cdot \mathbf{n}(\mathbf{U}_i) + \mathbb{F} \cdot \mathbf{n}(\mathbf{U}_j)\} - \frac{1}{2}|\mathbf{A}_{\mathbf{n}}(\mathbf{U}_i, \mathbf{U}_j)|(\mathbf{U}_j - \mathbf{U}_i) \\ &= \mathbb{F} \cdot \mathbf{n}(\mathbf{U}_j) - \mathbf{A}_{\mathbf{n}}^+(\mathbf{U}_i, \mathbf{U}_j)(\mathbf{U}_j - \mathbf{U}_i) \\ &\quad - \mathbb{F} \cdot \mathbf{n}(\mathbf{U}_i) + \mathbf{A}_{\mathbf{n}}^-(\mathbf{U}_i, \mathbf{U}_j)(\mathbf{U}_j - \mathbf{U}_i). \end{aligned}$$

It is well known that Roe's scheme may admit nonphysical (i.e., entropy violating) shocks at sonic points and, as in the one-dimensional case, some entropy correction is needed (see Chapter III, Section 3.1).

In the case of the gas dynamics equations, the numerical flux is

$$\begin{aligned} \mathbb{F} \cdot \mathbf{n} &= (\rho \mathbf{u} \cdot \mathbf{n}, \rho \mathbf{u}\mathbf{u} \cdot \mathbf{n} + p\mathbf{n}, (\rho e + p)\mathbf{u} \cdot \mathbf{n})^T \\ &= (\rho u_n, \rho u_n u + \rho \cos\theta, \rho u_n v + p \sin\theta, (\rho e + p)u_n)^T. \end{aligned}$$

For an ideal gas $p = (\gamma - 1)\rho\varepsilon$, defining the enthalpy by

$$H = \frac{(E+p)}{\rho} = \varepsilon + \frac{(u^2 + v^2)}{2} + \frac{p}{\rho},$$

the "parameter vectors" are (see Chapter III, Section 3.2)

$$w_1 = \rho^{1/2}, \quad w_2 = \rho^{1/2}u, \quad w_3 = \rho^{1/2}v, \quad w_4 = \rho^{1/2}H.$$

Following exactly the computations of Chapter III, Section 3, we get $\mathbf{A}_\omega(\mathbf{U}_L, \mathbf{U}_R) = \mathbf{A}_\omega(\overline{\mathbf{U}})$, where $\overline{u}, \overline{H}$ are given by Lemma 3.5 of Chapter III, Section 3, and

$$\overline{v} = \frac{(\rho_L^{1/2} v_L + \rho_R^{1/2} v_R)}{(\rho_L^{1/2} + \rho_R^{1/2})}$$

$$\overline{c} = \kappa\Big(\overline{H} - \frac{(\overline{u}^2 - \overline{v}^2)}{2}\Big) + \chi.$$

We then compute the eigenvectors and the coefficients in the basis of eigenvectors (more details are given in Section 4.3.3 below). Note that for the numerical flux we have

$$\text{if } \overline{\mathbf{u}} \cdot \boldsymbol{\omega} - \overline{c} > 0, \quad \mathbf{A}_\omega^-(\mathbf{U}_L, \mathbf{U}_R) = 0 \text{ and } \boldsymbol{\Phi}(\mathbf{U}_L, \mathbf{U}_R, \boldsymbol{\omega}) = \mathbb{F} \cdot \boldsymbol{\omega}(\mathbf{U}_L),$$
$$\text{if } \overline{\mathbf{u}} \cdot \boldsymbol{\omega} + \overline{c} < 0, \quad \mathbf{A}_\omega^+(\mathbf{U}_L, \mathbf{U}_R) = 0 \text{ and } \boldsymbol{\Phi}(\mathbf{U}_L, \mathbf{U}_R, \boldsymbol{\omega}) = \mathbb{F} \cdot \boldsymbol{\omega}(\mathbf{U}_R).$$

Lemma 4.4

Roe's scheme is invariant under rotation.

Proof. Let us check that (4.8) holds. Indeed, consider

$$\mathbf{A}_{\mathbf{e}_1}(\tilde{\mathbf{U}}_L, \tilde{\mathbf{U}}_R) = \mathbf{A}(\overline{\tilde{\mathbf{U}}}, \mathbf{e}_1),$$

where $\overline{\tilde{\mathbf{U}}}$ is the Roe's averaged state $\overline{\tilde{\mathbf{U}}} = \mathbf{U}(\tilde{\mathbf{W}}^*), \tilde{\mathbf{W}}^* = \frac{(\tilde{\mathbf{W}}_L + \tilde{\mathbf{W}}_R)}{2}$. A simple computation using the definition of the parameter vectors $\mathbf{W} = \rho^{1/2}(1, \mathbf{u}, H)^T$ yields

$$\tilde{\mathbf{W}}^* = \frac{\tilde{\mathbf{W}}_L + \tilde{\mathbf{W}}_R}{2} = R^{-1}\frac{\mathbf{W}_L + \mathbf{W}_R}{2} = R^{-1}\,\mathbf{W}^*$$

and

$$R^{-1}\overline{\mathbf{U}} = R^{-1}\mathbf{U}(\mathbf{W}^*) = \mathbf{U}(R^{-1}\mathbf{W}^*) = \mathbf{U}(\tilde{\mathbf{W}}^*).$$

Using (4.9b), it follows that if R is the rotation such that $R^{-1}\mathbf{n} = \mathbf{e}_1$ and $R^{-1}\mathbf{U} = \tilde{\mathbf{U}}$, we have

$$\mathbf{A}_\mathbf{n}(\mathbf{U}_L, \mathbf{U}_R) = \mathbf{A}(\overline{\mathbf{U}}, \mathbf{n}) = R\mathbf{A}(R^{-1}\overline{\mathbf{U}})R^{-1} = R\,\mathbf{A}_{\mathbf{e}_1}(\tilde{\mathbf{U}}_L, \tilde{\mathbf{U}}_R)R^{-1}$$

or (by (4.9c))

$$\mathbf{A}_\mathbf{n}(\mathbf{U}_L, \mathbf{U}_R)\mathbf{V} = R\,\mathbf{A}_{\mathbf{e}_1}(\tilde{\mathbf{U}}_L, \mathbf{U}_R)\tilde{\mathbf{V}},$$

which implies that (4.8) holds for Roe's scheme. □

Roe's scheme can be extended to the real gas case following the arguments of Chapter III, Section 3.3, to reactive flows (Dubroca and Morreuw 1992) and also to computing interface motion (Mulder, Osher and Sethian 1992).

However the numerical solutions stay sensitive to the chosen local triangulation. One way to overcome this problem is to develop a "grid independent wave model" which introduces an additional shear wave propagating in the normal direction (see Rumsey et al. 1993). We shall now develop a two-dimensional Roe's linearization. Other approaches for deriving "trully multidimensional schemes" can be found in Roe and Sidilkover (1992), Tamura and Fujii (1993), see Van Leer (1992) for a survey, Abgrall (1993).

4.3.2 Fully 2-dimensional Roe's linearization

Recently, some attempts have been made to construct a fully 2-D Roe linearization, which we present now (see Deconinck et al. 1993, Angrand and Lafon 1992). The point in presenting this particular method is to illustrate the fact that a 2-D linearization that respects the ideas of the one-dimensional case is not so easy construct. Given a triangle T with vertices $\mathbf{a}_i, \mathbf{a}_j, \mathbf{a}_k$, one constructs a Roe linearization matrix $\mathbf{A}_\omega(T) = \mathbf{A}_\omega(\mathbf{U}_i, \mathbf{U}_j, \mathbf{U}_k)$ which depends on the three states. This method is not "consistent" (see Abgrall 1994) for more precise statement) with the Euler system but guarantees hyperbolicity. Another approach which we shall not develop (see Abgrall 1994) derives jump relations and finds a linearization that respects these relations; the corresponding method is "consistent" but does not guarantee hyperbolicity.

One defines first for $\omega = \mathbf{e}_1$, $\mathbf{A}_{\mathbf{e}_1}(T) = \mathbf{A}_{\mathbf{e}_1}(\mathbf{U}_i, \mathbf{U}_j, \mathbf{U}_k)$ as the Jacobian $\mathbf{A} = \mathbf{f}'$ computed at the averaged state $\overline{\mathbf{U}} = \mathbf{U}(\mathbf{W}^\star)$, where now

$$\mathbf{W}^\star = \frac{1}{3}(\mathbf{W}_i + \mathbf{W}_j + \mathbf{W}_k)$$

is the parameter vector of the centroid of T; ones makes a similar definition for $\mathbf{B} = \mathbf{g}'$.

Indeed, let us recall the situation in the one-dimensional case: Given a parameter vector $\mathbf{W}$ (such that $\mathbf{U} = \mathbf{U}(\mathbf{W})$ and $\mathbf{f}(\mathbf{U}) = \mathbf{f} \circ \mathbf{U}(\mathbf{W})$ are quadratic), we wrote

$$\begin{aligned}\mathbf{U}_R - \mathbf{U}_L &= \mathbf{U}(\mathbf{W}_R) - \mathbf{U}(\mathbf{W}_L) \\ &= \int_0^1 \mathbf{U}'(\mathbf{W}_L + \theta(\mathbf{W}_R - \mathbf{W}_L))d\theta(\mathbf{W}_R - \mathbf{W}_L).\end{aligned}$$

Since $\mathbf{U}'(\mathbf{W})$ varies linearly, we have

$$\int_0^1 \mathbf{U}'(\mathbf{W}_L + \theta(\mathbf{W}_R - \mathbf{W}_L))d\theta = \mathbf{U}'(\mathbf{W}^*),$$

where

$$\mathbf{W}^* = \frac{(\mathbf{W}_R + \mathbf{W}_L)}{2};$$

similarly,

$$\begin{aligned}\mathbf{f}(\mathbf{U}_R) - \mathbf{f}(\mathbf{U}_L) &= \mathbf{f} \circ \mathbf{U}(\mathbf{W}_R) - \mathbf{f} \circ \mathbf{U}(\mathbf{W}_L) \\ &= \int_0^1 (\mathbf{f} \circ \mathbf{U})'(\mathbf{W}_L + \theta(\mathbf{W}_R - \mathbf{W}_L)) d\theta (\mathbf{W}_R - \mathbf{W}_L) \\ &= (\mathbf{f} \circ \mathbf{U})'(\mathbf{W}^\star)(\mathbf{W}_R - \mathbf{W}_L) \\ &= \mathbf{f}'(\mathbf{U}(\mathbf{W}^\star))\mathbf{U}'(\mathbf{W}^*)(\mathbf{W}_R - \mathbf{W}_L),\end{aligned}$$

and hence

$$\mathbf{f}(\mathbf{U}_R) - \mathbf{f}(\mathbf{U}_L) = \mathbf{f}'(\mathbf{U}(\mathbf{W}^*))(\mathbf{U}_R - \mathbf{U}_L),$$

which led us to define

$$\mathbf{A}(\mathbf{U}_L, \mathbf{U}_R) = \mathbf{f}'(\mathbf{U}(\mathbf{W}^*)).$$

The analog of this property (or (3.64), Chapter III, Section 3) in the two-dimensional case gives, if $L(\mathbf{W})$ is a linear function of $\mathbf{W}$ and $\mathbf{W}$ is linear in $\mathbf{x}$,

$$\frac{1}{|T|} \int_T L(\mathbf{W}(\mathbf{x})) d\mathbf{x} = L(\mathbf{W}^\star) = \frac{L(\mathbf{W}_i + \mathbf{W}_j + \mathbf{W}_k)}{3},$$

and thus if $\mathbf{W}$ is linear in $\mathbf{x}$

$$\frac{1}{|T|} \int_T \mathbf{U}'(\mathbf{W}) d\mathbf{x} = \mathbf{U}'(\mathbf{W}^\star) = \frac{\mathbf{U}'(\mathbf{W}_i + \mathbf{W}_j + \mathbf{W}_k)}{3}. \tag{4.25}$$

If we mimic the one-dimensional case, we are led to define

$$\mathbf{A}_{\mathbf{e}_1}(\mathbf{U}_i, \mathbf{U}_j, \mathbf{U}_k) = \mathbf{f}'(\mathbf{U}(\mathbf{W}^\star)), \tag{4.26a}$$

and similarly

$$\mathbf{A}_{\mathbf{e}_2}(\mathbf{U}_i, \mathbf{U}_j, \mathbf{U}_k) = \mathbf{g}'(\mathbf{U}(\mathbf{W}^\star)). \tag{4.26b}$$

Then, for any vector $\boldsymbol{\omega}$ in $\mathbb{R}^2$,
(4.26c)

$$\begin{aligned}\mathbf{A}_\omega(T) &= \mathbf{A}_\omega(\mathbf{U}_i, \mathbf{U}_j, \mathbf{U}_k) \\ &= \omega_1 \mathbf{A}_{\mathbf{e}_1}(\mathbf{U}_i, \mathbf{U}_j, \mathbf{U}_k) + \omega_2 \mathbf{A}_{\mathbf{e}_2}(\mathbf{U}_i, \mathbf{U}_j, \mathbf{U}_k) = \mathbf{A}(\overline{\mathbf{U}}, \boldsymbol{\omega}),\end{aligned}$$

with the notation (1.3), where $\overline{\mathbf{U}} = \mathbf{U}(\mathbf{W}^\star)$.

This linearization is consistent, since

$$\mathbf{A}_{\mathbf{e}_1}(\mathbf{U}, \mathbf{U}, \mathbf{U}) = \mathbf{A}(\mathbf{U}) = \mathbf{f}'(\mathbf{U}), \quad \mathbf{A}_{\mathbf{e}_2}(\mathbf{U}, \mathbf{U}, \mathbf{U}) = \mathbf{B}(\mathbf{U}) = \mathbf{g}'(\mathbf{U})$$

and for any $\boldsymbol{\omega}$, $\mathbf{A}_\omega(\mathbf{U}_i, \mathbf{U}_j, \mathbf{U}_k) = \mathbf{A}(\overline{\mathbf{U}}, \boldsymbol{\omega})$ has real eigenvalues and a complete set of eigenvectors. There remains to check the analog of the property (ii) of a Roe matrix. Since there is no obvious analog of the increment $\Delta \mathbf{U} = \mathbf{U}_R - \mathbf{U}_L$, we write instead

$$\begin{cases} \delta_{Tx}\mathbf{f} = \mathbf{A}_{\mathbf{e}_1}(\mathbf{U}_i, \mathbf{U}_j, \mathbf{U}_k)\delta_{Tx}\mathbf{U}, \\ \delta_{Ty}\mathbf{g} = \mathbf{A}_{\mathbf{e}_2}(\mathbf{U}_i, \mathbf{U}_j, \mathbf{U}_k)\delta_{Ty}\mathbf{U}, \end{cases} \tag{4.27}$$

for some "derivation rule" $\delta_T = (\delta_{Tx}, \delta_{Ty})$ directly inspired by the above computations and which we discuss now.

We can write the fact that $\mathbf{U}$ and $\mathbf{f}(\mathbf{U})$, $\mathbf{g}(\mathbf{U})$ are homogeneous quadratic functions of $\mathbf{W}$ in the following form:

$$\begin{aligned}
\mathbf{U}(\mathbf{W}) &= Q_{\mathbf{u}}(\mathbf{W}, \mathbf{W}) = \frac{1}{2}\mathbf{U}'(\mathbf{W})\mathbf{W}, \\
(\mathbf{f} \circ \mathbf{U})(\mathbf{W}) &= Q_{\mathbf{f}}(\mathbf{W}, \mathbf{W}) = \frac{1}{2}(\mathbf{f} \circ \mathbf{U})'(\mathbf{W})\mathbf{W}, \\
(\mathbf{g} \circ \mathbf{U})(\mathbf{W}) &= Q_{\mathbf{g}}(\mathbf{W}, \mathbf{W}) = \frac{1}{2}(\mathbf{g} \circ \mathbf{U})'(\mathbf{W})\mathbf{W},
\end{aligned}$$

where the Q's are bilinearly symmetric and $\mathbf{U}'$, $(\mathbf{f} \circ \mathbf{U})'$, and $(\mathbf{g} \circ \mathbf{U})'$ are matrices depending linearly on $\mathbf{W}$ (which are easily computed in the examples). Let us set for any quadratic function $\mathbf{h}(\mathbf{W}) = Q_{\mathbf{h}}(\mathbf{W}, \mathbf{W})$

$$\hat{\mathbf{h}}_T = \hat{\mathbf{h}}_T(\mathbf{W}) = Q_{\mathbf{h}}(\mathbf{W}^\star, \mathbf{W}). \tag{4.28}$$

Thus, freezing one of the variables, at the state $\mathbf{W}^\star$, we introduce the linearized variables

$$\hat{\mathbf{U}}_T = \hat{\mathbf{U}}(\mathbf{W}) = Q_{\mathbf{U}}(\mathbf{W}^\star, \mathbf{W}) = \frac{1}{2}\mathbf{U}'(\mathbf{W}^\star)\mathbf{W}, \tag{4.28a}$$

$$\hat{\mathbf{f}}_T = Q_{\mathbf{f}}(\mathbf{W}^\star, \mathbf{W}) = \frac{1}{2}(\mathbf{f} \circ \mathbf{U})'(\mathbf{W}^\star)\mathbf{W}, \tag{4.28b}$$

$$\hat{\mathbf{g}}_T = Q_{\mathbf{g}}(\mathbf{W}^\star, \mathbf{W}) = \frac{1}{2}(\mathbf{g} \circ \mathbf{U})'(\mathbf{W}^\star)\mathbf{W}, \tag{4.28c}$$

and define $\delta_T = (\delta_{Tx}, \delta_{Ty})$ by

$$\delta_T \mathbf{h} = 2\nabla\, \hat{\mathbf{h}}_T, \tag{4.29}$$

i.e.,

$$\delta_{Tx}\mathbf{h} = 2\,\frac{\partial \hat{\mathbf{h}}_T}{\partial x}, \quad \delta_{Ty}\mathbf{h} = 2\,\frac{\partial \hat{h}_T}{\partial y}.$$

Thus

$$\begin{aligned}
\delta_{Tx}\mathbf{U} &= 2\,\frac{\partial \hat{\mathbf{h}}_T}{\partial x}, \quad \delta_{Ty}\mathbf{U} = 2\,\frac{\partial \hat{U}_T}{\partial y}, \\
\delta_{Tx}\mathbf{f}(\mathbf{U}) &= 2\,\frac{\partial \hat{\mathbf{f}}}{\partial x}, \quad \delta_{Ty}\mathbf{g}(\mathbf{U}) = 2\,\frac{\partial \hat{\mathbf{g}}}{\partial y}.
\end{aligned}$$

Remark 4.6. The "2" on the right-hand side of (4.29) may seem funny but is natural since the functions are quadratic; if we consider the simple example $u(w) = w^2$, $\hat{u}(w) = w^\star w$, then $u'(w) = 2w$, and $2\hat{u}'(w) = 2w^\star$ approximates $u'(w)$ better than $\hat{u}'(w)$ does! □

The following result, together with the desired property (4.27) (analog of the relation on the increments (ii) for a Roe matrix), shows that δ_T appears as the mean of the gradient on the triangle.

Lemma 4.5

For a triangle T with vertices $\mathbf{a}_i$, $\mathbf{a}_j$, $\mathbf{a}_k$, define $\delta_T = (\delta_{Tx}, \delta_{Ty})$ by formulas (4.28) and (4.29) and assume that $\mathbf{W}$ is a linear function of $\mathbf{x}$ on T. Then, we have the identities

(4.30)

$$\delta_{Tx}\mathbf{U} = \frac{1}{|T|}\int_T \frac{\partial \mathbf{U}}{\partial x}\,d\mathbf{x}, \quad \delta_{Ty}\mathbf{U} = \frac{1}{|T|}\int_T \frac{\partial \mathbf{U}}{\partial y}\,d\mathbf{x},$$

$$\delta_{Tx}\mathbf{f}(\mathbf{U}) = \frac{1}{|T|}\int_T \frac{\partial}{\partial x}\mathbf{f}(\mathbf{U})\,d\mathbf{x}, \; \delta_{Ty}\mathbf{g}(\mathbf{U}) = \frac{1}{|T|}\int_T \frac{\partial}{\partial y}\mathbf{g}(\mathbf{U})\,d\mathbf{x},$$

and (4.27) holds, i.e.,

$$\delta_{Tx}\mathbf{f}(\mathbf{U}) = \mathbf{A}_{\mathbf{e}_1}(\mathbf{U}_i, \mathbf{U}_j, \mathbf{U}_k)\delta_{Tx}\mathbf{U}, \quad \delta_{Ty}\mathbf{f}(\mathbf{U}) = \mathbf{A}_{\mathbf{e}_2}(\mathbf{U}_i, \mathbf{U}_j, \mathbf{U}_k)\,\delta_{Ty}\mathbf{U},$$

where $A_{\mathbf{e}_1}(\mathbf{U}_i, \mathbf{U}_j, \mathbf{U}_k)$ and $A_{\mathbf{e}_2}(\mathbf{U}_i, \mathbf{U}_j, \mathbf{U}_k)$ are defined by (4.26).

Proof. Using (4.25), we write

$$\frac{\partial \hat{\mathbf{U}}_T}{\partial x} = \mathbf{U}'(\mathbf{W}^\star)\frac{\partial \mathbf{W}}{\partial x} = \frac{1}{|T|}\int_T \mathbf{U}'(\mathbf{W})d\mathbf{x}\,\frac{\partial \mathbf{W}}{\partial x},$$

and if we assume that $\mathbf{W}$ is a *linear* function of $\mathbf{x}$ on T, then

$$\int_T \mathbf{U}'(\mathbf{W})d\mathbf{x}\,\frac{\partial \mathbf{W}}{\partial x} = \int_T \mathbf{U}'(\mathbf{W})\frac{\partial \mathbf{W}}{\partial x}\,d\mathbf{x} = \int_T \frac{\partial \mathbf{U}}{\partial x}\,d\mathbf{x};$$

thus

$$\delta_{Tx}\mathbf{U}(\mathbf{W}) = \frac{1}{|T|}\int_T \frac{\partial \mathbf{U}}{\partial x}\,d\mathbf{x}.$$

Similarly,

$$\delta_{Ty}\mathbf{U}(\mathbf{W}) = \frac{\partial \hat{\mathbf{U}}(\mathbf{W})}{\partial y} = \mathbf{U}'(\mathbf{W}^\star)\frac{\partial \mathbf{W}}{\partial y} = \frac{1}{|T|}\int_T \frac{\partial \mathbf{U}}{\partial y}\,d\mathbf{x},$$

with analogous formulas for $\mathbf{f}$ and $\mathbf{g}$, which proves (4.30).

Now, we have

$$\begin{aligned}\delta_{Tx}\mathbf{f}(\mathbf{U}) &= (\mathbf{f}\circ\mathbf{U})'(\mathbf{W}^\star)\frac{\partial \mathbf{W}}{\partial x} = \mathbf{f}'(\mathbf{U}(\mathbf{W}^\star))\mathbf{U}'(\mathbf{W}^\star)\frac{\partial \mathbf{W}}{\partial x}\\ &= \mathbf{A}_{\mathbf{e}_1}(\mathbf{U}_i, \mathbf{U}_j, \mathbf{U}_k)\mathbf{U}'(\mathbf{W}^\star)\frac{\partial \mathbf{W}}{\partial \mathbf{x}} = \mathbf{A}_{\mathbf{e}_1}(\mathbf{U}_i, \mathbf{U}_j, \mathbf{U}_k)\frac{\partial \hat{\mathbf{U}}_T}{\partial x},\end{aligned}$$

and a similar formula for $\mathbf{g}$, which gives the desired result (4.27). □

Now, we use this linearization in the finite-volume method, where the cells are those of Example 4.2. The triangle $T = (\mathbf{a}_i, \mathbf{a}_j, \mathbf{a}_k)$ is divided into three quadrangular regions R_i, R_j, R_k, each belonging to a different control cell C_i, C_j, C_k and having a common center of mass. Consider a segment $e_{ij} = \mathbf{gm}_{ij}$ ($\mathbf{m}_{ij}$ is the midpoint of $\mathbf{a}_i\,\mathbf{a}_j$, $\mathbf{g}$ the barycenter of $(\mathbf{a}_i, \mathbf{a}_j, \mathbf{a}_k)$) separating R_i and R_j, which is thus part of the boundary between the dual cells C_i and C_j, and denote by $\mathbf{n}_{ij} = (n_{ij,x}, n_{ij,y})^T$ the outward unit normal to e_{ij} (pointing in the direction of R_j). Thus, we define the matrix $\mathbf{A}_{ij}(T) = \mathbf{A}_{\mathbf{n}_{ij}}(T)$ by (4.26),

$$\begin{aligned}\mathbf{A}_{ij}(T) &= \mathbf{A}_{\mathbf{n}_{ij}}(T)\\ &= n_{ij,x}\,\mathbf{A}_{\mathbf{e}_1}(\mathbf{U}_i, \mathbf{U}_j, \mathbf{U}_k) + n_{ij,y}\mathbf{A}_{\mathbf{e}_2}(\mathbf{U}_i, \mathbf{U}_j, \mathbf{U}_k) = \mathbf{A}(\overline{\mathbf{U}}, \mathbf{n}_{ij}),\end{aligned}$$

with the notation (1.3), where $\overline{\mathbf{U}} = \mathbf{U}(\mathbf{W}^\star)$, $\mathbf{W}^\star = \frac{(\mathbf{W}_i+\mathbf{W}_j+\mathbf{W}_k)}{3}$.

Consider a cell C_i that is the union of quadrangular regions R_i^T of type R_i. As in the general finite-volume method of Example 4.2, we integrate over $C_i = \cup R_i^T$ and, following Roe's method, we integrate, in fact, a linearized system. More precisely, here we integrate the linearized system in the "linearized" variables $\hat{\mathbf{U}}$, and on $R_i^T = R_i$ (dropping the dependence on T) Green's formula gives in particular a flux across the segment e_{ij}. Thus, as we have already explained for a general finite-volume method, we are led to approach the solution of the one-dimensional system

$$\frac{\partial \hat{\mathbf{U}}}{\partial t} + \mathbf{A}_{ij}(T)\frac{\partial \hat{\mathbf{U}}}{\partial \zeta} = 0, \quad \zeta = \mathbf{x}\cdot\mathbf{n}_{ij},$$

where

$$\tilde{\mathbf{U}} = \begin{cases}\hat{\mathbf{U}}_i & \zeta < 0,\\ \hat{\mathbf{U}}_j & \zeta > 0,\end{cases}$$

and with the linearized matrix $\mathbf{A}_{ij}(T)$ in the direction $\mathbf{n}_{ij}$ normal to e_{ij}. By Lemma 4.5 and the definition of $\mathbf{A}_{ij}(T) = n_{ij,x}\mathbf{A}_{\mathbf{e}_1}(\mathbf{U}_i, \mathbf{U}_j, \mathbf{U}_k) + n_{ij,y}\mathbf{A}_{\mathbf{e}_2}(\mathbf{U}_i, \mathbf{U}_j, \mathbf{U}_k) = \mathbf{A}(\overline{\mathbf{U}}, \mathbf{n}_{ij})$, we have

$$\begin{aligned}\mathbf{A}_{\mathbf{e}_1}(\mathbf{U}_i, \mathbf{U}_j, \mathbf{U}_k)n_{ij,x}\frac{\partial \hat{\mathbf{U}}}{\partial x} &+ \mathbf{A}_{\mathbf{e}_2}(\mathbf{U}_i, \mathbf{U}_j, \mathbf{U}_k)n_{ij,x}\frac{\partial \hat{\mathbf{U}}}{\partial y}\\ &= n_{ij,x}\frac{\partial \hat{\mathbf{f}}}{\partial x} + n_{ij,y}\frac{\partial \hat{\mathbf{g}}}{\partial y},\end{aligned}$$

and we approximate in an upwind way the one-dimensional system relative to the flux $\mathbb{F}\cdot\mathbf{n}_{ij}$, thus defining

$$\mathbf{\Phi}_T(\hat{\mathbf{U}}_i, \hat{\mathbf{U}}_j, \mathbf{n}_{ij}) = \frac{1}{2}\{\hat{\mathbb{F}}\cdot\mathbf{n}_{ij}(\mathbf{U}_i) + \hat{\mathbb{F}}\cdot\mathbf{n}_{ij}(\mathbf{U}_j)\} - \frac{1}{2}|\mathbf{A}_{ij}(T)|(\hat{\mathbf{U}}_j - \hat{\mathbf{U}}_i).$$

Thanks to Lemma 4.3,

$$\hat{\mathbb{F}}\cdot\mathbf{n}_{ij}(\mathbf{U}_i) - \hat{\mathbb{F}}\cdot\mathbf{n}_{ij}(\mathbf{U}_j) = \mathbf{A}_{ij}(T)(\hat{\mathbf{U}}_j - \hat{\mathbf{U}}_i).$$

Thus, we may also write

$$\begin{aligned}\mathbf{\Phi}_T(\hat{\mathbf{U}}_i, \hat{\mathbf{U}}_j, \mathbf{n}_{ij}) &= \hat{\mathbb{F}}\cdot\mathbf{n}_{ij}(\mathbf{U}_j) - \mathbf{A}_{ij}^+(T)(\hat{\mathbf{U}}_j - \hat{\mathbf{U}}_i)\\ &= \hat{\mathbb{F}}\cdot\mathbf{n}_{ij}(\mathbf{U}_i) + \mathbf{A}_{ij}^-(T)(\hat{\mathbf{U}}_j - \hat{\mathbf{U}}_i).\end{aligned}$$

The local increments in the linearized conservative variables $\hat{\mathbf{U}}$ are computed through

$$\hat{\mathbf{U}}_i - \hat{\mathbf{U}}_j = \mathbf{U}'(\mathbf{W}^\star)(\mathbf{W}_i - \mathbf{W}_j). \tag{4.31}$$

Among the various "entropy fix" we shall only detail the local Lax–Friedrichs scheme (at a sonic point, one turns back from the upwind flux to the local Lax–Friedrichs flux, see Chapter III, Section 3.1). The procedure is applied to each characteristic field indexed by K. Thus, denote by a_K^{ij} the eigenvalues and $\mathbf{r}_K^{ij}$ the eigenvectors of $\mathbf{A}_{ij}(T) = \mathbf{A}(\overline{\mathbf{U}}, \mathbf{n}_{ij})$, and $\mathbf{l}_K^{ij}$ the eigenvectors of $\mathbf{A}_{ij}^T$. We decompose the increment on the eigenbasis $(\mathbf{r}_K^{ij})$,

$$\hat{\mathbf{U}}_i - \hat{\mathbf{U}}_j = \sum_K \alpha_K^{ij} \mathbf{r}_K^{ij},$$

where

$$\alpha_K^{ij} = \mathbf{l}_K^{ij\;T}(\hat{\mathbf{U}}_i - \hat{\mathbf{U}}_j) = \mathbf{l}_K^{ij\;T}(\mathbf{U}'(\mathbf{W}^\star)(\mathbf{W}_i - \mathbf{W}_j)).$$

We shall also denote by a subscript K (i.e., by $\mathbf{V}_K$) the coefficient of a vector $\mathbf{V}$ on the eigenvector $\mathbf{r}_K^{ij}$ in the decomposition on the eigenbasis $(\mathbf{r}_K^{ij})$

$$\mathbf{V}_K = \mathbf{l}_K^{ij\;T}\mathbf{V}.$$

Then the Kth component ψ_K^{ij} of the LLF flux ψ^{ij} on $\mathbf{r}_K^{ij}$ is

$$\psi_K^{ij} = \frac{1}{2}\{\hat{\mathbb{F}} \cdot \mathbf{n}_{ij}(\mathbf{U}_i))_K + (\hat{\mathbb{F}} \cdot \mathbf{n}_{ij}(\mathbf{U}_j))_K\} - \sigma_K^{ij}\alpha_K^{ij},$$

where the coefficients σ_K^{ij} are defined by

$$\sigma_K^{ij} = \max_{i,j,k}\{a_K^{ij}(A_i), a_K^{ij}(A_j), a_K^{ij}(A_k)\}.$$

The $a_k^{ij}(A_i)$ denote the eigenvalues of the matrix $\mathbf{A}(\mathbf{U}(\mathbf{a}_i), \mathbf{n}_{ij})$, whereas the a_k^{ij} denote the eigenvalues of the matrix $\mathbf{A}_{ij} = \mathbf{A}(\mathbf{U}(\mathbf{W}^\star), \mathbf{n}_{ij})$. If the three signs of $a_k^{ij}(\mathbf{a}_i)$, $a_k^{ij}(\mathbf{a}_j)$, $a_k^{ij}(\mathbf{a}_k)$ relative to the three vertices of T are identical, we take for Φ_K^{ij} the Kth component of the upwind flow, i.e.,

$$\begin{aligned} &(\hat{\mathbb{F}} \cdot \mathbf{n}_{ij}(\mathbf{U}_i))_K \quad \text{if } a_K^{ij} \geq 0, \\ &(\hat{\mathbb{F}} \cdot \mathbf{n}_{ij}(\mathbf{U}_j))_K \quad \text{if } a_K^{ij} < 0. \end{aligned}$$

In the "sonic case", for an index K such that the three signs of $a_K^{ij}(\mathbf{a}_i)$, $a_K^{ij}(\mathbf{a}_j)$, $a_K^{ij}(\mathbf{a}_k)$ are not identical, we turn to the LLF flux and take for Φ_K^{ij} the Kth component ψ_K^{ij}; finally,

$$\mathbf{\Phi}_T(\hat{\mathbf{U}}_i, \hat{\mathbf{U}}_j, \mathbf{n}_{ij}) = \sum_K \Phi_K^{ij}\mathbf{r}_K^{ij}.$$

For an "optimal" choice of the coefficients σ_K^{ij}, we refer to Angrand and Lafon (1992).

4.3.3 Application to gas dynamics

In the case of the Euler system, the parameter vectors are

$$\begin{aligned}
w_1^\star &= \frac{1}{3}(\sqrt{\rho_i} + \sqrt{\rho_j} + \sqrt{\rho_k}),\\
w_2^\star &= \frac{1}{3}(\sqrt{\rho_i}\, u_i + \sqrt{\rho_j}\, u_j + \sqrt{\rho_k}\, u_k),\\
w_3^\star &= \frac{1}{3}(\sqrt{\rho_i}\, v_i + \sqrt{\rho_j}\, v_j + \sqrt{\rho_k}\, v_k),\\
w_4^\star &= \frac{1}{3}(\sqrt{\rho_i}\, H_i + \sqrt{\rho_j}\, H_j + \sqrt{\rho_k}\, H_k);
\end{aligned}$$

moreover;

$$\overline{u} = \frac{w_2^\star}{w_1^\star}, \quad \overline{v} = \frac{w_3^\star}{w_1^\star}, \quad \overline{H} = \frac{w_4^\star}{w_1^\star}.$$

The matrix $\mathbf{U}'(\mathbf{W})$ is easily computed for an ideal gas, following Lemma 3.2 in Chapter III):

$$\mathbf{U}'(\mathbf{W}) = \begin{pmatrix} 2w_1 & 0 & 0 & 0 \\ w_2 & w_1 & 0 & 0 \\ w_0 & 0 & w_1 & 0 \\ w_4/\gamma & w_2(\gamma-1)/\gamma & w_3(\gamma-1)/\gamma & w_1/\gamma \end{pmatrix}.$$

Note that the above matrix is triangular and involves only the density on the diagonal. We get similarly $(\mathbf{f}\circ\mathbf{U})'(\mathbf{W})$, $(\mathbf{g}\circ\mathbf{U})'(\mathbf{W})$, and

$$\mathbf{f}'(\mathbf{U}(\mathbf{W}^\star)) = (\mathbf{f}\circ\mathbf{U})'(\mathbf{W}^\star)\mathbf{U}'(\mathbf{W}^*)^{-1}.$$

As in the one-dimensional case, one can compute A and B in terms of u, v, H, and c. Setting

$$\kappa = \gamma - 1, \quad K = \frac{\kappa(u^2+v^2)}{2}, \quad c^2 = \kappa h = \kappa\Big(H - \frac{(u^2+v^2)}{2}\Big),$$

where $H = \frac{(E+p)}{\rho} = \varepsilon + \frac{(u^2+v^2)}{2} + \frac{p}{\rho}$, we get

$$\mathbf{A}(\mathbf{U}) = \begin{pmatrix} 0 & 1 & 0 & 0 \\ K-u^2 & (2-\kappa)u & -\kappa v & \kappa \\ -uv & v & u & 0 \\ u(K-H) & H-\kappa u^2 & -\kappa uv & (1+\kappa)u \end{pmatrix},$$

$$\mathbf{B}(\mathbf{U}) = \begin{pmatrix} 0 & 0 & 1 & 0 \\ -uv & v & u & 0 \\ K-v^2 & -\kappa u & (2-\kappa)v & \kappa \\ v(K-H) & -\kappa uv & H-\kappa v^2 & (1+\kappa)v \end{pmatrix},$$

which gives the matrix A_ω, whose eigenvalues are $\mathbf{u}\cdot\boldsymbol{\omega}\pm c$, and $\mathbf{u}\cdot\boldsymbol{\omega}$ (double eigenvalue). The eigenvectors are (see Section 2.3)

$$(4.32)\qquad \begin{cases} \mathbf{r}_1(\mathbf{U},\boldsymbol{\omega}) = (1, u-\omega_1 c, v-\omega_2 c, H-\mathbf{u}\cdot\boldsymbol{\omega}c)^T, \\ \mathbf{r}_4(\mathbf{U},\boldsymbol{\omega}) = (1, u+\omega_1 c, v+\omega_2 c, H+\mathbf{u}\cdot\boldsymbol{\omega}c)^T, \\ \mathbf{r}_2(\mathbf{U},\boldsymbol{\omega}) = (1, u, v, |\mathbf{u}|^2/2)^T, \\ \mathbf{r}_3(\mathbf{U},\boldsymbol{\omega}) = (0, -\omega_2, \omega_1, \mathbf{u}\cdot\boldsymbol{\omega}^\perp)^T. \end{cases}$$

In the decomposition of $\hat{\mathbf{U}}_i-\hat{\mathbf{U}}_j=\mathbf{U}'(\mathbf{W}^\star)(\mathbf{W}_i-\mathbf{W}_j)$, we have

$$\hat{\mathbf{U}}_i-\hat{\mathbf{U}}_j=(\Delta\rho,\Delta\rho u,\Delta\rho v,\Delta\rho e)=\sum_k \alpha_k^{ij}\mathbf{r}_k^{ij},$$

where

$$\begin{aligned} \Delta\rho &= 2w_1^\star\,\Delta w_1 = 2\sqrt{\rho^\star}\,\Delta\sqrt{\rho} \\ &= \alpha_1^{ij}+\alpha_2^{ij}+\alpha_4^{ij}, \\ \Delta\rho u &= w_2^\star\Delta w_1 + w_1^\star\Delta w_2 = (\sqrt{\rho}u)^\star\Delta\sqrt{\rho}+\sqrt{\rho^\star}\Delta(\sqrt{\rho}u) \\ &= \alpha_1^{ij}(u-\omega_1 c)+\alpha_2^{ij}u-\alpha_3^{ij}\omega_2+\alpha_4^{ij}(u-\omega_1 c), \\ \Delta\rho v &= w_3^\star\Delta w_1 + w_1^\star\Delta w_3 = (\sqrt{\rho}v)^\star\Delta\sqrt{\rho}+\sqrt{\rho^\star}\,\Delta(\sqrt{\rho}v) \\ &= \alpha_1^{ij}(v-\omega_2 c)+\alpha_2^{ij}v+\alpha_3^{ij}\omega_1+\alpha_4^{ij}(v-\omega_2 c), \\ \Delta\rho e &= \frac{w_4^\star}{\gamma}\Delta w_1 + w_2^\star\frac{(\gamma-1)}{\gamma}\Delta w_2 + w_3^\star\frac{(\gamma-1)}{\gamma}\Delta w_3 + \frac{w_1^\star}{\gamma}\Delta w_4 \\ &= \alpha_1^{ij}(H-\mathbf{u}\cdot\boldsymbol{\omega}c)+\alpha_2^{ij}\frac{|\mathbf{u}|^2}{2}+\alpha_3^{ij}\mathbf{u}\cdot\boldsymbol{\omega}^\perp+\alpha_4^{ij}(H+\mathbf{u}\cdot\boldsymbol{\omega}c), \end{aligned}$$

which gives a system in the α_k^{ij} (see Lemma 3.7, Chapter III). We find (with the notations of Chapter III, Section 3.2)

$$\begin{aligned} \alpha_1^{ij} &= \frac{1}{2\bar{c}^2}(\Delta p-\bar{c}\hat{m}(\rho)\Delta(\mathbf{u}\cdot\boldsymbol{\omega})), \\ \alpha_4^{ij} &= \frac{1}{2\bar{c}^2}(\Delta p+\bar{c}\hat{m}(\rho)\Delta(\mathbf{u}\cdot\boldsymbol{\omega})), \\ \alpha_2^{ij} &= \Delta\rho-\frac{\Delta p}{\bar{c}^2}, \\ \alpha_3^{ij} &= \hat{m}(\rho)\Delta(\mathbf{u}\cdot\boldsymbol{\omega}^\perp). \end{aligned}$$

The method can be extended to the real gas case following the arguments of Chapter III, Section 3.3.

4.3.4 The Osher scheme

We now follow the derivation of Chapter III, Section 4.2 and define the two-dimensional Osher flux by

$$(4.33)\qquad \boldsymbol{\Phi}(\mathbf{U}_i,\mathbf{U}_j,\mathbf{n}) = \frac{1}{2}\{\mathbb{F}\cdot\mathbf{n}(\mathbf{U}_i)+\mathbb{F}\cdot\mathbf{n}(\mathbf{U}_j)-\int_\Gamma |\mathbf{A}(\mathbf{w},\mathbf{n})|d\mathbf{w}\},$$

where Γ is a path consisting of portions $\Gamma_1 = \mathbf{U}_i\mathbf{U}_1^{ij}$, $\Gamma_2 = \mathbf{U}_1^{ij}\mathbf{U}_2^{ij}$,... of integral curves of the eigenvectors $\mathbf{r}_k$ in the state space, connecting $\mathbf{U}_i$ and $\mathbf{U}_j$, thus involving only two rarefaction or compression waves and a contact discontinuity. The intermediate states $\mathbf{U}_1^{ij}$, $\mathbf{U}_2^{ij}$,... can be computed by using the fact that the Riemann invariants are constant along the integral curves of $\mathbf{r}_k$.

In simple cases, we have a geometric interpretation (see G.R., Chapter III, Example 2.5). First, in the linear scalar case,

$$\frac{\partial u}{\partial t} + a\frac{\partial u}{\partial x} + b\frac{\partial u}{\partial y} = 0,$$

if we set $\mathbf{a} = (a,b)^T$, $\mathbb{F}(u) = \mathbf{a}u = (au, bu)^T$, $(\mathbb{F}\cdot\mathbf{n})(u) = \mathbf{a}\cdot\mathbf{n}\,u$, then (4.33) gives

$$\Phi(u, v, \mathbf{n}) = f^+(u) + f^-(v), \tag{4.34}$$

where

$$f^+(u) = u\ \max(\mathbf{a}\cdot\mathbf{n}, 0), \quad f^-(u) = u\ \min(\mathbf{a}\cdot\mathbf{n}), 0).$$

On a given edge of the cell C_i, with outward normal $\mathbf{n}$, we get that

$$\begin{aligned}
&\text{if } \mathbf{a}\cdot\mathbf{n} > 0, \text{ then } \Phi(u_i, u_j, \mathbf{n}) = f^+(u_i) = \mathbf{a}\cdot\mathbf{n}\,u_i,\\
&\text{if } \mathbf{a}\cdot\mathbf{n} > 0, \text{ then } \Phi(u_i, u_j, \mathbf{n}) = f^-(u_j) = \mathbf{a}\cdot\mathbf{n}\,u_j,
\end{aligned}$$

which is the upwind flux applied to the function $\mathbb{F}\cdot\mathbf{n}(u) = \mathbf{a}\cdot\mathbf{n}\,u$.

Also, in the simple nonlinear scalar case $f = g = \frac{u^2}{2}$, $(\mathbb{F}\cdot\mathbf{n})'(u) = (n_1 + n_2)\,\frac{u}{2}$, we can see that (4.34) holds, where

$$\begin{aligned}
f^+(u) &= u\ \max((\mathbb{F}\cdot\mathbf{n})'(u), 0) = (\mathbb{F}\cdot\mathbf{n})(\max(u, 0)) \text{ if } (n_1 + n_2) > 0,\\
f^-(u) &= u\ \min((\mathbb{F}\cdot\mathbf{n})'(u), 0) = (\mathbb{F}\cdot\mathbf{n})(\min(u, 0)) \text{ if } (n_1 + n_2) < 0.
\end{aligned}$$

More generally, we have similar results when $f = g$ is strictly convex, since $\mathbb{F} = f(\mathbf{e}_1 + \mathbf{e}_2)$, $\mathbb{F}\cdot\mathbf{n} = f(n_1 + n_2)$, and $(\mathbb{F}\cdot\mathbf{n})'$ is monotone.

In the case of the gas dynamics equations, we have two intermediate states and, using the expressions for the Riemann invariants (see Section 2.3), we write that

$\mathbf{U}_1 = \mathbf{U}_L$ is connected to $\mathbf{U}_2$, where $\mathbf{U}_2$ satisfies $\mathbf{u}_L\cdot\mathbf{n} + \ell_L = u_2\cdot\mathbf{n} + \ell_2$, $\mathbf{u}_L\cdot\mathbf{n}^\perp = \mathbf{u}_2\cdot\mathbf{n}^\perp$, $s_L = s_2$,

$\mathbf{U}_2$ is connected to $\mathbf{U}_3$ with $\mathbf{u}_3\cdot\mathbf{n} = \mathbf{u}_2\cdot\mathbf{n}$, $p_3 = p_2$,

$\mathbf{U}_3$ is connected to $\mathbf{U}_4 = \mathbf{U}_R$ with $\mathbf{u}_R\cdot\mathbf{n} - \ell_R = \mathbf{u}_3\cdot\mathbf{n} - \ell_3, \mathbf{u}_R\cdot\mathbf{n}^\perp = \mathbf{u}_3\cdot\mathbf{n}^\perp, s_R = s_3$.

Note that the tangential velocity $\mathbf{u}\cdot\mathbf{n}^\perp$ may jump across the contact discontinuity only. Then, the integral is split, so that

$$\begin{aligned}
\mathbf{\Phi}(\mathbf{U}_i, \mathbf{U}_j, \mathbf{n}) &= \mathbb{F}\cdot\mathbf{n}(\mathbf{U}_i) + \int_\Gamma \mathbf{A}^-(\mathbf{w}, \mathbf{n})d\mathbf{w}\\
&= \mathbb{F}\cdot\mathbf{n}(\mathbf{U}_j) + \int_\Gamma \mathbf{A}^+(\mathbf{w}, \mathbf{n})d\mathbf{w},
\end{aligned}$$

and one writes

$$\int_{\Gamma} \mathbf{A}^{+}(\mathbf{w},\mathbf{n})d\mathbf{w} = \sum_{k} \int_{\Gamma_k} \mathbf{A}^{+}(\mathbf{w},\mathbf{n})d\mathbf{w}.$$

In fact, it is simpler to use (4.8) and write the flux for $\mathbf{n} = \mathbf{e}_1$, $\mathbb{F}\cdot\mathbf{e}_1 = \mathbf{f}$, $\mathbf{A}(\mathbf{w},\mathbf{e}_1) = \mathbf{A}(\mathbf{w})$ since the flux is invariant under rotation. Indeed, by definition (see (1.9), Chapter III),

$$\mathbf{A}(\mathbf{U},\mathbf{n})^{\pm} = (\mathbf{T}\Lambda^{\pm}\mathbf{T}^{-1})(\mathbf{U},\mathbf{n}),$$

where $\Lambda(\mathbf{U},\mathbf{n})$ is the diagonal matrix of the eigenvalues of $\mathbf{A}(\mathbf{U},\mathbf{n})$, which are

$$a_1(\mathbf{U},\mathbf{n}) = \mathbf{u}\cdot\mathbf{n} - c, \quad a_2(\mathbf{U},\mathbf{n}) = a_3(\mathbf{U},\mathbf{n}) = \mathbf{u}\cdot\mathbf{n}, \quad a_4(\mathbf{U},\mathbf{n}) = \mathbf{u}\cdot\mathbf{n} + c,$$

and $\mathbf{T}(\mathbf{U},\mathbf{n})$ is the matrix whose columns are the corresponding eigenvectors $\mathbf{r}_k(\mathbf{U},\mathbf{n})$ given by (4.32). From (4.9), if R is the rotation such that $R^{-1}\mathbf{n} = \mathbf{e}_1$ and $R^{-1}\mathbf{U} = \tilde{\mathbf{U}}$, then

$$\mathbf{A}(\mathbf{U},\mathbf{n})\mathbf{r}_k(\mathbf{U},\mathbf{n}) = R\mathbf{A}(\tilde{\mathbf{U}})R^{-1}\mathbf{r}_k(\mathbf{U},\mathbf{n}) = a_k(\mathbf{U},\mathbf{n})\mathbf{r}_k(\mathbf{U},\mathbf{n}),$$

and we obtain

$$\mathbf{A}(\tilde{\mathbf{U}})R^{-1}\mathbf{r}_k(\mathbf{U},\mathbf{n}) = a_k(\mathbf{U},\mathbf{n})R^{-1}\mathbf{r}_k(\mathbf{U},\mathbf{n}),$$

and thus

$$a_k(\tilde{\mathbf{U}}) = a_k(\mathbf{U},\mathbf{n}), \quad \mathbf{r}_k(\tilde{\mathbf{U}}) = R^{-1}\mathbf{r}_k(\mathbf{U},\mathbf{n}),$$

which implies

$$\Lambda(\mathbf{U},\mathbf{n}) = \Lambda(\tilde{\mathbf{U}}), \quad \mathbf{T}(\tilde{\mathbf{U}}) = R^{-1}\mathbf{T}(\mathbf{U},\mathbf{n})$$

and

$$\mathbf{A}(\mathbf{U},\mathbf{n})^{\pm} = R\,\mathbf{T}(\tilde{\mathbf{U}})\Lambda^{\pm}(\tilde{\mathbf{U}})\mathbf{T}^{-1}(\tilde{\mathbf{U}})R = R\,\mathbf{A}^{\pm}(\tilde{\mathbf{U}})R^{-1}, \tag{4.35}$$

i.e., for any vector $\mathbf{w} \in \mathbb{R}^4$

$$\mathbf{A}(\mathbf{U},\mathbf{n})^{+}\mathbf{w} = R\,\mathbf{A}^{+}(\tilde{\mathbf{U}})R^{-1}\mathbf{w} = R\,\mathbf{A}^{+}(\tilde{\mathbf{U}})\tilde{\mathbf{w}}.$$

Thus

$$R\mathbf{\Phi}(\tilde{\mathbf{U}}_i,\tilde{\mathbf{U}}_j,\mathbf{e}_1) = R\{\mathbf{f}(\tilde{\mathbf{U}}_j) - \int_{\tilde{\Gamma}} \mathbf{A}^{+}(\mathbf{v})d\mathbf{v}\} = \mathbb{F}\cdot\mathbf{n}(\mathbf{U}_j) - \int_{\tilde{\Gamma}} R\mathbf{A}^{+}(\mathbf{v})d\mathbf{v}.$$

Now, making a change of variables in the integral, $\mathbf{v} = R^{-1}\mathbf{w}$,

$$\mathbf{A}(\mathbf{w},\mathbf{n})^{+} = R\,\mathbf{A}^{+}(\mathbf{v})R^{-1},$$

and noticing that the path $\tilde{\Gamma}$, consisting of portions $\tilde{\Gamma}_1$, $\tilde{\Gamma}_2$,... of integral curves of the eigenvectors $\mathbf{r}_k$ in the state space, connecting $\tilde{\mathbf{U}}_i$ and $\tilde{\mathbf{U}}_j$ is transformed into the path Γ connecting $\mathbf{U}_i$ and $\mathbf{U}_j$, we obtain

$$R\mathbf{\Phi}(\tilde{\mathbf{U}}_i,\tilde{\mathbf{U}}_j,\mathbf{e}_1) = \mathbb{F}\cdot\mathbf{n}(\mathbf{U}_j) - \int_{\Gamma} \mathbf{A}^{+}(\mathbf{w},\mathbf{n})d\mathbf{w} = \mathbf{\Phi}(\mathbf{U}_i,\mathbf{U}_j,\mathbf{n}).$$

The remaining computations then follow exactly the lines of Sections 4.2, 4.3 in Chapter III. In particular, each integral $\int_{\Gamma_k} \mathbf{A}^{\pm}(\mathbf{w})d\mathbf{w}$ appears as a difference of fluxes computed at the points $\mathbf{U}_L, \mathbf{U}_R, \mathbf{U}_2, \mathbf{U}_3$, and $\Phi(\mathbf{U}_i, \mathbf{U}_j, \mathbf{n})$ can be written as the sum $\Sigma(\varepsilon_i f(\mathbf{U}_i) + \overline{\varepsilon}_i f(\overline{\mathbf{U}}_i))$, where ε takes the value 0 or ± 1 according to the sign of the eigenvalues. The "sonic" states $\overline{\mathbf{U}}_1$, $\overline{\mathbf{U}}_4$ respectively are such that

$$\lambda_1 = \overline{u}_1 - c(u_1, s_L) = 0 \text{ and } \overline{u}_1 + \ell_1 = u_L + \ell_L, \overline{v}_1 = v_L,$$

$$\lambda_4 = \overline{u}_4 - c(u_4, s_R) = 0 \text{ and } \overline{u}_4 + \ell_4 = u_R + \ell_R, \overline{v}_4 = v_L.$$

One can similarly extend the method of Section 3.4 based on a shock-curve decomposition (for details see Mehlman 1991).

4.3.5 Flux-vector splitting

We now extend Steger and Warming's or van Leer's flux splitting. The flux $\mathbb{F} \cdot \boldsymbol{\omega}$ is split into

$$\mathbb{F} \cdot \boldsymbol{\omega} = \mathbf{F}_{\omega}^{+} + \mathbf{F}_{\omega}^{-},$$

in order that the split flux Jacobians $(\mathbf{F}_{\omega}^{+})'$ (resp. $(\mathbf{F}_{\omega}^{-})'$) have positive (resp. negative) eigenvalues. If the homogeneity property (see Chapter III, Section 5)

$$\mathbb{F} \cdot \omega(\mathbf{u}) = (\mathbb{F} \cdot \omega)'(\mathbf{u})\mathbf{u} = \mathbf{A}_{\omega}(\mathbf{u})\mathbf{u},$$

where

$$\mathbf{A}_{\omega}(\mathbf{u}) = (\mathbb{F} \cdot \omega)'(\mathbf{u}) = \mathbf{A}(\mathbf{u}, \boldsymbol{\omega}),$$

holds (which is easily verified for the ideal gas dynamics equations), following Steger and Warming, we can define

$$\mathbf{F}_{\omega}^{+}(\mathbf{u}) = \mathbf{A}_{\omega}^{+}(\mathbf{u})\mathbf{u}, \quad \mathbf{F}^{-}\omega(\mathbf{u}) = \mathbf{A}_{\omega}^{-}(\mathbf{u})\mathbf{u}$$

and

$$\boldsymbol{\Phi}(\mathbf{u}_i, \mathbf{u}_j, \boldsymbol{\omega}) = \mathbf{F}_{\omega}^{+}(\mathbf{u}_i) + \mathbf{F}_{\omega}^{-}(\mathbf{u}_j).$$

For the gas dynamics equations, as in the one-dimensional case, we can expand the flux $\mathbb{F} \cdot \boldsymbol{\omega}(\mathbf{u})$ in "fictitious" fluxes $\mathbf{F}_{\omega i}$ associated to each distinct eigenvalue

$$\mathbb{F} \cdot \boldsymbol{\omega} = \Sigma\, \mathbf{F}_{\omega i}$$

over $i = 1, 2, 4$, and then $\mathbf{F}_{\omega}^{+}(\mathbf{u})$ (resp. $\mathbf{F}_{\omega}^{-}(\mathbf{u})$) is the sum of the fluxes associated with the positive (resp. negative) eigenvalues (see Chapter III, Section 4.2). We get

$$\mathbf{F}_{\omega i} = \lambda_1 \frac{\rho}{2\gamma} \mathbf{r}_i, \quad i = 1, 4,$$

$$\mathbf{F}_{\omega 2} = \lambda_2 \frac{\rho(\gamma - 1)}{\gamma} \mathbf{r}_2(\mathbf{U}, \boldsymbol{\omega}),$$

where, with shorthand notations (see (4.32)),

$$\mathbf{r}_1(\mathbf{U},\boldsymbol{\omega}) = (1, \mathbf{u} - \omega c, H - \mathbf{u}\cdot\omega c)^T,$$
$$\mathbf{r}_4(\mathbf{U},\boldsymbol{\omega}) = (1, \mathbf{u} - \omega c, H - \mathbf{u}\cdot\omega c)^T,$$
$$\mathbf{r}_2(\mathbf{U},\boldsymbol{\omega}) = \Big(1, \mathbf{u}, \frac{|\mathbf{u}|^2}{2}\Big)^T.$$

If the normal speed $\mathbf{u}\cdot\boldsymbol{\omega}$ is supersonic, then

$$\mathbf{u}\cdot\boldsymbol{\omega} \geq c \Longrightarrow \mathbf{F}^+_{\omega}(\mathbf{U}) = \mathbb{F}\cdot\boldsymbol{\omega}(\mathbf{U}), \quad \mathbf{F}^-_{\omega}(\mathbf{U}) = 0,$$
$$\mathbf{u}\cdot\boldsymbol{\omega} \leq -c \Longrightarrow \mathbf{F}^+_{\omega}(\mathbf{U}) = 0, \quad \mathbf{F}^-_{\omega}(\mathbf{U}) = \mathbb{F}\cdot\boldsymbol{\omega}(\mathbf{U}).$$

Remark 4.7. In the case of the gas dynamics equations, the flux of Steger and Warming is invariant under rotation. By (4.35)

$$\mathbf{A}^{\pm}_{\mathbf{n}}(\mathbf{U}) = R\,\mathbf{A}^{\pm}(\tilde{\mathbf{U}})R^{-1},$$

i.e., for any vector $\mathbf{V} \in \mathbb{R}^4$,

$$\mathbf{A}^{\pm}_{\mathbf{n}}(\mathbf{U})\mathbf{V} = R\,\mathbf{A}^{\pm}(\tilde{\mathbf{U}})R^{-1}\mathbf{V} = R\,\mathbf{A}^{\pm}(\tilde{\mathbf{U}})\tilde{\mathbf{V}},$$

and

$$\mathbf{A}^{+}_{\mathbf{n}}(\mathbf{U}_i)\mathbf{U}_i + \mathbf{A}^{-}_{\mathbf{n}}(\mathbf{U}_j)\mathbf{U}_j = R\{\mathbf{A}^{+}(\tilde{\mathbf{U}}_i)\tilde{\mathbf{U}}_i + \mathbf{A}^{+}(\tilde{\mathbf{U}}_j)\tilde{\mathbf{U}}_j\},$$

so that (4.8) holds,

$$\boldsymbol{\Phi}(\mathbf{U}_i, \mathbf{U}_j, \mathbf{n}) = R\boldsymbol{\Phi}(\tilde{\mathbf{U}}_i, \tilde{\mathbf{U}}_j, \mathbf{e}_1) = R(\mathbf{f}^{+}(\tilde{\mathbf{U}}_i) + (\mathbf{f}^{-}(\tilde{\mathbf{U}}_j))$$

where $\mathbf{f}^+$ is the one-dimensional split flux associated to $\mathbf{f}$ ($\mathbf{f}^+ = \mathbf{F}^+_{\mathbf{e}_1}$ in the case of Steger–Warming splitting). Again, this property yields shorter computations. □

In the case of van Leer's flux splitting (see Chapter III, Section 4.3), we compute easily

$$\mathbf{f}^{\pm}(\mathbf{U}) = \Big(f_1^{\pm}, f_1^{\pm}\frac{(\gamma-1)u \pm 2c}{\gamma}, f_1^{\pm}v, f_1^{\pm}\frac{\{(\gamma-1)u \pm 2c\}^2}{2(\gamma^2-1)}\Big)^T,$$

where the mass flux $f_1^{\pm}$ is given by (see Chapter III, Section 4.3)

$$f_1^{\pm} = \pm\Big(\frac{\rho}{4c}\Big)(u \pm c)^2.$$

For $\mathbf{f}^{\pm}(\tilde{\mathbf{U}})$, we replace (u, v) by (u_n, u_τ) (for details, we refer to Steger and Warming 1981, van Leer 1982, Liu and Vinokur 1989, Fezoui and Steve 1988, Fezoui, Steve, and Selmin 1988, Liou, van Leer, and Shuen 1990). Van Leer's flux cannot preserve a steady contact discontinuity (see Mulder, Osher, and Sethian 1992) which is a drawback for the extension to the Navier–Stokes equations, since the shear layer may be interpreted as a layer of contact discontinuity. In that case, the Osher flux, which shares with van Leer's flux the property of beeing continuously differentiable (except at a stationary contact discontinuity (Larrouturou 1992) is to be preferred (see Koren 1988).

5 Second-order finite-volume schemes

5.1 MUSCL-type schemes

A "second-order" (in space) version can be obtained via a MUSCL (monotonic upstream schemes for conservation laws) approach introduced by van Leer (1979) that is now widely followed. It uses a piecewise linear reconstruction, instead of piecewise constant functions (see Fezoui 1985, Durlofsky, Engquist, and Osher 1992, Colella 1985, Rostand and Stoufflet 1988, Angrand and Lafon 1992, or piecewise constant functions in subcells (Perthame and Qiu 1994), together with a limitation procedure (on conservative, characteristic, or physical variables according to the chosen approach). For the time discretization, Runge–Kutta schemes (Shu and Osher 1988, Mulder, Osher, and Sethian 1992, Durlofsky, Engquist, and Osher 1992, Lallemand 1990, Angrand and Lafon 1992) allow a higher order of accuracy.

Let us sketch the main steps of the MUSCL method that was developed for one-dimensional systems in Chapter III, Section 6.1 (for the scalar one-dimensional case, see G.R., Chapter IV, Section 3). In formula (4.3), the first-order flux $\mathbf{\Phi}(\mathbf{u}_i, \mathbf{u}_j, \mathbf{n}_{ij})$, where $\mathbf{u}_i, \mathbf{u}_j$ are the constant values on each side of an edge $\Gamma_{ij} = \Omega_i \cap \Omega_j$, is now replaced by $\mathbf{\Phi}(\mathbf{u}_{ij}, \mathbf{u}_{ji}, \mathbf{n}_{ij})$, where $\mathbf{u}_{ij}, \mathbf{u}_{ji}$ are second-order approximations of the solution on each side of the edge $e = \Gamma_{ij}$. These second order approximations are computed through the following steps:
(i) prediction of the gradients $\boldsymbol{\nabla}\mathbf{u}_i = (\nabla_x \mathbf{u}_i, \nabla_y \mathbf{u}_i)$ in each cell (each $\nabla_x \mathbf{u}_i$ has p components; we shall also use the notation $\boldsymbol{\nabla}\mathbf{u}_i = (\nabla u_i)$, where u_i stands for some dependent variable and $\nabla u_i = (\nabla_x u_i, \nabla_y u_i) \in \mathbb{R}^2$);
(ii) linear extrapolation to define values $\mathbf{u}_{ij}, \mathbf{u}_{ji}$ on each side of the edge;
(iii) limitation procedure.

For the gas dynamics equations, stability considerations, for instance preservation of a maximum principle on the mass fractions (see Mehlmann 1991, Larrouturou 1991), the positivity of the pressure or temperature (see Perthame and Qiu 1994)... may impose the choice of dependent variables whose gradient is computed and limited: conservative $(\rho, \rho u, \rho v, \rho e)$, primitive (ρ, u, v, p) or physical (ρ, u, v, T), or entropy, characteristic variables (see Mulder and van Leer 1985, Arminjon et al. 1988). According to the scheme, the limitation procedure may thus be preceded by a computation of the components of the chosen vector and eventually followed by a return to the conservative variables. For instance, the velocity may be computed through $u_i^n = (\rho u)_i^n / \rho_i^n$, $v_i^n = (\rho v)_i^n / \rho_i^n$; one computes $(\rho, u, v)_i^{n+1}$ by a prediction-correction procedure, and then sets $(\rho u)_i^{n+1} = \rho_i^{n+1} u_i^{n+1}$, and so on.

Remark 5.1. In fact, in the van Leer–Hancock scheme (see G.R., Chapter IV, Section 3.2), the second-order approximation in space is linked to a prediction of the variables at time $t_n + \frac{\Delta t}{2}$. In the two-dimensional case, the solution is first updated at time $t_n + \frac{\Delta t}{2}$; first, one writes

$$\mathbf{u}_i^{n+1/2} \cong \mathbf{u}_i^n + \frac{\Delta t}{2} \left(\frac{\partial \mathbf{u}}{\partial t} \right)_i^n.$$

Then, one uses either the scheme

$$\left(\frac{\partial \mathbf{u}}{\partial t} \right)_n^i = \frac{1}{|\Omega_i|} \sum_j |\Gamma_{ij}| \mathbf{\Phi}(\mathbf{u}_i, \mathbf{u}_j, \mathbf{n}_{ij})$$

from section 4.1.1, and thus

$$\mathbf{u}_i^{n+1/2} \cong \mathbf{u}_i^n - \frac{\Delta t}{2} |\Omega_i| \sum_j |\Gamma_{ij}| \mathbf{\Phi}(\mathbf{u}_i, \mathbf{u}_j, \mathbf{n}_{ij}),$$

or else one can write (Dervieux and Vijayasundaram 1985, Fezoui 1985, Fezoui, Steve and Selmin 1988)

$$\left(\frac{\partial \mathbf{u}}{\partial t} \right)_i^n = -\{\mathbf{A}(\mathbf{u}_i^n) \nabla_x \mathbf{u}_i + \mathbf{B}(\mathbf{u}_i^n) \nabla_y \mathbf{u}_i)\}$$

which uses some prediction of the gradients.

Then, one defines $\mathbf{u}_{ij}^{n+1/2}$, $\mathbf{u}_{ji}^{n+1/2}$ on each side of the edge by interpolation and limitation, as we shall describe more precisely in some examples. □

For simplicity, we skip this prediction step so that the scheme reads

$$(5.1) \qquad |\Omega_i| \frac{(\mathbf{u}_i^{n+1} - \mathbf{u}_i^n)}{\Delta t} + \sum_{e \in \partial\Omega_i, e = \Gamma_{ij}} |e| \mathbf{\Phi}(\mathbf{u}_{ij}, \mathbf{u}_{ji}, \mathbf{n}_e) = \mathbf{0}.$$

The resulting sheme is only first order accurate in time.

Example 5.1. Rectangular mesh. (Example 4.4 revisited). We can define a linear function on $\Omega_{j,k} = (x_{j-1/2}, x_{j+1/2}) \times (y_{k-1/2}, y_{k+1/2})$ for any component u of $\mathbf{u}$,

$$u^n(x, y) = u_{j,k}^n + (x - x_j)\delta u_{j,k}^x + (y - y_k)\delta u_{j,k}^y, \ (x, y) \in \Omega_{j,k}.$$

In the case of a rectangular mesh, the prediction of the gradients $(\Delta_{j,k}^x, \delta_{j,k}^y)$ is rather straightforward. We may define a centered approximation of the gradients as in the one-dimensional case (see G.R., Chapter IV, Section 3.1)

$$\hat{\delta u}_{i,j}^x = \frac{(u_{i+1,j} - u_{i-1,j})}{2\Delta x}, \quad \hat{\delta u}_{i,j}^y = \frac{(u_{i+1,j} - u_{i-1,j})}{2\Delta y},$$

where to simplify we have chosen a uniform grid, and then limit componentwise (in x and y), for instance

$$\delta u_{i,j}^x = \text{minmod}\Big(\hat{\delta u}_{i,j}^x, \frac{(u_{i+1,j} - u_{i,j})}{\Delta x}, \frac{(u_{i,j} - u_{i-1,j})}{\Delta x} \Big)$$

(with the usual definition of the minmod function), and a similar formula for $\delta^y_{i,j}$. See also Jeng and Payne (1995) for another limiting procedure, Coirier and Powell (1995) for a related reconstruction. □

For a general unstructured mesh, step (i) is not so obvious. Let us give some examples of the prediction and limitation of the gradients.

5.1.1 MUSCL-type cell-centered schemes

We use the notations of Example 4.1. The control cell is a triangle T_i, and the center of the cell is the barycenter $\mathbf{g}_i$ of the triangle T_i; $\mathbf{m}_{ij}$ denotes the middle of the edge $\Gamma_{ij} = T_i \cap T_j$. We present in each step some different approaches.

(i) Prediction of the gradients

Approach 1: One can predict the gradients from the discretization of Green's formula (Deconinck et al. 1993, Dubois and Michaux 1990, see also W.K. Anderson 1994),

$$|T|\nabla u_i \cong \int_T \nabla u d\mathbf{x} = \int_{\partial T} u\, \mathbf{n} d\sigma \cong \sum_{e \subset \partial T} |e| u_e \mathbf{n}_e,$$

and on an edge $e = \Gamma_{ij}$ of ∂T, we take

$$u_e = \frac{u_i + u_j}{2}$$

or a convex combination of u_i, u_j,

$$u_e = (1 - \theta) u_i + \theta u_j,$$

where for instance $\theta = m_{ij} g_i / (m_{ij} g_i + m_{ij} g_j)$, or even the barycentric coordinate of the intersection of $\mathbf{g}_i \mathbf{g}_j$ with e.

Approach 2: Given the value $\mathbf{u}_i$ in T_i and the values $\mathbf{u}_j$, $\mathbf{u}_k$, $\mathbf{u}_\ell$ in the neighboring cells T_j, T_k, T_ℓ, we can use $\mathbf{u}_j - \mathbf{u}_i$ in order to compute the gradient in the direction $\mathbf{g}_i \mathbf{g}_j$, which gives a system of three equations to determine the two components $\nabla_x \mathbf{u}_i$, $\nabla_y \mathbf{u}_i$ solved by the least squares method. More precisely, let $\mathbf{d}u_i = (d_x, d_y)$ be any vector in $\mathbb{R}^2$, and let Δ^i_j denote the difference

$$\Delta^i_j = \Delta^i_j(d_x, d_y) \equiv u_j - u_i + \mathbf{d}u_i(\mathbf{g}_i - \mathbf{g}_j) = u_j - u_j^{\text{ext}}$$

between the mean value u_j in a neighboring mesh T_j and the extrapolated value

$$u_j^{\text{ext}} = u_i + \mathbf{d}u_i \cdot (\mathbf{g}_j - \mathbf{g}_i) = u_i + d_x (g_j - g_i)_x + d_y (g_j - g_i)_y$$

at the barycenter $\mathbf{g}_j$ of T_j, taking for the gradient the value $\mathbf{d}u_i = (d_x, d_y)$.

Then, define a prediction of the slope ∇u_i by solving the unconstrained (quadratic, convex) minimization problem (see Chevrier and Galley 1993, Benharbit 1992)

$$\nabla u_i \text{ is such that } \sum_j (\Delta^i_j)^2 (d_x, d_y) \text{ is minimum},$$

where the sum extends over all the indices j, k, ℓ of the neighboring cells. The gradients $\nabla_x u_i, \nabla_y u_i$ minimizing $\Sigma(\Delta_j^i)^2$ are a solution of a 2×2 linear system obtained by differentiating $\Sigma(\Delta_j^i)^2$ w.r.t. (d_x, d_y) and setting to zero the two partial derivatives

$$\frac{\partial}{\partial d_x} \sum_j (\Delta_j^i)^2 = 0, \quad \frac{\partial}{\partial d_y} \sum_j (\Delta_j^i)^2 = 0.$$

As already noted, this is done separately for each component u of the vector $\mathbf{u}$.

Approach 3: One does not explicitly compute the gradient but uses linear interpolation from the values u_i at the barycenter $\mathbf{g}_i$ of T_i and the values u_k, u_ℓ at the barycenters $\mathbf{g}_k, \mathbf{g}_\ell$ of two of the neighboring cells T_k, T_ℓ, in order to compute the value at the middle of the edge between T_i and the third neighboring cell T_j (see Figure 5.1). We get a piecewise linear function $p_{ik\ell}$ in T_i

$$p_{ik\ell} = u_i \psi_i + u_k \psi_k + u_\ell \psi_\ell,$$

where the ψ_k are the barycentric coordinates relative to the $\mathbf{g}_k$ (see Lin, Wu, and Chin 1993).
Then, we have simply

$$(5.2) \qquad (u_{ij})^{\text{pred}} = p_{i\ell k}(\mathbf{m}_{ij}) = u_i \Psi_i(\mathbf{m}_{ij}) + \mathbf{u}_k \boldsymbol{\psi}_k(\mathbf{m}_{ij}) + u_\ell \psi_\ell(\mathbf{m}_{ij}).$$

Approach 4: This approach uses one of the three preceding linear interpolants $p_{ik\ell}$, $p_{ij\ell}$, p_{ijk}, either the one that has a gradient with minimum norm or the one that has the greatest norm and that satisfies some limiting conditions as we shall discuss in (iii) (see Durlofsky, Engquist, and Osher 1992, X.-D. Liu 1993).
(ii) Linear extrapolation

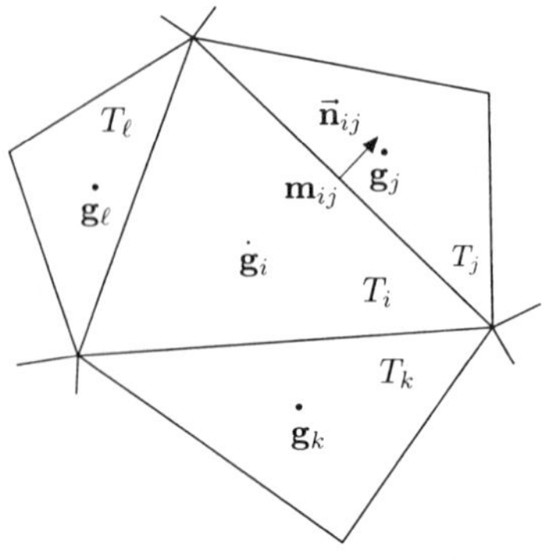

FIGURE 5.1. Neighboring cells (approach 3).

In approaches 1 or 2, we define the value at the middle of an edge by

$$(u_{ij})^{\text{pred}} = u_i + \nabla u_i \cdot (\mathbf{m}_{ij} - \mathbf{g}_i) = u_i + \nabla_x u_i (m_{ij} - g_i)_x + \nabla_y u_i (m_{ij} - g_i)_y,$$

and similarly

$$(u_{ji})^{\text{pred}} = u_j + \nabla u_j \cdot (\mathbf{m}_{ij} - \mathbf{g}_j),$$

where $\mathbf{m}_{ij} = (m_{ij\,x}, m_{ij\,y})$ denotes the coordinates of the middle of the edge $\Gamma_{ij} = T_i \cap T_j$.

It is known in the one-dimensional scalar case that the resulting scheme is not necessarily T.V.D, nor does it satisfy a discrete maximum principle.

(iii) Limitation procedure

Then $\mathbf{\nabla u}_i$ is limited, which corresponds to correcting the preceding stage and setting

$$\mathbf{u}_{ij} = \mathbf{u}_i + \boldsymbol{\alpha}_i \mathbf{\nabla u}_i \cdot (\mathbf{m}_{ij} - \mathbf{g}_i),$$

where the (diagonal matrix) limiter $\boldsymbol{\alpha}_i$ consists of a limiting factor α_i for each component u of $\mathbf{u}$,

$$u_{ij} = u_i + \alpha_i \nabla u_i \cdot (\mathbf{m}_{ij} - \mathbf{g}_i), \tag{5.3}$$

and α depends on the chosen criteria. In particular, if u_i is already a local extremum relative to its neighbors, then usually $\alpha_i = 0$.

For instance, one can simply limit in such a way that the value u_{ij} at the middle of the edge belongs to the interval between u_i and u_j. In the case of Approach 4, the limiting procedure does select the gradient among the three gradients $\nabla p_{ik\ell}$, $\nabla p_{ij\ell}$, ∇p_{ijk} as the one with greatest norm which meets this requirement (and if the gradient with second greatest norm does not satisfy this either, one takes the one with smallest norm).

Otherwise (see Mehlman 1991 in the case of unstructured quadrilaterals, Dubois and Michaux 1992, Dubois 1993) set

$$m_i = \min(u_j), \quad M_i = \max(u_j),$$

where the extrema are extended to the indices j of the neighboring cells, $j \neq i$. As already noted, if u_i is a local extremum, then $\alpha_i = 0$. If u_i belongs to $[m_i, M_i]$ the procedure requires that the value u_{ij} at the middle of the edges should satisfy

$$K \max_{j, u_j \leq u_i} (u_j - u_i) \leq u_{ij} - u_i \leq K \min_{j, u_j \leq u_i} (u_j - u_i), \tag{5.4}$$

where $0 \leq K \leq 1$, and α_i in (5.3) is the greatest possible value in $[0, 1]$ for which (5.4) holds. These are two-dimensional versions of the one-dimensional limiter (3.9) or (3.11) in G.R., Chapter IV, Section 3.1, which limits the gradient, and not each component in x and y.

In the case of Approach 2, another possibility considers a constrained minimization problem: $\Sigma_j (\Delta_j^i)^2 (d_x, d_y)$ is minimized on the "limited" set

(d_x, d_y) of vectors in $\mathbb{R}^2$ satisfying

$$- \text{ if } u_j - u_i \geq 0, \quad 0 \leq u_j^{\text{ext}} - u_i \leq u_j - u_i,$$
$$- \text{ if } u_j - u_i \leq 0, \quad u_j - u_i \leq u_j^{\text{ext}} - u_i \leq 0.$$

For details concerning the explicit computations, we refer to Buffard (1993).

In the case of Approach 3, the limiting procedure defines the values at the middle of the edge by

$$(5.5a) \qquad u_{ij} - u_i = \text{minmod}((u_{ij})^{\text{pred}} - u_i, K(u_j - u_i)),$$
$$(5.5b) \qquad u_{ji} - u_j = \text{minmod}((u_{ji})^{\text{pred}} - u_j, K(u_i - u_j)),$$

where $0 \leq K \leq \frac{1}{2}$, and we shall see later that the resulting scheme satisfies the maximium principle.

Remark 5.2. Let us interpret these limitations in the one-dimensional case. The values on each side at the "middle of the edge" are simply the values

$$u_{i+1/2-} = u_i + \frac{\delta_i}{2},$$
$$u_{i+1/2+} = u_{i+1} - \frac{\delta_{i+1}}{2},$$

(see G.R., Chapter IV, Section 3.1, 3.2), where δ_i is obtained from the prediction $\hat{\delta}_i$ by some limiter, for instance by the limiter minmod

$$\delta_i = \text{minmod}(|u_{i+1} - u_i|, \hat{\delta}_i, |u_i - u_{i-1}|).$$

Here, the function minmod is usually defined by (see G.R., Chapter IV, (2.28))

$$\text{minmod}(a_k) = \begin{cases} s \min(|a_k|) & \text{if the signs are identical, } s = sgn(a_k), \\ 0 & \text{otherwise.} \end{cases}$$

Since $\delta_i = 2(u_{i+1/2-} - u_i)$, we impose that the value $u_{i+1/2-}$ satisfies

$$2(u_{i+1/2-} - u_i) = \begin{cases} \min(u_{i+1} - u_i, u_i - u_{i-1}) \text{ if both increments } > 0, \\ \max(u_i - u_{i+1}, u_{i-1} - u_i) \text{ if both increments } < 0, \\ 0 \quad \text{otherwise,} \end{cases}$$

which leads to the above formula (5.3) with $K = \frac{1}{2}$ for limiting the 2-D gradients.

Approach 3 corresponds in the one-dimensional case to predicting first $u_{i+1/2,-}$ (resp. $u_{i+1/2,+}$) by extrapolation from u_{i-1}, u_i (resp. u_{i+1}, u_{i+2}) or equivalently to taking in G.R., formula (3.23), Chapter IV,

$$(u_{i+1/2,-})^{\text{pred}} - u_i = \frac{u_i - u_{i-1}}{2} = \frac{\Delta u_{i-1/2}}{2},$$
$$u_{i+1} - (u_{i+1/2,+})^{\text{pred}} = \frac{u_{i+2} - u_{i+1}}{2} = \frac{\Delta u_{i+3/2}}{2},$$

and then limiting by setting

$$(5.6a)\qquad \begin{aligned}(u_{i+1/2-})^{\lim} - u_i &= \operatorname{minmod}((u_{i+1/2,-})^{\text{pred}} - u_i, K(u_{i+1} - u_i)) \\ &= \operatorname{minmod}(\frac{\Delta u_{i-1/2}}{2}, K\Delta u_{i+1/2}),\end{aligned}$$

(5.6b)

$$\begin{aligned}u_{i+1} - (u_{i+1/2+})^{\ell im} &= \operatorname{minmod}(u_{i+1} - (u_{i+1/2,+})^{\text{pred}}, K(u_{i+1} - u_i)) \\ &= \operatorname{minmod}\Big(\frac{\Delta u_{i+3/2}}{2}, K\Delta u_{i+1/2}\Big)\end{aligned}$$

for some constant $K \leq 1$. For $K = \frac{1}{2}$, we recover the minmod limiter (see G.R., Chapter IV, (3.8)) which ensures that the scheme is T.V.D. and L^∞-stable. □

One can prove L^∞-stability for some of these two-dimensional schemes (see Benharbit 1992 in the case of Approach 2, Lin, Wu, and Chin 1993 for Approach 3, X.-D. Liu 1994 for Approach 4). For instance, see the following lemma.

Lemma 5.1

(Approach 3) Let u_{ij}, u_{ji} be defined by (5.2) and (5.5), and assume the following constraint on the triangulation:

$$\psi_q(\mathbf{m}_{ip}) \leq 0, \quad p, q \in \{j, k, \ell\}, \quad q \neq p.$$

Then, the scheme

$$|\Omega_i| \frac{(u_i^{n+1} - u_i^n)}{\Delta t} + \sum_{e \in \partial\Omega_j, e = \Gamma_{ij}} |e| \phi(u_{ij}, u_{ji}, \mathbf{n}_e) = 0,$$

where ϕ is the numerical flux associated to a 3-point monotone scheme by (4.6), satisfies

$$u_i^{n+1} = u_i^n + \{C_{ij}(u_j - u_i) + C_{ik}(u_k - u_i) + C_{i\ell}(u_\ell - u_i)\}$$

with nonnegative coefficients $C_{ij}, C_{ik}, C_{i\ell}$.

Proof. We consider first the one-dimensional case; we write

$$-\frac{(u_i^{n+1} - u_i^n)}{\lambda} = g_{i+1/2} - g_{i-1/2} = g_{i+1/2} - g(u_i, u_i) + g(u_i, u_i) - g_{i-1/2}$$

and

$$\begin{aligned}&g_{i+1/2} - g(u_i, u_i) = g(u_{i+1/2,-}, u_{i+1/2,+}) - g(u_i, u_i) \\ &= g(u_{i+1/2,-}, u_{i+1/2,+}) - g(u_i, u_{i+1/2,+}) + g(u_i, u_{i+1/2,+}) - g(u_i, u_i) \\ &= \frac{\partial g}{\partial u}(\xi_i)(u_{i+1/2,-} - u_i) + \frac{\partial g}{\partial v}(\eta_i)(u_{i+1/2,+} - u_i),\end{aligned}$$

where the derivatives are evaluated at some point and such that for a monotone scheme $\frac{\partial g}{\partial u}(\xi_i) \geq 0$, $\frac{\partial g}{\partial v}(\eta_i) \leq 0$. We have the prediction

$$(u_{i+1/2,-})^{\text{pred}} - u_i = \frac{(u_i - u_{i-1})}{2} = \frac{\Delta u_{i-1/2}}{2},$$

which, by (5.6a), we limit in such a way that

$$u_{i+1/2,-} - u_i = -\alpha_i(u_{i-1} - u_i),$$

where $\alpha_i = \min(\frac{1}{2}, \frac{k\Delta u_{i+1/2}}{\Delta u_{i-1/2}})$ if sgn $\Delta u_{i-1/2} =$ sgn $\Delta u_{i+1/2}$, and 0 otherwise. Also,

$$u_{i+1} - (u_{i+1/2,+})^{\text{pred}} = \frac{(u_{i+2} - u_{i+1})}{2} = \frac{\Delta u_{i+3/2}}{2},$$

and by (5.6b)

$$u_{i+1/2,+} - u_i = u_{i+1} - u_i - (u_{i+1} - u_{i+1/2,+}) = (1 - \beta_i)(u_{i+1} - u_i),$$

where $\beta_i = \min(k, \frac{\Delta u_{i+3/2}}{2\Delta u_{i+1/2}})$ if sgn $\Delta u_{i+3/2} =$ sgn $\Delta u_{i+1/2}$, and 0 otherwise.

Hence, we obtain

$$-\{g(u_{i+1/2,-}, u_{i+1/2,+}) - g(u_i, u_i)\} = c_i(u_{i-1} - u_i) + d_i(u_{i+1} - u_i)$$

with $c_i = \alpha_i \frac{\partial g}{\partial u}(\xi_i) \geq 0$, $d_i = -\frac{\partial g}{\partial v}(\eta_i)(1 - \beta_i) \geq 0$.

Similarly,

$$\begin{aligned} g_{i-1/2} - g(u_i, u_i) &= \frac{\partial g}{\partial u}(\xi_i')(u_{i-1/2,+} - u_i) + \frac{\partial g}{\partial v}(\eta_i')(u_{i-1/2,-} - u_i) \\ &= c_i'(u_i - u_{i-1}) + d_i'(u_i - u_{i+1}), \end{aligned}$$

where

$$d_i' = \alpha_i' \frac{\partial g}{\partial u}(\xi_i') \geq 0,$$

$$\alpha_i' = \min\Big(\frac{1}{2}, \frac{K\Delta u_{i-1/2}}{\Delta u_{i+1/2}}\Big) \text{ if sgn } \Delta u_{i-1/2} = \text{ sgn } \Delta u_{i+1/2}, \text{ 0 otherwise,}$$

$$c_i' = -\frac{\partial g}{\partial v}(\eta_i')(1 - \beta_i') \geq 0,$$

$$\beta_i' = \min\Big(K, \frac{\Delta u_{i-3/2}}{\Delta u_{i-1/2}}\Big) \text{ if sgn } \Delta u_{i-3/2} = \text{ sgn } \Delta u_{i-1/2}, \text{ 0 otherwise.}$$

Thus, setting $C_i = \lambda(c_i + c_i')$, $D_i = \lambda(d_i + d_i')$, we get

$$u_i^{n+1} - u_i^n = C_i(u_{i-1} - u_i) + D_i(u_{i+1} - u_i),$$

which can be written equivalently

$$u_i^{n+1} = u_i^n(1 - C_i - D_i)u_i + C_i u_{i-1} + D_i u_{i+1}, \quad \text{with } C_i \geq 0, D_i \geq 0.$$

Similarly, in the 2-D case, using the consistency property (4.5) and (4.14), we can write

$$\begin{aligned}\sum |e|\Phi(u_{ij}, u_{ji}, \mathbf{n_e}) &= \sum |e|\{\Phi(u_{ij}, u_{ji}, \mathbf{n_e}) - \Phi(u_i, u_i, \mathbf{n_e})\} \\ &= \sum |e|\{\Phi(u_{ij}, u_{ji}, \mathbf{n_e}) - \Phi(u_i, u_{ji}, \mathbf{n_e}) \\ &\qquad + \Phi(u_i, u_{ji}, \mathbf{n_e}) - \Phi(u_i, u_i, \mathbf{n_e})\} \\ &= \sum |e| \frac{\partial \Phi}{\partial u}(\xi_i)(u_{ij} - u_i) + \sum |e| \frac{\partial \Phi}{\partial v}(\eta_i)(u_{ji} - u_j + u_j - u_i),\end{aligned}$$

where $\frac{\partial \Phi}{\partial u} \geq 0$, $\frac{\partial \Phi}{\partial v} \leq 0$. In the the prediction step, we have

$$\begin{aligned}(u_{ij})^{\text{pred}} - u_i &= u_i \psi_i(\mathbf{m}_{ij}) + u_k \psi_k(\mathbf{m}_{ij}) + u_\ell \psi_\ell(m_{ij}) - u_i \\ &= (u_k - u_i)\psi_k(\mathbf{m}_{ij}) + (u_\ell - u_i)\psi_\ell(\mathbf{m}_{ij}),\end{aligned}$$

and we have assumed that the triangulation is such that the barycentric coordinates satisfy

$$\psi_k(\mathbf{m}_{ij}) \leq 0, \quad \psi_\ell(\mathbf{m}_{ij}) \leq 0.$$

We then use the formulas (5.5), and we conclude as in the one-dimensional case. □

Remark 5.3. The above decomposition of u_i^{n+1} is obviously to be compared with the incremental form of a numerical scheme (see Proposition 4.2 in Section 4.2.3 above), though the notations in the proof are slightly different. In the one-dimensional case, starting from a 3-point monotone scheme, we know that there exist unique incremental coefficients (see (1.2d) in Chapter III, Section 1)

$$\begin{aligned}C_{i+1/2} &= C(u_i, u_{i+1}) = \lambda \frac{g(u_i, u_i) - g(u_i, u_{i+1})}{u_{i+1} - u_i} = -\lambda \frac{\partial g}{\partial v}(\zeta_i), \\ D_{i+1/2} &= D(u_i, u_{i+1}) = \lambda \frac{g(u_{i+1}, u_{i+1}) - g(u_i, u_{i+1})}{u_{i+1} - u_i} = \frac{\partial g}{\partial u}(v_i),\end{aligned}$$

such that

$$u_i^{n+1} = u_i^n + C_{i+1/2}\Delta u_{i+1/2} - D_{i-1/2}\Delta u_{i-1/2},$$

and which, moreover, satisfy Harten's T.V.D. criteria

$$C_{i+1/2} \geq 0,\ D_{i+1/2} \geq 0,\ C_{i+1/2} + D_{i+1/2} < 1.$$

However, when substituting $u_{i+1/2,-}$, $u_{i+1/2,+}$ in the numerical flux g, the resulting scheme is 5-point and is not written in the same incremental form. We can use the fact that it is "essentially 3-point" to define incremental coefficients (given by G.R., formula (3.17) Chapter III), but it is not straightforward that the coeficients satisfy Harten's criteria. □

Of course, other linear reconstructions and limitations are also possible.

5.1.2 MUSCL-type cell vertex schemes

We now come to Example 4.2.

(i) Prediction of the gradients ∇u_i

Consider a triangle T with vertices $\mathbf{a}_i, \mathbf{a}_j, \mathbf{a}_k$. Given the values u_i, u_j, u_k of one component u of $\mathbf{u}$ at the vertices, one usually takes the $\mathbf{P}^1$-interpolate (i.e., the linear function p_T on T which takes the same values) and p'_T yields a constant approximate value ∇u_T of the gradient in T. More precisely, if the ψ_i are the $\mathbf{P}^1$-basis function (or barycentric coordinates), $\psi_i(\mathbf{a}_j) = \delta_{ij}$, then we take

$$p_T = u_i\psi_i + u_j\psi_j + u_k\psi_k,$$
$$\nabla u_T \equiv u_i\nabla\psi_i + u_j\nabla\psi_j + u_k\nabla\psi_k.$$

For ∇u_i we take a weighted average of the ∇u_T for all the triangles T having $\mathbf{a}_i$ as vertex. Hence (setting $\nabla u = \nabla u_T$ on T) we have

$$\nabla u_i = \frac{1}{|C_i|}\int_{C_i} \nabla u \, d\mathbf{x} = \frac{1}{|C_i|}\sum_{T/A_i \in T} \frac{|T|}{3}\nabla u_T.$$

(ii) Extrapolation at the middle of the edge

One simply sets

$$u_{ij} = u_i + \frac{1}{2}\nabla u_i \cdot (\mathbf{a}_j - \mathbf{a}_i),$$
$$u_{ji} = u_j - \frac{1}{2}\nabla u_j \cdot (\mathbf{a}_j - \mathbf{a}_i).$$

As earlier, the resulting scheme may generate oscillations, and one introduces some limitation.

(iii) The limitation procedure

The limitation can be achieved in different ways.

By limiting the gradients directly and choosing the value of least modulus on all the triangles T with vertex $\mathbf{a}_i$: for each component, we define

$$\nabla_x u_i^{\ell im} = \min_{T/A_i \in T} \bmod \nabla_x u_T, \quad \nabla_y u_i^{\ell im} = \min_{T/\mathbf{a}_i \in T} \bmod \nabla_y u_T$$

(∇_x, ∇_y may be replaced by the derivatives in the direction of the local gradients and the orthogonal direction, see Arminjon et al. 1989). Then, the extrapolation at the middle of the edge yields

$$u_{ij}^{\ell im} = u_i + \frac{1}{2}\nabla u_i^{\ell im} \cdot (\mathbf{a}_j - \mathbf{a}_i),$$
$$u_{ji}^{\ell im} = u_j + \frac{1}{2}\nabla u_j^{\ell im} \cdot (\mathbf{a}_j - \mathbf{a}_i).$$

Otherwise using another limiter (van Leer I 1972, Lallemand et al. 1987, Fezoui, Steve, and Silmin 1988),

$$\lim{}^I(a,b) = \frac{(a+b)(ab+|ab|+\varepsilon)}{(a^2+b^2+2\varepsilon)}$$

or (van Leer III 1977, Lallemand, Fezoui, and Perez 1987)

$$\lim^{III}(a,b) = \frac{(a+b+\varepsilon)(a+b)}{(a^2+b^2+\varepsilon^2)},$$

with $\varepsilon > 0$ small enough, one can set

$$u_{ij}^{\ell im} = u_i + \frac{1}{2}\lim(2\nabla u_i \cdot (\mathbf{a}_j - \mathbf{a}_i) - (u_j - u_i), u_j - u_i),$$

$$u_{ji}^{\ell im} = u_i - \frac{1}{2}\lim(2\nabla u_j \cdot (\mathbf{a}_j - \mathbf{a}_i) - (u_j - u_i), u_j - u_i).$$

In the one-dimensional case this limitation corresponds to replacing the increments $\Delta u_{j+1/2}$, $\Delta u_{j-1/2}$, which are involved in the averaging-limiting procedure (see the definition of $\hat{\delta}_j$ and δ_j in G.R., Chapter IV, Section 3.1), by $2\Delta u_i \cdot (\mathbf{a}_j - \mathbf{a}_i) - (u_j - u_i)$ and $(u_j - u_i)$.

Another procedure first defines gradients as follows (Billey et al. 1986, Rostand and Stoufflet 1988, Fezoui and Steve 1988, Arminjon et al. 1988). One considers for a given edge $\mathbf{a}_i\mathbf{a}_j$, the "upstream" and "downstream" triangles T_{ij} and T_{ji} in which the line $\mathbf{a}_i\mathbf{a}_j$ enters (i.e., $\mathbf{a}_i + \varepsilon(\mathbf{a}_i - \mathbf{a}_j) \in T_{ij}$ for small $\varepsilon > 0$ small enough (see Figure 5.2). Then, one sets

$$\nabla u_i = \nabla u_{T_{ij}}, \quad \nabla u_j = \nabla u_{T_{ji}}$$

and

$$u_{ij} = u_i + \frac{1}{2}\nabla u_i \cdot (\mathbf{a}_j - \mathbf{a}_i),$$

$$u_{ji} = u_i - \frac{1}{2}\nabla u_j \cdot (\mathbf{a}_j - \mathbf{a}_i),$$

or a variant

$$u_{ij} = u_i + \frac{1}{4}\{\nabla u_i \cdot (\mathbf{a}_j - \mathbf{a}_i) + u_j - u_i\},$$

$$u_{ji} = u_j - \frac{1}{4}\{\nabla u_j \cdot (\mathbf{a}_j - \mathbf{a}_i) + u_j - u_i\},$$

or one can even introduce an "upwinding" parameter k (Billey et al. 1986,

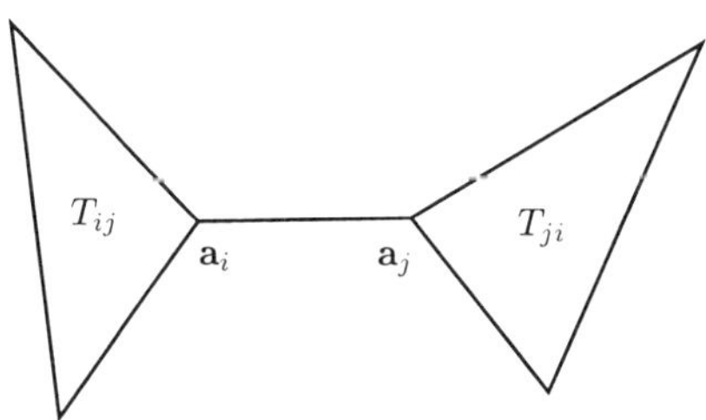

FIGURE 5.2. "Upstream" and "downstream" triangles.

Rostand and Stouflet 1988)

$$u_{ij} = u_i + \frac{1}{4}\{(1-k)\nabla u_i \cdot (\mathbf{a}_j - \mathbf{a}_i) + (1+k)(u_j - u_i)\},$$

$$u_{ji} = u_j - \frac{1}{4}\{(1-k)\nabla u_j \cdot (\mathbf{a}_j - \mathbf{a}_i) + (1+k)(u_j - u_i)\}.$$

This is followed by a limitation procedure; for instance,

$$u_{ij}^{\ell im} = u_i + \frac{1}{2}\ell(\nabla u_i \cdot (\mathbf{a}_j - \mathbf{a}_i), u_j - u_i),$$

where $\ell(a,b)$ is some limiter, for instance

$$\ell(a,b) = \frac{a(b^2+\varepsilon) + b(a^2+\varepsilon)}{a^2+b^2+2\varepsilon} \quad \text{if } \operatorname{sgn} a = \operatorname{sgn} b, \ 0 \text{ otherwise}.$$

(See Dervieux and Vijayasundaram 1983, Ciccoli et al. 1992, Arminjon et al. 1988, Radespiel and Kroll 1995 who precise the choice of ϵ for the van Albada limiter function). Or one can introduce a limiting factor and set

$$u_{ij}^{\ell im} = u_i + \frac{1}{4}s_{ij}\{(u_j - u_i) + \nabla u_i \cdot (\mathbf{a}_j - \mathbf{a}_i)\},$$

$$u_{ji}^{\ell im} = u_j - \frac{1}{4}s_{ji}\{(u_j - u_i) + \nabla u_j \cdot (\mathbf{a}_j - \mathbf{a}_i)\},$$

where

$$s_{ij} = s(\nabla u_i \cdot (\mathbf{a}_j - \mathbf{a}_i), u_j - u_i),$$

and $s(a,b)$ is van Albada's limiter (see Mulder and van Leer 1985, Jorgenson and Turkel 1993)

$$s(a,b) = \frac{2ab + \varepsilon^2}{(a^2+b^2+\varepsilon^2)}.$$

Remark 5.4. In the one-dimensional case, following the notations of G.R., Chapter IV, Remark 3.2, we had set

$$\delta_j = R(\theta_j)\hat{\delta}_j,$$

where R is a function of the ratio of consecutive increments $\theta_j = \frac{\Delta u_{j+1/2}}{\Delta u_{j-1/2}}$ (and $r_j = \frac{1}{\theta_j}$). To get the analogous formula in the 2-D case, we can set $a = \Delta u_{j-1/2}$, $b = \Delta u_{j+1/2}$ and take a function such that $R(\theta) = R(\frac{1}{\theta})$, and then set $s(a,b) = R(\frac{a}{b})$. For instance, Van Albada's limiter $R(\theta) = \frac{2\theta}{(1+\theta^2)}$ leads to the above formula for s. Note that in terms of flux limiters φ as defined in G.R., Chapter IV, Section 2, we have the relation $\varphi(r) = \frac{R(\theta)(1+\theta)}{2\theta}$ with $\theta = \frac{1}{r}$; then requiring that R satisfy $R(\theta) = R(\frac{1}{\theta})$ is equivalent to the "symetry property" $\varphi(\frac{1}{r}) = \frac{\varphi(r)}{r}$ for φ (see again G.R., Chapter IV, Remark 3.2). The particular Van Albada's limiter then gives $\varphi(r) = \frac{r^2+r}{r^2+1}$. □

Finally, one can also follow Davis' approach (see Arminjon et al. 1988, Arminjon and Dervieux 1993).

Again, the limiting procedure is applied on each dependent variable, conservative, primitive $\mathbf{W} = (\rho, u, v, p)$ (Fezoui, Steve, and Selmin 1988), or characteristic. In this last case, the characteristic variables are usually obtained by linearizing at the midpoint (see Chapter I, Remark 2.1), i.e., the limited variable is $T_{ij}^{-1}\mathbf{W}$, where the transformation matrix $\mathbf{T}$ that diagonalizes the Jacobian matrix is taken at the midvalue $\frac{(\mathbf{W}_i+\mathbf{W}_j)}{2}$ (see Fezoui and Steve 1988, Rostand and Stoufflet 1988, Arminjon et al. 1988, Mulder and van Leer 1985).

5.2 Other approaches

In fact, we shall only mention some of the most usual procedures.

Lax–Wendroff-type schemes were the first developed. These are usually centered predictor–corrector schemes on structured mesh such as the S_α^β-schemes (Lerat and Sidès 1982) or combined with a finite-element procedure such as the Richtmyer–Galerkin two-step schemes (Angrand et al. 1983, Billey et al. 1986, Arminjon and Dervieux 1993). Since their derivation is not so straightforward and not really linked to what precedes, we shall not detail it. Instead, we refer to the above-mentioned papers for details.

One can also use an E.N.O. schemes. These rely on a reconstruction procedure that gives a high-order accurate representation of the solution from given cell averages (see G.R., Chapter IV, Section 3.6) and usually involve Runge–Kutta methods for time integration. The reconstruction procedure is more easily achieved on structured mesh Casper and Atkins (1993) but can also be on unstructured mesh (Harten and Chakravarthy 1991, Shu and Osher 1989, Shu et al. 1992, Abgrall 1994, Angrand and Lafon 1992).

We can also mention different approaches, see Sanders and Li 1992, LeVeque 1988, Beam and Warming 1976, Cockburn , Hou, and Shu 1990, Berde and Borrel 1994 and Hansbo 1993, schemes based on switching strategies Harabetian and Pego 1993.

Notes

As we have already explained, there are few theoretical results concerning two-dimensional hyperbolic systems; besides the references already quoted in Remarks 1.2 and 1.3, we refer essentially to the book of Majda (1984); see also the texts of Jeffrey (1976), Hirsch (1990), Richtmyer and Morton (1967, Chapter 13) and Chan and Hsiao (1989) (which contains detailed computations concerning both the scalar two-dimensional case and steady gas dynamics); see those of Courant and Friedrichs (1976), J.D. Anderson (1982), Whitham (1974), Taniuti and Nishihara (1983), for a better understanding of physical phenomena; see also the papers of Lytton (1987), Loh and Hui (1990), Hui and Loh (1992), Koren (1988), Yang and Hsu

(1993), Glaz and Wardlaw (1985) (for computational aspects) and Glimm and Majda (1989).

For numerical approximation, this chapter is only a survey (by no means exhaustive) of a part of the huge literature on the subject. We mention the books of Godunov et al. (1979) and Hirsch (1990) for the approximation of gas dynamics by finite difference schemes; for finite-volume methods, we refer to Gallouët (1992). There are already so many references concerning numerical schemes in Sections 3 and 4 that we dare not add more than a few: the review article of Woodward and Colella (1984) and that of Vinokur (1989); Arminjon and Viallon (1995) for the extension of the Nessyahu–Tadmor scheme; Boukadida and Leroux (1994) for that of the Lax–Friedrichs scheme; for multigrid extensions, see Hemker and Spekreijse (1986), Koren and Hemker (1991), front-tracking methods Chern et al. (1986), Mao (1993), Davis (1992); for a Lagrangian approach Loh and Liou (1993), Dukowicz et al. (1989); positive schemes for linear advection can be found for instance in Hunsdorfer et al. and more details concerning computational aspects in W.K. Anderson (1994), Ajmani et al. (1994), Jiang and Forsyth (1995).

We have mainly considered the application to the compressible Euler equations; these methods are also used for incompressible flow (E and Shu 1994), Kelvin-Helmotz instabilities (Munz and Schmidt 1988), relativistic hydrodynamics (Schneider et al. 1993, Balsara 1994, Dolezal and Wong 1995), detonation waves (Clarke et al. 1993, Klein 1988, Quirk 1993) or solving other related problems such as magneto-hydrodynamics (Brio and Wu 1988, Tanaka 1994, Powell 1994, Dai and Woodward 1994, Zachary and Colella 1992, Cargo and Gallice 1995), shallow water equations (Alcrudo and Garcia-Navarro 1993), open-channel flows (Glaister 1994), petroleum reservoir simulation (Blunt and Rubin 1992), two-phase flow (Sainsaulieu 1995) through porous media (Durlofsky 1993).

V

An introduction to boundary conditions

The aim of this chapter is to introduce the unfamiliar reader to the topic of boundary conditions: we just want to give some insight into this question and do not pretend to give an exhaustive study. We recall first the main features of the initial boundary value problem (I.B.V.P.) before we present the numerical treatment of the question.

Considerations on characteristics show that one must be cautions about prescribing the solution on the boundary. In some particular cases, the boundary conditions can be found by physical considerations (such as a solid wall), but their derivation in the general case is not obvious. The problem of finding the "correct" boundary conditions, i.e., which lead to a well-posed problem, is difficult in general from both the theoretical and practical points of view (proof of the well-posedness, choice of the physical variables that can be prescribed).

The implementation of these boundary conditions is crucial in practice; however, it depends very much on the problem, and we shall give only some examples of the most usual situations. Moreover, it mostly remains a matter for the expert, whose know-how is seldom described in detail.

1 The initial boundary value problem in the linear case

It is well known that even the simple scalar I.B.V.P. in the "quarter plane"

$$\begin{cases} \dfrac{\partial u}{\partial t} + \dfrac{\partial}{\partial x} f(u) = 0, \quad x > 0,\ t > 0, \\ u(x,0) = u_0(x), \\ u(0,t) = g(t), \quad t > 0, \end{cases}$$

is ill-posed in general (which means that there may be no solution or one that does not depend in a continuous way on the initial and boundary data, or nonuniqueness). As soon as the initial condition $u(x,0) = u_0(x)$ is given, the function cannot be prescribed arbitrarily on the boundary, as we shall see below by considering the characteristics.

Many results concern the linear case and are then extended, mostly in a heuristic way, to the nonlinear case by "freezing" the Jacobian at a constant state, i.e., by linearizing around a constant state. That is why we begin by considering the linear case, and even the simplest scalar linear case.

1.1 Scalar advection equations

1.1.1 One-dimensional scalar advection equation

We consider first the problem

$$\begin{cases} \dfrac{\partial u}{\partial t} + a\dfrac{\partial u}{\partial x} = 0, & x > 0,\ t > 0, \\ u(x,0) = u_0(x), & x \geq 0. \end{cases} \tag{1.1}$$

In this problem, one sees (Figure 1.1) that if $a > 0$, the characteristics are leaving from the boundary $x = 0$, thus coming into the domain. Therefore, one needs to prescribe the solution on the boundary $x = 0$,

$$u(0,t) = g(t), \quad t > 0, \tag{1.2}$$

where g is some given function. If $M = (x,t)$ is any point in the domain $\mathbb{R}_+^\star \times \mathbb{R}_+^\star$, the value of u at M is then uniquely determined. The solution u of (1.1), (1.2) is then given by

$$\begin{cases} u(x,t) = u_0(x - at) & \text{if } x - at > 0, \\ u(x,t) = g\Big(t - \dfrac{x}{a}\Big) & \text{if } x - at < 0. \end{cases}$$

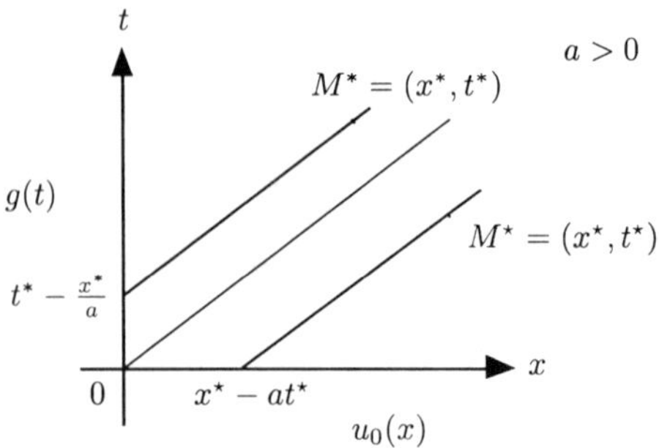

FIGURE 1.1. I.B.V.P. for the one-dimensional scalar advection equation.

The resulting solution is C^1 if the initial and boundary data are C^1 and satisfy the compatibility relations

$$u_0(0) = g(0), \quad u_0'(0) = -\frac{g'(0)}{a}.$$

If this does not hold, we have a weak solution satisfying the Rankine–Hugoniot jump condition on each side of the discontinuity $x = at$.

On the other hand, if $a < 0$, the characteristics are outgoing from the interior of the domain and impinging on the boundary; the information is thus carried from the given initial data u_0, and one cannot specify the solution on the boundary. The solution is

$$u(x,t) = u_0(x - at), \quad x \geq 0, \ t \geq 0,$$

and in particular

$$u(0,t) = u_0(-at).$$

Note that in the trivial case $a = 0$, the characteristics are vertical and no boundary condition is needed, as for outgoing characteristics.

Combining both cases enables one to treat the case of a bounded interval in space, for instance the strip $0 < x < 1$. The boundary condition that must be specified corresponds to incoming characteristics (see Figure 1.2)

$$\begin{aligned} u(0,t) &= g(t), \quad t > 0, \ \text{if } a > 0, \\ u(1,t) &= h(t), \quad t > 0, \ \text{if } a < 0. \end{aligned}$$

1.1.2 Two-dimensional scalar advection equation

The solution of the pure Cauchy problem

$$(1.3) \qquad \begin{cases} \dfrac{\partial u}{\partial t} + a\dfrac{\partial u}{\partial x} + b\dfrac{\partial u}{\partial y} = 0, & (x,y,t) \in \mathbb{R} \times \mathbb{R} \times \mathbb{R}_+, \\ u(x,y,0) = u_0(x,y), & x, y \in \mathbb{R} \end{cases}$$

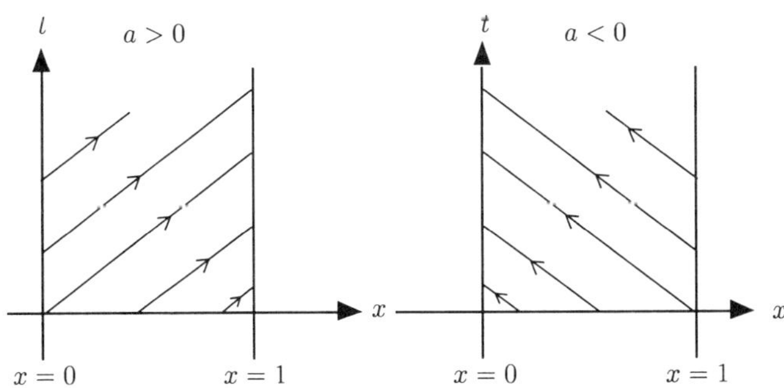

FIGURE 1.2. One-dimensional scalar advection equation in a strip.

is

$$u(x, y, t) = u_0(x - at, y - bt),$$

and is constant on the characteristic lines $x - at = \text{const.}$, $y - bt = \text{const.}$ (see Chapter IV, Section 1, Remark 1.2). The advection direction is $\mathbf{C} = (\mathbf{c}, 1)$, $\mathbf{c} = (a, b)^T$.

For an I.B.V.P., the independent variables (x, y, t) belong to a domain $Q = \mathcal{O} \times \mathbb{R}_+^\star$ of $\mathbb{R} \times \mathbb{R} \times \mathbb{R}_+$, with boundary Σ; Q is a cylinder and two different kinds of data given on the surface Σ:
(i) initial data on the set $\mathcal{O}$ of the plane $t = 0$,
(ii) boundary data on the remaining part Γ of Σ (Γ is the side of the cylinder): $\Gamma = \partial\mathcal{O} \times \mathbb{R}_+^\star$. On this surface Γ, $n_t = 0$, and one says that "the boundary of $\mathcal{O}$ is characteristic" at a point if

$$an_x + bn_y = \mathbf{c} \cdot \mathbf{n} = 0$$

at this point, where $\mathbf{n} = (n_x, n_y)^T$ is the outward normal to $\partial\mathcal{O}$ in the plane $t = 0$ (see Chapter IV, Section 1.2).

We first detail the "half-space problem," for which the computations are explicit. We have $\mathcal{O} = \{(x, y)/x > 0, y \in \mathbb{R}\}$ and $Q = \mathbb{R}_+^\star \times \mathbb{R} \times \mathbb{R}_+^\star$; therefore Σ is made of two half planes: $t = 0$, $x \geq 0$ on which the initial data are given and $\Gamma = \{x = 0, t > 0\}$. Let us follow the same arguments as in the one-dimensional case. Given any point $M^\star = (x^\star, y^\star, t^\star)$ in Q, one introduces the characteristic line through $M^\star : t \to (x = x(M^\star, t), y = y(M^\star, t))$, which is defined by

$$x - at = x^\star - at^\star, \quad y - bt = y^\star - bt^\star. \tag{1.4}$$

The solution is given by the value at the point where the characteristic line, on which u is constant, intersects the boundary Σ: if it intersects the plane $t = 0$, it is determined by the initial condition, whereas if it intersects the plane $x = 0$, it is given by the boundary data. The line (1.4) intersects the boundary $x = 0$ at a point corresponding to the time

$$t_0 = t^\star - \frac{x^\star}{a}.$$

(i) Assume first $a > 0$. If $x^\star > at^\star$, then $t_0 < 0$; but the line (1.4) intersects the boundary $t = 0$ at point $(x_0, y_0, 0)$, with $x_0 = x^\star - at^\star$, $y_0 = y^\star - bt^\star$, and the solution is given by the initial data

$$u(x^\star, y^\star, t^\star) = u(x^\star - at^\star, y^\star - bt^\star, 0) = u_0(x^\star - at^\star, y^\star - bt^\star).$$

Otherwise $0 \leq t_0 < t^\star$ if $0 \leq x^\star < at^\star$, and then the intersection point $M_0 = (0, y_0, t_0)$ belongs to $\Gamma = \{x = 0, t \geq 0\}$ and u is given by

$$u(x^\star, y^\star, t^\star) = u(0, y_0, t_0) = u\Big(0, y^\star - b(t^\star - t_0), t^\star - \frac{x^\star}{a}\Big),$$

which shows that one needs to prescribe the solution on the boundary $x = 0$ where the characteristics are "incoming,"

$$u(0, y, t) = g(y, t), \quad t > 0,$$

and

$$u(x^\star, y^\star, t^\star) = g\Big(y^\star - b\frac{x^\star}{a}, t^\star - \frac{x^\star}{a}\Big).$$

The solution u is then uniquely determined in the whole domain (see Figure 1.3).

(ii) If $a < 0$, the characteristic line intersects the boundary $x = 0$ at time $t_0 > T$, and the boundary $t = 0$ at $x_0 = x^\star - at^\star$, $y_0 = y^\star - bt^\star$. Therefore, one cannot specify the solution on the boundary $x = 0$; it is thoroughly determined by the initial data. In fact, the same thing occurs if $a = 0$, which means that if the boundary is characteristic, we consider it as part of the "outgoing" boundary.

In short, one has to prescribe the boundary data on the "incoming" part of $\Gamma : \Gamma_- = \partial\mathcal{O}_- \times \mathbb{R}^+$, where $\partial\mathcal{O}_- = \{(x, y) \in \partial\mathcal{O}, \mathbf{c}\cdot\mathbf{n} < 0\}$. In this particular case, the outward normal to $\partial\mathcal{O}$ is $\mathbf{n} = (-1, 0)^T$, and $\Gamma_- = \Gamma = \{x = 0, t > 0\}$ if $a > 0$ and is empty if $a < 0$.

More generally, consider now a bounded domain $\mathcal{O}$ of $\mathbb{R}^2$. In order to know whether the solution can indeed be defined at a point $(x^\star, y^\star, t^\star) \in Q = \mathcal{O} \times \mathbb{R}^{+\star}$, one draws as above the characteristic line (1.4) and looks for the intersection with the boundary Σ of Q. If the line remains in Q and intersects the plane $t = 0$ at a point $x_0 = x^\star - at^\star$, $y_0 = y^\star - bt^\star$ that lies inside $\mathcal{O}$, then, is determined by the initial condition

$$u(x^\star, y^\star, t^\star) = u_0(x_0, y_0) = u_0(x^\star - at^\star, y^\star - bt^\star).$$

In the other case, the line intersects Γ at some point (x_0, y_0, t_0) with time t_0 satisfying $0 \le t_0 < t^\star$. On the one hand, the points are on the

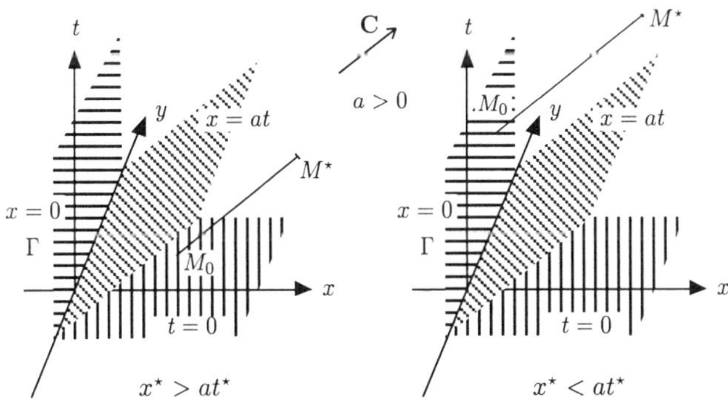

FIGURE 1.3. Two-dimensional scalar advection equation in a quarter of space.

same characteristic line, which yields

$$(x^\star - x_0, y^\star - y_0)^T = (t^\star - t_0)\mathbf{c}.$$

On the other hand, since $(x^\star, y^\star) \in \mathcal{O}$ and $(x_0, y_0) \in \partial\mathcal{O}$, we have (provided $\mathcal{O}$ is not characteristic at $m_0 = (x_0, y_0)$)

$$(x^\star - x_0,\ y^\star - y_0) \cdot \mathbf{n} < 0,$$

where $\mathbf{n}$ is the outward normal to $\mathcal{O}$ in the (x, y)-plane (see Figure 1.4).

Therefore, boundary data have to be prescribed on the part Γ_- of the boundary that corresponds to incoming characteristics

$$\begin{aligned} &\partial\mathcal{O}_- = \{(x, y) \in \partial\mathcal{O}, \mathbf{c} \cdot \mathbf{n}(x, y) < 0\}, \\ &u(\cdot, t) = g(\cdot, t) \text{ on } \partial\mathcal{O}_- \iff u = g \text{ on } \Gamma_- = \partial\mathcal{O}_- \times \mathbb{R}_+, \end{aligned}$$

and not on the part $\partial\mathcal{O}_+ = \{(x, y) \times \partial\mathcal{O}, \mathbf{c} \cdot \mathbf{n}(x, y) \geq 0\}$ where they are outgoing. Note that if $\mathcal{O}$ is characteristic at $m_0 = (x_0, y_0)$, it is easily seen that u cannot be specified on the corresponding part of Γ.

Denoting by $t_0 = t_0(M^\star)$ the time when the characteristic intersects Σ, $t_0 = \inf\{t \geq 0/(x(M^\star, t), y(M^\star, t), t) \in \overline{Q}\}$, and by (x_0, y_0) the coordinates of the intersection point, we have

$$\begin{aligned} u(x^\star, y^\star, t^\star) &= u_0(x_0, y_0) = u_0(x^\star - at^\star, y^\star - bt^\star) \\ &\qquad \text{if } t_0(x^\star, y^\star, t^\star) = 0, \\ u(x^\star, y^\star, t^\star) &= g(x_0, y_0, t_0) = g(x^\star - a(t^\star - t_0), y^\star - b(t^\star - x_0), t_0), \\ &\qquad \text{if } t_0(x^\star, y^\star, t^\star) > 0, \end{aligned}$$

since $t_0 > 0$ implies $x(M^\star, t_0)$, $y(M^\star, t_0) \in \partial\mathcal{O}_-$ and $(x(M^\star, t_0), y(M^\star, t_0), t_0) \in \Gamma_-$.

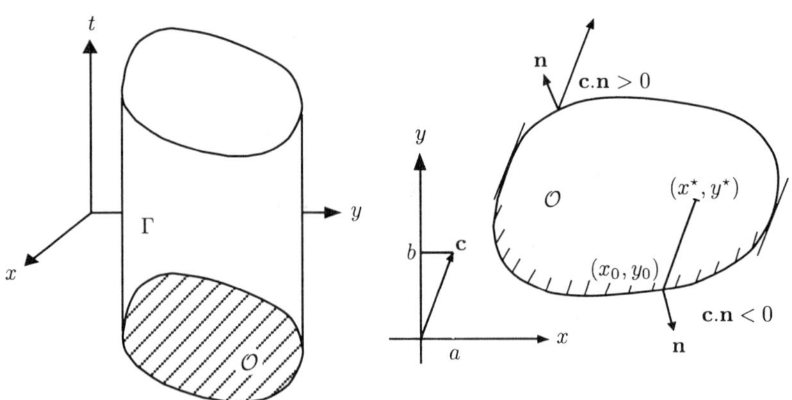

FIGURE 1.4. Boundary corresponding to incoming characteristics.

The resulting solution is C^1 if the data, which are assumed to be smooth enough, satisfy some compatibility relations. In particular, u is continuous on both sides of the surface S (corresponding to the line $x = at$ in the one-dimensional case, see Figure 1.1) defined by $S = \{M^\star/(x(M^\star, 0), y(M^\star, 0)) \in \partial\mathcal{O}_-\}$, where $(x(M^\star, 0), y(M^\star, 0))$ is the intersection of the backward characteristic (1.4) through $M^\star$ with the plane $t = 0$. S is an oblique cylinder (with base $\partial\mathcal{O}_-$) of direction along the vector $\mathbf{C} = (\mathbf{c}, 1)$. Therefore, the continuity of u across S supposes

$$u_0(x, y) = g(x, y, 0) \text{ on } \partial\mathcal{O}_-.$$

Otherwise, we get a weak solution: since the equation is linear, the Rankine–Hugoniot jump relation across S is simply (see the Introduction, Section 2, formula (2.8))

$$an_x + bn_y + n_t = 0,$$

where $(n_x, n_y, n_t)^T$ is a normal vector to S. By definition of S, this is indeed satisfied.

Remark 1.1. The arguments can easily be extended to a scalar problem with C^1 variable coefficients $\mathbf{c}(x, y, t) = (a(x, y, t), b(x, y, t))^T$ that are Lipschitz w.r.t space variables. Only the characteristics through $M^\star = (X^\star, t^\star)$, defined as the integral curves $t \to \mathbf{X}(t; \mathbf{X}^\star, t^\star) = (x(M^\star, t), y = y(M^\star, t))$ of

$$\begin{cases} \dfrac{d\mathbf{X}}{dt} = \mathbf{c}(\mathbf{X}(t), t), \\ \mathbf{X}(t^\star; \mathbf{X}^\star, t^\star) = \mathbf{X}^\star, \end{cases}$$

are no longer straight lines. One considers as above the "entrance" time $t_0 = t_0(\mathbf{X}^\star, t^\star)$ corresponding to the point where the characteristic enters the domain and then follows the same lines as above, depending on whether $t_0 = 0$ or $t_0 > 0$. The boundary data are still prescribed on the incoming part $\Gamma_- = \partial\mathcal{O}_- \times (0, T)$, $\partial\mathcal{O}_- = \{(x, y, t) \in \partial\mathcal{O} \times (0, T),\ \mathbf{c}(x, y, t) \cdot \mathbf{n}(x, y) < 0\}$,

$$u = g \text{ on } \Gamma_-.$$

We leave the technical details to the reader. □

1.2 *One-dimensional linear systems. Linearization*

Consider now a linear hyperbolic system in *diagonal* form,

$$\begin{cases} \dfrac{\partial \mathbf{u}}{\partial t} + \boldsymbol{\Lambda} \dfrac{\partial \mathbf{u}}{\partial x} = \mathbf{0}, \quad 0 < x < 1,\ t > 0, \\ \mathbf{u}(x, 0) = \mathbf{u}_0(x), \end{cases}$$

where $\mathbf{u} \in \mathbb{R}^p$, and $\boldsymbol{\Lambda}$ is a diagonal matrix with eigenvalues a_i, $1 \le i \le p$. At first, we assume that the a_i are nonvanishing: $a_i < 0$, $1 \le i \le p - q$, $a_i > 0$, $p - q + 1 \le i \le p$. One writes

$$\boldsymbol{\Lambda} = \boldsymbol{\Lambda}^I + \boldsymbol{\Lambda}^{II} = \boldsymbol{\Lambda}^+ + \boldsymbol{\Lambda}^-,$$

where, using the notations (1.9) in Chapter III, Section 1.1, the matrix $\boldsymbol{\Lambda}^I \equiv \operatorname{diag}(a_i^+) = \boldsymbol{\Lambda}^+$ (resp. $\boldsymbol{\Lambda}^{II} \equiv \operatorname{diag}(a_i^-) = \boldsymbol{\Lambda}^-$) has q positive (resp. $p - q$ negative) eigenvalues, and the corresponding partition for $\mathbf{u}$ is $\mathbf{u} = (\mathbf{u}^I, \mathbf{u}^{II})^T \in \mathbb{R}^q \times \mathbb{R}^{p-q}$ (incoming/outgoing dependent variables). The results of Section 1.2 lead to prescribing the following boundary conditions:

$$\mathbf{u}^I(0,t) = \mathbf{g}^I(t), \quad \mathbf{u}^{II}(1,t) = \mathbf{g}^{II}(t),$$

which means that one solves p uncoupled scalar equations. If one eigenvalue vanishes, the corresponding component of $\mathbf{u}$ is constant, determined by the initial data on $t = 0$. Thus, it should be considered as an outgoing variable, i.e., part of $\mathbf{u}^{II}$ on $x = 0$, and of $\mathbf{u}^I$ on $x = 1$.

A slight generalization consists in coupling these boundary conditions by introducing the already known components corresponding to the outgoing characteristics impinging on the boundary (which transport the information from the data given on the line $t = 0$). Thus, one can set

$$\text{(1.5)} \qquad \begin{cases} \mathbf{u}^I(0,t) = \mathbf{S}^I \mathbf{u}^{II}(0,t) + \mathbf{g}^I(t), \\ \mathbf{u}^{II}(1,t) = \mathbf{S}^{II} \mathbf{u}^I(1,t) + \mathbf{g}^{II}(t), \end{cases}$$

where $\mathbf{S}^I$ (resp. $\mathbf{S}^{II}$) is a $q \times (p-q)$ (resp. $(p-q) \times q$) matrix and the solution is still uniquely determined; (1.5) means that $\mathbf{u}^I$ is an affine function of $\mathbf{u}^{II}$ on $x = 0$ ("reflection of the outgoing waves"), and conversely $\mathbf{u}^{II}$ is an affine function of $\mathbf{u}^I$ at $x = 1$. In fact, this type of boundary condition leads in the case of linear systems with variable coefficients to the theory of well-posed systems in the sense of Kreiss (Kreiss 1970).

Now, for a *general linear* hyperbolic system with constant coefficients,

$$\text{(1.6)} \qquad \begin{cases} \dfrac{\partial \mathbf{u}}{\partial t} + \mathbf{A} \dfrac{\partial \mathbf{u}}{\partial x} = \mathbf{0}, \quad 0 < x < 1, \ t > 0, \\ \mathbf{u}(x,0) = \mathbf{u}_0(x), \end{cases}$$

one diagonalizes the matrix $\mathbf{A}$ (see Chapter III, Section 1.1)

$$\mathbf{A} = \mathbf{T} \boldsymbol{\Lambda} \mathbf{T}^{-1}.$$

We recall that $\mathbf{w} = \mathbf{T}^{-1}\mathbf{u}$ are the characteristic variables:

$$\text{(1.7)} \qquad \mathbf{u} = \sum_i \mathbf{w}_i \mathbf{r}_i, \quad \mathbf{w}_i = \mathbf{l}_i^T \mathbf{u},$$

where the $\mathbf{r}_i$ (resp. $\mathbf{l}_i$) are the eigenvectors of $\mathbf{A}$ (resp. $\mathbf{A}^T$). For ease of notation, we set p' = number of nonpositive eigenvalues of $\mathbf{A}$ ($a_i \le 0, 1 \le i \le p'$) and $q = p - p'$ = number of positive eigenvalues of $\mathbf{A}$ ($a_i >$

$0, p' + 1 \leq i \leq p$). Let the subscript I (resp. II) correspond to positive eigenvalues $a_i > 0$ (resp. negative $a_i \leq 0$) and set

$$\mathbf{w}^I = (w_{p'+1}, ..., w_p), \quad \mathbf{w}^{II} = (w_1, ..., w_{p'})^T.$$

Therefore, $\mathbf{w} = \mathbf{T}^{-1}\mathbf{u}$ is a solution of a decoupled system

$$\frac{\partial \mathbf{w}}{\partial t} + \boldsymbol{\Lambda}\frac{\partial \mathbf{w}}{\partial x} = \mathbf{0},$$

which is well-posed if the boundary conditions for $\mathbf{w} = (\mathbf{w}^I, \mathbf{w}^{II})^T \in \mathbb{R}^{p-p'} \times \mathbb{R}^{p'}$ can take the above form, i.e., are specified at $x = 0$ for $p' + 1 \leq i \leq p$,

(1.8a) $$\mathbf{w}^I(0, t) = \mathbf{g}^I(t),$$

where $\mathbf{g}^I(t)$ is a given $(p - q')$-component vector function.

If we consider the strip $0 < x < 1$, we set p'' = the number of negative eigenvalues of $\mathbf{A}$ ($a_i < 0$, for $1 \leq i \leq p''$), and at $x = 1$

$$\mathbf{w}^I = (w_{p''+1}, ..., w_p), \quad \mathbf{w}^{II} = (w_1, ..., w_{p''})^T,$$

the boundary data take the form

$$\mathbf{w}^{II}(1, t) = \mathbf{g}^{II}(t),$$

where $\mathbf{g}^{II}(t)$ is a given (p''-component vector) function; we can take more generally

(1.8b) $$\begin{cases} \mathbf{w}^I(0, t) = \mathbf{S}^I\mathbf{w}^{II}(0, t) + \mathbf{g}^I(t), \\ \mathbf{w}^{II}(1, t) = \mathbf{S}^{II}\mathbf{w}^I(1, t) + \mathbf{g}^{II}(t). \end{cases}$$

Now, if we are given boundary data at $x = 0$ in the form

(1.9) $$\mathbf{Eu}(0, t) = \mathbf{g}(t),$$

where $\mathbf{E}$ is a $N \times p$ matrix, and $\mathbf{g}$ is an N-component given function, we can ask whether the corresponding I.B.V.P. is well posed. In term of characteristic variables, the relation

$$\mathbf{ETw} = \mathbf{g}, \quad \text{at } x = 0$$

can be decomposed in blocks corresponding to the partition $\mathbf{w} = (\mathbf{w}^I, \mathbf{w}^{II})^T \in \mathbb{R}^q \times \mathbb{R}^{p-q}$ (q is the number of positive eigenvalues):

$$\mathbf{ETw} = (\mathbf{ET})^I\mathbf{w}^I + (\mathbf{ET})^{II}\mathbf{w}^{II} = \mathbf{g} \text{ at } x = 0,$$

where $(\mathbf{ET})^I$ (resp. $(\mathbf{ET})^{II}$) is a $N \times q$ (resp. $N \times (p - q)$) matrix. Since $\mathbf{w}^{II}(0, t)$ is given by the initial data, it reads

$$(\mathbf{ET})^I\mathbf{w}^I = \mathbf{g} - (\mathbf{ET})^{II}\mathbf{w}^{II}.$$

Hence, the problem is well posed iff one can compute $\mathbf{w}^I$, which supposes first that $N = q$, so that $(\mathbf{ET})^I$ is a square $q \times q$ matrix. Then, it is easily seen that

$$(\mathbf{ET})^I = \mathbf{E}(\mathbf{T})^I,$$

where the columns of $\mathbf{T}^I$ are the q eigenvectors of $\mathbf{A}$ corresponding to positive eigenvalues. The condition is therefore given in the following lemma.

Lemma 1.1

Consider the boundary condition (1.9) for the system (1.6). The resulting problem is well posed if $\mathbf{E}$ *is a* $q \times p$ *matrix such that* $\mathbf{E}(\mathbf{T})^I$ *is invertible, where* $\mathbf{T}^I$ *denotes the* $p \times q$ *matrix with columns the* q *eigenvectors of* $\mathbf{A}$ *corresponding to positive eigenvalues.*

We shall give below (Section 1.5) an example of such a situation in gas dynamics.

For a *nonlinear system*, one can linearize about a constant state ("freezing" theory) and apply the above procedure to the linearized system. We shall give another approach later.

In short, the number of boundary conditions should be equal to the number of characteristics pointing into the region, an argument that we have already used when introducing the Lax entropy condition (Chapter I, Section 5).

1.3 Multidimensional linear systems

Consider the "half-space" model I.B.V.P.

$$\frac{\partial \mathbf{u}}{\partial t} + \mathbf{A}\frac{\partial \mathbf{u}}{\partial x} + \sum_{j=1}^{d-1} \mathbf{B}_j \frac{\partial \mathbf{u}}{\partial y_j} = \mathbf{0}, \; x > 0, \; y \in \mathbb{R}^{d-1}, \; t > 0.$$

To simplify the presentation of the theory, which is already rather complicated, we shall mainly restrict ourselves to the two-dimensional constant coefficient case

$$(1.10a), \qquad \frac{\partial \mathbf{u}}{\partial t} + \mathbf{A}\frac{\partial \mathbf{u}}{\partial x} + \mathbf{B}\frac{\partial \mathbf{u}}{\partial y} = \mathbf{0}, \qquad x > 0, y \in \mathbb{R}, t > 0,$$

$$(1.10b), \qquad \mathbf{u}(x, y, 0) = \mathbf{u}_0(x, y),$$

$$(1.10c), \qquad \mathbf{Eu}(0, y, t) = \mathbf{g}(y, t),$$

where $\mathbf{E}$ is a $q \times p$ matrix and $\mathbf{g}$ a given q-component function. The system is hyperbolic if for any $(\xi, \eta) \in \mathbb{R}^2$, $\mathbf{A}\xi + \mathbf{B}\eta$ has real eigenvalues and is diagonalizable. We assume that the boundary $x = 0$ of the space domain is noncharacteristic, i.e., $\mathbf{A}$ is invertible (see Chapter IV, Section 1.2). Hence, with a possible change of variable, we can assume that $\mathbf{A}$ is diagonal $\mathbf{A} = \boldsymbol{\Lambda} = (\boldsymbol{\Lambda}^I, \boldsymbol{\Lambda}^{II})$. Following the considerations of the scalar case, a necessary condition for the I.B.V.P. to be well posed is that q boundary conditions are prescribed, where q = number of positive eivenvalues of $\boldsymbol{\Lambda}$, i.e., $\mathbf{u}^I(0, y, t) = \mathbf{g}^I(y, t)$ or $\mathbf{u}^I(0, y, t) = \mathbf{S}^I\mathbf{u}^{II}(0, y, t) + \mathbf{g}^I(y, t)$. But this is not sufficient, and there are examples of solutions of the corresponding I.B.V.P. with arbitrary growth in time (see Gustafsson and Kreiss 1979, Higdon 1986, Yee 1981), and more restrictions may be necessary.

In fact, the theory in the multidimensional case is not straightforward. We just want to sketch the main lines of the arguments because of their link to topics already developed such as Fourier modes, group velocity, and characteristic surfaces. This is only an introduction to the topic, and most details are skipped; for a precise study, we refer to Higdon (1981).

The system is assumed to be either strictly hyperbolic or symmetric hyperbolic; for simplicity, we again assume that that the boundary $x = 0$ of the space domain is noncharacteristic, i.e., $\mathbf{A}$ is invertible.

1.3.1 Uniform Kreiss condition (U.K.C.)

The theory developed by Kreiss relies on "normal mode analysis." Normal modes are elementary waves $\mathbf{u}(x, y, t) = \varphi(x)e^{i\eta y - st}$, $s \in \mathbb{C}$, which are introduced as follows. Applying a Fourier transform in the y-variable and a Laplace transform in the time variable t ($u \to L(u)(s) = \int_0^\infty e^{st}u(t)dt$, $\mathrm{Re}(s) < 0$) to the system (1.10 a) gives

$$\text{(1.11)}\qquad \begin{cases} -s\tilde{\mathbf{u}} + \mathbf{A}\dfrac{\partial \tilde{\mathbf{u}}}{\partial x} + i\eta\mathbf{B}\tilde{\mathbf{u}} = \mathbf{0}, \\ \tilde{\mathbf{u}} = \tilde{\mathbf{u}}(x, \eta, s) = \displaystyle\int_{\mathbb{R}}\int_0^\infty e^{-i\eta y}e^{st}\mathbf{u}(x, y, t)dy\, dt, \end{cases}$$

which can also be written

$$\frac{\partial \tilde{\mathbf{u}}}{\partial x} = \mathbf{A}^{-1}(s\mathbf{I} - i\eta\mathbf{B})\tilde{\mathbf{u}} = \mathbf{D}(\eta, s)\tilde{\mathbf{u}},$$

where we have set

$$\text{(1.12)}\qquad \mathbf{D}(\eta, s) = \mathbf{A}^{-1}(s\mathbf{I} - i\eta\mathbf{B}).$$

By the inverse transform, we see that the solution $\mathbf{u}$ is a "superposition" of elementary modes $\hat{\mathbf{u}}e^{\ell x}e^{i\eta y - st}$, where ℓ is an eigenvalue of $\mathbf{D}(\eta, s)$. We can immediately see that $\mathrm{Re}(\ell) \neq 0$.

Lemma 1.2

Assume $\mathrm{Re}(s) < 0$, *and* $\eta \in \mathbb{R}$. *The matrix* $\mathbf{D}(\eta, s)$ *defined by (1.12) has no purely imaginary eigenvalue.*

Proof. We write

$$\mathbf{D}(\eta, s)\varphi = i\xi\varphi \iff i(\xi\mathbf{A} + \eta\mathbf{B})\varphi = s\varphi.$$

Now, for $\mathbf{k} = (\xi, \eta)^T \in \mathbb{R}^2$, the matrix $\xi\mathbf{A} + \eta\mathbf{B}$ has real eigenvalues and a complete set of eigenvectors because the system is hyperbolic. Therefore

$$\xi \in \mathbb{R} \Longrightarrow s \text{ purely imaginary } (s = i\omega, \omega \in \mathbb{R}, \text{ and } \mathrm{Re} s = 0),$$

which gives the result. □

Thus, consider a normal mode $\mathbf{u}(x, y, t) = \varphi(x)e^{i\eta y - st}$, with $s \in \mathbb{C}$, which, once substituted in (1.10), is a solution of an ordinary differential equation for the amplitude function φ,

$$-s\varphi + \mathbf{A}\varphi' + i\eta\mathbf{B}\varphi = \mathbf{0},$$

or

$$\varphi' = \mathbf{A}^{-1}(s\mathbf{I} - i\eta\mathbf{B})\varphi = \mathbf{D}(\eta, s)\varphi.$$

The main idea in order to derive necessary conditions on the boundary data so that the problem is well posed is to exclude the cases that can lead to an ill-posed problem.

First, looking for particular normal modes (those giving rise to solutions of the form $\mathbf{u}_\alpha(x, y, t) = \varphi(\alpha x)e^{i\alpha\eta y - \alpha s t}$ that cannot satisfy an energy estimate) yields that the problem is ill posed if, for some η, the o.d.e. has an "eigenvalue" s with $\mathrm{Re}(s) < 0$.

Then, for those s with $\mathrm{Re}(s) < 0$ that yield an ill-posed problem, using the fact that the set of solutions of the o.d.e. is spanned by p independent solutions $\varphi_j(x) = \hat{\varphi}_j e^{\ell_j x}$, ℓ_j an eigenvalue, $\hat{\varphi}_j$ a corresponding eigenvector of $\mathbf{D}(\eta, s)$ (if ℓ_j is simple, otherwise $\varphi_j(x)$ is multiplied by a polynomial in x), one shows that q among them have finite norm, say $\varphi_1, \ldots, \varphi_q$. These functions φ_i correspond to the q eigenvalues with negative real part of the matrix $\mathbf{D}(\eta, s)$ (see Lemma 1.2).

We want, moreover, to prevent the possibility of such solutions satisfying the boundary conditions. Denoting by $\mathbf{u}_j = \varphi_j(x)e^{i\eta y - st} = \hat{\varphi}_j e^{\ell_j x + i\eta y - st}$ the corresponding solutions of (1.10 a), this eventually results in the necessary condition for well-posedness, which reads

$$\mathbf{E}[\mathbf{u}_1(0), \ldots, \mathbf{u}_q(0)] \text{ nonsingular}$$

(otherwise for $\mathbf{g} = \mathbf{0}$ in (1.10 c) we would have a solution of the I.B.V.P. that does not satisfy an energy estimate). This condition can be written equivalently

$$\mathbf{E}[\varphi_1(x = 0), \ldots, \varphi_q(x = 0)] \text{ nonsingular.}$$

Setting

$$\mathbf{N}(\eta, s) = \mathbf{E}[\varphi_1(0), \ldots, \varphi_q(0)], \tag{1.13}$$

$\mathbf{N}(\eta, s)$ is a square $q \times q$ matrix, and this in turn is equivalent to

$$\det \mathbf{N}(\eta, s) \neq 0, \quad \forall \eta \in \mathbb{R}, \forall s, \mathrm{Re}(s) < 0.$$

This is a necessary condition for well-posedness. The sufficient "uniform Kreiss condition" (U.K.C.) is written in a very similar way,

$$\det \overline{\mathbf{N}}(\eta, s) \geq \delta > 0, \quad \forall \eta \in \mathbb{R}, \forall s, \operatorname{Re}(s) < 0$$

(where $\overline{\mathbf{N}}$ is obtained as was $\mathbf{N}$, only after some normalization of the eigenfunctions), but the proof of sufficiency is not easy.

1.3.2 The characteristic variety

We would like to interpret the above U.K.C. condition in a more intuitive way involving "incoming" and "outgoing" notions. Following Higdon (1986), we introduce the following definition.

Definition 1.1

The characteristic variety for system (1.10 a) is the set of points $(\xi, \eta, \omega) \in \mathbb{R}^3$ *such that* $\det\,(-\omega \mathbf{I} + \xi \mathbf{A} + \eta \mathbf{B}) = 0$.

In view of Definition 1.2, Chapter IV, the characteristic variety is the set of (ξ, η, ω) such that the plane $\{(x, y, t), \omega t = \xi x + \eta y\}$ is characteristic. Therefore ω is an eigenvalue of the "principal symbol" $\xi \mathbf{A} + \eta \mathbf{B}$, which is real since the system is hyperbolic. Note that setting $\mathbf{n} = (-\omega, \mathbf{k})$, $\mathbf{k} = (\xi, \eta)^T$, we have

$$i\mathbf{A}(\mathbf{D}(\xi, i\omega) - i\xi \mathbf{I}) = -\omega \mathbf{I} + \eta \mathbf{B} + \xi \mathbf{A} = \mathbf{M}(\mathbf{n}),$$

where the matrix $\mathbf{M}$ is defined by (1.6), Chapter IV. Let us give an example.

Example 1.1. Consider the isentropic Euler equations (see Chapter IV, Section 1.1, Example 1.1; see also Section 2.4), linearized about a state (ρ, u, v); denoting by (ρ', u', v') the perturbations, they are a solution of the linear system

$$\begin{aligned}
\frac{\partial \rho'}{\partial t} + \rho \frac{\partial u'}{\partial x} + u \frac{\partial \rho'}{\partial x} + \rho \frac{\partial v'}{\partial y} + v \frac{\partial \rho'}{\partial y} &= R_1, \\
\frac{\partial u'}{\partial t} + u \frac{\partial u'}{\partial x} + \left(\frac{c^2}{\rho}\right) \frac{\partial \rho'}{\partial x} + v \frac{\partial u'}{\partial y} &= R_2, \\
\frac{\partial v'}{\partial t} + u \frac{\partial v'}{\partial x} + \left(\frac{c^2}{\rho}\right) \frac{\partial \rho'}{\partial y} + v \frac{\partial v'}{\partial y} &= R_3,
\end{aligned}$$

where the terms R_i do not contain derivatives of (ρ, u, v). This system can be symmetrized, defining the new dependent variables by $(u', v', (\frac{c}{\rho})\rho')$. The characteristic variety is the union of the plane

$$\omega = \mathbf{u} \cdot \mathbf{k} \quad \text{if } \mathbf{k} = (\xi, \eta)^T$$

and the cones C defined by

$$\omega = \mathbf{u} \cdot \mathbf{k} \pm c|\mathbf{k}|$$

(here $\mathbf{u} = (u, v)^T$ is the constant velocity of the state at which the system is linearized, c the speed of sound). □

Example 1.2. Maxwell's equations in empty space. Consider the system

(1.14*a*), $\dfrac{\partial}{\partial t}(\varepsilon_0 \mathbf{E}) - \dfrac{1}{\mu_0} \operatorname{curl} \quad \mathbf{B} = -J$ (Ampère's law),

(1.14*b*), $\dfrac{\partial \mathbf{B}}{\partial t} + \operatorname{curl} \quad \mathbf{E} = 0$ (Faraday's law),

(1.14*c*), $\operatorname{div} (\varepsilon_0 \mathbf{E}) = \rho$ (Poisson's or Gauss' law),

(1.14*d*) $\operatorname{div} \mathbf{B} = 0$,

where $\mathbf{B}$ denotes the magnetic field, $\mathbf{E}$ the electric field, ρ the charge density, and J the current density. In a vacuum, we assume constant permittivity ε_0 and permeability μ_0, and $c^2 = \frac{1}{\varepsilon_0 \mu_0}$ is the speed of light. If we take the divergence of equation (1.14 a), we get

$$\frac{\partial}{\partial t} \operatorname{div}(\varepsilon_0 \mathbf{E}) = -\operatorname{div} J,$$

so that with equation (1.14 c)

$$\frac{\partial \rho}{\partial t} + \operatorname{div} J = 0,$$

which is the equation of conservation of the total charge. If we take the divergence of (1.14 b), we get

$$\frac{\partial}{\partial t} (\operatorname{div} \mathbf{B}) = 0,$$

so that if we assume that

$$\operatorname{div} \mathbf{B}_0 = 0$$

and

$$\operatorname{div} (\varepsilon_0 \mathbf{E}_0) = \rho_0$$

at some initial time, (1.14 c) and (1.14 d) may be omitted. Then, the system simplifies to

$$\frac{\partial \mathbf{E}}{\partial t} - c^2 \operatorname{curl} \mathbf{B} = -\frac{1}{\varepsilon_0} J,$$
$$\frac{\partial \mathbf{B}}{\partial t} + \operatorname{curl} \mathbf{E} = 0.$$

(For more details concerning Maxwell's equation, see Dautray and Lions 1988). Setting

$$\mathbf{u} = (\mathbf{E}, c^2 \mathbf{B})^T,$$

introducing the matrices

$$\mathbf{D}_1 = \begin{pmatrix} 0 & 0 & 0 \\ 0 & 0 & c \\ 0 & -c & 0 \end{pmatrix}, \quad \mathbf{D}_2 = \begin{pmatrix} 0 & 0 & -c \\ 0 & 0 & 0 \\ c & 0 & 0 \end{pmatrix}, \quad \mathbf{D}_3 = \begin{pmatrix} 0 & c & 0 \\ -c & 0 & 0 \\ 0 & 0 & 0 \end{pmatrix},$$

and

$$\mathbf{A}_j = \begin{pmatrix} 0 & \mathbf{D}_j \\ \mathbf{D}_j^T & 0 \end{pmatrix}, \quad \mathbf{f} = \begin{pmatrix} -J/\varepsilon_0 \\ 0 \end{pmatrix},$$

the system is written

$$\frac{\partial \mathbf{u}}{\partial t} + \sum_{j=1}^{3} \mathbf{A}_j \frac{\partial \mathbf{u}}{\partial x_j} = \mathbf{f}.$$

The matrices $\mathbf{A}_j$ are symmetric, and for any $\mathbf{k} = (\xi, \boldsymbol{\eta}) \in \mathbb{R}^3$, the matrix $\xi \mathbf{A}_1 + \eta_2 \mathbf{A}_2 + \eta_3 \mathbf{A}_3$ has three real double eigenvalues, $\omega_{1,2} = -c|\mathbf{k}|$, $\omega_{3,4} = 0$, $\omega_{5,6} = c|\mathbf{k}|$. We see in this example that the assumption $\det \mathbf{A}_1 \neq 0$ is not always valid! The characteristic variety is made up of the horizontal plane $\omega = 0$ and the right circular cones $C : \omega = \pm c|\mathbf{k}|$ (propagation of electromagnetic waves). □

1.3.3 Group velocity and incoming/outgoing modes

The link between the characteristic variety and the above arguments introducing the U.K.C. appears in the following simple fact that we have already observed.

Lemma 1.3

The point $(\xi, \eta, \omega) \in \mathbb{R}^3$ lies on the characteristic variety for system (1.10a) iff $i\xi$ is a purely imaginary eigenvalue of the matrix $\mathbf{D}(\eta, i\omega)$ defined by (1.12).

Proof. Given any point $(\xi, \eta, \omega) \in \mathbb{R}^3$, we write as previously

$$-\omega \mathbf{I} + \xi \mathbf{A} + \eta \mathbf{B} = -\mathbf{A}(-\xi \mathbf{I} + \mathbf{A}^{-1}(\omega \mathbf{I} - \eta \mathbf{B}) = i\mathbf{A}(\mathbf{D}(\eta, i\omega) - i\xi \mathbf{I}).$$

Thus (ξ, η, ω) lies on the characteristic surface iff $i\xi$ is a purely imaginary eigenvalue of the matrix $\mathbf{D}(\eta, i\omega) = \mathbf{A}^{-1}(i\omega \mathbf{I} - i\xi \mathbf{B})$ introduced above. Note also that

$$\mathbf{D}(\eta, i\omega)\varphi = i\xi\varphi \iff (\xi \mathbf{A} + \eta \mathbf{B})\varphi - \omega\varphi,$$

which shows that the eigenvectors of $\mathbf{D}(\eta, i\omega)$ are those of $\xi \mathbf{A} + \eta \mathbf{B}$. □

The reason for introducing the characteristic variety results from the following elementary lemma.

Lemma 1.4

A plane wave of the form

$$\mathbf{u}(x, y, t) = \hat{\mathbf{u}} e^{i(\xi x + \eta y - \omega t)}$$

is a solution of (1.10a) iff the point $(\xi, \eta, \omega) \in \mathbb{R}^3$ lies on the characteristic variety.

Thus, by this lemma, the "characteristic variety" also describes the set of all wave numbers $\mathbf{k} = (\xi, \eta)^T$ and frequencies ω of plane wave solutions $\hat{\mathbf{u}}e^{i(\xi x+\eta y-\omega t)}$. The propagation velocity of such an individual plane wave is the phase velocity vector $\frac{\omega\mathbf{k}}{|\mathbf{k}|} = (\frac{\omega}{|\mathbf{k}|})(\xi, \eta)^T$, thus the wave is incoming (in the domain $x > 0$) if $\operatorname{sgn}(\omega\xi) \leq 0$. However, the direction in which a group of these waves (such as the superposition used in the inverse Fourier transform) propagates is directly linked to the group velocity

$$\gamma = \operatorname{grad} \omega = \left(\frac{\partial\omega}{\partial\xi}, \frac{\partial\omega}{\partial\eta}\right)^T$$

(see Chapter IV, Section 3.1.2). To illustrate this with a simple example, take the scalar case (1.3). The direction of propagation lies along bicharacteristics (1.4) with direction $\mathbf{c}$. The dispersion relation is $\omega = \mathbf{c} \cdot \mathbf{k}$, and the group velocity coincides with that of the bicharacteristics since $\gamma = \operatorname{grad} \omega = \mathbf{c}$, which points into or out of the domain $x > 0$ according to the sign of a (see Figure 1.3) (but the phase velocity vector of an individual plane wave can point in any desired direction). The characteristic variety is the plane $\omega = a\xi + b\eta = \mathbf{c} \cdot \mathbf{k}$.

We now focus on Example 1.1, of linearized isentropic Euler equations, which is representative of more general systems. On the one hand, the plane $\omega = \mathbf{u} \cdot \mathbf{k} = u\xi + v\eta$ (if $\xi = \frac{\omega - v\eta}{u}$, $i\xi$ is an eigenvalue of $\mathbf{D}(\eta, i\omega)$) corresponds to a constant group velocity $\mathbf{u}$ and thus to a translational motion. On the other hand, the cones correspond to group velocity $\gamma = \mathbf{u} \pm c\frac{\mathbf{k}}{|\mathbf{k}|}$ (propagation of sound waves). This can indeed be generalized in a rather simple way by using bicharacteristics, but we shall not go into details. Consider a cone $C(\omega = \mathbf{u} \cdot \mathbf{k} \pm c|\mathbf{k}|)$ and denote by Ξ the projection of this cone C onto the (η, ω)-plane (see Figure 1.5). When (η, ω) lies in the interior of Ξ, there are two corresponding points (ξ_j, η, ω), $j = 1, 2$, on

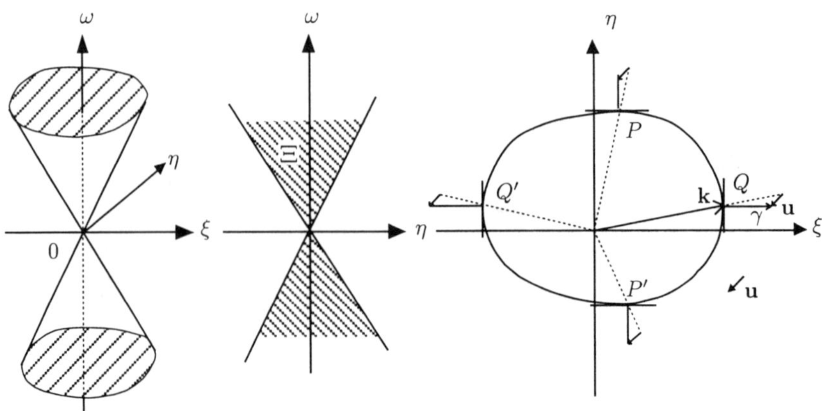

FIGURE 1.5. Characteristic cone and its projections.

the cone, where $\xi_j = \xi(\eta, \omega)$ is computed from

$$\omega = u\xi + v\eta \pm c(\xi^2 + \eta^2)^{1/2};$$

in that case $\mathbf{D}(\eta, i\omega)$ has two other imaginary eigenvalues $i\xi_j$. When (η, ω) approaches the boundary of Ξ, the corresponding points on the cone coalesce. Outside of Ξ, the eigenvalues are $\ell = i(\xi \pm i\rho) = \pm\rho + i\xi$ (this because $-\omega\mathbf{I} + \xi\mathbf{A} + \eta\mathbf{B}$ is real). When $\frac{\partial\omega}{\partial\eta} = 0$ (at points Q, Q' on Figure 1.5) the group velocity is tangent to the plane $y = 0$, corresponding to group velocity vectors grad ω pointing into or out of the domain (depending on whether $\frac{\partial\omega}{\partial\xi}$ is < 0 or > 0). On the other hand, when $\frac{\partial\omega}{\partial\eta} = 0$ (at points P, P'), the group velocity is tangent to the boundary $x = 0$. To understand the location of these points in Figure 1.5, note that the cross section of C has locally an equation of the form $\eta = \eta(\xi)$, and the points in the (ξ, η)-plane where $\frac{\partial\omega}{\partial\xi} = 0$ coincide with those where $\eta'(\xi) = 0$. Note also that in Figure 1.5 $\gamma = \mathbf{u} \pm c\frac{\mathbf{k}}{|\mathbf{k}|}$ is the sum of $\mathbf{u}$ and c times a unit vector $\frac{\mathbf{k}}{|\mathbf{k}|}$; the direction of $\mathbf{u}$ implies that in the present case $x = 0$ is an outflow boundary.

In short, the above considerations show that for (η, ω) in Ξ, each determination of $\xi = \xi(\eta, \omega)$ such that (ξ, η, ω) belongs to the cone C (i.e., such that $i\xi$ is an eigenvalue of $\mathbf{M}(\eta, i\omega)$) can be associated to a motion into ($\frac{\partial\omega}{\partial\xi} < 0$) or out of ($\frac{\partial\omega}{\partial\xi} > 0$) the spatial domain $x > 0$.

We have information on the behavior of the eigenvalues of $\mathbf{D}(\eta, s)$ as $\mathrm{Re} s = 0$. Now, we can think that in normal modes $\hat{\mathbf{u}}e^{\ell x + i\eta y - st}$, ℓ (which is an eigenvalue of $\mathbf{D}(\eta, s)$) and s (with $\mathrm{Re}(s) < 0$) are "perturbations" of $i\xi$ and $i\omega$ respectively:

$$\begin{aligned}
&i\xi \text{ is an eigenvalue of } \mathbf{D}(\eta, i\omega),\\
&\ell = \mathrm{Re}\ell + i\xi \ \text{ is an eigenvalue of } \mathbf{D}(\eta, s), \ \text{where } s = \mathrm{Re}(s) + i\omega.
\end{aligned}$$

We have $\ell = \ell(\eta, s)$ and, at least formally,

$$\frac{\partial\ell}{\partial s} = \frac{\partial\xi}{\partial\omega}.$$

One can then see, under some regularity assumption on the dependence of ℓ on s, that

$$\frac{\partial s}{\partial\ell} = \frac{\partial\omega}{\partial\xi}.$$

Therefore it appears that when s is "perturbed" so that $\mathrm{Re}(s) < 0$, the sign of $\mathrm{Re}(\ell)$ will indicate whether the mode corresponds to group velocity vectors pointing into ($\frac{\partial\omega}{\partial\xi} > 0$, thus $\mathrm{Re}(\ell) < 0$ since $\mathrm{Re}(s) < 0$) or out of ($\mathrm{Re}(\ell) > 0$, $\frac{\partial\omega}{\partial\xi} < 0$) the domain.

Informally speaking, identifying "incoming" and "outgoing" portions of the solution $\tilde{\mathbf{u}}$, obtained by means of Laplace and Fourier transform as a "superposition" of elementary modes $\hat{\mathbf{u}}e^{\ell x}e^{i\eta y - st}$, is linked to the sign of $\mathrm{Re}(\ell)$, ℓ an eigenvalue of $\mathbf{D}(\eta, s)$.

Now, there exists a matrix $\mathbf{Q} = \mathbf{Q}(\eta, s)$ such that $\mathbf{Q}^{-1}\mathbf{D}\mathbf{Q}(\eta, s)$ is made up of two Jordan blocks $\mathbf{D}_1(\eta, s)$ and $\mathbf{D}_2(\eta, s)$,

$$\mathbf{Q}\mathbf{D}\mathbf{Q}^{-1}(\eta, s) = \begin{pmatrix} \mathbf{D}_1 & \\ & \mathbf{D}_2 \end{pmatrix}, \tag{1.15a}$$

and such that the q eigenvalues ℓ of $\mathbf{D}_1$ (resp. $(p-q)$ eigenvalues of $\mathbf{D}_2$) have $\mathrm{Re}(\ell) < 0$ (resp. > 0). Let us set in (1.11)

$$\tilde{\mathbf{v}}(x, \eta, s) = \mathbf{Q}\tilde{\mathbf{u}} = (\tilde{\mathbf{v}}_1, \tilde{\mathbf{v}}_2)^T. \tag{1.15b}$$

Then, the differential equation (1.11) splits into

$$\begin{aligned} \frac{\partial \tilde{\mathbf{v}}_1}{\partial x} &= \mathbf{D}_1(\eta, s)\tilde{\mathbf{v}}_1, \\ \frac{\partial \tilde{\mathbf{v}}_2}{\partial x} &= \mathbf{D}_2(\eta, s)\tilde{\mathbf{v}}_2. \end{aligned}$$

The solutions $\tilde{\mathbf{v}}_j$, of the form $\hat{\mathbf{v}}_j \mathbf{e}^{\ell x}$, are associated to "incoming" (resp. "outgoing") waves since ℓ is an eigenvalue of $\mathbf{D}_1$ (resp. $\mathbf{D}_2$). Using the fact that $\mathbf{Q} = \mathbf{Q}(\eta, s)$ and that the Laplace–Fourier transform only acts on the function of the (y, t) variables, it happens that the boundary condition (1.10), $\mathbf{E}\mathbf{u}(0, y, t) = \mathbf{g}(y, t)$, becomes

$$\mathbf{E}\mathbf{Q}^{-1}\tilde{\mathbf{v}}(0, \eta, s) = \tilde{\mathbf{g}}(\eta, s).$$

We can also decompose the columns of $\mathbf{Q}^{-1}$ in two parts, $\mathbf{Q}^{-1} = (\mathbf{Q}_1, \mathbf{Q}_2)$, so that

$$\mathbf{Q}^{-1}\tilde{\mathbf{v}} = \mathbf{Q}_1\tilde{\mathbf{v}}_1 + \mathbf{Q}_2\tilde{\mathbf{v}}_2,$$

and the boundary conditions are transformed into

$$\mathbf{N}(\eta, s)\tilde{\mathbf{v}}_1 = -\mathbf{E}\mathbf{Q}_2\tilde{\mathbf{v}}_2 + \tilde{\mathbf{g}}$$

by definition (1.13) of $\mathbf{N}(\eta, s)$ (up to a linear change of dependent variables $\tilde{\mathbf{v}}_i$, which does not alter the class of "incoming" or "outgoing"). This hints that the uniform Kreiss condition can be interpreted as a uniform solvability condition, which enables one to solve for incoming dependent variables in terms of outgoing variables and boundary data.

Again, this is just an outline. In particular, tangential group velocities corresponding to waves moving tangent to the boundary, which we have completely skipped, are not so easily handled. For precise results and the nonconstant coefficient extension, we refer to Higdon (1986) and the references therein; see also Dutt (1990).

2 The nonlinear approach

2.1 Nonlinear equations

In this section, we introduce the reader to some results for a nonlinear system or equation, obtained without linearization. We shall not give any proof, and we refer to the cited papers for details.

Consider first the simple Burgers equation in the strip $0 < x < 1$, with boundary conditions given on $x = 0$ and $x = 1$. The problem is obviously more complex than in the linear case.

We take

$$\begin{cases} \dfrac{\partial u}{\partial t} + \dfrac{1}{2}\dfrac{\partial u^2}{\partial x} = 0, \\ u(x,0) = u_0(x), \\ u(0,t) = g(t), \\ u(1,t) = h(t). \end{cases}$$

According to the data, there may exist or not a weak solution satisfying the boundary data, and these data may be effectively necessary for the computation of the solution or not. Indeed, we leave as an exercise for the reader the computation of a solution corresponding to $u_0(x) = 0$, $g(t) = 1$, $h(t) = -1$; or $g(t) = -1$, $h(t) = 1$ and to $u_0(x) = 1$, $g(t) = -1$, $h(t) = 1$.

Kruzkov's existence and uniqueness result has been extended by Bardos LeRoux and Nedelec (1979) to the scalar multidimensional I.B.V.P. in a domain $\mathcal{O}$ of $\mathbb{R}^d$,

$$(2.1) \qquad \begin{cases} \dfrac{\partial u}{\partial t} + \displaystyle\sum_{i=1}^{d} \dfrac{\partial}{\partial x_i} f_i(u) = 0, \quad x \in \mathcal{O}, t \in (0,T), \\ u(\mathbf{x},0) = u_0(\mathbf{x}), \ \text{a.e. in } \mathcal{O}, \\ u(\mathbf{x},t) = 0, \ \text{on } \partial\mathcal{O} \times (0,T), \end{cases}$$

via the vanishing viscosity method (see the Introduction and G.R., Chapter II), by studying more closely the boundary integrals in the entropy inequality. We just mention their result, corresponding for simplicity to boundary conditions equal to zero. Note that the result supposes that one can define the trace of the entropy fluxes on the boundary $t = 0$ and on $\partial\mathcal{O} \times (0,T)$, which holds for instance for a function in $BV(\mathcal{O} \times (0,T))$ (the proof can be found in Giusti (1984); here, we assume that $\mathcal{O}$ is bounded and that u_0 belongs to $\mathbf{C}^2(\overline{\mathcal{O}})$.

Theorem 2.1

There exists a unique entropy solution u of (2.1), in the sense that u is the unique function in $BV(\mathcal{O} \times (0,T))$ satisfying for any test function

$\varphi \in \mathbf{C}_0^2(\overline{\mathcal{O}} \times [0,T[),\ \varphi \geq 0$

$$\int_0^T \int_{\mathcal{O}} \Big\{ |u-k| \frac{\partial \varphi}{\partial t} + \operatorname{sgn}(u-k) \sum_{i=1}^d (f_i(u) - f_i(k)) \frac{\partial \varphi}{\partial x_i} \Big\} d\mathbf{x}\, dt$$

$$+ \int_0^T \int_{\partial \mathcal{O}} \operatorname{sgn}(k) (\sum_{i=1}^d (f_i(0) - f_i(k)) \nu_i) \varphi(s,t) ds\, dt \geq 0,$$

where $\boldsymbol{\nu}$ is the unit outward normal to $\partial\mathcal{O}$ and

$$u(\mathbf{x}, 0) = u_0(\mathbf{x}) \ \textit{a.e. in } \mathcal{O}.$$

The solution is obtained as the limit of solutions of the following viscous equations (with $(\varepsilon > 0)$):

$$\begin{cases} \dfrac{\partial u_\varepsilon}{\partial t} + \displaystyle\sum_{i=1}^d \frac{\partial}{\partial x_i} f_i(u_\varepsilon) - \varepsilon \Delta u_\varepsilon = 0, & \mathbf{x} \in \mathcal{O}, t \in (0,T), \\ u_\varepsilon(\mathbf{x}, 0) = u_0(\mathbf{x}), \quad \mathbf{x} \in \mathcal{O}, \\ u_\varepsilon(\mathbf{x}, t) = 0, \quad \text{on } \partial\mathcal{O} \times (0,T). \end{cases}$$

The general case of nonzero boundary data g belonging to $\mathbf{C}^2(\partial\mathcal{O} \times (0,T))$ follows by translating the solution by $\tilde{g}$, where $\tilde{g}$ is an extension of g to $\mathcal{O} \times (0,T)$. The definition of a weak entropy solution in the theorem can be interpreted in the following terms. If the boundary data g are given, the value $u(x,t)$ the trace of u on the boundary at time t, must satisfy

$$(\operatorname{sgn}(u(x,t) - g(x,t)) \sum_{i=1}^d (f_i(u) - f_i(k)) \nu_i) \geq 0 \text{ a.e. on } \partial\mathcal{O} \times (0,T),$$

$\forall k \in I(u(x,t), g(x,t))$, where $I(a,b)$ denotes the interval $[\min(a,b), \max(a,b)]$. For a one-dimensional equation, this condition can be characterized in a rather simple way:

$$\frac{f(u) - f(k)}{u-k} \leq 0, \ \forall k \text{ between } u = u(0,t) \text{ and } g = g(t), \tag{2.2}$$

which says that the slope of the chord joining $(u, f(u))$ to $(k, f(k))$ is negative.

Example 2.1. Take again the above example of Burgers' equation with $u_0(x) = 1$, $g(t) = -1$, $h(t) = 1$. On the one hand, if we take the boundary data $g(t) = -1$ into account, obviously a difficulty appears at the boundary $x = 0$. On the other hand, if we do not consider the boundary data at $x = 0$, there is an infinity of weak continuous solutions. Indeed, consider for instance the functions:

$$u(x,t) = \begin{cases} \dfrac{(x+\alpha)}{t}, & x \le t-\alpha, \\ 1, \quad x \ge t-\alpha, \end{cases}$$

depending on a parameter $\alpha \in [0,1]$. For $t \le \alpha$, $u(x,t) = 1$ for all $x \ge 0$; for $t \ge \alpha$ the solution consists of a rarefaction and a constant state (see Figure 2.1). Let us study these solutions regarding the requirements (2.2) given by Theorem 2.1:
If $\alpha > 0$, for $t \le \alpha$, $I(u(0,t), g(t)) = [-1,1]$; for $t \ge \alpha$, $I(u(0,t), g(t)) = [-1, \frac{\alpha}{t}]$ and of course the boundary entropy condition (2.2) is not satisfied: for instance if $t \le \alpha$, $k \in [-1,1]$ (see Figure 2.1),

$$\frac{f(u(0,t) - f(k)}{1-k} > 0.$$

If $\alpha = 0$ the solution is such that the discontinuity at $x = 0$ indeed satisfies (2.2); $\forall k \in I(u(0,t), g(t)) = [-1,0]$,

$$\frac{f(u(0,t) - f(k)}{u-k} = \frac{f(0) - f(k)}{-k} \le 0.$$

It is thus the unique solution given by Theorem 2.1.

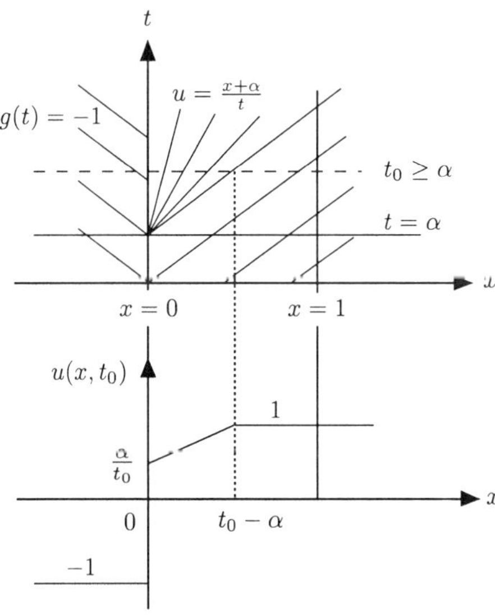

FIGURE 2.1. Example of I.B.V.P. for Burgers' equation.

If we consider the discontinuity connecting the state $u_L = -1 = g(t)$ to $u_R = u(0,t)$, the Rankine–Hugoniot condition gives

$$\sigma = \frac{f(u(0,t)) - f(g(t))}{u(0,t) - g(t)} = \frac{u(0,t) - 1}{2}.$$

Remembering Oleinik's entropy condition (see G.R., Chapter II, Lemma 6.1), that condition is obviously violated for any α, because $u(0,t) > g(t)$. However, for the solution given by the theorem, $\sigma < 0$ and the discontinuity curve leaves the domain immediately. □

2.2 Nonlinear system

The vanishing viscosity method has been used for one-dimensional systems

$$\frac{\partial \mathbf{u}}{\partial t} + \frac{\partial}{\partial x}\mathbf{f}(\mathbf{u}) = \mathbf{0}, \quad x > 0,\ t > 0, \tag{2.3a}$$

$$\mathbf{u}(x,0) = \mathbf{u}_0(x), \quad x > 0, \tag{2.3b}$$

in particular by Benabdallah and Serre (1987), Dubois and Le Floch (1988) and Gisclon and Serre (1994). The first authors define a set $\mathcal{E}(\mathbf{g})$ of admissible boundary values for which a boundary entropy inequality holds. The boundary condition is therefore written in the form $\mathbf{u}(0+,t) \in \mathcal{E}(\mathbf{g}(t))$, which stands in place of the strong nonhomogeneous Dirichlet condition $\mathbf{u} = \mathbf{g}$. Dubois and Le Floch also propose a second way of selecting admissible boundary conditions involving the resolution of Riemann problems. This leads to the introduction of another set of admissible boundary conditions $\mathbf{u}(0+,t) \in \mathcal{V}(\mathbf{g})$. The two kinds of boundary conditions are linked in general by

$$\mathbf{u} \in \mathcal{V}(\mathbf{g}) \Longrightarrow \mathbf{u} \in \mathcal{E}(\mathbf{g}),$$

and they coincide in some cases (scalar, linear systems, systems of the Temple class), but not always (see Benabdallah and Serre for a counterexample). Gisclon and Serre obtain a still different way of writing the boundary condition $\mathbf{u}(0,t) \in C(\mathbf{g}(t))$, with $C(\mathbf{g}) \subset \mathcal{E}(\mathbf{g})$. It is interesting to note that their "residual" boundary condition involves in general the viscosity matrix $\mathbf{B}$ of the perturbed system

$$\frac{\partial \mathbf{u}_\varepsilon}{\partial t} + \frac{\partial}{\partial x}\mathbf{f}(\mathbf{u}_\varepsilon) = \varepsilon \frac{\partial}{\partial x}\Big(\mathbf{B}(u_\varepsilon)\frac{\partial}{\partial x}\mathbf{u}^\varepsilon\Big).$$

Before entering into the description of these sets, we stress the point that, since very few results concerning the pure Cauchy problem are available (see the notes at the end of Chapter I), one cannot expect any general existence theorem concerning the I.B.V.P. (see however Li Ta-tsien 1994 for results in the case of a "reducible" 2×2 system and Li Ta-Tsien and Yu Wen-Ci 1985 for piecewise smooth solutions). The membership in $\mathcal{E}(\mathbf{g})$ or $\mathcal{V}(\mathbf{g})$ appears as a necessary condition to prove the existence of the corresponding

I.B.V.P. Only in some particular examples (either the scalar case, or linear systems, constant data, or p-system; see Benabdallah 1986, Dubroca and Gallice 1988) is this condition sufficient to prove existence and uniqueness. Other references can be found in Dubois and Le Floch (1988). Gisclon and Serre (1994) prove that their condition leads to a well, posed I.B.V.P., with convergence of the viscous solutions to the hyperbolic one.

Definition 2.1

For each state $\mathbf{g}$ *in* Ω, *the set* $\mathcal{E}(\mathbf{g})$ *of admissible values at the boundary is defined as the set* $\mathcal{E}(\mathbf{g})$ *of states* $\mathbf{u}$ *in* Ω *such that*

$$F(\mathbf{u}) - F(\mathbf{g}) - U'(\mathbf{g}) \cdot \{\mathbf{f}(\mathbf{u}) - \mathbf{f}(\mathbf{g})\} \leq 0 \tag{2.4}$$

for each entropy pair (U, F).

The "boundary condition" for the problem is then

$$\mathbf{u}(0+, t) \in \mathcal{E}(\mathbf{g}(t)), t > 0.$$

Let us now introduce another set, which is easier to deal with and directly applies to numerical approximation by finite-volume methods since it involves Riemann problems. Given a state $\mathbf{g} \in \Omega$, we consider the following set.

Definition 2.2

The set $\mathcal{V}(\mathbf{g})$ *of admissible values at the boundary is the set* $\mathcal{V}(\mathbf{g})$ *of states* $\mathbf{w}$ *in* Ω *such that*

$$\mathcal{V}(\mathbf{g}) = \{\mathbf{w}/\mathbf{w} = \mathbf{w}_R(0+; \mathbf{g}, \mathbf{u}_R), \mathbf{u}_R \in \Omega\}, \tag{2.5),}$$

where $\mathbf{w}_R$ *denotes the solution of the Riemann problem.*

We assume for simplicity that Ω is such that the Riemann problem has a solution (otherwise, we must add some restriction on $\mathbf{u}_R$). The corresponding boundary condition for the problem is then

$$\mathbf{u}(0+, t) \in \mathcal{V}(\mathbf{g}(t)), \quad t > 0.$$

Consider first the *scalar* convex case. If we assume that $\lim f(u) = +\infty$ as $u \to \pm\infty$, we can easily characterize $\mathcal{E}(g)$.

Proposition 2.1

Denote by $\overline{u}$ *the state where* f *is minimum. For a given* $g \neq \overline{u}$, *the set of "admissible" values* $\mathcal{E}(g)$ *is*

$$\mathcal{E}(g) = (-\infty, \overline{u}] \quad \textit{if } g \leq \overline{u},$$
$$\mathcal{E}(g) = (-\infty, \overline{g}] \cup \{g\} \quad \textit{if } g \geq \overline{u},$$

where $\overline{g}$ *denotes the solution* $\overline{g} \neq g$ *of* $f(\overline{g}) = f(g)$.

We also mention the fact that in the scalar (nonconvex) case, the two sets coincide.

Proposition 2.2

In the case of a scalar conservation law, for any g in Ω, the sets $\mathcal{V}(g)$ and $\mathcal{E}(g)$ coincide.

Example 2.1. (revisited). Let us illustrate the above results with Burgers' equation for which $\overline{u} = 0$, $\overline{g} = -g$. On the one hand, $g(t) = -1 \leq 0$, and the set of states u such that the chord has a negative slope is indeed $\mathcal{E}(g) = (-\infty, 0)$. On the other hand, if we look at all the limits $w_R(0+; -1, u_R)$, $u_R \in \Omega$ of the Riemann problems, we find :
(a) if $u_R \geq -1$, the solution of the Riemann problem is a rarefaction, and when u_R varies, $w_R(0_+; -1, u_R)$ lies in $[-1, 0]$,
(b) if $u_R < -1$, the solution of the Riemann problem is a shock, and when u_R varies, $w_R(0+; -1, u_R)$ lies in $(-\infty, -1[$, so that $\mathcal{V}(g) = (-\infty, -1] \cup [-1, 0] = (-\infty, 0]$ as expected.

Consider for instance the constant data $g \geq 0$. The proposition gives $\mathcal{V}(g) = (-\infty, -g] \cup \{g\}$. Let us interpret this result:
(i) If $u(0, t) < -g$, the trace of the solution u on the boundary $x = 0$ is "far" from the data; however, $u \in \mathcal{V}(g)$, and this outgoing wave $(u(0, t) \leq 0)$ is admissible: this is a truly nonlinear effect and could not have been obtained by linearization. The result is such that the solution of the Riemann problem (the discontinuity connecting the state $u_L = g$ to $u_R = u$ leaves the domain since it propagates with speed $\sigma = \frac{1}{2}(u + g) < 0$).
(ii) If u is "near" the data g, then the proposition yields that $u = g$: that could have been obtained by linearization. □

The characterization is also quite simple in the case of a strictly hyperbolic *linear system*; the sets $\mathcal{V}(\mathbf{g})$ and $\mathcal{E}(\mathbf{g})$ coincide, and we recover the formulation of Section 1.3.1. Denoting as previously by $p' = p - q$ the number of nonpositive eigenvalues of $\mathbf{A}$ ($a_i \leq 0$, $1 \leq i \leq p'$), we can state the following proposition.

Proposition 2.3

Assume that the system (2.3) is linear. For any $\mathbf{g}$ in $\mathbb{R}^n$,

$$\mathcal{V}(\mathbf{g}) = \mathcal{E}(\mathbf{g}) = \{\mathbf{u}, \ \exists(\alpha_i) \in \mathbb{R}^{p'}, \mathbf{u} = \mathbf{g} + \sum_{i=1}^{p'} \alpha_i \mathbf{r}_i\}. \tag{2.6}$$

Using Chapter I, Section 1, it is easy to check that $\mathcal{V}(\mathbf{g})$ is indeed $\{\mathbf{u} = \mathbf{g} + \Sigma_i \alpha_i \mathbf{r}_i\}$. For the other identity, we need the following lemma.

Lemma 2.1

For a strictly hyperbolic linear system, the pairs of entropy–entropy flux functions are given by

$$U(\mathbf{u}) = \sum_{i=1}^{p} \varphi_i(w_i), \quad F(\mathbf{u}) = \sum_{i=1}^{p} a_i \varphi_i(w_i)$$

for any arbitrary convex functions φ_i*, where* $\mathbf{w} = (w_i)$ *are the characteristic variables (1.7).*

This characterization of entropy–entropy flux pairs for a linear system is an easy consequence of Theorem 3.1 in Chapter I, Section 3 (which states that $U''\mathbf{A}$ is symmetric). For a detailed proof, we refer to Dubois and Le Floch (1988), Lemma 1.2.

Let us see that (2.6) gives another equivalent way of formulating (1.8) (i.e., $\mathbf{w}^I(0,t) = \mathbf{g}^I(t)$, where we recall that the subscript I (resp. II) corresponds to positive (resp. nonpositive) eigenvalues a_i, $\mathbf{w}^I = (w_{p'+1}, ..., w_p)^T$, $\mathbf{w}^{II} = (w_1, ..., w_{p'})^T$).

Indeed (2.6) says that $\mathbf{u}$ and $\mathbf{g}$ have the same components relative to the $p - p'$ positive eigenvalues (incoming characteristics).

If $\mathbf{g} \in \mathbb{R}^p$ is given and $\mathbf{u} \in \mathcal{V}(\mathbf{g})$, let $\mathbf{g}^I$ be the projection of $\mathbf{g}$ onto $\mathbf{r}_i$, $i = p' + 1, ..., p$, i.e.,

$$\mathbf{g} = \sum_{i=1}^{p} g_i \mathbf{r}_i, \quad \mathbf{g}^I = \sum_{i=p'+1}^{p} g_i \mathbf{r}_i.$$

Then (1.8) holds. Conversely, if $\mathbf{w}^I(0,t) = \mathbf{g}^I(t)$, then obviously $\mathbf{u} \in \mathcal{V}(\mathbf{g})$.

For a general *nonlinear system*, a first result regarding the well-posedness of the I.B.V.P. concerns the case where the initial data $\mathbf{u}_0(x) = \mathbf{u}_0$ and the boundary data $\mathbf{g}(t) = \mathbf{g}_0$ are constant; the I.B.V.P. is then defined by

$$\frac{\partial \mathbf{u}}{\partial t} + \frac{\partial}{\partial x}\mathbf{f}(\mathbf{u}) = \mathbf{0}, \quad x > 0,\ t > 0, \tag{2.7a}$$

$$\mathbf{u}(x,0) = \mathbf{u}_0, \quad x > 0, \tag{2.7b}$$

$$\mathbf{u}(0+,t) \in \mathcal{V}(\mathbf{g}_0), \quad t > 0. \tag{2.7c}$$

If the I.B.V.P. (2.7) is well posed, this means that $\mathbf{u}(x,t)$ has an extension as an entropy solution of a Riemann problem on $\mathbb{R} \times (0,\infty)$ in the "usual class" (discussed in Chapter I, Section 6). This extension is precisely $\mathbf{w}_R(\frac{x}{t}; \mathbf{g}_0, \mathbf{u}_0)$, and $\mathbf{u}(0,t)$ is the trace of this solution of the Riemann problem

$$\mathbf{u}(0,t) = \mathbf{w}_R(0+; \mathbf{g}_0, \mathbf{u}_0).$$

Theorem 2.2

Assuming that $\mathbf{u}_0$ *and* $\mathbf{g}_0$ *are constant, the problem (2.7) admits a unique entropy solution in the usual class of constant states separated by elementary waves.*

Example 2.2. $\mathcal{V}(\mathbf{g}_0)$ can be characterized in the particular case of the isentropic Euler system

$$\frac{\partial \rho}{\partial t} + \frac{\partial}{\partial x}(\rho u) = 0$$

$$\frac{\partial}{\partial t}(\rho u) + \frac{\partial}{\partial x}(\rho u^2 + p) = 0,$$

(see Chapter IV, Section 1, Example 1.1 with $v = 0$) where $p = k\rho^\gamma$, for which the rarefaction and shock curves can be explicitly computed, by studying closely the Riemann problem according to the position of $\mathbf{g}_0$ (sub- or supersonic, in- or outlet).

For instance, for supersonic inflow $\mathbf{g}_0 = (\overline{\rho}_0, \overline{q}_0 = \overline{\rho}_0\overline{u}_0)$, with $0 < \lambda_-(\mathbf{g}_0) = \overline{u}_0 - \overline{c}_0 < \lambda_+(\mathbf{g}_0) = \overline{u}_0 + \overline{c}_0$, one finds that $\mathbf{g}_0$ is the only supersonic inflow state ($u - c > 0$) in $\mathcal{V}(\mathbf{g}_0)$. Then, the boundary condition is $\mathbf{u}(0,t) = \mathbf{g}_0$, corresponding to the case where the linearization is valid. However, as we have already observed in the scalar case, a strong nonlinear behavior can be observed: $\mathcal{V}(\mathbf{g}_0)$ is reduced to $\{\mathbf{g}_0\}$ only near $\mathbf{g}_0$. Since the intersection of $\mathcal{V}(\mathbf{g}_0)$ with the subsonic states $\{\mathbf{u}, \quad -c \le u \le c\}$ is a part of the 1-shock curve $\mathcal{S}_1(\mathbf{g}_0)$, $\mathbf{u}(0,t)$ can be such that $\mathbf{g}_0$ is connected to $\mathbf{u}(0,t)$ by a 1-shock. Otherwise, since the intersection of $\mathcal{V}(\mathbf{g}_0)$ with the subsonic states $\{\mathbf{u}, \quad \lambda_-(\mathbf{u}) < \lambda_+(\mathbf{u}) < 0\}$ is locally of dimension 2, $\mathbf{u}(0,t)$ may even correspond to a supersonic outflow ($\mathbf{u}(0,t) = (\rho, \rho u)$ with $u + c < 0$). Again, we refer to Dubois and Le Floch (1988) for details. □

We also mention the recent result of Gisclon and Serre (1994) concerning a strictly hyperbolic system with nonzero eigenvalues. Since in the general case the definition of $C(g(t))$ needs further development, we refer to Gisclon (1994) for details.

3 Gas dynamics

Concerning physical problems and in particular gas dynamics, one has usually to distinguish between two types of boundary conditions:
actual boundary conditions: the boundary is that of the spatial (bounded) domain, which can be a fluid boundary, a solid boundary, or a free surface (a case we shall not consider);
artificial boundary conditions: when the spatial domain is unbounded (for instance, fluid flow in an exterior domain such as flow past an aerofoil, or in an interior infinite domain such as a channel), one limits the area of computation and introduces artificial boundaries. Then, arises the problem of specifing boundary data on this artificial boundary.

If we are given an external state $\mathbf{U}_\infty$ (uniform flow condition) at infinity, then, in view of the above analysis, one can think of two approaches. The usual one is linearization, which we shall detail later; one linearizes at this external state and "forces" some of the variables, say w, which is assigned the value w_∞ (for instance, supersonic inflow: $\mathbf{U} = \mathbf{U}_\infty$, supersonic outflow: no condition). This applies particularly to a stationary state computation, and the initial condition is then taken as $\mathbf{U}(\mathbf{x}, 0) = \mathbf{U}_\infty$.

One also can take into account nonlinear effects and compute the state at the boundary by solving Riemann problems.

One may derive absorbing (or radiation or nonreflecting) boundary conditions at the artificial boundary. The idea is that outgoing waves should be absorbed and not artificially reflected back into the flow from the computational boundary. We shall discuss some of these notions in Section 4.

Let us now detail the usual inflow and outflow boundary conditions obtained by linearization (see Oliger and Sundström 1978).

3.1 Fluid boundary (linearized approach)

The Euler system in one dimension linearized about a smooth state $\mathbf{U}_0$ is obtained by substituting $\mathbf{U} = \mathbf{U}_0 + \mathbf{U}'$ into the system (see Example 2.4, Chapter I) and neglecting terms of second order in $\mathbf{U}'$. We get, dropping the "prime,"

$$\frac{\partial \mathbf{U}}{\partial t} + \mathbf{A}(\mathbf{U}_0)\frac{\partial \mathbf{U}}{\partial x} = \mathbf{R}(\mathbf{U}, \mathbf{U}_0),$$

where the Jacobian matrix $\mathbf{A}$ is given in (3.34), Chapter III, and the source term $\mathbf{R}(\mathbf{U}, \mathbf{U}_0)$, which does not contain derivatives of $\mathbf{U}$, does not modify the boundary conditions, which are determined by the matrix $\mathbf{A}(\mathbf{U}_0)$.

The above considerations in Section 1.2 concern, a priori, the linearized characteristic variables (1.7), defined by $\mathbf{W} = \mathbf{T}^{-1}(\mathbf{U}_0)\mathbf{U}$, which satisfy the linear diagonal system

$$\frac{\partial \mathbf{W}}{\partial t} + \boldsymbol{\Lambda}(\mathbf{U}_0)\frac{\partial \mathbf{W}}{\partial x} = \mathbf{S}(\mathbf{W}, \mathbf{W}_0),$$

where $\boldsymbol{\Lambda}(\mathbf{U}_0) = \operatorname{diag}(u_0 - c_0, u_0, u_0 + c_0)$ and $c = \frac{\partial p}{\partial \rho}(\rho, s)$.

We decompose $\mathbf{W} = (\mathbf{w}^I, \mathbf{w}^{II})^T \in \mathbb{R}^q \times \mathbb{R}^{3-q}$, in incoming/outgoing variables, and one has to specify $\mathbf{w}^I$ at $x = 0$.

In practice, it is more convenient to prescribe "physical," i.e., measurable, variables such as pressure, velocity, or temperature. Thus, one can be interested in specifying as many conservative $\mathbf{u}^I$ or nonconservative variables $\mathbf{v}^I$. One has then to check that the corresponding boundary conditions on $\mathbf{W}$ lead to a well-posed problem. For instance, in the case $\mathbf{V}' = (\rho, u, p)^T$, $\mathbf{V}'$ is a solution of (with the notations of Chapter I, Remark 2.2)

$$\frac{\partial \mathbf{V}'}{\partial t} + \mathbf{B}'(\mathbf{V}_0')\frac{\partial \mathbf{V}'}{\partial x} = \mathbf{R}'(\mathbf{V}', \mathbf{V}_0'),$$

where $\mathbf{T}'^{-1}\mathbf{B}'\mathbf{T}' = \boldsymbol{\Lambda}$; setting

$$\mathbf{W}(\mathbf{V}') = \mathbf{T}'^{-1}(\mathbf{V}_0')\mathbf{V}',$$

the relation between $\mathbf{V}'$ and $\mathbf{W}(\mathbf{V}')$ is

$$\begin{aligned} \mathbf{W}(\mathbf{V}') &= \left(-\frac{\rho_0 u}{2c_0} + \frac{p}{2c_0^2}, \rho - \frac{p}{2c_0^2}, \frac{\rho_0 u}{2c_0} + \frac{p}{2c_0^2}\right)^T \\ &= \frac{1}{2c_0^2}(-\rho_0 c_0 u + p, 2(\rho c_0^2 - p), \rho_0 c_0 u + p)^T. \end{aligned} \tag{3.1}$$

Remark 3.1. If we take instead $\mathbf{V} = (\rho, u, s)^T$, $\mathbf{V}$ is a solution of (with the notations of Chapter I, Remark 2.1)

$$\frac{\partial \mathbf{V}}{\partial t} + \mathbf{B}(\mathbf{V}_0)\frac{\partial \mathbf{V}}{\partial x} = \mathbf{r}(\mathbf{V}, \mathbf{V}_0),$$

with $\mathbf{T}^{-1}\mathbf{B}\mathbf{T} = \boldsymbol{\Lambda}$; setting

$$\mathbf{W}(\mathbf{V}) = \mathbf{T}^{-1}(\mathbf{V}_0)\mathbf{V},$$

the relation between $\mathbf{V}$ and $\mathbf{W}(\mathbf{V})$ is

$$\mathbf{W}(\mathbf{V}) = \Big(\frac{\rho}{2\rho_0} - \frac{u}{2c_0} + \frac{p_s^0 s}{2\rho_0 c_0^2}, -\frac{s}{c_0^2}, \frac{\rho}{2\rho_0} + \frac{u}{2c_0} + \frac{p_s^0 s}{2\rho_0 c_0^2}\Big)^T.$$

If we introduce a linearized pressure p,

$$p = p_s^0 s + p_\rho^0 \rho = p_s^0 s + c_0^2 \rho,$$

we can write

$$\mathbf{W}(\mathbf{V}) = \frac{1}{2\rho_0 c_0^2}(p - \rho_0 c_0 u, -2\rho_0 s, p + \rho_0 c_0 u)^T.$$

We notice that the first and last components of $\mathbf{W}(\mathbf{V}')$ and $\mathbf{W}(\mathbf{V})$ coincide (up to a constant, which corresponds to the fact that we have not normalized the eigenvectors) but not the second one, though the equation $\frac{\partial w_2}{\partial t} + u_0 \frac{\partial w_2}{\partial x} = 0$ is of course satisfied by both $w_2(\mathbf{V}')$ and $w_2(\mathbf{V})$ □

Then, if we have a partition $\mathbf{V} = (\mathbf{v}^I, \mathbf{v}^{II}) \in \mathbb{R}^N \times \mathbb{R}^{p-N}$, the relation $\mathbf{W}(\mathbf{V}) = \mathbf{T}^{-1}(\mathbf{V}_0)\mathbf{V}$ can be written in matrix form,

$$\begin{pmatrix} \mathbf{w}^I \\ \mathbf{w}^{II} \end{pmatrix} = \begin{pmatrix} \mathbf{a}_I & \mathbf{a}_{II} \\ \mathbf{b}_I & \mathbf{b}_{II} \end{pmatrix} \begin{pmatrix} \mathbf{v}^I \\ \mathbf{v}^{II} \end{pmatrix},$$

with matrices that are rectangular ($\mathbf{a}_{II}, \mathbf{b}_I$) or square ($\mathbf{a}_I, \mathbf{b}_{II}$) and possibly empty, and this last relation yields

$$\begin{aligned} \mathbf{w}^I &= \mathbf{a}_I \mathbf{v}^I + \mathbf{a}_{II}\mathbf{v}^{II}, \\ \mathbf{w}^{II} &= \mathbf{b}_I \mathbf{v}^I + \mathbf{b}_{II}\mathbf{v}^{II}. \end{aligned}$$

Assume that the boundary data are "$\mathbf{v}^I$ prescribed." The condition to get admissible boundary conditions is first that the partition of $\mathbf{W}$ corresponds to incoming/outgoing variables. Hence, we must have $N = q$. Next, we want for $\mathbf{w}^I$ an expression of the form

$$\mathbf{w}^I(0, t) = \mathbf{S}^I \mathbf{w}^{II}(0, t) + \mathbf{g}^I(t);$$

this condition supposes that either $\mathbf{b}_{II}$ is invertible, in which case

$$\mathbf{w}^I = \mathbf{a}_{II}\mathbf{b}_{II}^{-1}\mathbf{w}^{II} + (\mathbf{a}_I - \mathbf{a}_{II}\mathbf{b}_{II}^{-1}\mathbf{b}_I)\mathbf{v}^I,$$

or $\mathbf{b}_{II}$ is empty and $\mathbf{w}^I = \mathbf{a}_I \mathbf{v}^I$. This should be checked in all the particular examples.

For a one-dimensional problem, assuming that the boundary is on the left of the domain (for instance, the domain is $x > 0$), we have four possible cases. We exclude the case $u_0 = 0$, which corresponds to a rigid wall boundary and will be considered later.
(1) Supersonic inflow, $u_0 > c_0$, $q = 3$, all three eigenvalues are positive: three boundary conditions, which means that the whole state must be prescribed.
(2) Subsonic inflow, $q = 2$ positive eigenvalues, $c_0 > u_0 > 0 > -c_0$: two conditions. A precise study (Oliger and Sundström 1978) gives that, in conservative variables, one can impose any pair from a wong $(\rho, \rho u, \rho e)$ or even any two linear combinations of $(\rho, \rho u, \rho e)$. In primitive variables, one can prescribe (ρ, u) or (ρ, p) but not the pair (u, p). Indeed, setting $\mathbf{v}^I = (u, p)$, $\mathbf{v}^{II} = \rho$, (3.1) gives $\mathbf{b}^{II} = 0$ and yields an ill-posed problem.
(3) Subsonic outflow, $q = 1$ positive eigenvalue, $0 > u_0 > -c_0$: one condition, which can be the density, the pressure or the velocity (in fact any combination of $(\rho, \rho u, \rho e)$ or (ρ, u, p)).
(4) Supersonic outflow $u_0 < -c_0$, $q = 0$: no conditions.

For a two-dimensional problem, in the domain, say, $x > 0$, $y \in \mathbb{R}$, $t > 0$, taking into account the fact that two eigenvalues collapse, the corresponding inflow–outflow problems lead to 4, 3, 1, 0 prescribed boundary conditions.

3.2 Solid or rigid wall boundary

The usual "slip boundary condition" is prescribed:

$$\mathbf{u} \cdot \mathbf{n} = 0, \tag{3.2}$$

which means that the flow does not cross the boundary but may move tangentially.

Remark 3.2. In the case of Maxwell's system (Example 1.2), the corresponding boundary condition would be

$$\mathbf{E} \times \mathbf{n} = 0, \tag{3.3}$$

the notation $\mathbf{E} \times \mathbf{n}$ denotes the vector $(E_y n_z - E_z n_y, E_z n_x - E_x n_z, E_x n_y - E_y n_x)^T$. Condition (3.3) means that the boundary is a perfect conductor. Then Faraday's law gives

$$\mathbf{B} \cdot \mathbf{n} = \mathbf{B}_0 \cdot \mathbf{n}$$

(see Dautray and Lions 1988, Chapter I, Part A, §4, Section 2.4.3). □

4 Absorbing boundary conditions

In general, a boundary condition is called "exact" if the boundary is "transparent," i.e., the (approximate) solution, in the finite domain with artificial boundary obtained with this boundary condition, coincides with the exact solution in the unbounded domain. The boundary condition is called "absorbing" if it yields the decreasing with time of some energy function. One also speaks of "radiation" or "nonreflecting" boundary conditions if they allow the wave motion to pass through the boundary of the domain without generating reflections back into the interior, or at least with a reduced amount of spurious reflection, but allow true physical reflections (see Hagstrom and Hariharan 1988; we refer to Givoli 1991 for a review of the problem).

If one uses normal mode analysis, one can derive exact or "perfectly absorbing" boundary conditions at normal incidence (or in dimension one) by imposing boundary conditions that annihilate the outgoing waves (or prevent the generation of incoming waves). Otherwise, one minimizes the amplitude of waves reflected from the artificial boundary (but the reflection coefficients depend on the incidence). Let us illustrate these ideas with a very simple example.

Example 4.1. Consider the wave equation which is often taken as the simplest example for illustrating absorbing boundary conditions:

$$\frac{\partial^2 \rho}{\partial t^2} - c^2 \frac{\partial^2 \rho}{\partial x^2} = 0, \quad c > 0.$$

The solution of this equation is a function of the form $f(x - ct) + g(x + ct)$, i.e., two waves traveling to the right (resp. to the left) with constant speed c (resp. $-c$). If we put an artificial boundary at $x = 0$ (domain $x > 0$), we want to let the wave $g(x + ct)$ that travels to the left leave the domain, and we do not want to let a wave enter the domain; thus, we do exclude waves of the form $f(x - ct)$. The perfectly absorbing condition is the Sommerfeld condition (see Givoli and Cohen 1995)

$$\frac{\partial \rho}{\partial t} - c \frac{\partial \rho}{\partial x} = 0 \text{ on } x = 0, \quad t > 0. \tag{4.1}$$

For an artificial boundary at $x = 1$ (domain $x < 1$), it would be

$$\frac{\partial \rho}{\partial t} + c \frac{\partial \rho}{\partial x} = 0 \text{ on } x = 1, \quad t > 0.$$

The motivation to study this equation here can be found in Chapter IV, Remark 2.4. The wave equation was obtained from the linear system

$$\frac{\partial \rho}{\partial t} + \frac{\partial}{\partial x}(\rho_0 u) = 0,$$
$$\frac{\partial}{\partial t}(\rho_0 u) + c_0^2 \frac{\partial \rho}{\partial x} = 0,$$

which we write

$$\frac{\partial \mathbf{u}}{\partial t} + \mathbf{A}\frac{\partial \mathbf{u}}{\partial x} = 0;$$

it is hyperbolic, with eigenvalues $-c_0$, c_0 and corresponding eigenvectors $\mathbf{r}_1 = (-1, c_0)^T$ and $\mathbf{r}_2 = (1, c_0)^T$, which give the columns of $\mathbf{T}$. The characteristic variables are $\mathbf{w} = \mathbf{T}^{-1}\mathbf{u}$, i.e., $w_1 = -c_0\rho + \rho_0 u$ and $w_2 = c_0\rho + \rho_0 u$, which thus satisfy a decoupled system

$$\frac{\partial w_1}{\partial t} - c_0\frac{\partial w_1}{\partial x} = 0,$$
$$\frac{\partial w_2}{\partial t} + c_0\frac{\partial w_2}{\partial x} = 0.$$

We can annihilate the rightward traveling waves by setting

$$w_2 = c_0\rho + \rho_0 u = 0. \tag{4.2}$$

Condition (4.2) implies (4.1) (with $c \equiv c_0$) since then

$$\frac{\partial \rho}{\partial t} - c_0\frac{\partial \rho}{\partial x} = \frac{\partial \rho}{\partial t} - c_0\Big(\frac{-\rho_0}{c_0}\Big)\frac{\partial u}{\partial x} = 0.$$

Conversely, if (4.1) is satisfied,

$$\frac{\partial w_2}{\partial t} = c_0\frac{\partial \rho}{\partial t} + \rho_0\frac{\partial u}{\partial t} = c_0^2\frac{\partial \rho}{\partial x} + \rho_0\frac{\partial u}{\partial t} = 0;$$

similarly, $\frac{\partial w_2}{\partial x} = 0$, and $w_2 = 0$ up to a constant.

Let us now use normal mode analysis as an introduction to the two-dimensional case that follows. We note that a mode $e^{ik(x+c_0t)}$, with phase velocity $-c_0$, travels out of the domain $x > 0$. We want to prevent waves of the form $e^{ik(x-c_0t)}$ from entering. By Fourier transform, if we define

$$\hat{\mathbf{u}}(x,\omega) = \int e^{+i\omega t}u(x,t)dt$$

(the + sign in $e^{+i\omega t}$ is taken here only for convenience), the linear system becomes

$$\frac{\partial \hat{\mathbf{u}}}{\partial x} = i\omega\mathbf{A}^{-1}\hat{\mathbf{u}}(x,\omega),$$

and diagonalizing $\mathbf{A}^{-1} = \mathbf{T}\Lambda^{-1}\mathbf{T}^{-1}$, setting $\hat{\mathbf{v}} = T^{-1}\hat{\mathbf{u}}$, it decouples into

$$\frac{\partial \hat{v}_1}{\partial x} = i\Big(\frac{\omega}{c_0}\Big)\hat{v}_1(x,\omega),$$
$$\frac{\partial \hat{v}_2}{\partial x} = -i\Big(\frac{\omega}{c_0}\Big)\hat{v}_2(x,\omega).$$

If we set

$$\hat{v}_1 = 0 \quad \text{at } x = 0, \tag{4.3}$$

(here the subscript 1 corresponds to the positive eigenvalue $+c$), there will be no incoming wave, since then

$$\mathbf{u} = \int e^{-i\omega t}\hat{\mathbf{u}}(x,\omega)d\omega = \int e^{-i\omega t}\mathbf{T}\hat{\mathbf{v}}(x,\omega)d\omega$$
$$= \int e^{-i\omega t - i\omega x/c_0}\hat{v}_2(0,\omega)\mathbf{r}_2 d\omega$$

is only a function of $x + c_0 t$. Thus, these approaches give the same result. They provide a perfectly absorbing condition in one dimension or at normal incidence in several dimensions since a wave $e^{ik(x+ct)}$ can be regarded as a wave in several dimensions traveling at normal incidence to the boundary $x = 0$. □

Example 1.2. (revisited). The same approach applies to the (homogeneous) Maxwell system in one dimension. Assuming slab symmetry, the variables depend only on z and t,

$$\mathbf{E} = \mathbf{E}(z,t), \quad \mathbf{B} = \mathbf{B}(z,t)$$

(Figure 4.1). E_z and B_z are independent of (z,t) (no propagation) and $(E_x, E_y), (B_x, B_y)$ satisfy (1.14 a),

$$\frac{\partial E_x}{\partial t} + c^2 \frac{\partial B_y}{\partial z} = 0,$$
$$\frac{\partial E_y}{\partial t} + c^2 \frac{\partial B_x}{\partial z} = 0,$$

together with (1.14 b)

$$\frac{\partial B_y}{\partial t} + \frac{\partial E_x}{\partial z} = 0,$$
$$\frac{\partial B_x}{\partial t} + \frac{\partial E_y}{\partial z} = 0.$$

The system can also be written in the characteristic form of two waves

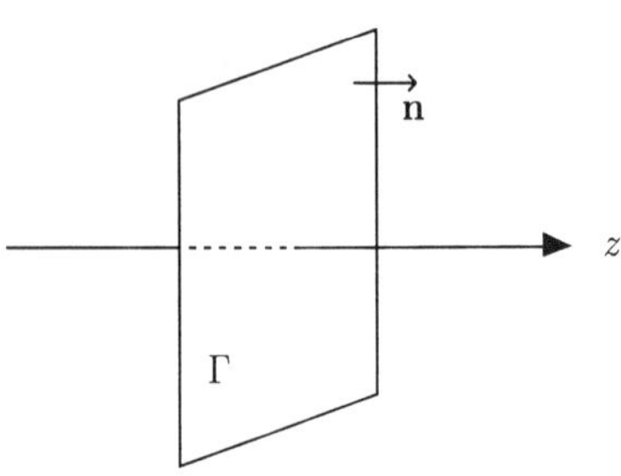

FIGURE 4.1. Boundary for Maxwell system with slab symetry.

propagating with characteristic velocity $\pm c$,

$$\frac{\partial}{\partial t}\left\{\begin{pmatrix} E_x \\ E_y \end{pmatrix} + c\begin{pmatrix} B_y \\ -B_x \end{pmatrix}\right\} + c\frac{\partial}{\partial z}\left\{\begin{pmatrix} E_x \\ E_y \end{pmatrix} + c\begin{pmatrix} B_y \\ -B_x \end{pmatrix}\right\} = \mathbf{0},$$

$$\frac{\partial}{\partial t}\left\{\begin{pmatrix} E_x \\ E_y \end{pmatrix} - c\begin{pmatrix} B_y \\ -B_x \end{pmatrix}\right\} - c\frac{\partial}{\partial z}\left\{\begin{pmatrix} E_x \\ E_y \end{pmatrix} - c\begin{pmatrix} B_y \\ -B_x \end{pmatrix}\right\} = \mathbf{0}.$$

Therefore, the incoming (resp. outgoing) wave corresponds to

$$\begin{pmatrix} E_x \\ E_y \end{pmatrix} - c\begin{pmatrix} B_y \\ -B_x \end{pmatrix} \quad \left(\text{resp. } \begin{pmatrix} E_x \\ E_y \end{pmatrix} + c\begin{pmatrix} B_y \\ -B_x \end{pmatrix}\right) = \mathbf{0}.$$

The perfectly absorbing boundary condition corresponds to no incoming wave (no reflected wave at the artificial boundary),

$$\begin{pmatrix} E_x \\ E_y \end{pmatrix} - c\begin{pmatrix} B_y \\ -B_x \end{pmatrix} = \mathbf{0} \Longleftrightarrow (\mathbf{E} - c\mathbf{B} \times \mathbf{n}) \times \mathbf{n} = \mathbf{0}$$

on the artificial boundary, which are called the Silver–Müller boundary conditions. When applied more generally, they yield perfect absorption for plane waves at normal incidence. We can also admit a given incoming wave

$$(\mathbf{E} - c\mathbf{B} \times \mathbf{n}) \times \mathbf{n} = \mathbf{e} \times \mathbf{n} \text{ on the artificial boundary,}$$

or equivalently

$$\left(\mathbf{B} + \left(\frac{1}{c}\right)\mathbf{E} \times \mathbf{n}\right) \times \mathbf{n} = \mathbf{b} \times \mathbf{n}.$$

See Cioni et al. (1993) for the numerical approximation. □

Consider now a linear, strictly hyperbolic system with constant coefficients, in dimension $d = 2$, in the whole space

$$\frac{\partial \mathbf{u}}{\partial t} + \mathbf{A}\frac{\partial \mathbf{u}}{\partial x} + \mathbf{B}\frac{\partial \mathbf{u}}{\partial y} = \mathbf{0};$$

for example, consider the case of linearized isentropic Euler equations (Example 1.1). We set the artificial boundary at $x = 0$ and assume that it is noncharacteristic; the interior of the computational domain corresponds to $x > 0$ and we thus recover problem (1.10 a). Applying a Fourier transform w.r.t. to y, t yields

$$\frac{\partial \hat{\mathbf{u}}}{\partial x} = \mathbf{A}^{-1}(i\omega\mathbf{I} - i\eta\mathbf{B})\hat{\mathbf{u}} = \mathbf{D}(\eta, i\omega)\hat{\mathbf{u}},$$

where $\hat{\mathbf{u}} = \hat{\mathbf{u}}(x, \eta, \omega) = \iint e^{-i\eta y}e^{+i\omega t}u(x, y, t)dy\,dt$, and following arguments similar to those of Section 1.3.3, we can derive "perfectly absorbing" boundary conditions by annihilating the wave entering the domain. Recall that $\mathbf{D}(0, i) = i\mathbf{A}^{-1}$ has distinct (purely imaginary) eigenvalues and for $|\frac{\eta}{\omega}| + |\omega|$ small enough $\mathbf{D}(\eta, i\omega)$ has also distinct purely imaginary eigenvalues (we have seen that it is the case if (η, ω) lies in Ξ, see Figure 1.5). $\mathbf{D}(\eta, i\omega)$ is diagonalizable in this neighborhood, which geometrically corresponds to near normal incidence; indeed, in the present situation (half-space problem), the pair $(\eta, \omega) = (0, 1)$ corresponds to normal incidence. Therefore, the eigenvalues of $\mathbf{D}(\eta, i\omega)$ are of the form $i\xi_k$, where $\xi = \xi(\eta, \omega)$ is real. Following (1.15), let us set

$$\hat{\mathbf{v}}(x, \eta, \omega) = \mathbf{Q}(\eta, i\omega), \qquad \hat{\mathbf{u}} = (\hat{\mathbf{v}}_1, \hat{\mathbf{v}}_2)^T,$$

where $\mathbf{Q}^{-1}(\eta, i\omega)$ is the matrix with columns the eigenvectors of $\mathbf{D}(\eta, i\omega)$ ($\mathbf{Q}$ is the matrix with rows the eigenvectors $\mathbf{l}_i^T$, $\mathbf{l}_i$ the eigenvectors of $\mathbf{D}^T$ and $\mathbf{QDQ}^{-1}(\eta, i\omega)$ is diagonal), and the eigenvalues of $\mathbf{D}_1$ (resp. $\mathbf{D}_2$) are such that sgn $\xi(\frac{\eta}{\omega}, 1) \geq 0$ (resp. ≤ 0). The system satisfied by $\hat{\mathbf{u}}$ is decoupled,

$$\frac{\partial \hat{\mathbf{v}}_i}{\partial x} = \mathbf{D}_i(\eta, i\omega)\hat{\mathbf{v}}_i,$$

and thus

$$\hat{v}_{1k}(x, \eta, \omega) = \exp(i\xi(\eta, \omega))\hat{v}_{1k}(0, \eta, \omega),$$

and $\hat{\mathbf{v}}_1$ corresponds to the incoming waves $\xi_k(\frac{\eta}{\omega}, 1) \geq 0$. Indeed the phase velocity of a wave $e^{+i\eta y - i\omega t + i\xi x}$, which is involved in the inverse Fourier transform $\mathbf{u} = \iint e^{+i\eta y - i\omega t}\hat{\mathbf{u}}(x, \eta, \omega) d\eta\, d\omega$, is the vector $\frac{(\omega\xi, \omega\eta)^T}{|\mathbf{k}|^2}$, and it enters the domain if sgn $\omega\xi \geq 0$ and thus is an incoming wave if sgn $(\omega\xi(\eta, \omega))$ = sgn $\xi(\frac{\eta}{\omega}, 1) \geq 0$. Then, the condition

$$\hat{\mathbf{v}}_1 = 0 \text{ at } x = 0 \tag{4.4}$$

is the "perfectly absorbing" boundary condition. It can be equivalently written

$$\pi_q \mathbf{Q}\hat{\mathbf{u}} = \mathbf{0},$$

where π_q denotes the projection

$$\pi_q(v_1, ..., v_p) = (v_1, ..., v_q)$$

on the coordinates related to the "incoming" eigenvalues of $\mathbf{D}(\eta, i\omega)$, i.e., as we have just seen, related to eigenvalues with positive imaginary part $\xi_k(\frac{\eta}{\omega}, 1) \geq 0$.

However, since the Fourier transform is used and $\mathbf{Q}$ is not constant (it depends on η, ω), this provides global conditions for $\mathbf{u}$. Therefore, local conditions (boundary values of the incoming variables $\mathbf{u}^I$ are given by a function of $\mathbf{u}^{II}$ at the same place and time) should be derived by some approximation w.r.t. $\frac{\eta}{\omega}$. Let $\mathbf{P}^{-1}(\eta, \omega)$ denote the matrix with columns the eigenvectors of $\mathbf{D}(\frac{\eta}{\omega}, i) = (\frac{1}{\omega})\mathbf{D}(\eta, i\omega)$ ($\mathbf{Q}^{-1}(\eta, \omega)$ denoted the matrix with columns the eigenvectors of $\mathbf{D}(\eta, i\omega)$). We assume that we can write near normal incidence (for $|\frac{\eta}{\omega}| + |\omega|$ small enough)

$$\mathbf{P}(\eta, \omega) = \mathbf{V}(0, 1) + \frac{\eta}{\omega}\frac{\partial}{\partial \eta}\mathbf{P}(0, 1) + O\Big(\frac{\eta}{\omega}\Big)^2,$$

where $\frac{\partial}{\partial\eta}\mathbf{P}$ denotes the derivative w.r.t. the first variable. The condition (4.4) yields

$$\pi_q\Big\{\mathbf{P}(0,1)+\frac{\eta}{\omega}\frac{\partial}{\partial\eta}\mathbf{P}(0,1)+O\Big(\frac{\eta}{\omega}\Big)^2\Big\}\hat{\mathbf{u}}=\mathbf{0}.$$

We get then first-order approximating boundary conditions

$$\pi_q^0\mathbf{P}(0,1)\hat{\mathbf{u}}=\mathbf{0}\quad\text{at } x=0,$$

which by inverse transform gives

$$\pi_q^0\mathbf{P}(0,1)\mathbf{u}=\mathbf{P}_I(0,1)\mathbf{u}=\mathbf{0}.$$

Here $\mathbf{P}_I$ denotes the matrix with lines $\mathbf{l}_i^T$ corresponding to the q eigenvectors of $\mathbf{A}^{-1}$, i.e., of $\mathbf{A}$ associated to positive eigenvalues.

Remark 4.1. We have previously denoted $\mathbf{P}(0,1)$ by $\mathbf{T}^{-1}$ (for instance in (1.8), Chapter III). Some elementary algebra gives

$$\mathbf{u}=\sum v_i\mathbf{r}_i=\mathbf{P}^{-1}\mathbf{v}=\mathbf{T}\mathbf{v}\ (\text{where } v_i=\mathbf{l}_i^T\mathbf{u})\Longrightarrow(v_1,...,v_q)=\mathbf{P}_I\mathbf{u},$$

or

$$\mathbf{l}_i^T\mathbf{u}=\mathbf{0}\text{ at } x=0.$$

$\mathbf{P}_I\mathbf{u}$ is the projection on the space spanned by the eigenvectors corresponding to "incoming" eigenvalues of $\mathbf{A}^{-1}$, i.e., to positive eigenvalues of $\mathbf{A}$. This is usually denoted as above with a projection operator, say π_q^0, which coincides with π_q if we assume that the eigenvalues do not vanish in the neighborhood. □

This means exactly that the characteristic variables corresponding to incoming characteristics are set to zero. We can also admit a given incoming wave, "$\mathbf{P}_I(0,1)\mathbf{u}$ given at $x=0$", and we recover the boundary condition (1.8).

A second-order approximation is

$$\pi_q^0\Big\{\omega\mathbf{P}(0,1)+\eta\frac{\partial}{\partial\eta}\mathbf{P}(0,1)\Big\}\hat{\mathbf{u}}=\mathbf{0},$$

to which we apply again the inverse transform; now the variables η,ω in the symbol correspond by inverse transform to partial differential operators, and we get

$$\pi_q^0\Big(\mathbf{P}(0,1)\frac{\partial u}{\partial t}+\Big(\frac{\partial}{\partial\eta}\mathbf{P}(0,1)\Big)\frac{\partial u}{\partial y}\Big)=0\text{ at } x=0.$$

Again, this is just an outline, and for more details concerning the derivation of these conditions we refer to Engquist and Majda (1977) and Kröner (1991). The first authors apply the theory to the particular example of the linearized shallow water equation, whereas the last author studies the linearized Euler system in dimension $d=2$ ($p=4$) and derives precisely the corresponding first-order and second-order absorbing boundary conditions by computing explicitly $\mathbf{V}(\eta, \omega)$. For instance, a subsonic outflow requires one condition, which often reduces to fixing the pressure at the boundary. However, it can reflect pressure disturbance back into the computational domain, and a first-order approximation of the nonreflecting boundary condition is $p - \overline{\rho c} u = 0$ ($p - \rho c$ is the incoming characteristic variable see Remark 1.2). See also Higdon , Gustafsson and Ferm (1979), Jiang and Wong (1990), Hagstrom and Hariharan (1988), Rudy and Strikwerda (1980), Bayliss and Turkel (1982).

Remark 4.2. In fact, Engquist and Majda do not work with the Fourier transform but with the corresponding differential operators. The matrix $\mathbf{D}(\eta, -i\omega) = -\mathbf{A}^{-1}(i\omega\mathbf{I} + i\eta\mathbf{B})$ (which they denote by $\mathbf{M}(\eta, \omega)$) corresponds to the symbol of the differential operator $\mathbf{A}^{-1}(\frac{\partial}{\partial t}) + \mathbf{A}^{-1}\mathbf{B}(\frac{\partial}{\partial y})$ that results from rewriting (1.10a) as

$$\frac{\partial \mathbf{u}}{\partial x} + \mathbf{A}^{-1}\Big(\frac{\partial \mathbf{u}}{\partial t}\Big) + \mathbf{A}^{-1}\mathbf{B}\Big(\frac{\partial \mathbf{u}}{\partial y}\Big) = \mathbf{0},$$

and then the matrices $\mathbf{D}_i$ in (1.15) are the symbols of differential operators, more precisely pseudo-differential operators, since the matrices depend on x, y. We refer to Engquist and Majda (1977), Section 2 for more details. □

The approach followed by Hedstrom (1979) and Thompson (1987) for a nonlinear hyperbolic system relies on characteristics. In one dimension (see Chapter I, Section 5), we have written the system in characteristic form,

$$\mathbf{l}_i^T(\mathbf{u})\Big\{\frac{\partial \mathbf{u}}{\partial t} + \lambda_i(\mathbf{u})\frac{\partial \mathbf{u}}{\partial x}\Big\} = 0.$$

Hedstrom's nonreflecting boundary condition can be written

$$\mathbf{l}_i^T(\mathbf{u})\frac{\partial \mathbf{u}}{\partial t} = 0, \text{ for all } i \text{ such that } \lambda_i(\mathbf{u}) > 0,$$

i.e., the amplitude of an incoming wave is constant with time at the boundary. If the system is linear, it means exactly that the characteristic variables corresponding to incoming waves are constant. In the nonlinear case, he shows that if there are only simple waves going out, this condition gives no wave coming into the domain fom the boundary $x = 0$. Otherwise, the strength of the reflected shock is of order 3. In the previous example of subsonic outflow, it gives $\frac{\partial p}{\partial t} - \rho c\frac{\partial u}{\partial t} = 0$. Such a boundary condition is found for instance in Cambier, Escande and Veuillot (1986).

Thompson has extended this condition to the multidimensional case

$$\frac{\partial \mathbf{u}}{\partial t} + \frac{\partial}{\partial x}\mathbf{f}(\mathbf{u}) + \frac{\partial}{\partial y}\mathbf{g}(\mathbf{u}) = \mathbf{0}. \tag{4.5}$$

Considering as previously a boundary $x = 0$ of the domain $x > 0$ and defining

$$\mathbf{L}_i(\mathbf{u}) = \mathbf{l}_i^T(\mathbf{u})\lambda_i(\mathbf{u})\frac{\partial \mathbf{u}}{\partial x},$$

where $\lambda_i(\mathbf{u}) = \lambda_i(\mathbf{u}, \mathbf{e}_1)$ are the eigenvalues of $\mathbf{A} = \mathbf{f}'$ and $\mathbf{l}_i^T(\mathbf{u})$ the eigenvectors of $\mathbf{A}^T$, the characteristic equations of the system (4.4) projected on the normal $\mathbf{e}_1$ to the boundary (see (1.5), Chapter IV , Section 1) are

$$\mathbf{l}_i^T(\mathbf{u})\frac{\partial \mathbf{u}}{\partial t} + \mathbf{L}_i(\mathbf{u}) = -\mathbf{l}_i^T(\mathbf{u})\mathbf{B}(\mathbf{u})\frac{\partial \mathbf{u}}{\partial y}. \tag{4.6}$$

The terms on the right-hand side contain derivatives in the direction transverse to the boundary and may be evaluated from values in the interior, as well as the terms corresponding to outgoing waves. There remains to specify "nonreflecting" boundary conditions that determine the values of $\mathbf{L}_i(\mathbf{u})$ for incoming waves ($\lambda_i(\mathbf{u}) > 0$) by requiring as above that the amplitude of an incoming wave remain constant in time. For subsonic outflow, the condition can be written $\frac{\partial p}{\partial t} - \rho c \frac{\partial u}{\partial t} = 0$, where u now represents the normal component of the velocity; we refer to Thompson (1987) for details, see also Vanajakshi et al. (1989), Poinsot and Lele (1992), Baum et al. (1995), Sun et al. (1995).

5 Numerical treatment

Concerning the numerical treatment, we have to distinguish between the given "analytical" boundary conditions that follow from the considerations of the preceding sections, and the "numerical" conditions required for computation: if a variable cannot be prescribed, its value may be needed in the computation, and eventually one must describe the way to compute it. For example, we have just seen that for subsonic inflow, two conditions, for instance (ρ, u), are specified, but one must still specify the value of p. Now, there are many ways of deriving a boundary treatment, so we just give some examples and the reader is asked to refer to the papers cited for details.

5.1 Finite difference schemes

For a finite difference scheme, in the boundary condition (1.8), $\mathbf{w}_0^{I,n} \sim \mathbf{w}^I(0, t_n)$ is prescribed (given by $\mathbf{g}^I(t_n)$), whereas the nonspecified value $\mathbf{w}_0^{II,n} \sim \mathbf{w}^{II}(0, t_n)$ must be computed from the values or from the differential equation in the interior of the domain: the value $\mathbf{w}^{II}(0, t_n)$ is therefore approximated either by extrapolation or by a finite difference discretization. Note that "$\mathbf{w}^I(0, t_n)$ is prescribed" means that the information for $\mathbf{w}^I(0, t)$ cannot be extracted solely from the differential equations in the interior of the domain as for $\mathbf{w}^{II}(0, t)$. If $\mathbf{g}^I(t)$ is not exactly known, as can be the case in some practical situations where no data are available, it must still be determined by additional information or else instability occurs (see Gustafsson 1982, Tadmor 1983).

Example 5.1. Consider a 3-point (linear) conservative scheme for approximating the simple advection equation (1.1) on the strip $0 < x < 1$. The boundary condition (see Section 1.1.1) is $u(0,t) = g(t)$ if $a > 0$, and no boundary condition at $x = 1$. In the formula

$$u_j^{n+1} = u_j^n - \lambda\{\varphi(u_j^n, u_{j+1}^n) - \varphi(u_{j-1}^n, u_j^n)\},$$

where $0 \le j \le N$, $N+1 = \frac{1}{\Delta x}$, $u_0^n = g(n\Delta t)$ is given by the boundary condition, whereas the value u_{N+1}^n is needed and should be computed from the interior values.

The simplest zeroth-order (locally first-order) extrapolation is

$$u_{N+1}^n = u_N^n \text{ (horizontal)},$$

and a first-order (or linear) extrapolation is

$$u_{N+1}^n = 2u_N^n - u_{N-1}^n.$$

Other possible oblique extrapolations are

$$\begin{aligned} u_{N+1}^n &= u_N^{n-1} \text{ (zeroth order) or} \\ u_{N+1}^n &= 2u_N^{n-1} - u_{N-1}^{n-2} \text{ (first order).} \end{aligned}$$

Otherwise, we can use the upwind scheme (first order)

$$u_{N+1}^{n+1} = u_{N+1}^n - \lambda c(u_{N+1}^n - u_N^n),$$

one can also use implicit schemes (Beam, Warming, and Yee 1982). The order of accuracy is studied as always by means of Taylor expansion.

If one uses a 5-point scheme, one still sets $u_0^n = g(t_n)$, and u_{-1}^n will be computed from the Taylor expansion w.r.t x, where the space derivatives are replaced by time derivatives of g, thanks to the equation (1.1),

$$\frac{\partial u}{\partial x}(0, t_n) = -\left(\frac{1}{a}\right) g'(t_n),$$

and so on, which gives for instance

$$u_{-1}^n = g(t_n) + \Delta x \frac{g'(t_n)}{c}.$$

One then needs u_{N+1}^n and u_{N+2}^n. If one takes for u_{N+1}^n and u_{N+2}^n the same formula, i.e., the same coefficients in the extrapolation or difference scheme, for example

$$\begin{aligned} u_{N+1}^n &= 2u_N^n - u_{N-1}^n \text{ and} \\ u_{N+2}^n &= 2u_{N+1}^n - u_N^n, \end{aligned}$$

the boundary conditions are called "translatory". □

Example 5.2. For a linear system, a usual approach introduces "compatibility relations". The boundary condition (1.8b)

$$\mathbf{w}^I(0,t) = \mathbf{S}^I \mathbf{w}^{II}(0,t) + \mathbf{g}^I(t)$$

provides q relations at the boundary. The $p-q$ other relations for $\mathbf{w}^{II}$ are obtained from the discretization of the $p-q$ differential equations in the characteristic outgoing variables

$$\frac{\partial \mathbf{w}^{II}}{\partial t} + \boldsymbol{\Lambda}_{II} \frac{\partial \mathbf{w}^{II}}{\partial x} = \mathbf{0}.$$

As above in the scalar case, one uses an upwind scheme for the spatial derivative and then computes $\mathbf{w}^{II}(0, t_{n+1})$ from $\mathbf{w}(0, t_n)$, where $\mathbf{w}^{II}(0, t_n)$ is given by the scheme and $\mathbf{w}^{I}(0, t_n)$ by $\mathbf{S}^{I}\ \mathbf{w}^{II}(0, t_n) + \mathbf{g}^{I}(t_n)$. □

Example 5.3. In two dimensions, for instance the linearized Euler system, one also uses the characteristic equations of the system projected on the outward normal $\boldsymbol{\nu}$ to the boundary (see Chapter IV, Section 1, (1.8))

$$\mathbf{l}_k^T(\mathbf{u}, \boldsymbol{\nu})\Big\{ \frac{\partial \mathbf{u}}{\partial t} + \lambda_k(\mathbf{u}, \boldsymbol{\nu}) \frac{\partial \mathbf{u}}{\partial \nu})\Big\} = S_k,$$

where the right-hand side S_k involves derivatives in the tangential direction $\boldsymbol{\nu}^{\perp}$ only. For the Euler system that is invariant by rotation, it is equivalent to consider a boundary of the type $x = 0$. One keeps only the equations corresponding to outgoing characteristics for the one-dimensional projected system (i.e., to positive eigenvalues). The equations are discretized, using the same upwind scheme as in the interior and without taking into account the boundary conditions, using extrapolations to estimate the derivatives or modifying the scheme so as to involve only interior points; the term S_k may be considered as known from values inside the domain. This gives a value noted, say, $\mathbf{u}^{\star}$; then $\mathbf{u}^{n+1}$ satisfies the "compatibility relations," which are

$$\mathbf{l}_k^T(\mathbf{u}^{\star}, \boldsymbol{\nu})\{\mathbf{u}^{\star} - \mathbf{u}^{n+1}\} = 0, \text{ if } \lambda_k(\mathbf{u}^{\star}, \boldsymbol{\nu}) \geq 0;$$

the other relations needed in order to compute $\mathbf{u}^{n+1}$ are given by the prescribed boundary data if $\lambda_k(\mathbf{u}^{\star}, \boldsymbol{\nu}) \leq 0$. The expression is often linearized by taking $\mathbf{l}_k^T(\mathbf{u}^n, \boldsymbol{\nu})$ instead of $\mathbf{l}_k^T(\mathbf{u}^{\star}, \boldsymbol{\nu})$. This approach is well suited for subdomain computations (Veuillot and Cambier 1983).

Note that if we take for $\mathbf{u}$ the primitive variables (ρ, u_ν, u_ξ, p), the expressions for $\mathbf{l}_i^T(\mathbf{u}, \boldsymbol{\nu})$ are very simple (see Cambier et al. 1985, 1986).

The same idea is used by Hagstrom and Hariharan (1988) for the numerical treatment of their "reflecting" boundary conditions. □

For what concerns stability, the relevance of these different boundary schemes results from the G.K.S. theory (for Gustafsson, Kreiss, Oliger). See Gustafsson (1982), Gustafsson and Kreiss (1972), Goldberg and Tadmor (1987) for convenient stability criteria; Trefethen (1984, 1985), Beam, Warming, and Yee (1982), Sod (1987), Daru and Lerat (1983). However, this stability analysis does not extend to the nonlinear case, and linear extrapolation may be unsatisfactory (Kamowitz 1988).

In fact, let us see that the finite-volume method offers a very simple treatment of the boundary via fluxes, which does not require the explicit computation of the nonprescribed variables.

5.2 Finite-volume approach

Assume that the boundary of the spatial domain $\mathcal{O}$ is made up of boundaries of the finite volumes (one uses a "body-fitted" grid; for another approach, the so-called Cartesian grid method, see Pember et al. (1995) for instance). Following the arguments of Chapter IV, Section 4.1.1, we have to integrate the system over a boundary element, say Ω_i, and thus to compute a flux through the boundary of Ω_i. We have already taken care of the parts of $\partial\Omega_i$ adjacent to another cell, and there remains to approximate the flux through an edge e that is on the boundary of $\mathcal{O}$ (Figure 5.1). We denote this flux by $\Phi(\mathbf{V}_i, \mathbf{V}_\infty)$ and describe its computation according to the different cases. The notation $\mathbf{V}_\infty$ is not significant: it refers to a state in the far field, which in fact will not always be the case.

5.2.1 Solid wall boundary

Taking into account the slip boundary condition $\mathbf{u} \cdot \mathbf{n} = 0$, we see that the only contribution to the exact flux comes from the pressure

$$\mathbb{F} \cdot \mathbf{n} = (0, p_b\mathbf{n}, 0)^T.$$

This pressure p_b at the body can be estimated differently according to the chosen method. In the case of a (i) node based scheme (cell vertex), set $p_b = p_i$. In the case of a (ii) center based scheme, given a state $\mathbf{V}_i$ in a

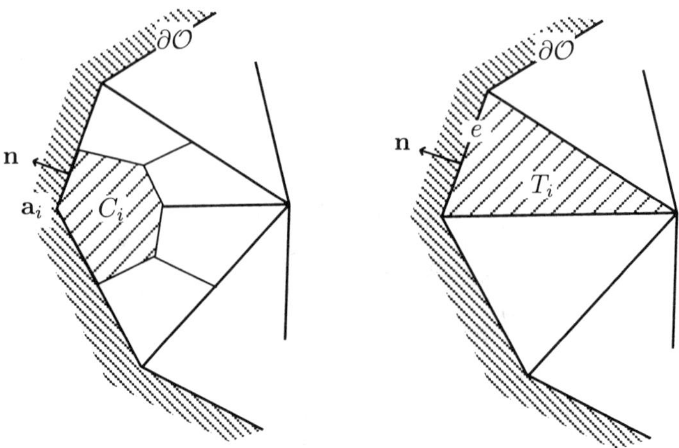

FIGURE 5.1. Boundary cell in the cell vertex and cell center approaches.

cell next to the boundary one, introduces a mirror state $\tilde{\mathbf{V}}_i$ with the same $\rho_i, p_i, \mathbf{u}_{\tau i} = \mathbf{u}_i \cdot \mathbf{n}^\perp$, only opposite normal velocity $\tilde{u}_n = -\mathbf{u} \cdot \mathbf{n}$, and then one solves the corresponding Riemann problem in an exact or approximate way (it can be viewed as an external state in a fictitious cell opposite to the edge e)

$$\Phi(\mathbf{V}_i, \mathbf{V}_\infty) \equiv \Phi(\mathbf{V}_i, \tilde{\mathbf{V}}_i). \tag{5.1}$$

The exact solution at $\xi = 0$ of the one-dimensional Riemann problem corresponding to $\mathbf{U}_L = \mathbf{U} = (\rho, -u, p)$ and $\mathbf{U}_R = \tilde{\mathbf{U}} = (\rho, u, p)$, with for instance $u > 0$ (and u not too large), is a 1-rarefaction connecting $\mathbf{U}$ to $\mathbf{U}^\star = (\rho_I, 0, p^\star)$, followed by a 2-contact discontinuity on $x = 0$ connecting to $(\rho_{II}, 0, p^\star)$, and a 3-rarefaction connecting to $\tilde{\mathbf{U}}$, (see Chapter II, Section 3). The one-dimensional Godunov flux gives

$$\mathbf{g}^{\mathrm{God}}(\mathbf{U}, \tilde{\mathbf{U}}) = (0, p^\star, 0)^T.$$

The expressions with Roe's scheme are also very simple, since Roe's average state is easily computed (see Chapter III, Section 3),

$$\overline{H} = H, \quad \overline{u} = 0, \quad \overline{c} = \kappa\Big(H - \frac{u^2}{2}\Big) = c, \quad \overline{\rho} = \rho.$$

Roe's flux gives

$$\mathbf{g}^{\mathrm{Roe}}(\mathbf{U}, \tilde{\mathbf{U}}) = (0, p_b, 0)^T,$$

where

$$p_b = p + \rho u^2 + \rho c u.$$

Thus, if on the right-hand side of (5.1) one uses Godunov's flux, we get for the boundary flux $\Phi^{\mathrm{God}}(\mathbf{V}_i, \tilde{\mathbf{V}}_i) = (0, p^\star \mathbf{n}, 0)^T$, where $p^\star$ is the pressure "on the boundary," i.e., the pressure of the above-mentioned intermediate state in the solution of the Riemann problem.

Now, if one uses Roe's flux, an easy computation (see Chapter IV, Section 4.3.1) shows that Roe's intermediate state is such that

$$\overline{\rho}_i = \rho_i, \quad \overline{H}_i = H_i, \quad \overline{u}_{\tau i} = \overline{\mathbf{u}}_i \cdot \mathbf{n}^\perp = u_{\tau i}, \quad \overline{u}_{ni} = \overline{\mathbf{u}}_i \cdot \mathbf{n} = 0,$$

$$\overline{c}_i = \kappa\Big(H_i - \frac{u_{\tau i}^2}{2}\Big),$$

and then one computes Roe's flux (using formula (3.11 b), (3.32), and (3.44) in Chapter III, which in the present case are very simple because $\Delta\rho = \Delta p = 0$); this yields

$$\Phi^{Roe}(\mathbf{V}_i, \tilde{\mathbf{V}}_i) = (0, p_b \mathbf{n}, 0)^T,$$

with

$$p_b = p_i + \rho_i u_{ni}^2 + \rho_i \overline{c}_i u_{ni}.$$

We might also use other schemes.

5.2.2 Fluid boundary

Be aware that $\mathbf{n}$ is the outward unit normal to e in Ω_i and thus in the case of a fluid boundary, in the above one-dimensional linearized study, $-\mathbf{n}$ gives the direction $x > 0$, inflow corresponds to $\mathbf{u} \cdot \mathbf{n} < 0$, outflow to $\mathbf{u} \cdot \mathbf{n} > 0$, and the prescribed boundary conditions to negative eigenvalues. The state $\mathbf{V}_\infty$ is a state in the far field, which represents the flow in a fictitious cell adjacent to the boundary that "satisfies the linearized boundary conditions" in the sense that for a subsonic inflow (for instance, stagnant gas taken from a reservoir expanding in a nozzle), total enthalpy and physical entropy of $\mathbf{V}_\infty$ are specified, for a subsonic outflow (for instance, flow in a tunnel exhausting into the atmosphere) the pressure is prescribed, for supersonic inflow (external flow), the whole state is given, whereas for supersonic outflow no condition is required on $\mathbf{V}_\infty$. The reason for specifying enthalpy H for a subsonic inflow is that H is constant along streamlines for steady adiabatic flow; thus, for a flow originating from a reservoir of common total enthalpy, the total enthalpy remains constant throughout the complete flowfield (see J.D. Anderson 1982, Chapter 6.4). Other problems may lead one to choose other quantities (velocity, temperature, pressure).

A very natural approach consists of Steger and Warming's flux-splitting formula (see Chapter IV, Section 4.3.5), slightly modified in the sense that the matrix is computed at the same state, $\mathbf{V}_i$, i.e.,

$$\Phi(\mathbf{V}_i, \mathbf{V}_\infty) = \mathbf{A}_n^+(\mathbf{V}_i)\mathbf{V}_i + \mathbf{A}_n^-(\mathbf{V}_\infty)\mathbf{V}_\infty$$

is replaced by

$$\Phi(\mathbf{V}_i, \mathbf{V}_\infty) = \mathbf{A}_n^+(\mathbf{V}_i)\mathbf{V}_i + \mathbf{A}_n^-(\mathbf{V}_i)\mathbf{V}_\infty,$$

which uses known information since, roughly speaking, negative eigenvalues correspond to prescribed boundary data. Note that for a supersonic inflow, $\mathbf{A}_\mathbf{n}^+ = 0$ and thus one "forces" in a weak sense $\mathbf{V}$ to take the value $\mathbf{V}_\infty$ (only the flux is prescribed, not the state, but the four dependent variables are involved), whereas for a supersonic outflow, $\mathbf{A}_\mathbf{n}^- = 0$,

$$\Phi(\mathbf{V}_i, \mathbf{V}_\infty) = \mathbf{A}_\mathbf{n}^+(\mathbf{V}_i)\mathbf{V}_i, \tag{5.2}$$

which corresponds to an extrapolation.

This linearized approach is valid for "weak" nonlinear effects, not for "strong" nonlinear ones. Assume for instance that in the last case of a supersonic outflow ($a_4(\mathbf{V}_\infty, -\mathbf{n}) = u_{-n\infty} + c_\infty = -u_{n\infty} + c_\infty \leq 0$), the internal state is subsonic ($a_4(\mathbf{V}_i, -\mathbf{n}) = u_{-ni} + c_i \geq 0$, $a_{2,3}(\mathbf{V}_i, -\mathbf{n}) \leq 0$) and thus does not correspond to a supersonic outflow. Then, one expects the extrapolation (5.2) to be unstable, and instead one introduces an intermediate "boundary" state. This state is obtained by again solving a Riemann problem following the approach of Osher and Chakravarthy (1983) and Section 2.2 above. It consists in solving a Riemann problem between an unknown left state $\mathbf{V}_0$ and the right state $\mathbf{V}_i$, so that we obtain a state on the boundary $\mathbf{V}_\infty = \lim \mathbf{w}_R(\xi; \mathbf{V}_0, \mathbf{V}_i)$ as $\xi \to 0$. It can also be implemented by introducing a fictitious cell on the other side of the boundary, where the state is $\mathbf{V}_0$.

More precisely, according to the (linearized) boundary conditions, the "state" $\mathbf{V}_\infty$ is only characterized by the fact that it belongs to some variety $\mathcal{V}$ with codimension q = number of specified conditions = number of positive eigenvalues. One solves the "Riemann problem" between this variety $\mathcal{V}$ (or rather the adherence $\overline{\mathcal{V}}$) and $\mathbf{V}_i$, i.e., one looks for a state $\mathbf{V}_0$ in $\overline{\mathcal{V}}$ and $q-1$ intermediate states such that $\mathbf{V}_0$ can be connected to $\mathbf{V}_i$ by a succession of at most q waves of the families corresponding to positive eigenvalues; then one sets

$$\Phi(\mathbf{V}_i, \mathbf{V}_\infty) \equiv \Phi(\mathbf{V}_i, \mathbf{V}_0).$$

On the right-hand side, the flux may be that of Godunov, Osher, and Mehlmann (Chapter III, Section 3.4). In the case of Godunov's flux, it means that we solve exactly a one-dimensional Riemann problem where the right state $\mathbf{V}_i$ is given, and the left state $\mathbf{V}_0$ is unknown but belongs to $\overline{\mathcal{V}}$, and the resulting flux through $x = 0$ provides the boundary flux.

For outflow and weak nonlinearities ($\mathbf{V}_i$ is "near" $\mathcal{V}$), the wave solutions of the Riemann problem are all in the same quadrant $\frac{x}{t} > 0$, so that there is only one state $\mathbf{V}_0$ in the quadrant $\frac{x}{t} \leq 0$, and one can speak of a boundary state $\mathbf{V}_0$.

In the above example (supersonic outflow with subsonic internal state Figure 5.2), one looks for one ($q = 0$) supersonic (or sonic) state $\mathbf{V}_0$ that can be connected to $\mathbf{V}_i$ by a 4-wave. Since $a_4(\mathbf{V}_0, -\mathbf{n}) = -u_{n0} + c_0 \leq 0 \leq a_4(\mathbf{V}_i, -\mathbf{n}) = -u_{-ni} + c_i$, we find that this wave is a 4-rarefaction (see Chapter II, Section 3, Lemma 3.1), and for $\mathbf{V}_0$ we get the unique sonic state connected to $\mathbf{V}_i$ by a 4-rarefaction ($\mathbf{V}_0$ is obtained by writing that the state is sonic $-u_{n0} + c_0 = 0$, and that the 4-Riemann invariants $\mathbf{u} \cdot \mathbf{n} - \ell$, $\mathbf{u} \cdot \mathbf{n}^\perp$ and s are constant across the 4-wave). Then, if Godunov's flux is

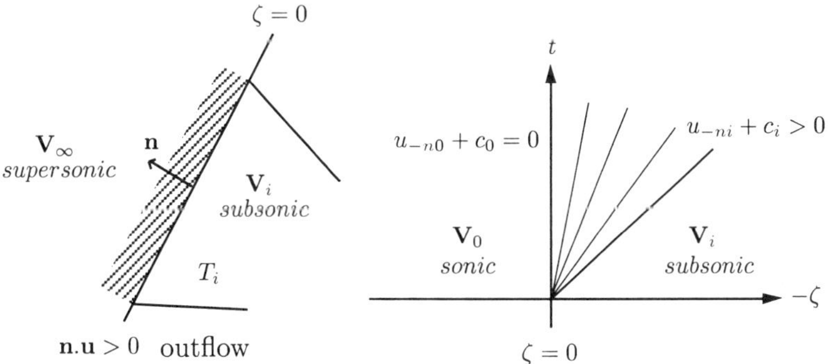

FIGURE 5.2. Supersonic outflow with subsonic internal state.

used,

$$\Phi(\mathbf{V}_i, \mathbf{V}_\infty) \equiv \mathbb{F} \cdot \mathbf{n}(\mathbf{V}_0).$$

For a subsonic outflow $a_{2,3}(\mathbf{V}_\infty, -\mathbf{n}) \leq 0 \leq a_4(\mathbf{V}_\infty, -\mathbf{n})$, the variety is $\mathcal{V} = \{\mathbf{U} = (\rho, \mathbf{u}, p) \ / \ p = p_\infty\}$. In the weak nonlinear case, $a_4(\mathbf{V}_i, -\mathbf{n}) \geq 0$, and we look for a state $\mathbf{V}_0$ with given pressure connected to $\mathbf{V}_i$ by a 4-rarefaction; again, the three 4-Riemann invariants are constant across the 4-simple wave, so that $\mathbf{V}_0$ is completely determined. Otherwise, $\mathbf{V}_i$ corresponds to a supersonic outflow $a_4(\mathbf{V}_i, -\mathbf{n}) \leq 0$, and the 4-wave can be a shock. The state $\mathbf{V}_0$ is obtained by taking in the (u,p)-plane the intersection of the 4-shock curve through $\mathbf{V}_i$ with the horizontal line $\{p = p_\infty\}$.

For details concerning the other cases and using Osher's scheme, see Dubois (1987, 1988, 1993).

Notes

We have already mentioned in each section of this chapter the contributions to which we refer (in particular, Higdon 1986, Yee 1987, Gustafsson et al. 1972, Engquist and Majda 1977, Hedstrom 1979, Thompson 1986, Bardos et al. 1979, Dubois and Le Floch 1988, Gisclon and Serre 1994, Cambier et al. (1985, 1986), to which we can add a nice (unpublished) report by Wu; see also Serre's book (1996) for theoretical results, and the papers of Sablé-Tougeron (1993), Asakura (1994); Yee (1981), Karni (1988), for the numerical treatment. The reader will find in the above papers many other important references that we did not quote. There are also more details concerning many parts of this chapter in the book of Hirsch (1990). Boundary conditions for reactive flows can be found in Baum et al. (1995), for magnetohydrodynamic flows in Sun et al. (1995), both following Thompson's approach.

Bibliography

This bibliography, though it is rather long, does not pretend to be exhaustive. An exhaustive survey indeed seems impossible, so important is the recent development of the subject from both the theoretical and numerical points of view. Moreover, any omissions are without prejudice.

We have not cited again all the references of our first book (in particular all those concerning only the scalar case), though many of them are still topical. We have quoted many more references in this book than in the first volume and also added some others in the notes at the end of each chapter. We mention a reference in the text mostly when a proof is not self-contained or when we have approached a topic only in an example or a remark. The contribution of many of the following authors to the subject is considerable, and we have made extensive use of some of the references cited, even if they are not always quoted in the text.

References

Abgrall, R. Généralisation du schéma de Roe pour le calcul d'écoulements de gaz à concentrations variables, *La Recherche Aérospatiale* 6 (1988), 31–43.

Preliminary results on an extrension of Roe's approximate Riemann solver to nonequilibrium flows, INRIA Research Report 987 (1989), INRIA Rocquencourt, 78153 Le Chesnay, France.

An extension of Roe's upwind scheme to algebraic equilibrium real gas models, *Comput. Fluids* 19 (2) (1991), 71–182.

Design of an Essentially Non-Oscillatory reconstruction procedure on finite-element type meshes, ICASE Report No 91–84 (1991), ICASE NASA Langley Research Center, Hampton, VA.

On Essentially Non-Oscillatory schemes on unstructured meshes: analysis and implementation, *J. Comp. Phys.* 114 (1994), 45–58.

A genuinely multidimensional Riemann solver, INRIA Research Report 1859 (1993), INRIA Rocquencourt, 78153 Le Chesnay, France.

Approximation du problème de Riemann vraiment multidimensionnel des équations d'Euler par une méthode de type Roe, *C.R. Acad. Sci. Paris* 319

Série I (1994), I : La linéarisation, 499–504; II : Solution du problème de Riemann approché, 625–629.
Approximation of the multidimensional Riemann problem in Compressible Fluid Mechanics by a Roe-type method, INRIA Research Report 2343 (1994), INRIA Rocquencourt, 78153 Le Chesnay, France (subm. to *SIAM J. Numer. Anal.*).

Abgrall, R., L. Fezoui, and J. Talandier. An extension of Osher's Riemann solver for chemical and vibrational nonequilibrium gas flows, *Int. J. Numer. Methods Engrg.* 14 (1992), 935–960.

Abgrall, R. and J.-L. Montagné. Généralisation du schéma d'Osher pour le calcul d'écoulements de mélanges de gaz à concentrations variables et de gaz réels, *La Recherche Aérospatiale*, 4 (1989), 1–13.

Adamczewski, M., J.-F. Colombeau, and A.-Y. Le Roux. Convergence of numerical schemes involving powers of the Dirac delta function, *J. Math. Anal. Applic.* 145 (1990), 172–185.

Ajmani, K., W.F. Ng, and M-S. Liou. Preconditionned conjugate gradient methods for the Navier–Stokes equations, *J. Comp. Phys.* 110 (1994), 68–81.

Alcrudo, F. and P. Garcia-Navarro. A high resolution Godunov-type scheme in finite volumes for the 2d shallow-water equations, *Int. J. Numer. Methods Fluids* 16 (1993), 489–505.

Alinhac, S. Existence d'ondes de raréfaction pour des systèmes quasi - linéaires hyperboliques multidimensionnels, *Comm. Part. Diff. Equations* 14 (2) (1989), 173–230.

Anderson, J.D. *Modern compressible flow with historical perspective*, McGraw-Hill series in mechanical engineering, McGraw-Hill, New York, 1982.

Anderson, W.K. A grid generation and flow solution method for the Euler equations on unstructured grids, *J. Comp. Phys.* 110 (1994), 23–38.

Angrand, F., V. Boulard, A. Dervieux, J. Périaux, and G. Vijayasundaram. Triangular finite element methods for the Euler equations, in *Computing Methods in Applied Sciences and Engineering, VI*, Proceedings of the Sixth International Symposium on Computing Methods in Applied Sciences and Engineering, France (1983), R. Glowinski, J.L. Lions (Eds.), Elsevier Science Publishers B.V. (North-Holland) INRIA (1984), 535–551.

Angrand, F. and F.C. Lafon. Flux formulation using a fully 2D approximate Roe Riemann solver, in *Nonlinear Hyperbolic Problems-Theoretical, Applied and Computational Aspects* Proceedings of the Fourth International Conference on Hyperbolic Problems, Taormina 1992, A. Donato and F. Oliveri (Eds.), Notes on Numerical Fluid Mechanics 43, Vieweg (1993), 15–22. et Rapport CEA (submitted to *J. Comp. Phys.*).

Arminjon, P. and A. Dervieux. Construction of TVD-like artificial viscosities on two-dimensional arbitrary FEM grids, *J. Comp. Phys.* 106 (1993), 176–198.

Arminjon, P., A. Dervieux, L. Fezoui, H. Steve, and B. Stoufflet. Non-oscillatory schemes for multidimensional Euler calculations with unstrutured grids, in *Nonlinear Hyperbolic Equations-Theory, Computation Methods, and Applications* (Proceedings of the Second International Conference on Nonlinear Hyperbolic Problems, Aachen 1988), J. Ballmann and R. Jeltsch (Eds.), Notes on Numerical Fluid Mechanics 24, Vieweg, Braunschweig (1989), 31–42.

Arminjon, P. and M.C. Viallon. Généralisation du schéma de Nessyahu–Tadmor pour une équation hyperbolique à deux dimensions d'espace, *C. R. Acad. Sci. Paris*, 320, série I (1995), 85–88.

Asakura, F. The initial boundary value problem for hyperbolic conservation laws, in *Hyperbolic problems: theory, numerics, applications*, Proceedings of the Fifth International Conference on Hyperbolic Problems, StonyBrook 1994, J. Glimm et al. (Eds.), World Scientific, Singapore (1996), 278–383.

Azevedo, A.V. and D. Marchesin. Multiple viscous profile Riemann solutions in mixed elliptic-hyperbolic models for flow in porous media, in *Nonlinear Evolution Equations that Change Type*, B.L. Keyfitz and M. Shearer (Eds.), IMA Volumes in Mathematics and its Applications 27, Springer-Verlag, New York (1991), 1–17.

Ball, J. A version of the fundamental theorem for Young measures, in *Proceedings of conference on partial differential equations and continuum models of phase transitions*, M. Rascle, D. Serre, and M. Slemrod (Eds.), Lecture Notes in Physics 359, Springer, Berlin (1989), 3–16.

Ballmann, J. and R. Jeltsch Editors. *Nonlinear Hyperbolic Equations - Theory, Computation Methods, and Applications*, Proceedings of the Second International Conference on Nonlinear Hyperbolic Problems, Aachen 1988, Notes on Numerical Fluid Mechanics 24, Vieweg, Braunschweig (1989).

Balsara, D.S. Riemann solver for relativistic hydrodynamics, *J. Comp. Phys.* 114 (1994), 284-297.

Baraille, D.R., G. Bourdin, F. Dubois, and A.-Y. LeRoux. Une version à pas fractionnaires du schéma de Godunov pour l'hydrodynamique, *C. R. Acad. Sci. Paris*, 314, Série I (1992), 147–152.

Bardos, C. Introduction aux problèmes hyperboliques non linéaires, in *Fluid dynamics - Varenna (1982)*, Lecture Notes in Mathematics 1047, Springer-Verlag, Berlin (1984), 1-74.

Une interprétation des relations existant entre les équations de Boltzmann, de Navier–Stokes et d'Euler à l'aide de l'entropie, *Matematica Aplicada e Comp.* 6 (1987), 97–117.

Different approach for the relation between the kinetic and the macroscopic equations, in *Nonlinear Hyperbolic Problems, Proceedings St. Etienne 1986*, C. Carasso, P.-A. Raviart, D. Serre (Eds.), Lecture Notes in Mathematics 1270, Springer-Verlag, Berlin (1987), 308–323.

Bardos, C., A.-Y. LeRoux, and J.-C. Nedelec. First order quasilinear equations with boundary conditions, *Comm. Part. Diff. Equations* 4 (9) (1979), 1017–1034.

Baum, M., T. Poinsot, and D. Thévenin. Accurate boundary conditions for multicomponent reactive flows, *J. Comp. Phys.* 116 (1995), 247–261.

Bayliss, A. and E. Turkel. Far field boundary conditions for compressible flows, *J. Comp. Phys.* 48 (1982), 182–199.

Beam, R.M. and R.F. Warming. An implicit finite-difference algorithm for hyperbolic systems in conservation-law form, *J. Comp. Phys.* 22 (1976), 87–110.

Beam, R.M., R.F. Warming, and H.C. Yee. Stability analysis of numerical boundary conditions and implicit difference approximations for hyperbolic equations, *J. Comp. Phys.* 48 (1982), 200–222.

Bell, J.B., P. Colella, and J.A. Trangenstein. Higher order Godunov methods for general hyperbolic systems of conservation laws, *J. Comp. Phys.* 82 (1989), 362–397.

Bell, J.B., C.N. Dawson, and G.R. Shubin. An unsplit higher order Godunov method for scalar conservation laws in multiple dimension, *J. Comp. Phys.* 74 (1988), 1–24.

Benabdallahh, A. Le p-système dans un intervalle, *C. R. Acad. Sci. Paris* 303, Sér. I, (1986), 123–126.

Benabdallahh, A. and D. Serre. Problèmes aux limites pour des systèmes hyperboliques non linéaires de deux équations à une dimenson d'espace, *C. R. Acad. Sci. Paris* 305, Sér. I, (1987), 677–680.

Ben-Artzi, M. The generalized Riemann problem for reactive flows, *J. Comp. Phys.* 81 (1989), 70–101.

Ben-Artzi, M. and A. Birman. Computation of reactive duct flows in external fields, *J. Comp. Phys.* 86 (1990), 225–255.

Ben-Artzi, M. and J. Falcovitz. A second order Godunov-type scheme for compressible fluid dynamics, *J. Comp. Phys.* 55 (1984), 1–32.

Benharbit, S. *Sur la théorie et l'approximation numérique des problèmes hyperboliques non linéaires. Application à la dynamique des gaz compressible*, Doctoral dissertation, Thesis Université J. Fourier, Grenoble, France (1992).

Benharbit, S., A. Chalabi, and J.P. Vila. Numerical viscosity and convergence of finite volume methods for conservation laws with boundary conditions, in *Nonlinear Hyperbolic Problems-Theoretical, Applied and Computational Aspects*, Proceedings of the Fourth International Conference on Hyperbolic Problems, Taormina 1992, A. Donato and F. Oliveri (Eds.), Notes on Numerical Fluid Mechanics 43, Vieweg, Braunschweig (1993), 48–55, and *SIAM J. Numer. Anal.* 32, 3 (1995), 775–796.

Benzoni-Gavage, S. and D. Serre. Compensated compactness for a class of hyperbolic systems of p conservation laws with $p \geq 3$, in *Progress in partial differential equations: the Metz surveys 2*, M. Chipot (Ed.), Pitman Research Notes in Mathematics Series 296, Longman Scientific & Technical, Harlow, UK. (1993), 3–11.

Berde B. and M. Borrel. Comparison of high-order Godunov-type schemes for the Euler equations on irregular meshes, *Eccomas-94* (Proceedings of the European conference, Stuttgart) (1994).

Bereux, F. and L. Sainsaulieu. Un schéma numérique de type Roe pour les systèmes hyperboliques avec relaxation, *C. R. Acad. Sci. Paris* 320 (1995), 379–384.

Berger, F. and J.-F. Colombeau. Numerical solutions of one-pressure models in multifluid flows, *SIAM J. Numer. Anal.* 32 (1995), 1139–1154.

Bermudez, A. and M.E. Vazquez. Upwind methods for hyperbolic conservation laws with source terms, *Comp. Fluids* 23 (1994), 1049–1071.

Beux, F., S. Lanteri, A. Dervieux, and B. Larrouturou. Upwind stabilization of Navier–Stokes solvers, INRIA Research Report 1885 (1993), INRIA Rocquencourt, 78153 Le Chesnay, France.

Billet, G. Finite difference scheme with dissipation control joined to a generalization of van Leer flux splitting, in *Nonlinear Hyperbolic Equations - Theory, Computation Methods, and Applications* (Proceedings of the Second International Conference on Nonlinear Hyperbolic Problems, Aachen 1988), J. Ballmann and R. Jeltsch (Eds.), Notes on Numerical Fluid Mechanics 24, Vieweg, Braunschweig (1989), 11–20.

Billey, V., J. Périaux, P. Perrier, and B. Stoufflet. 2-D and 3-D Euler computations with finite element methods in aerodynamics, in *Nonlinear Hyperbolic Problems, Proceedings St. Etienne 1986*, C. Carasso, P.-A. Raviart, D. Serre (Eds.), Lecture Notes in Mathematics 1270, Springer-Verlag, Berlin (1987), 64–81.

Blunt, M. and B. Rubin. Implicit flux limiting schemes for petroleum reservoir simulation, *J. Comp. Phys.* 102 (1992), 194–210.

Boillat, G. Chocs caractéristiques, *C.R. Acad. Sci. Paris* 274, Série A (1972), 1018–1021.

Böing, H., K. Werner, and J. Jackisch. Construction of the entropy solution of hyperbolic conservation laws by a geometrical interpretation of the conservation principle, *J. Comp. Phys.* 95 (1991), 40–58.

Bouchut, F., C. Bourdarias, and B. Perthame. Un exemple de méthode MUSCL satisfaisant toutes les inégalités d'entropie numériques, *C. R. Acad. Sci. Paris* 317 (1993), 619–624.

Boukadida, T. and A.-Y. LeRoux. A new version of the two-dimensional Lax–Friedrichs scheme, *Math. Comp.* 63 (1994), 541–553.

Bourdel, F., Ph. Delorme, and P.A. Mazet. Convexity in hyperbolic problems. Application to a discontinuous Galerkin method for the resolution of the polydimensional Euler equations, in *Nonlinear Hyperbolic Equations - Theory, Computation Methods, and Applications* (Proceedings of the Second International Conference on Nonlinear Hyperbolic Problems, Aachen 1988), J. Ballmann and R. Jeltsch (Eds.), Notes on Numerical Fluid Mechanics 24, Vieweg, Braunschweig (1989), 31–42.

Bourgeade, A., Ph. Le Floch, and P. A. Raviart. An asymptotic expansion for the solution of the generalized Riemann problem. Part 2 : applicaton to the equations of gas dynamics, *Ann. Inst. H. Poincaré, Nonlinear analysis* 6 (6) (1989), 437–480.

Bourgeat, A. and J. Koebbe. Minimization of grid orientation effects in simulation of oil recovery processes through the use of an unsplit higher order scheme, Preprint, Publications de l'équipe d'Analyse numérique de Lyon-St Etienne 139 (1992), France.

Brenier, Y. Averaged multivalued solutions for scalar conservation laws, *SIAM J. Numer. Anal.* 21 (1984), 1013–1037.

Systèmes hyperboliques de lois de conservation, cours de DEA d'Analyse Numérique (1992-93), Université Pierre et Marie Curie, Paris (France).

Brenier, Y. and S. Osher. Approximate Riemann solvers and numerical flux functions *SIAM J. Numer. Anal.* 23 (1986), 259–273.

Bressan, A. and R.M. Colombo. The semi-group generated by 2×2 conservation laws, *Arch. Rat. Mech. Anal.* 133 (1995), 1–75.

Brio, M. Admissibility conditions for weak solutions of nonstrictly hyperbolic systems, in *Nonlinear Hyperbolic Equations - Theory, Computation Methods, and Applications* (Proceedings of the Second International Conference on Nonlinear Hyperbolic Problems, Aachen 1988), J. Ballmann and R. Jeltsch (Eds.), Notes on Numerical Fluid Mechanics 24, Vieweg, Braunschweig (1989), 43–50.

Brio, M. and C. C. Wu. An upwind differencing scheme for the equations of ideal magnetohydrodynamics, *J. Comp. Phys.* 75 (1988), 400–422.

Buffard, T. *Analyse de quelques méthodes de volumes finis non structurés pour la résolution des équations d'Euler*, Doctoral dissertation, Thesis Université Paris 6, France (1993).

Buffard, T. and J. M. Hérard. Un schéma conservatif à pas fractionnaires pour résoudre les équations d'Euler en maillage non structuré, *C. R. Acad. Sci. Paris* 316 (1993), 575–582.

Euler solvers and fractional step methods, in *Finite elements in fluids, New trends and applications*, Proceedings of the eighth International Conference on Finite Elements in fluids, Barcelona, Spain (1993), K. Morgan et al. (Eds.), CIMNE, Pineridge Press (1993), 319–328.

Bukiet, B. Application of front tracking to two-dimensional curved detonation front, *SIAM J. Sci. Statist. Comput.* 9 (1988), 80–99.

Cambier, L., F. Dusson, and J.-P. Veuillot. Méthode multi-domaines pour les équations d'Euler. Applications pour des sous-domaines avec recouvrement, *La Recherche Aérospatiale* 3 (1985), 181–188.

Cambier, L., B. Escande, and J.-P. Veuillot. Calculs d'écoulements internes à grand nombre de Reynolds par résolution numérique des équations de Navier–Stokes, *La Recherche Aérospatiale* 6 (1986), 415–432.

Canic S. and B.J. Plohr. Shock wave admissibility for quadratic conservation laws, *J. Diff. Equations* 118 (1995), 293–335.

Carasso, C., P.-A. Raviart, and D. Serre (Editors). Nonlinear Hyperbolic Problems, Proceedings St. Etienne 1986, Lecture Notes in Mathematics 1270, Springer-Verlag, Berlin (1987).

Cargo, P. and G. Gallice. Un solveur de Roe pour les équations de la magnétohydrodynamique, *C. R. Acad. Sci. Paris* 320, Série I (1995), 1269–1272.

Casper, J. and H.L. Atkins. A finite-volume high-order ENO scheme for two-dimensional hyperbolic systems, *J. Comp. Phys.* 106 (1993), 62–76.

Cauret, J.J., J.F. Colombeau, and A.-Y. LeRoux. Discontinuous generalized solutions of nonlinear nonconservative hyperbolic equations, *J. Math. Anal. Applications* 139 (1989), 552–573.

Cercignani, C. *The Boltzmann Equation and its Applications*, Applied Mathematical Sciences 67, Springer-Verlag, New York (1988).

Cercignani, C., R. Illner, and M. Pulvirenti. *The mathematical theory of dilute gases*, Applied Mathematical Sciences 106, Springer-Verlag, New York (1994).

Chakravarthy, S.R. and S. Osher. Computing with high resolution upwind schemes for hyperbolic equations, in *Large-scale computations in Fluid Mechanics*, Proceedings of AMS/SIAM Summer Seminar, La Jolla, CA (1983), Lectures in Applied Mathematics 22, Part 1, B. Engquist, S. Osher and R.C.J. Sommerville (Eds.), AMS, Providence, RI, (1985), 57–86.

Chalabi, A. Stable upwind schemes for hyperbolic conservation laws with source terms, *IMA J. Numer. Anal.* 12 (1992), 217–241.

Approximation de lois de conservation non homogènes via une approximation affine du flux, *C. R. Acad. Sci. Paris* 317 (1993), 1079–1082.

Chalabi, A. and J. P. Vila. Eclatement d'opérateur, méthode à pas fractionnaires et condition d'entropie pour les systèmes hyperboliques non linéaires, Internal report (1992), Université Paul Sabatier Toulouse (France).

Champier, S. and T. Gallouët. Convergence d'un schéma décentré amont sur un maillage triangulaire pour un système hyperbolique linéaire, *M^2AN, Modélisation mathématique et analyse numérique* 26 (1992), 835–853.

Champier, S., T. Gallouët, and R. Herbin. Convergence of an upstream finite-volume scheme for a nonlinear hyperbolic equation on a triangular mesh, *Numerische Mathematik* 66 (1993), 139–158.

Chang, T. and L. Hsiao. *The Riemann problem and interaction of waves in gas dynamics*, Pitman Monographs and Surveys in Pure and Applied Mathematics 41, Longman Scientific and Technical, Harlow, UK. 1989.

Chang, C.L. and C.L. Merkle. The relation between flux vector splitting and parabolized schemes, *J. Comp. Phys.* 80 (1989), 344–361.

Chargy, D., R. Abgrall, L. Fezoui, and B. Larrouturou. Comparisons of several upwind schemes for multicomponent inviscid flows, INRIA Research Report 1253 (1990), INRIA Rocquencourt, 78153 Le Chesnay, France. Conservative numerical schemes for multicomponent inviscid flows, *Recherche Aérospatiale (English edition)* 1992-2 (1992), 61–79.

Charrier, P., B. Dubroca, and L. Flandrin. Un solveur de Riemann approché pour l'étude d'écoulements hypersoniques bidimensionnels, *C. R. Acad. Sci. Paris* 317 , Sér. I, (1993), 1083–1086.

Charrier, P. and B. Tessieras. On front-tracking methods applied to hyperbolic systems of nonlinear conservation laws, *SIAM J. Numer. Anal.* 23 (1986), 461–472.

Chen, G.-Q. The method of quasi-decoupling for discontinuous solutions of conservation laws, *Arch. Rat. Mech. Anal.* 121 (1992), 131–185.

Chen, G.-Q. and J. Glimm. Shock capturing and global solutions to the compressible Euler equations with geometrical structure, in *Hyperbolic problems: theory, numerics, applications*, Proceedings of the Fifth International Conference on Hyperbolic Problems, StonyBrook 1994, J. Glimm et al. (Eds.), World Scientific, Singapore (1996), 101–109.

Chen, G.-Q. and P.G. Le Floch. Entropy flux-splittings for hyperbolic conservation laws. Part I: General framework, *Comm. Pure Appl. Math.* 48(7) (1995), 691–729.

Chen, G.-Q., C.D. Levermore, and T.P. Liu. Hyperbolic conservation laws with stiff relaxation terms and entropy, *Comm. Pure. Appl. Math.* 47 (6) (1994), 787–830.

Chen, G.Q. and J.G. Liu. Convergence of second order schemes for isentropic gas dynamics, *Math. Comp.* 61 (1993), 607–627.

Chen, G.Q. and D. H. Wagner. Large time, weak solutions to reacting Euler equations, in *Nonlinear Hyperbolic Problems-Theoretical, Applied and Computational Aspects*, Proceedings of the Fourth International Conference on Hyperbolic Problems, Taormina 1992, A. Donato and F. Oliveri (Eds.), Notes on Numerical Fluid Mechanics 43, Vieweg, Braunschweig, (1993), 144–149.

Chéret, R. *La Détonation des Explosifs Condensés*, Tome 1, Série scientifique, Collection CEA, Masson Paris (1988).

Chern, I.L., J. Glimm, O. McBryan, B. Plohr, and S. Yaniv. Front tracking for gas dynamics, *J. Comp. Phys.* 62 (1986), 83–110.

Chévrier, P. and H. Galley. A van Leer finite volume scheme for the Euler equations on unstructured meshes, *M^2AN, Modélisation mathématique et analyse numérique*, 27 (2) (1993), 183–201.

Chorin, A.J. Random choice solution of hyperbolic systems, *J. Comp. Phys.* 22 (1976), 517–533.

Random choice methods with application to reacting gas flow, *J. Comp. Phys.* 25 (1977), 253–272.

Chorin, A.J., T.R. Hughes, M.F. McCracken, and J.E. Marsden. Product formulas and numerical algorithms, *Comm. Pure Appl. Math.* 31 (1978), 205–256.

Chorin, A.J. and J.E. Marsden. *A Mathematical Introduction to Fluid Mechanics*, Texts in Applied Mathematics 4, Springer-Verlag, New York, Third Edition, 1993.

Chuey, K.N., C.C. Conley, and J.A. Smoller. Positively invariant regions of nonlinear diffusion equations, *Indiana Univ. Math. J.* 26 (2) (1977), 373–392.

Ciccoli, M.-C., L. Fezoui, and J.A. Désidéri. Efficient methods for inviscid nonequilibrium hypersonic flow fields, *Recherche Aérospatiale* (1992) 1, 37–52.

Cioni, J.P., L. Fezoui, and H. Steve. A parallel time-domain Maxwell solver using upwind schemes and triangular meshes, INRIA Research Report 1867 (1993), INRIA Sophia-Antipolis, France.

Clarke, J. F., S. Karni, J.J. Quirk, P.L. Roe, L.G. Simmonds, and E.F. Toro. Numerical computation for two-dimensional unsteady detonation waves in high energy solids, *J. Comp. Phys.* 106 (1993), 215–233.

Cockburn, B. The quasi-monotone schemes for scalar conservation laws, III, *SIAM J. Numer. Anal.* 27 (1990), 259–276.

Cockburn, B., F. Coquel, and Ph. Le Floch. An error estimate for finite volume methods for multidimensional conservation laws, *Math. Comp.* 63 (1994), 77–103.

Convergence of the finite volume method for multidimensional conservation laws, *SIAM J. Numer. Anal.* 32 (3) (1995), 687–706.

Cockburn, B., S. Hou, and C.-W. Shu. The Runge–Kutta local projection discontinuous Galerkin finite element method for conservation laws IV: the multidimensional case, *Math. Comp.* 54 (1990), 545–581.

Cockburn, B. and C.-W. Shu. Nonlinearly stable compact schemes for shock calculations, *SIAM J. Numer. Anal.* 31(3) (1994), 607–627.

Coirier, W.J. and K.G. Powell. An accuracy assessment of Cartesian-mesh approaches for the Euler equations, *J. Comp. Phys.* 117 (1995), 121–131.

Cole, J. D. On a quasi-linear parabolic equation occuring in aerodynamics, *Quarterly Appl. Math.* 9 (1951), 225–236.

Colella, P. A direct Eulerian MUSCL scheme for gas dynamics, *SIAM J. Sci. Statist. Comput.* 6 (1985), 104–117.

Multidimensional upwind methods for hyperbolic conservation laws, *J. Comp. Phys.* 87 (1990), 171–200.

Colella, P. and H. Glaz. Efficient solution algorithms for the Riemann problem for real gases, *J. Comp. Phys.* 59 (1985), 264–289.

Colella, P., A. Majda, and V. Roytburd. Theoretical and numerical structure for reacting shock waves, *SIAM J. Sci. Statist. Comput.* 7 (1986), 1059–1080.

Colella, P. and P.R. Woodward. The Piecewise Parabolic Method (PPM) for gas-dynamical simulations, *J. Comp. Phys.* 54 (1984), 174–201.

Collins, J.P., P. Colella, and H.M. Glaz. An implicit-explicit Eulerian Godunov scheme for compressible flow, *J. Comp. Phys.* 116 (1995), 195–211.

Colombeau, J.F., A.-Y. Le Roux, A. Noussair, and B. Perrot. Microscopic profiles of shock waves and ambiguities in multiplications of distributions, *SIAM J. Numer. Anal.* 26 (1989), 871–883.

Coquel, F. and Ph. Le Floch. Convergence of finite difference schemes for conservation laws in several space dimensions: the corrected antidiffusive flux approach, *Math. Comp.* 57 (195) (1991), 169–210.

Convergence of finite difference schemes for conservation laws in several space dimensions: a general theory, *SIAM J. Numer. Anal.* 30 (1993), 675–700.

An entropy satisfying MUSCL scheme for systems of conservation laws, École Polytechnique, CMAP Internal Report 322, 91128 Palaiseau, France (1995), to appear in *Numer. Math.*.

Coquel, F. and M.-S. Liou. Stable and low diffusive hybrid upwind splitting methods, in *Proceedings of the First European CFD Conference, Brussels (1992)*, C. Hirsh and E. Onate (Eds.), Elsevier, Amsterdam (1992), 9–16.

A reliable and efficient upwind scheme by a field by field decomposition - Hybrid upwind splitting (HUS), preprint (1994).

Coron, F. and B. Perthame. Numerical passage from kinetic to fluid equations, *SIAM J. Numer. Anal.* 28 (1991), 26–42.

Courant, R. and K.O. Friedrichs. *Supersonic Flow and Shock Waves*, Applied Mathematical Sciences 21, Springer-Verlag, New York (1976).

Crandall, M.G. and A. Majda. Monotone difference approximations for scalar conservation laws, *Math. Comp.* 34 (1980), 1–21.

The method of fractional steps for conservation laws, *Numer. Math.* 34 (1980), 285–314.

Croisille, J.-P. *Contribution à l'étude théorique et à l'approximation par éléments finis du système hyperbolique de la dynamique des gaz multidimensionnelle et multiespèces*, Doctoral dissertation, Thèse de l'Université Pierre et Marie Curie-Paris 6 (1990).

Croisille, J.-P. and P. Delorme. Kinetic symmetrizations and pressure laws for the Euler equations, Université de Paris-Sud Mathématiques Report 91-67, 91405 Orsay, France (1991).

Croisille, J.-P. and P. Villedieu. Entropies de Lax pour les équations d'Euler en déséquilibre thermochimique, *C. R. Acad. Sci. Paris* 318 (1994), 723–727.

Dafermos, C.M. The entropy rate admissibility criteria for solutions of hyperbolic conservation laws, *J. Diff. Equations* 14 (1973), 265–298.

Hyperbolic Systems of Conservation Laws, in *Systems of Nonlinear Partial Differential Equations*, J.M. Ball (Ed.), NATO ASI Series, Reidel, Dordrecht (1983), 25–70.

Admissible wave fans in nonlinear hyperbolic systems, *Arch. Rat. Mech. Anal.* 106 (1989), 243–260.

Generalized characteristics in hyperbolic systems of conservation laws, *Arch. Rat. Mech. Anal.* 107 (1989), 127–155.

Equivalence of referential and spatial fields equations in continuum physics, in *Nonlinear Hyperbolic Problems - Theoretical, Applied and Computational*

Aspects, Proceedings of the Fourth International Conference on Hyperbolic Problems, Taormina 1992, A. Donato and F. Oliveri (Eds.), Notes on Numerical Fluid Mechanics 43, Vieweg, Braunschweig (1993), 179–183.

Dai, W. and P.R. Woodward. An approximate Riemann solver for ideal magnetohydrodynamics, *J. Comp. Phys.* 111 (1994), 354–372.

A simple Riemann solver and high-order Godunov schemes for hyperbolic conservation laws, *J. Comp. Phys.* 121 (1995), 51–65.

Dal Maso, Ph. Le Floch, and F. Murat. Definition and weak stability of nonconservative products, *J. Math. Pures Appl.* 74 (1995), 483–458.

Daru, V. and A. Lerat. Analysis of an implicit Euler solver, in *Numerical methods for the Euler equations of fluid dynamics*, Proceedings of the INRIA workshop, Rocquencourt, France (1983), F. Angrand et al. (Eds.), SIAM, Philadelphia (1985), 246–280.

Dautray, R. and J.-L. Lions. *Analyse mathématique et calcul numérique pour les sciences et les techniques, Vol. 1 : Modèles physiques*, INSTN-CEA Collection Enseignement, Masson, Paris, 1988.

Davis, S.F. A rotationally biased upwind difference scheme for the Euler equations, *J. Comp. Phys.* 56 (1984), 65–92.

TVD finite difference schemes and artificial viscosity, ICASE Report N° 84-20 (1984), ICASE NASA Langley Research Center, Hampton, VA.

Simplified second-order Godunov-type methods, *SIAM J. Sci. Statist. Comput.* 9 (1988), 445–473.

An interface tracking method for hyperbolic systems of conservation laws, *Applied Numerical Mathematics* 10 (1992), 447–472, North-Holland, Amsterdam.

Dawson, C. Godunov-mixed methods for advection-diffusion equation in multidimensions, *SIAM J. Numer. Anal.* 30, 25 (1994), 1315–1332.

Deconinck, H., P.L. Roe, and R. Struijs. A multidimensional generalization of Roe's flux difference splitter for the Euler equations, *Computers Fluids*, 22 (1993), 215–222.

Dervieux, A. and G. Vijayasundaram. On numerical schemes for solving Euler equations of gas dynamics, in *Numerical methods for the Euler equations of fluid dynamics, Proceedings of the INRIA workshop, Rocquencourt, France (1983)*, F. Angrand et al. (Eds.), SIAM, Philadelphia (1985), 121–144.

Deshpande, S. M. A second-order accurate kinetic-theory-based method for inviscid compressible flows, NASA Technical paper, 2613 (1986).

Kinetic theory based new upwind methods for inviscid compressible flows, AIAA 86-0275, 24th Aerospace Sciences Meeting, Reno (1986).

Desideri, J.A., A. Goudjo, and V. Selmin. Third-order numerical schemes for hyperbolic problems, INRIA Research Report 607 (1987), INRIA Rocquencourt, 78153 Le Chesnay, France.

De Vuyst, A. *Schémas non-conservatifs et schémas cintétiques pour la simulation numérique d'écoulements hypersoniques non visqueux en déséquilibre thermochimique*, Doctoral dissertation, Thesis Université Paris 6, France (1994).

A new implicit second-order scheme based on a kinetic interpretation for solving the Euler equations, in Conference on numerical methods for fluid dynamics, University of Reading U.K. (April 1992), Oxford University Press, Oxford (1992), 425–443.

Upwinding process of the characteristic variables for Roe's scheme. A macroscopic or kinetic approach according to the nature of the characteristic fields, *C. R. Acad. Sci. Paris* 320 , Sér. I, (1995), 743–748.

Boltzmann-type schemes and derivatives for the numerical solution of the steady nonequilibrium compressible Euler equations, to appear in *Enumath 1995*, Paris 1995.

DiPerna, R. The structure of solutions to hyperbolic conservation laws, in *Nonlinear analysis and mechanics: Heriot-Watt Symposium, Vol. IV*, R.J. Knops (Ed.). Research Notes in Mathematics 39, Pitman, London (1979), 1–16.

Convergence of approximate solutions to conservation laws, *Arch. Rat. Mech. Anal.* 82 (1983), 27–70.

Convergence of the viscosity method for isentropic gas dynamics, *Comm. Math. Phys.* 91 (1983), 1–30.

Measure valued solutions to conservation laws, *Arch. Rat. Mech. Anal.* 88 (1985), 223–270.

Dolezal, A. and S.S.M. Wong. Relativistic hydrodynamics and essentially nonoscillatory shock capturing scheme, *J. Comp. Phys.* 120 (1995), 266–277.

Don, W.S. and C.B. Quillen. Numerical simulation of shock-cylinder interactions: I. Resolution, *J. Comp. Phys.* 122 (1995), 244–265.

Dubois, F. Boundary conditions and the Osher scheme for the Euler equations of gas dynamics, École Polytechnique, CMAP Internal Report, 91128 Palaiseau, France (1987).

Conditions aux limites fortement nonlinéaires pour les équations d'Euler, Notes de cours l'École CEA-EDF-INRIA sur les problèmes nonlinéaires appliqués nov-déc 1988, INRIA Rocquencourt, 78153 Le Chesnay, France.

Code Euler 3D implicite en hypersonique et supersonique élévé. Evaluation du flux d'Osher pour l'air à l'équilibre chimique, Aérospatiale Report 41747, (1989).

Concavité de l'entropie thermostatique et convexité de l'entropie mathématique au sens de Lax, *La Recherche Aérospatiale* 3 (1990), 77–80.

Résolution numérique des équations d'Euler multidimensionnelles, cours de DEA, École Polytechnique, 91128 Palaiseau, France (1993).

Dubois, F. and P. Le Floch. Boundary conditions for nonlinear hyperbolic systems of conservation laws, *J. Diff. Equations* 71 (1988), 93–122.

Dubois, F. and G. Mehlman. A non-parametrized entropy correction for Roe's approximate Riemann solver, in *Computing Methods in Applied Sciences and Engineering*, Proceedings of the tenth International Conference on Computing Methods in Applied Sciences and Engineering, France (1992), R. Glowinski (Ed.), Nova Science Publishers, Inc, New York, 479–488, and *Numer. Math.* 73 (2) (1996), 169–208.

Dubois, F. and G. Michaux. Solution of the Euler equations around a double ellipsoidal shape using unstructured meshes and including real gas effects, in *Proceedings of the Workshop on Hypersonic flows and reentry problems, Antibes France (1990)*, Désidéri-Glowinski-Périaux (Eds.), Springer-Verlag, Berlin, vol. 2 (1992), 358–373.

Dubroca, B. and G. Gallice. Problème mixte pour un système de lois de conservation monodimensionnel, *C. R. Acad. Sci. Paris* 306 , Sér. I, (1988), 317–320.

Dubroca, B. and J.P. Morreeuw. An extension of Roe's approximate Riemann solver for the approximation of Navier–Stokes equations in chemical-nonequilibrium cases, in *Computing Methods in Applied Sciences and Engineering*, Proceedings of the tenth International Conference on Computing Methods in Applied Sciences and Engineering, France (1992), R. Glowinski (Ed.), Nova Science Publishers, Inc, New York, 345–372.

Dukowicz, J. and A. Dvinsky. Approximate factorization as a high order splitting for the implicit incompressible flow equations, *J. Comp. Phys.* 102 (1992), 336–347.

Dukowicz, J., M.C. Cline, and F. L. Addessio. A general topology Godunov method, *J. Comp. Phys.* 82 (1989), 29–63.

Durlofsky, L. A triangle based mixed finite element-finite volume technique for modeling two phase flow through porous media, *J. Comp. Phys.* 105 (1993), 252–266.

Durlofsky, L., B. Engquist, and S. Osher. Triangle based adaptive stencils for the solution of hyperbolic conservation laws, *J. Comp. Phys.* 98 (1992), 64–73.

Dutt, P. Stable Boundary Conditions and Difference Schemes for Navier–Stokes Equations, *SIAM J. Numer. Anal.* 25 (1988), 245–267.

Spectral methods for initial boundary value problems - An alternative approach, *SIAM J. Numer. Anal.* 27 (1990), 885–903.

E, W. Homogenization of scalar conservation laws with oscillatory forcing terms, *SIAM J. Appl. Math.* 52 (4) (1992), 959–972.

E, W. and C.W. Shu. A numerical resolution study of high order essentially non-oscillatory schemes applied to incompressible flow *J. Comp. Phys.* 110 (1994), 39–46.

Einfeldt, B. On Godunov-type methods for gas dynamics, *SIAM J. Numer. Anal.* 25 (1988), 294–318.

Einfeldt, B., C.D. Munz, P.L. Roe, and B. Sjögreen. On Godunov-type methods near low densities, *J. Comp. Phys.* 92 (1991), 273–295.

Embid, P., J. Hunter, and A. Majda. Simplified asymptotic equations for the transition to detonation in reactive granular material, *SIAM J. Appl. Math.* 52(5) (1992), 1199–1237.

Engquist, B. and A. Majda. Absorbing boundary conditions for the numerical simulation of waves, *Math. Comp.* 31 (1977), 629–651.

Engquist, B. and S. Osher. Stable and entropy satisfying approximations for transonic flow calculations, *Math Comp.* 34 (1980), 45–75.

Estivalezes, J.L. and P. Villedieu. High order positivity preserving schemes for the compressible Euler equations, subm. to *SIAM J. Numer. Anal.*

Eymard, R. and T. Gallouët. Convergence d'un schéma de type éléments finis - volumes finis pour un système formé d'une équation elliptique et d'une équation hyperbolique, *M^2AN, Modélisation mathématique et analyse numérique*, 27 (7) (1993), 843–862.

Falcovitz, J. and M. Ben-Artzi. Recent developments of the GRP Method, Institute of Mathematics internal Report 30, Hebrew University, Jerusalem 91904 Israel (1992-93).

Fan, H. and J.K. Hale. Large-time behavior in inhomogeneous conservation laws, *Arch. Rat. Mech. Anal.* 125 (1993), 201–216.

Fernandez, G. and B. Larrouturou. Hyperbolic schemes for multi-component Euler equations, in *Nonlinear Hyperbolic Equations - Theory, Computation Methods, and Applications* (Proceedings of the Second International Conference on Nonlinear Hyperbolic Problems, Aachen 1988), J. Ballmann and R. Jeltsch (Eds.), Notes on Numerical Fluid Mechanics 24, Vieweg, Braunschweig (1989), 128–138.

Fey, M. Decomposition of the multidimensional Euler equations into advection equations, SAM Research Report 95-14, ETH Zürich, Switzerland (1995).

Fey, M., R. Jeltsch, and S. Müller. The influence of a source term, an example: chemically reacting hypersonic flows, in *Nonlinear Hyperbolic Problems - Theoretical, Applied and Computational Aspects*, Proceedings of the Fourth International Conference on Hyperbolic Problems, Taormina 1992, A. Donato and F. Oliveri (Eds.), Notes on Numerical Fluid Mechanics 43, Vieweg, Braunschweig (1993), 235–245.

Fezoui, L. Résolution des équations d'Euler par un schéma de van Leer en éléments finis, INRIA Research Report 358 (1985), INRIA Rocquencourt, 78153 Le Chesnay, France.

Fezoui, L., S. Lanteri, B. Larrouturou, and C. Olivier. Résolution numérique des équations de Navier–Stokes pour un fluide compressible en maillage triangulaire, INRIA Research Report 1033 (1989), INRIA Rocquencourt, 78153 Le Chesnay, France.

Fezoui, L. and H. Steve. Décomposition de flux de van Leer en éléments finis, INRIA Research Report 830 (1988), INRIA Rocquencourt, 78153 Le Chesnay, France.

Fezoui, L., H. Steve, and V. Selmin. Simulation numérique d'écoulements compressibles 3-D par un schéma décentré en maillage non structuré, INRIA Research Report 825 (1988), INRIA Rocquencourt, 78153 Le Chesnay, France.

Fezoui, L. and B. Stoufflet. A class of implicit upwind schemes for Euler simulations with unstructured meshes, *J. Comp. Phys.* 84 (1989), 174–206.

Fickett, W. and W.C. Davis. *Detonation*, University of California Press, Berkeley-Los Angeles-London, 1979.

Freistühler, H. A standard model for generic rotational degeneracy, in *Nonlinear Hyperbolic Equations - Theory, Computation Methods, and Applications* (Proceedings of the Second International Conference on Nonlinear Hyperbolic Problems, Aachen 1988), J. Ballmann and R. Jeltsch (Eds.), Notes on Numerical Fluid Mechanics 24, Vieweg, Braunschweig (1989), 149–158.

Instability of vanishing viscosity approximation to hyperbolic systems of conservation laws with rotational invariance, *J. Diff. Equations* 87 (1990), 205–226.

On the Cauchy problem for a class of hyperbolic systems of conservation laws, *J. Diff. Equations* 112 (1994), 170–178.

Freistühler, H. and E.B. Pitman. A numerical study of a rotationally degenerate hyperbolic system. Part I. The Riemann problem, *J. Comp. Phys.* 100 (1992), 306–321.

A numerical study of a rotationally degenerate hyperbolic system. Part II. The Cauchy problem, *SIAM J. Numer. Anal.* 32 (3) (1995), 741–753.

Friedrichs, K.O. and P.D. Lax. Systems of conservation laws with a convex extension, *Proc. Nat. Acad. Sci. U.S.A.* 68 (1971), 1686–1688.

Gallice, G. Matrices de Roe pour des lois de conservation générales sous forme eulérienne ou lagrangienne. Application à la dynamique des gaz et à la MHD, *C. R. Acad. Sci. Paris* 321, Série I (1995), 1069–1072.

Gallouët, T. An introduction to finite volume methods, cours CEA-EDF-INRIA (France) (1992), to appear in *Finite volume methods* by R. Eymard, T. Gallouët, and R. Herbin, Handbook of Numerical Analysis, P. G. Ciarlet and J.-L. Lions (Eds.), North-Holland, Amsterdam.

Garcia-Navarro, P., M.E. Hubbard, and A. Priestley. Genuinely multidimensional upwinding for the 2d shallow water equations, *J. Comp. Phys.* 121 (1995), 79–93.

Gasser, I. and P. Szmolyan A geometric singular perturbation analysis of detonation and deflagration waves, *SIAM J. Math. Anal.* 24 (4) (1993), 968–986.

Gilquin, H., J. Laurens, and C. Rosier. Multi-dimensional Riemann problems for linear hyperbolic systems, in *Nonlinear Hyperbolic Problems - Theoretical, Applied and Computational Aspects*, Proceedings of the Fourth International Conference on Hyperbolic Problems, Taormina 1992, A. Donato and F. Oliveri (Eds.), Notes on Numerical Fluid Mechanics 43, Vieweg, Braunschweig (1993), 276–290.

Gisclon, M. Étude des conditions aux limites pour des système strictement hyperboliques, *via* l'approximation parabolique, Thesis Université Claude Bernard - Lyon I (France) (1994).

Gisclon, M. and D. Serre. Étude des conditions aux limites pour un système strictement hyperbolique via l'approximation parabolique, *C. R. Acad. Sci. Paris*, 319, Série I (1994), 377–382.

Giusti, E. *Minimal surfaces and functions of bounded variation*, Monograph in Mathematics, vol 80, Birkhauser, Boston (1984).

Givoli, D. Non-reflecting boundary conditions, Review article, *J. Comp. Phys.* 94 (1991), 1–29.

Givoli, D. and D. Cohen. Non-reflecting boundary conditions based on Kirchhoff-type formulae, *J. Comp. Phys.* 117 (1995), 102–113.

Glaister, P. An approximate linearized Riemann solver for the Euler equations for real gases, *J. Comp. Phys.* 74 (1988), 382–408.

An approximate linearized Riemann solver for the three-dimensional Euler equations for real gases using operator splitting, *J. Comp. Phys.* 77 (1988), 361–383.

Flux difference splitting for open-channel flows, *Int. J. Numer. Methods Fluids* 16 (1993), 629–654.

A weak formulation of Roe's approximate Riemann solver applied to 'barotropic' flows, *Comp. Math. Appl.* 27 (1994), 87–90.

A weak formulation of Roe's approximate Riemann solver applied to the St. Venant equations, *J. Comp. Phys.* 116 (1995), 189–191.

Glaz, H.M. Self-similar shock reflection in two space dimensions, in *Multidimensional Hyperbolic Problems and Computations*, Proceedings of the IMA workshop (1989), The IMA Volumes in Mathematics and its applications 29, J. Glimm, A. J. Majda (Eds.), Springer-Verlag, New York (1991), 70–88.

Glaz, H.M. and A. Wardlaw. A high-order Godunov scheme for steady supersonic gas dynamics, *J. Comp. Phys.* 58 (1985), 157–187.

Glimm, J. Solutions in the large for nonlinear hyperbolic systems of equations, *Comm. Pure Appl. Math.* 18 (1965), 697–715.

Nonuniqueness of solutions of Riemann problems, in *Nonlinear Hyperbolic Equations - Theory, Computation Methods, and Applications* (Proceedings of the Second International Conference on Nonlinear Hyperbolic Problems, Aachen 1988), J. Ballmann and R. Jeltsch (Eds.), Notes on Numerical Fluid Mechanics 24, Vieweg, Braunschweig (1989), 169–178.

Glimm, J., M. J. Graham, J.W. Grove, and J.B. Plohr (Editors). *Hyperbolic problems: theory, numerics, applications*, Proceedings of the Fifth International Conference on Hyperbolic Problems, StonyBrook 1994, World Scientific, Singapore (1996).

Glimm, J., C. Klingenberg, O. McBryan, B. Plohr, D. Sharp, and S. Yaniv. Front tracking and two dimensional Riemann problems, *Adv. Appl. Math.* 6 (1985), 259–290.

Glimm, J., B. Lindquist, and Q. Zhang. Front tracking, oil reservoirs, engineering scale problems and mass conservation, in *Multidimensional Hyperbolic Problems and Computations*, Proceedings of the IMA workshop (1989), The IMA Volumes in Mathematics and its applications 29, J. Glimm, A.J. Majda (Eds.), Springer-Verlag, New York (1991), 123–139.

Glimm, J. and A.J. Majda (Editors). *Multidimensional Hyperbolic Problems and Computations*, Proceedings of the IMA workshop on Multidimensional Hyperbolic Problems and Computations (1989), IMA Volumes in Mathematics and its Applications 29, Springer-Verlag, New York, 1991.

Godlewski E. and P.-A. Raviart. *Hyperbolic systems of conservation laws*, Mathématiques et Applications, Ellipses, Paris, 1991.

Godunov, S.K. A difference scheme for numerical computation of discontinuous solutions of equations of fluid dynamics, *Mat. Sb.* 47(89) (1959), 271–306.

Lois de conservation et intégrales d'énergie des équations hyperboliques, in *Nonlinear Hyperbolic Problems, Proceedings St. Etienne 1986*, C. Carasso, P.-A. Raviart, D. Serre (Eds.), Lecture Notes in Mathematics 1270, Springer-Verlag, Berlin (1987), 135–149.

Godunov, S.K., A. Zabrodine, M. Ivanov, A. Kraïko, and G. Prokopov. *Résolution numérique des problèmes multidimensionnels de la dynamique des gaz*, traduit du russe, Editions Mir, Moscou, 1979.

Goldberg, M. and E. Tadmor. New stability criteria for difference approximations of hyperbolic initial-boundary value problems, in *Large-scale computations in Fluid Mechanics*, Proceedings of AMS/SIAM Summer Seminar, La Jolla, CA (1983), Part 1, B. Engquist, S. Osher and R.C.J. Sommerville (Eds.), Lectures in Applied Mathematics 22, AMS, Providence, RI (1985), 177–192, also *Math. Comp.* 48, 178 (1987), 503–520.

Convenient stability criteria for difference approximations of hyperbolic initial-boundary value problems, *Math. Comp.* 48 (1987), 503–520.

Simple stability criteria for difference approximations of hyperbolic initial-boundary value problems, in *Nonlinear Hyperbolic Equations - Theory, Computation Methods, and Applications* (Proceedings of the Second International Conference on Nonlinear Hyperbolic Problems, Aachen 1988), J. Ballmann and R. Jeltsch (Eds.), Notes on Numerical Fluid Mechanics 24, Vieweg, Braunschweig (1989), 179–185.

Goodman, J.B. and R.J. LeVeque. On the accuracy of stable schemes for 2D scalar conservation laws, *Math. Comp.* 45 (1988), 503–520.

Goodman, J.B. and Z. Xin. Viscous limits for piecewise smooth solutions to systems of conservation laws, *Arch. Rat. Mech. Anal.* 121 (1992), 235–265.

Greenberg, J.M. and A.-Y. LeRoux. A well balanced scheme for the numerical processing of source terms in hyperbolic equations, subm. to *SIAM J. Numer. Anal.*.

Griffiths, D.F., A.M. Stuart, and Y.C. Yee. Numerical wave propagation in an advection equation with a nonlinear source term, *SIAM J. Numer. Anal.* 29 (1992), 1244–1260.

Grossman, B. and P. Cinella. Flux-split algorithms for flows with non-equilibrium chemistry and vibrational relaxation, *J. Comp. Phys.* 88 (1990), 131–168.

Guillard, H. Mixed element volume methods in computational fluid dynamics, von Karman Institute for Fluid Dynamics Lecture Series 1995-02.

Gustafsson, B. The choice of numerical boundary conditions for hyperbolic systems, *J. Comp. Phys.* 48 (1982), 270–283.

Gustafsson, B. and L. Ferm. Far field boundary conditions for steady state solutions to hyperbolic systems, in *Nonlinear Hyperbolic Problems, Proceedings St. Etienne 1986*, C. Carasso P.-A. Raviart D. Serre (Eds.), Lecture Notes in Mathematics 1270, Springer-Verlag, Berlin (1987), 238–252.

Far field boundary conditions for time-dependent hyperbolic systems, *SIAM J. Sci. Statist. Comput.* 9 (1988), 812–848.

Gustafsson, B. and H.-O. Kreiss. Boundary conditions for time-dependent problems with an artificial boundary, *J. Comp. Phys.* 30 (1979), 333–351.

Gustafsson, B., H.-O. Kreiss, and A. Sundström. Stability theory of difference approximations for mixed initial boundary value problems, II, *Math. Comp.*, 26 (1972), 649–686.

Hagstrom, T and S.I. Hariharan. Accurate boundary conditions for exterior problems in gas dynamics, *Math. Comp.* 51 (1988), 581–597.

Hall, M.S. A comparison of first and second order rezoned and Lagrangian Godunov solutions, *J. Comp. Phys.* 90 (1990), 458–485.

Hannapel, R., T. Hauser, and R. Friedrich. A comparison of ENO and TVD schemes for the computation of shock-turbulence interaction, *J. Comp. Phys.* 121 (1995), 176–184.

Hansbo, P. Explicit streamline diffusion finite element methods for compressible Euler equations in conservation variables, *J. Comp. Phys.* 109 (1993), 274–88.

Aspects of conservation in finite element flow computations, *Comp. Meth. Appl. Mech. and Engrg.* 117 (1994), 423–437.

Harabetian, E. A numerical method for computing viscous shock layers, in *Nonlinear Hyperbolic Equations - Theory, Computation Methods, and Applications* (Proceedings of the Second International Conference on Nonlinear Hyperbolic Problems, Aachen 1988), J. Ballmann and R. Jeltsch (Eds.), Notes on Numerical Fluid Mechanics 24, Vieweg, Braunschweig (1989), 220–229.

A numerical method for viscous perturbations of hyperbolic conservation laws, *SIAM J. Numer. Anal.* 27 (1990), 870–884.

A subcell resolution method for viscous systems of conservation laws, *J. Comp. Phys.* 103 (1992), 350–358.

Harabetian, E. and R. Pego. Nonconservative hybrid shock capturing schemes, *J. Comp. Phys.* 105 (1993), 1–13.

Harten, A. On the symmetric form of systems of conservation laws with entropy, *J. Comp. Phys.* 49 (1983), 151–164.

High resolution schemes for hyperbolic conservation laws, *J. Comp. Phys.* 49 (1983), 357–393.

On a class of high resolution total-variation-stable finite difference schemes, *SIAM J. Numer. Anal.* 21 (1984), 1–23.

Preliminary results on the extension of ENO schemes to two-dimensional problems, in *Nonlinear Hyperbolic Problems, Proceedings St. Etienne 1986*, C. Carasso, P.-A. Raviart, D. Serre (Eds.), Lecture Notes in Mathematics 1270, Springer-Verlag, Berlin (1987), 23–40.

Harten, A. and S. Chakravarthy. Multidimensional ENO schemes for general geometries, ICASE Report No 91-76 (1991), ICASE NASA Langley Research Center, Hampton, VA.

Harten, A. and J.M. Hyman. Self-adjusting grid methods for one-dimensional hyperbolic conservation laws, *J. Comp. Phys.* 50 (1983), 235–269.

Harten, A., J.M. Hyman, and P.D. Lax. On finite difference approximations and entropy conditions for shocks, *Comm. Pure Appl. Math.* 29 (1976), 297–322.

Harten, A., P.D. Lax, and B. van Leer. On upstream differencing and Godunov-type schemes for hyperbolic conservation laws, *SIAM Rev.* 25 (1983), 35–61.

Hattori, H. The Riemann Problem for a van der Waals fluid with entropy rate admissibility criterion. Isothermal case, *Arch. Rat. Mech. Anal.* 92 (1986), 247–263. Non isothermal case, *J. Diff. Equations* 65 (1986), 158–174.

Hedstrom, G.W. Nonreflecting boundary conditions for nonlinear hyperbolic systems, *J. Comp. Phys.* 30 (1979), 222–237.

Heibig, A. Existence et unicité des solutions pour certains systèmes de lois de conservation, *C. R. Acad. Sci. Paris* 311 (1990), 861–866.

Existence and uniqueness of solutions for some hyperbolic systems of conservation laws, *Arch. Rat. Mech. Anal.* 126 (1994), 79–101.

Hemker, P.W. and S. Spekreijse. Multigrid and Osher's scheme for efficient solution of the steady Euler equations, *Appl. Numer. Math.* 2 (1986), 475–493.

Henshaw, W.D. A scheme for the numerical solution of hyperbolic systems of conservation laws, *J. Comp. Phys.* 68 (1987), 25–47.

Higdon, R.L. Initial-boundary value problems for linear hyperbolic systems, *SIAM Rev.* 28 (1986), 177–217.

Hilditch, J. and P. Colella. A front tracking method for compressible flames in one dimension, *SIAM J. Sci. Statist. Comput.* 16 (4) (1995), 755–772.

Hirsch, C. *Numerical computation of internal and external flows, vol 2: Computational methods for inviscid and viscous flows*, John Wiley & Sons, Chichester, 1990, reprinted 1995.

Hoff, D. Invariant regions for systems of conservation laws, *Trans. Amer. Math. Soc.* 289 (2) (1985), 591–610.

Holden, L. On the strict hyperbolicity of the Buckley–Leverett equations for three-phase flow, in *Nonlinear Evolution Equations that Change Type*, IMA Volumes in Mathematics and its Applications 27, B.L. Keyfitz and M. Shearer (Eds.), Springer-Verlag, New York (1991), 79–88.

Holden, H. and L. Holden. On some recent results for an explicit conservation law of mixed type in one dimension, in *Nonlinear Hyperbolic Equations - Theory, Computation Methods, and Applications* (Proceedings of the Second International Conference on Nonlinear Hyperbolic Problems, Aachen 1988), J. Ballmann and R. Jeltsch (Eds.), Notes on Numerical Fluid Mechanics 24, Vieweg, Braunschweig (1989), 238–245.

Holden, H., L. Holden, and N.H. Risebro. Some qualitative properties of 2×2 systems of conservation laws, in *Nonlinear Evolution Equations that Change Type*, IMA Volumes in Mathematics and its Applications 27, B.L. Keyfitz and M. Shearer (Eds.), Springer-Verlag, New York (1991), 67–78.

Holden, H. and N.H. Risebro. A mathematical model of traffic flow on a network of unidirectional roads, *SIAM J. Math. Anal.* 26 (4) (1995), 999–1017.

Holt, M. *Numerical Methods in Fluid Dynamics*, Springer Series in Computational Physics, Springer-Verlag, Berlin 1977.

Hopf, E. The partial differential equation $u_t + uu_x = \mu u_{xx}$, *Comm. Pure Appl. Math.* 3 (3) (1950), 201–230.

Hou, T. and P.G. Le Floch. Why nonconservative schemes converge to wrong solutions: error analysis, *Math. of Comp.* 62 (1993), 497–530.

Hsiao, L. Qualitative behaviour of solutions for Riemann problems of conservation laws of mixed type, in *Nonlinear Hyperbolic Equations - Theory, Computation Methods, and Applications* (Proceedings of the Second International Conference on Nonlinear Hyperbolic Problems, Aachen 1988), J. Ballmann and R. Jeltsch (Eds.), Notes on Numerical Fluid Mechanics 24, Vieweg, Braunschweig (1989), 246–256.

Admissibility criteria and admissible weak solutions of Riemann problems for conservation laws of mixed type: a summary, in *Nonlinear Evolution Equations that Change Type*, IMA Volumes in Mathematics and its Applications 27, B.L. Keyfitz and M. Shearer (Eds.), Springer-Verlag, New York (1991), 85–88.

Hugues, T.J.R., M. Mallet, Y. Taki, T.E. Tezduyar, and R. Zanutta. A one-dimensional shock capturing finite element method and multi-dimensional generalization, in *Numerical methods for the Euler equations of fluid dynamics*, Proceedings of the INRIA workshop, Rocquencourt, France (1983), F. Angrand et al. (Eds.), SIAM, Philadelphia (1985), 371–408.

Hui, W.H. and C.Y. Loh. A new Lagrangian method for steady supersonic flow computation II. Slip line resolution, *J. Comp. Phys.* 103 (1992), 450–464.

III. Strong shocks, Note, *J. Comp. Phys.* 103 (1992), 465–471.

Hundsdorfer, W., B. Koren, M. van Loon, and J.G. Verwer. A positive finite-difference advection scheme, *J. Comp. Phys.* 117 (1995), 35–46.

Huynh, H.T. A piecewie-parabolic dual-mesh method for the Euler equations, *AIAA* 95-1739-CP, 12th AIAA Computational Fluid Dynamics Conference (San Diego 1995) 1054-1066.

Accurate upwind schemes for the Euler equations, *SIAM J. Numer. Anal.* 32 (5) (1995), 1565–1618.

Isaacson, E., D. Marchesin, and B.J. Plohr. The structure of the Riemann solution for non-strictly hyperbolic conservation laws, in *Nonlinear Hyperbolic Equations - Theory, Computation Methods, and Applications* (Proceedings of the Second International Conference on Nonlinear Hyperbolic Problems, Aachen

1988), J. Ballmann and R. Jeltsch (Eds.), Notes on Numerical Fluid Mechanics 24, Vieweg, Braunschweig (1989), 269–278.

Isaacson, E. and B. Temple. Analysis of a singular hyperbolic system of conservation law, *J. Diff. Equations* 65 (1986), 250–268.

Nonlinear resonance in systems of conservation laws, *SIAM J. Appl. Math.* 52 (2) (1992), 1260–1278.

Convergence of a 2 × 2 Godunov method for a general resonant nonlinear balance law, *SIAM J. Appl. Math.* 55 (1995), 625–640.

Jameson, A. Numerical solution of the Euler equations for compressible inviscid fluids, in *Numerical methods for the Euler equations of fluid dynamics, Proceedings of the INRIA workshop, Rocquencourt, France (1983)*, F. Angrand et al. (Eds.), SIAM, Philadelphia (1985), 199–245.

Jeffrey, A. *Quasilinear hyperbolic systems and waves*, Research Notes in Mathematics 5, Pitman, London (1976).

Jeffrey, A. and T. Taniuti. *Non-linear wave propagation*, Mathematics in Science and Engineering 9, Academic Press, New York 1964.

Jeng, Y.N. and J.L. Chen. Truncation error analysis of the finite volume method for a model steady convective equation, *J. Comp. Phys.* 100 (1992), 64–76.

Jeng, Y.N. and U.J. Payne. An adaptative TVD limiter, *J. Comp. Phys.* 118 (1995), 229–241.

Jiang, H. and P.A. Forsyth. Robust linear and nonlinear strategies for solution of the transonic Euler equations, *Comp. Fluids* 24 (7) (1995), 753–770.

Jiang, H. and Y.S. Wong. Absorbing boundary conditions for second-order hyperbolic equations, *J. Comp. Phys.* 88 (1990), 205–231.

Jin, B.X. On an essentially conservative scheme for hyperbolic conservation laws, *J. Comp. Phys.* 112 (1994), 308–315.

Jin, S. Runge–Kutta methods for hyperbolic conservation laws with stiff relaxation terms, *J. Comp. Phys.* 122 (1995), 51–67.

Jin, S. and Z. Xin. The relaxation schemes for systems of conservation laws in arbitrary space dimensions, *Comm. Pure Appl. Math.* 48 (3) (1995), 235–276.

John, F. Formation of singularities in one-dimensional nonlinear wave propagation, *Comm. Pure Appl. Math.* 27 (3) (1974), 377–405.

Partial Differential Equations, Applied Mathematical Sciences 1, Springer-Verlag, New York (1982).

Jorgenson, P. and E. Turkel. Central difference TVD schemes for time dependent and steady state problems, *J. Comp. Phys.* 107 (1993), 297–308.

Kamowitz, D. Some observations on boundary conditions for numerical conservation laws, ICASE Report No 88-67 (1988), ICASE NASA Langley Research Center, Hampton, VA.

Kan, P.T. and G.-Q. Chen. Global solutions to hyperbolic conservation laws with umbilic degeneracy, in *Hyperbolic problems: theory, numerics, applications*, Proceedings of the Fifth International Conference on Hyperbolic Problems, StonyBrook 1994, J. Glimm et al. (Eds.), World Scientific, Singapore (1996), 368–374.

Kaniel, S. A kinetic model for the compressible flow equations, *Indiana Univ. Math. J.* 37 (1988), 537–563.

Karni, S. Far field boundaries and their numerical treatment: an unconventional approach, *Numerical methods for fluid dynamics III*, Proceedings of the conference on Numerical methods for fluid dynamics (held in Oxford, March 1988), K.W. Morton and M.J. Baines (Eds.); IMA Conference Series 17, Clarendon Press, Oxford, 1988, 307–512.

Viscous shock profiles and primitive formulations, *SIAM J. Numer. Anal.* 29 (1992), 1592–1609.

Multicomponent flow calculations by a consistent primitive algorithm, *J. Comp. Phys.* 112 (1994), 31–43.

Kato, T. The Cauchy problem for quasi-linear symmetric hyperbolic systems, *Arch. Rat. Mech. Anal.* 58 (1975), 181–205.

Kawashima, S. and A. Matsumura. Stability of shock profiles in viscoelasticity with non-convex constitutive relations, *Comm. Pure Appl. Math.* 47 (12) (1994), 1547–1569.

Kevorkian, J., J. Yu, and L. Wang. Weakly nonlinear waves for a class of linearly unstable hyperbolic conservation laws with source terms, *SIAM J. Appl. Math.* 55(2) (1995), 446–484.

Keyfitz, B.L. Some elementary connections among nonstrictly hyperbolic conservation laws, in *Nonstrictly Hyperbolic Conservation Laws*, Proceedings of an AMS Special Session, Anaheim CA (1985), B.L. Keyfitz and H.C. Kranzer (Eds.), Contemporary Mathematics 60, AMS, Providence, RI (1987), 79–88.

A survey of nonstrictly hyperbolic conservation laws, in *Nonlinear Hyperbolic Problems, Proceedings St. Etienne 1986*, C. Carasso, P.-A. Raviart and D. Serre (Eds.), Lecture Notes in Mathematics 1270, Springer-Verlag, Berlin (1987), 152–162.

Keyfitz, B.L. and H.C. Kranzer (Editors). *Nonstrictly Hyperbolic Conservation Laws*, Proceedings of an AMS Special Session, Anaheim, CA (1985), Contemporary Mathematics 60, AMS, Providence, RI (1987).

Keyfitz, B.L. and H.C. Kranzer. A system of non-strictly hyperbolic conservation laws arising in elasticity theory, *Arch. Rat. Mech. Anal.* 72 (1980), 219–241.

Spaces of weighted measures for conservation laws with singular shock solutions, *J. Diff. Equations* 118 (1995), 420–451.

Keyfitz, B.L. and M. Shearer (Editors). *Nonlinear Evolution Equations that Change Type*, IMA Volumes in Mathematics and its Applications 27, Springer-Verlag, New York, 1991.

Khanfir, R. *Approximation volumes finis de type cinétique du système hyperbolique de la MHD idéale compressible à pression isotrope*, Doctoral dissertation, Thesis Université Paris-Sud, Orsay, France (1995).

Khobalatte, B. and B. Perthame. A maximum principle on the entropy and minimal limitations for kinetic schemes, *Math. Comp.* 62 (1994), 119–132.

Kim, C.A. and A. Jameson. Flux limited dissipation schemes for high speed unsteady flows, *AIAA* 95-1738-CP, 12th AIAA Computational Fluid Dynamics Conference (San Diego, 1995), 1040–1053.

Klein, R. Detonation initiation due to shock wave-boundary interactions, in *Nonlinear Hyperbolic Equations - Theory, Computation Methods, and Applications* (Proceedings of the Second International Conference on Nonlinear Hyperbolic Problems, Aachen 1988), J. Ballmann and R. Jeltsch (Eds.), Notes on Numerical Fluid Mechanics 24, Vieweg, Braunschweig (1989), 279–288.

Klingenberg, C. and Y.G. Lu. Existence of solutions to resonant systems of conservation laws, in *Hyperbolic problems: theory, numerics, applications*, Proceedings of the Fifth International Conference on Hyperbolic Problems, StonyBrook 1994, J. Glimm et al. Eds., World Scientific, Singapore (1996), 383–374.

Klingenberg, C. and S. Osher. Nonconvex scalar conservation laws in one and two space dimensions, in *Nonlinear Hyperbolic Equations - Theory, Computation Methods, and Applications* (Proceedings of the Second International Conference on Nonlinear Hyperbolic Problems, Aachen 1988), J. Ballmann and R. Jeltsch (Eds.), Notes on Numerical Fluid Mechanics 24, Vieweg, Braunschweig (1989), 289–299.

Klingenstein, P. Hyperbolic conservation laws with source terms: errors of the shock location, SAM Research Report 94-07, ETH Zürich, Switzerland (1994).

Koren, B. Multigrid and defect correction for the steady Navier–Stokes equations. Application to aerodynamics, CWI Tract 74 (Centrum voor Wiskunde en Informatica).

Upwind schemes for the Navier–Stokes equations, in *Nonlinear Hyperbolic Equations - Theory, Computation Methods, and Applications* (Proceedings of the Second International Conference on Nonlinear Hyperbolic Problems, Aachen 1988), J. Ballmann and R. Jeltsch (Eds.), Notes on Numerical Fluid Mechanics 24, Vieweg, Braunschweig (1989), 300–309.

Koren, B. and P.W. Hemker. Damped, direction dependent multigrid for hypersonic flow computations, *Applied Numerical Mathematics* 7 (1991), 309–328, North-Holland.

Koren, B. and S. Spekreijse. Multigrid and defect correction for the efficient solution of the steady Euler equations, in *Research in Numerical Fluid Mechanics (1988), Proceedings of the 25th Meeting of the Dutch Association for Numerical Fluid Mechanics, Delft (1986)*, P. Wesseling (Ed.), Notes on Numerical Fluid Mechanics 17, Vieweg (1987), 87–100.

Kranzer, H.C. and B.L. Keyfitz. A strictly hyperbolic system of conservation laws admitting singular shocks, in *Nonlinear Evolution Equations that Change Type*, IMA Volumes in Mathematics and its Applications 27, B.L. Keyfitz and M. Shearer (Eds.), Springer-Verlag, New York (1991), 107–125.

Spaces of weighted measures for conservation laws with singular shock solutions, *J. Diff. Equations* 118 (1995), 420–451.

Kreiss, H.-O. On Difference Approximations of the Dissipative Type for Hyperbolic Differential Equations, *Comm. Pure Appl. Math.* 17 (1964), 335–353.

Stability theory for difference approximations of mixed initial boundary value problems. I, *Math. Comp.* 22 (1968), 703–714.

Kröner, D. Absorbing boundary conditions for the linearized Euler equations in 2-D, *Math. Comp.* 57 (195) (1991), 153–167.

Kröner, D. and M. Rokyta. Convergence of upwind finite volume schemes for scalar conservation laws in two dimensions, *SIAM J. Numer. Anal.* 31, 2 (1994), 324–343.

Kružkov, S. First-order quasilinear equations in several independent variables, *Mat. Sb.* 123 (1970), 228–255, English translation in *Math. USSR. Sb.* 10 (1970), 217–243.

Kumbaro, A. *Modélisation, analyse mathématique et numérique des modèles bi-fluides d'écoulement diphasique*, Doctoral dissertation, Thesis Université Paris-Sud, Orsay, France (1992).

Kunhardt E.E. and C. Wu. Towards a more accurate Flux Corrected Transport algorithm, *J. Comp. Phys.* 68 (1987), 127–150.

Lallemand, M.H. Dissipative properties of Runge–Kutta schemes with upwind spatial approximation for the Euler equations, INRIA Research Report 1173 (1990), INRIA Rocquencourt, 78153 Le Chesnay, France.

Lallemand, M.H., F. Fezoui, and E. Perez. Un schéma multigrille en éléments finis décentré pour les équations d'Euler, INRIA Research Report 602 (1987), INRIA Rocquencourt, 78153 Le Chesnay, France.

Larrouturou, B. Recent progress in reactive flow computations, in *Computing methods in applied sciences and engineering (Proceedings of the ninth conference on computing methods in applied sciences and engineering, INRIA, Paris 1990)*, R. Glowinski and A. Lichnewsky (Eds.), SIAM, Philadelphia (1990), 249–272.

How to preserve the mass fractions positivitiy when computing compressible multicomponent flows, *J. Comp. Phys.* 92 (1991), 273–295.

On upwind approximations of multi-dimensional multi-species flows, in *Proceedings of the first European CFD Conference*, Ch. Hirsh, J. Périaux and E. Onate (Eds.), Elsevier, Amsterdam (1992), 117–126.

Modélisation physique, numérique et mathématique des phénomènes de propagation de flammes, in *Recent Advances in Combustion Modelling*, Series on Advances in Mathematics for Applied Sciences, vol. 6, World Scientific, Singapore (1991).

Larrouturou, B. and L. Fezoui. On the equations of multi-component perfect or real gas inviscid flow, in *Nonlinear Hyperbolic Problems*, (Proceedings of the Second International Conference on Nonlinear Hyperbolic Problems, Aachen 1988), Carasso et al. (Eds.), Lecture Notes in Mathematics 1402, Springer-Verlag, Berlin (1989).

Lax, P.D. Shock waves and entropy, in *Contributions to nonlinear functional analysis*, E.A. Zarantonello (Ed.), Academic Press, New York (1971), 603–634.

Hyperbolic systems of conservation laws and the mathematical theory of shock waves. Conf. Board. Math. Sci. Regional Conference Series in Applied Mathematics 11, SIAM, Philadelphia, 1972.

Lax, P. and B. Wendroff. Systems of conservation laws, *Comm. Pure Appl. Math.* 13 (1960), 217–237.

Difference schemes for hyperbolic equations with high order of accuracy, *Comm. Pure Appl. Math.* 17 (1964), 381–398.

Le Floch, Ph. Entropy weak solutions to nonlinear hyperbolic systems in nonconservation form, in *Nonlinear Hyperbolic Equations - Theory, Computation Methods, and Applications* (Proceedings of the Second International Conference on Nonlinear Hyperbolic Problems, Aachen 1988), J. Ballmann and R. Jeltsch (Eds.), Notes on Numerical Fluid Mechanics 24, Vieweg, Braunschweig (1989), 362–373, and *Comm. Part. Diff. Equations* 13 (6) (1988), 669–127.

An existence and uniqueness result for two nonstrictly hyperbolic systems, in *Nonlinear Evolution Equations that Change Type*, IMA Volumes in

Mathematics and its Applications 27, B.L. Keyfitz and M. Shearer (Eds.), Springer-Verlag, New York (1991), 126–138.

Le Floch, Ph. and Li Ta-tsien. Un développement asymptotique défini globalement en temps pour la solution du problème de Riemann généralisé, *C. R. Acad. Sci. Paris* 309, Série I (1989), 807–810.

Le Floch, Ph. and J.G. Liu. Discrete entropy and monotonicity criteria for hyperbolic conservation laws, *C. R. Acad. Sci. Paris* 319, Série I (1994), 881–886.

Le Floch, Ph. and T.P. Liu. Existence theory for nonlinear hyperbolic systems in nonconservative form, CMA, École Polytechnnique internal report 254 (1992), 91128 Palaiseau, France, to appear in Forum Mathematicum.

Le Floch, Ph. and P.-A. Raviart. An asymptotic expansion for the solution of the generalized Riemann problem. Part 1: general theory, *Ann. Inst. H. Poincaré, Nonlinear analysis* 5 (2) (1988), 179–207.

Le Floch, Ph. and Z. Xin. Uniqueness via the adjoint problems for systems of conservation laws, *Comm. Pure Appl. Math.* 46 (11) (1993), 1489–1534.

Le Gruyer, E. and A.-Y. LeRoux. A two-dimensional Lagrange-Euler technique for gas dynamics, in *Numerical methods for the Euler equations of fluid dynamics*, Proceedings of the INRIA workshop, Rocquencourt, France (1983), F. Angrand et al. (Eds.), SIAM, Philadelphia (1985), 176–196.

Lerat, A. *Sur le calcul des solutions faibles des systèmes hyperboliques de lois de conservation à l'aide de schémas aux différences*, Doctoral dissertation, Publication ONERA 1981-1 (1981), B.P. 72, 92322 Chatillon Cedex.

Propriété d'homogénéité et décomposition des flux en dynamique des gaz, *J. de Mécanique théorique et appliquée* 2 (2) (1983), 185–213.

Difference methods for hyperbolic problems with emphasis on space-centered approximations, Computational Fluid Dynamics, von Karman Institute for Fluid Dynamics, Lecture Series 1990-03.

Lerat, A. and J. Sidès. A new finite-volume method for the Euler equations with application to transonic flow, in *Numerical Methods in Aeronautical fluid dynamics*, P.L. Roe Ed., Academic Press, New York (1982), 245–288.

LeRoux, A.-Y. *Approximation de quelques problèmes hyperboliques non linéaires*, Doctoral dissertation, Thèse d'état de l'Université de Rennes (1979), 35042 Rennes Cedex, France.

LeVeque, R.J. A large time step generalization of Godunov's method for systems of conservation laws, *SIAM J. Numer. Anal.* 22 (1985), 1051–1073.

Second order accuracy of Brenier's time-discrete method for nonlinear systems of conservation laws *SIAM J. Numer. Anal.* 25 (1988), 1–7.

High resolution finite volume methods on arbitrary grids via wave propagation, *J. Comp. Phys.* 78 (1988), 36–63.

Numerical methods for conservation laws, Lectures in Mathematics, ETH Zürich, Birkhäuser, Basel (1990).

LeVeque, R.J. and J.B. Goodman. TVD Schemes in one and two space dimensions, in *Large-scale computations in Fluid Mechanics*, Proceedings of AMS/SIAM Summer Seminar, La Jolla, CA (1983), Lectures in Applied Mathematics 22, Part 1, B. Engquist, S. Osher and R.C.J. Sommerville (Eds.), AMS, Providence, RI (1985), 51-62.

LeVeque, R.J. and W. Jinghua. A linear hyperbolic system with stiff source term, in *Nonlinear Hyperbolic Problems-Theoretical, Applied and Computational Aspects*, Proceedings of the Fourth International Conference on Hyperbolic Problems, Taormina 1992, A. Donato and F. Oliveri (Eds.), Notes on Numerical Fluid Mechanics 43, Vieweg, Braunschweig (1993), 401–408.

LeVeque, R.J. and K.-M. Shyue. One-dimensional front tracking based on high resolution wave propagation methods, *SIAM J. Sci. Comput.* 16 (1995), 348–379.

Two-dimensional front tracking based on high resolution wave propagation methods, *J. Comp. Phys.* 123 (1996), 354–368.

LeVeque, R.J. and B. Temple. Convergence of Godunov's method for a class of 2×2 conservation laws, *Trans. Am. Math. Soc.* 288 (1985), 115–123.

LeVeque, R.J. and L.N. Trefethen. On the resolvent condition in the Kreiss matrix theorem, ICASE Report No 172177 (1983), ICASE NASA Langley Research Center, Hampton, VA.

LeVeque, R.J. and R. Walder. Grid alignment effects and rotated methods for computing complex flows in astrophysics, ETH Research Report 91-09, CH. 8092 Zürich, Switzerland (1991).

LeVeque, R.J. and H.C. Yee. A study of numerical methods for hyperbolic conservation laws with stiff source terms, *J. Comp. Phys.* 86 (1990), 187–210.

Levy, D.W., K.G. Powell, and B. van Leer. Use of a rotated Riemann solver for the two-dimensional Euler equations, *J. Comp. Phys.* 106 (1993), 201–214.

Liang, S.M. and J.J. Chan. An improved upwind scheme for the Euler equations, *J. Comp. Phys.* 84 (1989), 461–473.

Lighthill, J. *Waves in fluids*, Cambridge University Press (1978)

Lin, H.C. Dissipation additions to flux-difference splitting, *J. Comp. Phys.* 117 (1995), 20-27.

Lin S.-Y., T.-M. Wu, and Y.-S. Chin. Upwind finite-volume method with a triangular mesh for conservation laws, *J. Comp. Phys.* 107 (1993), 324–337.

Lindquist, W.B. The scalar Riemann problem in two spatial dimensions: piecewise smoothness of solutions and its breakdown, *SIAM J. Math. Anal.* 17, (1986) 1178–1197 and in *Nonstrictly Hyperbolic Conservation Laws*, Proceedings of an AMS Special Session, Anaheim, CA (1985), B.L. Keyfitz and H.C. Kranzer (Eds.), Contemporary Mathematics 60, AMS, Providence, RI (1987), 79–88.

Construction of solutions for two-dimensional Riemann problems, *Comp. Math. Appl.* 12A (1986), 615–630.

Lindquist, W.B. (Editor). *Current progress in hyperbolic systems: Riemann problems and computations*, Proceedings of a summer research conference (July 1988), Contemporary Mathematics 100, AMS, Providence, RI (1989).

Lions, P.-L. On kinetic equations, in *Proceedings of the International Congress of Mathematicians, Kyoto, Japan, 1990, Vol. II*, The Mathematical Society of Japan 1991, Springer-Verlag, Tokyo,1173–1185.

Lions, P.-L., B. Perthame, and E. Tadmor. A kinetic formulation of multidimensional scalar conservation laws and related equations, *J. Am. Math. Soc.* 7 (1) (1994), 169–191.

Kinetic formulation of the isentropic gas dynamics and *p*-systems, *Comm. Math. Phys.* 163 (1994), 415–431.

Lions, P.-L. and E. Souganidis. Convergence of MUSCL and filtered schemes for scalar conservation laws and Hamilton–Jacobi equations, *Numerische Mathematik* 69 (4) (1995), 441–470.

Liou M.S., C.J. Steffen, Jr. A new flux-splitting scheme, *J. Comp. Phys.* 107 (1993), 23–39.

Liou M.S., B. van Leer and J.S. Shuen. Splitting of inviscid fluxes for real gases, *J. Comp. Phys.* 87 (1990), 1–24.

Li Ta-tsien. *Global classical solutions for quasilinear hyperbolic systems*, Research in applied mathematics, Masson-Wiley, Paris-Chichester (1994).

Li Ta-tsien and Yu Wen-ci. *Boundary value problems for quasilinear hyperbolic systems*, Duke University mathematical Series V, Durham, 1985.

Liu, J.G. and Z. Xin. Nonlinear stability of discrete shocks for systems of conservation laws, *Arch. Rat. Mech. Anal.* 125 (1993), 217–256.

Liu, T.P. The Riemann problem for general systems of conservation laws, *J. Diff. Equations* 56 (1975), 218-234.

The entropy condition and the admissibility of shocks, *Arch. Rat. Mech. Anal.* 53 (1976), 78–88.

Solutions in the large for the equations of nonisentropic gas dynamics, *Indiana Univ. Math. J.* 26 (1977), 147–177.

Admissible solutions of hyperbolic conservation laws, *A.M.S. Memoirs* 30 (240) (1981).

Liu, T.-P. and Z. Xin. Overcompressive shock wave, in *Nonlinear Evolution Equations that Change Type*, IMA Volumes in Mathematics and its Applications 27, B.L. Keyfitz and M. Shearer (Eds.), Springer-Verlag, New York (1991), 139–145.

Liu, T.-P. and L.-A. Ying. Nonlinear stability of strong detonations for a viscous combustion model, *SIAM J. Math. Anal.* 26 (3) (1995), 519–528.

Liu, X.B. A maximum principle satisfying modification of triangle based adaptive stencils for the solution of scalar hyperbolic conservation laws, *SIAM J. Numer. Anal.* 30 (1993), 701–716.

Liu X.D., S. Osher, and T. Chan. Weighted essentially non-oscillatory schemes, *J. Comp. Phys.* 115 (1994), 200–212.

Liu, Y. and M. Vinokur. Nonequilibrium flow computations. An analysis of numerical formulations of conservation laws, *J. Comp. Phys.* 83 (1989), 373–397.

Loh, C.Y. and W.H. Hui. A new Lagrangian method for steady supersonic flow computation I. Godunov scheme, *J. Comp. Phys.* 89 (1990), 207–240.

Loh, C.Y. and M.S. Liou. Lagrangian solution of supersonic real gas flow, *J. Comp. Phys.* 104 (1993), 150-161.

Lucier, B.J. Error bounds for the methods of Glimm, Godunov and LeVeque, *SIAM J. Numer. Anal.* 22 (1985), 1074–1081.

Lytton, C.C. Solution of the Euler equations for transonic flow over a lifting aerofoil - The Bernoulli formulation (Roe–Lytton method), *J. Comp. Phys.* 73 (1987), 395–431.

Macrossan, M.N. The equilibrium flux method for the calculation of flows with nonequilibrium chemical reactions, *J. Comp. Phys.* 80 (1989), 204–231.

Majda, A. A qualitative model for dynamic combustion, *SIAM J. Appl. Math.* 41 (1) (1981), 70-93.

Compressible fluid flow and systems of conservation laws in several space variables, Appl. Math. Science 53, Springer-Verlag, New York (1984).

Nonlinear geomatric optics for hyperbolic systems of conservation laws, in On perspective and open problems in multi-dimensional conservation laws, in *Oscillation Theory, Computation, and Methods of Compensated compactness*, IMA Volumes in Mathematics and its Applications 2, C. Dafermos et al. (Eds.), Springer-Verlag, New York (1986), 115–165.

Multidimensional Hyperbolic Problems and Computations, Proceedings of the IMA workshop (1989), The IMA Volumes in Mathematics and its applications 29, J. Glimm, A.J. Majda (Eds.), Springer-Verlag, New York (1991), 217–238.

Majda, A. and S. Osher. Numerical viscosity and the entropy condition, *Comm. Pure. Appl. Math.* 32 (1979), 797–838.

Majda, A. and R.L. Pego. Stable viscosity matrices for systems of conservation laws, *J. Diff. Equations* 56 (1985), 229–262.

Majda, A. and J. Ralston. Discrete shock profiles for systems of conservation laws, *Comm. Pure. Appl. Math.* 32 (1979), 445–482.

Mandal, J.C. and S.M. Deshpande. Higher order accurate kinetic flux vector splitting method for Euler equations, in *Nonlinear Hyperbolic Equations - Theory, Computation Methods, and Applications* (Proceedings of the Second International Conference on Nonlinear Hyperbolic Problems, Aachen 1988), J. Ballmann and R. Jeltsch (Eds.), Notes on Numerical Fluid Mechanics 24, Vieweg, Braunschweig (1989), 384–392.

Kinetic flux vector splitting for Euler equations, *Comp. Fluids* 23 (1994), 447–478.

Mao, D.K. A treatment of discontinuities for finite difference methods, *J. Comp. Phys.* 103 (1992), 359–369.

A treatment of discontinuities for finite difference methods in the two-dimensional case, *J. Comp. Phys.* 104 (1993), 377–397.

Marshall, G. and B. Plohr. A random choice method for two-dimensional steady supersonic shock wave diffraction problems, *J. Comp. Phys.* 56 (1992), 410–427.

Mavripilis, D.J. Adaptive mesh generation for viscous flows using Delaunay triangulation, *J. Comp. Phys.* 90 (1990), 271–291.

Unstructured and adaptative mesh generation for high Reynolds number viscous flow, in *Numerical grid generation in computational fluid dynamics and related fields*, Proc. of the third Int. Conference, Barcelona, Spain, June 1991, A.S.-Arcilla et al. (Eds.), Elsevier Science Publishers B.V., North Holland, Amsterdam (1991), 79–91.

Unstructured mesh algorithms for aerodynamic calculations, ICASE Report 92-35 (1992), ICASE NASA Langley Research Center, Hampton, VA.

An advancing front Delaunay triangulation algorithm designed for robustness, *J. Comp. Phys.* 117 (1995), 90–101.

Mazet, P.A. Sur une formulation variationnelle des systèmes hyperboliques conservatifs, *La Recherche Aérospatiale* 5 (1983), 121–129.

Mazet, P.A. and F. Bourdel. Sur une formulation variationnelle entropique des systèmes hyperboliques conservatifs : cas pluridimensionnel, *La Recherche Aérospatiale* 5 (1985), 369–378.

Mehlman, G. *Étude de quelques problèmes liés aux écoulements en déséquilibre chimique*, Doctoral dissertation, Thesis École Polytechnique (1991), Palaiseau, France.

Menikoff, R. Analogies between Riemann problem for 1-D fluid dynamics and 2-D steady supersonic flow, in *Current progress in hyperbolic systems: Riemann problems and computations*, Proceedings of a summer research conference (July 1988), Contemporary Mathematics 100, AMS, Providence, RI (1989).

Menikoff, R. and B.J. Plohr. The Riemann problem for fluid flow of real materials, *Rev. Mod. Phy.* 61 (1) (1989), 75–130.

Menne, S., C. Weiland, D. D'Ambrosio, and M. Pandolfi. Comparison of real gas simulations using different numerical methods, *Comp. Fluids* 24 (3) (1995), 189–208.

Metivier, G. Interaction de deux chocs pour un système de deux lois de conservation en dimension deux d'espace, *Trans. Am. Math. Soc.* 296 (1986), 431–479.
Stability of multi-dimensional weak shocks, in *Multidimensional Hyperbolic Problems and Computations*, Proceedings of the IMA workshop (1989), The IMA Volumes in Mathematics and its applications 29, J. Glimm, A.J. Majda (Eds.), Springer-Verlag, New York (1991), 239–250.

Montagné, J.-L. Étude de schémas numériques décentrés en dynamique des gaz bidimensionnelle, *La Recherche Aérospatiale* 5 (1984), 323–338.
Utilisation d'un schéma décentré pour la simulation d'écoulements non visqueux de gaz réel à l'équilibre, *La Recherche Aérospatiale* 6 (1986), 433–441.

Montagné, J.-L., H.C. Yee, and M. Vinokur. Comparative study of high-resolution shock-capturing schemes for a real gas, in *Proceedings of the seventh GAMM-Conference on Numerical Methods in Fluid Mechanics (1988)*, M. Deville (Ed.), Notes on Numerical Fluid Mechanics 20, Vieweg, Braunschweig (1988), 219–228.

Morton, K.W. and P.K. Sweby. A comparison of flux limited difference methods and characteristic Galerkin methods for shock modelling, *J. Comp. Phys.* 73 (1987), 203–229.

Mulder, W., S. Osher, and J. Sethian. Computing interface motion in compressible gas dynamics, *J. Comp. Phys.* 100 (1992), 209–228.

Mulder, W. and B. van Leer. Experiments with implicit upwind methods for the Euler equations, *J. Comp. Phys.* 59 (1985), 232-246.

Müller, E. Flux vector splitting for the Euler equations for real gases, *J. Comp. Phys.* 79 (1988), 227–230.

Munz, C.-D. On Godunov-type schemes for Lagrangian gas dynamics, *SIAM J. Numer. Anal.* 31 (1994), 17–42.

Munz, C.-D. and L. Schmidt. Numerical simulations of compressible hydrodynamic instabilities with high resolution schemes, in *Nonlinear Hyperbolic Equations - Theory, Computation Methods, and Applications* (Proceedings of the Second International Conference on Nonlinear Hyperbolic Problems, Aachen 1988), J. Ballmann and R. Jeltsch (Eds.), Notes on Numerical Fluid Mechanics 24, Vieweg, Braunschweig (1989), 456–465.

Murat, F. Compacité par compensation, *Ann. Scuola Norm. Sup. Pisa* 5 (1978), 489-507.

Nessyahu, H. and E. Tadmor. Non-oscillatory central differencing for hyperbolic conservation laws, *J. Comp. Phys.* 87 (1990), 408–463.

The convergence rate of approximate solutions for nonlinear scalar conservation laws, *SIAM J. Numer. Anal.* 29 (6) (1992), 1505–1519.

Nessyahu, H., E. Tadmor, and T. Tassa. The convergence rate of Godunov-type schemes, *SIAM J. Numer. Anal.* 31 (1) (1994), 1–16.

N'konga, B. and H. Guillard. Godunov type method on non-structured meshes for three-dimensional moving boundary problems, *Comp. Methods Appl. Mech. Engrg.* 113 (1994), 183–204.

Nishida, T. and J. Smoller. Convergence of finite difference approximations to nonlinear parabolic systems, in *Nonlinear partial differential equations*, Proceedings of a conference held in Durham (1982), Smoller J.A. (Ed.), Contemporary Mathematics 17, AMS, Providence, RI (1980).

Noelle, S. Hyperbolic systems of conservation laws, the Weyl equation, and multidimensional upwinding, *J. Comp. Phys.* 115 (1994), 22–26.

Oleinik, O. Discontinuous solutions of nonlinear differential equation, *Usp. Mat. Nauk. (NS)* 12 (1957), 3-73. English translation in *Am. Math. Soc. Trans., Ser. 2* 26 (1963), 95–172.

Oliger, J. and A. Sundström. Theoretical and practical aspects of some initial boundary value problems in fluid dynamics, *SIAM J. Appl. Math.* 35 (1978), 419–446.

Oran, E.S. and J.P. Boris. *Numerical simulation of reactive flow*, Elsevier, New York (1987).

Osher, S. Riemann solvers, the entropy condition and difference approximations, *SIAM J. Numer. Anal.* 21 (1984), 217–235.

Convergence of generalized MUSCL schemes, *SIAM J. Numer. Anal.* 22 (1985), 947–961.

Osher, S. and S. Chakravarthy. Upwind schemes and boundary conditions with applications to Euler equations in general geometries, *J. Comp. Phys.* 50 (1983), 447–481.

Osher, S. and R. Sanders. Numerical approximations to nonlinear conservation laws with locally varying time and space grids, *Math. Comp.* 41 (1983), 321–336.

Osher, S. and F. Solomon. Upwind difference schemes for hyperbolic systems of conservation laws, *Math. Comp.* 38 (1982), 339–374.

Osher, S. and E. Tadmor. On the convergence of difference approximations to scalar conservation laws, *Math. Comp.* 50 (1988), 19–51.

Pego, R.L. Phase transitions in one-dimensional nonlinear viscoelasticity: admissibility and stability, *Arch. Rat. Mech. Anal.* 97 (1987), 353–394.

Pember, R.B., J.B. Bell, P. Colella, W.Y. Crutchfield, and M.L. Welcome. An adaptive cartesian grid method for unsteady compressible flow in irregular regions, *J. Comp. Phys.* 120 (1995), 278–304.

Peng, Y.-J. Solutions faibles globales pour l'équation d'Euler d'un fluide compressible avec de grandes données initiales, *Comm. Part. Diff. Equations* 17 (1992), 161–187.

Perthame, B. Global existence of solutions to the BGK model of Boltzmann equations, *J. Diff. Equations* 81 (1989), 191–205.

Boltzmann type schemes for gas dynamics and the entropy property, *SIAM J. Numer. Anal.* 27 (6) (1990), 1405–1421.

Second order Boltzmann schemes for compressible Euler equations in one and

two space variables, *SIAM J. Numer. Anal.* 29 (1) (1992), 1–19.
Convergence of N-schemes for linear advection equations, *SIAM Symposium on trends in applications of mathematics to mechanics* (Rodrigues Ed.), Lisbon July 1994.
Introduction to the collision models in Boltzmann's theory, to appear in *Fluides et collisions*, P.-A. Raviart (Ed.), Research in Applied Mathematics, Wiley-Masson, Paris-New York (1996).

Perthame, B. and M. Pulvirenti. Weighted L^∞ bounds and uniqueness for the Boltzmann BGK model, *Arch. Rat. Mech. Anal.* 125 (1993), 289–295.

Perthame, B. and Y. Qiu. A variant of van Leer's method for multidimensional systems of conservation laws, *J. Comp. Phys.* 112 (1994), 370–381.

Perthame, B., Y. Qiu, and B. Stoufflet. Sur la convergence des schémas fluctuation-splitting pour l'advection et leur utilisation en dynamique des gaz, *C. R. Acad. Sci. Paris* 319 , Série I (1994), 283–288.
Kinetic discretization of gas dynamics using fluctuation-splitting, in *Hyperbolic problems: theory, numerics, applications*, Proceedings of the Fifth International Conference on Hyperbolic Problems, StonyBrook 1994, J. Glimm et al. Eds., World Scientific, Singapore (1996), 207–216.

Perthame, B. and C.-W. Shu. On positivity preserving finite volume schemes for Euler equations, *Numerische Mathematik* 73 (1) (1996), 119–130.

Perthame, B. and E. Tadmor. A kinetic equation with kinetic entropy functions for scalar conservation laws, *Comm. Math. Phys.* 136 (1991), 501–517.

Peyret, R. and T.D. Taylor. *Computational Methods for Fluid Flow*, Springer Series in Computational Physics, Springer-Verlag, New York (1983).

Pike, J. Grid adaptative algorithms for the solution of the Euler equations on irregular grids, *J. Comp. Phys.* 71 (1987), 194–223.

Poinsot, T.J. and S.K. Lele. Boundary conditions for direct simulations of compressible viscous flows, *J. Comp. Phys.* 101 (1992), 104–129.

Powell K.G. An approximate Riemann solver for magnetohydrodynamics, ICASE Report No 94-24 (1994), ICASE NASA Langley Research Center, Hampton, VA.

Prendertgast, K.H. and K. Xu. Numerical hydrodynamics from gas-kinetic theory, *J. Comp. Phys.* 109 (1993), 53–66.

Pullin, D. Direct simulation methods for compressible inviscid ideal gas flow, *J. Comp. Phys.* 34 (1980), 231–244.

Quirk, J. Godunov-type schemes applied to detonation flows, ICASE Report No. 93-15 (1993), ICASE NASA Langley Research Center, Hampton, VA.
A contribution to the great Riemann solver debate, *Int. J. Numer. Methods Fluids* 18 (1994), 555–574.

Radespiel, R. and N. Kroll. Accurate flux-vector splitting for shocks and shear layers, *J. Comp. Phys.* 121 (1995), 66–78.

Rascle, M. Convergence of approximate solutions to some systems of conservative laws: a conjecture on the product of Riemann invariants, in *Oscillation Theory, Computation, and Methods of Compensated compactness*, IMA Volumes in Mathematics and its Applications 2, C. Dafermos et al. (Eds.), Springer-Verlag, New York (1986), 85–88.

Rauch, J. BV estimates fail for most quasilinear hyperbolic systems in dimension greater than one, *Comm. Math. Phys.* 106 (1986), 484–489.

Raviart, P.A. and L. Sainsaulieu. Mathematical and numerical modelling of two-phase flows, in *Computing Methods in Applied Sciences and Engineering*, Proceedings of the tenth International Conference on Computing Methods in Applied Sciences and Engineering, France (1992), R. Glowinski (Ed.), Nova Science Publishers, Inc., New York, 119–132.

Reitz, R.D. One-dimensional compressible gas dynamics calculations using the Boltzmann equation, *J. Comp. Phys.* 42 (1981), 108–123.

Richtmyer, R.D. and K.W. Morton. *Difference Methods for Initial-Value Pro:-blems*, Interscience, New York (1967).

Rider, W.J. A review of approximate Riemann solvers with Godunov's method in Lagrangian coordinates, *Comp. Fluids*, 23 (1994), 397–413.

Roache, P.J. *Computational fluid dynamics*, Hermosa Publishers, Albuquerque, NM, 1972.

Roberts, T.W. The behavior of flux difference splitting schemes near slowly moving shock waves, *J. Comp. Phys.* 90 (1990), 141–160.

Roe, P.L. Approximate Riemann solvers, parameter vectors and difference schemes, *J. Comp. Phys.* 43 (1981), 357–372.

Upwind schemes using various formulations of the Euler equations, in *Numerical Methods for the Euler Equations of fluid Dynamics*, Proceedings of the INRIA workshop, Rocquencourt, France (1983), F. Angrand et al. (Eds.), SIAM, Philadelphia (1985).

Some Contributions to the Modelling of Discontinuous Flows, in *Large-scale computations in Fluid Mechanics*, Proceedings of AMS/SIAM Summer Seminar, La Jolla, USA (1983), B. Engquist et al. (Eds.), Part 2, Lectures in Applied Mathematics 22, AMS, Providence, RI (1985), 163–193.

A basis of upwind differencing of the two-dimensional unsteady Euler equations, in *Numerical methods for fluid dynamics II*, Proceedings of the conference on Numerical methods for fluid dynamics (held in Reading, April 1985), K.W. Morton and M.J. Baines (Eds.); IMA Conference Series, Clarendon Press, Oxford (1986), 55–80.

Discrete models for the numerical analysis of time-dependent multidimensional gas dynamics, *J. Comp. Phys.* 63 (1986), 458–476.

Characteristic based schemes for the Euler equations, *Annu. Rev. Fluid Mech.* 18 (1986), 337–365.

Upwind differencing schemes for hyperbolic conservation laws with source terms, in *Nonlinear Hyperbolic Problems, Proceedings St. Etienne 1986*, C. Carasso P-A. Raviart D. Serre (Eds.), Lecture Notes in Mathematics 1270, Springer-Verlag, Berlin (1987), 41–51.

Remote boundary conditions for unsteady multidimensional aerodynamic computations, ICASE Report No. 86-75 (1986), ICASE NASA Langley Research Center, Hampton, VA.

Discontinuous solutions to hyperbolic systems under operator splitting, ICASE Report No. 87-64 (1987), ICASE NASA Langley Research Center, Hampton, VA.

Sonic flux formulae, *SIAM J. Sci. Statist. Comp.* 13 (1992), 611–630.

Linear bicharacteristic schemes without dissipation, ICASE Report No. 94-65 (1994), ICASE NASA Langley Research Center, Hampton, VA.

Local reduction of certain wave operators to one-dimensional form, ICASE

Report No. 94-66 (1994), ICASE NASA Langley Research Center, Hampton, VA.

Roe, P.L. and J. Pike. Efficient Construction and Utilisation of Approximate Riemann Solutions, in *Computing Methods in Applied Sciences and Engineering, VI*, Proceedings of the Sixth International Symposium on Computing Methods in Applied Sciences and Engineering, France (1983), R. Glowinski, J.L. Lions (Eds.), Elsevier Science Publishers B.V., North-Holland, Amsterdam, INRIA (1984), 499–518.

Roe, P.L. and D. Sidilkover. Optimum positive linear schemes for advection in two and three dimensions, *SIAM J. Numer. Anal.* 29 (1992), 1542–1568.

Roe, P.L., R. Struijs, and H. Deconinck. A conservative linearisation of the multi-dimensional Euler equations, to appear in *J. Comp. Physics.*

Rostand, P. and B. Stoufflet. TVD schemes to compute compressible viscous flows on unstructured meshes, in *Nonlinear Hyperbolic Equations - Theory, Computation Methods, and Applications* (Proceedings of the Second International Conference on Nonlinear Hyperbolic Problems, Aachen 1988), J. Ballmann and R. Jeltsch (Eds.), Notes on Numerical Fluid Mechanics 24, Vieweg, Braunschweig (1989), 510–520.

Rubino, B. On the vanishing viscosity approximation to the Cauchy problem for a 2×2 system of conservation laws, *Ann. Inst. Henri Poincaré* 10 (1993), 627–656.

Convergence of approximate solutions of the Cauchy problem for a 2×2 nonstrictly hyperbolic system of conservation laws, *Nonlinear Hyperbolic Problems-Theoretical, Applied and Computational Aspects*, Proceedings of the Fourth International Conference on Hyperbolic Problems, Taormina 1992, A. Donato and F. Oliveri (Eds.), Notes on Numerical Fluid Mechanics 43, Vieweg, Braunschweig (1993), 487–494.

Rudy D.H. and J.C. Strikwerda. A nonrefecting outflow boundary condition for subsonic Navier–Stokes calculations, *J. Comp. Phys.* 36 (1980), 35–70.

Rumsey, C., B. van Leer, and P.L. Roe. A multidimensional flux function with applications to the Euler and Navier–Stokes equations, *J. Comp. Phys.* 105 (1993), 306–323.

Sablé-Tougeron, M. Méthode de Glimm et problème mixte, *Ann. Inst. H. Poincaré, Nonlinear analysis* 10 (1993), 423–443.

Sainsaulieu, L. *Modélisation, analyse mathématique et numérique d'écoulements diphasiques constitués d'un brouillard de gouttes*, Doctoral dissertation, Thesis École Polytechnique (Palaiseau), France (1991).

Travelling waves solutions of convection-diffusion systems in nonconservation form, CERMICS Report 93-15 (1993), CERMICS-ENPC, 93167 Noisy le Grand, France, to appear in *SIAM J. Math. Anal.*

Finite volume approximation of two-phase-fluid flows based on an approximate Roe-type Riemann solver, *J. Comp. Phys.* 121 (1995), 1–28.

Salari, K. and S. Steinberg. Flux-Corrected transport in a moving grid, *J. Comp. Phys.* 111 (1994), 24–32.

Saltzman, J. An unsplit 3d upwind method for hyperbolic conservation laws, *J. Comp. Phys.* 115 (1994), 153–168.

Sanders, R. On convergence of monotone finite difference schemes with variable spatial differencing, *Math. Comp.* 40 (1983), 91–106.

Finite difference techniques for nonlinear hyperbolic conservation laws, in *Large-scale computations in Fluid Mechanics*, Proceedings of AMS/SIAM Summer Seminar, La Jolla, CA (1983), Lectures in Applied Mathematics 22, Part 1, B. Engquist, S. Osher, and R.C.J. Sommerville (Eds.), AMS, Providence, RI (1985), 209–220.

Sanders, R. and C.P. Li. A variation nonexpansive central differencing scheme for nonlinear hyperbolic conservation laws, in *Computing Methods in Applied Sciences and Engineering*, Proceedings of the tenth International Conference on Computing Methods in Applied Sciences and Engineering, France (1992), R. Glowinski (Ed.), Nova Science Publishers, Inc., New York (1992), 511–526.

Sanders, R. and A. Weiser. A high resolution staggered mesh approach for nonlinear hyperbolic systems of conservation laws, *J. Comp. Phys.* 101 (1992), 314–329.

Saurel, R., M. Larini, and J.-C. Loraud. Exact and approximate Riemann solvers for real gases, *J. Comp. Phys.* 112 (1994), 126–137.

Schaeffer, D.G. and M. Shearer. The classification of 2×2 systems of non-strictly hyperbolic conservation laws with application to oil recovery, *Comm. Pure. Appl. Math.* 40(1987), 141–178.

Loss of hyperbolicity in yield vertex plasticity models under nonproportional loading, in *Nonlinear Evolution Equations that Change Type*, IMA Volumes in Mathematics and its Applications 27, B.L. Keyfitz and M. Shearer (Eds.), Springer-Verlag, New York (1991), 192–217.

Schneider, V., U. Katscher, D.H. Rischke, B. Waldhauser, J.A. Maruhn, and C.-D. Munz. New algorithms for ultra-relativistic numerical hydrodynamics, *J. Comp. Phys.* 105 (1993), 92–107.

Schochet, S. Sufficient conditions for local existence via Glimm's scheme for large BV data, *J. Diff. Equations* 89 (1991), 317–354.

Schochet, S. and E. Tadmor. Regularized Chapman–Enskog expansion for scalar conservation laws, *Arch. Rat. Mech. Anal.* 119 (1992), 95–107.

Schroll, H.J., A. Tveito, and R. Winther. A system of conservation laws with a relaxation term, in *Hyperbolic problems: theory, numerics, applications*, Proceedings of the Fifth International Conference on Hyperbolic Problems, StonyBrook 1994, J. Glimm et al. Eds., World Scientific, Singapore (1996), 431–439.

Schulz-Rinne, C. Classification of the Riemann problem for two-dimensional gas dynamics, *SIAM J. Math. Anal.* 24 (1) (1993) 76–88.

Schulz-Rinne, C., J.P. Collins, and H.M. Glaz. Numerical solution of the Riemann problem for two-dimensional gas dynamics, *SIAM J. Sci. Comp.* 14 (6) (1993).

Selmin, V. and L. Quartapelle. A unified approach to build artificial dissipation operators for finite element and finite volume discretisation, in *Finite elements in fluids, New trends and applications*, Proceedings of the eigth International Conference on Finite Elements in fluids, Barcelona, Spain (1993), K. Morgan et al. (Eds.), CIMNE, Pineridge Press (1993), Part II, 1329–1341.

Serre, D. La compacité par compensation pour les sytèmes non linéaires de deux équations à une dimension d'espace, *J. Math. Pures et Appl.* 65 (1987), 423–468.

Solutions à variations bornées pour certains systèmes hyperboliques de lois de conservation, *J. Diff. Equations* 68 (1987), 137–168.

Problèmes de Riemann singuliers, Preprint, Publications de l'équipe d'Analyse numérique de Lyon-St Etienne 81 (1992), France.

Richness and the classification of quasilinear hyperbolic systems, in *Multi-dimensional Hyperbolic Problems and Computations*, Proceedings of the IMA workshop (1989), The IMA Volumes in Mathematics and its applications 29, J. Glimm, A.J. Majda (Eds.), Springer-Verlag, New York (1991), 315–333.

La stabilité d'une méthode de pas fractionnaires pour la dynamique des gaz, Report UMPA 86 (1992), ENS Lyon, France.

Systèmes de lois de conservation, tome I et II, Diderot, Paris (1996).

Shearer, M. Loss of strict hyperbolicity of the Buckley–Leverett equations for three phase flow in a porous medium, in *Numerical simulation in oil recovery*, Proceedings of the IMA workshop (1986-87), The IMA Volumes in Mathematics and its applications 11, M.F. Wheeler (Ed.), Springer-Verlag, New York (1988), 263–283.

Shearer, M., D.G. Schaeffer, D. Marchesin, and P.L. Paes-Leme. Solution of the Riemann problem for a prototype 2×2 system of non-strictly hyperbolic conservation laws, *Arch. Rat. Mech. Anal.* 97 (1987), 299–320.

Shearer, M. and S. Schecter. Undercompressive shocks in systems of conservation laws, in *Nonlinear Evolution Equations that Change Type*, IMA Volumes in Mathematics and its Applications 27, B.L. Keyfitz and M. Shearer (Eds.), Springer-Verlag, New York (1991), 218–231.

Sheng, W.C. and D. Tan. Weak deflagration solutions to the simplest combustion model, *J. Diff. Equations* 107 (1994), 207–230.

Shu, C.W. T.V.B. high-order schemes for conservation laws, *Math. Comp.* 49 (1987) 105–121.

A numerical method for systems of conservation laws of mixed type admitting hyperbolic flux splitting, *J. Comp. Phys.* 100 (1992), 424–429.

Shu, C.W. and S. Osher. Efficient implementation of Essentially Non-Oscillatory shock-capturing schemes, I, *J. Comp. Phys.* 77 (1988), 439–471. II, *J. Comp. Phys.* 83 (1989), 32–78.

Shu, C.W., T. Zang, G. Erlebacher, D. Whitaker, and S. Osher. High-order ENO schemes applied to two- and three-dimensional compressible flow, *Appl. Num. Math.* 9 (1992), 45–71.

Shuen, J.S., M.-S. Liou, and B. van Leer. Inviscid flux-splitting algorithms for real gases with non-equilibrium chemistry, *J. Comp. Phys.* 90 (1990), 371–395.

Sidilkover, D. A genuinely multidimensional upwind scheme and efficient multigrid solver for the compressible Euler equations, ICASE Report 94-84 (1994), ICASE NASA Langley Research Center, Hampton, VA, and in *Hyperbolic problems: theory, numerics, applications*, Proceedings of the Fifth International Conference on Hyperbolic Problems, StonyBrook 1994, J. Glimm et al. Eds., World Scientific, Singapore (1996), 447–455.

Sidilkover, D. and P.L. Roe. Unification of some advection schemes in two dimensions, ICASE Report No 95-10 (1987), ICASE NASA Langley Research Center, Hampton, VA.

Smith, R. The Riemann problem in gas dynamics, *Trans. Am. Math. Soc.* 249, 1 (1979), 1–50.

Smoller, J. *Shock Waves and Reaction-Diffusion Equations*, Grundlehren der mathematischen Wissenschaften 258, Springer-Verlag, New York (Second Edition 1994).

Smoller J.A. (Editor). *Nonlinear partial differential equations*, Proceedings of a conference held in Durham (1982), Contemporary Mathematics 17, AMS, Providence, RI (1980).

Smoller, J.A., J.B. Temple, and Z.P. Xin. Instability of rarefaction shocks in systems of conservation laws, *Arch. Rat. Mech. Anal.* 112 (1990), 63–81.

Sod, G.A. A survey of several finite difference methods for systems of nonlinear hyperbolic conservation laws, *J. Comp. Phys.* 27 (1978), 1–31.

Numerical Methods in fluid dynamics: Initial and boundary value problems, Cambridge University Press, Cambridge 1987.

Song, Y. and T. Tang. Dispersion and group velocity in numerical schemes for three-dimensional hydrodynamic equations, *J. Comp. Phys.* 105 (1993), 72–82.

Spekreijse, S. Multigrid solution of monotone second-order discretization of hyperbolic conservation laws, *Math. Comp.* 49 (1987), 135–156.

Steger, J. and R.F. Warming. Flux vector splitting of the inviscid gas dynamics equation with application to finite difference methods, *J. Comp. Phys.* 40 (1981), 263–293.

Stewart, H.B. and B. Wendroff. Two phase flow: models and methods, *J. Comp. Phys.* 56 (1984), 363–409.

Stoufflet, B. Implicit finite element methods for the Euler equations, in *Numerical methods for the Euler equations of fluid dynamics*, Proceedings of the INRIA workshop, Rocquencourt, France (1983), F. Angrand et al. (Eds.), SIAM Philadelphia (1985), 409–434.

Sun, M.T., S.T. Wu, and M. Dryer. On the time-dependent numerical boundary conditions of magnetohydrodynamic flows, *J. Comp. Phys.* 116 (1995), 330–342.

Swanson, R.C. and E. Turkel. On central-difference and upwind schemes, *J. Comp. Phys.* 101 (1992), 292–306.

Sweby, P.K. High resolution schemes using flux limiters for hyperbolic conservation laws, *SIAM J. Numer. Anal.* 21 (1944), 995–1011.

"TVD" schemes for inhomogenous conservation laws, in *Nonlinear Hyperbolic Equations - Theory, Computation Methods, and Applications* (Proceedings of the Second International Conference on Nonlinear Hyperbolic Problems, Aachen 1988), J. Ballmann and R. Jeltsch (Eds.), Notes on Numerical Fluid Mechanics 24, Vieweg, Braunschweig (1989), 599–607.

Szepessy, A. Measure valued solutions to scalar conservation laws with boundary conditions, *Arch. Rat. Mech. Anal.* (1989), 181–193.

An existence result for scalar conservation laws using measure valued solutions, *Comm. Part. Diff. Equations* 14 (10) (1989), 1329–1350.

Convergence of a streamline diffusion finite element method for scalar conservation laws with boundary condition, *M^2AN, Modélisation mathématique et analyse numérique*, 25 (1991), 749–782.

Tadmor, E. The equivalence of L^2-stability, the resolvent condition and strict H-stability, *Linear algebra and its applications* 41 (1981), 151–159.

The unconditional instability of inflow-dependent boundary conditions in difference approximations of hyperbolic systems, *Math. Comp.* 41 (164) (1983),

309–319.
A minimum entropy principle in the gas dynamics equations, ICASE Report 86-33, (1986), or in Advances in numerical and applied mathematics, J.C. South and M.Y. Hussaini (Eds.), ICASE Report 86-18, (1986), 100–118, ICASE NASA Langley Research Center, Hampton, VA.
Entropy functions for symmetric systems of conservation laws, *J. Math. Anal. Appl.* 122 (1987), 355–359.
The numerical viscosity of entropy stable schemes for systems of conservation laws, I, *Math. Comp.* 49 (1987), 91–103.
The numerical viscosity of entropy stable schemes for systems of conservation laws, II, ICASE Report, ICASE NASA Langley Research Center, Hampton, VA.
Tadmor, E. and T. Tassa. On the piecewise smoothness of entropy solutions to scalar conservation laws, *Comm. Part. Diff. Equations* 18 (9&10) (1993), 1631–1652.
Tamura, Y. and K. Fujii. A multi-dimensional upwind scheme for the Euler equations on structured grid, *Computers Fluids*, 22 (1993), 125–138.
Tan, D. and T. Zhang. Riemann problem for self-similar ZND-Model in gas dynamical combustion, *J. Diff. Equations* 95 (1992), 331–369.
Two-dimensional Riemann problem for a hyperbolic system of nonlinear conservation laws, I. Four-J cases, *J. Diff. Equations* 111 (1994), 203–254.
Tan, D., T. Zhang, and Y. Zheng. Delta-shock waves as limits of vanishing viscosity for hyperbolic systems of conservation laws, *J. Diff. Equations* 112 (1994), 1–32.
Tanaka, T. Finite volume TVD scheme on an unstructured grid system for three-dimensional M.H.D. simulation of inhomogeneous sytems including strong background potential fields, *J. Comp. Phys.* 111 (1994), 381–389.
Tang, T. and Z.H. Teng. Error bounds for fractional step methods for conservation laws with source terms, *SIAM J. Numer. Anal.* 32 (1995), 110–127.
Taniuti, T. and K. Nishihara. *Nonlinear waves*, Pitman Monographs and Studies in Mathematics 15, Pitman, London (1983).
Tartar, L. Compensated compactness and applications to partial differential equations, in *Nonlinear analysis and mechanics: Heriot-Watt Symposium, Vol. IV*, R.J. Knops (Ed.), Research Notes in Mathematics 39, Pitman London (1979), 136–192.
The compensated compactness method applied to systems of conservation laws, in *Systems of Nonlinear Partial Differential Equations*, J.M. Ball (Ed.), NATO ASI Series, Reidel, Dordrecht (1983), 263–285.
Une introduction à la théorie mathématique des systèmes hyperboliques de lois de conservation, Report 682 (1989), Istituto di Analisi del CNR, 27100 Pavia, Italy.
Taylor, M. *Partial Differential Equations*, Applied Mathematical Sciences 117, Springer-Verlag, New York (1996).
Temple, B. Global solution of the Cauchy problem for a class of 2 × 2 nonstrictly hyperbolic conservation laws, *Adv. Appl. Math.* 3 (1982), 335–375.
Systems of conservation laws with invariant submanifolds, *Trans. Am. Math. Soc.* 280 (1983), 781–795.

The L^1-norm distinguishes the strictly hyperbolic from a non-strictly hyperbolic theory of the initial value problem for systems of conservation laws, in *Nonlinear Hyperbolic Equations - Theory, Computation Methods, and Applications* (Proceedings of the Second International Conference on Nonlinear Hyperbolic Problems, Aachen 1988), J. Ballmann and R. Jeltsch (Eds.), Notes on Numerical Fluid Mechanics 24, Vieweg, Braunschweig (1989), 608–616.

Teng, Z.-H. On the accuracy of fractional step methods for conservation laws in two dimensions, *SIAM J. Numer. Anal.* 31 (1994), 43–63.

Teng, Z.-H., A.J. Chorin, and T.-P. Liu. Riemann problems for reacting gas with applications to transition, *SIAM J. Appl. Math.* 42 (5) (1982), 964–981.

Thompson, K.W. Time dependent boundary conditions for hyperbolic systems, *J. Comp. Phys.* 68 (1987), 1–24.

Time dependent boundary conditions for hyperbolic systems, II, *J. Comp. Phys.* 89 (1990), 439–461.

Thouvenin, J. Les mécanismes élémentaires de la détonique, unpublished.

Toro, E.F. A weighted average flux method for hyperbolic conservation laws, *Proc. R. Soc. London* A 423 (1989), 401–418.

Riemann-solver adaptation for gas dynamics, in *Computing Methods in Applied Sciences and Engineering*, Proceedings of the tenth International Conference on Computing Methods in Applied Sciences and Engineering, France (1992), R. Glowinski (Ed.), Nova Science Publishers, Inc, New York (1992), 489–500.

Toumi, I. A weak formulation of Roe's approximate Riemann solver, *J. Comp. Phys.* 102 (1992), 360–373.

Toumi, I. and A. Kumbaro. An approximate linearized Riemann solver for a two-fluid model, *J. Comp. Phys.* 124 (1996), 286–300.

Trefethen, L.N. Group velocity in finite difference schemes, *SIAM Rev.* 24 (1982), 113–136.

Stability of hyperbolic finite-difference models with one or two boundaries, in *Large-scale computations in Fluid Mechanics*, Proceedings of AMS/SIAM Summer Seminar, La Jolla, USA (1983), B. Engquist et al. (Eds.), Part 2, Lectures in Applied Mathematics 22, AMS, Providence, RI (1985), 311–326.

Instability of difference models for hyperbolic initial boundary value problems, *Comm. Pure Appl. Math.* 37 (1984), 329–367.

Turkel, E. Acceleration to a steady state for the Euler equations, in *Numerical methods for the Euler equations of fluid dynamics*, Proceedings of the INRIA workshop, Rocquencourt, France (1983), F. Angrand et al. (Eds.), SIAM, Philadelphia (1985), 281–311.

Accuracy of schemes with nonuniform meshes for compressible fluid flows, ICASE Report 85-43 (1985), ICASE NASA Langley Research Center, Hampton, VA.

Tveito, A. and R. Winther. Existence uniqueness and continuous dependence for a system of hyperbolic conservation laws modeling polymer flooding, *SIAM J. Math. Anal.* 22 (1991), 905–933.

An error estimate for a finite difference scheme approximating a hyperbolic system of conservation laws, *SIAM J. Numer. Anal.* 30 (2) (1993), 401–424.

The solution of nonstrictly hyperbolic conservation laws may be hard to compute, *SIAM J. Sci. Comput.* 16 (2) (1995), 320–329.

Vallet, M.G., F. Hecht, and B. Mantel. Anisotropic control of mesh generation based upon a Voronoi type method, in *Numerical grid generation in computational fluid dynamics and related fields*, Proc. of the third Int. Conference, Barcelona, Spain, June 1991, A.S. Arcilla et al. (Eds.), Elsevier Science Publishers B.V., North-Holland, Amsterdam (1991), 93–103.

Vanajakshi, T.C., K.W. Thompson, and D.C. Black. Boundary value problems in magnetohydrodynamics (and fluid dynamics). I. Radiation boundary condition, *J. Comp. Phys.* 84 (1989), 343–359 .

Van Leer, B. Towards the Ultimate Conservative Difference Scheme: I. The quest of monotonicity in *Proceedings of the third international conference on numerical methods in fluid mechanics (1972).* Lecture Notes in Physics 18, Springer-Verlag, Berlin (1973), 163–168.

II. Monotonicity and conservation combined in a second-order scheme, *J. Comp. Phys.* 14 (1974), 361–370.

III. Upstream-centered finite-difference schemes for ideal compressible flow, *J. Comp. Phys.* 23 (1977), 263–275.

IV. A new approach to numerical convection, *J. Comp. Phys.* 23 (1977), 276–279.

V. A second-order sequel to Godunov's method, *J. Comp. Phys.* 32 (1979), 101–136.

Flux-vector splitting for the Euler equations, ICASE Report No 82-30 (1982), ICASE NASA Langley Research Center, Hampton, VA. and *Proceedings of the eighth International Conference on Numerical Methods in Fluid Dynamics, Aachen (1982)*, E. Krause (Ed.), Lecture Notes in Physics 170, Springer-Verlag, Berlin (1982), 507–512.

Computational methods for ideal compressible flow, ICASE Report 172180 (1983), ICASE NASA Langley Research Center, Hampton, VA.

Multidimensional explicit difference schemes for hyperbolic conservation laws, in *Computing Methods in Applied Sciences and Engineering, VI*, Proceedings of the Sixth International Symposium on Computing Methods in Applied Sciences and Engineering, France (1983), R. Glowinski, J.L. Lions (Eds.), Elsevier Science Publishers B.V., North-Holland, Amsterdam, INRIA (1984), 493–497.

Upwind difference methods for aerodynamic problems governed by the Euler equations, in *Large-scale computations in Fluid Mechanics*, Proceedings of AMS/SIAM Summer Seminar, La Jolla, CA (1983), B. Engquist et al. (Eds.), Lectures in Applied Mathematics 22, AMS, Providence, RI (1985), 327–336.

On the relation between upwind differencing schemes of Godunov, Engquist–Osher and Roe, *SIAM J. Sci. Statist. Comp.* 5 (1984), 1–20.

Progress in multi-dimensional upwind differencing, ICASE Report 92-43 (1992), ICASE NASA Langley Research Center, Hampton, VA.

Van Leer, B. and W. Mulder. Relaxation methods for hyperbolic conservation laws, in *Numerical methods for the Euler equations of fluid dynamics*, Proceedings of the INRIA workshop, Rocquencourt, France (1983), F. Angrand et al. (Eds.), SIAM, Philadelphia (1985), 312–333.

Venkatakrishnan, V. Convergence to steady state solutions of the Euler equations on unstructured grids with limiters, *J. Comp. Phys.* 118 (1995), 120–130.

Veuillot, L. and J.-P. Cambier. A sub-domain approach for the computation of compressible inviscid flows, in *Numerical methods for the Euler equations of fluid dynamics*, Proceedings of the INRIA workshop, Rocquencourt, France (1983), F. Angrand et al. (Eds.), SIAM Philadelphia (1985), 470–489.

Vila, J.-P. Convergence and error estimates in finite volume schemes for general multidimensional scalar conservation laws I. Explicite monotone schemes, *RAIRO Math. Model. Numer. Anal.* 28 (1994), 267–295.

Vijayasundaram, G. Transonic flow simulations using an upstream centered scheme of Godunov in finite elements, Review article, *J. Comp. Phys.* 63 (1986), 416–433.

Villedieu, P. *Approximation de type cinétique du système hyperbolique de la dynamique des gaz hors équilibre thermochimique*, Doctoral dissertation, Thesis Université Paul Sabatier (Toulouse), France (1994).

Vinokur, M. An analysis of finite-difference and finite-volume formulations of conservation laws, Review article, *J. Comp. Phys.* 81 (1989), 1–52.

Vinokur, M. and and J.-L. Montagné. Generalized flux-vector splitting and Roe average for an equilibrium real gas, *J. Comp. Phys.* 89 (1990), 276–300.

Wagner, D.H. The Riemann problem in two space dimensions for a single conservation law, *SIAM J. Math. Anal.* 14 (1983), 534–559.

Equivalence of the Euler and Lagrangian equations of gas dynamics for weak solutions, *J. Diff. Equations* 68 (1987), 118–136.

Conservation laws, coordinate transformations, and differential forms, in *Hyperbolic problems: theory, numerics, applications*, Proceedings of the Fifth International Conference on Hyperbolic Problems, StonyBrook 1994, J. Glimm et al. Eds., World Scientific, Singapore (1996), 471–477.

Wang, J. and G. Warnecke. On entropy consistency of large time-step schemes, I: The Godunov and Glimm schemes, II: Approximate Riemann solvers, *SIAM J. Numer. Anal.* 30 (1993), 1229–1251, 1252–1267.

Wang, J. and G.F. Widhopf. A high-resolution TVD finite volume scheme for the Euler equations in conservation form, *J. Comp. Phys.* 84 (1989), 145–173.

Warming, R.F. and R.M. Beam. On the construction and application of implicit factored schemes for conservation laws, *SIAM-AMS Proceedings* 11 (1978), 85–129.

Warming, R.F., R.M. Beam, and B.J. Hyett. Diagonalization and simultaneous symmetrization of the gas dynamic matrices, *Math. Comp.* 29 (1975), 1037–1045.

Warnecke, G. Admissibility of solutions to the Riemann problem for systems of mixed type, in *Nonlinear Evolution Equations that Change Type*, IMA Volumes in Mathematics and its Applications 27, B.L. Keyfitz and M. Shearer (Eds.), Springer-Verlag, New York (1991), 258–284.

Wendroff, B. The Riemann problem for materials with nonconvex equations of state I: Isentropic flow, *J. Math. Anal. Appl.* 38 (1972), 454–466.

The Riemann problem for materials with nonconvex equations of state II: General flow, *J. Math. Anal. Appl.* 38 (1972), 640–658.

Wendroff, B. and A.B. White. Some supraconvergent schemes for hyperbolic equations on irregular grids, in *Nonlinear Hyperbolic Equations - Theory, Computation Methods, and Applications* (Proceedings of the Second International Conference on Nonlinear Hyperbolic Problems, Aachen 1988), J. Ballmann and R. Jeltsch (Eds.), Notes on Numerical Fluid Mechanics 24, Vieweg, Braunschweig (1989), 671–677.

Whitham, G.B. *Linear and Nonlinear Waves*, Pure and applied mathematics, Wiley-Interscience, New York (1974).

Williams, F.A. *Combustion Theory*, Benjamin/Cummings, Menlo Park Publishing company, Menlo Park, CA, 1985.

Woodward, P.R. and P. Colella. The numerical simulation of two-dimensional fluid flow with strong shocks, Review article, *J. Comp. Phys.* 31 (1984), 115–173.

Wu Z.N. Conditions aux limites pour un problème hyperbolique, Internal Report (unpublished), Sinumef, ENSAM, 75013 Paris (France).

Xu, K. and A. Jameson. Gas-kinetic relaxation (BGK-type) schemes for the compressible Euler equations, *AIAA* 95-1736-CP, 12th AIAA Computational Fluid Dynamics Conference (San Diego 1995) 1012—1019.

Xu, K., L. Martinelli, and A. Jameson. Gas-kinetic finite volume methods, flux-vector splitting and artificial diffusion *J. Comp. Phys.* 120 (1995), 48–65.

Xu, K. and K.H. Prendergast. Numerical Navier–Stokes solutions from gas-kinetic theory, *J. Comp. Phys.* 114 (1994), 9–17.

Yang, H. An artificial compression method for ENO schemes: the slope modification method, *J. Comp. Phys.* 89 (1990), 125–160.

Yang, H.Q. and C.-A. Hsu. Numerical experiments with nonoscillatory schemes using Eulerian and new Lagrangian formulations, *Computers Fluids*, 22 (1993), 163–177.

Yang, H.Q. and A.J. Przekwas. A comparative study of advanced shock-capturing schemes applied to Burgers' equation, *J. Comp. Phys.* 102 (1992), 139–159.

Yang, J.Y. and C.A. Hsu. Numerical experiments with nonoscillatory schemes using Eulerian and new Lagrangian formulations, *Comp. Fluids* 22 (1993), 163–177.

Yang, J.Y. and J.C. Huang. Rarefied flow computations using nonlinear model Boltzmann equations, *J. Comp. Phys.* 120 (1995), 323–339.

Yee, H.C. Numerical approximation of boundary conditions with applications to inviscid equations of gas dynamics, NASA Technical memorandum 81285 (1981).

Construction of explicit and implicit symmetric TVD schemes and their applications, *J. Comp. Phys.* 68 (1987), 151–179.

Yee, H.C., G.H. Klopfer, and J.-L. Montagné. High-resolution shock-capturing schemes for inviscid and viscous hypersonic flows, *J. Comp. Phys.* 88 (1990), 31–61.

Yee, H.C., R.F. Warming, and A. Harten. Implicit Total Variation Diminishing (TVD) schemes for steady-state calculations, *J. Comp. Phys.* 57 (1985), 327–360.

Application of TVD schemes for the Euler equations of gas dynamics, in *Large-scale computations in Fluid Mechanics*, Proceedings of AMS/SIAM Summer Seminar, La Jolla, CA (1983), Lectures in Applied Mathematics 22, Part 1, B. Engquist, S. Osher, and R.C.J. Sommerville (Eds.), AMS, Providence, RI (1985), 357–377.

Young, R. and B. Temple. Solutions to the Euler equations with large data, in *Hyperbolic problems: theory, numerics, applications*, Proceedings of the Fifth International Conference on Hyperbolic Problems, StonyBrook 1994, J. Glimm et al. Eds., World Scientific, Singapore (1996), 258–267.

Zachary, A.L. and P. Colella. A higher-order Godunov method for the equations of ideal magnetohydrodynamics, *J. Comp. Phys.* 99 (1992), 341–347.

Zachary, A.L., A. Malagoli, and P. Colella. A higher-order Godunov method for multidimensional ideal magnetohydrodynamics, *SIAM J. Sci. Comp.* 15 (1994), 263–284.

Zalesak, S.T. Fully multidimensional Flux-Corrected Transport algorithms for fluids, *J. Comp. Phys.* 31 (1979), 335–362.

A preliminary comparison of modern shock-capturing schemes: Linear advection, in *Advances in Computer Methods for Partial Differential Equations-VI*, R. Vichnevetsky and R.S. Stepleman (Eds.), Publ. IMACS (1987).

Zelmanse, A. *Formulation cintétique et schémas de Boltzmann pour le calcul numérique en mécanique des fluides*, Doctoral dissertation, Thesis Université Paris 13, France (1995).

Zhang, T. and Y. Zheng. Two-dimensional Riemann problem for a single conservation law, *Trans. Am. Math. Soc.* 312 (2) (1989), 589–619.

Conjecture on the structure of solutions of the Riemann problem for two-dimensional gas dynamics systems, *SIAM J. Math. Anal.* 21 (1990), 593–630.

Zhong, X., T.Y. Hou and P.G. Le Floch. Computational methods for propagating phase boundaries, *J. Comp. Phys.* 124 (1996), 192–216.

Index

Applied Mathematical Sciences

(continued from page ii)

61. *Sattinger/Weaver:* Lie Groups and Algebras with Applications to Physics, Geometry, and Mechanics.
62. *LaSalle:* The Stability and Control of Discrete Processes.
63. *Grasman:* Asymptotic Methods of Relaxation Oscillations and Applications.
64. *Hsu:* Cell-to-Cell Mapping: A Method of Global Analysis for Nonlinear Systems.
65. *Rand/Armbruster:* Perturbation Methods, Bifurcation Theory and Computer Algebra.
66. *Hlavácek/Haslinger/Necasl/Lovísek:* Solution of Variational Inequalities in Mechanics.
67. *Cercignani:* The Boltzmann Equation and Its Applications.
68. *Temam:* Infinite Dimensional Dynamical Systems in Mechanics and Physics.
69. *Golubitsky/Stewart/Schaeffer:* Singularities and Groups in Bifurcation Theory, Vol. II.
70. *Constantin/Foias/Nicolaenko/Temam:* Integral Manifolds and Inertial Manifolds for Dissipative Partial Differential Equations.
71. *Catlin:* Estimation, Control, and the Discrete Kalman Filter.
72. *Lochak/Meunier:* Multiphase Averaging for Classical Systems.
73. *Wiggins:* Global Bifurcations and Chaos.
74. *Mawhin/Willem:* Critical Point Theory and Hamiltonian Systems.
75. *Abraham/Marsden/Ratiu:* Manifolds, Tensor Analysis, and Applications, 2nd ed.
76. *Lagerstrom:* Matched Asymptotic Expansions: Ideas and Techniques.
77. *Aldous:* Probability Approximations via the Poisson Clumping Heuristic.
78. *Dacorogna:* Direct Methods in the Calculus of Variations.
79. *Hernández-Lerma:* Adaptive Markov Processes.
80. *Lawden:* Elliptic Functions and Applications.
81. *Bluman/Kumei:* Symmetries and Differential Equations.
82. *Kress:* Linear Integral Equations.
83. *Bebernes/Eberly:* Mathematical Problems from Combustion Theory.
84. *Joseph:* Fluid Dynamics of Viscoelastic Fluids.
85. *Yang:* Wave Packets and Their Bifurcations in Geophysical Fluid Dynamics.
86. *Dendrinos/Sonis:* Chaos and Socio-Spatial Dynamics.
87. *Weder:* Spectral and Scattering Theory for Wave Propagation in Perturbed Stratified Media.
88. *Bogaevski/Povzner:* Algebraic Methods in Nonlinear Perturbation Theory.
89. *O'Malley:* Singular Perturbation Methods for Ordinary Differential Equations.
90. *Meyer/Hall:* Introduction to Hamiltonian Dynamical Systems and the N-body Problem.
91. *Straughan:* The Energy Method, Stability, and Nonlinear Convection.
92. *Naber:* The Geometry of Minkowski Spacetime.
93. *Colton/Kress:* Inverse Acoustic and Electromagnetic Scattering Theory.
94. *Hoppensteadt:* Analysis and Simulation of Chaotic Systems.
95. *Hackbusch:* Iterative Solution of Large Sparse Systems of Equations.
96. *Marchioro/Pulvirenti:* Mathematical Theory of Incompressible Nonviscous Fluids.
97. *Lasota/Mackey:* Chaos, Fractals, and Noise: Stochastic Aspects of Dynamics, 2nd ed.
98. *de Boor/Höllig/Riemenschneider:* Box Splines.
99. *Hale/Lunel:* Introduction to Functional Differential Equations.
100. *Sirovich (ed):* Trends and Perspectives in Applied Mathematics.
101. *Nusse/Yorke:* Dynamics: Numerical Explorations.
102. *Chossat/Iooss:* The Couette-Taylor Problem.
103. *Chorin:* Vorticity and Turbulence.
104. *Farkas:* Periodic Motions.
105. *Wiggins:* Normally Hyperbolic Invariant Manifolds in Dynamical Systems.
106. *Cercignani/Illner/Pulvirenti:* The Mathematical Theory of Dilute Gases.
107. *Antman:* Nonlinear Problems of Elasticity.
108. *Zeidler:* Applied Functional Analysis: Applications to Mathematical Physics.
109. *Zeidler:* Applied Functional Analysis: Main Principles and Their Applications.
110. *Diekmann/van Gils/Verduyn Lunel/Walther:* Delay Equations: Functional-, Complex-, and Nonlinear Analysis.
111. *Visintin:* Differential Models of Hysteresis.
112. *Kuznetsov:* Elements of Applied Bifurcation Theory.
113. *Hislop/Sigal:* Introduction to Spectral Theory: With Applications to Schrödinger Operators.
114. *Kevorkian/Cole:* Multiple Scale and Singular Perturbation Methods.
115. *Taylor:* Partial Differential Equations I, Basic Theory.
116. *Taylor:* Partial Differential Equations II, Qualitative Studies of Linear Equations.
117. *Taylor:* Partial Differential Equations III, Nonlinear Equations.
118. *Godlewski/Raviart:* Numerical Approximation of Hyperbolic Systems of Conservation Laws.
119. *Wu:* Theory and Applications of Partial Functional Differential Equations.
120. *Kirsch:* An Introduction to the Mathematical Theory of Inverse Problems.